TABLE 3.6 Summary of Discrete Compounding Formulas with Di... [barcode: D1194537]

Flow Type	Factor Notation	Formula	Excel Command	Cash Flow Diagram
SINGLE	Compound amount $(F/P, i, N)$	$F = P(1 + i)^N$	$= \text{FV}(i, N, 0, P)$	
	Present worth $(P/F, i, N)$	$P = F(1 + i)^{-N}$	$= \text{PV}(i, N, 0, F)$	
EQUAL PAYMENT SERIES	Compound amount $(F/A, i, N)$	$F = A\left[\dfrac{(1 + i)^N - 1}{i}\right]$	$= \text{FV}(i, N, A)$	
	Sinking fund $(A/F, i, N)$	$A = F\left[\dfrac{i}{(1 + i)^N - 1}\right]$	$= \text{PMT}(i, N, 0, F)$	
	Present worth $(P/A, i, N)$	$P = A\left[\dfrac{(1 + i)^N - 1}{i(1 + i)^N}\right]$	$= \text{PV}(i, N, A)$	
	Capital recovery $(A/P, i, N)$	$A = P\left[\dfrac{i(1 + i)^N}{(1 + i)^N - 1}\right]$	$= \text{PMT}(i, N, P)$	
GRADIENT SERIES	Linear gradient Present worth $(P/G, i, N)$	$P = G\left[\dfrac{(1 + i)^N - iN - 1}{i^2(1 + i)^N}\right]$		
	Equal-Payment Conversion factor $(A/G, i, N)$	$A = G\left[\dfrac{(1 + i)^N - iN - 1}{i[(1 + i)^N - 1]}\right]$		
	Geometric gradient Present worth $(P/A_1, g, i, N)$	$P = \left[\begin{array}{l} A_1\left[\dfrac{1 - (1 + g)^N(1 + i)^{-N}}{i - g}\right] \\[2mm] A_1\left(\dfrac{N}{1 + i}\right) \text{ (if } i = g) \end{array}\right]$		

Summary of Project Analysis Methods

Analysis Method	Description	Single Project Evaluation	Mutually Exclusive Projects	
			Revenue Projects	**Service Projects**
Payback period PP	A method for determining when in a project's history it breaks even. Management sets the benchmark PP°.	$PP < PP°$		
Discounted payback period DPP	A variation of payback period when factors in the time value of money. Management sets the benchmark PP*.	$DPP < PP*$		
Present worth PW(*i*)	A method which translates a project's cash flows into an equivalent net present value	$PW(i) > 0$	Select the one with the largest PW	Select the one with the least cost in PW
Future worth FW(*i*)	A variation of the PW: a project's cash flows are translated into an equivalent net future value	$FW(i) > 0$	Select the one with the largest FW	Select the one with the least cost in FW
Capitalized equivalent CE(*i*)	A variation of the PW: a perpetual or very long-lived project that generates a constant stream of annual net cash flow	$CE(i) > 0$	Select the one with the largest CE	Select the one with the least cost in CE
Annual equivalence AE(*i*)	A variation of the PW: a project's cash flows are translated into an annual equivalent sum	$AE(i) > 0$	Select the one with the largest AE	Select the one with the least AEC
Internal rate of return IRR	A relative percentage method which measures the yield as a percentage of investment over the life of a project: The IRR must exceed the minimum required rate of return (MARR).	$IRR > MARR$	Incremental analysis: If $IRR_{A2-A1} > MARR$, select the higher cost investment project, A2.	
Benefit-cost ratio BC(*i*)	A relative measure to evaluate public projects by finding the ratio of the equivalent benefits over the equivalent costs	$BC(i) > 1$	Incremental analysis: If $BC(i)_{A2-A1} > 1$, select the higher cost investment project, A2.	
Profitability index PI(*i*)	A relative measure to evaluate projects in terms of net equivalent benefits generated per dollar invested	$PI(i) > 0$	Incremental analysis: If $PI(i)_{A2-A1} > 1$, select the higher cost investment project, A2.	

MyEngineeringLab™

Right now, in your course, there are young men and women whose engineering achievements could revolutionize, improve, and sustain future generations.

Don't Let Them Get Away.

Contemporary Engineering Economics, Sixth Edition, together with MyEngineeringLab, is a complete solution for providing an engaging in-class experience that will inspire your students to stay in engineering, while also giving them the practice and scaffolding they need to keep up and be successful in the course.

Learn more at **myengineeringlab.com**

CONTEMPORARY ENGINEERING ECONOMICS

Sixth Edition

Global Edition

Sixth Edition

Global Edition

CONTEMPORARY ENGINEERING ECONOMICS

Chan S. Park

Department of Industrial
and Systems Engineering
Auburn University

Boston   Columbus   Indianapolis   New York   San Francisco   Hoboken
Amsterdam   Cape Town   Dubai   London   Madrid   Milan   Munich   Paris   Montréal   Toronto   Delhi   Mexico
City   São Paulo   Sydney   Hong Kong   Seoul   Singapore   Taipei   Tokyo

Vice President and Editorial Director, ECS: *Marcia J. Horton*
Executive Editor: *Holly Stark*
Field Marketing Manager: *Demetrius Hall*
Senior Product Marketing Manager: *Bram van Kempen*
Marketing Assistant: *Jon Bryant*
Senior Managing Editor: *Scott Disanno*
Production Project Manager: *Rose Kernan*
Program Manager: *Erin Ault*
Senior Digital Producer: *Felipe Gonzalez*
Global HE Director of Vendor Sourcing and Procurement: *Diane Hynes*
Senior Acquisitions Editor, Global Edition: *Sandhya Ghoshal*
Associate Project Editor, Global Edition: *Sinjita Basu*
Media Production Manager, Global Edition: *Vikram Kumar*
Senior Manufacturing Controller, Production, Global Edition: *Trudy Kimber*
Director of Operations: *Nick Sklitsis*
Operations Specialist: *Maura Zaldivar-Garcia*
Full-Service Project Management: *Laserwords Pvt. Ltd.*
Cover Photo Source: *Shutterstock*
Cover Printer: *Ashford Colour Press*

Pearson Education Limited

Edinburgh Gate

Harlow

Essex CM20 2JE

England

and Associated Companies throughout the world

Visit us on the World Wide Web at:

www.pearsonglobaleditions.com

© Pearson Education Limited 2016

ISBN 10: 1-292-10909-2

ISBN 13: 978-1-292-10909-1

British Library Cataloguing-in-Publication Data

A catalogue record for this book is available from the British Library.

10 9 8 7 6 5

20 19 18 17

Typeset in 10.5/12, Times LT Pro by Laserwords Pvt. Ltd.

Printed by Ashford Colour Press Ltd. in the United Kingdom.

For Sophie and Alexander

Sixth Edition

Global Edition

CONTEMPORARY ENGINEERING ECONOMICS

Chan S. Park

Department of Industrial
and Systems Engineering
Auburn University

PEARSON

Boston   Columbus   Indianapolis   New York   San Francisco   Hoboken
Amsterdam   Cape Town   Dubai   London   Madrid   Milan   Munich   Paris   Montréal   Toronto   Delhi   Mexico
City   São Paulo   Sydney   Hong Kong   Seoul   Singapore   Taipei   Tokyo

Vice President and Editorial Director, ECS: *Marcia J. Horton*
Executive Editor: *Holly Stark*
Field Marketing Manager: *Demetrius Hall*
Senior Product Marketing Manager: *Bram van Kempen*
Marketing Assistant: *Jon Bryant*
Senior Managing Editor: *Scott Disanno*
Production Project Manager: *Rose Kernan*
Program Manager: *Erin Ault*
Senior Digital Producer: *Felipe Gonzalez*
Global HE Director of Vendor Sourcing and Procurement: *Diane Hynes*
Senior Acquisitions Editor, Global Edition: *Sandhya Ghoshal*
Associate Project Editor, Global Edition: *Sinjita Basu*
Media Production Manager, Global Edition: *Vikram Kumar*
Senior Manufacturing Controller, Production, Global Edition: *Trudy Kimber*
Director of Operations: *Nick Sklitsis*
Operations Specialist: *Maura Zaldivar-Garcia*
Full-Service Project Management: *Laserwords Pvt. Ltd.*
Cover Photo Source: *Shutterstock*
Cover Printer: *Ashford Colour Press*

Pearson Education Limited
Edinburgh Gate
Harlow
Essex CM20 2JE
England

and Associated Companies throughout the world

Visit us on the World Wide Web at:

www.pearsonglobaleditions.com

ISBN 10: 1-292-10909-2

ISBN 13: 978-1-292-10909-1

British Library Cataloguing-in-Publication Data

A catalogue record for this book is available from the British Library.

10 9 8 7 6 5

20 19 18 17

Typeset in 10.5/12, Times LT Pro by Laserwords Pvt. Ltd.

Printed by Ashford Colour Press Ltd. in the United Kingdom.

For Sophie and Alexander

CONTENTS

Preface   21

PART I BASICS OF FINANCIAL DECISIONS   31

Chapter 1   Engineering Economic Decisions   32

1.1   Role of Engineers in Business   33
      1.1.1   Types of Business Organization   34
      1.1.2   Engineering Economic Decisions   35
      1.1.3   Personal Economic Decisions   36
      1.1.4   Economic Decisions Versus Design Decisions   36
1.2   What Makes the Engineering Economic Decision Difficult?   37
1.3   Large-Scale Engineering Projects   38
      1.3.1   Are Tesla's Plans for a Giant Battery Factory Realistic?   38
      1.3.2   Impact of Engineering Projects on Financial Statements   40
1.4   Common Types of Strategic Engineering Economic Decisions   40
      1.4.1   Equipment or Process Selection   41
      1.4.2   Equipment Replacement   41
      1.4.3   New Product or Product Expansion   42
      1.4.4   Cost Reduction   42
      1.4.5   Improvement in Service or Quality   42
1.5   Fundamental Principles of Engineering Economics   42
      Summary   45
      Short Case Studies   45

Chapter 2   Accounting and Financial Decision Making   46

2.1   Accounting: The Basis of Decision Making   48
2.2   Financial Status for Businesses   49
      2.2.1   The Balance Sheet   51
      2.2.2   The Income Statement   56
      2.2.3   The Cash Flow Statement   57
2.3   Using Ratios to Make Business Decisions   64
      2.3.1   Debt Management Analysis   6
      2.3.2   Liquidity Analysis
      2.3.3   Asset Management Analysis
      2.3.4   Profitability Analysis

2.3.5   Market Value Analysis   72
2.3.6   Limitations of Financial Ratios in Business Decisions   73
Summary   76
Problems   76
Short Case Studies   83

Chapter 3   Interest Rate and Economic Equivalence   84

3.1   **Interest: The Cost of Money**   85
   3.1.1   The Time Value of Money   86
   3.1.2   Elements of Transactions Involving Interest   87
   3.1.3   Methods of Calculating Interest   91
3.2   **Economic Equivalence**   94
   3.2.1   Definition and Simple Calculations   94
   3.2.2   Equivalence Calculations: General Principles   96
3.3   **Development of Formulas for Equivalence Calculations**   101
   3.3.1   The Five Types of Cash Flows   101
   3.3.2   Single-Cash-Flow Formulas   103
   3.3.3   Equal-Payment Series   111
   3.3.4   Linear-Gradient Series   123
   3.3.5   Geometric Gradient Series   129
   3.3.6   Irregular (Mixed) Payment Series   135
3.4   **Unconventional Equivalence Calculations**   141
   3.4.1   Composite Cash Flows   141
   3.4.2   Determining an Interest Rate to Establish Economic Equivalence   147
   3.4.3   Unconventional Regularity in Cash Flow Pattern   149
Summary   150
Problems   151
Short Case Studies   160

Chapter 4   Understanding Money and Its Management   162

4.1   **Nominal and Effective Interest Rates**   163
   4.1.1   Nominal Interest Rates   164
   4.1.2   Effective Annual Interest Rates   164
   4.1.3   Effective Interest Rates per Payment Period   167
   4.1.4   Continuous Compounding   169

4.2   **Equivalence Calculations with Effective Interest Rates**   **171**
    4.2.1   When Payment Period is Equal to Compounding Period   171
    4.2.2   Compounding Occurs at a Different Rate than That at Which
        Payments are Made   172
    4.2.4   Compounding is Less Frequent than Payments   176
4.3   **Equivalence Calculations with Continuous Compounding**   **180**
    4.3.1   Discrete-Payment Transactions with Continuous
        Compounding   180
    4.3.2   Continuous-Funds Flow with Continuous Compounding   182
4.4   **Changing Interest Rates**   **187**
    4.4.1   Single Sums of Money   187
    4.4.2   Series of Cash Flows   189
4.5   **Debt Management**   **190**
    4.5.1   Commercial Loans   190
    4.5.2   Loan versus Lease Financing   198
    4.5.3   Home Mortgage   202
4.6   **Investing in Financial Assets**   **209**
    4.6.1   Investment Basics   209
    4.6.2   How to Determine Your Expected Return   209
    4.6.3   Investing in Bonds   212
    Summary   **220**
    Problems   **221**
    Short Case Studies   **230**

PART 2 EVALUATION OF BUSINESS AND ENGINEERING ASSETS   233

Chapter 5   Present-Worth Analysis   234

5.1   **Describing Project Cash Flows**   **236**
    5.1.1   Loan versus Project Cash Flows   236
    5.1.2   Independent versus Mutually Exclusive Investment Projects   239
5.2   **Initial Project Screening Method**   **240**
    5.2.1   Payback Period: The Time It Takes to Pay Back   240
    5.2.2   Benefits and Flaws of Payback Screening   243
    5.2.3   Discounted Payback Period   243
    5.2.4   Where Do We Go From Here?   244

5.3	**Discounted Cash Flow Analysis**	**245**
	5.3.1 Net-Present-Worth Criterion	245
	5.3.2 Meaning of Net Present Worth	249
	5.3.3 Basis for Selecting the MARR	252
5.4	**Variations of Present-Worth Analysis**	**253**
	5.4.1 Future-Worth Analysis	253
	5.4.2 Capitalized Equivalent Method	256
5.5	**Comparing Mutually Exclusive Alternatives**	**261**
	5.5.1 Meaning of Mutually Exclusive and "Do Nothing"	261
	5.5.2 Service Projects versus Revenue Projects	262
	5.5.3 Application of Investment Criteria	262
	5.5.4 Scale of Investment	263
	5.5.5 Analysis Period	265
	5.5.6 Analysis Period Matches Project Lives	266
	5.5.7 Analysis Period Differs from Project Lives	269
	5.5.8 Analysis Period is Not Specified	276
	Summary	**279**
	Problems	**280**
	Short Case Studies	**293**

Chapter 6   Annual Equivalent-Worth Analysis   294

6.1	**Annual Equivalent-Worth Criterion**	**296**
	6.1.1 Fundamental Decision Rule	296
	6.1.2 Annual-Worth Calculation with Repeating Cash Flow Cycles	298
	6.1.3 Comparing Mutually Exclusive Alternatives	300
6.2	**Capital Costs Versus Operating Costs**	**303**
6.3	**Applying Annual-Worth Analysis**	**306**
	6.3.1 Benefits of AE Analysis	306
	6.3.2 Unit Profit or Cost Calculation	306
	6.3.3 Make-or-Buy Decision—Outsourcing Decisions	308
	6.3.4 Pricing the Use of an Asset	311
6.4	**Life-Cycle Cost Analysis**	**312**
6.5	**Design Economics**	**320**
	Summary	**331**
	Problems	**331**
	Short Case Studies	**342**

Chapter 7 Rate-of-Return Analysis 346

7.1	**Rate of Return**	**348**
	7.1.1 Return on Investment	348
	7.1.2 Return on Invested Capital	350
7.2	**Methods for Finding the Rate of Return**	**351**
	7.2.1 Simple versus Nonsimple Investments	351
	7.2.2 Predicting Multiple $i*$s	353
	7.2.3 Computational Methods	356
7.3	**Internal-Rate-of-Return Criterion**	**363**
	7.3.1 Relationship to PW Analysis	363
	7.3.2 Net-Investment Test: Pure versus Mixed Investments	363
	7.3.3 Decision Rule for Pure Investments	366
	7.3.4 Decision Rule for Mixed Investments	368
	7.3.5 Modified Internal Rate of Return (MIRR)	377
7.4	**Mutually Exclusive Alternatives**	**379**
	7.4.1 Flaws in Project Ranking by IRR	379
	7.4.2 Incremental Investment Analysis	380
	7.4.3 Handling Unequal Service Lives	387
	Summary	**390**
	Problems	**391**
	Short Case Studies	**405**

PART 3 ANALYSIS OF PROJECT CASH FLOWS 409

Chapter 8 Cost Concepts Relevant to Decision Making 410

8.1	**General Cost Terms**	**412**
	8.1.1 Manufacturing Costs	413
	8.1.2 Nonmanufacturing Costs	414
8.2	**Classifying Costs for Financial Statements**	**415**
	8.2.1 Period Costs	415
	8.2.2 Product Costs	415
8.3	**Cost Classification for Predicting Cost Behavior**	**418**
	8.3.1 Volume Index	419
	8.3.2 Cost Behaviors	419
	8.3.3 Cost–Volume–Profit Analysis	424

8.4	**Future Costs for Business Decisions**	**429**
	8.4.1 Differential Cost and Revenue	429
	8.4.2 Opportunity Cost	433
	8.4.3 Sunk Costs	435
	8.4.4 Marginal Cost	435
8.5	**Estimating Profit from Operation**	**441**
	8.5.1 Calculation of Operating Income	441
	8.5.2 Annual Sales Budget for a Manufacturing Business	441
	8.5.3 Preparing the Annual Production Budget	442
	8.5.4 Preparing the Cost-of-Goods-Sold Budget	444
	8.5.5 Preparing the Nonmanufacturing Cost Budget	445
	8.5.6 Putting It All Together: The Budgeted Income Statement	447
	8.5.7 Looking Ahead	449
	Summary	**449**
	Problems	**450**
	Short Case Studies	**453**

Chapter 9   Depreciation and Corporate Taxes   456

9.1	**Asset Depreciation**	**458**
	9.1.1 Economic Depreciation	459
	9.1.2 Accounting Depreciation	460
9.2	**Factors Inherent in Asset Depreciation**	**460**
	9.2.1 Depreciable Property	460
	9.2.2 Cost Basis	461
	9.2.3 Useful Life and Salvage Value	463
	9.2.4 Depreciation Methods: Book and Tax Depreciation	464
9.3	**Book Depreciation Methods**	**464**
	9.3.1 Straight-Line Method	465
	9.3.2 Declining Balance Method	466
	9.3.3 Units-of-Production Method	473
9.4	**Tax Depreciation Methods**	**474**
	9.4.1 MACRS Depreciation	474
	9.4.2 MACRS Depreciation Rules	476
9.5	**Depletion**	**481**
	9.5.1 Cost Depletion	482
	9.5.2 Percentage Depletion	482
9.6	**Repairs or Improvements Made to Depreciable Assets**	**485**
	9.6.1 Revision of Book Depreciation	485
	9.6.2 Revision of Tax Depreciation	485

9.7	**Corporate Taxes**	**487**
	9.7.1 Income Taxes on Operating Income	487
9.8	**Tax Treatment of Gains or Losses on Depreciable Assets**	**490**
	9.8.1 Disposal of a MACRS Property	490
	9.8.2 Calculations of Gains and Losses on MACRS Property	490
9.9	**Income Tax Rate to Be Used in Economic Analysis**	**496**
	9.9.1 Incremental Income Tax Rate	496
	9.9.2 Consideration of State Income Taxes	499
9.10	**The Need For Cash Flow in Engineering Economic Analysis**	**500**
	9.10.1 Net Income versus Net Cash Flow	501
	9.10.2 Treatment of Noncash Expenses	501
	Summary	**504**
	Problems	**506**
	Short Case Studies	**514**

Chapter 10 Developing Project Cash Flows 516

10.1	**Cost–Benefit Estimation for Engineering Projects**	**518**
	10.1.1 Simple Projects	518
	10.1.2 Complex Projects	519
10.2	**Incremental Cash Flows**	**520**
	10.2.1 Elements of Cash Outflows	520
	10.2.2 Elements of Cash Inflows	522
	10.2.3 Classification of Cash Flow Elements	522
10.3	**Developing Cash Flow Statements**	**524**
	10.3.1 When Projects Require Only Operating and Investing Activities	524
	10.3.2 When Projects Require Working-Capital Investments	528
	10.3.3 When Projects are Financed with Borrowed Funds	533
	10.3.4 When Projects Result in Negative Taxable Income	535
	10.3.5 When Projects Require Multiple Assets	539
10.4	**Generalized Cash-Flow Approach**	**543**
	10.4.1 Setting up Net Cash-Flow Equations	543
	10.4.2 Presenting Cash Flows in Compact Tabular Formats	544
	10.4.3 Lease-or-Buy Decision	547
	Summary	**553**
	Problems	**554**
	Short Case Studies	**563**

PART 4 HANDLING RISK AND UNCERTAINTY   567

Chapter 11   Inflation and Its Impact on Project Cash Flows   568

11.1 Meaning and Measure of Inflation	**570**
11.1.1 Measuring Inflation	570
11.1.2 Actual versus Constant Dollars	575
11.2 Equivalence Calculations Under Inflation	**578**
11.2.1 Market and Inflation-Free Interest Rates	578
11.2.2 Constant-Dollar Analysis	578
11.2.3 Actual-Dollar Analysis	579
11.2.4 Mixed-Dollar Analysis	583
11.3 Effects of Inflation on Project Cash Flows	**583**
11.3.1 Multiple Inflation Rates	587
11.3.2 Effects of Borrowed Funds Under Inflation	589
11.4 Rate-of-Return Analysis Under Inflation	**592**
11.4.1 Effects of Inflation on Return on Investment	592
11.4.2 Effects of Inflation on Working Capital	596
Summary	**598**
Problems	**600**
Short Case Studies	**605**

Chapter 12   Project Risk and Uncertainty   608

12.1 Origins of Project Risk	**610**
12.2 Methods of Describing Project Risk	**610**
12.2.1 Sensitivity (What-if) Analysis	611
12.2.2 Break-Even Analysis	616
12.2.3 Scenario Analysis	620
12.3 Probability Concepts for Investment Decisions	**622**
12.3.1 Assessment of Probabilities	622
12.3.2 Summary of Probabilistic Information	627
12.3.3 Joint and Conditional Probabilities	630
12.3.4 Covariance and Coefficient of Correlation	632
12.4 Probability Distribution of NPW	**634**
12.4.1 Procedure for Developing an NPW Distribution	634
12.4.2 Aggregating Risk over Time	639
12.4.3 Decision Rules for Comparing Mutually Exclusive Risky Alternatives	645

12.5   Risk Simulation   **649**
12.5.1 Computer Simulation   649
12.5.2 Model Building   650
12.5.3 Monte Carlo Sampling   654
12.5.4 Simulation Output Analysis   659
12.5.5 Risk Simulation with Oracle Crystal Ball   661

12.6   Decision Trees and Sequential Investment Decisions   **664**
12.6.1 Structuring a Decision-Tree Diagram   665
12.6.2 Worth of Obtaining Additional Information   669
12.6.3 Decision Making after Having Imperfect Information   673

Summary   **678**
Problems   **679**
Short Case Studies   **689**

Chapter 13   Real-Options Analysis   692

13.1   Risk Management: Financial Options   **693**
13.1.1 Features of Financial Options   694
13.1.2 Buy Call Options When You Expect the Price to Go Up   695
13.1.3 Buy Put Options When You Expect the Price to Go Down   697

13.2   Option Strategies   **699**
13.2.1 Buying Calls to Reduce Capital That is at Risk   699
13.2.2 Protective Puts as a Hedge   702

13.3   Option Pricing   **705**
13.3.1 Replicating-Portfolio Approach with a Call Option   705
13.3.2 Risk-Free Financing Approach   708
13.3.3 Risk-Neutral Probability Approach   709
13.3.4 Put-Option Valuation   711
13.3.5 Two-Period Binomial Lattice Option Valuation   712
13.3.6 Multiperiod Binomial Lattice Model   713
13.3.7 Black–Scholes Option Model   716

13.4   Real-Options Analysis   **718**
13.4.1 How is Real Options Analysis Different?   718
13.4.2 A Conceptual Framework for Real Options
        in Engineering Economics   719

13.5   Simple Real-Option Models   **724**
13.5.1 Option to Defer Investment   724
13.5.2 Patent and License Valuation   727
13.5.3 Growth Option—Option to Expand   728

13.5.4 Scale-Up Option | 730
13.5.5 Compound Options | 733

13.6   **Estimating Volatility at the Project Level** | **739**
13.6.1 Mathematical Relationship between σ and σ_T | 739
13.6.2 Estimating V_T Distribution | 740

Summary | **746**
Problems | **747**
Short Case Studies | **751**

PART 5 SPECIAL TOPICS IN ENGINEERING ECONOMICS   755

Chapter 14   Replacement Decisions   756

14.1   **Replacement Analysis Fundamentals** | **757**
14.1.1 Basic Concepts and Terminology | 758
14.1.2 Opportunity Cost Approach to Comparing Defender and Challenger | 760

14.2   **Economic Service Life** | **763**

14.3   **Replacement Analysis when the Required Service is Long** | **768**
14.3.1 Required Assumptions and Decision Frameworks | 769
14.3.2 Replacement Strategies under the Infinite Planning Horizon | 771
14.3.3 Replacement Strategies under the Finite Planning Horizon | 776
14.3.4 Consideration of Technological Change | 780

14.4   **Replacement Analysis with Tax Considerations** | **780**

Summary | **794**
Problems | **795**
Short Case Studies | **805**

Chapter 15   Capital-Budgeting Decisions   810

15.1   **Methods of Financing** | **812**
15.1.1 Equity Financing | 812
15.1.2 Debt Financing | 814
15.1.3 Capital Structure | 816

15.2   **Cost of Capital** | **820**
15.2.1 Cost of Equity | 821
15.2.2 Cost of Debt | 826
15.2.3 Calculating the Cost of Capital | 827

15.3 Choice of Minimum Attractive Rate of Return **829**
15.3.1 Choice of MARR when Project Financing is Known 829
15.3.2 Choice of MARR when Project Financing is Unknown 831
15.3.3 Choice of MARR under Capital Rationing 833

15.4 Capital Budgeting **837**
15.4.1 Evaluation of Multiple Investment Alternatives 837
15.4.2 Formulation of Mutually Exclusive Alternatives 838
15.4.3 Capital-Budgeting Decisions with Limited Budgets 839

Summary **847**
Problems **848**
Short Case Studies **852**

Chapter 16   Economic Analysis in the Service Sector   858

16.1 What Is The Service Sector? **860**
16.1.1 Characteristics of the Service Sector 860
16.1.2 Difficulty of Pricing Service 861

16.2 Economic Analysis in The Public Sector **862**
16.2.1 What is Benefit–Cost Analysis? 863
16.2.2 Framework of Benefit–Cost Analysis 863
16.2.3 Valuation of Benefits and Costs 864
16.2.4 Quantifying Benefits and Costs 866
16.2.5 Difficulties Inherent in Public Project Analysis 871

16.3 Benefit–Cost Ratios **872**
16.3.1 Definition of Benefit–Cost Ratio 872
16.3.2 Profitability Index (Net B/C Ratio) 875
16.3.2 Relationship Among B/C Ratio, Profitability Index, and NPW 876
16.3.4 Comparing Mutually Exclusive Alternatives: Incremental
Analysis 877

16.4 Analysis of Public Projects Based on Cost-Effectiveness **881**
16.4.1 Cost-Effectiveness Studies in the Public Sector 881
16.4.2 A Cost-Effectiveness Case Study 882

16.5 Economic Analysis in Health-Care Service **890**
16.5.1 Economic Evaluation Tools 890
16.5.2 Cost–Effectiveness Analysis in the Healthcare Sector 891
16.5.3 Cost-Utility Analysis 896

Summary **899**
Problems **900**
Short Case Studies **904**

Appendix A   Fundamentals of Engineering
Review Questions   911

Appendix B   Interest Factors for Discrete Compounding   931

Appendix C   Values of the Standard Normal
Distribution Function   961

Index   965

PREFACE

What is "Contemporary" About Engineering Economics?

Decisions made during the engineering design phase of product development determine the majority of the costs associated with the manufacturing of that product (some say that this value may be as high as 85%). As design and manufacturing processes become more complex, engineers are making decisions that involve money more than ever before. Thus, the competent and successful engineer in the twenty-first century must have an improved understanding of the principles of science, engineering, and economics, coupled with relevant design experience. Increasingly, in the new world economy, successful businesses will rely on engineers with such expertise.

Economic and design issues are inextricably linked in the product/service life cycle. Therefore, one of my strongest motivations for writing this text was to bring the realities of economics and engineering design into the classroom and to help students integrate these issues when contemplating many engineering decisions. Of course, my underlying motivation for writing this book was not simply to address contemporary needs, but to address as well the ageless goal of all educators: to help students to learn. Thus, thoroughness, clarity, and accuracy of presentation of essential engineering economics were my aim at every stage in the development of the text.

New to the Sixth Edition

Much of the content has been streamlined to provide materials in depth and to reflect the challenges in contemporary engineering economics. Some of the highlighted changes are as follows:

- All the chapter opening vignettes—a trademark of *Contemporary Engineering Economics*—have been updated or completely replaced with more current and thought-provoking issues. Selection of vignettes reflects the important segment of global economy in terms of variety and scope of business as well. With more than 80% of the total GDP (Gross Domestic Product) in the United States provided by the service sector, engineers work on various economic decision problems in the service sector as well. For this reason, many engineering economic decision problems from the service sector are presented in this sixth edition.
- Excel spreadsheet modeling techniques are incorporated into various economic decision problems to provide many "what-if" solutions to key decision problems.
- About 20% of end-of-chapter problems are either new or revised. There are a total of 618 end-of-chapter problems and 65 short case-study questions. There are also 196 fully worked-out examples and 40 carefully selected and fully worked out Fundamentals of Engineering Exam Review Questions in Appendix A.

Chapter Opening Vignettes				
Chapters	**Vignettes**	**Company**	**Sector**	**Industry**
1	• Electric vehicles	Tesla	Consumer Goods	Auto Manufacturers
2	• Communication chips	Broadcom	Technology	Semiconductor—Integrated Circuits
3	• Powerball—Lottery winning	Cindy and Mark Hill	Services	Lottery
4	• Financing home mortgage	Personal Finance	Financial	Banking/Housing
5	• Football stadium expansion	University of Colorado	Services	Sports
6	• Industrial robots	Delta	Industrial	Manufacturing
7	• Investment in antique car	Personal	Personal	Automobile
8	• iPhone manufacturing	Apple	Consumer Goods	Electronic Equipment
9	• Airline baggage handling	Delta Airlines	Services	Airlines
10	• Aircraft manufacturing	Eclipse	Industrial Goods	Aerospace
11	• Big Mac index	Personal	Services	Restaurants
12	• Aluminum auto body	Alcoa	Basic Materials	Aluminum
13	• Insurance	Personal	Services	Travel
14	• Replacing absorption chiller	UCSF Medical Center	Healthcare	Hospitals
15	• Capital budgeting	Laredo Petroleum	Energy	Oil drilling
16	• Auto inspection program	State of Pennsylvania	Public	Government

• Some other specific changes in each chapter are summarized as follows:

Chapters	
1	• Revised Section 1.3 by providing one of contemporary issues—electric vehicle and battery manufacturing.
2	• Replaced all financial analyses (including financial ratios) based on the financial statements by Broadcom Corporation.
	• Provided two chapter examples and solutions to improve the understanding of financial analysis.
3	• Redesigned all Excel worksheets to take advantage of its financial functions in solving various economic equivalence problems.

Chapters	
4	• Revised Section 4.3.2 to enhance the understanding of continuous-funds flow with continuous compounding. • Revised Section 4.6.3 to reflect the current bond market.
5	• Revised all Excel worksheets. • Streamlined the presentation.
6	• Revised Section 6.3.3 with a new make-buy example. • Introduced a new example of HVAC retrofit life-cycle-costing analysis.
7	• Created a new section (7.3.5) on modified internal rate of return.
8	• Streamlined the presentation. • Updated all data related to cost of owning and operating a vehicle.
9	• Updated tax information. • Updated all Excel worksheets of generating depreciation schedules.
10	• Revised all cash flow statement tables by using Excel.
11	• Updated all data related to consumer price index as well as other cost data to reflect the current trend in inflation as well as deflation in various economic sectors. • Revised all cash flow statements by using Excel.
12	• Revised Excel worksheet related to sensitivity analysis.
13	• Revised all financial options examples by providing many graphical illustrations to explain complex conceptual financial as well as real option problems. • Extended Example 13.14 on how to estimate project volatility.
14	• Created a new graphical chart (Figure 14.8) to facilitate the understanding of overall replacement strategies under infinite planning horizon.
15	• Created a new figure (Figure 15.1) to illustrate the capital structure of a typical firm. • Extended Section 15.4.3 to include an example on how to find the optimal capital budget if projects cannot be accepted in part (Example 15.12).
16	• Streamlined the presentation. • Provide a new detailed vehicle inspection program on cost-benefit analysis. • Added a new section (16.5.3) on cost-utility analysis to improve the pedagogical aspect of healthcare decisions.

Overview of the Text

Although it contains little advanced math and few truly difficult concepts, the introductory engineering economics course is often curiously challenging for the sophomores, juniors, and seniors. There are several likely explanations for this difficulty.

- The course is the student's first analytical consideration of money (a resource with which he or she may have had little direct control beyond paying for tuition, housing, food, and textbooks).

- The emphasis on theory may obscure the fact that the course aims, among other things, to develop a very practical set of analytical tools for measuring project worth. This is unfortunate since, at one time or another, virtually every engineer—not to mention every individual—is responsible for the wise allocation of limited financial resources.
- The mixture of industrial, civil, mechanical, electrical, and manufacturing engineering students, as well as other undergraduates who take the course, often fail to "see themselves" using in the skills the course and text are intended to foster. This is perhaps less true for industrial engineering students for whom many texts take as their primary audience. But other disciplines are often motivationally shortchanged by a text's lack of applications that appeal directly to their students.

Goal of the Text

This text aims not only to provide sound and comprehensive coverage of the concepts of engineering economic but also aims to address the difficulties of students as outlined previously, all of which have their basis in inattentiveness to the practical concerns of engineering economics. More specifically, this text has the following chief goals:

- To build a thorough understanding of the theoretical and conceptual basis upon which the practice of financial project analysis is built.
- To satisfy the very practical needs of the engineer toward making informed financial decisions when acting as a team member or project manager for an engineering project.
- To incorporate all critical decision-making tools—including the most contemporary, computer-oriented ones that engineers bring to the task of making informed financial decisions.
- To appeal to the full range of engineering disciplines for which this course is often required: industrial, civil, mechanical, electrical, computer, aerospace, chemical, and manufacturing engineering, as well as engineering technology.

Prerequisites

The text is intended for undergraduate engineering students at the sophomore level or above. The only mathematical background required is elementary calculus. For Chapters 12 and 13, a first course in probability or statistics is helpful but not necessary, since the treatment of basic topics there is essentially self-contained.

Taking Advantage of the Internet

The integration of computer use is another important feature of *Contemporary Engineering Economics*. Students have greater access to and familiarity with the various spreadsheet tools and instructors have greater inclination either to treat these topics explicitly in the course or to encourage students to experiment independently.

A remaining concern is that the use of computers will undermine true understanding of course concepts. This text does not promote the use of trivial spreadsheet applications as a replacement for genuine understanding of and skill in applying traditional solution methods. Rather, it focuses on the computer's productivity-enhancing benefits for complex project cash flow development and analysis. For spreadsheet coverage, the emphasis is on demonstrating a chapter concept that embodies some complexity that can be much more efficiently resolved on a computer than by traditional long-hand solutions.

MyEngineeringLab™

- MyEngineeringLab is now available with *Contemporary Engineering Economics*, Sixth Edition and provides a powerful homework and test manager which lets instructors create, import, and manage online homework assignments, quizzes, and tests that are automatically graded. You can choose from a wide range of assignment options, including time limits, proctoring, and maximum number of attempts allowed. The bottom line: MyEngineeringLab means less time grading and more time teaching.
- Algorithmic-generated homework assignments, quizzes, and tests that directly correlate to the textbook.
- Automatic grading that tracks students' results.
- Learning Objectives mapped to ABET outcomes provide comprehensive reporting tools. If adopted, access to MyEngineeringLab can be bundled with the book or purchased separately.

Resources for Instructors and Students

- MyEngineeringLab, myengineeringlab.com, which is also available as MyEngineeringLab with Pearson eText, a complete online version of the book. It allows highlighting, note taking, and search capabilities.
- Excel files of selected example problems from the text as well as end-of-chapter problems.
- Instructor's Solutions Manual in both WORD and PDF versions.
- PowerPoint lecture notes.

Acknowledgments

This book reflects the efforts of a great many individuals over a number of years. In particular, I would like to recognize the following individuals, whose reviews and comments on prior editions have contributed to this edition. Once again, I would like to thank each of them:

Kamran Abedini, *California Polytechnic—Pomona*
James Alloway, *Syracuse University*
Mehar Arora, *U. Wisconsin—Stout*
Joel Arthur, *California State University—Chico*
Robert Baker, *University of Arizona*
Robert Barrett, *Cooper Union and Pratt Institute*
Tom Barta, *Iowa State University*
Charles Bartholomew, *Widener University*
Richard Bernhard, *North Carolina State University*
Bopaya Bidanda, *University of Pittsburgh*

James Buck, *University of Iowa*
Philip Cady, *The Pennsylvania State University*
Tom Carmichal, *Southern College of Technology*
Jeya Chandra, *The Pennsylvania State University*
Max C. Deibert, *Montana State University*
Stuart E. Dreyfus, *University of California–Berkeley*
Philip A. Farrington, *University of Alabama at Huntsville*
W.J. Foley, *RPI*
Jane Fraser, *University of Southern Colorado*
Terry L Friesz, *Penn State University*
Anil K. Goyal, *RPI*
R. Michael Harnett, *Kansa State University*
Bruce Hartsough, *University of California–Davis*
Carl Hass, *University of Texas–Austin*
John Held, *Kansas State University*
T. Allen Henry, *University of Alabama*
R.C. Hodgson, *University of Notre Dame*
Scott Iverson, *University of Washington*
Peter Jackson, *Cornell University*
Philip Johnson, *University of Minnesota*
Harold Josephs, *Lawrence Tech*
Henry Kallsen, *University of Alabama*
Alla Kammerdiner, *Arizona State University*
W.J. Kennedy, *Clemson University*
Oh Keytack, *University of Toledo*
Wayne Knabach, *South Dakota State University*
Bahattin Koc, *University of Buffalo*
Stephen Kreta, *California Maritime Academy*
John Krogman, *University of Wisconsin–Platteville*
Dennis Kroll, *Bradley University*
Michael Kyte, *University of Idaho*
Gene Lee, *University of Central Florida*
William Lesso, *University of Texas–Austin*
Martin Lipinski, *Memphis State University*
Robert Lundquist, *Ohio State University*
Richard Lyles, *Michigan State University*
Gerald T. Mackulak, *Arizona State University*
Abu S. Masud, *The Wichita State University*
Sue McNeil, *Carnegie-Mellon University*
James Milligan, *University of Idaho*
Richard Minesinger, *University of Massachusetts–Lowell*
Gary Moynihan, *The University of Alabama*
Kumar Muthuraman, *University of Texas*
James S. Noble, *University of Missouri–Columbia*
Michael L. Nobs, *Washington University–St. Louis*
Kurt Norlin, *Laurel Tech Integrated Publishing Solutions*
Peter O'Grady, *University of Iowa*
Wayne Parker, *Mississippi State University*

Global Edition Contributors and Reviewers

Pearson wishes to thank and acknowledge the following people for their work on the Global Edition:

Contributor and Reviewer

Anupam De, *National Institute of Technology, Durgapur*

Contributor

Abhik Kumar Mukherjee, *University of Burdwan*

Reviewers

Pradip Banerjee, *Indian Institute of Management, Indore*
Soheli Ghose, *St. Xavier's College, Kolkata*
Gagari Chakrabarti, *Presidency University, Kolkata*

Elizabeth Pate-Cornell, *Stanford University*
Cecil Peterson, *GMI*
George Prueitt, *U.S. Naval Postgraduate School*
J.K. Rao, *California State University–Long Beach*
Susan Richards, *GMI*
Bruce A. Reichert, *Kansas State University*
Mark Roberts, *Michigan Tech*
John Roth, *Vanderbilt University*
Stan Settle, *University of Southern California*
Paul L. Schillings, *Montana State University*
Bill Shaner, *Colorado State University*
Fred Sheets, *California Polytechnic—Pomona*
Dean Shup, *University of Cincinnati*
David Sly, *Iowa State University*
Milton Smith, *Texas Tech*
Stephen V. Smith, *Drexel University*
David C. Slaughter, *University of California–Davis*
Charles Stavridge, *FAMU/FSU*
Junius Storry, *South Dakota State University*
Frank E. Stratton, *San Diego State University*
George Stukhart, *Texas A&M University*
Donna Summers, *University of Dayton*
Joe Tanchoco, *Purdue University*
Deborah Thurston, *University of Illinois at Urbana-Champaign*
Lt. Col. James Treharne, *U.S. Army*
L. Jackson Turaville, *Tennessee Technological University*
Theo De Winter, *Boston University*
Yoo Yang, *Cal Poly State University*

Special Acknowledgment

Personally, I wish to thank Professor Stan Settle of University of Southern California for his inputs to the sixth edition with a detailed list of suggestions for improvement. My special thanks are due to Kyongsun Kim, who served as an accuracy checker for many solutions to the end-of-chapter problems. Her technical knowledge as well as pointed comments improved the solutions manual in many directions. I would also like to thank Erin Ault, Program Manager at Pearson, who assumed responsibility for the overall project and Rose Kernan, my production editor at RPK Editorial Services, Inc., who oversaw the entire book production.

CHAN S. PARK
AUBURN, ALABAMA

CONTEMPORARY ENGINEERING ECONOMICS

Sixth Edition

Global Edition

BASICS OF FINANCIAL DECISIONS

Engineering Economic Decisions

Are Electric Cars Finally the Next Big Thing? Tesla Motors was founded in 2003 by a group of engineers and venture capitalists. Tesla designs, develops, manufactures, and sells premium electric vehicles (EVs) and advanced electric vehicle powertrain components by order only. Tesla's business plan recognizes that innovative technology is often very expensive and that the very rich are usually the first people to adopt it. Once prices come down, the technology can move down into

Source: Alexkava/Shutterstock

the market. That's why Tesla's first car is a high-end sports car only made in limited numbers. In its 10 years since founding, Tesla has launched both a high-end limited edition "Tesla Roadster" and its "Model S" production car, and introduced "Model X," a sport utility vehicle with seating for seven adults in 2015. Despite a public controversy about its limited driving range before recharging, the Model S had received the coveted Car of the Year Award and earned the highest rating that Consumer Reports ever gave to a car, saying that "The mere fact the Tesla Model S exists at all is a testament to innovation and entrepreneurship, the very qualities that once made the American automobile industry the largest, richest, and most powerful in the world."[1] While some of its most visible EV competitors went bankrupt or halted

---

[1] Angus MacKenzie, "2013 Motor Trend Car of the Year: Tesla Model S," *MotorTrend*, January 2013.

production, Tesla became a darling of many investors and Wall Street analysts. Tesla's goal is to be a mass manufacturer of electric cars.

The story of how the Tesla founders got motivated to develop a series of luxury electric cars and eventually transformed their invention to a multibillion-dollar business is a typical one. Companies such as Google, Facebook, and Microsoft all produce computer-related products and have market values of several hundred billion dollars. These companies were all started by highly motivated young college students. One thing that is also common to all these successful businesses is that they have capable and imaginative engineers who constantly generate good ideas for capital investment, execute them well, and obtain good results. You might wonder what kind of role these engineers play in making such business decisions. In other words, what specific tasks are assigned to these engineers, and what tools and techniques are available to them for making such capital investment decisions? We answer these questions and explore related issues throughout this text.

CHAPTER LEARNING OBJECTIVES

After completing this chapter, you should understand the following concepts:

- The role of engineers in business.
- Types of business organization.
- The nature and types of engineering economic decisions.
- What makes the engineering economic decisions difficult.
- How a typical engineering project idea evolves in business.
- Fundamental principles of engineering economics.

1.1 Role of Engineers in Business

Facebook, Google, and Microsoft produce computer products and have a market value of several hundred billion dollars each, as stated earlier. These companies were all started by young college students with technical backgrounds. When they went into the computer business, these students initially organized their companies as proprietorships. As the businesses grew, they became partnerships and were eventually converted to corporations. This chapter begins by introducing the three primary forms of business organization and briefly discusses the role of engineers in business.

1.1.1 Types of Business Organization

As an engineer, you should understand the nature of the business organization with which you are associated. This section will present some basic information about the type of organization you should choose should you decide to go into business for yourself. The three legal forms of business, each having certain advantages and disadvantages, are proprietorships, partnerships, and corporations.

Proprietorships

A **proprietorship** is a business owned by one individual. This person is responsible for the firm's policies, owns all its assets, and is personally liable for its debts. A proprietorship has two major advantages. First, it can be formed easily and inexpensively. No legal and organizational requirements are associated with setting up a proprietorship, and organizational costs are therefore virtually nil. Second, the earnings of a proprietorship are taxed at the owner's personal tax rate, which may be lower than the rate at which corporate income is taxed. Apart from personal liability considerations, the major disadvantage of a proprietorship is that it cannot issue stocks and bonds, making it difficult to raise capital for any business expansion.

Partnerships

A **partnership** is similar to a proprietorship, except that it has more than one owner. Most partnerships are established by a written contract between the partners. The contract normally specifies salaries, contributions to capital, and the distribution of profits and losses. A partnership has many advantages, among which are its low cost and ease of formation. Because more than one person makes contributions, a partnership typically has a larger amount of capital available for business use. Since the personal assets of all the partners stand behind the business, a partnership can borrow money more easily from a bank. Each partner pays only personal income tax on his or her share of a partnership's taxable income.

On the negative side, under partnership law, each partner is liable for a business's debts. This means that the partners must risk all their personal assets—even those not invested in the business. And while each partner is responsible for his or her portion of the debts in the event of bankruptcy, if any partners cannot meet their pro rata claims, the remaining partners must take over the unresolved claims. Finally, a partnership has a limited life, insofar as it must be dissolved and reorganized if one of the partners quits.

Corporations

A **corporation** is a legal entity created under provincial or federal law. It is separate from its owners and managers. This separation gives the corporation four major advantages:

1. It can raise capital from a large number of investors by issuing stocks and bonds.
2. It permits easy transfer of ownership interest by trading shares of stock.
3. It allows limited liability—personal liability is limited to the amount of the individual's investment in the business.
4. It is taxed differently than proprietorships and partnerships, and under certain conditions, the tax laws favor corporations.

On the negative side, it is expensive to establish a corporation. Furthermore, a corporation is subject to numerous governmental requirements and regulations.

As a firm grows, it may need to change its legal form, because the form of a business affects the extent to which it has control of its own operations and its ability to acquire funds. The legal form of an organization also affects the risk borne by its owners in case of bankruptcy and the manner in which the firm is taxed. Apple Computer, for example, started out as a two-man garage operation. As the business grew, the owners felt constricted by this form of organization: It was difficult to raise capital for business expansion; they felt that the risk of bankruptcy was too high to bear; and as their business income grew, their tax burden grew as well. Eventually, they found it necessary to convert the partnership into a corporation. With a market value of close to $700 billion in 2014, it is the largest corporation in the United States.

In the United States, the overwhelming majority of business firms are proprietorships, followed by corporations and partnerships. However, in terms of total business volume (dollars of sales), the quantity of business transacted by proprietorships and partnerships is several times less than that of corporations. Since most business is conducted by corporations, this text will generally address economic decisions encountered in that form of ownership.

1.1.2 Engineering Economic Decisions

What role do engineers play within a firm? What specific tasks are assigned to the engineering staff, and what tools and techniques are available to it to improve a firm's profits? Engineers are called upon to participate in a variety of decisions, ranging from manufacturing, through marketing, to financing decisions. We will restrict our focus, however, to various economic decisions related to engineering projects. We refer to these decisions as **engineering economic decisions**.

In manufacturing, engineering is involved in every detail of a product's production, from conceptual design to shipping. In fact, engineering decisions account for the majority (some say 85%) of product costs. Engineers must consider the effective use of capital assets such as buildings and machinery. One of the engineer's primary tasks is to plan for the acquisition of equipment (**capital expenditure**) that will enable the firm to design and produce products economically.

With the purchase of any fixed asset—equipment, for instance—we need to estimate the profits (more precisely, cash flows) that the asset will generate during its period of service. In other words, we have to make capital expenditure decisions based on predictions about the future. Suppose, for example, you are considering the purchase of a deburring machine to meet the anticipated demand for hubs and sleeves used in the production of gear couplings. You expect the machine to last 10 years. This decision thus involves an implicit 10-year sales forecast for the gear couplings, which means that a long waiting period will be required before you will know whether the purchase was justified.

An inaccurate estimate of the need for assets can have serious consequences. If you invest too much in assets, you incur unnecessarily heavy expenses. Spending too little on fixed assets, however, is also harmful, for then the firm's equipment may be too obsolete to produce products competitively, and without an adequate capacity, you may lose a portion of your market share to rival firms. Regaining lost customers involves heavy marketing expenses and may even require price reductions or significant product improvements, both of which are costly.

1.1.3 Personal Economic Decisions

In the same way that an engineer can play a role in the effective utilization of corporate financial assets, each of us is responsible for managing our personal financial affairs. After we have paid for nondiscretionary or essential needs, such as housing, food, clothing, and transportation, any remaining money is available for discretionary expenditures on items such as entertainment, travel, and investment. For money we choose to invest, we want to maximize the economic benefit at some acceptable risk. The investment choices are virtually unlimited and include savings accounts, guaranteed investment certificates, stocks, bonds, mutual funds, registered retirement savings plans, rental properties, land, business ownership, and more.

How do you choose? The analysis of one's personal investment opportunities utilizes the same techniques that are used for engineering economic decisions. Again, the challenge is predicting the performance of an investment into the future. Choosing wisely can be very rewarding, while choosing poorly can be disastrous. Some investors in the energy stock Enron who sold prior to the fraud investigation became millionaires. Others, who did not sell, lost everything. A wise investment strategy is a strategy that manages risk by diversifying investments. With such an approach, you have a number of different investments ranging from very low to very high risk and are in a variety of business sectors. Since you do not have all your money in one place, the risk of losing everything is significantly reduced. (We discuss some of these important issues in Chapters 12 and 13.)

In this text, we will consider many types of investments—personal investments as well as business investments. The focus, however, will be on evaluating engineering projects on the basis of their economic desirability and on dealing with investment situations that a typical firm or a public institution faces.

1.1.4 Economic Decisions Versus Design Decisions

Economic decisions differ in a fundamental way from the types of decisions typically encountered in engineering design. In a design situation, the engineer utilizes known physical properties, the principles of chemistry and physics, engineering design correlations, and engineering judgment to arrive at a workable and optimal design. If the judgment is sound, the calculations are done correctly, and we ignore technological advances, the design is time invariant. In other words, if the engineering design to meet a particular need is done today, next year, or in five years' time, the final design would not change significantly.

In considering economic decisions, the measurement of investment attractiveness, which is the subject of this text, is relatively straightforward. However, the information required in such evaluations always involves predicting or forecasting product sales, product selling prices, and various costs over some future time frame—five years, 10 years, 25 years, etc.

All such forecasts have two things in common. First, they are never completely accurate compared with the actual values realized at future times. Second, a prediction or forecast made today is likely to be different from one made at some point in the future. It is this ever-changing view of the future that can make it necessary to revisit and even change previous economic decisions. Thus, unlike engineering design, the conclusions reached through economic evaluation are not necessarily time invariant. Economic decisions have to be based on the best information available at the time of the decision and a thorough understanding of the uncertainties in the forecasted data.

1.2 What Makes the Engineering Economic Decision Difficult?

The economic decisions that engineers make in business differ very little from the financial decisions made by individuals, except for the scale of the concern. For example, everyone who experienced the Great Blackout of 2003 remembers where they were when it happened that summer day. The blackout, which cut power to much of the Northeastern and Midwestern United States, as well as parts of Canada, brought home the reality that the electrical grid in the United States was outdated.[2] Updating the grid will not be cheap—estimates range as high as $2 trillion—but the massive effort will also present huge opportunities for U.S. manufacturers, with a market that could reach $1 trillion. The race is on to capitalize on smart-grid technologies, which would include building new power plants, transmission lines, and focus on conservation. (See Figure 1.1.)

Figure 1.1 A helicopter lowers towers for high-voltage power lines into place. Many say the country needs to build more of these lines to move renewable power and become more efficient.

[2]Frank Andorka, "Powering Up: The Smart Grid's Next Steps," *Industry Week*, April 2011.

Obviously, this level of engineering decision by electric power companies is far more complex and more significant than a business decision about when to introduce a new product. Projects of this nature involve large sums of money over long periods of time, and it is difficult to estimate the magnitude of economic benefits in any precise manner. Even if we decide to rebuild the electric grid systems, should we build in incremental steps, or should we build to withstand a demand to occur 20 years from now? Even if we can justify the project on economic reasoning, how to finance the project is another issue. Any engineering economic decision pertaining to this type of a large-scale project will be extremely difficult to make.

1.3 Large-Scale Engineering Projects

In the development of any product, a company's engineers are called upon to translate an idea into reality. A firm's growth and development depend largely upon a constant flow of ideas for new products, and for the firm to remain competitive, it has to make existing products better or produce them at a lower cost. We will present an example of how a large-scale engineering project evolves and what types of financial decisions have to be considered in the process of executing such a project.[3]

1.3.1 Are Tesla's Plans for a Giant Battery Factory Realistic?

Tesla Motors introduced the world's first luxury electrical vehicles whose engines cut air pollution to zero and boosted operating efficiency to significant levels. Tesla, in short, wanted to launch and dominate a new "green" era for automobiles and plans to build one of the world's largest factories of any kind in the U.S. But it wouldn't build its electric cars there—it would make the batteries to power them. The plant, slated for completion by 2017 at a cost of as much as $5 billion, would be able to turn out more lithium-ion batteries than all the battery factories in the world today. Tesla finally broke ground in June of 2014 on the site in Reno, Nevada, and expects to start producing batteries at the plant by 2017. It says the scale will help drive the cost of batteries down, in turn helping to make a mass-manufacturing within reach.

How Economical is Tesla's Plan?

The biggest question remaining about the mass production of the electric vehicle concerned its battery production cost. Costs would need to come down for Tesla's electric cars to be competitive around the world, where gasoline prices were stable or even declining. Economies of scale would help as production volumes increase, but further advances in engineering also would be essential. With the initial engineering specification, Tesla has designed the powerpacks and their associated circuitry, each of which contains up to 7,000 standard lithium-ion cells of the sort found in laptops. The firm is said to buy more of these sorts of cell than all the world's computer-makers combined.

---

[3]"Elon Musk's Tesla Picks Nevada to Host Battery Gigafactory," *Scientific American*, September 5, 2014. This article presents various economic and financial issues associated with locating the battery plant in Nevada. Some of the performance is from Tesla Motors Corporation.

However, in the case of the Tesla products, if a significant reduction in production cost never materializes, demand might remain insufficient to justify the investment in the battery factory.

1.3.2 Impact of Engineering Projects on Financial Statements

Engineers must understand the business environment in which a company's major business decisions are made. It is important for an engineering project to generate profits, but it also must strengthen the firm's overall financial position. How do we measure Tesla's success in the giant battery factory project? Will enough electric car models be sold, for example, to keep the green-engineering business as Tesla's major source of profits? While the giant battery project would provide revolutionary energy storage system for the company's customers, the bottom line is its financial performance over the long run.

Regardless of a business' form, each company has to produce basic financial statements at the end of each operating cycle (typically a year). These financial statements provide the basis for future investment analysis. In practice, we seldom make investment decisions solely on the basis of an estimate of a project's profitability, because we must also consider the overall impact of the investment on the financial strength and position of the company.

Suppose that you were the president of the Tesla Corporation. Suppose further that you even hold some shares in the company, making you one of the company's many owners. What objectives would you set for the company? While all firms are in business in hopes of making a profit, what determines the market value of a company are not profits per se, but cash flow. It is, after all, available cash that determines the future investments and growth of the firm. Therefore, one of your objectives should be to increase the company's value to its owners (including yourself) as much as possible. To some extent, the market price of your company's stock represents the value of your company.

Many factors affect your company's market value: present and expected future earnings, the timing and duration of those earnings, and the risks associated with them. Certainly, any successful investment decision will increase a company's market value. Stock price can be a good indicator of your company's financial health, and may also reflect the market's attitude about how well your company is managed for the benefit of its owners.

Any successful investment decision on the giant battery factory's scale will tend to increase a firm's stock prices in the marketplace and promote long-term success. Thus, in making a large-scale engineering project decision, we must consider its possible effect on the firm's market value. (In Chapter 2, we discuss the financial statements in detail and show how to use them in our investment decision making.)

1.4 Common Types of Strategic Engineering Economic Decisions

The story of how the Tesla Corporation successfully introduced a new product and became the market leader in the advanced electric car market is typical: Someone had a good idea, executed it well, and obtained good results. Project ideas such as the giant

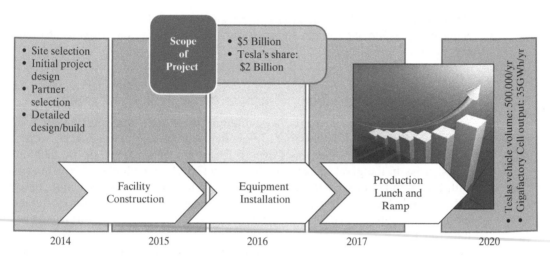

Figure 1.2 Projected Timeline of Tesla's Gigafactory.
Source: "Assault on Batteries," *The Economist*, June 14, 2014.

Tesla argues that its battery packs, including their power-management and cooling systems, currently cost less than $300 a kilowatt-hour (kWh) of storage capacity: about half the costs of its rivals.

The gigafactory, which will eventually turn out batteries for 500,000 vehicles, should cut their costs by another 30%; two-thirds of that saving will come from scale alone with the rest due to improved manufacturing technology. When costs drop below $200 a kWh, battery-powered cars start to become competitive with conventional ones without government subsidies. The gigafactory could bring Tesla close to that.

What is the Business Risk?

Although engineers at Tesla claim that they would be able to cut its current battery co drastically, many financial analysts are skeptical as raw materials account for 70% of price of a lithium battery. This would make the scope for savings limited, and even factory does turn out many cheap battery cells, that may not be enough. Technical key to increasing range and performance is to improve the efficiency, size, and the electronics that manage the power, along with overall vehicle weight. Tesla have the same advantages in these areas as it has with its batteries.

At a cost of $5 billion, which Tesla will share with Panasonic of Japan battery supplier, and other partners, the gigafactory is a big gamble. Also, demand stalls, the question is what we do with the huge output of cheap b is a lot of cost that can be removed at larger scales of battery manufactu about the capacity utilization. A battery plant that is not running will cos

Despite Tesla management's decision to build the giant battery fac analysts were still uncertain whether there would be enough demand petitors, including U.S. automakers, just did not see how Tesla coul mies of scale needed to produce electric cars at a profit. The pri design, however, is that the electric vehicle could cut auto pollut is a feature that could be very appealing at a time when governn are becoming more rigorous and consumer interest in the env

However, in the case of the Tesla products, if a significant reduction in production cost never materializes, demand might remain insufficient to justify the investment in the battery factory.

1.3.2 Impact of Engineering Projects on Financial Statements

Engineers must understand the business environment in which a company's major business decisions are made. It is important for an engineering project to generate profits, but it also must strengthen the firm's overall financial position. How do we measure Tesla's success in the giant battery factory project? Will enough electric car models be sold, for example, to keep the green-engineering business as Tesla's major source of profits? While the giant battery project would provide revolutionary energy storage system for the company's customers, the bottom line is its financial performance over the long run.

Regardless of a business' form, each company has to produce basic financial statements at the end of each operating cycle (typically a year). These financial statements provide the basis for future investment analysis. In practice, we seldom make investment decisions solely on the basis of an estimate of a project's profitability, because we must also consider the overall impact of the investment on the financial strength and position of the company.

Suppose that you were the president of the Tesla Corporation. Suppose further that you even hold some shares in the company, making you one of the company's many owners. What objectives would you set for the company? While all firms are in business in hopes of making a profit, what determines the market value of a company are not profits per se, but cash flow. It is, after all, available cash that determines the future investments and growth of the firm. Therefore, one of your objectives should be to increase the company's value to its owners (including yourself) as much as possible. To some extent, the market price of your company's stock represents the value of your company.

Many factors affect your company's market value: present and expected future earnings, the timing and duration of those earnings, and the risks associated with them. Certainly, any successful investment decision will increase a company's market value. Stock price can be a good indicator of your company's financial health, and may also reflect the market's attitude about how well your company is managed for the benefit of its owners.

Any successful investment decision on the giant battery factory's scale will tend to increase a firm's stock prices in the marketplace and promote long-term success. Thus, in making a large-scale engineering project decision, we must consider its possible effect on the firm's market value. (In Chapter 2, we discuss the financial statements in detail and show how to use them in our investment decision making.)

1.4 Common Types of Strategic Engineering Economic Decisions

The story of how the Tesla Corporation successfully introduced a new product and became the market leader in the advanced electric car market is typical: Someone had a good idea, executed it well, and obtained good results. Project ideas such as the giant

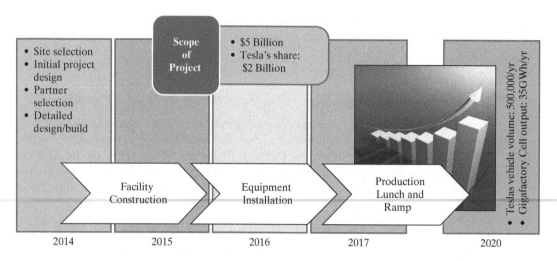

Figure 1.2 Projected Timeline of Tesla's Gigafactory.
Source: "Assault on Batteries," *The Economist*, June 14, 2014.

Tesla argues that its battery packs, including their power-management and cooling systems, currently cost less than $300 a kilowatt-hour (kWh) of storage capacity: about half the costs of its rivals.

The gigafactory, which will eventually turn out batteries for 500,000 vehicles, should cut their costs by another 30%; two-thirds of that saving will come from scale alone with the rest due to improved manufacturing technology. When costs drop below $200 a kWh, battery-powered cars start to become competitive with conventional ones without government subsidies. The gigafactory could bring Tesla close to that.

What is the Business Risk?

Although engineers at Tesla claim that they would be able to cut its current battery costs drastically, many financial analysts are skeptical as raw materials account for 70% of the price of a lithium battery. This would make the scope for savings limited, and even if the factory does turn out many cheap battery cells, that may not be enough. Technically, the key to increasing range and performance is to improve the efficiency, size, and price of the electronics that manage the power, along with overall vehicle weight. Tesla does not have the same advantages in these areas as it has with its batteries.

At a cost of $5 billion, which Tesla will share with Panasonic of Japan, its current battery supplier, and other partners, the gigafactory is a big gamble. Also, if electric-car demand stalls, the question is what we do with the huge output of cheap batteries. There is a lot of cost that can be removed at larger scales of battery manufacturing, but it's all about the capacity utilization. A battery plant that is not running will cost Tesla a fortune.

Despite Tesla management's decision to build the giant battery factory, the financial analysts were still uncertain whether there would be enough demand. Furthermore, competitors, including U.S. automakers, just did not see how Tesla could achieve the economies of scale needed to produce electric cars at a profit. The primary advantage of the design, however, is that the electric vehicle could cut auto pollution to a zero level. This is a feature that could be very appealing at a time when government air-quality standards are becoming more rigorous and consumer interest in the environment is getting strong.

battery factory can originate from many different levels in an organization. Since some ideas will be good, while others will not, we need to establish procedures for screening projects.

Many large companies have a specialized project analysis division that actively searches for new ideas, projects, and ventures. Once project ideas are identified, they are typically classified as (1) equipment or process selection, (2) equipment replacement, (3) new product or product expansion, (4) cost reduction, or (5) improvement in service or quality. This classification scheme allows management to address key questions: Can the existing plant, for example, be used to attain the new production levels? Does the firm have the knowledge and skill to undertake the new investment? Does the new proposal warrant the recruitment of new technical personnel? The answers to these questions help firms screen out proposals that are not feasible, given a company's resources.

The giant battery project represents a fairly complex engineering decision that required the approval of top executives and the board of directors. Virtually all big businesses face investment decisions of this magnitude at some time. In general, the larger the investment, the more detailed is the analysis required to support the expenditure. For example, expenditures aimed at increasing the output of existing products or at manufacturing a new product would invariably require a very detailed economic justification. Final decisions on new products, as well as marketing decisions, are generally made at a high level within the company. By contrast, a decision to repair damaged equipment can be made at a lower level. The five classifications of project ideas are as follows.

1.4.1 Equipment or Process Selection

This class of engineering decision problems involves selecting the best course of action out of several that meet a project's requirements. For example, which of several proposed items of equipment shall we purchase for a given purpose? The choice often hinges on which item is expected to generate the largest savings (or the largest return on the investment). For example, the choice of material will dictate the manufacturing process for the body panels in the automobile. Many factors will affect the ultimate choice of the material, and engineers should consider all major cost elements, such as the cost of machinery and equipment, tooling, labor, and material. Other factors may include press and assembly, production and engineered scrap, the number of dies and tools, and the cycle times for various processes.

1.4.2 Equipment Replacement

This category of investment decisions involves considering the expenditure necessary to replace worn-out or obsolete equipment. For example, a company may purchase 10 large presses, expecting them to produce stamped metal parts for 10 years. After five years, however, it may become necessary to produce the parts in plastic, which would require retiring the presses early and purchasing plastic molding machines. Similarly, a company may find that, for competitive reasons, larger and more accurate parts are required, making the purchased machines become obsolete earlier than expected.

Suppose, for example, that a firm is using a lathe that was purchased 12 years ago to produce pump shafts. As the production engineer in charge of this product, you expect demand to continue into the foreseeable future. However, the lathe has begun to show its age: It has broken frequently during the last two years and has finally stopped operating altogether. Now you have to decide whether to replace or repair it. If you expect a more efficient lathe to be available in the next one or two years, you might repair the old lathe instead of replacing it. The major issue is whether you should make the considerable investment in a new lathe now or later. As an added complication, if demand for your product begins to decline, you may have to conduct an economic analysis to determine whether declining profits from the project offset the cost of a new lathe.

1.4.3 New Product or Product Expansion

Investments in this category increase company revenues if output is increased. One common type of expansion decision includes decisions about expenditures aimed at increasing the output of existing production or distribution facilities. In these situations, we are basically asking, "Shall we build or otherwise acquire a new facility?" The expected future cash inflows in this investment category are the profits from the goods and services produced in the new facility. A second type of expenditure decision includes considering expenditures necessary to produce a new product or to expand into a new geographic area. These projects normally require large sums of money over long periods.

1.4.4 Cost Reduction

A cost-reduction project is a project that attempts to lower a firm's operating costs. Typically, we need to consider whether a company should buy equipment to perform an operation currently done manually or spend money now in order to save more money later. The expected future cash inflows on this investment are savings resulting from lower operating costs.

1.4.5 Improvement in Service or Quality

Most of the examples in the previous sections were related to economic decisions in the manufacturing sector. The decision techniques we develop in this book are also applicable to various economic decisions related to improving services or quality of product. We will provide several economic decision problems from the service sector throughout the text.

1.5 Fundamental Principles of Engineering Economics

This book is focused on the principles and procedures engineers use to make sound economic decisions. To the first-time student of engineering economics, anything related to money matters may seem quite strange when compared to other engineering subjects. However, the decision logic involved in solving problems in this domain is quite similar to that employed in any other engineering subject. There are fundamental principles to follow in engineering economics that unite the concepts and techniques presented in this text, thereby allowing us to focus on the logic underlying the practice of engineering economics.

- **Principle 1: A dollar earned today is worth more than a dollar earned in the future.** A fundamental concept in engineering economics is that money has a time value associated with it (Figure 1.3). Because we can earn interest on money received today, it is better to receive money earlier than later. This concept will be the basic foundation for all engineering project evaluation.

Figure 1.3 Time value of money.

- **Principle 2: All that counts are the differences among alternatives.** An economic decision should be based on the *differences* among the alternatives considered (Figure 1.4). All that is common is irrelevant to the decision. Certainly, any economic decision is no better than the alternatives being considered. Thus, an economic decision should be based on the objective of making the best use of limited resources. Whenever a choice is made, something is given up. The opportunity cost of a choice is the value of the best alternative given up.

Comparing Buy versus Lease
Whatever you decide, you need to spend the same amount of money on fuel and maintenance

Option	Monthly Fuel Cost	Monthly Maintenance	Cash Outlay at Signing	Monthly Payment	Salvage Value at End of Year 3
Buy	$960	$550	$6,500	$350	$9,000
Lease	$960	$550	$2,400	$550	0

Irrelevant items in decision making

Figure 1.4 Differential analysis.

- **Principle 3: Marginal revenue must exceed marginal cost.** Effective decision making requires comparing the additional costs of alternatives with the additional benefits (Figure 1.5). Each decision alternative must be justified on its own economic merits before being compared with other alternatives. Any increased economic activity must be justified on the basis of the fundamental economic principle that marginal revenue must exceed marginal cost. Here, *marginal revenue* means the additional revenue made possible by increasing the activity by one unit (or small unit). *Marginal cost* has an analogous definition. Productive resources—the natural resources, human resources, and capital goods available to make goods and services—are limited. Therefore, people cannot have all the goods and services they want; as a result, they must choose some things and give up others.

To justify your action, marginal revenue must exceed marginal cost

Figure 1.5 Marginal analysis.

- **Principle 4: Additional risk is not taken without the expected additional return.** For delaying consumption, investors demand a minimum return that must be greater than the anticipated rate of inflation or any perceived risk (Figure 1.6). If they didn't receive enough to compensate for anticipated inflation and the perceived investment risk, investors would purchase whatever goods they desired ahead of time or invest in assets that would provide a sufficient return to compensate for any loss from inflation or potential risk. The preceding four principles are as much statements of common sense as they are theoretical precepts. These principles provide the logic behind what is to follow. We build on them and attempt to draw out their implications for decision-making. As we continue, keep in mind that, while the topics being treated may change from chapter to chapter, the logic driving our treatment of them is constant and rooted in the four fundamental principles.

Expected returns from bonds and stocks are normally higher than the expected return from a savings account

Investment Class	Potential Risk	Expected Return
Savings account (cash)	Low/None	1.5%
Bond (debt)	Moderate	4.8%
Stock (equity)	High	11.5%

Figure 1.6 Risk and return trade off.

SUMMARY

- This chapter has given us an overview of a variety of engineering economic problems that commonly are found in the business world. We examined the role and the increasing importance of engineers in the firm, as evidenced by the development of innovative electric vehicles at Tesla. Commonly, engineers are called upon to participate in a variety of strategic business decisions ranging from product design to marketing.

- The term **engineering economic decision** refers to any investment decision related to an engineering project. The facet of an economic decision that is of most interest from an engineer's point of view is the evaluation of costs and benefits associated with making a capital investment.

- The five main types of engineering economic decisions are (1) equipment or process selection, (2) equipment replacement, (3) new product or product expansion, (4) cost reduction, and (5) improvement in service or quality.

- The factors of **time** and **uncertainty** are the defining aspects of any investment project.

- The four fundamental principles that must be applied in all engineering economic decisions are (1) the time value of money, (2) differential (incremental) cost and revenue, (3) marginal cost and revenue, and (4) the trade-off between risk and reward.

Short Case Studies

ST1.1 Review the contents of *The Wall Street Journal* for the past three months. Then, identify and categorize the types of investment decisions appearing in the journal according to the types of strategic economics decisions discussed in the text.

ST1.2 Work in small groups and brainstorm ideas about how a common appliance, device, or tool could be redesigned to improve it in some way. Identify the steps involved and the economic factors that you would need to consider prior to making a decision to manufacture the redesigned product. A detailed design and actual cost estimates are not required.

Some items which could be considered for this redesign exercise are a shopping cart, telephone, can opener, screwdriver, etc.

ST1.3 Many oil price forecasts in the early 2000s indicated that the price of oil in the year 2007 would not exceed $50 per barrel. What is the price of oil today? Why are these prices so difficult to predict? Imagine what the consequences would be if you used these optimistic estimates in your economic analysis in your early project undertaking. What would be some practical ways to consider this type of variation in economic analysis?

2

Accounting and Financial Decision Making

Broadcom's Expansion to China   Based in Irvine, CA, Broadcom is engaged in designing and marketing semiconductor components of network voice, video, and data traffic for various applications. Broadcom Corporation (BRCM) is looking to expand its sales in China by making chips that support a wider range of handsets, including those that run on the network of the world's largest telecom carrier, China Mobile Limited.

Broadcom's strategy is a bet that demand will continue to rise for low-cost smartphones that work on China Mobile's third-generation network. China Mobile's network is based on a technology rarely used outside of China. However, the company's more than 760 million subscribers have created demand for phones, and in turn chips, that support that technology. With more than 500 million smartphone users, China is the world's largest smartphone market.

Suppose you are a chief engineer who designed the chip to be used in the China's smartphone market and also participated in the Broadcom's strategic decision to expand into China. How would you determine the financial success of your project? Or, as an investor, if you want to explore investing in BRCM stock, what information would you go by? You would certainly prefer that BRCM

Source: FrankRamspott/Getty Images

have a record of accomplishment of profitable operations, earning a profit (net income) year after year. The company would need a steady stream of cash coming in and a manageable level of debt. How would you determine whether the company met these criteria? Investors commonly use the financial statements contained in the annual report as a starting point in forming expectations about future levels of earnings and about the firm's riskiness.

Before making any financial decision, it is good to understand an elementary aspect of your financial situation—one that you'll also need for retirement planning, estate planning, and, more generally, to get an answer to the question, "How am I doing?" It is called your **net worth**. If you want to invest $10,000 in BRCM stocks, how would that affect your net worth? You may need this information for your own financial planning, but it is routinely required whenever you have to borrow a large sum of money from a financial institution. For example, when you are buying a home, you need to apply for a mortgage. Invariably, the bank will ask you to submit your net-worth statement as a part of loan processing. Your net-worth statement is a snapshot of where you stand financially at a given point in time. You do that by adding your assets—such as cash, investments, and pension plans—in one column and your liabilities—or debts—in the other. Then subtract your liabilities from your assets to find your net worth. In other words, your net worth is what you would be left with if you sold everything and paid off all you owe. The bank will determine how creditworthy you are by examining your net worth. In a similar way, a corporation prepares the same kind of information for its financial planning or to report its financial health to stockholders or investors. The reporting document is known as the financial statements.

CHAPTER LEARNING OBJECTIVES

After completing this chapter, you should understand the following concepts:

- The role of accounting in economic decisions.
- Four types of financial statements prepared for investors and regulators.

■ How to read the balance sheet statement.

■ How to use the income statement to manage a business.

■ The sources and uses of cash in business operation.

■ How to conduct the ratio analysis and what the numbers really mean.

2.1 Accounting: The Basis of Decision Making

We need financial information when we are making business decisions. Virtually all businesses keep accounting records to aid in making decisions. As illustrated in Figure 2.1, accounting is the information system that measures business activities, processes the resulting information into reports, and communicates the results to decision makers. For this reason, we call accounting "the language of business." The better you understand this language, the better you can manage your financial well-being, and the better your financial decisions will be.

The uses of accounting information are many and varied.

• **Business managers** use accounting information to set goals for their organizations, to evaluate progress toward those goals, and to take corrective actions if necessary. Decisions based on accounting information may include which building or equipment to purchase, how much merchandise to keep on hand as inventory, and how much cash to borrow.

• **Investors and creditors** provide the money a business needs to begin operations. To decide whether to help start a new venture, potential investors evaluate what income they can expect on their investment. Such an evaluation involves analyzing the financial statements of the business. Before making a loan, banks determine the borrower's

Figure 2.1 The accounting system, which illustrates the flow of information.

ability to meet scheduled payments. This kind of evaluation includes a projection of future operations and revenue based on accounting information.

An essential product of an accounting information system is a series of financial statements that allows people to make informed decisions. For business use, financial statements are the documents that report financial information about a business entity to decision makers. They tell us how a business is performing and where it stands financially. Our purpose is not to present the bookkeeping aspects of accounting, but to acquaint you with financial statements and to give you the basic information you need to make sound engineering economic decisions through the remainder of the text.

2.2 Financial Status for Businesses

Of the various reports corporations issue to their stockholders, the annual report is by far the most important, containing basic financial statements as well as management's opinion of the past year's operations and the firm's future prospects. What would managers and investors want to know about a company at the end of the fiscal year? The following are four basic questions that managers or investors are likely to ask:

- What is the company's financial position at the end of the fiscal period?
- How well did the company operate during the fiscal period?
- On what did the company decide to use its profits?
- How much cash did the company generate and spend during the fiscal period?

As illustrated in Figure 2.2, the answer to each of these questions is provided by one of the following financial statements: the balance sheet statement, the income statement, the statement of retained earnings, and the cash flow statement. The fiscal year (or operating cycle) can be any 12-month term but is usually January 1 through December 31 of a calendar year.

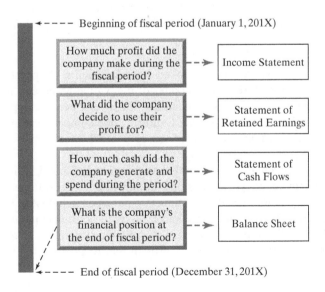

Figure 2.2   Information reported on the financial statements.

Figure 2.3 Summary of major factors affecting stock prices.

As mentioned in Section 1.1.2, one of the primary responsibilities of engineers in business is to plan for the acquisition of equipment (capital expenditure) that will enable the firm to design and produce products economically. This type of planning will require an estimation of the savings and costs associated with the acquisition of equipment and the degree of risk associated with execution of the project. Such estimation will affect the business's **bottom line** (profitability), which will eventually affect the firm's stock price in the marketplace. Therefore, engineers should understand the various financial statements in order to communicate with upper management regarding the status of a project's profitability. The situation is summarized in Figure 2.3.

Broadcom Corporation

For illustration purposes, we use data taken from Broadcom Corporation (BRCM),[1] a leading designer, manufacturer, and marketer of innovative wired and wireless communications. Broadcom (BRCM) was founded in 1991 by engineers and managed by experienced industry leaders. Based in Irvine, California, the company estimates that 99.98% of Internet traffic crosses at least one Broadcom chip—in the home, in the hand, and across the network.

The company's revenue in 2013 totaled approximately $8.305 billion. During fiscal 2013, BRCM maintained its position as one of the world's largest fabless semiconductor companies,[2] with a performance that continued to outpace the industry. In the company's 2013 annual report, management painted an even more optimistic picture for the future, stating that BRCM will continue to invest in information systems, research, development, and engineering activities to support its growth and to provide for new competitive products.

---

[1]Corporate Profile, BRCM, http://www.broadcom.com.
[2]Fabless manufacturing refers to the business methodology of outsourcing the manufacturing of silicon wafers.

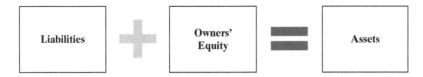

Figure 2.4 Accounting equation.

What can individual investors make of all this from the financial information contained in the annual report? Actually, they can make quite a bit. As you will see, investors use the information contained in an annual report to form expectations about future earnings and dividends. Therefore, the annual report is certainly of great interest to investors.

2.2.1 The Balance Sheet

What is the company's financial position at the end of the reporting period? We find the answer to this question in the company's **balance sheet statement**. A company's balance sheet, sometimes called its **statement of financial position**, reports three main categories of items: assets, liabilities, and stockholders' equity. This statement is based on the most fundamental tool of accounting: the accounting equation.

Accounting Equation

The **accounting equation** shows the relationship among assets, liabilities, and owners' equity, as in Figure 2.4. Every business transaction, no matter how simple or complex, can be expressed in terms of its effect on the accounting equation. Regardless of whether a business grows or contracts, the equality between the assets and the claims against the assets is always maintained. In other words, any change in the amount of total assets is necessarily accompanied by an equal change on the other side of the equation—that is, by an increase or decrease in either the liabilities or the owners' equity.

The general format used in reporting a typical balance sheet statement is shown in Figure 2.5. The dollar amount shown in the assets portion of the balance sheet represents how much the company owns at the time it issues the report. Assets are arranged in order of liquidity. Current assets are so critical that they are separately broken out and totaled. They are what will hold the business afloat for the next year. The claims against assets are of two types: liabilities and stockholders' equity. The liabilities of a company indicate where the company obtained the funds to acquire its assets and to operate the business.

Figure 2.5 Using the four quadrants of the balance sheet.

Liability is money the company owes. Stockholders' equity is that portion of the assets of a company that is provided by the investors (owners). Therefore, stockholders' equity is the liability of a company to its owners.

Assets

The asset items are listed in the order of their "liquidity" or the length of time it takes to convert them to cash. The most liquid assets appear at the top of the list; the least liquid assets appear at the bottom. Because cash is the most liquid of all assets, it is always listed first.

- **Current assets** can be converted to cash or its equivalent in less than one year. Current assets generally include three major accounts.
  1. The first is *cash*. A firm typically has a cash account at a bank to provide for the funds needed to conduct day-to-day business. Although assets are always stated in terms of dollars, only cash represents actual money. Cash-equivalent items are also listed and include marketable securities and short-term investments.
  2. The second account is *accounts receivable*—money that is owed the firm but that has not yet been received. For example, when a company receives an order from a smartphone manufacturer, the company will send an invoice along with the shipment to the manufacturer. Then the unpaid bill immediately falls into the accounts receivable category. When the bill is paid, it will be deducted from the accounts receivable account and placed into the cash category. A typical firm will have a 30- to 45-day accounts receivable, depending on the frequency of its bills and the payment terms for customers.
  3. The third account is *inventories*, which show the dollar amount that a company has invested in raw materials, work in process, and finished goods available for sale.
- **Long-term (fixed) assets** are relatively permanent and take time to convert into cash. Fixed assets reflect the amount of money a company paid for its plant and equipment when it acquired those assets. The most common fixed asset is the physical investment in the business, such as land, buildings,[3] factory machinery, office equipment, and automobiles. With the exception of land, most fixed assets have a limited useful life. For example, buildings and equipment are used up over a period of years. Each year, a portion of the usefulness of these assets expires, and a portion of their total cost should be recognized as a depreciation expense. The term *depreciation* denotes the accounting process for this gradual conversion of fixed assets into expenses. *Property, plant and equipment*, and *net* thus represent the current book value of these assets after deducting depreciation expenses.
- Finally, **other assets** include investments made in other companies and intangible assets such as goodwill, copyrights, franchises, and so forth. Goodwill appears on the balance sheet only when operating business is purchased in its entirety. Goodwill indicates any additional amount paid for the business above the fair market value of the business. (Here, the fair market value is defined as the price that a buyer is willing to pay when the business is offered for sale.)

---

[3]Land and buildings are commonly called **real assets** to distinguish them from equipment and machinery.

Liabilities

Liabilities are arranged in order of payment, the most pressing at the top of the list, the least pressing at the bottom. Like current assets, current liabilities are so critical that they are separately broken out and totaled. They are what will be paid out during the next year.

- **Current liabilities** are those debts that must be paid in the near future (normally, within one year). The major current liabilities include *accounts and notes payable* within a year. Also included are accrued expenses (wages, salaries, interest, rent, taxes, etc., which are owed but not yet due for payment) and advance payments and deposits from customers.
- **Long-term and other liabilities** include *long-term liabilities*, such as bonds, mortgages, and long-term notes that are due and payable more than one year in the future.

Stockholders' Equity (Owners' Net Worth)

Stockholders' equity represents the amount that is available to the owners after all other debts have been paid. Generally, stockholders' equity consists of preferred and common stock, treasury stock, capital surplus, and retained earnings. Preferred stock is a hybrid between common stock and debt. In case the company goes bankrupt, it must pay its preferred stockholders after its debtors, but before its common stockholders. Preferred dividend is fixed, so preferred stockholders do not benefit if the company's earnings grow. In fact, many firms do not use preferred stock. The common stockholders' equity is a residual.

- **Common (capital) stock** is the aggregate par value of the company's stock issued. Companies rarely issue stocks at a discount (i.e., at an amount below the stated par). Normally, corporations set the par value sufficiently low enough so that, in practice, stock is usually sold at a premium.
- **Paid-in capital** (capital surplus) is the amount of money received from the sale of stock that is over and above the par value of the stock. Outstanding stock is the number of shares issued that actually are held by the public.
- **Treasury stock** is often created when shares of a company are initially issued. In this case, not all shares are issued to the public, as some are kept in the company's treasury to be used to create extra cash should it be needed. If the corporation buys back part of its own issued stock on the open market, that stock is also listed as *treasury stock* on the balance sheet.
- **Retained earnings** represent the cumulative net income of the firm since its inception, less the total dividends that have been paid to stockholders. In other words, retained earnings indicate the amount of assets that have been financed by plowing profits back into the business. Therefore, retained earnings belong to the stockholders.

What to Read from BRCM's Balance Sheet Statement

As shown in Table 2.1, the first half of BRCM's year-end 2013 and 2012 balance sheets lists the firm's assets, while the remainder shows the liabilities and equity or claims against those assets. Note that all figures are in millions.

- BRCM had total assets of $3,915 and obtained the bulk of the funds it used to buy these assets:
  1. by buying on credit from its suppliers (accounts payable);
  2. by borrowing from financial institutions (notes payable and long-term bonds);

TABLE 2.1 Consolidated Statements of Financial Position for BRCM, Ltd.

Period End Date	2013 12/31/2013	2012 12/31/2012	Changes in Account
Assets			
Cash and Cash Equivalent	$ 1,657.0	$ 1,617.0	+40
Short-term Marketable Securities	775.0	757.0	+18
Account Receivables	795.0	740.0	+55
Inventories	525.0	527.0	−2
Prepaid Assets and Others	163.0	140.0	+23
Total Current Assets	**$ 3,915.0**	**$ 3,781.0**	**+134**
Property, Plant, and Equipment, net	593.0	485.0	+108
Goodwill and Other Intangible Assets	4,937.0	5,512.0	−575
Long-term Investments	1,939.0	1,348.0	+591
Other Non-Current Assets	111.0	82.0	+29
Total Non-Current Assets	**$ 7,580.0**	**$ 7,427.0**	**+153**
Total Assets	**$11,495.0**	**$11,208.0**	**+287**
Liabilities and Shareholders' Equity			
Account Payables	1,475.0	1,360.0	+115
Current Portion of Long-term Debt	0.0	300.0	−300
Deferred Revenue	21.0	22.0	−1
Total Current Liabilities	**$ 1,496.0**	**$ 1,682.0**	**−186**
Long-term Debt	1,394.0	1,393.0	+1
Other Long-Term Liabilities	234.0	294.0	−60
Total Non-Current Liabilities	**$ 1,628.0**	**$ 1,687.0**	**−59**
Total Liabilities	**$ 3,124.0**	**$ 3,369.0**	**−245**
Common (Capital) Stock	0.0	0.0	0
Retained Earnings	−4,107.0	−4,531.0	+424
Additional Paid in Capital	12,475.0	12,403.0	+72
Other Equity	3.0	−33.0	+36
Total Equity	**$ 8,371.0**	**$ 7,839.0**	**+532**
Total Liabilities and Equity	**$11,495.0**	**$11,208.0**	**+287**

Source: Data from BRMC 2013 Annual Report

3. by issuing common stock to investors;

4. by plowing earnings into the business, as reflected in the retained earnings entry.

- BRCM had a very strong cash position in the amount of $1,657, which is sufficient amount to cover all current liabilities if they are due on the same date.
- BRCM had the total current assets of $3,915 and the total current liabilities of $1,496. The difference is known as the **working capital**.

$$\text{Working capital} = \text{Current assets} - \text{Current liabilities}$$

The working capital indicates whether a company has enough short-term assets to cover its short-term debt.

- BRCM had a total long-term debt of $1,394 at the end of the fiscal year, consisting of several bonds issued in previous years. According to BRCM's 2013 Annual Report, BRCM repaid its outstanding long-term debt of $300 in 2013 and interest expense on long-term debt for the year was $30.
- BRCM issued about 581 million shares of common stock with a par value of $0.0001, so the book value of common stock is about $58,100, or $0.0058 million. Since it is such small number, it is shown as "−" in the balance sheet.
- BRCM had 581 shares of common stock outstanding. Investors actually provided the company with a total capital of $11,495. However, BRCM has retained the current as well as previous earnings of $4,107 since it was incorporated. BRCM also held $12,475 worth of paid-in-capital, which was raised through the current as well as previous stock offerings. The company did not issue any preferred stock. The combined net stockholders' equity was $8,371, and these equities belong to BRCM's common stockholders.

$$\text{Assets} - \text{Liabilities} - \text{Preferred stock} = \text{Common stockholders' equity}$$
$$\$11,495 - \$3,124 - \$0 = \$8,371$$

- On the average, stockholders have a total investment of $14.41 per share ($8,371/581 shares) in the company. The $14.41 figure is known as the stock's *book value*. In December of 2013, the stock traded in the general range from $26 to $29 per share. Note that this market price is quite different from the stock's book value. Many factors affect the market price, the most important one being how investors expect the company to perform in the future. Certainly, the company's innovative chip products have had a major influence on the market value of its stock.
- Repurchasing its own stock is equivalent to investing the firm's idle cash from operation in the stock market. BRCM could have bought another company's stock, such as IBM or Microsoft stock, with the money, but if BRCM bought their own stock, it means that BRCM liked its own stock better than any other stocks on the market.
- BRCM had negative retained earnings for two years in a row. When a company records a loss, this too is recorded in retained earnings. If the amount of the loss exceeds the amount of profit previously recorded in the retained earnings account, then a company is said to have *negative retained earnings*.

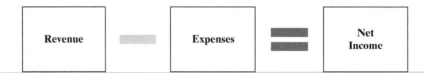

Figure 2.6 Net income-accounting profit.

2.2.2 The Income Statement

The second financial report is the **income statement**, which indicates whether the company is making or losing money during a stated *period*, usually a year. Most businesses prepare quarterly and monthly income statements as well. The company's accounting period refers to the period covered by an income statement.

Reporting Format

Just like the accounting equation, Figure 2.6 shows a simple net income equation. Typical items that are itemized in the income statement are as follows:

- **Revenue** is the income from goods sold and services rendered during a given accounting period.
- **Net revenue** represents gross sales, less any sales return and allowances.
- Shown on the next several lines are the expenses and costs of doing business as deductions from revenue. The largest expense for a typical manufacturing firm is the expense it incurs in making a product (such as labor, materials, and overhead), called the **cost of revenue** (Cost of sales or cost of goods sold).
- Net revenue less the cost of revenue gives the **gross margin**.
- Next, we subtract any other operating expenses from the gross margin. These other operating expenses are expenses associated with paying interest, leasing machinery or equipment, sales, marketing, and administration. This results in the operating income, or **operating margin**.
- Finally, we determine the **net income** (or net profit) by subtracting the income taxes from the taxable income. Net income is also commonly known as *accounting income*, or **net margin**.

Earnings per Share

Another important piece of financial information provided in the income statement is the **earnings per share** (EPS).[4] In simple situations, we compute the EPS by dividing the available earnings to common stockholders by the number of shares of common stock outstanding. Stockholders and potential investors want to know what their share of profits is, not just the total dollar amount. The presentation of profits on a per share basis allows the stockholders to relate earnings to what they paid for a share of stock. Naturally, companies want to report a higher EPS to their investors as a means of summarizing how well they managed their businesses for the benefits of the owners.

---

[4]In reporting EPS, the firm is required to distinguish between "basic EPS" and "diluted EPS." Basic EPS is the net income of the firm divided by the number of shares of common stock outstanding. By contrast, the diluted EPS includes all common stock equivalents (convertible bonds, preferred stock, warrants, and rights) along with common stock. Therefore, diluted EPS will usually be less than basic EPS.

Retained Earnings

As a supplement to the income statement, many corporations also report their retained earnings during the accounting period. When a corporation makes some profits, it has to decide what to do with those profits. The corporation may decide to pay out some of the profits as dividends to its stockholders. Alternatively, it may retain the remaining profits in the business in order to finance expansion or support other business activities.

When the corporation declares dividends, preferred stock has priority over common stock. Preferred stock pays a stated dividend, much like the interest payment on bonds. The dividend is not a legal liability until the board of directors has declared it. However, many corporations view the dividend payments to preferred stockholders as a liability. Therefore, "available for common stockholders" reflects the net earnings of the corporation, less the preferred stock dividends. When preferred and common stock dividends are subtracted from net income, the remainder is retained earnings (profits) for the year. As mentioned previously, these retained earnings are reinvested into the business.

What to Read from the BRCM Income Statement

For BRCM, the accounting period begins on January 1 and ends on December 31 of the following year. Table 2.2 gives the 2013 and 2012 income statements for BRCM. Clearly, it was not a good year as it had to report the declining net income even though it was still a profitable year.

BRCM's net revenue was $8,305 in 2013, compared with $8,006 in 2012, a gain of 3.74%. However, profits from operations (operating income) decreased 30.18% to $472, and net income was down 41.02% to $424.

BRCM issued no preferred stock, so there is no required cash dividend. Therefore, the entire net income of $424 could be available as cash dividends to the common stockholders. Normally, BRCM could pay it out as dividends from these earnings to the common stockholders. Instead, BRCM retained this income fully to reduce the accrued losses and decided to pay out cash dividends of $0.44 per share ($254) from their cash reserve. As shown in Table 2.1, the beginning balance of the retained earnings was −$4,531. Therefore, the total loss of retained earnings reduced to −$4,107.

Earnings per common share dropped at a faster pace in 2013, to $0.74, a decrease of 42.63% over 2012 (Table 2.2), but it also paid a larger amount of cash dividends ($0.44 per share in 2013, and $0.40 per share in 2012). Probably, BRCM's management thought their worst years were behind and needed to make the investors happy by paying cash dividends out of their current cash reserve.

2.2.3 The Cash Flow Statement

The income statement explained in the previous section indicates only whether the company was making or losing money during the reporting period. Therefore, the emphasis was on determining the net income (profits) of the firm for supporting its operating activities. However, the income statement ignores two other important business activities for the period: financing and investing activities. Therefore, we need another financial statement—the cash flow statement, which details how the company generated the cash it received and how the company used that cash during the reporting period.

TABLE 2.2 Consolidated Statements of Income (in millions, except per share amounts) for BRCM, Ltd.

	2013	2012
Period End Date	12/31/2013	12/31/2012
Total Revenue	**$ 8,305.0**	**$ 8,006.0**
Cost of Revenue	4,088.0	4,027.0
Gross Profit	**$ 4,217.0**	**$ 3,979.0**
Selling, General and Administrative	706.0	696.0
Research and Development	2,486.0	2,318.0
Depreciation, Amortization and Depletion	57.0	113.0
Special Income/Charges	496.0	176.0
Operating Expenses	**$ 3,745.0**	**$ 3,303.0**
Operating Income	**$ 472.0**	**$ 676.0**
Net Interest Income	−30.0	−30.0
Other Income/Expense, Net	3.0	10.0
Income before Income Taxes	**$ 445.0**	**$ 656.0**
Provision for Income Tax	21.0	−63.0
Net Income	**$ 424.0**	**$ 719.0**
Dividend Per Share	0.44	0.40
Tax Rate	0.047191	0.096037
Basic earnings per share	$ 0.74	$ 1.29
Diluted earnings per share	$ 0.73	$ 1.25
Ordinary shares outstanding	581	569

Source: Data from BRMC 2013 Annual Report

Sources and Uses of Cash

The difference between the sources (inflows) and uses (outflows) of cash represents the net cash flow during the reporting period. This is a very important piece of information, because investors determine the value of an asset (or, indeed, of a whole firm) by the cash flows it generates. Certainly, a firm's net income is important, but cash flows are even more important, particularly because the company needs cash to pay dividends and to purchase the assets required to continue its operations. As mentioned in the previous section, the goal of the firm should be to maximize the market value of the firm (or stock price). Since the value of any asset depends on the cash flows produced by the asset, managers want to maximize the cash flows available to investors over the long run. Therefore, we should make investment decisions on the basis of cash flows rather than profits. For such investment decisions, it is necessary to convert profits (as determined in the income statement) to cash flows. Here are the guidelines for distinguishing between a source and a use of cash:

- A company generates cash by reducing an asset or by increasing a liability.
- A company also uses cash to increase an asset account or to reduce a liability account.
- The net cash flow is the difference between the sources and uses of cash.

Reporting Format

In preparing the cash flow statement, many companies identify the sources and uses of cash according to the types of business activities. Basically a cash flow statement simply expands and rearranges the sources and uses statement, placing each source or use into one of three broad categories: (1) cash flows from operating activities, (2) cash flows from investing activities, and (3) cash flows from financing activities.

1. **Cash Flow from Operating Activities.** We start with the net change in operating cash flows from the income statement. Here, operating cash flows represent those cash flows related to production and the sales of goods or services. First, we add all noncash charges, such as depreciation, amortization, and deferred charges, back to net income. Although we may charge these items against current income as expenses, they do not involve an actual cash outflow. Then we add the changes in current assets and liabilities to net income, recognizing that some sales did not increase cash because customers had not yet paid. Also changes in other current assets and liabilities, such as inventories and prepaid expenses, appear here because the accountant, following the matching principle, ignored these cash flows when calculating net income for that operating period. We can lump together all these changes in current assets and current liabilities under the heading of "Changes in **working capital**.[5]

$$\text{Cash flow from operating activities} = \text{Net income} + \text{Noncash items} \\ + \text{Changes in working capital}.$$

2. **Cash Flow from Investing Activities.** Once we determine the operating cash flows, we consider any cash flow transactions related to investment activities, which include purchasing new fixed assets (cash outflow), reselling old equipment (cash inflow), and buying and selling financial assets.

3. **Cash Flow from Financing Activities.** Finally, we detail cash transactions related to financing any capital used in business. For example, the company could borrow more money or sell more stock, resulting in cash inflows. Paying off existing debt will result in cash outflows.

By summarizing cash inflows and outflows from three activities for a given accounting period, we obtain the net change in the cash flow position of the company.

What to Read from BRCM's Cash Flow Statement

Table 2.3 is BRCM's statement of cash flows, as it would appear in the company's annual report. Note that all figures are in millions.

The cash flow categories and their values for BRCM in 2013 are as follows:

1. Cash flows from operating activities: $1,785 (source).
2. Cash flows from investing activities: −$996 (use).
3. Cash flows from financing activities: −$749 (use).

And the net cash generated during the reporting period is $40. BCRM's operating activities provided $1,785 in cash. This was primarily the result of net income of $424 and

---

[5]The difference between the change in current assets (excluding the cash and cash equivalents) and the change in current liabilities is the **net change in net working capital**. If this change is positive, then additional financing is needed to fund the increase in current assets. This will further reduce the cash flow from the operating activities.

TABLE 2.3 Consolidated Statements of Cash Flows (in millions) for BRCM, Ltd.

	2013	2012
Period End Date	12/31/2013	12/31/2012
Operating Activities:		
Net Income	$ 424.0	$ 719.0
Depreciation and Amortization	401.0	445.0
Other Non-cash Items	$ 1,027.0	$ 615.0
Changes in Working Capital	−67.0	152.0
Cash Flow from Operating Activities	**$ 1,785.0**	**$ 1,931.0**
Investing Activities:		
Purchase/Sale of Property, Plant, Equip: Net	−228.0	−244.0
Purchase/Sale of Business, Net	−142.0	−3,582.0
Purchase/Sale of Investments, Net	−626.0	−970.0
Cash Flow from Investing Activities	**$ −996.0**	**$ −4,796.0**
Financing Activities:		
Issuance/Payments of Debt, Net	−300.0	492.0
Issuance/Payments of Common Stock, Net	−65.0	278.0
Cash Dividends Paid	−254.0	−224.0
Other Financing Changes, Net	−130.0	−210.0
Cash Flow from Financing Activities	**$ −749.0**	**$ 336.0**
Cash, Equivalents, Start of Period	1,617.0	4,146.0
Cash, Equivalents, End of Period	1,657.0	1,617.0
Change in Cash	**$ 40.0**	**$ −2,529.0**

Source: Data from BRMC 2013 Annual Report

net non-cash operating expenses of $1,428, offset in part by changes in current assets and liabilities of $67. Clearly, the net cash generated from the operating activities ($1,785) is quite different from the net income earned ($424) for the same period. Why is the cash position so different from the net income earned? The main reason for the difference lies in the accrual-basis accounting principle used by the BRCM Corporation. In **accrual-basis accounting**, an accountant recognizes the impact of a business event as it occurs. When the business performs a service, makes a sale, or incurs an expense, the accountant enters the transaction into the books, regardless of whether cash has or has not been received or paid. For example, an increase in account receivables of $795 − $740 = $55 during 2013 represents the amount of total sales on credit (Table 2.1). Since the $55 figure was included in the total sales in determining the net income, we need to subtract it to determine the company's true cash position. After making similar adjustments to all other current assets and current liabilities, BRCM reported the net changes under the name of "Changes in Working Capital" in the amount of $67. Factoring in non-cash outlay items, the net cash provided from operating activities is $1,785.

- With regards to investment activities, BRCM used $996 in cash in 2013, which was primarily the result of $611 in net purchases of marketable securities, $228 of capital equipment purchases mostly to support the company's research and development efforts, $142 in net cash paid for the business acquisition, and $15 of purchases of strategic investments which is listed under "Purchase/Sale of Investment."
- Financing activities used $749 in cash; $65 of which was primarily the result of $597 in repurchases of the firm's Class A common stock, offset in part by $532 in proceeds received from issuances of common stock, the repayment of long-term debt of $300, dividend paid of $254, and $130 in other business transactions, Normally most firms would pay cash dividends out of earnings but BRCM paid out of the proceeds from stock sales.
- Together, the three types of activities generated a total cash inflow of $40. With the initial cash balance of $1,617, the ending cash balance thus increased to $1,785. This same amount denotes the change in BRCM's cash position, as shown in the cash accounts in the balance sheet in Table 2.1.

EXAMPLE 2.1 Estimating Cash Flows from Operations

A comparative balance sheet for J&M, Inc. containing data for the two reporting periods is given in Table 2.4, and the reported income statement during 2015 is also shown in Table 2.5. All figures (except the tax rate) are in thousands. The following additional information is available about the company's activities during 2015, the current year.

- Cash dividends declared and paid to the preferred stockholders totaled $600.
- Cash dividends paid to the common shareholders totaled $9,400.
- J&M purchased $9,000 worth of equipment.
- J&M paid off $2,000 long-term debt.
- The number of outstanding preferred and common stockholders was 100 and 10,000 respectively.
- The effective tax rate is 28.77%.

TABLE 2.4 Comparative Balance Sheet for J&M, Inc.

J&M, Inc.			
Comparative Balance Sheet Statement			
December 31, 2015, and 2014			
	2015	**2014**	**Changes**
Assets			
Current assets:			
Cash and cash equivalents	$ 8,500	$ 6,100	2,400
Short-term marketable securities	3,000	5,000	(2,000)

TABLE 2.4 (*continued*)

J&M, Inc. Comparative Balance Sheet Statement December 31, 2015, and 2014			
	2015	**2014**	**Changes**
Accounts receivable	23,700	19,500	4,200
Inventory	37,700	39,800	(2,100)
Prepaid expenses	2,000	1,500	500
Deferred charges	2,500	3,000	(500)
Total current assets	**77,400**	**74,900**	2,500
Long-term Assets:			–
Property and equipment	154,000	145,000	9,000
Less accumulated depreciation	(70,000)	(50,000)	(20,000)
Total assets	**161,400**	**169,900**	(8,500)
Liabilities and Shareholders' Equity			
Current liabilities:			
Accounts payable	10,000	26,000	(16,000)
Wages payable	16,000	15,000	1,000
Accrued taxes	2,000	3,500	(1,500)
Total current liabilities	**28,000**	**44,500**	(16,500)
Other liabilities			
Long-term debt	30,000	32,000	(2,000)
Total liabilities	**58,000**	**76,500**	(18,500)
Shareholders' equity			
Preferred stock	10,000	10,000	
Common stock	40,000	40,000	
Additional paid-in capital	11,000	11,000	
Retained earnings	42,400	32,400	10,000
Total owners' equity	**103,400**	**93,400**	10,000
Total liabilities and equity	**161,400**	**169,900**	(8,500)

TABLE 2.5   The Income Statement for J&M, Inc.

J&M, Inc.	
The Income Statement	
(For Year Ended December 31, 2015)	
Total revenue	$ 300,000
Cost of revenue	188,000
Gross profit	112,000
Selling, general and administrative	44,720
Depreciation	20,000
Lease payment	14,000
Operating income	33,280
Interest expense, net	(5,200)
Income before income taxes	28,080
Provision for income taxes	(8,080)
Net income	$ 20,000

Based on the information provided, estimate the cash flows from three activities in 2015.

SOLUTION

Given: Financial data as given in Tables 2.4 and 2.5

Find: (a) cash flows from operating activities, (b) cash flows from investing activities, and (c) cash flows from financing activities.

(a) Cash flows from operating activities: Note that we can estimate the cash flows from operating activities with the following equation:

$$\text{Cash flow from operating activities} = \text{Net income} + \text{Noncash items}$$
$$(\text{Depreciation}) + \text{Changes in working capital}.$$

	Source/Use		Cash Flow
Net income			$ 20,000
Adjustments to net income			
Depreciation			20,000
Changes in current assets and liabilities:			
Decrease in short-term marketable securities	Source	2,000	
Increase in accounts receivables	Use	(4,200)	
Decrease in inventory	Source	2,100	

(continued)

	Source/Use		Cash Flow
increase in prepaid expenses	Use	(500)	
Decrease in deferred charges	Source	500	
Net changes in current assets	**Use**		(100)
Decrease in accounts payable	Use	(16,000)	
Increase in wages payable	Source	1,000	
Decrease in accrued taxes	Use	(1,500)	
Net changes in current liabilities	**Use**		(16,500)
Changes in working capital			(16,600)
Cash flows from operating activities			**$ 23,400**

(b) Cash flows from investing activities:

$$\text{Purchased equipment: } \$9,000 \text{ (use)}$$

(c) Cash flows from financing activities:

$$\text{Reduction in long-term debt: } \$2,000 \text{ (use)}$$
$$\text{Cash dividends: } \$10,000 \text{ (use)}$$

Net cash flows in 2015 = $23,400 − $9,000 − $2,000 − $10,000 = **$2,400**

COMMENTS: This cash amount is exactly the same as the increase of $2,400 in the cash account reported in the J&M's balance sheet.

2.3 Using Ratios to Make Business Decisions

As we have seen in BRCM's financial statements, the purpose of accounting is to provide information for decision-making. Accounting information tells what happened at a particular point in time. In that sense, financial statements are essentially historical documents. However, most users of financial statements are concerned about what will happen in the future. For example:

- Stockholders are concerned with future earnings and dividends.
- Creditors are concerned with the company's ability to repay its debts.
- Managers are concerned with the company's ability to finance future expansion.
- Engineers are concerned with planning actions that will influence the future course of business events.

Although financial statements are historical documents, they can still provide valuable information bearing on all of these concerns. An important part of financial analysis is the calculation and interpretation of various financial ratios. In this section, we consider some of the ratios that analysts typically use in attempting to predict the future course of events in business organizations. We may group these ratios into five categories (debt management, liquidity, asset management, profitability, and market trend) as outlined in Figure 2.7. In all financial ratio calculations, we will use the 2013 financial statements for BRCM.

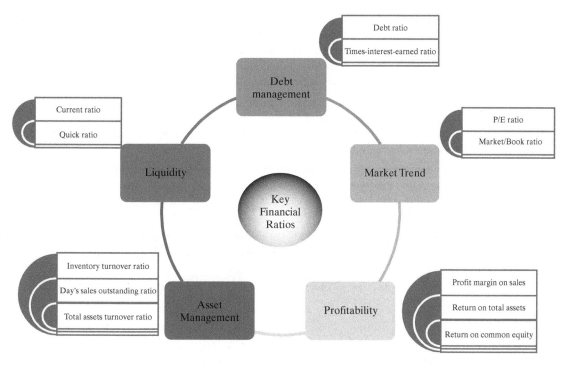

Figure 2.7 Types of financial ratios used in evaluating a firm's financial health.

2.3.1 Debt Management Analysis

All businesses need assets to operate. To acquire assets, the firm must raise capital. When the firm finances its long-term needs externally, it may obtain funds from the capital markets. Capital comes in two forms: **debt** and **equity**. Debt capital is capital borrowed from financial institutions. Equity capital is capital obtained from the owners of the company.

The basic methods of financing a company's debt are through bank loans and the sale of bonds. For example, suppose a firm needs $10,000 to purchase a computer. In this situation, the firm would borrow money from a bank and repay the loan, together with the interest specified, in a few years. This kind of financing is known as *short-term debt financing*. Now suppose that the firm needs $100 million for a new product development project. Normally, it would be very expensive (or require a substantial mortgage) to borrow the money directly from a bank. In this situation, the company would go public to borrow money on a long-term basis. When investors lend capital to a company and the company consents to repay the loan at an agreed-upon interest rate, the investor is the creditor of the corporation. The document that records the nature of the arrangement between the issuing company and the investor is called a **bond**. Raising capital by issuing a bond is called *long-term debt financing*.

Similarly, there are different types of equity capital. For example, the equity of a proprietorship represents the money provided by the owner. For a corporation, equity capital comes in two forms: *preferred stock* and *common stock*. Investors provide capital to a corporation, and the company agrees to endow the investor with fractional ownership in the corporation. Preferred stock pays a stated *dividend*, much like the interest payment on bonds. However, the dividend is not a legal liability until the company declares it. Preferred stockholders have preference over common stockholders as regards the receipt of dividends if the company has to liquidate its assets. We can examine the extent to which a company uses debt financing (or financial leverage) in the operation of its business if we:

- Check the balance sheet to determine the extent to which borrowed funds have been used to finance assets; and
- Review the income statement to see the extent to which fixed charges (interests) are covered by operating profits.

Two essential indicators of a business's ability to pay its long-term liabilities are the *debt ratio* and the *times-interest-earned ratio*, as shown in Figure 2.8.

Debt Ratio

The relationship between total liabilities and total assets, generally called the **debt ratio**, tells us the proportion of the company's assets that it has financed with debt:

$$\text{Debt ratio} = \frac{\text{Total debt}}{\text{Total assets}}$$

$$= \frac{\$3,124}{\$11,495} = 27.17\%$$

Total debt includes both current liabilities and long-term debt. If the debt ratio is unity, then the company has used debt to finance all of its assets. As of December 31, 2013, BRCM's debt ratio was 27.17%; this means that its creditors have supplied close to 28% of the firm's total financing. Certainly, most creditors prefer low debt ratios, because the lower the ratio, the greater is the cushion against creditors' losses in case of liquidation. If a company seeking financing already has large liabilities, then additional debt payments may be too much for the business to handle. For such a highly leveraged company, creditors generally charge higher interest rates on new borrowing to help protect themselves.

Times-Interest-Earned Ratio

The most common measure of the ability of a company's operations to provide protection to the long-term creditor is the times-interest-earned ratio. We find this ratio by dividing earnings before interest and income taxes (EBIT) by the yearly interest charges that must

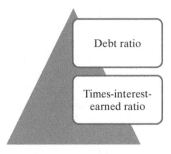

Debt Management

Ratios that show how a firm uses debt financing and its ability to meet debt repayment obligations

Debt ratio

Times-interest-earned ratio

Figure 2.8   Debt management ratios.

be met. BRCM issued $1,394 worth of senior notes and long-term bonds. This results in $30 in interest expenses[6] in 2013:

$$\text{Times-interest-earned ratio} = \frac{\text{EBIT}}{\text{Interest expense}}$$

$$= \frac{\$445 + \$30}{\$30} = 15.83 \text{ times}$$

The times-interest-earned ratio measures the extent to which operating income can decline before the firm is unable to meet its annual interest costs. Failure to meet this obligation can bring legal action by the firm's creditors, possibly resulting in the company's bankruptcy. Note that we use the earnings before interest and income taxes, rather than net income, in the numerator. Because BRCM must pay interest with pretax dollars, BRCM's ability to pay current interest is not affected by income taxes. Only those earnings remaining after all interest charges are subject to income taxes. For BRCM, the times-interest-earned ratio for 2013 would be 15.83 times. This ratio is relatively low compared with the rest of the industry's 54.91 times during the same operating period.

2.3.2 Liquidity Analysis

If you were one of the many suppliers to BRCM, your primary concern would be whether BRCM will be able to pay off its debts as they come due over the next year or so. Short-term creditors want to be repaid on time; therefore they focus on BRCM's cash flows and on its working capital, as these are the company's primary sources of cash in the near future. The excess of current assets over current liabilities is known as **working capital**, a figure that indicates the extent to which current assets can be converted to cash to meet current obligations. Therefore, we view a firm's net working capital as a measure of its *liquidity* position. In general, the larger the working capital, the better able the business is to pay its debt. As shown in Figure 2.9, two ratios are related to liquidity analysis.

Liquidity Analysis

Ratios that show the relationship of a firm's cash and other assets to its current liabilities

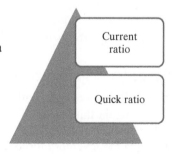

Figure 2.9   Ratios related to liquidity analysis.

<hr />

[6]Unless the interest expenses are itemized in the income statement, you will find them in the firm's annual report.

Current Ratio

We calculate the **current ratio** by dividing current assets by current liabilities:

$$\text{Current ratio} = \frac{\text{Current assets}}{\text{Current liabilities}}$$

$$= \frac{\$3,915}{\$1,496} = 2.62 \text{ times}$$

If a company is getting into financial difficulty, it begins paying its bills (accounts payable) more slowly, borrowing from its bank, and so on. If current liabilities are rising faster than current assets, the current ratio will fall, and that could spell trouble. What is an acceptable current ratio? The answer depends on the nature of the industry. The general rule of thumb calls for a current ratio of 2 to 1. This rule, of course, is subject to many exceptions, depending heavily on the composition of the assets involved.

Quick (Acid-Test) Ratio

The quick ratio tells us whether a company could pay all of its current liabilities if they came due immediately. We calculate the quick ratio by deducting inventories from current assets and then dividing the remainder by current liabilities:

$$\text{Quick ratio} = \frac{\text{Current assets} - \text{Inventories}}{\text{Current liabilities}}$$

$$= \frac{\$3,915 - \$525}{\$1,496} = 2.27 \text{ times}$$

The quick ratio measures how well a company can meet its obligations without having to liquidate or depend too heavily on its inventory. Inventories are typically the least liquid of a firm's current assets; hence, they are the assets on which losses are most likely to occur in the case of liquidation. These ratios indicate that BRCM's liquidity position is relatively strong, as it has carried very little inventory in its current assets (only $525 out of $3,915 of current assets, or 13.41%). We often compare against industry average figures and should note at this point that an industry average is not an absolute number that all firms should strive to maintain. In fact, some very well managed firms will be above the average, while other good firms will be below it. However, if we find that a firm's ratios are quite different from the average for its industry, we should examine the reason for the difference.

2.3.3 Asset Management Analysis

The ability to sell inventory and collect accounts receivables is fundamental to business success. Therefore, the third group of ratios measures how effectively the firm is managing its assets. As shown in Figure 2.10, we will review three ratios related to a firm's asset management: (1) the inventory turnover ratio, (2) the day's sales outstanding ratio, and (3) the total asset turnover ratio. The purpose of these ratios is to answer this question: Does the total amount of each type of asset, as reported on the balance sheet, seem reasonable in view of current and projected sales levels? The acquisition of any asset requires the use of funds. On the one hand, if a firm has too many assets, its cost of capital will be too high; hence, its profits will be depressed. On the other hand, if assets are too low, the firm is likely to lose profitable sales.

Asset Management

A set of ratios that measure how effectively a firm is managing its assets

Figure 2.10 Ratios related to asset management.

Inventory Turnover

The inventory turnover ratio measures how many times the company sold and replaced its inventory over a specific period—for example, during the year. We compute the ratio by dividing sales by the average level of inventories on hand. We compute the average inventory figure by taking the average of the beginning and ending inventory figures. Since BRCM has a beginning inventory figure of $527 and an ending inventory figure of $525, its average inventory for the year would be $526, or ($527 + $525)/2. We then compute BRCM's inventory turnover for 2013 as

$$\text{Inventory turnover ratio} = \frac{\text{Sales}}{\text{Average inventory balance}}$$

$$= \frac{\$8,305}{\$526} = 15.79 \text{ times}$$

As a rough approximation, BRCM was able to sell and restock its inventory 15.79 times per year. BRCM's turnover of 15.79 times is much faster than its competitor QCOM (Qualcomm, Inc.) at 7.91 times during the same reporting period. This suggests that QCOM is holding excess stocks of inventory; excess stocks are, of course, unproductive, and they represent an investment with a low or zero rate of return.

Day's Sales Outstanding (Accounts Receivable Turnover)

The day's sales outstanding (DSO) is a rough measure of how many times a company's accounts (or trade) receivable have been turned into cash during the year. We determine this ratio, also called the **average collection period**, by dividing accounts receivable by average sales per day. In other words, the DSO indicates the average length of time the firm must wait after making a sale before receiving cash. For BRCM,

$$\text{DSO} = \frac{\text{Receivables}}{\text{Average sales per day}} = \frac{\text{Receivables}}{\text{Annual sales}/365}$$

$$= \frac{\$795}{\$8,305/365} = 34.94 \text{ days}$$

Thus, on average, it takes BRCM 34.94 days to collect on a credit sale. During the same period, QCOM's average collection period was 12.52 days. Whether the average of 34.94 days taken to collect an account is good or bad depends on the credit terms BRCM is offering its customers. If the credit terms are 30 days, we can say that BRCM's customers, on average, are not paying their bills on time. In order to improve their working-capital position, most customers tend to withhold payment for as long as the credit terms will allow and may even go over a few days. The long collection period may signal either that customers are in financial trouble or that the company manages its credit poorly.

Total Assets Turnover

The total assets turnover ratio measures how effectively the firm uses its total assets in generating its revenues. It is the ratio of sales to all the firm's assets:

$$\text{Total assets turnover ratio} = \frac{\text{Sales}}{\text{Total assets}}$$

$$= \frac{\$8,305}{\$11,495} = 0.7225 \text{ times}$$

BRCM's ratio of 0.7225 times, compared with QCOM's 0.54, is almost 34% faster, indicating that BRCM is using its total assets about 34% more intensively than QCOM is. In fact, BRCM's total investment in plant and equipment is about 20% of QCOM's. If we view BRCM's ratio as the industry average, we can say that QCOM has too much investment in inventory, plant, and equipment compared to the size of sale.

2.3.4 Profitability Analysis

One of the most important goals for any business is to earn a profit. The ratios examined thus far provide useful clues about the effectiveness of a firm's operations, but the profitability ratios show the combined effects of liquidity, asset management, and debt on operating results. Therefore, ratios that measure profitability play a large role in decision-making. As shown in Figure 2.11, three ratios are related to profitability analysis.

Profit Margin on Sales

We calculate the profit margin on sales by dividing net income by sales. This ratio indicates the profit per dollar of sales:

$$\text{Profit margin on sales} = \frac{\text{Net income available to common stockholders}}{\text{Sales}}$$

$$= \frac{\$424}{\$8,305} = 5.11\%$$

Profitability Analysis

A set of ratios that show the combined effects of liquidity, asset management, and debt on operating results

Figure 2.11 Ratios related to profitability analysis.

Thus, BRCM's profit margin is equivalent to 5.11 cents for each sales dollar generated. BRCM's profit margin is much smaller than QCOM's profit margin of 26.9%, indicating that, although QCOM's sales are about 300% more than BRCM's revenue during the same operating period, QCOM's operation is equally more efficient than BRCM's. BRCM's lower profit margin is also a result of its carrying a higher level of debt. Recall that net income is income after taxes. Therefore, if two firms have identical operations in the sense that their sales, operating costs, and earnings before income tax are the same, but if one company uses more debt than the other, it will have higher interest charges. Those interest charges will pull net income down, and since sales are constant, the result will be a relatively low profit margin.

Return on Total Assets

The return on total assets—or simply, return on assets (ROA)—measures a company's success in using its assets to earn a profit. The ratio of net income to total assets measures the return on total assets after taxes:

$$\text{Return on total assets} = \frac{\text{Net income} + \text{interest expense} (1 - \text{tax rate})}{\text{Average total assets}}$$

$$= \frac{\$424 + \$30(1 - 0.047191)}{(\$11,208 + \$11,495)/2} = 4\%$$

Adding interest expenses back to net income results in an adjusted earnings figure that shows what earnings would have been if the assets had been acquired solely by selling shares of stock. (Note that BRCM's effective tax rate was 4.72% in 2013.) With this adjustment, we may be able to compare the return on total assets for companies with differing amounts of debt. Again, BRCM's 4% return on assets is well below the 14.49% for QCOM. This high return for QCOM results from (1) the company's high basic earning power and (2) its low use of debt, both of which cause its net income to be relatively high.

Return on Common Equity

Another popular measure of profitability is rate of return on common equity. This ratio shows the relationship between net income and common stockholders' investment in the company—that is, how much income is earned for every $1 invested by the common stockholders. To compute the return on common equity, we first subtract preferred dividends from net income, yielding the net income available to common stockholders. With no preferred stocks for BRCM, there were no preferred cash dividends. We then divide this net income available to common stockholders by the average common stockholders' equity during the year. We compute average common equity by using the beginning and ending balances. At the beginning of fiscal year 2013, BRCM's common equity balance was $7,839; at the end of fiscal year 2013, the balance was $8,371. The average balance is then simply $8,105, and we have

$$\text{Return on common equity} = \frac{\text{Net income available to common stockholders}}{\text{Average common equity}}$$

$$= \frac{\$424}{\$8,105} = 5.23\%$$

The rate of return on common equity for BRCM was 5.23% during 2013. Over the same period, QCOM's return on common equity amounted to 18.23%, a better performance relative to the computer industry (14.92% in 2013) in general, but BRCM had a poor year again.

To learn more about what management can do to increase the return on common equity, or ROE, we may rewrite the ROE in terms of the following three components:

$$\text{ROE} = \frac{\text{Net income available to common stockholders}}{\text{Average stockholders' equity}}$$

$$= \frac{\text{Net income}}{\text{Sales}} \times \frac{\text{Sales}}{\text{Assets}} \times \frac{\text{Assets}}{\text{Average stockholders' equity}}$$

The three principal components can be described as the profit margin, asset turnover, and financial leverage, respectively, so that

$$\text{ROE} = (\text{Profit margin}) \times (\text{Asset turnover}) \times (\text{Financial leverage})$$

$$= (5.11\%) \times (0.7225) \times \left(\frac{\$11{,}495}{\$8{,}105} \right)$$

$$= 5.23\%$$

This expression tells us that management has only three key ratios for controlling a company's ROE: (1) the earnings from sales (the profit margin); (2) the revenue generated from each dollar of assets employed (asset turnover); and (3) the amount of equity used to finance the assets in the operation of the business (financial leverage).

2.3.5 Market Value Analysis

When you purchase a company's stock, what are your primary factors in valuing the stock? In general, investors purchase stock to earn a return on their investment. This return consists of two parts: (1) gains (or losses) from selling the stock at a price that differs from the investors' purchase price; and (2) dividends—the periodic distributions of profits to stockholders. As shown in Figure 2.12, the market value ratios, such as the price-to-earnings ratio and the market-to-book ratio, relate the firm's stock price to its earnings and book value per share, respectively. These ratios give management an indication of what investors think of the company's past performance and future prospects. If the firm's asset

Market Value Analysis

A set of ratios that relate the firm's stock price to its earnings and book value per share

Figure 2.12 Ratios related to market value analysis.

and debt management is sound and its profit is rising, then its market value ratios will be high, and its stock price will probably be as high as can be expected.

Price-to-Earnings Ratio

The price-to-earnings (*P/E*) ratio shows how much investors are willing to pay per dollar of reported profits. BRCM's stock sold for $29.64 at the end of the fiscal period (December 31, 2013), so with an EPS of $0.74, its *P/E* ratio was 40.05:

$$P/E \text{ ratio} = \frac{\text{Price per share}}{\text{Earnings per share}}$$

$$= \frac{\$29.64}{\$0.74} = 40.05$$

That is, the stock was selling for about 40 times its current earnings per share. In general, *P/E* ratios are higher for firms with high growth prospects, other things held constant, but they are lower for firms with lower expected earnings. BRCM's expected annual increase in operating earnings is 14.6% over the next 3 to 5 years. Since BRCM's ratio is greater than 7.55%, the average for other computer industry firms, this suggests that investors value BRCM's stock more highly than most as having excellent growth prospects, believing that their worst years are behind. However, all stocks with high *P/E* ratios carry high risk whenever the expected growths fail to materialize. Any slight earnings disappointment tends to punish the market price significantly.

Book Value per Share

Another ratio frequently used in assessing the well-being of the common stockholders is the book value per share, which measures the amount that would be distributed to holders of each share of common stock if all assets were sold at their balance sheet carrying amounts and if all creditors were paid off. We compute the book value per share for BRCM's common stock as follows:

$$\text{Book value per share} = \frac{\text{Total stockholders' equity} - \text{preferred stock}}{\text{Average shares outstanding}}$$

$$= \frac{\$8,371 - \$0}{(569 + 581)/2} = \$14.56$$

If we compare this book value with the market price of $29.64 at the end of BRCM's fiscal period (December 31, 2013), then we may say that the stock appears to be overpriced. Once again, though, market prices reflect expectations about future earnings and dividends, whereas book value largely reflects the results of events that occurred in the past. Therefore, the market value of a stock tends to exceed its book value.

2.3.6 Limitations of Financial Ratios in Business Decisions

Business decisions are made in a world of uncertainty. As useful as ratios are, they have limitations. We can draw an analogy between their use in decision making and a physician's use of a thermometer. A reading of 102°F indicates that something is wrong with the patient, but the temperature alone does not indicate what the problem is or how to cure it. In other words, ratio analysis is useful, but analysts should be aware of ever-changing market conditions and make adjustments as necessary. It is also difficult to generalize about whether a particular ratio is "good" or "bad." For example, a high current ratio may indicate a strong liquidity position, which is good, but holding too much cash in a bank account (which will increase the current ratio) may not be the best utilization of funds. Ratio analysis based on any one year may not represent the true business condition. It is important to analyze trends in various financial ratios, as well as their absolute levels, for trends give clues as to whether the financial situation is likely to improve or deteriorate. To do a **trend analysis**, one simply plots a ratio over time. As a typical engineering student, your judgment in interpreting a set of financial ratios is understandably weak at this point, but it will improve as you encounter many facets of business decisions in the real world. Again, accounting is a language of business, and as you speak it more often, it can provide useful insights into a firm's operations.

EXAMPLE 2.2 Ratio Analysis for J&M, Inc.

Reconsider the J&M's financial statements given in Tables 2.4 and 2.5 and other data given in Example 2.1. Compute the following values for 2015.

(a) The amount of working capital.
(b) The debt to equity ratio.
(c) The time-interest-earned ratio.
(d) The current ratio.
(e) The quick ratio.
(f) The inventory-turnover ratio.
(g) The days-sales-outstanding ratio.
(h) The total asset turnover ratio.
(i) The profit margin.
(j) The return on total assets.
(k) The return on common equity.

SOLUTION

Given: Financial statements given in Tables 2.4 and 2.5, and other financial data provided in Example 2.1, tax rate = 28.77%.
Find: Compute the financial ratios.

(a) The amount of working capital:

$$\text{Working capital} = \text{Current assets} - \text{Current liabilities}$$

$$= \$77,400 - \$28,000$$

$$= \$49,400$$

(b) The debt ratio:

$$\text{Debt ratio} = \frac{\text{Total debt}}{\text{Total assets}} = \frac{\$58,000}{\$161,400} = 35.94\%$$

(c) The time-interest-earned ratio:

$$\text{Times-interest-earned ratio} = \frac{\text{EBIT}}{\text{Interest expense}} = \frac{\$33,280}{\$5,200} = 6.40 \text{ times}$$

(d) The current ratio:

$$\text{Current ratio} = \frac{\text{Current assets}}{\text{Current liabilities}} = \frac{\$77,400}{\$28,000} = 2.76 \text{ times}$$

(e) The quick ratio:

$$\text{Quick ratio} = \frac{\text{Current assets} - \text{Inventory}}{\text{Current liabilities}} = \frac{\$77,400 - \$37,700}{\$28,000}$$

$$= 1.42 \text{ times}$$

(f) The inventory-turnover ratio:

$$\text{Inventory Turnover} = \frac{\text{Sales}}{\text{Average Inventory}} = \frac{\$300,000}{(\$39,800 + \$37,700)/2}$$

$$= 7.74 \text{ times}$$

The typical item sits in inventory almost 1.55 months (12 months/7.74) or 46.51 days before being sold.

(g) The days-sales-outstanding ratio:

$$\text{DSO} = \frac{\text{A/R}}{\text{Average sales per day}} = \frac{\$23,700}{\$300,000/365} = 28.84 \text{ days}$$

(h) The total asset turnover ratio:

$$\text{Total Asset Turnover} = \frac{\text{Sales}}{\text{Total Assets}} = \frac{\$300,000}{\$161,400} = 1.86 \text{ times}$$

(i) The net profit margin:

$$\text{Net Margin Ratio} = \frac{\text{Net Income (\$)}}{\text{Net Sales}} = \frac{\$20,000}{\$300,000} = 6.67\%$$

(j) The return on total assets:

$$\text{Return on Total Assets} = \frac{\text{Net Income} + \text{interest expenses} (1 - \text{tax rate})}{\text{Average total assets}}$$

$$= \frac{\$20,000 + \$5,200 (1 - 0.2877)}{(\$161,400 + \$169,900)/2}$$

$$= 14.31\%$$

(k) The return on common equity.

$$\text{Return on Equity} = \frac{\text{Net Income} - \text{Cash Dividend to Preferred Stockholders}}{\text{Average Common Equity}}$$

$$= \frac{\$20,000 - \$600}{(\$93,400 + \$83,400)/2}$$

$$= 21.95\%$$

SUMMARY

The primary purposes of this chapter were to (1) describe the basic financial statements, (2) present some background information on cash flows and corporate profitability, and (3) discuss techniques used by investors and managers to analyze financial statements. The following are some concepts we covered.

- Before making any major financial decisions, it is important to understand their impact on your net worth. Your net-worth statement is a snapshot of where you stand financially at a given point in time.

- The three basic financial statements contained in the annual report are the balance sheet, the income statement, and the statement of cash flows. Investors use the information provided in these statements to form expectations about future levels of earnings and dividends and about the firm's risk-taking behavior.

- A firm's balance sheet shows a snapshot of a firm's financial position at a particular point in time through three categories: (1) assets the firm owns, (2) liabilities the firm owes, and (3) owners' equity (assets less liabilities).

- A firm's income statement reports the results of operations over a period of time and shows earnings per share as its "bottom line." The main items are (1) revenues and gains, (2) expenses and losses, and (3) net income or net loss (revenue less expenses).

- A firm's statement of cash flows reports the impact of operating, investing, and financing activities on cash flows over an accounting period.

- The purpose of calculating a set of financial ratios is twofold: (1) to examine the relative strengths and weaknesses of a company compared with those of other companies in the same industry; and (2) to learn whether the company's position has been improving or deteriorating over time.

- Liquidity ratios show the relationship of a firm's current assets to its current liabilities and thus its ability to meet maturing debts. Two commonly used liquidity ratios are the current ratio and the quick (acid-test) ratio.

- Asset management ratios measure how effectively a firm is managing its assets. Some of the major ratios are inventory turnover, fixed assets turnover, and total assets turnover.

- Debt management ratios reveal (1) the extent to which a firm is financed with debt and (2) the firm's likelihood of defaulting on its debt obligations. In this category are the debt ratio and the times-interest-earned ratio.

- Profitability ratios show the combined effects of liquidity, asset management, and debt management policies on operating results. Profitability ratios include the profit margin on sales, the return on total assets, and the return on common equity.

- Market value ratios relate the firm's stock price to its earnings and book value per share, and they give management an indication of what investors think of the company's past performance and future prospects. Market value ratios include the price-to-earnings ratio and the book value per share.

- Trend analysis, in which one plots a ratio over time, is important, because it reveals whether the firm's ratios are improving or deteriorating over time.

PROBLEMS

Financial Statements

2.1 Consider the balance-sheet entries for War Eagle Corporation in Table P2.1.

(a) Compute the firm's
   Current assets: $_____

Current liabilities: $_____
Working capital: $_____
Shareholders' equity: $_____

TABLE P2.1 Balance Sheet Statement as of December 31, 2015

Assets		
Cash		$ 150,000
Marketable securities		200,000
Accounts receivable		150,000
Inventories		50,000
Prepaid taxes and insurance		30,000
Manufacturing plant at cost	$ 600,000	
Less accumulated depreciation	100,000	
Net fixed assets		500,000
Goodwill		20,000
Liabilities and Shareholders' Equity		
Notes payable		50,000
Accounts payable		100,000
Income taxes payable		80,000
Long-term mortgage bonds		400,000
Preferred stock, 6%, $100 par value (1,000 shares)		100,000
Common stock, $15 par value (10,000 shares)		150,000
Capital surplus		150,000
Retained earnings		70,000

(b) If the firm had a net income of $500,000 after taxes, what are the earnings per share?

(c) When the firm issued its common stock, what was the market price of the stock per share?

2.2 A chemical processing firm is planning on adding a duplicate polyethylene plant at another location. The financial information for the first project year is shown in Table P2.2.

(a) Compute the working-capital requirement during the project period.

(b) What is the taxable income during the project period?

(c) What is the net income during the project period?

(d) Compute the net cash flow from the project during the first year.

Financial Ratio Analysis

2.3 The data in Table P2.3 are available for two companies, A and B, all stated in millions.

(a) Calculate each company's return on equity (ROE) and return on total assets (ROA).

(b) Why Company B's ROE so much higher than Company A's? Does this mean Company B is a better company? Why or why not?

(c) If Companies A and B were combined (merged), what would be the impact on the results on ROE? Under what conditions would such a combination make sense?

TABLE P2.2 Financial Information for First Project Year

Sales		$ 1,500,000
Manufacturing costs		
Direct materials	$ 150,000	
Direct labor	200,000	
Overhead	100,000	
Depreciation	200,000	
Operating expenses		150,000
Equipment purchase		400,000
Borrowing to finance equipment		200,000
Increase in inventories		100,000
Decrease in accounts receivable		20,000
Increase in wages payable		30,000
Decrease in notes payable		40,000
Income taxes		272,000
Interest payment on financing		20,000

TABLE P2.3

Description	Company A	Company B
Earnings before interest and taxes	$ 300	$ 560
Interest expenses	20	160
Earnings before tax	280	400
Taxes at 40%	112	160
Earnings after tax (Net income)	168	240
Debt	$ 200	$1,600
Equity	800	400

2.4 Table P2.4 shows financial statements for Apple Computer Corporation. The closing stock price for Apple was $68.11 on September 30, 2013. The average number of outstanding shares was 6.03 billion. On the basis of the financial data presented, compute the various financial ratios and make an informed analysis of Apple's financial health.

(a) Debt ratio
(b) Times-interest-earned ratio
(c) Current ratio
(d) Quick (acid-test) ratio
(e) Inventory-turnover ratio
(f) Day's-sales-outstanding
(g) Total-assets-turnover ratio
(h) Profit margin on sales
(i) Return on total assets
(j) Return on common equity
(k) Price/earnings ratio
(l) Book value per share

TABLE P2.4 Financial Statements for Apple Computer (All numbers in thousands)

THE BALANCE SHEET STATEMENT		
Period Ending	**Sep 28, 2013**	**Sep 29, 2012**
Cash And Cash Equivalents	$ 14,259,000	$ 10,746,000
Short Term Investments	26,287,000	18,383,000
Net Receivables	24,094,000	21,275,000
Inventory	1,764,000	791,000
Other Current Assets	6,882,000	6,458,000
Total Current Assets	**$ 73,286,000**	**$ 57,653,000**
Long Term Investments	106,215,000	92,122,000
Property Plant and Equipment	16,597,000	15,452,000
Goodwill	1,577,000	1,135,000
Intangible Assets	4,179,000	4,224,000
Other Assets	5,146,000	5,478,000
Total Assets	**$207,000,000**	**$176,064,000**
Accounts Payable	36,223,000	32,589,000
Short/Current Long Term Debt	–	–
Other Current Liabilities	7,435,000	5,953,000
Total Current Liabilities	**$ 43,658,000**	**$ 38,542,000**
Long Term Debt	16,960,000	–
Other Liabilities	20,208,000	16,664,000
Deferred Long Term Liability Charges	2,625,000	2,648,000
Total Liabilities	**$ 83,451,000**	**$ 57,854,000**
Common Stock	19,764,000	16,422,000
Retained Earnings	104,256,000	101,289,000
Treasury Stock	–	–
Capital Surplus	–	–
Other Stockholder Equity	(471,000)	499,000
Total Stockholder Equity	**$123,549,000**	**$118,210,000**
Total Liabilities and Stockholders' Equity	**$207,000,000**	**$176,064,000**
Income Statement		
Total Revenue	**$170,910,000**	**$156,508,000**
Cost of Revenue	106,606,000	87,846,000
Gross Profit	**$ 64,304,000**	**$ 68,662,000**

(continued)

TABLE P2.4 *(continued)*

THE BALANCE SHEET STATEMENT

Period Ending	Sep 28, 2013	Sep 29, 2012
Research Development	4,475,000	3,381,000
Selling General and Administrative	10,830,000	10,040,000
Operating Income or Loss	**$ 48,999,000**	**$ 55,241,000**
Total Other Income/Expenses Net	1,156,000	522,000
Earnings Before Interest And Taxes	50,155,000	55,763,000
Interest Expense	–	–
Income Before Tax	50,155,000	55,763,000
Income Tax Expense	13,118,000	14,030,000
Minority Interest	–	–
Net Income From Continuing Ops	37,037,000	41,733,000
Net Income	**$ 37,037,000**	**$ 41,733,000**
Preferred Stock And Other Adjustments	–	–
Net Income Applicable To Common Shares	**$ 37,037,000**	**$ 41,733,000**

Source: Data from Apple Reports Fourth Quarter Results, Apple, October 28, 2013.

2.5 Table P2.5 on Page 54 summarizes the financial conditions for Intel Corporation (INTC), a manufacturer of various computer-processing chips for fiscal year 2013. Compute the various financial ratios and interpret the firm's financial health during fiscal year 2013. The closing stock price was $25.50 on December 31, 2013 and the average number of outstanding shares was 4.98 billion.

(a) Debt ratio
(b) Times-interest-earned ratio
(c) Current ratio
(d) Quick (acid-test) ratio
(e) Inventory-turnover ratio
(f) Days-sales-outstanding
(g) Total-assets-turnover ratio
(h) Profit margin on sales
(i) Return on total assets (with a tax rate of 40%)
(j) Return on common equity
(k) Price/earnings ratio
(l) Book value per share

2.6 R.C. had earnings per share of $12 in year 2015, and it paid a $6 dividend. Book value per share at year's end was $80. During the same period, the total retained earnings increased by $42 million. R.C. has no preferred stock, and no new common stock was issued during the year. If R.C.'s year-end debt (which equals its total liabilities) was $240 million, what was the company's year-end debt ratio?

2.7 If Company B uses more debt than Company A and both companies have identical operations in terms of sales, operating costs, etc., which of the following statements is *true*?

(a) Company A has a higher profit margin on sales than Company B.
(b) Company A has a lower profit margin on sales than Company B.
(c) Both companies have identical profit margins on sales.
(d) Company A will definitely have a higher current ratio.

TABLE P2.5 Financial Statements for Intel Co.

BALANCE SHEET		
Period Ending	**Dec 28, 2013**	**Dec 29, 2012**
Cash And Cash Equivalents	$ 5,674,000	$ 8,478,000
Short Term Investments	14,413,000	9,684,000
Net Receivables	6,176,000	5,950,000
Inventory	4,172,000	4,734,000
Other Current Assets	1,649,000	2,512,000
Total Current Assets	**$ 32,084,000**	**$ 31,358,000**
Long Term Investments	7,694,000	4,917,000
Property Plant and Equipment	31,428,000	27,983,000
Goodwill	10,513,000	9,710,000
Intangible Assets	5,150,000	6,235,000
Other Assets	5,489,000	4,148,000
Total Assets	**$ 92,358,000**	**$ 84,351,000**
Accounts Payable	11,191,000	10,654,000
Short/Current Long Term Debt	281,000	312,000
Other Current Liabilities	2,096,000	1,932,000
Total Current Liabilities	**$ 13,568,000**	**$ 12,898,000**
Long Term Debt	13,165,000	13,136,000
Other Liabilities	2,972,000	3,702,000
Deferred Long Term Liability Charges	4,397,000	3,412,000
Total Liabilities	**$ 34,102,000**	**$ 33,148,000**
Common Stock	21,536,000	19,464,000
Retained Earnings	35,477,000	32,138,000
Other Stockholder Equity	1,243,000	(399,000)
Total Stockholder Equity	**$ 58,256,000**	**$ 51,203,000**
Total Liabilities and Stockholders' Equity	**$ 92,358,000**	**$ 84,351,000**
Income Statement		
Total Revenue	**$170,910,000**	**$156,508,000**
Cost of Revenue	106,606,000	87,846,000
Gross Profit	**$ 64,304,000**	**$ 68,662,000**
Research Development	4,475,000	3,381,000
Selling General and Administrative	10,830,000	10,040,000
Operating Income or Loss	**$ 48,999,000**	**$ 55,241,000**

(continued)

TABLE P2.5 (*continued*)

BALANCE SHEET		
Period Ending	**Dec 28, 2013**	**Dec 29, 2012**
Total Other Income/Expenses Net	1,156,000	522,000
Earnings Before Interest And Taxes	50,155,000	55,763,000
Interest Expense	–	–
Income Before Tax	50,155,000	55,763,000
Income Tax Expense	13,118,000	14,030,000
Net Income From Continuing Ops	37,037,000	41,733,000
Net Income Applicable To Common Shares	**$ 37,037,000**	**$ 41,733,000**

Source: Data from Intel Corporation

2.8 You are trying to assess the well-being of the common stockholders of a company. Which of the following ratios would help you make such an assessment?
(a) Debt ratio
(b) Current ratio
(c) Book value per share
(d) Total asset turnover

2.9 Which of the following statements is *correct*?
(a) The quickest way to determine whether the firm has too much debt is to calculate the Times-interest-earned ratio.
(b) The best rule of thumb for determining the firm's liquidity is to calculate the current ratio.
(c) From an investor's point of view, the price-to-earnings ratio is a good indicator of whether or not a firm is generating an acceptable return to the investor.
(d) The operating margin is determined by subtracting all operating and non-operating expenses from the gross margin.

2.10 Consider the following financial data for North-gate Corporation ($ in millions):
• Cash and marketable securities, $100
• Total fixed assets, $280
• Annual sales, $1,200
• Net income, $418
• Inventory, $180
• Current liabilities, $134
• Current ratio, 3.2

• Average correction period, 45 days
• Average common equity, $550
On the basis of these financial data, determine the firm's *return on (common) equity*.
(a) 114.6%
(b) 71.6%
(c) 76%
(d) 30%

2.11 Consider B&B Company's financial data as follows: (unit: millions of dollars except ratio figures):
• Cash and marketable securities $5,000
• Fixed assets $16,500
• Sales $50,200
• Net income $4,100
• Inventory $8,750
• Current ratio   2.8
• Average collection period   73 days
• Average common equity $25,000

Fill in the blanks by calculating the appropriate accounting entries in the balance or the income statement.

Category	Financial Data
Accounts Receivable	
Current Assets	
Long-Term Debt	
Total Asset Turnover	

2.12 The following table summarizes some of key financial data for Copeland Corporation (unit: thousand dollars except ratio figures):

Item	Value
Cash and marketable securities	$10,000
Fixed assets	$90,000
Sales	$200,000
Net income	$15,000
Inventory	$150,000
Current ratio	4.20
Average collection period	91.25 days
Long-term debt	$200,000

(a) Determine the accounts receivables.
(b) Determine the amount of current liabilities.
(c) Calculate the amount of long-term debt.
(d) Calculate the return on common equity.

2.13 Consider Fisher & Company's financial data as follows: (unit: millions of dollars except ratio figures):

Item	Value
Cash and marketable securities	$100
Fixed assets	$280
Sales	$1,200
Net income	$358
Inventory	$180
Current ratio	3.2
Average collection period	45 days
Average common equity	$500

(a) Find Fisher's accounts receivable.
(b) Calculate the amount of current assets.
(c) Determine the amount of current liabilities.
(d) Determine the amount of total assets.
(e) Calculate the amount of the long-term debt.
(f) Calculate the profit margin.
(g) Calculate the Return on Common Equity

Short Case Studies

ST2.1 Consider the two companies Google and Yahoo, which compete with each other in the mobile advertising sector. Google enjoys strong relationships among businesses and has a dominant role in online search. Get these companies' most recent annual reports from their websites, and answer the following questions.

(a) On the basis of the most recent financial statements, comment on each company's financial performance in the following areas:
  • Asset management
  • Liquidity
  • Debt management
  • Profitability
  • Market trend

(b) Check the current stock prices for both companies. The stock ticker symbol is GOOG for Google and YHOO for Yahoo. Based on your analysis in (a), which company would you bet your money on and why?

ST2.2 Compare Monsanto Company (MON) and E.I. du Pont de Nemours & Company (DD) using a thorough financial ratios analysis.

(a) For each company, compute all the ratios listed in Figure 2.7 (i.e., debt management, liquidity, asset management, market trend, and profitability) for the fiscal year chosen.

(b) Compare and contrast the two companies, using the ratios you calculated from part (a).

(c) Carefully read and summarize the "risk management" or "hedging" practices described in the financial statements of each company.

(d) If you were a mutual fund manager and could invest in only one of these companies, which one would you select and why? Be sure to justify your answer by using your results from parts (a), (b), and (c).

ST2.3 Read the Coca-Cola Company's Annual Report (most recent one) and calculate the various financial ratios defined in Figure 2.7. Then give your assessment of the company's general financial condition compared with the one of Pepsi Company.

CHAPTER 3

Interest Rate and Economic Equivalence

Dearborn Couple Claim Missouri's Largest Jackpot—$293.75 Million Every once in a while, a lottery jackpot becomes so large that it attracts national attention. Cindy and Mark Hill posed with an oversized check for $293,750,000 at the press conference after winning their half of the $588 million Powerball jackpot on November 28, 2012. Mark and Cindy are common people—Mark works for a meatpacking plant and Cindy is a clerical worker.

According to lottery officials, the odds of winning the jackpot were approximately 1 in 175 million. The winning ticket had the numbers 5–16–22–23–29 and Mega Ball number 6. According to lottery rules they could elect for the lump sum cash payment of $192.37 million (or exactly

Source: Edw./Shutterstock

$192,373,928) immediately, or an average annuity payment of $9.79 million ($9,791,667 to be exact)—either sum would go a long way in the farming town of 500 where they live. The Hills decided to take their money in a lump sum of $192.37 million, which, after state and federal taxes, would amount to $132 million.

Now put yourself in the Hills' position. You might wonder why the value of the lump sum payment, $192.37 million paid immediately, is so much lower than the $293.75 million paid in 30 installments. Isn't receiving the $293.75 million overall a lot better than receiving $192.37 million now?

First, most people familiar with investments would tell the Hills that receiving a lump amount of $192.37 million today is likely to prove a far better deal than receiving $293.75 million for 30 years, even if they live long enough to collect the entire annual payments. This reasoning involves the principles we will discuss in this chapter, namely, the interest operation and the **time value of money**. A dollar in hand today is worth more than one that will be paid to you in the future.

In engineering economics, the principles discussed in this chapter are regarded as the underpinning for nearly all project investment analysis. This is because we always need to account for the effect of interest operating on sums of cash over time. Interest formulas allow us to place different cash flows received at different times in the same time frame and to compare them. As will become apparent, almost our entire study of engineering economics is built on the principles introduced in this chapter.

CHAPTER LEARNING OBJECTIVES

After completing this chapter, you should understand the following concepts:

- The time value of money.
- The difference between simple interest and compound interest.
- The meaning of economic equivalence and why we need it in economic analysis.
- How to compare two different money series by means of the concept of economic equivalence.
- The interest operation and the types of interest formulas used to facilitate the calculation of economic equivalence.

3.1 Interest: The Cost of Money

Most of us are familiar in a general way with the concept of interest. We know that money left in a savings account earns interest so that the balance over time is greater than the sum of the deposits. We also know that borrowing to buy a car means repaying an amount over time; that amount includes interest, and it is therefore greater than the amount borrowed. What may be unfamiliar to us is the idea that, in the financial world, money itself is a commodity and, like other goods that are bought and sold, money costs money.

The cost of money is established and measured by a **market interest rate**, a percentage that is periodically applied and added to an amount (or varying amounts) of money over a specified length of time. When money is borrowed, the interest paid is the charge to the borrower for the use of the lender's property; when money is lent or invested, the interest earned is the lender's gain from providing a service to another (Figure 3.1). **Interest**, then, may be defined as the cost of having money available for use. In this section, we examine how interest operates in a free-market economy, and we establish a basis for understanding the more complex interest relationships that follow later on in the chapter.

3.1.1 The Time Value of Money

The "time value of money" seems like a sophisticated concept, yet it is a concept that you grapple with every day. Suppose you have two options; (1) receiving $1,000 cash now, or (2) receiving $1,090 one year from now. Which option would you prefer? Here, we assume that there is no risk of not receiving the money a year from now. Ask yourself what you would do if you decide to receive $1,000 now. If you don't need money now for any other purpose, you may deposit the money in a bank. If the bank's interest rate is 10%, then your balance would grow to $1,100. Certainly waiting a year to receive $1,090 does not make sense.

Suppose that you could expect a 4% inflation during your deposit period. Inflation is simply the loss of purchasing power. That is, as prices increase, it would require more money to purchase the same amount of goods and services. (We will devote a full chapter on this topic in Chapter 11.) In the previous example, the receipt of $1,090 one year from now is only worth $1,048 (= $1,090/1.04) if you expect a 4% inflation. Conceptually, the rate at which you earn interest from your investment should be higher than the inflation rate to make any economic sense of the delayed consumption, as your purchasing power will continue to decrease over time. In order to make up this future loss in purchasing power, your earning interest rate should be sufficiently larger than the anticipated inflation rate.

Considering both Earning and Purchasing Power

What the example above illustrates is that we must connect the "earning power" and the "purchasing power" to the concept of time. After all, time, like money, is a finite resource. There are only 24 hours in a day, so time has to be budgeted, too. When we deal with large amounts of money, long periods of time, or high interest rates, the change in the value of a sum of money over time becomes extremely significant. For example, at a current annual interest rate of 10%, $1 million will earn $100,000 in interest in one year; thus, to wait one year to receive $1 million clearly involves a significant sacrifice. When deciding among

Charge or Cost to Borrower

Interest Rate
8%

Profit or Earning to Lender

Figure 3.1 The meaning of *interest rate* to the lender (bank) and to the borrower.

alternative proposals, we must take into account the operation of interest and the time value of money in order to make valid comparisons of different amounts at various times.

The way interest operates reflects the fact that money has a time value. This is why amounts of interest depend on lengths of time; interest rates, for example, are typically given in terms of a percentage per year. We may define the principle of the time value of money as follows: The economic value of a sum depends on when it is received. Because money has both **earning power** as well as **purchasing power** over time, as shown in Figure 3.2 (it can be put to work, earning more money for its owner), a dollar received today has a greater value than a dollar received at some future time.

Market Interest Rate and Cash Flow Transactions in Actual Dollars

When lending or borrowing, interest rates are quoted by financial institutions in the marketplace; those interest rates reflect the desired amounts to be earned, as well as any protection from loss in the future purchasing power of money because of inflation. We will assume that, unless otherwise mentioned, *the interest rate used in this book reflects the market interest rate,* which takes into account the earning power as well as the effect of inflation perceived in the marketplace. We will also assume that all cash flow transactions are given in terms of **actual dollars** with the effect of inflation, if any, reflected in the amount. (If we want to know the true desired earnings in isolation from inflation, we can determine the real interest rate. We consider this issue in Chapter 11. The earning power of money and its loss of value because of inflation are calculated by different analytical techniques.)

3.1.2 Elements of Transactions Involving Interest

Many types of transactions (e.g., borrowing or investing money, or purchasing machinery on credit) involve interest, but certain elements are common to all of these types of transactions.

- An initial amount of money in transactions involving debt or investments is called the **principal**.
- The **interest rate** measures the cost or price of money and is expressed as a percentage per period of time.

Figure 3.2 The time value of money. This is a two-edged sword whereby earning grows, but purchasing power decreases as time goes by due to inflation.

- A period of time, called the **interest period**, determines how frequently interest is calculated.
- A specified length of time marks the duration of the transaction and thereby establishes a certain **number of interest periods**.
- A **plan for receipts or disbursements** yields a particular cash flow pattern over a specified length of time. (For example, we might have a series of equal monthly payments that repay a loan.)
- A **future amount of money** results from the cumulative effects of the interest rate over a number of interest periods.

For the purposes of calculation, these elements are represented by the following variables.

A_n = A discrete payment or receipt occurring at the end of some interest period.

i = The interest rate per interest period.

N = The total number of interest periods.

P = A sum of money at a time chosen as time zero (now) for purposes of analysis; sometimes referred to as the **present value** or **present worth**.

F = A future sum of money at the end of the analysis period.

A = An end-of-period payment or receipt in a uniform series that continues for N periods. This is a special situation where $A_1 = A_2 = \cdots = A_N$.

V_n = An equivalent sum of money at the end of a specified period n that considers the effect of the time value of money. Note that $V_0 = P$ and $V_N = F$.

Because we make frequent use of these symbols in this text, it is important that you become familiar with them. Note, for example, the distinction between A, A_n and A_N. The symbol A_n refers to a specific payment or receipt, at the end of period n, in any series of payments. A_N is the final payment in such a series, because N refers to the total number of interest periods. A refers to any series of cash flows in which all payments or receipts are equal.

Example of an Interest Transaction

As an example of how the elements we have just defined are used in a particular situation, let us suppose that an electronics manufacturing company buys a machine for $25,000 and borrows $20,000 from a bank at a 9% annual interest rate. In addition, the company pays a $200 loan origination fee when the loan commences. The bank offers two repayment plans as follows:

- **Plan 1.** The principal amount P is $20,000, and the interest rate i is 9%. The interest period is one year, and the duration of the transaction is five years, which means there are five interest periods ($N = 5$). As mentioned earlier, whereas one year is a common interest period, interest is frequently calculated at other intervals: monthly, quarterly, or semiannually, for instance. For this reason, we used the term **period** rather than **year** when we defined the preceding list of variables. The receipts and disbursements planned over the duration of this transaction yield a cash flow pattern of five equal payments A of $5,141.85 each, paid at year's end during years 1 through 5. (You will have to accept these amounts on faith for now—the next section presents the formula

TABLE 3.1   **Repayment Plans for Example Given in Text (for $N = 5$ years and $i = 9\%$)**

End of Year	Receipts	Payment Plan	
		Plan 1	Plan 2
Year 0	$20,000.00	$ 200.00	$ 200.00
Year 1		5,141.85	1,800.00
Year 2		5,141.85	1,800.00
Year 3		5,141.85	1,800.00
Year 4		5,141.85	1,800.00
Year 5		5,141.85	21,800.00

Note: An origination fee is an up-front fee charged by a lender for processing a new loan. You actually borrow $19,800 with the origination fee of $200, but you pay back on the basis of $20,000.

used to arrive at the amount of these equal payments, given the other elements of the problem.)
- **Plan 2.**   This plan has most of the elements of Plan 1, except that the partial payment of principal is not allowed. Instead, only interest is paid each year and the principal is paid in a lump sum when the loan matures.

These two payment plans are summarized in Table 3.1.

Cash Flow Diagrams

Problems involving the time value of money can be conveniently represented in graphic form with a cash flow diagram (Figure 3.3). **Cash flow diagrams** represent time by a horizontal line marked off with the number of interest periods specified. The cash flows over time are represented by arrows at relevant periods: Upward arrows denote positive flows

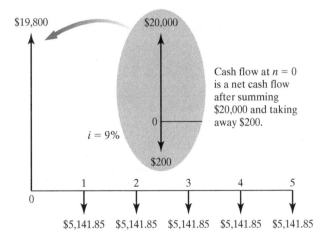

Figure 3.3   A cash flow diagram for Plan 1 of the loan repayment example summarized in Table 3.1.

(receipts) and downward arrows negative flows (disbursements). Note, too, that the arrows actually represent **net cash flows**: Two or more receipts or disbursements made at the same time are summed and shown as a single arrow. For example, $20,000 received during the same period as a $200 payment would be recorded as an upward arrow of $19,800. Also, the lengths of the arrows can suggest the relative values of particular cash flows.

Cash flow diagrams function in a manner similar to free-body diagrams or circuit diagrams, which most engineers frequently use: Cash flow diagrams give a convenient summary of all the important elements of a problem in a graphical form to determine whether the statement of the problem has been converted into its appropriate parameters. This text frequently uses this graphic tool, and you are strongly encouraged to develop the habit of using well-labeled cash flow diagrams as a means to identify and summarize pertinent information in a cash flow problem. Similarly, a table such as Table 3.1 can help you organize information in another summary format.

End-of-Period Convention

In practice, cash flows can occur at the beginning or in the middle of an interest period—or indeed, at practically any point in time. One of the simplifying assumptions we make in engineering economic analysis is the **end-of-period convention**, which is the practice of placing all cash flow transactions at the end of an interest period. (See Figure 3.4.) This assumption relieves us of the responsibility of dealing with the effects of interest within an interest period, which would greatly complicate our calculations.

Like many of the simplifying assumptions and estimates we make in modeling engineering economic problems, the end-of-period convention inevitably leads to some discrepancies between our model and real-world results. Suppose, for example, that $100,000 is deposited during the first month of the year in an account with an interest period of one year and an interest rate of 10% per year. In such a case, the difference of 1 month would cause an interest income loss of $10,000. This is because, under the end-of-period convention, the $100,000 deposit made during the interest period is viewed as if the deposit were made at the end of the year, as opposed to 11 months earlier.

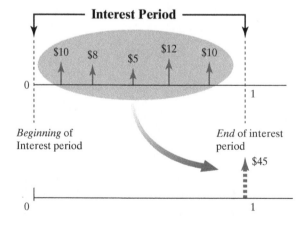

Figure 3.4 Any cash flows occurring during the interest period are summed to a single amount and placed at the end of the interest period.

This example gives you a sense of why financial institutions choose interest periods that are less than one year, even though they usually quote their rate as an annual percentage. (We will discuss this issue in Chapter 4.)

Armed with an understanding of the basic elements involved in interest problems, we can now begin to look at the details of calculating interest.

3.1.3 Methods of Calculating Interest

Money can be lent and repaid in many ways, and, equally, money can earn interest in many different ways. Usually, however, at the end of each interest period, the interest earned on the principal amount is calculated according to a specified interest rate. The two computational schemes for calculating this earned interest are said to yield either **simple interest** or **compound interest**. Engineering economic analysis uses the compound-interest scheme almost exclusively.

Simple Interest

Simple interest is interest earned on only the principal amount during each interest period. In other words, with simple interest, the interest earned during each interest period does not earn additional interest in the remaining periods, *even though you do not withdraw it*.

In general, for a deposit of P dollars at a simple interest rate of i for N periods, the total earned interest would be

$$I = (iP)N \tag{3.1}$$

The total amount available at the end of N periods thus would be

$$F = P + I = P(1 + iN) \tag{3.2}$$

Simple interest is commonly used with add-on loans or bonds, as we will see in Chapter 4.

Compound Interest

Under a compound-interest scheme, the interest earned in each period is calculated on the basis of the total amount at the end of the previous period. This total amount includes the original principal plus the accumulated interest that has been left in the account. In this case, you are, in effect, increasing the deposit amount by the amount of interest earned. In general, if you deposited (invested) P dollars at interest rate i, you would have $P + iP = P(1 + i)$ dollars at the end of one period. If the entire amount (principal and interest) is reinvested at the same rate i for another period, at the end of the second period you would have

$$P(1 + i) + i[P(1 + i)] = P(1 + i)(1 + i)$$
$$= P(1 + i)^2$$

Continuing, we see that the balance after the third period is

$$P(1 + i)^2 + i[P(1 + i)^2] = P(1 + i)^3$$

This interest-earning process repeats, and after N periods, the total accumulated value (balance) F will grow to

$$F = P(1 + i)^N \tag{3.3}$$

Recall that under the simple-interest scheme, you earn interest only on the principal amount at the end of each interest period. Under the compounding scheme, you earn interest on the principal, as well as interest on interest. From Eq. (3.3), the total interest earned over N periods is

$$I = F - P = P[(1 + i)^N - 1] \tag{3.4}$$

Compared with the simple-interest scheme, the additional interest earned with compound interest is

$$\Delta I = P[(1 + i)^N - 1] - (iP)N \tag{3.5}$$

$$= P[(1 + i)^N - (1 + iN)] \tag{3.6}$$

As either i or N becomes large, the difference in interest earnings also becomes large, so the effect of compounding is further pronounced. Note that when $N = 1$, compound interest is the same as simple interest. *Unless otherwise mentioned, all interest rates mentioned in this text refer to compound interest.*

To illustrate the difference between the compound interest and the simple interest, let's consider two examples. Example 3.1 illustrates the computational aspects of the interest operations. Example 3.2 illustrates the extraordinary power of compound interest.

EXAMPLE 3.1 Simple Interest versus Compound Interest

Suppose you deposit $1,000 in a bank savings account that pays interest at a rate of 10% compounded annually. Assume that you don't withdraw the interest earned at the end of each period (one year), but let it accumulate. How much would you have at the end of year 3?

SOLUTION
Given: $P = \$1,000$, $N = 3$ years, and $i = 10\%$ per year.
Find: F.

Applying Eq. (3.3) to our three-year, 10% case, we obtain

- Simple interest:

$$F = \$1,000[1 + 0.10(3)] = \$1,300$$

- Compound interest:

$$F = \$1,000(1 + 0.10)^3 = \$1,331$$

The total interest earned is $331, which is $31 more than was accumulated under the simple-interest method. We can keep track of the interest accrual process more precisely as follows.

Period	Amount at Beginning of Interest Period	Interest Earned for Period	Amount at End of Interest Period
1	$ 1,000	$ 1,000(0.10)	$ 1,100
2	1,100	1,100(0.10)	1,210
3	1,210	1,210(0.10)	1,331

COMMENTS: At the end of the first year, you would have $1,000, plus $100 in interest, or a total of $1,100. In effect, at the beginning of the second year, you would be depositing $1,100, rather than $1,000. Thus, at the end of the second year, the interest earned would be 0.10($1,100) = $110, and the balance would be $1,100 + $110 = $1,210. This is the amount you would be depositing at the beginning of the third year, and the interest earned for that period would be 0.10($1,210) = $121. With a beginning principal amount of $1,210 plus the $121 interest, the total balance would be $1,331 at the end of year 3. The sequence of the compounding process is illustrated in Figure 3.5.

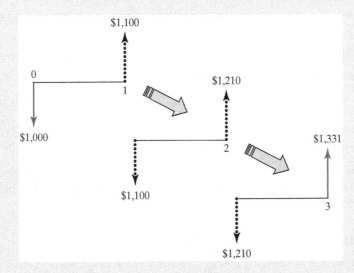

Figure 3.5   The process of computing the balance when $1,000 at 10% is deposited for three years (Example 3.1).

EXAMPLE 3.2 $24 for Manhattan a Good Deal?

Peter Minuit of the Dutch West India Company purchased the island of Manhattan from Native Americans on May 26, 1626 for goods valued at 60 Dutch guilders, which was estimated to be the equivalent of $24 (or $680 today). Instead of purchasing the island, if Minuit had invested the $24 in a savings account that earned 8% interest, how much would it be worth in 2015?

SOLUTION

Given: $P = \$24$, $i = 8\%$ per year, and $N = 389$ years.
Find: F, based on (a) 8% simple interest, and (b) 8% compound interest.

(a) With 8% simple interest,

$$F = \$24[1 + (0.08)(389)] = \$771$$

(b) With 8% compound interest,

$$F = \$24(1 + 0.08)^{389} = \$241{,}019{,}469{,}461{,}115$$

COMMENTS: The significance of compound interest is obvious in this example. Many of us can hardly comprehend the magnitude of $241 trillion. In 2015, the total population in the United States was estimated to be around 320 million. If the money were distributed equally among the population, each individual would receive $753,186. Certainly, there is no way of knowing exactly how much Manhattan Island is worth today, but most real-estate experts would agree that the value of the island is nowhere near $241 trillion. (Note that the U.S. Domestic Gross Product for 2013 was merely $15.68 trillion. The total value of all New York City property was assessed at $914.8 billion for the 2015 fiscal year.)

3.2 Economic Equivalence

The observation that money has a time value leads us to two important questions:

- How do we measure and compare various cash payments received at different points in time?
- How do we know, for example, whether we would prefer to have $20,000 today and $50,000 ten years from now or $8,000 each year for the next ten years?

In this section, we describe the basic analytical techniques for making these comparisons. Then, in Section 3.3, we will use these techniques to develop a series of formulas that can greatly simplify our calculations.

3.2.1 Definition and Simple Calculations

The central question in deciding among alternative cash flows involves comparing their economic worth. This would be a simple matter if, in the comparison, we did not need to consider the time value of money: We could simply add the individual payments within a cash flow, treating receipts as positive cash flows and payments (disbursements) as negative cash flows. The fact that money has a time value, however, makes our calculations more complicated. We need to know more than just the size of a payment in order to determine its economic effect completely. In fact, as we will see in this section, we need to know several things.

- The size of the payment.
- The type of the payment: Is it a receipt or a disbursement?
- The timing of the payment: When is it made?
- The interest rate in operation during the period under consideration.

To assess the economic value of a series of payments, we must consider the value of each payment individually.

Calculations for determining the economic effects of one or more cash flows are based on the concept of economic equivalence. **Economic equivalence** exists between cash flows that have the same economic effect and could therefore be traded for one another in the financial marketplace, which we assume to exist.

Economic equivalence refers to the fact that a cash flow—whether a single payment or a series of payments—can be converted to an *equivalent* cash flow at any point in time. For example, we could find the equivalent future value F of a present amount P at interest rate i at period n; or we could determine the equivalent present value P of N equal payments A.

We can easily extend the preceding strict concept of equivalence to include the comparison of alternatives. For example, we could compare the value of two proposals by finding the equivalent value of each at *any common point in time*. If financial proposals that appear to be quite different turn out to have the same monetary value, then we can be *economically indifferent* to choosing between them: In terms of economic effect, one would be an even exchange for the other, so no reason exists to prefer one over the other in terms of their economic value.

A way to see the concepts of equivalence and economic indifference at work in the real world is to note the variety of payment plans offered by lending institutions for consumer loans. Recall Table 3.1 where we developed two different repayment plans for a loan of $20,000 for five years at 9% interest. You will notice, perhaps to your surprise, that the two plans require significantly different repayment patterns and different total amounts of repayment. However, because money has a time value, these plans are equivalent, and economically, the bank is indifferent to a consumer's choice of plan. We will now discuss how such equivalence relationships are established.

Equivalence Calculations: A Simple Example

We can view equivalence calculations as an application of the compound-interest relationships we developed in Section 3.1. Suppose that we invest $1,000 at 12% annual interest for five years. The formula developed for calculating compound interest, $F = P(1 + i)^N$ (Eq. 3.3), expresses the equivalence between some present amount P and a future amount F, for a given interest rate i and a number of interest periods N. Therefore, at the end of the investment period, our sums grow to

$$\$1,000(1 + 0.12)^5 = \$1,762.34$$

Thus, we can say that at 12% interest, $1,000 received now is equivalent to $1,762.34 received in five years and that we could trade $1,000 now for the promise of receiving $1,762.34 in five years. Example 3.3 further demonstrates the application of this basic technique.

EXAMPLE 3.3 Equivalence

Suppose you are offered the alternative of

- Option 1: receiving $3,000 at the end of five years.
- Option 2: receiving P dollars today.

What value of P would make you indifferent to your choice between P dollars today and the promise of $3,000 at the end of five years? Here you assume that if you go with Option 2, you would deposit the money in an account that pays 8% interest because you have no current need for the money. You further assume that there is no question that the $3,000 will be paid in full (no risk).

STRATEGY: Our job is to determine the present amount that is economically equivalent to $3,000 in five years, given the earning potential of 8% per year. The "indifference" ascribed to you refers to economic indifference; that is, in a marketplace where 8% is the applicable interest rate, you could trade either cash flow for the other.

SOLUTION

Given: $F = \$3,000$, $N = 5$ years, and $i = 8\%$ per year.
Find: P.

Equation:
Eq. (3.3), $F = P(1 + i)^N$.

Rearranging terms to solve for P gives

$$P = \frac{F}{(1 + i)^N}$$

Substituting yields

$$P = \frac{\$3,000}{(1 + 0.08)^5} = \$2,042$$

COMMENTS: In this example, it is clear that if P is anything less than $\$2,042$, you would prefer the promise of $\$3,000$ in five years to P dollars today; if P is greater than $\$2,042$, you would prefer P.

3.2.2 Equivalence Calculations: General Principles

In spite of their numerical simplicity, the examples we have developed reflect several important general principles, which we will now explore.

Principle 1: Equivalence Calculations Made to Compare Alternatives Require a Common Time Basis

When we travel to foreign countries, we often need to convert U.S. dollars to equivalent local currencies via prevailing foreign exchange rates; we must also convert cash flows to a common basis to compare their value. One aspect of this basis is the choice of a single point in time at which to make our calculations. In Example 3.3, if we had been given the magnitude of each cash flow and had been asked to determine whether they were equivalent, we could have chosen any reference point and used the compound interest formula to find the value of each cash flow at that point.

When selecting a point in time at which to compare the value of alternative cash flows, we commonly use either the present time, which yields what is called the **present worth** of the cash flows, or some point in the future, which yields their **future worth**. The choice of the point in time often depends on the circumstances surrounding a particular decision, or it may be chosen for convenience. For instance, if the present worth is known for the first two of three alternatives, all three may be compared simply by calculating the present worth of the third. In Example 3.3, as you can readily see, the choice of $n = 0$ or $n = 5$ would make our problem simpler, because we need to make only one set of calculations: At 8% interest, either convert $\$2,042$ at time 0 to its equivalent value at time 5, or convert $\$3,000$ at time 5 to its equivalent value at time 0. To see how to choose a different reference point, take a look at Example 3.4.

EXAMPLE 3.4 Equivalent Cash Flows Are Equivalent at Any Common Point in Time

In Example 3.3, we determined that, given an interest rate of 8% per year, receiving $2,042 today is equivalent to receiving $3,000 in five years. Are these cash flows also equivalent at the end of year 3?

STRATEGY: This problem is summarized in Figure 3.6. The solution consists of solving two equivalence problems.

(a) What is the future value of $2,042 after three years at 8% interest?

(b) Given the sum of $3,000 after five years and an interest rate of 8%, what is the equivalent sum after three years?

SOLUTION

Given:

(a) $P = \$2,042$; $i = 8\%$ per year; $N = 3$ years.

(b) $F = \$3,000$; $i = 8\%$ per year; $N = 5 - 3 = 2$ years.

Find:

(a) V_3;

(b) V_3. Are these two values equal?

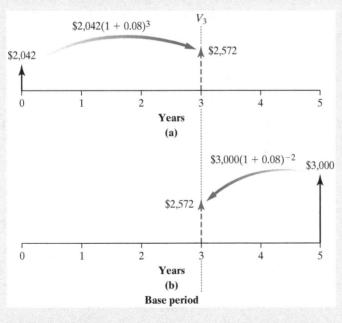

Figure 3.6   Selection of a base period for an equivalence calculation (Example 3.4).

Equation:

(a) $F = P(1 + i)^N$.

(b) $P = F(1 + i)^{-N}$.

Notation: The usual terminology of F and P is confusing in this example, since the cash flow at $n = 3$ is considered a future sum in part (a) of the solution and a past cash flow in part (b) of the solution. To simplify matters, we are free to arbitrarily designate a reference point $n = 3$ and understand that it need not be the present. Therefore, we assign the equivalent cash flow at $n = 3$ to a single variable, V_3.

(a) The equivalent worth of $2,042 after three years is

$$V_3 = 2,042(1 + 0.08)^3$$
$$= \$2,572$$

(b) The equivalent worth of the sum $3,000 two years earlier is

$$V_3 = F(1 + i)^{-N}$$
$$= \$3,000(1 + 0.08)^{-2}$$
$$= \$2,572$$

(Note that $N = 2$ because that is the number of periods during which discounting is calculated in order to arrive back at year 3.)

(c) While our solution doesn't strictly prove that the two cash flows are equivalent at any time, they will be equivalent at any time as long as we use an interest rate of 8%.

COMMENTS: If we were looking for an equivalent amount at $n = 2$, instead of $n = 0$, we may solve the following

$$V_2 = \frac{\$3,000}{(1 + 0.08)^{5-3}} = \frac{\$3,000}{(1 + 0.08)^2} = \$2,381$$

Similarly, we can calculate the equivalent values of other times as shown in Figure 3.7.

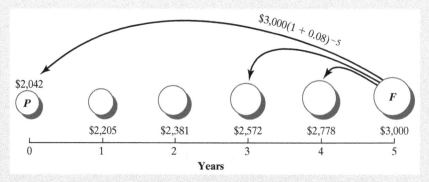

Figure 3.7 Various dollar amounts that will be economically equivalent to $3,000 in five years, given an interest rate of 8% (Example 3.4).

Principle 2: Equivalence Depends on Interest Rate

The equivalence between two cash flows is a function of the amount and timing of individual cash flows and the interest rate or rates that operate on those flows. This principle is easy to grasp in relation to our previous Example 3.3.

EXAMPLE 3.5 Changing the Interest Rate Destroys Equivalence

In Example 3.3, we determined that, given an interest rate of 8% per year, receiving $2,042 today is equivalent to receiving $3,000 in five years. Are these cash flows also equivalent at an interest rate of 10%? If not, which option is more economical?

SOLUTION

Given: $P = \$2,042$, $i = 10\%$ per year, and $N = 5$ years.
Find: F and is it equal to $3,000?

We first determine the base period under which an equivalence value is computed. Since we can select any period as the base period, let's select $N = 5$. We then need to calculate the equivalent value of $2,042 today five years from now.

$$F = \$2,042(1 + 0.10)^5 = \$3,289.$$

Since this amount is greater than $3,000, the change in interest rate breaks the equivalence between the two cash flows. As you may have already guessed, at a lower interest rate, P must be higher to be equivalent to the future amount. For example, at $i = 4\%$, $P = \$2,466$.

Principle 3: Equivalence Calculations May Require the Conversion of Multiple Payment Cash Flows to a Single Cash Flow

In all the examples presented thus far, we have limited ourselves to the simplest case of converting a single payment at one time to an equivalent single payment at another time. Part of the task of comparing alternative cash flow series involves moving each individual cash flow in the series to the same single point in time and summing these values to yield a single equivalent cash flow. We perform such a calculation in Example 3.6.

EXAMPLE 3.6 Equivalence Calculations with Multiple Payments

Recall the repayment plans in Table 3.1. Determine whether these repayment plans are equivalent to the bank with a lending rate of 9%.

STRATEGY: Notice that under Plan 1, we will pay a total of $25,909.25, whereas under Plan 2, we will pay a total of $29,200. Before concluding that we prefer Plan 1, remember that a comparison of the two cash flows is based on a *combination of payment amounts and the timing of those payments*. To make our comparison, we must compare the equivalent value of each option at a single point in time. It is simplest to convert the cash flow pattern of both plans to a single value at $n = 5$. To do this, we must convert the six disbursements of each plan to their respective equivalent values at $n = 5$. At that point, since they share a time in common, we can simply sum them in order to compare them (Figure 3.8).

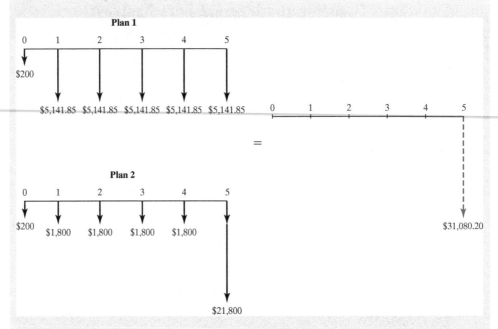

Figure 3.8 Loan repayment schedules for two options (Example 3.6).

SOLUTION

Given: Loan repayment plans; $i = 9\%$ per year and $N = 5$

Find: A single future value F of the cash flows at the end of year 5.

Equation: $F = P(1 + i)^N$, applied to each disbursement in the cash flow diagram, where N is the number of periods during which interest is in effect, and n is the period number (i.e., for year 1, $n = 1$). We determine the value of F by finding the number of interest periods applicable for each payment. By converting the cash flow in Plan 2 to a single future payment at year 5, we can compare Plans 1 and 2.

- Plan 1

$$\begin{aligned}
F_5 \text{ for } \$200 \text{ paid at } n = 0: && \$200(1 + 0.09)^{5-0} &= \$ \quad 307.72 \\
F_5 \text{ for } \$5{,}141.85 \text{ paid at } n = 1: && \$5{,}141.85(1 + 0.09)^{5-1} &= \$ \ 7{,}258.14 \\
F_5 \text{ for } \$5{,}141.85 \text{ paid at } n = 2: && \$5{,}141.85(1 + 0.09)^{5-2} &= \$ \ 6{,}658.84 \\
F_5 \text{ for } \$5{,}141.85 \text{ paid at } n = 3: && \$5{,}141.85(1 + 0.09)^{5-3} &= \$ \ 6{,}109.03 \\
F_5 \text{ for } \$5{,}141.85 \text{ paid at } n = 4: && \$5{,}141.85(1 + 0.09)^{5-4} &= \$ \ 5{,}604.62 \\
F_5 \text{ for } \$5{,}141.85 \text{ paid at } n = 5: && \$5{,}141.85(1 + 0.09)^{5-5} &= \underline{\$ \ 5{,}141.85} \\
&& & \ \$31{,}080.20
\end{aligned}$$

- Plan 2

$$
\begin{aligned}
F_5 \text{ for } \$200 \text{ paid at } n = 0: && \$200(1 + 0.09)^{5-0} &= \$307.72 \\
F_5 \text{ for } \$1{,}800 \text{ paid at } n = 1: && \$1{,}800(1 + 0.09)^{5-1} &= \$\ 2{,}540.85 \\
F_5 \text{ for } \$1{,}800 \text{ paid at } n = 2: && \$1{,}800(1 + 0.09)^{5-2} &= \$\ 2{,}331.05 \\
F_5 \text{ for } \$1{,}800 \text{ paid at } n = 3: && \$1{,}800(1 + 0.09)^{5-3} &= \$\ 2{,}138.58 \\
F_5 \text{ for } \$1{,}800 \text{ paid at } n = 4: && \$1{,}800(1 + 0.09)^{5-4} &= \$\ 1{,}962.00 \\
F_5 \text{ for } \$1{,}800 \text{ paid at } n = 5: && \$21{,}800(1 + 0.09)^{5-5} &= \$21{,}800.00 \\
&& &\ \overline{\$31{,}080.20}
\end{aligned}
$$

We see that the two repayment schedules are equivalent. Thus, the bank would be economically indifferent to a borrower's choice between the two plans. Note that the final payment in each plan does not accrue any compound interest.

3.3 Development of Formulas for Equivalence Calculations

Now that we have established some working assumptions and notations and a preliminary understanding of the concept of equivalence, we will develop a series of interest formulas for use in more complex comparisons of cash flows.

As we begin to compare series of cash flows instead of single payments, the required analysis becomes more complicated. We will classify five major categories of cash flow transactions, develop interest formulas for them, and present several working examples of each type.

3.3.1 The Five Types of Cash Flows

Whenever we identify patterns in cash flow transactions, we may use those patterns to develop concise expressions for computing either the present or future worth of the series. For this purpose, we will classify cash flow transactions into five categories: (1) a single cash flow, (2) a uniform series, (3) a linear-gradient series, (4) a geometric-gradient series, and (5) an irregular series. To simplify the description of various interest formulas, we will use the following notation.

1. **Single Cash Flow.** The simplest case involves the equivalence of a single, present amount and its future worth. Thus, the single-cash-flow formulas deal with only two amounts: a single, present amount P and its future worth F (Figure 3.9a). You have already seen the derivation of one formula for this situation in Section 3.1.3, which gave us Eq. (3.3):

$$F = P(1 + i)^N$$

2. **Equal-Payment (Uniform) Series.** Probably the most familiar category includes transactions arranged as a series of equal cash flows at regular intervals, known as a

Figure 3.9 Five types of cash flows: (a) Single cash flows, (b) equal (uniform) payment series, (c) linear-gradient series, (d) geometric-gradient series, and (e) irregular payment series.

uniform series (or **equal-payment series**) (Figure 3.9b). For example, this category describes the cash flows of the common installment loan contract, which arranges the repayment of a loan in equal periodic installments (*A*). The uniform series formulas deal with the equivalence relations *P*, *F*, and *A*.

3. **Linear-Gradient Series.** While many transactions involve series of cash flows, the amounts are not always uniform; they may, however, vary in some regular way. One common pattern of variation occurs when each cash flow in a series increases (or decreases) by a fixed amount (Figure 3.9c). A five-year loan repayment plan might specify, for example, a series of annual payments that increase by $500 each year. We call this type of cash flow pattern a **linear-gradient series** because its cash

flow diagram produces an ascending (or descending) straight line, as you will see in Section 3.3.5. In addition to using P, F, and A, the formulas employed in such problems involve a *constant amount G* of the change in each cash flow.

4. **Geometric-Gradient Series.**   Another kind of gradient series is formed when the series in a cash flow is determined not by some fixed amount, like $500, but by some fixed *rate* expressed as a percentage. For example, in a five-year financial plan for a project, the cost of a particular raw material might be budgeted to increase at a rate of 4% per year. The curving gradient in the diagram of such a series suggests its name: a **geometric-gradient series** (Figure 3.9d). In the formulas dealing with such series, the rate of change is represented by a lowercase g.

5. **Irregular (Mixed) Series.**   Finally, a series of cash flows may be irregular, in that it does not exhibit a regular overall pattern (Figure 3.9e). Even in such a series, however, one or more of the patterns already identified may appear over segments of time in the total length of the series. The cash flows may be equal, for example, for five consecutive periods in a 10-period series. When such patterns appear, the formulas for dealing with them may be applied and their results included in calculating an equivalent value for the entire series.

3.3.2 Single-Cash-Flow Formulas

We begin our coverage of interest formulas by considering the simplest of cash flows: single cash flows.

Compound Amount Factor

Given a present sum P invested for N interest periods at interest rate i, what sum will have accumulated at the end of the N periods? You probably noticed right away that this description matches the case we first encountered in describing compound interest. To solve for F (the future sum), we use Eq. (3.3):

$$F = P(1 + i)^N$$

Because of its origin in the compound-interest calculation, the factor $(1 + i)^N$ is known as the **compound-amount factor**. Like the concept of equivalence, this factor is one of the foundations of engineering economic analysis. Given the compound-amount factor, all the other important interest formulas can be derived.

This process of finding F is often called the **compounding process**. The cash flow transaction is illustrated in Figure 3.10. (Note the time-scale convention: The first period begins at $n = 0$ and ends at $n = 1$.) If a calculator is handy, it is easy enough to calculate $(1 + i)^N$ directly.

Interest Tables

Interest formulas, such as the one developed in Eq. (3.3), $F = P(1 + i)^N$, allow us to substitute known values from a particular situation into the equation and to solve for the unknown. As we will see in later sections, more complex formulas will require even more involved calculations. To simplify the process, tables of compound-interest factors were developed, and these tables allow us to find the appropriate factor for a given interest rate and the number of interest periods. Even with hand calculators, it is still often convenient to use such tables, and they are included in this text in Appendix B. Take some time now

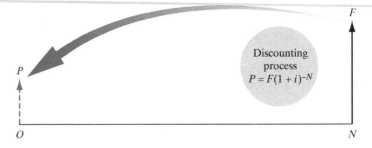

Figure 3.10 Equivalence relation between P and F.

to become familiar with their arrangement and, if you can, locate the compound-interest factor for the example just presented, in which we know P. Remember that to find F, we need to know the factor by which to multiply $20,000 when the interest rate i is 12% and the number of periods is 5:

$$F = \$20,000\underbrace{(1 + 0.12)^5}_{1.7623} = \$35,246$$

TABLE 3.2 A Typical Compound Interest Table (12%)

	Single Payment		Equal Payment Series				Gradient Series		
	Compound Amount Factor	Present Worth Factor	Compound Amount Factor	Sinking Fund Factor	Present Worth Factor	Capital Recovery Factor	Gradient Uniform Series	Gradient Present Worth	
N	$(F/P, i, N)$	$(P/F, i, N)$	$(F/A, i, N)$	$(A/F, i, N)$	$(P/A, i, N)$	$(A/P, i, N)$	$(A/G, i, N)$	$(P/G, i, N)$	N
1	1.1200	0.8929	1.0000	1.0000	0.8929	1.1200	0.0000	0.0000	1
2	1.2544	0.7972	2.1200	0.4717	1.6901	0.5917	0.4717	0.7972	2
3	1.4049	0.7118	3.3744	0.2963	2.4018	0.4163	0.9246	2.2208	3
4	1.5735	0.6355	4.7793	0.2092	3.0373	0.3292	1.3589	4.1273	4
5	1.7623	0.5674	6.3528	0.1574	3.6048	0.2774	1.7746	6.3970	5

Factor Notation

As we continue to develop interest formulas in the rest of this chapter, we will express the resulting compound-interest factors in a conventional notation that can be substituted in a formula to indicate precisely which table factor to use in solving an equation. In the preceding example, for instance, the formula derived as Eq. (3.3) is $F = P(1 + i)^N$. In ordinary language this tells us that in order to determine what future amount F is equivalent to a present amount P, we need to multiply P by a factor $(1 + i)^N$. We may also express that factor in functional notation as $(F/P, i, N)$, which is read as "Find F, given $P, i,$ and N." This is known as the **single-payment compound-amount factor**. When we incorporate the table factor into the formula, it is expressed as

$$F = P(1 + i)^N = P(F/P, i, N)$$

Thus, in the preceding example, where we had $F = \$20,000(1.12)^5$, we can write $F = \$20,000(F/P, 12\%, 5)$. The table factor tells us to use the 12% interest table and find the factor in the F/P column for $N = 5$. Because using the interest tables is often the easiest way to solve an equation, this factor notation is included for each of the formulas derived in the sections that follow.

EXAMPLE 3.7 Single Amounts: Find *F*, Given *i*, *N*, and *P*

If you had \$2,000 now and invested it at 10% interest compounded annually, how much would it be worth in eight years?

SOLUTION

Given: $P = \$2,000, i = 10\%$ per year, and $N = 8$ years (Figure 3.11).
Find: F.

We can solve this problem in any of three ways.

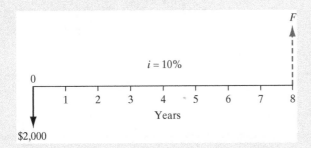

$2,000

Figure 3.11   A cash flow diagram from the investor's point of view (Example 3.7).

1. **Using a calculator.**   You can simply use a calculator to evaluate the $(1 + i)^N$ factor (financial calculators are pre-programmed to solve most future-value problems):

$$F = \$2,000(1 + 0.10)^8$$
$$= \$4,287.18$$

2. **Using compound-interest tables.** The interest tables can be used to locate the compound-amount factor for $i = 10\%$ and $N = 8$. From the 10% interest factor table in Appendix B, we obtain

$$F = \$2,000(F/P, 10\%, 8) = \$2,000(2.1436) = \$4,287.20$$

This is essentially identical to the value obtained by the direct evaluation of the single-cash-flow compound-amount factor. The slight difference is due to rounding errors in the table.

3. **Using Excel.** Excel provides financial functions to evaluate various interest formulas, where the future-worth calculation can be done using the financial function such as `FV(i,N,0,-P)`. Figure 3.12 illustrates the process of finding the future worth.

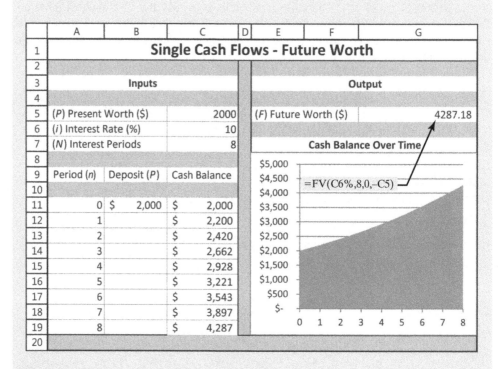

Figure 3.12 Excel: `FV` function.

Present-Worth Factor

Finding the present worth of a future sum is simply the reverse of compounding and is known as the **discounting process**. In Eq. (3.3), we can see that if we were to find a present sum P, given a future sum F, we simply solve for P:

$$P = F\left[\frac{1}{(1+i)^N}\right] = F(P/F, i, N) \tag{3.7}$$

The factor $(1 + i)^{-N}$ is known as the **single-payment present-worth factor** and is designated $(P/F, i, N)$. Tables have been constructed for P/F factors and for various values of i and N. The interest rate i and the P/F factor are also referred to as the **discount rate** and **discounting factor**, respectively.

EXAMPLE 3.8 Single Amounts: Find *P*, Given *F*, *i*, and *N*

Suppose that $1,000 is to be received in five years. At an annual interest rate of 12%, what is the present worth of this amount?

SOLUTION
Given: $F = \$1,000$, $i = 12\%$ per year, and $N = 5$ years.
Find: P.

$$P = \$1,000(1 + 0.12)^{-5} = \$1,000(0.56743) = \$567.43.$$

Using a calculator may be the best way to make this simple calculation. To have $1,000 in your savings account at the end of five years, you must deposit $567.43 now.

We can also use the interest tables to find that

$$P = \$1,000 \overbrace{(P/F, 12\%, 5)}^{(0.56743)} = \$567.43.$$

Again, you could use a financial calculator or a computer to find the present worth. With Excel, the present-worth can be calculated using the financial function $PV(i, N, 0, -F)$, as shown in Figure 3.13.

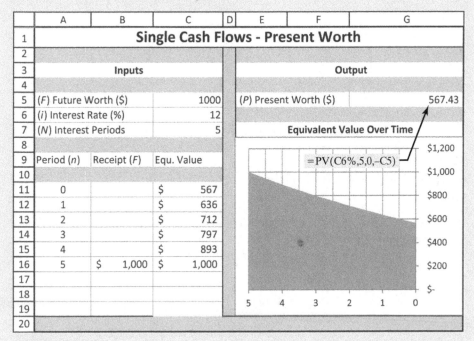

	A	B	C	D	E	F	G
1			**Single Cash Flows - Present Worth**				
2							
3		**Inputs**				**Output**	
4							
5	(F) Future Worth ($)		1000		(P) Present Worth ($)		567.43
6	(i) Interest Rate (%)		12				
7	(N) Interest Periods		5		**Equivalent Value Over Time**		
8							$1,200
9	Period (n)	Receipt (F)	Equ. Value		=PV(C6%,5,0,–C5)		
10							$1,000
11	0		$ 567				$800
12	1		$ 636				
13	2		$ 712				$600
14	3		$ 797				
15	4		$ 893				$400
16	5	$ 1,000	$ 1,000				$200
17							
18							$-
19				5 4 3 2 1 0			
20							

Figure 3.13 Excel: PV function.

Solving for Time and Interest Rates

At this point, you should realize that the compounding and discounting processes are reciprocals of one another and that we have been dealing with one equation in two forms:

$$\text{Future-value form: } F = P(1 + i)^N$$

$$\text{Present-value form: } P = F(1 + i)^{-N}$$

There are four variables in these equations: P, F, N, and i. If you know the values of any three, you can find the value of the fourth. Thus far, we have always given you the interest rate i and the number of years N, plus either P or F. In many situations, though, you will need to solve for i or N, as we discuss next.

EXAMPLE 3.9 Solving for *i*

Suppose you buy a share of stock for $20 and sell it for $40. Then your profit is $20. If that happens within a year, your rate of return (interest rate earned on your investment) is an impressive 100% . If it takes five years, what would be the compound annual rate of return on your investment? (See Figure 3.14.)

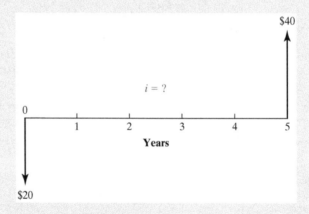

Figure 3.14   Cash flow diagram (Example 3.9).

SOLUTION

Given: $P = \$20$, $F = \$40$, and $N = 5$.
Find: i.

Here, we know P, F, and N, but we do not know i, the interest rate you will earn on your investment. This type of rate of return is easy to calculate, because you make only a one-time lump-sum investment. Problems such as this are solved as follows.

$$F = P(1 + i)^N$$

$$\$40 = \$20(1 + i)^5; \text{ solve for } i$$

$$i = 2^{1/5} - 1$$

$$= 14.87\%$$

- **Using Excel.**   The most practical approach is to use either a financial calculator or an electronic spreadsheet, such as Excel. A financial function such as RATE($N, 0, P, F$) allows us to calculate an unknown interest rate. The precise command statement would be as shown in Figure 3.15. Note that, in Excel format, we enter the present value (P) as a negative number, indicating a cash outflow.

	A	B	C	D	E	F	G
1		**Single Cash Flows - Compound Interest Rate**					
2							
3		**Inputs**				**Output**	
4							
5	(P) Present Worth ($)		20		(i) Interest Rate (%)		14.87%
6	(F) Future Worth		40				
7	(N) Interest Periods		5		**Cash Balance Over Time**		
8							
9	Period (n)	Deposit (P)	Cash Balance				
10							
11	0	$ 20	$ 20				
12	1		$ 23				
13	2		$ 26				
14	3		$ 30				
15	4		$ 35				
16	5		$ 40				
17							
18							
19							
20							

Chart: Cash Balance Over Time with vertical axis from $- to $45 and horizontal axis 0 to 5. Callout: =RATE(C7,0,–C5,C6)

Figure 3.15   Excel: RATE function

EXAMPLE 3.10 Single Amounts: Find *N*, Given *P*, *F*, and *i*

You have just purchased 100 shares of Facebook stock at $60 per share. You will sell the stock when its market price has doubled. If you expect the stock price to increase 20% per year, how long do you anticipate waiting before selling the stock (Figure 3.16)?

Figure 3.16   Cash flow diagram.

SOLUTION

Given: $P = \$6,000$, $F = \$12,000$, and $i = 20\%$ per year.

Find: N (years).

Using the single-payment compound-amount factor, we write

$$F = P(1 + i)^N = P(F/P, i, N)$$

$$\$12,000 = \$6,000(1 + 0.20)^N = \$6,000(F/P, 20\%, N)$$

$$2 = (1.20)^N = (F/P, 20\%, N)$$

Again, we could use a calculator or Excel to find N.

1. **Using a Calculator.** Solving for N gives

$$\log 2 = N \log 1.20$$

or

$$N = \frac{\log 2}{\log 1.20}$$

$$= 3.80 \approx 4 \text{ years}$$

2. **Using Excel.** Within Excel, the financial function NPER(i,0,P,F) computes the number of compounding periods it will take an investment (P) to grow to a future value (F), earning a fixed interest rate (i) per compounding period. In our example, the Excel command would look like Figure 3.17.

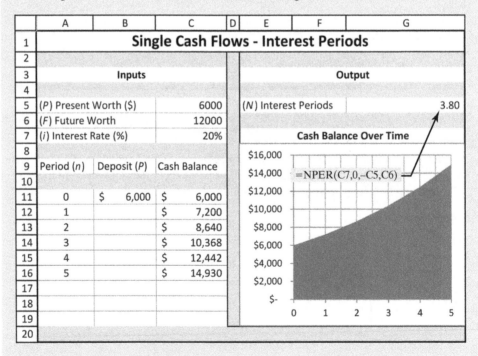

Figure 3.17 Excel: NPER function

COMMENTS: A very handy rule of thumb, called the Rule of 72, estimates approximately how long it will take for a sum of money to double. The rule states that, to find the time it takes for a present sum of money to grow by a factor of two, we divide 72 by the interest rate. In our example, the interest rate is 20%. Therefore, the Rule of 72 indicates 72/20 = 3.60, or roughly four years, for a sum to double. This is, in fact, relatively close to our exact solution. Figure 3.18 illustrates the approximate number of years required to double an investment at various interest rates using the Rule of 72.

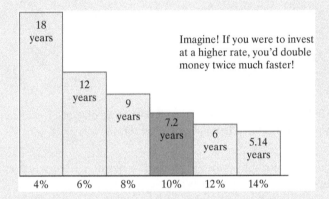

Figure 3.18 Number of years required to double an initial investment at various interest rates.

3.3.3 Equal-Payment Series

A common cash flow transaction involves a series of disbursements or receipts. Familiar examples of series payments are payment of installments on car loans and home mortgage payments. Payments on car loans and home mortgages typically involve identical sums to be paid at regular intervals. We can find the present worth of a stream of future cash flows always by summing the present worth of each of the individual cash flows. However, if cash flow regularities are present within the stream, then the use of shortcuts, such as finding the present worth of a uniform series, may be possible.

Compound-Amount Factor: Find *F*, Given *A*, *i*, and *N*

Suppose we are interested in the future amount F of a fund to which we contribute A dollars each period and on which we earn interest at a rate of i per period. The contributions are made at the end of each of N equal periods. These transactions are graphically illustrated in Figure 3.19. Looking at this diagram, we see that, if an amount A is invested at the end of each period for N periods, the total amount F that can be withdrawn at the end of the N periods will be the sum of the compound amounts of the individual deposits.

As shown in Figure 3.19, the A dollars we put into the fund at the end of the first period will be worth $A(1 + i)^{N-1}$ at the end of N periods. The A dollars we put into the fund at the end of the second period will be worth $A(1 + i)^{N-2}$, and so forth. Finally, the

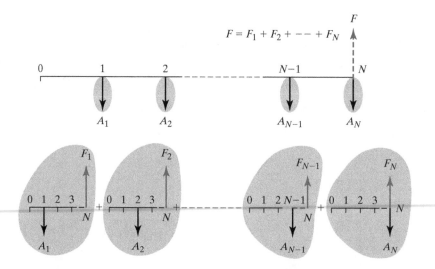

Figure 3.19 The future worth of a cash flow series obtained by summing the future-worth figures of each of the individual flows.

last A dollars that we contribute at the end of the Nth period will be worth exactly A dollars at that time. This means that there exists a series of the form

$$F = A(1 + i)^{N-1} + A(1 + i)^{N-2} + \cdots + A(1 + i) + A$$

or expressed alternatively,

$$F = A + A(1 + i) + A(1 + i)^2 + \cdots + A(1 + i)^{N-1} \tag{3.8}$$

To find a closed form expression to the problem, we may multiply Eq. (3.8) by $(1 + i)$

$$(1 + i)F = A(1 + i) + A(1 + i)^2 + \cdots + A(1 + i)^N \tag{3.9}$$

Subtracting Eq. (3.8) from Eq. (3.9) to eliminate common terms gives us

$$F(1 + i) - F = -A + A(1 + i)^N$$

Solving for F yields

$$F = A\left[\frac{(1 + i)^N - 1}{i}\right] = A(F/A, i, N) \tag{3.10}$$

The bracketed term in Eq. (3.10) is called the **equal-payment series compound-amount factor**, or the **uniform series compound-amount factor**; its factor notation is $(F/A, i, N)$. This interest factor has been calculated for various combinations of i and N in the interest tables.

EXAMPLE 3.11 Equal-Payment Series: Find *F*, Given *i*, *A*, and *N*

Suppose you make an annual contribution of $3,000 to your savings account at the end of each year for 10 years. If the account earns 7% interest annually, how much can be withdrawn at the end of 10 years (Figure 3.20)?

SOLUTION

Given: $A = \$3{,}000$, $N = 10$ years, and $i = 7\%$ per year.

Find: F.

$$F = \$3{,}000(F/A, 7\%, 10)$$

$$= \$3{,}000(13.8164)$$

$$= \$41{,}449.20$$

To obtain the future value of the annuity with the use of Excel, we may use the following financial command: $=\text{FV}(7\%,10,-3000)$, as shown in Figure 3.21.

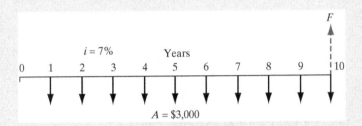

Figure 3.20 Cash flow diagram (Example 3.11).

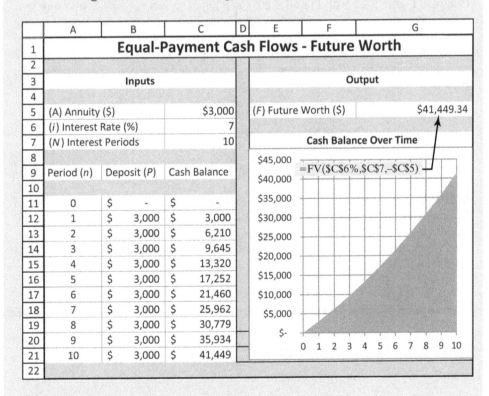

Figure 3.21 Excel worksheet to find future worth (F).

EXAMPLE 3.12 Handling Time Shifts in a Uniform Series

In Example 3.11, the first deposit of the 10-deposit series was made at the end of period 1 and the remaining nine deposits were made at the end of each following period. Suppose that all deposits were made at the *beginning* of each period instead. How would you compute the balance at the end of period 10?

SOLUTION

Given: Cash flow as shown in Figure 3.22, and $i = 7\%$ per year.
Find: F_{10}.

Figure 3.22 Cash flow diagram (Example 3.12).

Compare Figure 3.22 with Figure 3.20. Each payment has been shifted to one year earlier; thus, each payment would be compounded for one extra year. Note that with the end-of-year deposit, the ending balance (F) was \$41,449.20. With the beginning-of-year deposit, the same balance accumulates by the end of period 9. This balance can earn interest for one additional year. Therefore, we can easily calculate the resulting balance as

$$F_{10} = \$41{,}449.20(1.07) = \$44{,}350.64$$

The annuity due can be easily evaluated with the following financial command available on Excel:

$$= \texttt{FV}(7\% \ ,10,-3000,0,1)$$

COMMENTS: Another way to determine the ending balance is to compare the two cash flow patterns. By adding the \$3,000 deposit at period 0 to the original cash flow and subtracting the \$3,000 deposit at the end of period 10, we obtain the second cash flow. Therefore, the ending balance can be found by making the following adjustment to the \$41,449.20:

$$F_{10} = \$41{,}449.20 + \$3{,}000(F/P, 7\%, 10) - \$3{,}000 = \$44{,}350.64.$$

Sinking-Fund Factor: Find *A*, Given *F*, *i*, and *N*

If we solve Eq. (3.10) for *A*, we obtain

$$A = F\left[\frac{i}{(1 + i)^N - 1}\right] = F(A/F, i, N) \tag{3.11}$$

The term within the brackets is called the **equal-payment series sinking-fund factor**, or **sinking-fund factor**, and is referred to by the notation (A/F, i, N). A sinking fund is an interest-bearing account into which a fixed sum is deposited each interest period; it is commonly established for the purpose of replacing fixed assets or retiring corporate bonds.

EXAMPLE 3.13 The Cost of Waiting in Building a Retirement Fund

It is human nature to put things off, and say, "I will start saving next year." But when it comes to financial matters, the cost of procrastination can be staggering. Suppose that you are making $5,000 annual contributions to your retirement funds at the age of 20, and you earn a 7% return. Assuming that you plan to retire at age 65, you'll have $1,428,747 saved at retirement.

$$F = \$5,000(F/A, 7\%, 45) = \$1,428,747$$

Now if you want to save the same amount ($1,428,747) at retirement, what are the required annual contributions as a function of starting age?

SOLUTION
Given: $F = \$1,428,747$, $N = 65 - n$, and $i = 7\%$ per year.
Find: A.

The required annual contribution as a function of starting age n can be determined as

$$A_n = \$1,428,747(A/F, 7\%, 65 - n)$$

For example, if you wait even three years (or age 23), your annual contributions would increase to nearly $6,195 to save that same amount by age 65. And if you were to wait until age 40, you'd have to contribute nearly $22,589 a year! Figure 3.23 illustrates the general trend of required annual contributions and Table 3.3 illustrates the cost of procrastination at each age.

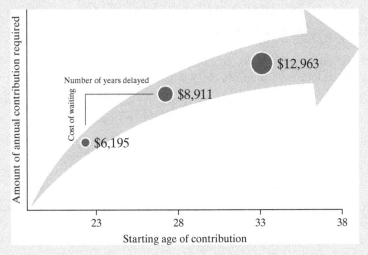

Figure 3.23 The cost of waiting in building a retirement fund.

TABLE 3.3 Required Annual Contributions as a Function of Starting Age to Build Retirement Funds in the Amount of $1,428,747

	A	B	C	D	E
1					
2			Annual	Investment	Final
3	Start	End	Contribution	Return	Value
4					
5	18	65	$4,339.73	7%	$1,428,747
6	19	65	$4,657.66	7%	$1,428,747
7	20	65	$5,000.00	7%	$1,428,747
8	21	65	$5,368.79	7%	$1,428,747
9	22	65	$5,766.27	7%	$1,428,747
10	23	65	$6,194.91	7%	$1,428,747
11	24	65	$6,657.42	7%	$1,428,747
12	25	65	$7,156.79	7%	$1,428,747
13	26	65	$7,696.32	7%	$1,428,747
14	27	65	$8,279.66	7%	$1,428,747
15	28	65	$8,910.88	7%	$1,428,747
16	29	65	$9,594.48	7%	$1,428,747
17	30	65	$10,335.49	7%	$1,428,747
18	31	65	$11,139.56	7%	$1,428,747
19	32	65	$12,012.99	7%	$1,428,747
20	33	65	$12,962.90	7%	$1,428,747
21	34	65	$13,997.30	7%	$1,428,747
22	35	65	$15,125.29	7%	$1,428,747
23	36	65	$16,357.22	7%	$1,428,747
24	37	65	$17,704.92	7%	$1,428,747
25	38	65	$19,181.97	7%	$1,428,747
26	39	65	$20,804.02	7%	$1,428,747
27	40	65	$22,589.22	7%	$1,428,747
28					

EXAMPLE 3.14 Comparison of Three Different Investment Plans

Consider three investment plans for an individual who just celebrated her 24th birthday at an annual interest rate of 8% (Figure 3.24):

Plan A. Invest $2,000 per year for the first 10 years of your career. At the end of 10 years, make no further investments, but reinvest the amount accumulated at the end of 10 years for the next 31 years.

Plan B. Do nothing for the first 10 years. Then start investing $2,000 per year for the next 31 years.

Plan C. Invest $2,000 per year for the entire 41 years.

Note that all investments are made at the birthday of each year; the first deposit will be made on the 25th birthday ($n = 1$), and you want to calculate the balance on the 65th birthday ($n = 41$).

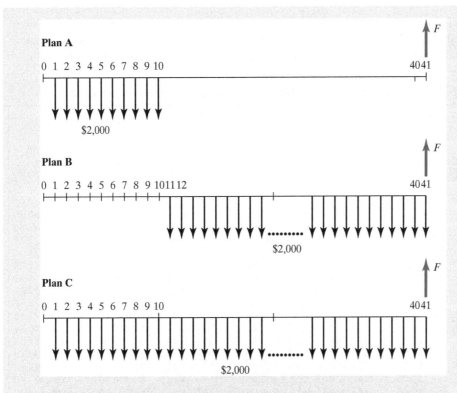

Figure 3.24 Cash flow diagrams for three investment plans (Example 3.14).

SOLUTION

Given: Three different deposit scenarios with $i = 8\%$ and $N = 41$ years.

Find: Balance at the end of 41 years (or on the 65th birthday).

Plan A:

$$F_{65} = \underbrace{\$2,000(F/A, 8\%, 10)}_{\$28,973.12}(F/P, 8\%, 31)$$

Balance at the end of 10 years

$$= \$314,870.34$$

Plan B:

$$F_{65} = \underbrace{\$2,000(F/A, 8\%, 31)}_{\$246,691.74}$$

$$= \$246,691.74$$

Plan C:

$$F_{65} = \underbrace{\$2,000(F/A, 8\%, 41)}_{\$561,562.08}$$

$$= \$561,562.08$$

To illustrate your balance changes at the end of each year, you may want to construct a tableau such as the one shown in Table 3.4. Note that, due to rounding errors, the final balance figures can be slightly off from those calculated by interest formulas.

TABLE 3.4 Three Investment Plans - How the Timing of Contributions Affects the Wealth

	A	B	C	D	E	F	G	H
1								
2			Plan A			Plan B		Plan C
3				Year-End		Year-End		Year-End
4	Age	Year	Contribution	Value	Contribution	Value	Contribution	Value
5								
6	24	0						
7	25	1	$2,000.00	$2,000.00	$0.00	$0.00	$2,000.00	$2,000.00
8	26	2	$2,000.00	$4,160.00	$0.00	$0.00	$2,000.00	$4,160.00
9	27	3	$2,000.00	$6,492.80	$0.00	$0.00	$2,000.00	$6,492.80
10	28	4	$2,000.00	$9,012.22	$0.00	$0.00	$2,000.00	$9,012.22
11	29	5	$2,000.00	$11,733.20	$0.00	$0.00	$2,000.00	$11,733.20
12	30	6	$2,000.00	$14,671.86	$0.00	$0.00	$2,000.00	$14,671.86
13	31	7	$2,000.00	$17,845.61	$0.00	$0.00	$2,000.00	$17,845.61
14	32	8	$2,000.00	$21,273.26	$0.00	$0.00	$2,000.00	$21,273.26
15	33	9	$2,000.00	$24,975.12	$0.00	$0.00	$2,000.00	$24,975.12
16	34	10	$2,000.00	$28,973.12	$0.00	$0.00	$2,000.00	$28,973.12
17	35	11	$0.00	$31,290.97	$2,000.00	$2,000.00	$2,000.00	$33,290.97
18	36	12	$0.00	$33,794.25	$2,000.00	$4,160.00	$2,000.00	$37,954.25
19	37	13	$0.00	$36,497.79	$2,000.00	$6,492.80	$2,000.00	$42,990.59
20	38	14	$0.00	$39,417.62	$2,000.00	$9,012.22	$2,000.00	$48,429.84
21	39	15	$0.00	$42,571.03	$2,000.00	$11,733.20	$2,000.00	$54,304.23
22	40	16	$0.00	$45,976.71	$2,000.00	$14,671.86	$2,000.00	$60,648.57
23	41	17	$0.00	$49,654.84	$2,000.00	$17,845.61	$2,000.00	$67,500.45
24	42	18	$0.00	$53,627.23	$2,000.00	$21,273.26	$2,000.00	$74,900.49
25	43	19	$0.00	$57,917.41	$2,000.00	$24,975.12	$2,000.00	$82,892.53
26	44	20	$0.00	$62,550.80	$2,000.00	$28,973.12	$2,000.00	$91,523.93
27	45	21	$0.00	$67,554.87	$2,000.00	$33,290.97	$2,000.00	$100,845.84
28	46	22	$0.00	$72,959.26	$2,000.00	$37,954.25	$2,000.00	$110,913.51
29	47	23	$0.00	$78,796.00	$2,000.00	$42,990.59	$2,000.00	$121,786.59
30	48	24	$0.00	$85,099.68	$2,000.00	$48,429.84	$2,000.00	$133,529.52
31	49	25	$0.00	$91,907.65	$2,000.00	$54,304.23	$2,000.00	$146,211.88
32	50	26	$0.00	$99,260.26	$2,000.00	$60,648.57	$2,000.00	$159,908.83
33	51	27	$0.00	$107,201.09	$2,000.00	$67,500.45	$2,000.00	$174,701.54
34	52	28	$0.00	$115,777.17	$2,000.00	$74,900.49	$2,000.00	$190,677.66
35	53	29	$0.00	$125,039.35	$2,000.00	$82,892.53	$2,000.00	$207,931.87
36	54	30	$0.00	$135,042.49	$2,000.00	$91,523.93	$2,000.00	$226,566.42
37	55	31	$0.00	$145,845.89	$2,000.00	$100,845.84	$2,000.00	$246,691.74
38	56	32	$0.00	$157,513.56	$2,000.00	$110,913.51	$2,000.00	$268,427.07
39	57	33	$0.00	$170,114.65	$2,000.00	$121,786.59	$2,000.00	$291,901.24
40	58	34	$0.00	$183,723.82	$2,000.00	$133,529.52	$2,000.00	$317,253.34
41	59	35	$0.00	$198,421.73	$2,000.00	$146,211.88	$2,000.00	$344,633.61
42	60	36	$0.00	$214,295.47	$2,000.00	$159,908.83	$2,000.00	$374,204.30
43	61	37	$0.00	$231,439.10	$2,000.00	$174,701.54	$2,000.00	$406,140.64
44	62	38	$0.00	$249,954.23	$2,000.00	$190,677.66	$2,000.00	$440,631.89
45	63	39	$0.00	$269,950.57	$2,000.00	$207,931.87	$2,000.00	$477,882.44
46	64	40	$0.00	$291,546.62	$2,000.00	$226,566.42	$2,000.00	$518,113.04
47	65	41	$0.00	**$314,870.34**	$2,000.00	**$246,691.74**	$2,000.00	**$561,562.08**
48								
49	Note that all contributions are made at the birthdate of each year.							
50	First contribution is made at the 25th birthdate for both Plan A and Plan C.							

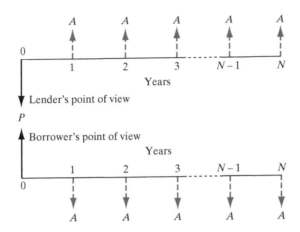

Figure 3.25 A repayment series for a loan by two different viewpoints.

Capital Recovery Factor (Annuity Factor): Find A, Given P, i, and N

We can determine the amount of a periodic payment A if we know P, i, and N. Figure 3.25 illustrates this situation. To relate P to A, recall the relationship between P and F in Eq. (3.3): $F = P(1 + i)^N$. Replacing F in Eq. (3.11) by $P(1 + i)^N$, we get

$$A = P(1 + i)^N \left[\frac{i}{(1 + i)^N - 1} \right]$$

or

$$A = P \left[\frac{i(1 + i)^N}{(1 + i)^N - 1} \right] = P(A/P, i, N) \tag{3.12}$$

Now we have an equation for determining the value of the series of end-of-period payments A when the present sum P is known. The portion within the brackets is called the **equal-payment series capital-recovery factor**, or simply **capital-recovery factor**, which is designated $(A/P, i, N)$. In finance, this A/P factor is referred to as the **annuity factor** and indicates a series of payments of a fixed, or constant, amount for a specified number of periods.

EXAMPLE 3.15 Equal-Payment Series: Find A, Given P, i, and N

BioGen Company, a small biotechnology firm, has borrowed $250,000 to purchase laboratory equipment for gene splicing. The loan carries an interest rate of 8% per year and is to be repaid in equal installments over the next six years. Compute the amount of the annual installments (Figure 3.26).

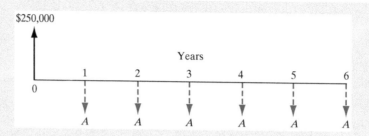

Figure 3.26 A loan cash flow diagram from BioGen's point of view.

SOLUTION

Given: $P = \$250,000$, $i = 8\%$ per year, and $N = 6$ years.
Find: A.

$$A = \$250,000(A/P, 8\%, 6)$$
$$= \$250,000(0.2163)$$
$$= \$54,075$$

Here is an Excel solution (Figure 3.27) using an annuity fuction command:

$$= \mathbf{PMT}(\mathtt{i,\ N,\ P})$$

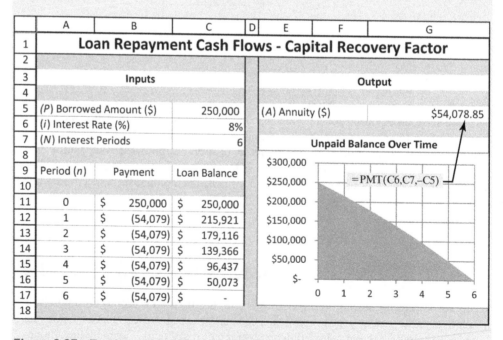

Figure 3.27 Excel: PMT function.

EXAMPLE 3.16 Deferred Loan Repayments

In Example 3.15, suppose that BioGen wants to negotiate with the bank to defer the first loan repayment until the end of year 2 (but still desires to pay in six equal installments at 8% interest). If the bank wishes to earn the same profit, what should be the annual installment, also known as **deferred annuity** (Figure 3.28)?

SOLUTION

Given: $P = \$250,000$, $i = 8\%$ per year, and $N = 6$ years, but the first payment occurs at the end of year 2.

Find: A.

By deferring the loan for a year, the bank will add the interest accrued during the first year to the principal. In other words, we need to find the equivalent worth P' of $250,000 at the end of year 1:

$$P' = \$250,000(F/P, 8\%, 1)$$

$$= \$270,000$$

In fact, BioGen is borrowing $270,000 for six years. To retire the loan with six equal installments, the deferred equal annual payment on P' will be

$$A' = \$270,000(A/P, 8\%, 6)$$

$$= \$58,401$$

By deferring the first payment for one year, BioGen needs to make additional payments of $4,326 in each year.

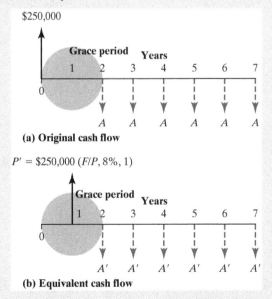

(a) Original cash flow

(b) Equivalent cash flow

Figure 3.28 A deferred loan cash flow diagram from BioGen's point of view (Example 3.16).

Present-Worth Factor: Find *P*, Given *A*, *i*, and *N*

What would you have to invest now in order to withdraw *A* dollars at the end of each of the next *N* periods? In answering this question, we face just the opposite of the equal-payment series capital-recovery factor situation: *A* is known, but *P* has to be determined. With the capital recovery factor given in Eq. (3.12), solving for *P* gives us

$$P = A\left[\frac{(1 + i)^N - 1}{i(1 + i)^N}\right] = A(P/A, i, N) \tag{3.13}$$

The bracketed term is referred to as the **equal-payment series present-worth factor** and is designated (*P/A*, *i*, *N*).

EXAMPLE 3.17 Equal-Payment Series: Find *P*, Given *A*, *i*, and *N*

Let us revisit Cindy and Mark Hill's lottery problem, introduced in the chapter opening.

Determine whether or not the couple's decision was wise to accept the lump sum in the amount of $192,373,928 as opposed to receiving $9,791,667 each year over 30 years at an interest rate of 5%. (See Figure 3.29.)

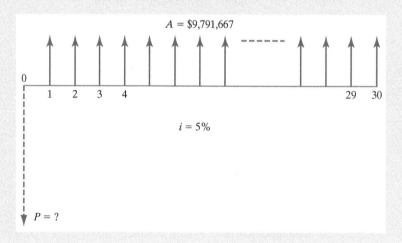

Figure 3.29 A cash flow series for Cindy and Mark Hill's lottery winning.

SOLUTION

Given: *i* = 5% per year, *A* = $9,791,667, and *N* = 30 years.
Find: *P*.

- Using interest factor:

$$P = \$9,791,667(P/A, 5\%, 30) = \$9,791,667(15.37245103)$$
$$= \$150,521,916$$

- Using Excel:

$$= \text{PV}(5\%, 30, -9791667) = \$150,521,921$$

COMMENTS: Clearly, we can tell the couple that giving up $9,791,667 a year for 30 years to receive $192,373,928 today is a winning proposition even if they can earn only a 5% return on their investment.

3.3.4 Linear-Gradient Series

Engineers frequently encounter situations involving periodic payments that increase or decrease by a constant amount (G) from period to period. These situations occur often enough to warrant the use of special equivalence factors that relate the arithmetic gradient to other cash flows. Figure 3.30 illustrates a **strict gradient series**, $A_n = (n - 1)G$. Note that the origin of the series is at the end of the first period with a zero value. The gradient G can be either positive or negative.

- If $G > 0$, the series is referred to as an *increasing* gradient series.
- If $G < 0$, it is a *decreasing* gradient series.

Unfortunately, the strict form of the increasing or decreasing gradient series does not correspond with the form that most engineering economic problems take. A typical problem involving a linear gradient includes an initial payment during period 1 that increases by G during some number of interest periods, a situation illustrated in Figure 3.31. This contrasts with the strict form illustrated in Figure 3.30, in which no payment is made during period 1 and the gradient is added to the previous payment beginning in period 2.

Gradient Series as Composite Series

In order to utilize the strict gradient series to solve typical problems, we must view cash flows, as shown in Figure 3.31, as a **composite series**, or a set of two cash flows. Each cash flow corresponds to a form that we can recognize and easily solve: a uniform series of N payments of amount A_1 and a gradient series of increments of constant amount G. The need to view cash flows that involve linear-gradient series as composites of two series is very important in solving problems, as we will now see.

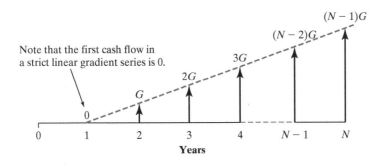

Figure 3.30 A cash flow diagram for a strict gradient series.

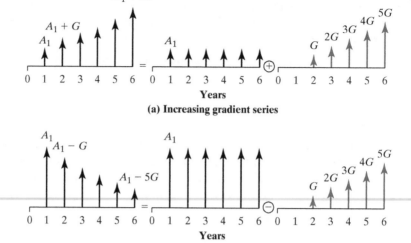

Figure 3.31 Two types of linear-gradient series as composites of a uniform series of N payments of A_1 and the gradient series of increments of constant amount G.

Present-Worth Factor: Linear Gradient: Find *P*, Given *G*, *N*, and *i*

How much would you have to deposit now to withdraw the gradient amounts specified in Figure 3.30? To find an expression for the present amount P, we apply the single-payment present-worth factor to each term of the series and obtain

$$P = 0 + \frac{G}{(1 + i)^2} + \frac{2G}{(1 + i)^3} + \cdots + \frac{(N - 1)G}{(1 + i)^N}$$

or

$$P = \sum_{n=1}^{N}(n - 1)G(1 + i)^{-n} \tag{3.14}$$

Letting $G = a$ and $1/(1 + i) = x$ yields

$$P = 0 + ax^2 + 2ax^3 + \cdots + (N - 1)ax^N$$
$$= ax[0 + x + 2x^2 + \cdots + (N - 1)x^{N-1}] \tag{3.15}$$

Since an arithmetic–geometric series $\{0, x, 2x^2, \ldots, (N - 1)x^{N-1}\}$ has the finite sum

$$0 + x + 2x^2 + \cdots + (N - 1)x^{N-1} = x\left[\frac{1 - Nx^{N-1} + (N - 1)x^N}{(1 - x)^2}\right]$$

we can rewrite Eq. (3.15) as

$$P = ax^2\left[\frac{1 - Nx^{N-1} + (N - 1)x^N}{(1 - x)^2}\right] \tag{3.16}$$

Replacing the original values for A and x, we obtain

$$P = G\left[\frac{(1 + i)^N - iN - 1}{i^2(1 + i)^N}\right] = G(P/G, i, N) \qquad (3.17)$$

The resulting factor in brackets is called the **gradient-series present-worth factor**, which we denote as $(P/G, i, N)$.

EXAMPLE 3.18   Linear Gradient: Find P, Given A_1, G, i, and N

A textile mill has just purchased a lift truck that has a useful life of five years. The engineer estimates that maintenance costs for the truck during the first year will be $1,000. As the truck ages, maintenance costs are expected to increase at a rate of $250 per year over the remaining life. Assume that the maintenance costs occur at the end of each year. The firm wants to set up a maintenance account that earns 12% annual interest. All future maintenance expenses will be paid out of this account. How much does the firm have to deposit in the account now?

SOLUTION

Given: $A_1 = \$1,000$, $G = \$250$, $i = 12\%$ per year, and $N = 5$ years.
Find: P.

Asking how much the firm has to deposit now is equivalent to asking what the equivalent present worth for this maintenance expenditure is if 12% interest is used. The cash flow may be broken into two components, as shown in Figure 3.32.

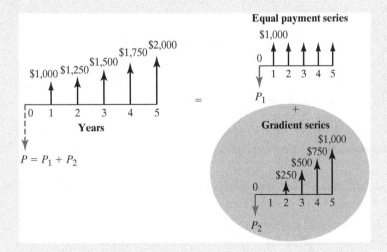

Figure 3.32   Cash flow diagram with a gradient series (Example 3.18).

The first component is an equal-payment series (A_1), and the second is a linear-gradient series (G). We have

$$P = P_1 + P_2$$

$$P = A_1(P/A, 12\%, 5) + G(P/G, 12\%, 5)$$

$$= \$1,000(3.6048) + \$250(6.397)$$

$$= \$5,204$$

Note that the value of N in the gradient factor is 5, not 4. This is because, by definition of the series, the first gradient value begins at period 2.

COMMENTS: As a check, we can compute the present worth of the cash flow by using the (P/F, 12%, n) factors, as in Figure 3.33.

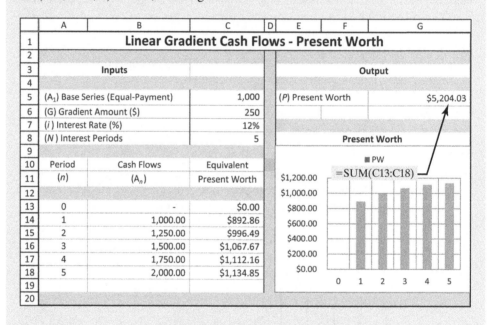

Figure 3.33   Excel worksheet to calculate the required deposit amount (P).

The slight difference is caused by a rounding error.

Gradient-to-Equal-Payment Series Conversion Factor: Find A, Given G, i, and N

We can obtain an equal payment series equivalent to the gradient series, as depicted in Figure 3.34, by substituting Eq. (3.17) for P into Eq. (3.12) to obtain

$$A = G\left[\frac{(1 + i)^N - iN - 1}{i[(1 + i)^N - 1]}\right] = G(A/G, i, N) \tag{3.18}$$

where the resulting factor in brackets is referred to as the **gradient-to-equal-payment series conversion factor** and is designated (A/G, i, N).

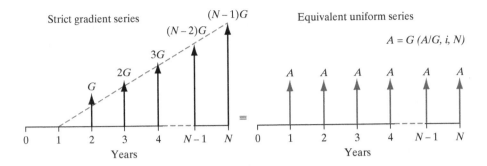

Figure 3.34   Conversion of a gradient series to its equivalent uniform series.

EXAMPLE 3.19 Linear Gradient: Find *A*, Given *A*₁, *G*, *i*, and *N*

John and Barbara have just opened two savings accounts at their credit union. The accounts earn 10% annual interest. John wants to deposit $1,000 in his account at the end of the first year and increase this amount by $300 for each of the next five years. Barbara wants to deposit an equal amount each year for the next six years. What should be the size of Barbara's annual deposit so that the two accounts will have equal balances at the end of six years (Figure 3.35)?

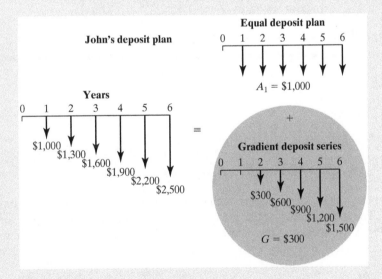

Figure 3.35   John's deposit series viewed as a combination of uniform and gradient series (Example 3.19).

SOLUTION

Given: $A_1 = \$1,000$, $G = \$300$, $i = 10\%$, and $N = 6$.
Find: A.

Since we use the end-of-period convention unless otherwise stated, this series begins at the end of the first year and the last contribution occurs at the end of the sixth year. We can separate the constant portion of $1,000 from the series, leaving the gradient series of 0, 0, 300, 600, ..., 1,500.

To find the equal payment series beginning at the end of year 1 and ending at year 6 that would have the same present worth as that of the gradient series, we may proceed as follows:

$$A = \$1,000 + \$300(A/G, 10\%, 6)$$
$$= \$1,000 + \$300(2.2236)$$
$$= \$1,667.08$$

Barbara's annual contribution should be $1,667.08.

COMMENTS: Alternatively, we can compute Barbara's annual deposit by first computing the equivalent present worth of John's deposits and then finding the equivalent uniform annual amount. The present worth of this combined series is

$$P = \$1,000(P/A, 10\%, 6) + \$300(P/G, 10\%, 6)$$
$$= \$1,000(4.3553) + \$300(9.6842)$$
$$= \$7,260.56$$

The equivalent uniform deposit is

$$A = \$7,260.56(A/P, 10\%, 6) = \$1,667.02$$

(The slight difference in cents is caused by a rounding error.)

EXAMPLE 3.20 Declining Linear Gradient: Find F, Given A_1, G, i, and N

Suppose that you make a series of annual deposits into a bank account that pays 10% interest. The initial deposit at the end of the first year is $1,200. The deposit amounts decline by $200 in each of the next four years. How much would you have immediately after the fifth deposit?

SOLUTION
Given: Cash flow shown in Figure 3.36, $i = 10\%$ per year, and $N = 5$ years.
Find: F.

The cash flow includes a decreasing gradient series. Recall that we derived the linear-gradient factors for an increasing gradient series. For a decreasing gradient series, the solution is most easily obtained by separating the flow into two components: a uniform series and an increasing gradient that is *subtracted* from the uniform series (Figure 3.36). The future value is

$$F = F_1 - F_2$$
$$= A_1(F/A, 10\%, 5) - \$200(P/G, 10\%, 5)(F/P, 10\%, 5)$$
$$= \$1,200(6.1051) - \$200(6.8618)(1.6105)$$
$$= \$5,115.93$$

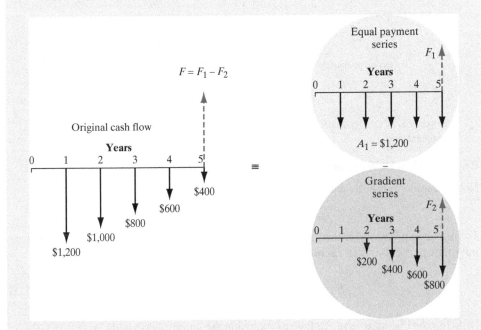

Figure 3.36 Recomposition of a gradient series into an equal-payment series and a strict gradient series.

3.3.5 Geometric Gradient Series

Many engineering economic problems—particularly those relating to construction costs or operating costs—involve cash flows that increase or decrease over time, not by a constant amount (as with a linear gradient), but rather by a constant percentage (a **geometric gradient**). This kind of cash flow is called **compound growth**. Price changes caused by inflation are a good example of a geometric-gradient series. If we use g to designate the percentage change in a payment from one period to the next, the magnitude of the nth payment, A_n, is related to the first payment A_1 by the formula

$$A_n = A_1(1 + g)^{n-1}, n = 1, 2, \ldots, N \tag{3.19}$$

The variable g can take either a positive or a negative sign, depending on the type of cash flow. If $g > 0$, the series will increase, and if $g < 0$, the series will decrease. Figure 3.37 illustrates the cash flow diagram for this situation.

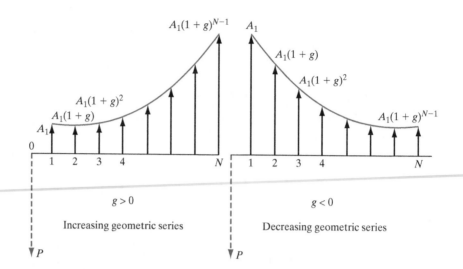

Figure 3.37 A geometrically increasing or decreasing gradient series at a constant rate g.

Present-Worth Factor: Find P, Given A_1, g, i, and N

Notice that the present worth of any cash flow A_n at interest rate i is

$$P_n = A_n(1 + i)^{-n} = A_1(1 + g)^{n-1}(1 + i)^{-n}$$

To find an expression for the present amount P for the entire series, we apply the **single-payment present-worth factor** to each term of the series:

$$P = \sum_{n=1}^{N} A_1(1 + g)^{n-1}(1 + i)^{-n} \tag{3.20}$$

Bringing the constant term $A_1(1 + g)^{-1}$ outside the summation yields

$$P = \frac{A_1}{(1 + g)} \sum_{n=1}^{N} \left[\frac{1 + g}{1 + i}\right]^n \tag{3.21}$$

Let $a = \dfrac{A_1}{1 + g}$ and $x = \dfrac{1 + g}{1 + i}$. Then rewrite Eq. (3.21) as

$$P = a(x + x^2 + x^3 + \cdots + x^N) \tag{3.22}$$

Since the summation in Eq. (3.22) represents the first N terms of a geometric series, we may obtain the closed-form expression as follows. First, multiply Eq. (3.22) by x to get

$$xP = a(x^2 + x^3 + x^4 + \cdots + x^{N+1}) \tag{3.23}$$

Then, subtract Eq. (3.23) from Eq. (3.22):

$$P - xP = a(x - x^{N+1})$$
$$P(1 - x) = a(x - x^{N+1})$$
$$P = \frac{a(x - x^{N+1})}{1 - x}, (x \neq 1). \tag{3.24}$$

If we replace the original values for a and x, we obtain

$$
P = \begin{cases} \dfrac{A_1}{i - g}\left[1 - \left(\dfrac{1 + g}{1 + i} \right)^{N} \right] & \text{if } i \neq g \\[2ex] \dfrac{NA_1}{1 + i} & \text{if } i = g \end{cases} \tag{3.25}
$$

Or in terms of engineering economy factor notation, we have

$$ P = A_1(P/A_1, g, i, N) $$

The factor within brackets is called the **geometric-gradient series present-worth factor** and is designated $(P/A_1, g, i, N)$. In the special case where $i = g$, Eq. (3.21) becomes $P = [NA_1/(1 + i)]$.

EXAMPLE 3.21 Geometric Gradients: Find P, Given A_1, g, i, and N

Ansell, Inc., a medical device manufacturer, uses compressed air in solenoids and pressure switches in its machines to control various mechanical movements. Over the years, the manufacturing floor has changed layouts numerous times. With each new layout, more piping was added to the compressed-air delivery system to accommodate new locations of manufacturing machines. None of the extra, unused old pipe was capped or removed; thus, the current compressed-air delivery system is inefficient and fraught with leaks. Because of the leaks, the compressor is expected to run 70% of the time that the plant will be in operation during the upcoming year. This will require 260 kW of electricity at a rate of $0.05/kWh for 6,000 hours. (The plant runs 250 days a year, 24 hours per day.)

- If Ansell continues to operate the current air delivery system, the compressor run time will increase by 7% per year for the next four years because of ever-worsening leaks. (After five years, the current system will not be able to meet the plant's compressed-air requirement, so it will have to be replaced.)
- If Ansell decides to replace all of the old piping now, it will cost $28,570.

The compressor will still run the same number of days; however, it will run 23% less (or will have $70\%(1 - 0.23) = 53.9\%$ usage during the day) because of the reduced air-pressure loss. If Ansell's interest rate is 12%, is the machine worth fixing now?

SOLUTION

Given: Current power consumption, $g = 7\%$, $i = 12\%$, and $N = 5$ years.

Find: A_1 and P.

Step 1. We need to calculate the cost of power consumption of the current piping system during the first year:

Power cost = % of day operating
        × days operating per year
        × hours per day
        × kW × $/kWh

$$= (70\%) \times (250 \text{ days/year}) \times (24 \text{ hours/day})$$
$$\times (260 \text{ kW}) \times (0.05/\text{kWh})$$
$$= \$54,600$$

Step 2. Each year, the annual power cost will increase at the rate of 7% over the previous year's cost. The anticipated power cost over the five-year period is summarized in Figure 3.38. The equivalent, present lump-sum cost at 12% for this geometric-gradient series is

$$P_{\text{Old}} = \$54,600(P/A_1, 7\%, 12\%, 5)$$

$$= \frac{\$54,600}{0.12 - 0.07}\left[1 - \left(\frac{1 + 0.07}{1 + 0.12}\right)^5\right]$$

$$= \$222,937$$

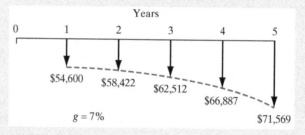

Figure 3.38 Annual power cost series if repair is not performed.

Step 3. If Ansell replaces the current compressed-air system with the new one, the annual power cost will be 23% less during the first year and will remain at that level over the next five years. The equivalent present lump-sum cost at 12% is then

$$P_{\text{New}} = \$54,600(1 - 0.23)(P/A, 12\%, 5)$$

$$= \$42,042(3.6048)$$

$$= \$151,552$$

Step 4. The net cost for not replacing the old system now is $71,385 (= $222,937 − $151,552).

Since the new system costs only $28,570, the replacement should be made now.

COMMENTS: In this example, we assumed that the cost of removing the old system was included in the cost of installing the new system. If the removed system has some salvage value, replacing it will result in even greater savings. We will consider many types of replacement issues in Chapter 14.

EXAMPLE 3.22 Geometric Gradients: Find A_1, Given F, g, i, and N

Jimmy Carpenter, a self-employed individual, is opening a retirement account at a bank. His goal is to accumulate $1,000,000 in the account by the time he retires from work in 20 years time. A local bank is willing to open a retirement account that pays 8% interest compounded annually throughout the 20 years. Jimmy expects that his annual income will increase 6% yearly during his working career. He wishes to start with a deposit at the end of year 1 (A_1) and increase the deposit at a rate of 6% each year thereafter. What should be the size of his first deposit (A_1)? The first deposit will occur at the end of year 1, and subsequent deposits will be made at the end of each year. The last deposit will be made at the end of year 20.

SOLUTION

Given: $F = \$1,000,000$, $g = 6\%$ per year, $i = 8\%$ per year, and $N = 20$ years.
Find: A_1 as in Figure 3.39.
We have

$$F = \overbrace{A_1(P/A_1, 6\%, 8\%, 20)}^{\substack{\text{Equivalent present worth of} \\ \text{the total deposit}}}(F/P, 8\%, 20)$$

$$= \frac{A_1}{0.08 - 0.06}\left[1 - \left(\frac{1 + 0.06}{1 + 0.08}\right)^{20}\right](F/P, 8\%, 20)$$

$$= A_1(72.6911)$$

Solving for A_1 yields

$$A_1 = \$1,000,000/72.6911 = \$13,757$$

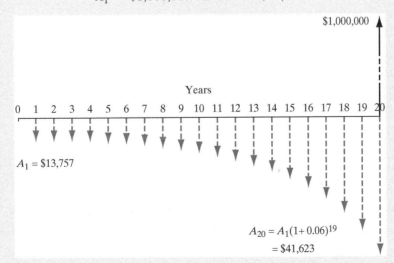

Figure 3.39 Jimmy Carpenter's retirement plan (Example 3.22).

COMMENTS: This example demonstrates the process of finding the initial deposit amount (A_1) analytically. With Excel, we can easily find the solution by using the `Goal Seek` function.[1] In Figure 3.40, the cell that contains the formula that you want to settle is called the **`Set cell ($C$30)`**. The value you want the formula to change to is called the **`To value (1000000)`** and the part of the formula that you wish to change is called the **`By changing cell ($B$2, initial deposit amount)`**. The **`Set cell`** *must* always contain a formula or a function, whereas the **`By changing cell`** must contain a value only, not a formula or function. Upon clicking on "OK," Excel will find the value (A_1) to be $13,756.85, which is the same as the one obtained by solving the equivalence formula.

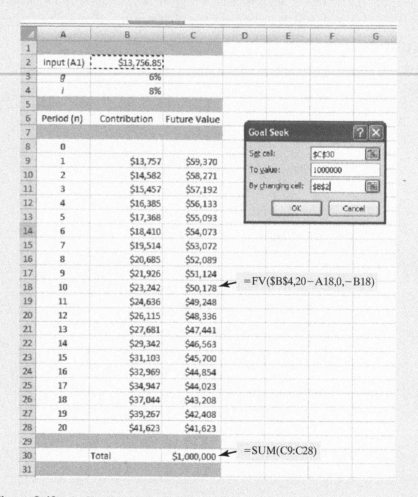

Figure 3.40 An Excel worksheet to demonstrate the use of a `Goal Seek` function.

[1]`Goal Seek` can be used when you know the result of a formula, but not the input value required by the formula to decide the result. You can change the value of a specified cell until the formula that is dependent on the changed cell returns the result you want. `Goal Seek` is found under the DATA/"What If Analysis" menu.

3.3.6 Irregular (Mixed) Payment Series

If there is no clear pattern over the series, we call the transaction an uneven cash flow series. We can find the present worth of any uneven stream of payments by calculating the present value of each individual payment and summing the results. Once the present worth is found, we can make other equivalence calculations (e.g., future worth can be calculated by using the interest factors developed in the previous section).

EXAMPLE 3.23 Present Values of an Uneven Series by Decomposition Into Single Payments

Wilson Technology, a growing machine shop, wishes to set aside money now to meet the anticipated future expenses for its production department. The company can earn 10% on a lump sum deposited now, and it wishes to withdraw the money from this account in the following increments.

> **Year 1:** $25,000 to purchase a machine and software designed for automation.
> **Year 2:** $3,000 to purchase additional hardware to accommodate anticipated growth in use of the system.
> **Year 3:** No expenses.
> **Year 4:** $5,000 to purchase software upgrades.

How much money must be deposited now to cover the anticipated payments over the next four years? Assume each withdrawal and payment to occur at the end of each year.

STRATEGY: This problem is equivalent to asking what value of P would make you indifferent in your choice between P dollars today and the future expense stream ($25,000, $3,000, $0, and $5,000). One way to deal with an uneven series of cash flows is to calculate the equivalent present value of each single cash flow and to sum the present values to find P. In other words, the cash flow is broken into three parts, as shown in Figure 3.41.

SOLUTION
Given: Uneven cash flow in Figure 3.41 with $i = 10\%$ per year.
Find: P.

$$P = \$25,000(P/F, 10\%, 1) + \$3,000(P/F, 10\%, 2)$$
$$+ \$5,000(P/F, 10\%, 4)$$
$$= \$28,622$$

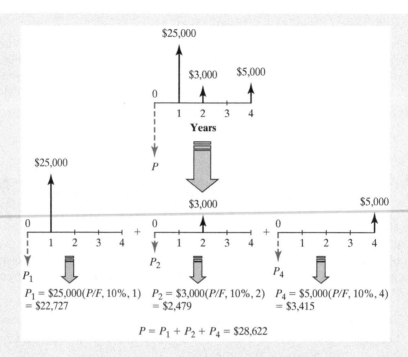

Figure 3.41 Decomposition of uneven cash flow series (Example 3.23).

COMMENTS: To see if $28,622 is indeed sufficient, let's calculate the balance at the end of each year. If you deposit $28,622 now, it will grow to (1.10)($28,622), or $31,484, at the end of year 1. From this balance, you pay out $25,000. The remaining balance, $6,484, will again grow to (1.10)($6,484), or $7,132, at the end of year 2. Now you make the second payment ($3,000) out of this balance, which will leave you with only $4,132 at the end of year 2. Since no payment occurs in year 3, the balance will grow to $(1.10)^2$($4,132), or $5,000, at the end of year 4. The final withdrawal in the amount of $5,000 will deplete the balance completely.

EXAMPLE 3.24 Miguel Cabrera's $292 Million Contract

On March 27, 2014, Miguel Cabrera became one of the richest players in the National Baseball League by agreeing to extend his 8-year contract with the Detroit Tigers for $248 million. Cabrera's extension averages $31 million per season for eight seasons, making his the biggest contract ever paid to a baseball player. When added to the two years remaining on his current contract for $22 million per season, that is a grand total of $292 million. The annual payment schedule looks like Table 3.5.

TABLE 3.5   Cabrera's $292 Million Contract

End of Season	Current Contract	Extended Contract	Net Amount
2013			
2014	$22,000,000		$22,000,000
2015	$22,000,000		$22,000,000
2016		$31,000,000	$31,000,000
2017		$31,000,000	$31,000,000
2018		$31,000,000	$31,000,000
2019		$31,000,000	$31,000,000
2020		$31,000,000	$31,000,000
2021		$31,000,000	$31,000,000
2022		$31,000,000	$31,000,000
2023		$31,000,000	$31,000,000
Total	$44,000,000	$248,000,000	$292,000,000

How much is Cabrera's contract actually worth at the time of signing? Assume that Cabrera's interest rate is 6% per year and payments are made at the end of each season. Here we assume that the signing was taken place at the end of 2013 season or the beginning of 2014 season.

SOLUTION
Given: Payment series given in Figure 3.42, with $i = 6\%$ per year.
Find: P.

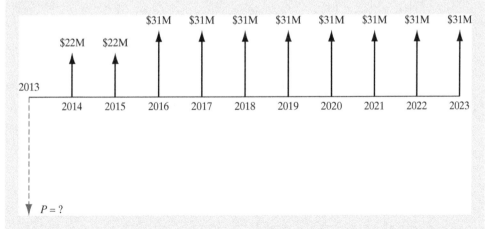

Figure 3.42   Miguel Cabrera's baseball contract with Detroit Tigers (Example 3.24).

Actual worth of the contract at the time of signing:

- **Method 1:** Summing the present values of individual cash flows:

$$
\begin{aligned}
P_{\text{contract}} &= \$22{,}000{,}000(P/F, 6\%, 1) + \$22{,}000{,}000(P/F, 6\%, 2) \\
&\quad + \$31{,}000{,}000(P/F, 6\%, 3) + \cdots + \$31{,}000{,}000(P/F, 6\%, 10) \\
&= \$211{,}662{,}165
\end{aligned}
$$

- **Method 2:** Using the multiple interest factors:

$$
P_{\text{contract}} = \overbrace{\$22{,}000{,}000(P/A, 6\%, 2)}^{\text{Equivalent present worth of the old contract}} + \underbrace{\$31{,}000{,}000(P/A, 6\%, 8)(P/F, 6\%, 2)}_{\substack{\text{Equivalent present worth of} \\ \text{the extended contract at } n = 2}}
$$

$$
= \$211{,}662{,}165
$$

Note that the actual worth of the contract is much less than the published figure of $292 million.

EXAMPLE 3.25 Future Values of an Uneven Series with Varying Interest Rates

Suppose that you have a savings account with your federal credit union. By looking at the history of the account, you learned the interest rate in each period during the last five years was as shown in Figure 3.43. Show how the credit union calculated your balance at the end of year 5.

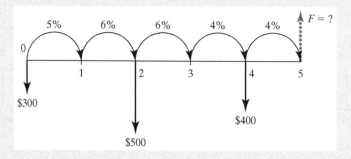

Figure 3.43 A future-value calculation with changing interest rates.

SOLUTION

Given: Deposit series shown in Figure 3.43 with varying interest rates; $N = 5$ years.

Find: Balance at the end of year 5 (F_5).

- Contribution of $300 at $n = 0$ toward F_5:

$$\underbrace{300\underbrace{\overbrace{(F/P, 5\%, 1)}^{\text{Balance at } n = 1}(F/P, 6\%, 2)}_{\text{Balance at } n = 3}(F/P, 4\%, 2)}_{\text{Balance at } n = 5} = \$382.82$$

- Contribution of $500 at $n = 2$ toward F_5:

$$\underbrace{500\overbrace{(F/P, 6\%, 1)}^{\text{Balance at } n = 3}(F/P, 4\%, 2)}_{\text{Balance at } n = 5} = \$573.25$$

- Contribution of $400 at $n = 4$ toward F_5:

$$\$400(F/P, 4\%, 1) = \$416$$

- Total balance at $n = 5$:

$$F_5 = \$382.82 + \$573.25 + \$416.00 = \$1,372.06$$

Table 3.6 on the next page summarizes the interest formulas developed in this section and the cash flow situations in which they should be used. Recall that these formulas are applicable only to situations where the interest (compounding) period is the same as the payment period (e.g., annual compounding with annual payment). Also, we present some useful Excel financial commands in the table.

TABLE 3.6 Summary of Discrete Compounding Formulas with Discrete Payments

Flow Type	Factor Notation	Formula	Excel Command	Cash Flow Diagram
S I N G L E	Compound Amount $(F/P, i, N)$	$F = P(1 + i)^N$	= FV(i,N,0,P)	
	Present Worth $(P/F, i, N)$	$P = F(1 + i)^{-N}$	= PV(i,N,0,F)	
E Q U A L	Compound Amount $(F/A, i, N)$	$F = A\left[\dfrac{(1 + i)^N - 1}{i}\right]$	= FV(i,N,A)	
P A Y M E N T	Sinking fund $(A/F, i, N)$	$A = F\left[\dfrac{i}{(1 + i)^N - 1}\right]$	= PMT(i,N,0,F)	
S E R I E S	Present worth $(P/A, i, N)$	$P = A\left[\dfrac{(1 + i)^N - 1}{i(1 + i)^N}\right]$	= PV(i,N,A)	
	Capital recovery $(A/P, i, N)$	$A = P\left[\dfrac{i(1 + i)^N}{(1 + i)^N - 1}\right]$	= PMT(i,N,P)	
G R A D I E N T	Linear gradient Present worth $(P/G, i, N)$	$P = G\left[\dfrac{(1 + i)^N - iN - 1}{i^2(1 + i)^N}\right]$		
	Equal-payment Conversion factor $(A/G, i, N)$	$A = G\left[\dfrac{(1 + i)^N - iN - 1}{i[(1 + i)^N - 1]}\right]$		
S E R I E S	Geometric gradient Present worth $(P/A_1, g, i, N)$	$P = \left[\begin{array}{l} A_1\left[\dfrac{1 - (1 + g)^N(1 + i)^{-N}}{i - g}\right] \\ A_1\left(\dfrac{N}{1 + i}\right) \text{(if } i = g\text{)} \end{array}\right]$		

3.4 Unconventional Equivalence Calculations

In the preceding section, we occasionally presented two or more methods of attacking problems, even though we had standard interest factor equations by which to solve them. It is important that you become adept at examining problems from unusual angles and that you seek out unconventional solution methods, because not all cash flow problems conform to the neat patterns for which we have discovered and developed equations. Two categories of problems that demand unconventional treatment are composite (mixed) cash flows and problems in which we must determine the interest rate implicit in a financial contract. We will begin this section by examining instances of composite cash flows.

3.4.1 Composite Cash Flows

Although many financial decisions do involve constant or systematic changes in cash flows, others contain several components of cash flows that do not exhibit an overall pattern. Consequently, it is necessary to expand our analysis to deal with such mixed types of cash flows. Also, note that in some instances we may want to find the equivalent value of a stream of payments at some point other than the present (year 0). In this situation, we proceed as before, but compound and discount to some other point in time—say, year 2, rather than year 0.

To illustrate, consider the cash flow stream shown in Figure 3.44. We may view this cash flow series in two ways.

- **Situation 1.** If you make four $100 annual deposits in your savings account, which earns 10% annual interest, what equal annual amount (*A*) can be withdrawn over four subsequent years? This is basically a savings plan where an individual makes a series

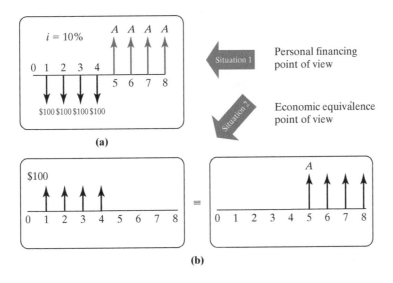

(a)

(b)

Figure 3.44 Equivalence calculations based on two different points of view.

of $100 deposits over four years, and then starts to withdraw money from this savings account (Figure 3.44a).
- **Situation 2.** What value of A would make the two cash flow transactions equivalent at an interest rate of 10%? (Figure 3.44b).

Method 1. Select the base period at $n = 0$. Then we calculate the equivalent lump-sum amount of four deposits in the amount of $100 at period 0, and the equivalent lump-sum amount of four withdrawals at period 0. The solution procedure involves the following steps.

- **Equivalent lump-sum deposit:**

$$P_{\text{deposits}} = \$100(P/A, 10\%, 4)$$
$$= \$316.99$$

- **Equivalent lump-sum withdrawals:** Find the equivalent worth of an "A" equal payment series at year 4 and then bring this equivalent worth at year 0 again:

$$P_{\text{withdrawals}} = \underbrace{A(P/A, 10\%, 4)}_{V_1}(P/F, 10\%, 4)$$
$$= 2.1651A$$

- **Find the unknown (A):** These two amounts must be the same, so we equate them and solve for unknown A.

$$2.1651A = \$316.99$$
$$A = \$146.41$$

A pictorial view of this computational process is given in Figure 3.45.

Method 2. In computing the equivalent worth of the equal payment series components, we may select the base period at $n = 4$.

- **Equivalent lump-sum deposits at $n = 4$:**

$$V_{4, \text{deposits}} = \$100(F/A, 10\%, 4)$$
$$= \$464.10$$

- **Equivalent lump-sum withdrawals at $n = 4$:**

$$V_{4, \text{withdrawals}} = A(P/A, 10\%, 4)$$
$$= 3.1699A$$

- **Find the unknown A:**

$$3.1699A = \$464.10$$
$$A = \$146.41$$

By selecting the base period at $n = 4$, we may calculate the unknown A more efficiently with the lesser number of interest factors, as shown in Figure 3.46.

Method 1: Establish economic equivalence
at period 0

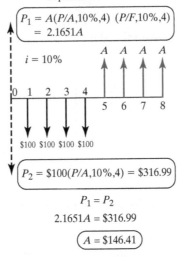

$P_1 = A(P/A,10\%,4) \ (P/F,10\%,4)$
$\quad = 2.1651A$

$i = 10\%$

$A \quad A \quad A \quad A$

0 1 2 3 4

5 6 7 8

\$100 \$100 \$100 \$100

$P_2 = \$100(P/A,10\%,4) = \$316.99$

$P_1 = P_2$

$2.1651A = \$316.99$

$A = \$146.41$

Figure 3.45   Establishing economic equivalence by selecting base period at $n = 0$.

Method 2: Establish economic equivalence
at period 4

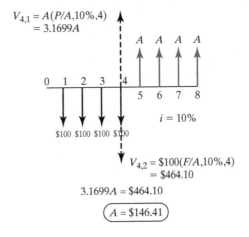

$V_{4,1} = A(P/A,10\%,4)$
$\quad = 3.1699A$

$A \quad A \quad A \quad A$

0 1 2 3 4

5 6 7 8

$i = 10\%$

\$100 \$100 \$100 \$100

$V_{4,2} = \$100(F/A,10\%,4)$
$\quad = \$464.10$

$3.1699A = \$464.10$

$A = \$146.41$

Figure 3.46   Establishing economic equivalence by selecting base period at $n = 4$.

EXAMPLE 3.26 Cash Flows with Subpatterns

The two cash flows in Figure 3.47 are equivalent at an interest rate of 12% compounded annually. Determine the unknown value C.

Figure 3.47   Equivalence calculation (Example 3.26).

SOLUTION

Given: Cash flows as in Figure 3.47; $i = 12\%$ per year.
Find: C.

- **Method 1.** Compute the present worth of each cash flow at time 0:

$$P_1 = \$100(P/A, 12\%, 2) + \$300(P/A, 12\%, 3)(P/F, 12\%, 2)$$

$$= \$743.42;$$

$$P_2 = C(P/A, 12\%, 5) - C(P/F, 12\%, 3)$$

$$= 2.8930C$$

Since the two flows are equivalent, $P_1 = P_2$ and we have

$$743.42 = 2.8930C$$

Solving for C, we obtain $C = \$256.97$.

- **Method 2.** We may select a time point other than 0 for comparison. The best choice of a base period is determined largely by the cash flow patterns. Obviously, we want to select a base period that requires the minimum number of interest factors for the equivalence calculation. Cash flow 1 represents a combined series of two equal payment cash flows, whereas cash flow 2 can be viewed as an equal payment series with the third payment missing. For cash flow 1, computing the equivalent worth at period 5 will require only two interest factors:

$$V_{5,1} = \$100(F/A, 12\%, 5) + \$200(F/A, 12\%, 3)$$

$$= \$1,310.16$$

For cash flow 2, computing the equivalent worth of the equal payment series at period 5 will also require two interest factors:

$$V_{5,2} = C(F/A, 12\%, 5) - C(F/P, 12\%, 2)$$

$$= 5.0984C$$

Therefore, the equivalence would be obtained by letting $V_{5,1} = V_{5,2}$:

$$\$1,310.16 = 5.0984C$$

Solving for C yields $C = \$256.97$, which the same result is obtained from Method 1. The alternative solution of shifting the time point of comparison will require only four interest factors, whereas Method 1 requires five interest factors.

EXAMPLE 3.27 Establishing a College Fund

A couple with a newborn daughter wants to save for their child's college expenses in advance. The couple can establish a college fund that pays 7% annual interest. Assuming that the child enters college at age 18, the parents estimate that an amount of \$40,000 per year (actual dollars) will be required to support the child's college expenses for four years. Determine the equal annual amounts the couple must save until they send their child to college. (Assume that the first deposit will be made on the child's first birthday and the last deposit on the child's 18th birthday. The first withdrawal will be made at the beginning of the freshman year, which also is the child's 18th birthday.)

Figure 3.48 Establishing a college fund (Example 3.27). Note that the \$40,000 figure represents the actual anticipated expenditures, considering future inflation.

SOLUTION

Given: Deposit and withdrawal series shown in Figure 3.48; $i = 7\%$ per year.
Find: Unknown annual deposit amount (X).

- **Method 1.** Establish economic equivalence at period 0.
  Step 1: Find the equivalent single lump-sum deposit now:

$$P_{\text{Deposit}} = X(P/A, 7\%, 18)$$
$$= 10.0591X$$

Step 2: Find the equivalent single lump-sum withdrawal now:

$$P_{\text{Withdrawal}} = \$40,000(P/A, 7\%, 4)(P/F, 7\%, 17)$$

$$= \$42,892$$

Step 3: Since the two amounts are equivalent, by equating $P_{\text{Deposit}} = P_{\text{Withdrawal}}$ we obtain X.

$$10.0591X = \$42,892$$

$$X = \$4,264$$

- **Method 2.** Establish the economic equivalence at the child's 18th birthday.
  Step 1: Find the accumulated deposit balance on the child's 18th birthday.

$$V_{18} = X(F/A, 7\%, 18)$$

$$= 33.9990X$$

Step 2: Find the equivalent lump-sum withdrawal on the child's 18th birthday:

$$V_{18} = \$40,000 + \$40,000(P/A, 7\%, 3)$$

$$= \$144,972$$

Step 3: Since the two amounts must be the same, we obtain

$$33.9990X = \$144,972$$

$$X = \$4,264$$

The computational steps are summarized in Figure 3.49. In general, the second method is the more efficient way to obtain an equivalence solution to this type of decision problem.

COMMENTS: To verify that the annual deposits of $4,264 over 18 years would be sufficient to meet the child's college expenses, we can calculate the actual year-by-year balances: With the 18 annual deposits of $4,264, the balance on the child's 18th birthday is

$$\$4,264(F/A, 7\%, 18) = \$144,972$$

From this balance, the couple will make four annual tuition payments:

Year N	Beginning Balance	Interest Earned	Tuition Payment	Ending Balance
Freshman	$144,972	$ 0	$40,000	$104,972
Sophomore	104,972	7,348	40,000	72,320
Junior	72,320	5,062	40,000	37,382
Senior	37,382	2,618	40,000	0

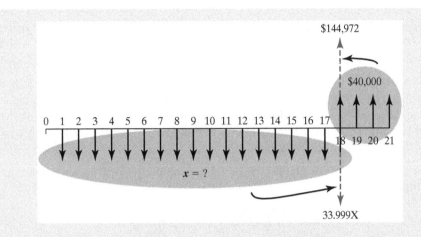

Figure 3.49   An alternative equivalence calculation (Example 3.27).

3.4.2 Determining an Interest Rate to Establish Economic Equivalence

Thus far, we have assumed that, in equivalence calculations, a typical interest rate is given and invariant. Now we can use the same interest formulas that we developed earlier to determine interest rates that are explicit in equivalence problems. For most commercial loans, interest rates are already specified in the contract. However, when making some investments in financial assets, such as stocks, you may want to know the rate of growth (or rate of return) at which your asset is appreciating over the years. (This kind of calculation is the basis of rate-of-return analysis, which is covered in Chapter 7.) Although we can use interest tables to find the rate that is implicit in single payments and annuities, it is more difficult to find the rate that is implicit in an uneven series of payments. In such cases, a trial-and-error procedure or computer software may be used. To illustrate, consider Example 3.28.

EXAMPLE 3.28   Calculating an Unknown Interest Rate

Reconsider Example 3.17. At this point, we may be interested in knowing at just what rate of return the couple's deal (receiving $9,791,667 a year) would in fact make sense.

 Option 1: Take a lump sum payment in the amount of $192,373,928 now.
 Option 2: Take the 30-installment option.

At what interest rate on the couple's savings (or earning opportunity) will make these two options equivalent?

SOLUTION
Given: $P = \$192{,}373{,}928$, $N = 30$, and $A = \$9{,}791{,}667$.
Find: i at which the two options are equivalent.

In this problem, you are looking for the interest rate that caused the *P/A* factor to equal $(P/A, i, 30) = (\$192,373,928/\$9,791,667) = 19.6467$. Since we are dealing with an annuity, we could proceed as follows.

- To use the interest tables, first recognize that $\$192,373,928 = \$9,791,667 \times (P/A, i, 30)$ or $(P/A, i, 30) = 19.6467$. Look up 19.6467 or a close value in Appendix B. In the *P/A* column with $N = 30$ in the 3% interest table, you will find that $(P/A, 3\%, 30) = 19.6004$. If you look up the 2% interest table, you find that $(P/A, 2\%, 30) = 22.3965$.

When using the interest tables, we need to eventually resort to a trial-and-error method.

Interest Rate	(P/A, i, 30)
2%	22.3965
?	19.6467
3%	19.6004

As a result, we find that the interest rate is between 2% and 3% and may be approximated by **linear interpolation**, as shown in Figure 3.50:

$$i = 2\% + (3\% - 2\%)\left[\frac{22.3965 - 19.6467}{22.3965 - 19.6004}\right]$$

$$= 2\% + 1\%\left[\frac{2.7498}{2.7961}\right]$$

$$= 2.9834\%$$

Figure 3.50 Linear interpolation to find an unknown interest rate.

- To use Excel's financial command, you simply evaluate the following command to solve the unknown interest rate problem for an annuity:

$$= \texttt{RATE}\,(N,A,P,F,\textit{type},\textit{guess})$$

$$= \texttt{RATE}\,(30,9791667,-192373928,0,0,3\%)$$

$$= 2.98177\%$$

It is more than likely that the couple will find a financial investment which provides a rate of return higher than 3% in the marketplace.

COMMENTS: This example demonstrates that finding an interest rate is an iterative process that is more complicated and generally less precise than finding an equivalent worth at a known interest rate. Since computers and financial calculators can speed the process of finding unknown interest rates, such tools are highly recommended for these types of problem solving.

3.4.3 Unconventional Regularity in Cash Flow Pattern

Sometimes we may need to deal with the situation where there is some regularity in a cash flow pattern, but we can't directly use the standard interest factors to calculate the equivalent value. For example, you may need to paint the house every five years during your ownership. The painting expense occurs every five years, instead of every year, but as you look at the five-year cycle, it is a kind of uniform series as well. To illustrate, consider Example 3.29.

EXAMPLE 3.29 Equivalence Calculations for a Skipping Pattern in Cash Flows

Suppose that you purchased industrial equipment that requires special software to run the machine. The licensing agreement calls for a $20,000 payment immediately and $10,000 every other year to renew and get any update on the software. If the machine is to be kept for 12 years, determine the total equivalent cost in present worth at an interest rate of 10%.

STRATEGY: Since the cash flow occurs every other year, let's find out the equivalent compound interest rate that covers the two-year period. In our example, if interest is compounded 10% annually, the equivalent interest rate for a two-year period is

$$i = (1 + 0.10)(1 + 0.10) - 1 = 1.21 - 1 = 21\%$$

We then interpret the problem as having a $10,000 uniform series over six periods.

SOLUTION

Given: Payment series shown in Figure 3.51 and $i = 10\%$.
Find: P.

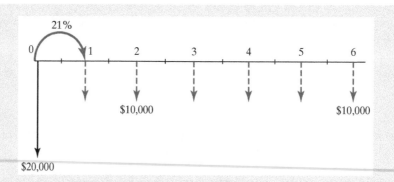

Figure 3.51   A cash flow diagram with a skipping cash flow pattern.

The present-worth calculation is

$$P = \$20,000 + \$10,000(P/A, 21\%, 6)$$
$$= \$52,446$$

SUMMARY

- Money has a time value because it can earn more money over time. A number of terms involving the time value of money were introduced in this chapter:

  Interest is the cost of money. More specifically, it is a cost to the borrower and an earning to the lender, above and beyond the initial sum borrowed or loaned.

  Interest rate is a percentage periodically applied to a sum of money to determine the amount of interest to be added to that sum.

  Simple interest is the practice of charging an interest rate only to an initial sum.

  Compound interest is the practice of charging an interest rate to an initial sum *and* to any previously accumulated interest that has not been withdrawn from the initial sum. Compound interest is by far the most commonly used system in the real world.

  Economic equivalence exists between individual cash flows and/or patterns of cash flows that have the same value. Even though the amounts and timing of the cash flows may differ, the appropriate interest rate makes them equal.

- The compound-interest formula, perhaps the single most important equation in this text, is

  $$F = P(1 + i)^N$$

  where P is a present sum, i is the interest rate, N is the number of periods for which interest is compounded, and F is the resulting future sum. All other important interest formulas are derived from this one.

- **Cash flow diagrams** are visual representations of cash inflows and outflows along a timeline. They are particularly useful for helping us detect which of the following five patterns of cash flow is represented by a particular problem.

  1. **Single payment.** A single present or future cash flow.

  2. **Uniform series.** A series of flows of equal amounts at regular intervals.

  3. **Linear-gradient series.** A series of flows increasing or decreasing by a fixed amount at regular intervals.

4. Geometric-gradient series. A series of flows increasing or decreasing by a fixed percentage at regular intervals.

5. Irregular series. A series of flows exhibiting no overall pattern. However, patterns might be detected for portions of the series.

■ **Cash flow patterns** are significant because they allow us to develop **interest formulas**, which streamline the solution of equivalence problems. Table 3.6 summarizes the important interest formulas that form the foundation for all other analyses you will conduct in engineering economic analysis.

PROBLEMS

Types of Interest

3.1 If you open a savings account that earns 6.5% simple interest per year, what is the minimum number of years you must wait to triple your balance? Suppose you open another account that earns 6% interest compounded yearly. How many years will it take now to triple your balance?

3.2 Compare the interest earned by $15,000 for ten years at 8% simple interest with that earned by the same amount for ten years at 8% compounded annually.

3.3 You are considering investing $5,400 at 9% per year simple interest for five years or investing the $5,400 at an interest rate of 8% compounded annually for five years. Which option is better?

3.4 You are about to borrow $25,000 from a bank at an interest rate of 10% compounded annually. You are required to make three equal annual repayments in the amount of $10,052.87 per year, with the first repayment occurring at the end of year 1. Show the interest payment and principal payment in each year.

Equivalence Concept

3.5 Suppose you have the alternative of receiving either $24,000 at the end of four years or P dollars today. Currently, you have no need for money, so you could deposit the P dollars in a bank that pays 4% interest. What value of P would make you indifferent in your choice between P dollars today and the promise of $24,000 at the end of four years?

3.6 Suppose that you are obtaining a personal loan from your uncle in the amount of $40,000 (now) to be repaid in three years to cover some of your college expenses. If your uncle usually earns 8% interest (annually) on his money, which is invested in various sources, what minimum lump-sum payment three years from now would make your uncle happy?

3.7 If you deposited $200 now ($n = 0$) and $500 three years from now ($n = 3$) in a savings account that pays 10% annual interest, how much would you have at the end of year 12?

3.8 At what interest rate are these two transactions in Figure P3.8 equivalent?

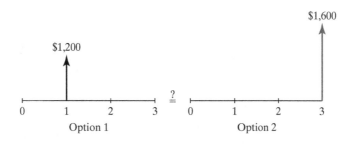

Figure P3.8

Single Payments (Use of F/P or P/F Factors)

3.9 The average price of a new home is $520,000. If new home prices are increasing at a rate of 10% per year, how much will a new home cost in 6 years?

3.10 What will be the amount accumulated by each of these present investments?

(a) $4,000 in 6 years at 7% compounded annually.

(b) $6,750 in 15 years at 9% compounded annually.

(c) $10,500 in 25 years at 6% compounded annually.

(d) $15,000 in 8 years at 7.5% compounded annually.

3.11 You are interested in buying a piece of real property that could be worth $300,000 in 8 years. Assuming that your money is worth 10%, how much would you be willing to pay for the property?

3.12 What is the present worth of these future payments?

(a) $25,500 eight years from now at 12% compounded annually.

(b) $58,000 twelve years from now at 4% compounded annually.

(c) $25,000 nine years from now at 6% compounded annually.

(d) $35,000 four years from now at 9% compounded annually.

3.13 For an interest rate of 12% compounded annually, determine the following:

(a) How much can be lent now if $13,000 will be repaid at the end of five years?

(b) How much will be required in four years to repay a $35,000 loan received now?

3.14 How many years will it take an investment to double if the interest rate is 6% compounded annually?

3.15 You bought 300 shares of General Electric (GE) stock at $7,800 on January 2, 2015. Your intention is to keep the stock until it doubles in value. If you expect 6% annual growth for GE stock, how many years do you anticipate holding onto the stock? Compare your answer with the solution obtained by the Rule of 72 (discussed in Example 3.10).

3.16 On October 1, 1970, Walmart first offered 300,000 shares of its common stock to the public at a price of $16.50 per share. Since that time, Walmart has had 11 two-for-one stock splits. So if you purchased 100 shares in 1970, you would have 204,800 shares as of September 30, 2014. The closing price of the stock on that day was $77.50. What is the compound growth of the Walmart stock over 44 years? (This is equivalent to asking what kind of interest rate you have to earn to match the same growth from your savings account.)

3.17 Warren Buffett's Berkshire Hathaway Company went public in 1965. The public offering price was $18 per share. The stock was trading at $190,500 on July 1, 2014, and the market value of the firm was $304.8 billion.

(a) What is the firm's average annual compound growth rate over last 49 years?

(b) If Mr. Buffett's company continues to grow at the historical growth rate, what will be his company's total market value when he reaches the age of 100? (Buffett celebrated his 83th birthday in 2014.)

Uneven Payment Series

3.18 Consider the following sequence of deposits and withdrawals over a period of four years. If you earn 10% interest on your deposits, how much will you able to withdraw at the end of four years?

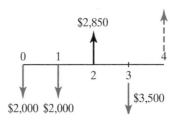

Figure P3.18

3.19 If you desire to withdraw the amounts given in Table P3.19 over the next five years from a savings account that earns 8% interest compounded annually, how much do you need to deposit now?

TABLE P3.19

N	Amount
2	$13,000
3	25,000
4	38,000
5	46,000

3.20 If $2,200 is invested now, $2,800 three years from now, and $3,500 five years from now at an interest rate of 6.5% compounded annually, what will be the total amount in 8 years?

3.21 A local newspaper headline blared, "Bo Smith Signed for $30 Million." A reading of the article revealed that on April 1, 2014, Bo Smith, the former record-breaking running back from Football University, signed a $30 million package with the Dallas Rangers. The terms of the contract were $3 million immediately, $2.4 million per year for the first five years (with the first payment after one year) and $3 million per year for the next five years (with the first payment at year 6). If Bo's interest rate is 8% per year, what would his contract be worth at the time he signs it?

3.22 How much invested now at 7% would be just sufficient to provide three payments, with the first payment in the amount of $10,000 occurring two years hence, then $7,500 four years hence, and finally $4,000 eight years hence?

Equal Payment Series

3.23 What is the future worth of a series of equal year-end deposits of $5,000 for 12 years in a savings account that earns 8% annual interest if the following were true?

(a) All deposits are made at the *end* of each year.
(b) All deposits are made at the *beginning* of each year.

3.24 You are paying into a mutual fund that earns 7% compound interest. If you are making an annual contribution of $14,000, how much will be in the funds in 25 years?

3.25 What is the future worth of the following series of payments?

(a) $9,000 at the end of each year for 6 years at 11.25% compounded annually.
(b) $4,500 at the end of each year for 15 years at 4.75% compounded annually.
(c) $8,000 at the end of each year for 18 years at 6.45% compounded annually.
(d) $6,200 at the end of each year for 20 years at 8.5% compounded annually.

3.26 What equal annual series of payments must be paid into a sinking fund to accumulate the following amounts?

(a) $54,000 in 12 years at 7% compounded annually.
(b) $38,000 in 15 years at 8% compounded annually.
(c) $25,000 in 8 years at 6% compounded annually.
(d) $13,000 in 10 years at 10% compounded annually.

3.27 You want to save money from your business operation to replace a truck that has been used in delivery. The truck will be replaced after 12 years from now and the replacement cost would be about $60,000. If you earn 5% interest on your savings, how much must you deposit at the end of each year to meet the needs?

3.28 Part of the income that a machine generates is put into a sinking fund to replace the machine when it wears out. If $2,400 is deposited annually at 5% interest, how many years must the machine be kept before a new machine costing $45,000 can be purchased?

3.29 A no-load (commission-free) mutual fund has grown at a rate of 12% compounded annually since its beginning. If it is anticipated that it will continue to grow at that rate, how much must be invested every year so that $26,000 will be accumulated at the end of seven years?

3.30 To help you reach a $16,000 goal 8 years from now, your father offers to give you $4,000 now. You plan to get a part-time job and make five additional deposits, one at the end of each year. (The first deposit is made at the end of the first year.) If all your money is deposited in a bank that pays 6% interest, how large must your annual deposit be?

3.31 What equal annual payment series is required to repay the following present amounts?

(a) $14,000 in five years at 7.5% interest compounded annually.
(b) $7,500 in seven years at 6.25% interest compounded annually.
(c) $5,000 in six years at 5% interest compounded annually.
(d) $20,000 in 18 years at 4.75% interest compounded annually.

3.32 From the interest tables in Appendix B, determine the values of the following factors by interpolation and compare your results with those obtained from evaluating the *A/P* and *P/A* interest formulas.

(a) The capital-recovery factor for 38 periods at 6.25% interest.

(b) The equal-payment series present-worth factor for 85 periods at 9.25% interest.

3.33 You have borrowed $60,000 at an interest rate of 10.5%. Equal payments will be made over a five-year period. (The first payment will be made at the end of the first year.) What will the annual payment be, and what will the interest payment be for the fourth year?

3.34 You borrowed $18,000 from a bank with the agreement that you repay the balance in 12 equal annual installments at an interest rate of 9.5%. What should be the required annual payment size?

3.35 What is the present worth of the following series of payments?

(a) $1,500 at the end of each year for 10 years at 7.25% compounded annually.

(b) $4,900 at the end of each year for seven years at 9.75% compounded annually.

(c) $1,000 at the end of each year for 15 years at 8.2% compounded annually.

(d) $19,500 at the end of each year for eight years at 7.5% compounded annually.

3.36 You are considering buying a piece of industrial equipment to automate a part of your production process. This automation will save labor costs by as much as $42,000 per year over 12 years. The equipment costs $270,000. Should you purchase the equipment if your interest rate is 10%?

3.37 Recently, Matt Ryan, an NFL quarterback, agreed to a five-year, $103.75 million contract extension with the Atlanta Falcons, which made him one of the highest-paid players in professional football history. The contract consisted of a $30 million signing bonus and average annual salary of $20.75 million. The $30 million signing bonus was pro-rated over the course of the contract, in such that $2 million was paid upfront and remaining balance over the five-year contract period. The detail breakdown of the contract is as follows:

Year	Base	Signing Bonus
2013	$ 2,000,000	$7,600,000
2014	9,500,000	5,600,000
2015	11,500,000	5,600,000
2016	15,750,000	5,600,000
2017	15,750,000	5,600,000
2018	19,250,000	

(a) How much was the quarterback's contract actually worth at the time of signing (year 2013)? Assume that his interest rate is 6%.

(b) For the signing bonus portion, suppose that the quarterback was allowed to take either the pro-rated payment option as just described, or a lump-sum payment option in the amount of $25 million at the time he signed the contract. Should he have taken the lump-sum option instead of the pro-rated one?

Linear-Gradient Series

3.38 An individual deposits an annual bonus into a savings account that pays 5% interest compounded annually. The size of the bonus increases by $6,000 each year, and the initial bonus amount was $30,000. Determine how much will be in the account immediately after the fifth deposit.

3.39 Six annual deposits in the amounts of $12,000, $10,000, $8,000, $6,000, $4,000, and $2,000, in that order, are made into a fund that pays interest at a rate of 7% compounded annually. Determine the amount in the fund immediately after the sixth deposit.

3.40 Compute the value of P in the accompanying cash flow diagram, assuming that $i = 8\%$.

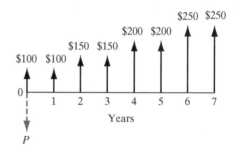

Figure P3.40

3.41 What is the equal payment series for 12 years that is equivalent to a payment series of $25,000 at the end of the first year, decreasing by $1,200 each year over 12 years? Interest is 7% compounded annually.

3.42 How much do you have to deposit now in your savings account that earns an 8% annual interest if you want to withdraw the annual payment series in the figure below?

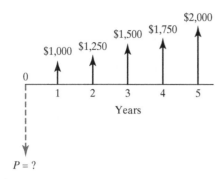

Figure P3.42

Geometric-Gradient Series

3.43 Anderson Manufacturing Co., a small fabricator of plastics, needs to purchase an extrusion molding machine for $150,000. Anderson will borrow money from a bank at an interest rate of 7% over five years. Anderson expects its product sales to be slow during the first year, but to increase subsequently at an annual rate of 10%. Anderson therefore arranges with the bank to pay off the loan on a "balloon scale," which results in the lowest payment at the end of the first year and each subsequent payment being just 10% higher over the previous one. Determine the five annual payments.

3.44 Joe's starting salary as a mechanical engineer is around $75,000. Joe is planning to place a total of 12% of his salary each year in the mutual fund. Joe expects a 6% salary increase each year for the next 28 years of employment. If the mutual fund will average 8% annual return over the course of his career, what can Joe expect at retirement?

3.45 Suppose that an oil well is expected to produce 150,000 barrels of oil during its first year in production. However, its subsequent production (yield) is expected to decrease by 10% over the previous year's production. The oil well has a proven reserve of 1,500,000 barrels.

(a) (a) Suppose that the price of oil is expected to be $64 per barrel for the next several years. What would be the present worth of the anticipated revenue stream at an interest rate of 12% compounded annually over the next seven years?

(b) (b) Suppose that the price of oil is expected to start at $64 per barrel during the first year, but to increase at the rate of 5% over the previous year's price. What would be the present worth of the anticipated revenue stream at an interest

rate of 12% compounded annually over the next seven years?

(c) Consider part (b) again. After three years' production, you decide to sell the oil well. What would be a fair price?

3.46 A city engineer has estimated the annual toll revenues from a newly proposed highway construction over 25 years as follows:

$$A_n = (2,500,000)(n)(1.08)^{n-1}$$
$$n = 1, 2, \ldots, 25$$

To validate the bond, the engineer was asked to present the estimated total present value of toll revenue at an interest rate of 8%. Assuming annual compounding, find the present value of the estimated toll revenue.

3.47 What is the amount of 15 equal annual deposits that can provide five annual withdrawals? A first withdrawal of $22,000 is made at the end of year 16 and subsequent withdrawals increase at the rate of 9% per year over the previous year's withdrawal. Determine the amounts from the following rates.

(a) The interest rate is 8.4% compounded annually.

(b) The interest rate is 6.3% compounded annually.

Various Interest Factor Relationships

3.48 By using only those factors given in interest tables, find the values of the factors that follow, which are not given in your tables. Show the relationship between the factors by using factor notation, and calculate the value of the factor. Then compare the solution you obtained by using the factor formulas with a direct calculation of the factor values.

For example, $(F/P, 8\%, 38) =$
$(F/P, 8\%, 30) (F/P, 8\%, 8) = 18.6253$.

(a) $(P/F, 8\%, 67)$

(b) $(A/P, 8\%, 42)$

(c) $(P/A, 8\%, 135)$

3.49 Prove the following relationships among the following interest factors.

(a) $(F/P, i, N) = i(F/A, i, N) + 1$

(b) $(P/F, i, N) = 1 - (P/A, i, N)i$

(c) $(A/F, i, N) = (A/P, i, N) - i$

(d) $(A/P, i, N) = i/[1 - (P/F, i, N)]$

(e) $(P/A, i, N \to \infty) = 1/i$

(f) $(A/P, i, N \to \infty) = i$

(g) $(P/G, i, N \to \infty) = 1/i^2$

Equivalence Calculations

3.50 Find the present worth of the cash receipts where $i = 10\%$ compounded annually with only four interest factors.

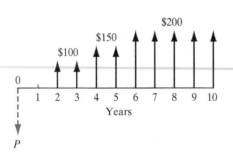

Figure P3.50

3.51 Find the equivalent present worth of the cash receipts where $i = 9\%$. In other words, how much do you have to deposit now (with the second deposit in the amount of $300 at the end of the first year) so that you will be able to withdraw $300 at the end of second year, $180 at the end of third year, and so forth if the bank pays you an 9% annual interest on your balance?

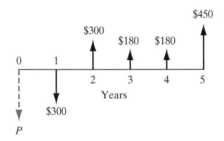

Figure P3.51

3.52 What value of A makes the two annual cash flows equivalent at 14% interest compounded annually?

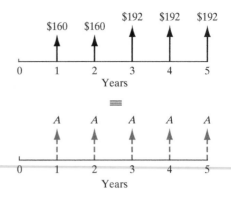

Figure P3.52

3.53 The two cash flow transactions shown in the accompanying cash flow diagram are said to be equivalent at 9% interest compounded annually. Find the unknown value of X that satisfies the equivalence.

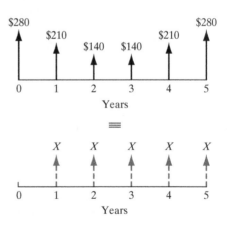

Figure P3.53

3.54 Solve for the present worth of this cash flow using at most three interest factors at 12% interest compounded annually.

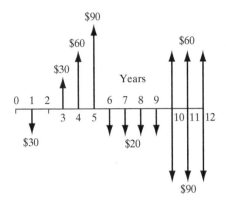

Figure P3.54

3.55 From the accompanying cash flow diagram, find the value of C that will establish the economic equivalence between the deposit series and the withdrawal series at an interest rate of 10% compounded annually.

(a) $2,082

(b) $1,375

(c) $1,506

(d) $1,388

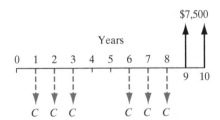

Figure P3.55

3.56 The following equation describes the conversion of a cash flow into an equivalent equal payment series with $N = 10$:

$$A = [625 + 40(A/G, 8\%, 7)]$$
$$\times (P/A, 8\%, 7)(A/P, 8\%, 10)$$
$$+ [440(F/A, 8\%, 3) - 500](A/F, 8\%, 10)$$
$$= \$629.04$$

Reconstruct the original cash flow diagram.

3.57 Consider the cash flow shown in the accompanying diagram. What value of C makes the inflow series equivalent to the outflow series at an interest rate of 9%?

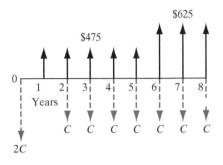

Figure P3.57

3.58 Find the value of X so that the two cash flows shown in the diagram are equivalent for an interest rate of 9%.

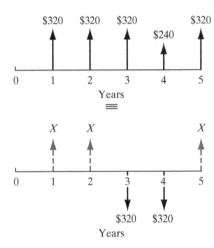

Figure P3.58

3.59 What single amount at the end of the sixth year is equivalent to a uniform annual series of $7,500 per year for 12 years if the interest rate is 10% compounded annually?

3.60 From the following list, identify all the correct equations used in computing either the future worth (F) or equivalent present worth (P) for the cash flow shown at $i = 12\%$.

(a) $F = R(F/A, 12\%, 6)$

(b) $F = R(F/A, 12\%, 5) + R$

(c) $F = R(F/A, 12\%, 5) + R(F/P, 12\%, 5)$

(d) $F = 2R(F/A, 12\%, 3)$

(e) $P = R + R(P/A, 12\%, 3) + R(P/A, 12\%, 2)$
$(P/F, 12\%, 3)$

(f) $P = R(P/A, 12\%, 6) - R$

(g) $P = R + R(P/A, 12\%, 5)$

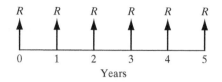

Figure P3.60

3.61 On the day his baby is born, a father decides to establish a savings account for the child's college education. Any money that is put into the account will earn an interest rate of 9% compounded annually. The father will make a series of annual deposits in equal amounts on each of his child's birthdays from the 1st through the 18th, so that the child can make five annual withdrawals from the account in the amount of $45,000 on each birthday. Assuming that the first withdrawal will be made on the child's 18th birthday, which of the following equations are correctly used to calculate the required annual deposit?

(a) $A = \$45,000(F/A, 9\%, 5)$
$\times (P/F, 9\%, 22)(A/P, 9\%, 18)$

(b) $A = \$45,000(P/A, 9\%, 18)$
$\times (F/P, 9\%, 22)(A/F, 9\%, 5)$

(c) $A = \$45,000[(P/F, 9\%, 18) + (P/F, 9\%, 19)$
$+ (P/F, 9\%, 20) + (P/F, 9\%, 21)$
$+ (P/F, 9\%, 22)](A/P, 9\%. 18)$

(d) $A = [\$45,000(P/A, 9\%, 4)$
$+ \$45,000](A/F, 9\%, 18)$

(e) $A = (\$45,000 \times 5)/18$

3.62 Find the equivalent equal payment series (A) using an A/G factor such that the two cash flows are equivalent at 10% compounded annually.

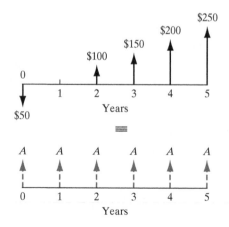

Figure P3.62

TABLE P3.63

Year End	Payment
0	$500
1–5	$1,000

3.63 Consider the following cash flow.

In computing F at the end of year 5 at an interest rate of 12%, which of the following equations is *incorrect*?

(a) $F = \$1,000(F/A, 12\%, 5) - \$500(F/P, 12\%, 5)$

(b) $F = \$500(F/A, 12\%, 6) + \$500(F/A, 12\%, 5)$

(c) $F = [\$500 + \$1,000(P/A, 12\%, 5)]$
$\times (F/P, 12\%, 5)$

(d) $F = [\$500(A/P, 12\%, 5) + \$1,000]$
$\times (F/A, 12\%, 5)$

3.64 Consider the cash flow series given. In computing the equivalent worth at $n = 4$, which of the following equations is *incorrect*?

(a) $V_4 = [\$100(P/A, i, 6) - \$100(P/F, i, 4)]$
$(F/P, i, 4)$

(b) $V_4 = \$100(F/A, i, 3) + \$100(P/A, i, 2)$

(c) $V_4 = \$100(F/A, i, 4) - \$100 + \$100(P/A, i, 2)$

(d) $V_4 = [\$100(F/A, i, 6) - \$100(F/P, i, 2)]$
$(P/F, i, 2)$

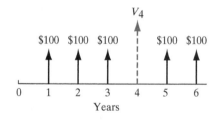

Figure P3.64

3.65 Henry Cisco is planning to make two deposits: $25,000 now and $30,000 at the end of year 6. He wants to withdraw C at the end of each year for the first six years and (C + $1,000) each year for the next six years. Determine the value of C if the deposits earn 10% interest compounded annually.

(a) $7,711

(b) $5,794

(c) $6,934

(d) $6,522

Solving for an Unknown Interest Rate or Unknown Interest Periods

3.66 At what rate of interest compounded annually will an investment double in five years?

3.67 Determine the interest rate (i) that makes the pairs of cash flows shown economically equivalent.

Figure P3.67

3.68 You have $15,000 available for investment in stock. You are looking for a growth stock whose value can grow to $40,000 over five years. What kind of growth rate are you looking for?

3.69 How long will it take to save $1 million if you invest $2,000 each year at 6%?

3.70 You may have already won $2 million! Just peel the game piece off the Instant Winner Sweepstakes ticket, and mail it to us along with your order for subscriptions to your two favorite magazines. As a grand-prize winner, you may choose between a $1 million cash prize paid immediately or $100,000 per year for 20 years—that's $2 million! Suppose that, instead of receiving one lump sum of $1 million, you decide to accept the 20 annual installments of $100,000. If you are like most jackpot winners, you will be tempted to spend your winnings to improve your lifestyle during the first several years. Only after you get this type of spending "out of your system" will you save later sums for investment purposes. Suppose that you are considering the following two options.

Option 1: You save your winnings for the first seven years and then spend every cent of the winnings in the remaining 13 years.

Option 2: You do the reverse, spending for seven years and then saving for 13 years.

If you can save winnings at 7% interest, how much would you have at the end of 20 years, and what interest rate on your savings will make these two options equivalent?

3.71 Read the following letter from a magazine publisher.

> Dear Parent:
>
> Currently your *Growing Child/Growing Parent* subscription will expire with your 24-month issue. To renew on an annual basis until your child reaches 72 months would cost you a total of $63.84 ($15.96 per year). We feel it is so important for you to continue receiving this material until the 72nd month that we offer you an opportunity to renew now for $57.12. Not only is this a savings of 10% off the regular rate, but it is an excellent inflation hedge for you against increasing rates in the future. Please act now by sending $57.12.

(a) If your money is worth 6% per year, determine whether this offer can be of any value.

(b) What rate of interest would make you indifferent between the two renewal options?

3.72 Fairmont Textile has a plant in which employees have been having trouble with carpal tunnel syndrome (CTS, an inflammation of the nerves that pass through the carpal tunnel, a tight space at the base of the palm), resulting from long-term repetitive activities, such as years of sewing. It seems as if 15 of the employees working in this facility developed signs of CTS over the last five years. DeepSouth, the company's insurance firm, has been increasing Fairmont's liability insurance steadily because of this problem. DeepSouth is willing to lower the insurance premiums to $16,000 a year (from the current $30,000 a year) for the next five years if Fairmont implements an acceptable CTS-prevention program that includes making the employees aware of CTS and how to reduce the chances of it developing. What would be the maximum amount that Fairmont should invest in the program to make it worthwhile? The firm's interest rate is 12% compounded annually.

Short Case Studies

ST3.1 A judge denied a 94-year-old woman's attempt to force the Massachusetts Lottery Commission to pay her entire $5.6 million winnings up front on the grounds that she otherwise won't live long enough to collect it all. The ruling means that the commission can pay Louise Outing, a retired waitress, in installments over 20 years. After an initial gross payment of $283,770, Outing would be paid 19 annual gross checks of $280,000. That's about $197,000 after taxes. Lottery Executive Director Joseph Sullivan said all players are held to the same rules, which are printed on the back of Megabucks tickets. Lottery winners are allowed to "assign" their winnings to a state-approved financial company that makes the full payment—but only in return for a percentage of the total winnings. Outing, who won a Megabucks drawing in September, has seven grandchildren, nine great-grandchildren, and six great-great-grandchildren. "I'd like to get it and do what I want with it," she said. "I'm not going to live 20 years. I'll be 95 in March."

(a) Suppose that she were able to find an investor who was willing to buy her lottery ticket for $2 million. Recall that after an initial gross payment of $283,770, Outing would be paid 19 annual gross checks of $280,000. If she could invest her money at 8% interest, what would be the fair amount to trade her 19 future lottery receipts? (Note that she already cashed in $283,770 after winning the lottery, so she is giving up 19 future lottery checks in the amount of $280,000.)

(b) What is the rate of return for this investor who is willing to buy Ms. Outing's lottery ticket?

ST3.2 Yuma, Arizona resident Rosalind Setchfield won $1.3 million in an Arizona lottery drawing, to be paid in 20 annual installments of $65,277. However, her husband, a construction worker, suffered serious injuries when a crane dropped a heavy beam on him. The couple's medical expenses and debt mounted. Six years later, in early 2015, a prize broker from Singer Asset Finance Co. telephoned Mrs. Setchfield with a promising offer. Singer would immediately give her $140,000 for one-half of each of her next nine prize checks ($\frac{1}{2} \times 9 \times \$65,277 = \$293,746$), an amount close to 48% of the $293,746 she had coming over that period. A big lump sum had obvious appeal to Mrs. Setchfield at that time, and she ended up signing a contract with Singer. Did she make the right decision? Table ST3.2 gives the details of the two options:

TABLE ST3.2

Year	Installment	Year	Installment	Reduced Payment
2008	$65,277	2015	$65,277	$32,639
2009	65,277	2016	65,277	32,639
2010	65,277	2017	65,277	32,639
2011	65,277	2018	65,277	32,639
2012	65,277	2019	65,277	32,639
2013	65,277	2020	65,277	32,639
2014	65,277	2021	65,277	32,639
		2022	65,277	32,639
		2023	65,277	32,639
		2024	65,277	
		2025	65,277	
		2026	65,277	
		2027	65,277	

ST3.3 It always makes the papers when a well-known sports star signs a multi-million dollar contract, (such as the payment schedule for a football player who has signed a 10-year, $130 million contract extension that guarantees him an NFL-record $37 million in bonuses.) The contract also comes with two categories of signing bonuses.

The initial signing bonus ($30 million) can be pro-rated only over the first six years of the contract. Then the second roster bonus of $7,002,000 can be pro-rated over the final nine seasons of the contract. With the salary and signing bonus paid at the beginning of each season, the net annual payment schedule looks like Table ST3.3.

(a) How much is the player's contract actually worth at the time of signing, at an interest rate of 6%?

(b) For the initial signing bonus and the second roster bonus, suppose that the ball club allows the player to take either the pro-rated payment option as just described ($30 million over five years) or a lump-sum payment option in the amount of $23 million at the time he signs the contract. Should the player take the lump-sum option instead of pro-rated one?

TABLE ST3.3

Beginning of Season	Base Salary	Pro-rated Signing Bonus	Total Annual Payment
1	$ 600,000	$5,000,000	$ 5,600,000
2	1,400,000	5,000,000 + 778,000	7,178,000
3	6,000,000	5,000,000 + 778,000	11,778,000
4	7,000,000	5,000,000 + 778,000	12,778,000
5	9,000,000	5,000,000 + 778,000	14,778,000
6	10,500,000	5,000,000 + 778,000	16,278,000
7	13,500,000	778,000	14,278,000
8	13,000,000	778,000	13,778,000
9	15,000,000	778,000	15,778,000
10	17,000,000	778,000	17,778,000

Understanding Money and Its Management

How Would You Finance Your Dream Home? At some point in your life, if you are buying a home, you would face a decision on how to finance your mortgage. Your choices would more than likely be between a fixed-rate and an adjustable-rate mortgage (ARM). How would you decide? Home mortgage loans come in and out of fashion for a few reasons. When rates on fixed loans are perceived to be low, borrowers tend to stay away from ARMs. When the difference between fixed and adjustable rates is small, people again tend to shun ARMs. Adjustable-rate loans are considered riskier than fixed-rate mortgages because the monthly payment can jump higher, if market interest rates increase. Here is an example of how much that can cost you. If you took out a 30-year mortgage in January 2003, the

Adjusting the Math

A homeowner using an ARM over the past 10 years would have saved more than $23,000

| ARM total over 10 years **$119,181** | **$23,479** |
| Fixed-rate total over 10 years **$142,660** | |

Note: One-year adjustable-rate mortgage amortized over 30 years

Source: Simon Constable, "Why Flexible-Rate Mortgages Make Sense," *The Wall Street Journal*, November 10, 2013. The Wall Street Journal by News Corporation; Dow Jones & Co Reproduced with permission of DOW JONES COMPANY in the format Republish in a book via Copyright Clearance Center.

average fixed rate was 5.92%. Ten years of interest and principal payments on a $200,000 mortgage would have cost you $142,660. But if you went with a one-year ARM, which kicked off at 3.99%, after resetting each year the total cost would have been $119,181. That's a savings of $23,479. How would you calculate all these savings? Of course, we have no way of knowing exactly how the market interest rates vary in the future, but we could predict the floating-rate range when the ARM would be preferred over the fixed-rate mortgage by simulating the worst- and best-case scenarios.

I n this chapter, we will consider several concepts crucial to managing money. In Chapter 3, we examined how time affects the value of money, and we developed various interest formulas for that purpose. Using these basic formulas, we will now extend the concept of equivalence to determine interest rates that are implicit in many financial contracts, by introducing several examples in the area of loan transactions. For example, many commercial loans require that interest compound more frequently than once a year—such as monthly or quarterly. To consider the effect of more frequent compounding, we must begin with an understanding of the concepts of nominal and effective interest.

CHAPTER LEARNING OBJECTIVES

After completing this chapter, you should understand the following concepts:

- The difference between the nominal interest rate and the effective interest rate.
- The procedure for computing the effective interest rate, based on a payment period.
- How commercial loans and mortgages are structured in terms of interest and principal payments.
- The basics of investing in financial assets.

4.1 Nominal and Effective Interest Rates

In all the examples in Chapter 3, the implicit assumption was that payments are received *once a year* (or *annually*) or at the end of interest period. However, some of the most familiar financial transactions involve payments that are not based on one annual payment—for example, monthly mortgage payments and quarterly earnings on savings accounts. Thus, if we are to compare different cash flows with different compounding periods, we need to evaluate them on a common basis. This need has led to the development of the concepts of the **nominal interest rate** and the **effective interest rate**.

4.1.1 Nominal Interest Rates

Take a closer look at the billing statement from any of your credit cards. Or if you financed a new car recently, examine the loan contract. You will typically find the interest that the bank charges on your unpaid balance. Even if a financial institution uses a unit of time other than one year—say, one month or one quarter (e.g., when calculating interest payments)—the institution usually quotes the interest rate on an *annual basis*. Many banks, for example, state the interest arrangement for credit cards in this way:

APR 18% compounded monthly

This statement simply means that each month the bank will charge 1.5% interest on an unpaid balance. We say that 18% is the **nominal interest rate** or **annual percentage rate (APR)**, and the compounding frequency is monthly (12). As shown in Figure 4.1, to obtain the interest rate per compounding period, we divide, for example, 18% by 12 to get 1.5% per month.

Although the annual percentage rate (APR) is commonly used by financial institutions and is familiar to many customers, the APR does not explain precisely the amount of interest that will accumulate in a year. To explain the true effect of more frequent compounding on annual interest amounts, we will introduce the term **effective interest rate**, commonly known as **annual effective yield** or **annual percentage yield (APY)**.

4.1.2 Effective Annual Interest Rates

The **effective annual interest rate** is the one rate that truly represents the interest earned in one year. For instance, in our credit card example, the bank will charge 1.5% interest on any unpaid balance at the end of each month. Therefore, the 1.5% rate represents the effective interest rate per month. On a yearly basis, you are looking for a cumulative rate—1.5% each month for 12 months. This cumulative rate predicts the actual interest payment on your outstanding credit card balance.

Suppose you deposit $10,000 in a savings account that pays you at an interest rate of 9% compounded quarterly. Here, 9% represents the nominal interest rate, and the interest rate per quarter is 2.25% (9%/4). Table 4.1 is an example of how interest is compounded when it is paid quarterly.

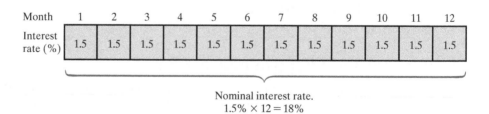

Figure 4.1 The nominal interest rate is determined by summing the individual interest rates per period.

TABLE 4.1 Interest Calculation with Quarterly Compounding

End of Period	Base Amount	Interest Earned 2.25% × (Base Amount)	New Base
First Quarter	$10,000.00	2.25% × $10,000.00 = $225.00	$10,225.00
Second Quarter	10,255.00	2.25% × $10,255.00 = $230.06	10,455.06
Third Quarter	10,455.06	2.25% × $10,455.06 = $235.24	10,690.30
Fourth Quarter	10,690.30	2.25% × $10,690.30 = $240.53	10,930.83

Clearly, you are earning more than 9% on your original deposit. In fact, you are earning 9.3083% ($930.83/$10,000). We could calculate the total annual interest payment for a principal amount of $10,000 with the formula given in Eq. (3.3). If $P = \$10,000$, $i = 2.25\%$, and $N = 4$, we obtain

$$F = P(1 + i)^N$$
$$= 10,000(1 + 0.0225)^4$$
$$= \$10,930.83$$

The implication is that, for each dollar deposited, you are earning an equivalent annual interest of 9.308 cents. In terms of an effective annual interest rate (i_a), the interest payment can be rewritten as a percentage of the principal amount:

$$i_a = \frac{930.83}{10,000} = 0.093083, \text{ or } 9.3083\%$$

In other words, earning 2.25% interest per quarter for four quarters is equivalent to earning 9.3083% interest just one time each year.

Table 4.2 shows effective interest rates at various compounding intervals for 4% to 12% APR. As you can see, depending on the frequency of compounding, the effective interest

TABLE 4.2 Nominal and Effective Interest Rates with Different Compounding Periods

	Effective Interest Rates				
Nominal Rate	Compounding Annually	Compounding Semiannually	Compounding Quarterly	Compounding Monthly	Compounding Daily
4%	4.00%	4.04%	4.06%	4.07%	4.08%
5	5.00	5.06	5.09	5.12	5.13
6	6.00	6.09	6.14	6.17	6.18
7	7.00	7.12	7.19	7.23	7.25
8	8.00	8.16	8.24	8.30	8.33
9	9.00	9.20	9.31	9.38	9.42
10	10.00	10.25	10.38	10.47	10.52
11	11.00	11.30	11.46	11.57	11.62
12	12.00	12.36	12.55	12.68	12.74

earned or paid by the borrower can differ significantly from the APR. Therefore, truth-in-lending laws require that financial institutions quote both the nominal interest rate and the compounding frequency (i.e., the effective interest) when you deposit or borrow money.

Certainly, more frequent compounding increases the amount of interest paid over a year at the same nominal interest rate. Assuming that the nominal interest rate is r, and M compounding periods occur during the year, we can calculate the effective annual interest rate or annual percentage yield (APY)

$$i_a = \left(1 + \frac{r}{M}\right)^M - 1 \tag{4.1}$$

When $M = 1$, we have the special case of annual compounding. Substituting $M = 1$ in Eq. (4.1) reduces it to $i_a = r$. That is, when compounding takes place once annually, the effective interest is equal to the nominal interest. Thus, in the examples in Chapter 3 in which only annual interest was considered, we were (by definition) using effective interest rates.

EXAMPLE 4.1 Determining a Compounding Period

Ally Bank offers several types of certificate of deposit (CD) on May 15, 2015.

Product CD Type	APY (%)*
3-month CD	0.30
6-month CD	0.60
1-year CD	0.95
3-year CD	1.20
5-year CD	1.60

*Annual percentage yield = effective annual interest rate (i_a).

In the table, no mention is made of specific APRs, but interest compounded daily. (a) Find the APR assumed for the five-year CD offered. (b) Find the total balance for a deposit amount of $10,000 at the end of five years.

SOLUTION
Given: i_a (APY) $= 1.6\%$, $P = \$10,000$, $M = 365$, and $N = 5$ years.
Find: APR (r)

(a) The effective annual interest rate (yield) is 1.6%. Using Eq. (4.1), we obtain the expression

$$0.016 = \left(1 + \frac{r}{365}\right)^{365} - 1$$

or

$$1.016 = \left(1 + \frac{r}{365}\right)^{365}$$

Solving for r, we obtain

$$\ln 1.016 = 365 \ln\left(1 + \frac{r}{365}\right)$$

$$\frac{0.015873}{365} = 0.000043 = \ln\left(1 + \frac{r}{365}\right)$$

$$e^{0.000043} = 1 + \frac{r}{365}$$

$$r = 1.5874\%$$

Thus, the five-year CD earns 1.5874% interest compounded daily. Normally, if the CD is not cashed at maturity, it will be renewed automatically at the original interest rate. Similarly, we can find the APRs for the other CDs.

(b) With APR of 1.5874% compounded daily, your CD earns an effective annual interest (APY) of 1.6%. The balance after 5 years would be

$$F = P(1 + i_a)^N$$

$$= 10,000(1 + 0.016)^5$$

$$= \$10,826$$

COMMENTS: Whenever there is a small difference between r and i_a, the effect of more frequent compounding is less noticeable.

4.1.3 Effective Interest Rates per Payment Period

We can generalize the result of Eq. (4.1) to compute the effective interest rate for periods of *any duration*. As you will see later, the effective interest rate is usually computed on the basis of the payment (transaction) period. For example, if cash flow transactions occur quarterly, but interest is compounded monthly, we may wish to calculate the effective interest rate on a quarterly basis. To do this, we may redefine Eq. (4.1) as

$$i = \left(1 + \frac{r}{M}\right)^C - 1$$

$$= \left(1 + \frac{r}{CK}\right)^C - 1 \tag{4.2}$$

where

M = number of interest periods per year

C = the number of interest periods per payment period

K = the number of payment periods per year

Note that $M = CK$ in Eq. (4.2). For the special case of annual payments with annual compounding, we obtain $i = i_a$ with $C = M$ and $K = 1$.

EXAMPLE 4.2 Effective Rate per Payment Period

Suppose that you make quarterly deposits in a savings account that earns 9% interest compounded monthly. Compute the effective interest rate (a) per quarter and (b) per year.

SOLUTION

Given: $r = 9\%$, $C =$ three interest periods per quarter, $K =$ four quarterly payments per year, and $M = 12$ interest periods per year.

Find: i.

Using Eq. (4.2), we compute the effective interest rate per quarter as

(a)
$$i = \left(1 + \frac{0.09}{4 \times 3}\right)^3 - 1$$
$$= 2.27\%$$

(b)
$$i_a = (1 + 0.027)^4 - 1$$
$$= 9.38\%$$

COMMENTS: Figure 4.2 illustrates the relationship between the nominal and effective interest rates per payment period.

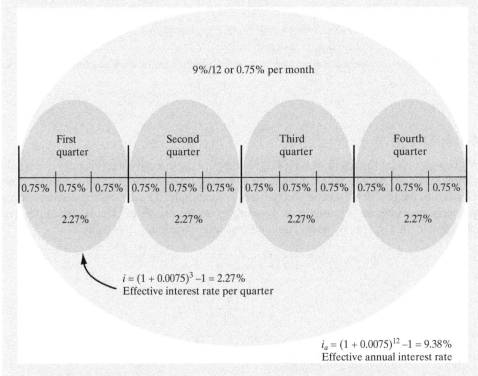

$$i = (1 + 0.0075)^3 - 1 = 2.27\%$$
Effective interest rate per quarter

$$i_a = (1 + 0.0075)^{12} - 1 = 9.38\%$$
Effective annual interest rate

Figure 4.2 Functional relationships among r, i, and i_a, where interest is calculated based on 9% compounded monthly and payments occur quarterly (Example 4.2).

4.1.4 Continuous Compounding

To be competitive in the financial market or to entice potential depositors, some financial institutions offer continuous compounding. As the number of compounding periods (M) becomes very large, the interest rate per compounding period (r/M) becomes very small. As M approaches infinity and r/M approaches zero, we approximate the situation of **continuous compounding**.

By taking limits on both sides of Eq. (4.2), we obtain the effective interest rate per payment period as

$$i = \lim_{CK \to \infty} \left[\left(1 + \frac{r}{CK} \right)^C - 1 \right]$$

$$= \lim_{CK \to \infty} \left[\left(1 + \frac{r}{CK} \right)^{CK} \right]^{1/K} - 1$$

$$= (e^r)^{1/K} - 1$$

Therefore, the effective interest rate per payment period is

$$i = e^{r/K} - 1 \tag{4.3}$$

To calculate the effective annual interest rate for continuous compounding, we set K equal to one, resulting in

$$i_a = e^r - 1 \tag{4.4}$$

As an example, the effective annual interest rate for a nominal interest rate of 12% compounded continuously is $i_a = e^{0.12} - 1 = 12.7497\%$.

EXAMPLE 4.3 Calculating an Effective Interest Rate with Quarterly Payment

Find the effective interest rate per quarter at a nominal rate of 8% compounded (a) quarterly, (b) monthly, (c) weekly, (d) daily, and (e) continuously.

SOLUTION

Given: $r = 8\%$, M, C, and $K = 4$ quarterly payments per year.
Find: i.

(a) Quarterly compounding:

$r = 8\%$, $M = 4$, $C = 1$ interest period per quarter, and $K = 4$ payments per year. Then

$$i = \left(1 + \frac{0.08}{4} \right)^1 - 1 = 2.00\%$$

(b) Monthly compounding: $r = 8\%$, $M = 12$, $C = 3$ interest periods per quarter, and $K = 4$ payments per year. Then

$$i = \left(1 + \frac{0.08}{12}\right)^3 - 1 = 2.013\%$$

(c) Weekly compounding: $r = 8\%$, $M = 52$, $C = 13$ interest periods per quarter, and $K = 4$ payments per year. Then

$$i = \left(1 + \frac{0.08}{52}\right)^{13} - 1 = 2.0186\%$$

(d) Daily compounding: $r = 8\%$, $M = 365$, $C = 91.25$ days per quarter, and $K = 4$. Then

$$i = \left(1 + \frac{0.08}{365}\right)^{91.25} - 1 = 2.0199\%$$

(e) Continuous compounding: $r = 8\%$, $M \to \infty$, $C \to \infty$, and $K = 4$. Then, from Eq. (4.3),

$$i = e^{0.08/4} - 1 = 2.0201\%$$

COMMENTS: Note that the difference between daily compounding and continuous compounding is often negligible. Many banks offer continuous compounding to entice deposit customers, but the extra benefits are relatively small.

4.2 Equivalence Calculations with Effective Interest Rates

All the examples in Chapter 3 assumed annual payments and annual compounding. However, a number of situations involve cash flows that occur at intervals that are not the same as the compounding intervals often used in practice. Whenever payment and compounding periods differ from each other, *one or the other must be transformed so that both conform to the same unit of time*. For example, if payments occur quarterly and compounding occurs monthly, the most logical procedure is to calculate the effective interest rate per quarter. By contrast, if payments occur monthly and compounding occurs quarterly, we may be able to find the equivalent monthly interest rate. The bottom line is that, to proceed with equivalency analysis, the compounding and payment periods must be in the same order.

4.2.1 When Payment Period is Equal to Compounding Period

Whenever the compounding and payment periods are equal ($M = K$), whether the interest is compounded annually or at some other interval, the following solution method can be used.

Step 1. Identify the number of compounding periods (M) per year.

Step 2. Compute the effective interest rate per payment period—that is, using Eq. (4.2). Then with $C = 1$ and $K = M$, we have

$$i = \frac{r}{M}$$

Step 3. Determine the number of compounding periods:

$$N = M \times (\text{number of years})$$

EXAMPLE 4.4 Calculating Auto Loan Payments

Suppose you want to buy a car. You have surveyed the dealers' newspaper advertisements, and the following one has caught your attention:

College Graduate Special: New 2015 Chevy Cruze Turbo with automatic transmission, A/C, power package, and cruise control.

Manufacturer's Suggested Retail Price (MSRP)	$20,870
Dealer's discount	$1,143
Manufacturer rebate	$800
College graduate cash	$500
Sale price:	$18,427

Price and payment is plus tax, title, and customer service fee, with approved credit for 72 months at 6.25% APR.

You can afford to make a down payment of $3,427 (and taxes and insurance as well), so the net amount to be financed is $15,000. What would the monthly payment be (Figure 4.3)?

Figure 4.3   A car loan cash transaction (Example 4.4).

DISCUSSION: The advertisement does not specify a compounding period, but in automobile financing, the interest and the payment periods are almost always monthly. Thus, the 6.25% APR means 6.25% compounded monthly.

SOLUTION

Given: $P = \$15{,}000$, $r = 6.25\%$ per year, $K = 12$ payments per year, $N = 72$ months and $M = 12$ interest periods per year.

Find: A.

In this situation, we can easily compute the monthly payment with Eq. (3.12):

$$i = 6.25\%/12 = 0.5208\% \text{ per month}$$

$$N = (12)(6) = 72 \text{ months}$$

$$A = \$15{,}000(A/P, 0.5208\%, 72) = \$250.37$$

4.2.2 Compounding Occurs at a Different Rate than That at Which Payments are Made

We will consider two situations: (1) compounding is more frequent than payments, and (2) compounding is less frequent than payments.

Compounding is More Frequent than Payments

The computational procedure for establishing economic equivalence is as follows.

Step 1. Identify the number of compounding periods per year (M), the number of payment periods per year (K), and the number of interest periods per payment period (C).

Step 2. Compute the effective interest rate per payment period:

• For discrete compounding, compute

$$i = \left(1 + \frac{r}{M}\right)^{C} - 1$$

- For continuous compounding, compute

$$i = e^{r/K} - 1$$

Step 3. Find the total number of payment periods:

$$N = K \times \text{(number of years)}$$

Step 4. Use i and N in the appropriate formulas in Table 3.6.

EXAMPLE 4.5 Compounding Occurs More Frequently than Payments Are Made (Discrete-Compounding Case)

Suppose you make equal quarterly deposits of $1,500 into a fund that pays interest at a rate of 6% compounded monthly, as shown in Figure 4.4. Find the balance at the end of year 2.

SOLUTION

Given: $A = \$1,500$ per quarter, $r = 6\%$ per year, $M = 12$ compounding periods per year, and $N = 8$ quarters.
Find: F.

We follow the aforementioned procedure for non-comparable compounding and payment periods.

Step 1. Identify the parameter values for M, K, and C, where

$$M = 12 \text{ compounding periods per year}$$

$$K = 4 \text{ payment periods per year}$$

$$C = 3 \text{ interest periods per payment period (quarter)}$$

Step 2. Use Eq. (4.2) to compute the effective interest:

$$i = \left(1 + \frac{0.06}{12}\right)^3 - 1$$

$$= 1.5075\% \text{ per quarter}$$

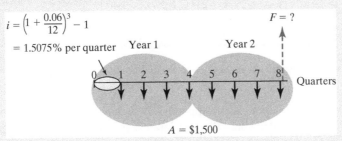

Figure 4.4 Quarterly deposits with monthly compounding (Example 4.5).

Step 3. Find the total number of payment periods, N:

$$N = K(\text{number of years}) = 4(2) = 8 \text{ quarters}$$

Step 4. Use i and N in the appropriate equivalence formulas:

$$F = \$1,500(F/A, 1.5075\%, 8) = \$12,652.61$$

Figure 4.5 illustrates the process of obtaining the future worth of the quarterly payment series in Excel format.

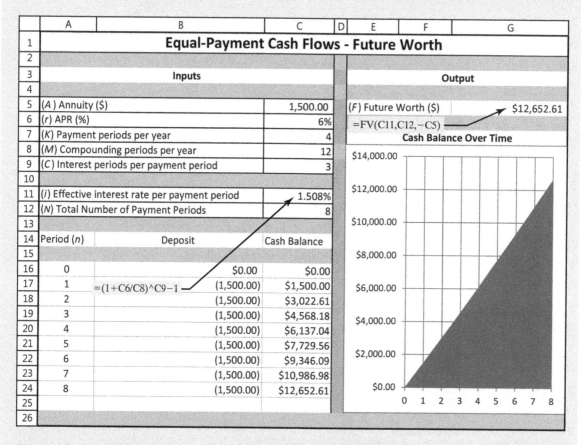

Figure 4.5   An Excel worksheet to illustrate the process of accumulating the future worth.

COMMENT: No 1.5075% interest table appears in Appendix B, but F can still be evaluated using $F = \$1,500\,(A/F, 0.5\%, 3)(F/A, 0.5\%, 24)$, where the first interest factor finds its equivalent monthly payment and the second interest factor converts the monthly payment series to an equivalent lump-sum future payment.

EXAMPLE 4.6 Compounding Occurs More Frequently than Payments are Made (Continuous-Compounding Case)

A series of equal quarterly receipts of $500 extends over a period of five years, as shown in Figure 4.6. What is the present worth of this quarterly payment series at 8% interest compounded continuously?

DISCUSSION: A question that is equivalent to the preceding one is "How much do you need to deposit now in a savings account that earns 8% interest compounded continuously so that you can withdraw $500 at the end of each quarter for five years?" Since the payments are quarterly, we need to compute i per quarter for the equivalence calculations:

$$i = e^{r/K} - 1 = e^{0.08/4} - 1$$

$$= 2.02\% \text{ per quarter}$$

$$N = (4 \text{ payment periods per year})(5 \text{ years})$$

$$= 20 \text{ quarterly periods}$$

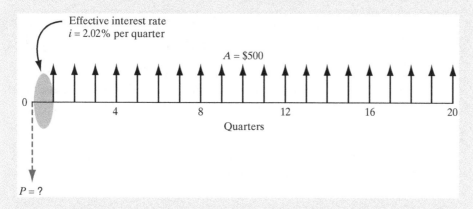

Figure 4.6 A present-worth calculation for an equal payment series with an interest rate of 8% compounded continuously (Example 4.6).

SOLUTION

Given: $i = 2.02\%$ per quarter, $N = 20$ quarters, and $A = \$500$ per quarter.
Find: P.

Using the $(P/A, i, N)$ factor with $i = 2.02\%$ and $N = 20$, we find that

$$P = \$500(P/A, 2.02\%, 20)$$

$$= \$500(16.3197)$$

$$= \$8,159.85$$

4.2.4 Compounding is Less Frequent than Payments

The next two examples contain identical parameters for savings situations in which compounding occurs less frequently than payments. However, two different underlying assumptions govern how interest is calculated.

- **Case 1:** We assume that, whenever a deposit is made, it starts to earn interest. As a result, in Example 4.7 we transform the compounding period to conform to the payment period.
- **Case 2:** We assume that the deposits made within a quarter do not earn interest until the end of that quarter. In Example 4.8 we lump several payments together to match the compounding period.

In the real world, which assumption is applicable depends on the transactions and the financial institutions involved. The accounting methods used by many firms record cash transactions that occur within a compounding period as if they had occurred at the end of that period. For example, when cash flows occur daily, but the compounding period is monthly, the cash flows within each month are summed (ignoring interest) and treated as a single payment on which interest is calculated.

Note: *In this text, we assume that whenever the time point of a cash flow is specified, one cannot move it to another time point without considering the time value of money (i.e., the practice demonstrated in Example 4.7 should be followed).*

EXAMPLE 4.7 Compounding is Less Frequent than Payments: Effective Interest Rate per Payment Period

Suppose you make $500 monthly deposits to a tax-deferred retirement plan that pays interest at a rate of 10% compounded quarterly. Compute the balance at the end of 10 years. Interest is accrued on deposits made during the compounding period.

SOLUTION

Given: $r = 10\%$ per year, $M = 4$ quarterly compounding periods per year, $K = 12$ payment periods per year, $A = \$500$ per month, and $N = 120$ months.
Find: i, F.

As in the case of Example 4.5, the procedure for non-comparable compounding and payment periods is followed.

Step 1. The parameter values for M, K, and C are

$$M = 4 \text{ compounding periods per year}$$

$$K = 12 \text{ payment periods per year}$$

$$C = \frac{1}{3} \text{ interest period per payment period}$$

Step 2. As shown in Figure 4.8, the effective interest rate per payment period is calculated with Eq. (4.2):

$$i = 0.826\% \text{ per month}$$

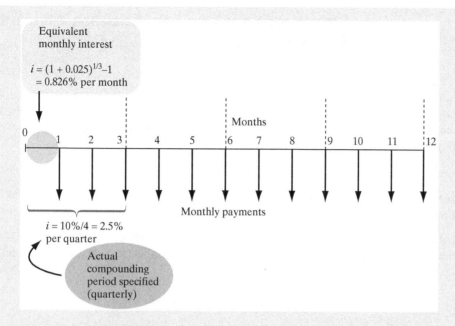

Figure 4.7 Calculation of equivalent monthly interest when the quarterly interest rate is specified (Example 4.7).

Step 3. Find N:

$$N = (12)(10) = 120 \text{ payment periods}$$

Step 4. Use i and N in the appropriate equivalence formulas (Figure 4.8):

$$F = \$500(F/A, 0.826\%, 120)$$

$$= \$101,907.89$$

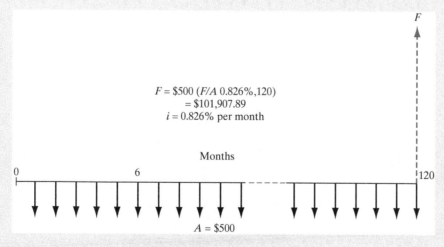

Figure 4.8 Cash flow diagram (Example 4.7).

EXAMPLE 4.8 Compounding is Less Frequent than Payment: Summing Cash Flows to the End of the Compounding Period

Some financial institutions will not pay interest on funds deposited between the interest periods. To illustrate, consider Example 4.7 again. Suppose that money deposited during a quarter (the compounding period) will not earn any interest (Figure 4.9). Compute what F will be at the end of 10 years.

SOLUTION

Given: Same as for Example 4.7; however, no interest on flow during the interest period.
Find: F.

In this case, the three monthly deposits during each quarterly period will be placed at the end of each quarter. Then the payment period coincides with the interest period, and we have

$$i = \frac{10\%}{4} = 2.5\% \text{ per quarter}$$

$$A = 3(\$500) = \$1,500 \text{ per quarter}$$

$$N = 4(10) = 40 \text{ payment periods}$$

$$F = \$1,500(F/A, 2.5\%, 40) = \$101,103.83$$

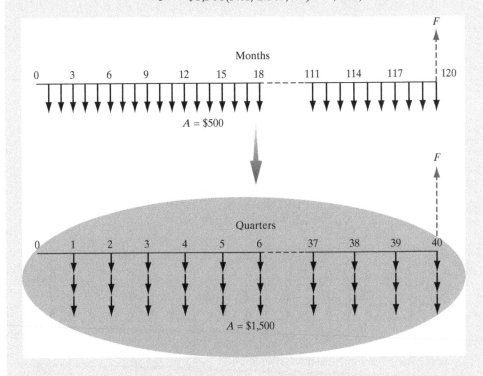

Figure 4.9 Transformed cash flow diagram created by summing monthly cash flows to the end of the quarterly compounding period (Example 4.8).

COMMENTS: In Example 4.8, the balance will be $804.06 less than in Example 4.7, a fact that is consistent with our understanding that increasing the frequency of compounding increases the future value of money. Some financial institutions follow the practice illustrated in Example 4.7. As an investor, you should reasonably ask yourself whether it makes sense to make deposits in an interest-bearing account more frequently than interest is paid. In the interim between interest compoundings, you may be tying up your funds prematurely and forgoing other opportunities to earn interest.

Figure 4.10 is a decision chart that allows you to sum up how you can proceed to find the effective interest rate per payment period, given the various possible compounding and interest arrangements.

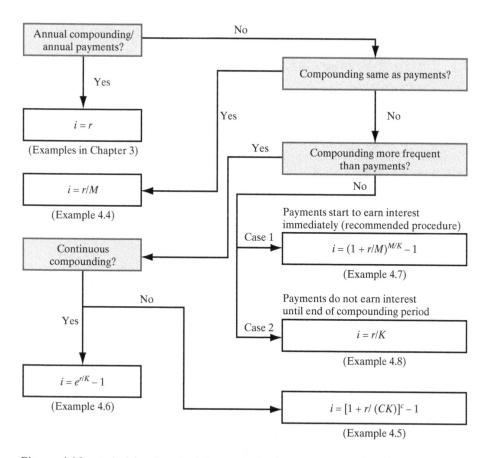

Figure 4.10 A decision flowchart demonstrating how to compute the effective interest rate i per payment period.

4.3 Equivalence Calculations with Continuous Compounding

As we have seen so far, interest can be compounded annually, semiannually, monthly, or even continuously. Discrete compounding is appropriate for many financial transactions; mortgages, bonds, and installment loans, which require payments or receipts at discrete times, are good examples. In most businesses, however, transactions occur continuously throughout the year. In these circumstances, we may describe the financial transactions as having a continuous flow of money for which continuous compounding and discounting are more realistic. This section illustrates how one establishes economic equivalence between cash flows under continuous compounding.

Continuous cash flows represent situations in which money flows continuously and at a known rate throughout a given period. In business, many daily cash-flow transactions can be viewed as continuous. An advantage of the continuous-flow approach is that it more closely models the realities of business transactions. Typical examples are costs for labor, for carrying inventory, and for operating and maintaining equipment. Others include capital improvement projects that conserve energy or water or that process steam. Savings on these projects can occur continuously.

4.3.1 Discrete-Payment Transactions with Continuous Compounding

First we will illustrate how single-payment formulas for continuous compounding and discounting are derived. Suppose that you invested P dollars at a nominal rate of $r\%$ interest for N years. If interest is compounded continuously, the effective annual interest is $i = e^r - 1$. The future value of the investment at the end of N years is obtained with the F/P factor by substituting $e^r - 1$ for i:

$$
\begin{aligned}
F &= P(1 + i)^N \\
&= P(1 + e^r - 1)^N \\
&= Pe^{rN}
\end{aligned}
$$

This implies that $1 invested now at an interest rate of $r\%$ compounded continuously accumulates to e^{rN} dollars at the end of N years. Correspondingly, the present value of F due N years from now and discounted continuously at an interest rate of $r\%$ is equal to

$$
P = Fe^{-rN}
$$

We can say that the present value of $1 due N years from now and discounted continuously at an annual interest rate of $r\%$ is equal to e^{-rN} dollars. Similarly, we can develop interest formulas for discrete payments with continuous compounding as in Table 4.3.

TABLE 4.3 Summary of Selected Continuous Compounding Formulas with Discrete Payments

Flow Type	Find	Given	Factor Notation	Formula
Compound-Amount	F	P	$(F/P, r, N)$	$F = P(e^{rN})$
Present-Worth	P	F	$(P/F, r, N)$	$P = F(e^{-rN})$
Compound-Amount	F	A	$(F/A, r, N)$	$F = A\left[\dfrac{(e^r)^N - 1}{e^r - 1}\right]$
Sinking-Fund	A	F	$(A/F, r, N)$	$A = F\left[\dfrac{e^r - 1}{(e^r)^N - 1}\right]$
Present-Worth	P	A	$(P/A, r, N)$	$P = A\left[\dfrac{1 - e^{-rN}}{e^r - 1}\right]$
Capital Recovery	A	P	$(A/P, r, N)$	$A = P\left[\dfrac{e^r - 1}{1 - e^{-rN}}\right]$

EXAMPLE 4.9 Loan Repayment Under Continuous Compounding

You need $30,000 immediately in order to make a down payment on a new home. Suppose that you can borrow the money from your insurance company. You will be required to repay the loan in equal amounts made every 6 months over the next five years. The nominal interest rate being charged is 8% compounded continuously. What is the amount of each payment?

SOLUTION

Given: $P = \$30,000$, $r = 8\%$ per year, and $N = 5$ years (or 10 semiannual periods).
Find: A.

Note that 8% APR is stated over a year, so the APR over a 6-month period is simply $8\%/2 = 4\%$.

$$i = e^{r/K} - 1 = e^{0.08/2} - 1 = 4.0811\% \text{ per semiannual period.}$$

$$N = 5 \times 2 = 10$$

$$A = (A/P, 4.0811\%, 10)$$

$$= \$3,713.68$$

Or, using the formula in Table 4.3,

$$A = \$30,000\left[\frac{e^{0.04} - 1}{1 - e^{-0.04(10)}}\right]$$

$$= \$30,000(0.123789)$$

$$= \$3,713,67$$

4.3.2 Continuous-Funds Flow with Continuous Compounding

Suppose that an investment's future cash flow per unit of time (e.g., per year) can be expressed by a continuous function ($f(t)$) that can take any shape. Suppose also that the investment promises to generate cash of $f(t)\Delta t$ dollars between t and $t + \Delta t$, where t is a point in the time interval $0 \leq t \leq N$ (Figure 4.11). If the nominal interest rate is a constant r during this interval, the present value of the cash stream is given approximately by the expression

$$[f(t)\Delta t]e^{-rt}$$

where e^{-rt} is the discounting factor that converts future dollars into present dollars. With the project's life extending from 0 to N, we take the summation over all subperiods (compounding periods) in the interval from 0 to N. As the interval is divided into smaller and smaller segments (i.e., as Δt approaches zero), we obtain the expression for the present value by the integral

$$P = \int_0^N f(t)e^{-rt}dt \tag{4.5}$$

Similarly, the expression for the future value of the cash flow stream is given by the equation

$$F = Pe^{rN} = \int_0^N f(t)e^{r(N-t)}\,dt \tag{4.6}$$

where $e^{r(N-t)}$ is the compounding factor that converts present dollars into future dollars. It is important to observe that the time unit is the *year*, because the effective interest rate is expressed in terms of a year. Therefore, all time units in equivalence calculations must be converted into years. Table 4.4 summarizes some typical continuous cash functions that can facilitate equivalence calculations.[1]

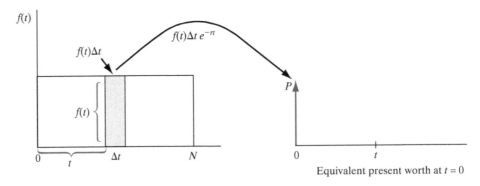

Figure 4.11 Finding an equivalent present worth of a continuous-flow payment function $f(t)$ at a nominal rate of $r\%$.

[1]Chan S. Park and Gunter P. Sharp-Bette, *Advanced Engineering Economics*, New York: John Wiley & Sons, 1990. (Reprinted by permission of John Wiley & Sons, Inc.), p. 62

TABLE 4.4 **Summary of Interest Factors for Typical Continuous Cash Flows with Continuous Compounding**

Type of Cash Flow	Cash Flow Function	Parameters Find	Parameters Given	Algebraic Notation	Factor Notation
Uniform (step)	$f(t) = \bar{A}$	P	$\bar{A}$	$\bar{A}\left[\dfrac{e^{rN} - 1}{re^{rN}}\right]$	$(P/\bar{A}, r, N)$
		$\bar{A}$	P	$P\left[\dfrac{re^{rN}}{e^{rN} - 1}\right]$	$(\bar{A}/P, r, N)$
		F	$\bar{A}$	$\bar{A}\left[\dfrac{e^{rN} - 1}{r}\right]$	$(F/\bar{A}, r, N)$
		$\bar{A}$	F	$F\left[\dfrac{r}{e^{rN} - 1}\right]$	$(\bar{A}/P, r, N)$
Gradient (ramp)	$f(t) = Gt$	P	G	$\dfrac{G}{r^2}(1 - e^{-rN}) - \dfrac{G}{r}(Ne^{-rN})$	
Decay	$f(t) = ce^{-jt}$ $\quad j^t = $ decay rate with time	P	c, j	$\dfrac{c}{r + j}(1 - e^{-(r+j)N})$	

Source: Chan S. Park and Gunter P. Sharp-Bette, *Advanced Engineering Economics*, New York: John Wiley & Sons, 1990. (Reprinted by permission of John Wiley & Sons, Inc.), p. 62.

EXAMPLE 4.10 Comparison of Daily Flows and Daily Compounding with Continuous Flows and Continuous Compounding

Consider a situation in which money flows daily. Suppose you own a retail shop and generate $200 cash each day. You establish a special business account and deposit your daily cash flows in an account for 15 months. The account earns an interest rate of 6%. Compare the accumulated cash values at the end of 15 months, assuming

(a) Daily compounding

(b) Continuous compounding

SOLUTION

(a) Daily compounding.
Given: A = \$200 per day, r = 6% per year, M = 365 compounding periods per year, and N = 455 days.
Find: F.

Assuming that there are 455 days in the 15-month period, we find that

$$i = 6\%/365$$

$$= 0.01644\% \text{ per day}$$

$$N = 455 \text{ days}$$

The balance at the end of 15 months will be

$$F = \$200(F/A, 0.01644\%, 455)$$

$$= \$200(472.4095)$$

$$= \$94,482$$

(b) Continuous compounding.
Given: $\overline{A}$ = \$73,000 per year, r = 6% per year, compounded continuously, and N = 1.25 years.
Find: F.

Now we approximate this discrete cash flow series by a uniform continuous cash flow function as shown in Figure 4.12. In this situation, an amount flows at the rate of $\overline{A}$ per year for N years.

Note: *Our time unit is one year.* Thus, a 15-month period is 1.25 years. Then the cash flow function is expressed as

$$f(t) = \overline{A}, 0 \leq t \leq 1.25$$

$$= \$200(365)$$

$$= \$73,000 \text{ per year}$$

Substituting these values back into Eq. (4.6) yields

$$F = \int_0^{1.25} 73,000 e^{0.06(1.25-t)} \, dt$$

$$= \$73,000 \left[\frac{e^{0.075} - 1}{0.06} \right]$$

$$= \$94,759$$

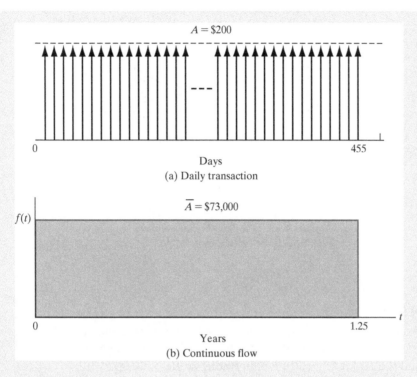

Figure 4.12   Comparison between daily transaction and continuous-funds flow transaction (Example 4.10).

The factor in the bracket is known as the **funds flow compound-amount factor** and is designated $(F/\overline{A}, r, N)$ as shown in Table 4.4. Notice that the difference between the two methods is only $277 (less than 0.3%).

COMMENTS: As shown in this example, the differences between discrete daily compounding and continuous compounding have no practical significance in most cases. Consequently, as a mathematical convenience, instead of assuming that money flows in discrete increments at the end of each day, we could assume that money flows continuously at a uniform rate during the period in question. This type of cash flow assumption is common practice in the chemical industry.

EXAMPLE 4.11   A Ramp Cash Flow Function with Continuous Compounding

Consider a situation in which money flows according to the diagram shown below (Figure 4.13). Here, a cash flow increases uniformly over two years to accumulate $10,000.

(a) What is the expression for the present worth at 10% interest compounded continuously?

(b) Suppose that the rate of operating expenditure is essentially constant at $2,000 per year. Determine the present worth of operating profit over a life of 2 years.

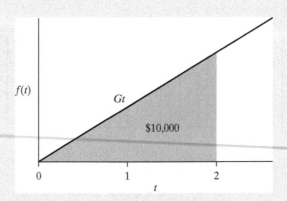

Figure 4.13 Finding the present worth of a ramp cash flow function (Example 4.11).

SOLUTION

Given: $C = \$10,000$ over 2 years, $r = 6\%$ per year, compounded continuously, $N = 2$ years, and the rate of operating expenditure = $\$2,000$ per year.

Find: (a) P of revenue; (b) P of operating profit.

(a) The ramp cash flow function in Figure 4.13 shows the total amount to be received over two years, (i.e., the area), but the slope of the ramp function is not given. Therefore, our first task is to determine the slope G. Since the area of the ramp function is given, we can find the unknown slope of the function as

$$\int_0^2 Gt \cdot dt = \$10,000$$

$$G\frac{t^2}{2}\bigg|_0^2 = 2G = \$10,000$$

$$G = \$5,000$$

Substituting these values back into the formula in Table 4.4 yields

$$P = \int_0^2 5{,}000t \cdot e^{-0.06t}\, dt$$

$$= \frac{5{,}000}{0.06^2}(1 - e^{-0.06(2)}) - \frac{5{,}000}{0.06}(2e^{-0.06(2)})$$

$$= \$9{,}234.88$$

(b) The present worth of operating profit over a life of two years is

$$P = \int_0^2 (5000t - 2000)e^{-0.06t}\, dt$$

$$= \$9,234.88 - 2000 \int_0^2 e^{-0.06t}\, dt$$

$$= \$9,234.88 - 2000 \left[\frac{e^{0.06(2)} - 1}{0.06e^{0.06(2)}} \right]$$

$$= \$9,234.88 - \$3,769.32$$

$$= \$5,465.56$$

4.4 Changing Interest Rates

Up to this point, we have assumed a constant interest rate in our equivalence calculations. When an equivalence calculation extends over several years, more than one interest rate may be applicable to properly account for the time value of money. That is to say, over time, interest rates available in the financial marketplace fluctuate, and a financial institution that is committed to a long-term loan may find itself in the position of losing the opportunity to earn higher interest because some of its holdings are tied up in a lower interest loan. The financial institution may attempt to protect itself from such lost earning opportunities by building gradually increasing interest rates into a long-term loan at the outset. Adjustable-rate mortgage (ARM) loans are perhaps the most common examples of variable interest rates. In this section, we will consider variable interest rates in both single payments and a series of cash flows.

4.4.1 Single Sums of Money

To illustrate the mathematical operations involved in computing equivalence under changing interest rates, first consider the investment of a single sum of money, P, in a savings account for N interest periods. If i_n denotes the interest rate appropriate during period n, then the future-worth equivalent for a single sum of money can be expressed as

$$F = P(1 + i_1)(1 + i_2) \cdots (1 + i_{N-1})(1 + i_N) \tag{4.7}$$

and solving for P yields the inverse relation

$$P = F[(1 + i_1)(1 + i_2) \cdots (1 + i_{N-1})(1 + i_N)]^{-1} \tag{4.8}$$

EXAMPLE 4.12 Changing Interest Rates with a Lump-Sum Amount

Suppose you deposit $2,000 in an individual retirement account (IRA) that pays interest at 6% compounded monthly for the first two years and 9% compounded monthly for the next three years. Determine the balance at the end of five years (Figure 4.14).

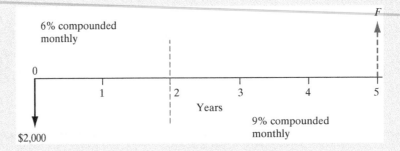

Figure 4.14 Future worth calculation with changing interest rates (Example 4.12).

SOLUTION

Given: $P = \$2,000$, $r = 6\%$ per year for first two years, 9% per year for last three years, $M = 4$ compounding periods per year, and $N = 60$ months.

Find: F.

We will compute the value of F in two steps. First we will compute the balance B_2 in the account at the end of two years. With 6% compounded monthly, we have

$$i = 6\%/12 = 0.5\%$$

$$N = 12(2) = 24 \text{ months}$$

$$B_2 = \$2,000(F/P, 0.5\%, 24)$$

$$= \$2,000(1.12716)$$

$$= \$2,254$$

Since the fund is not withdrawn but is reinvested at 9% compounded monthly, as a second step, we compute the final balance as follows.

$$i = 9\%/12 = 0.75\%$$

$$N = 12(3) = 36 \text{ months}$$

$$F = B_2(F/P, 0.75\%, 36)$$

$$= \$2,254(1.3086)$$

$$= \$2,950$$

4.4.2 Series of Cash Flows

The phenomenon of changing interest rates easily can be extended to a series of cash flows. In this case, the present worth of a series of cash flows can be represented as

$$P = A_1(1 + i_1)^{-1} + A_2[(1 + i_1)^{-1} (1 + i_2)^{-1}] + \dots$$
$$+ A_N[(1 + i_1)^{-1}(1 + i_2)^{-1} \dots (1 + i_N)^{-1}] \tag{4.9}$$

The future worth of a series of cash flows is given by the inverse of Eq. (4.9):

$$F = A_1[(1 + i_2)(1 + i_3) \dots (1 + i_N)]$$
$$+ A_2[(1 + i_3)(1 + i_4) \dots (1 + i_N)] + \dots + A_N \tag{4.10}$$

The uniform series equivalent is obtained in two steps. First, the present-worth equivalent of the series is found from Eq. (4.9). Then A is obtained after establishing the following equivalence equation.

$$P = A(1 + i_1)^{-1} + A[(1 + i_1)^{-1} (1 + i_2)^{-1}] + \dots$$
$$+ A[(1 + i_1)^{-1} (1 + i_2)^{-1} \dots (1 + i_N)^{-1}] \tag{4.11}$$

EXAMPLE 4.13 Changing Interest Rates with Uneven Cash Flow Series

Consider the cash flow in Figure 4.15 with the interest rates indicated, and determine the uniform series equivalent of the cash flow series.

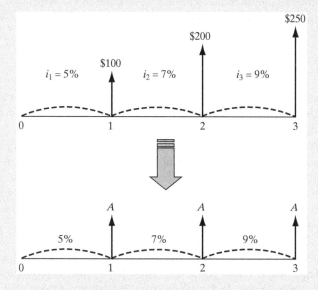

Figure 4.15 Equivalence calculation with changing interest rates (Example 4.13).

DISCUSSION: In this problem and many others, the easiest approach involves collapsing the original flow into a single equivalent amount, for example at time zero, and then converting the single amount into the final desired form.

SOLUTION

Given: Cash flows and interest rates as shown in Figure 4.15; $N = 3$.
Find: A.

Using Eq. (4.9), we find the present worth:

$$P = \$100(P/F, 5\%, 1) + \$200(P/F, 5\%, 1)(P/F, 7\%, 1)$$
$$+ \$250(P/F, 5\%, 1)(P/F, 7\%, 1)(P/F, 9\%, 1)$$
$$= \$477.41$$

Then we obtain the uniform series equivalent as

$$\$477.41 = A(P/F, 5\%, 1) + A(P/F, 5\%, 1)(P/F, 7\%, 1)$$
$$+ A(P/F, 5\%, 1)(P/F, 7\%, 1)(P/F, 9\%, 1)$$
$$= 2.6591A$$
$$A = \$179.54$$

4.5 Debt Management

Credit card debt and commercial loans are among the most significant financial transactions involving interest. Many types of loans are available, but here we will focus on those most frequently used by individuals and in business.

4.5.1 Commercial Loans

One of the most important applications of compound interest involves loans that are paid in **installments** over time. If the loan is to be repaid in equal periodic amounts (weekly, monthly, quarterly, or annually), it is said to be an **amortized loan**.

Examples of installment loans include automobile loans, loans for appliances, home mortgage loans, and the majority of business debts other than very short-term loans. Most commercial loans have interest that is compounded monthly. With an auto loan, a local bank or a dealer advances you the money to pay for the car, and you repay the principal plus interest in monthly installments, usually over a period of three to five years. The car is your collateral. If you don't keep up with your payments, the lender can repossess the car and keep all the payments you have made.

Two things determine what borrowing will cost you: the finance charge and the length of the loan. The cheapest loan is not the one with the lowest payments or even the one with the lowest interest rate. Instead, you have to look at (1) the total cost of borrowing, which depends on the interest rate plus fees, and (2) the term or length of time it takes you to repay the loan. While you probably cannot influence the rate or the fees, you may be able to arrange for a shorter term.

- **The annual percentage rate (APR)** is set by lenders, who are required to tell you what a loan will actually cost per year (expressed as an APR). Some lenders charge lower interest, but add high fees; others do the reverse. Combining the fees with a year of interest charges to give you the true annual interest rate, the APR allows you to compare these two kinds of loans on equal terms.
- **Fees** are the expenses the lender will charge to lend the money. The application fee covers processing expenses. Attorney fees pay the lender's attorney. Credit search fees cover researching your credit history. Origination fees cover administrative costs and sometimes appraisal fees. All these fees add up very quickly and can substantially increase the cost of your loan.
- **Finance charges** are the cost of borrowing. For most loans, they include all the interest, fees, service charges, points, credit-related insurance premiums, and any other charges.
- **The periodic interest rate** is the interest the lender will charge on the amount you borrow. If a lender also charges fees, the periodic interest rate will not be the true interest rate.
- **The term of your loan** is crucial in determining its cost. Shorter terms mean squeezing larger amounts into fewer payments. However, they also mean paying interest for fewer years, saving a lot of money.

Amortized Installment Loans

So far, we have considered many instances of amortized loans in which we calculated present or future values of the loans or the amounts of the installment payments. An additional aspect of amortized loans that will be of great interest to us is calculating the amount of interest versus the portion of the principal that is paid off in each installment. As we will explore more fully in Chapter 10, the interest paid on a loan is an important element in calculating taxable income and has repercussions for both personal and business loan transactions. For now, we will focus on several methods of calculating interest and principal paid at any point in the life of the loan.

In calculating the size of a monthly installment, lending institutions may use two types of schemes. The first is the conventional amortized loan, based on the compound-interest method, and the second is the **add-on loan**, based on the simple-interest concept. We explain each method in what follows, but it should be understood that the amortized loan is the most common in various commercial lending. The add-on loan is common in financing appliances as well as furniture.

In a typical amortized loan, the amount of interest owed for a specified period is calculated on the basis of the remaining balance on the loan at the beginning of the period. A set of formulas has been developed to compute the remaining loan balance, interest payment, and principal payment for a specified period. Suppose we borrow an amount P at an interest rate i and agree to repay this principal sum P (including interest) in equal payments A over N periods. Then the size of the payment is $A = P(A/P, i, N)$, and each payment is divided into an amount that is interest and a remaining amount that goes toward paying the principal.

Let

$$B_n = \text{Remaining balance at the end of period } n, \text{ with } B_0 = P$$

$$I_n = \text{Interest payment in period } n, \text{ where } I_n = B_{n-1}i$$

$$P_n = \text{Principal payment in period } n$$

Then each payment can be defined as

$$A = P_n + I_n \tag{4.12}$$

The interest and principal payments for an amortized loan can be determined in several ways; two are presented here. No clear-cut reason is available to prefer one method over the other. Method 1, however, may be easier to adopt when the computational process is automated through a spreadsheet application, whereas method 2 may be more suitable for obtaining a quick solution when a period is specified. You should become comfortable with at least one of these methods; pick the one that comes most naturally to you.

Method 1: Tabular Method. The first method is tabular. The interest charge for a given period is computed progressively on the basis of the remaining balance at the beginning of that period. Example 4.14 illustrates the process of creating a loan repayment schedule based on an iterative approach.

EXAMPLE 4.14 Loan Balance, Principal, and Interest: Tabular Method

Suppose you secure a home improvement loan in the amount of $5,000 from a local bank. The loan officer computes your monthly payment as

$$\text{Contract amount} = \$5,000$$

$$\text{Contract period} = 24 \text{ months}$$

$$\text{Annual percentage rate} = 12\%$$

$$\text{Monthly installments} = \$235.37$$

Figure 4.16 is the cash flow diagram. Construct the loan payment schedule by showing the remaining balance, interest payment, and principal payment at the end of each period over the life of the loan.

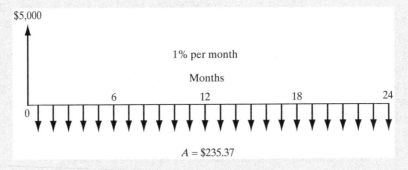

Figure 4.16 Cash flow diagram of the home improvement loan with an APR of 12% (Example 4.14).

SOLUTION

Given: $P = \$5,000$, $A = \$235.37$ per month, $r = 12\%$ per year, $M = 12$ compounding periods per year, and $N = 24$ months.

Find: B_n and I_n for $n = 1$ to 24.

We can easily see how the bank calculated the monthly payment of $235.37. Since the effective interest rate per payment period on this loan is 1% per month, we establish the following equivalence relationship:

$$\$235.37(P/A, 1\%, 24) = \$235.37(21.2434) = \$5,000$$

The loan payment schedule can be constructed as in Table 4.5. The interest due at $n = 1$ is \$50.00; 1% of the \$5,000 outstanding during the first month. The remaining \$185.37 is applied to the principal, reducing the amount outstanding in the second month to \$4,814.63. The interest due in the second month is 1% of \$4,814.63 (or \$48.15), leaving \$187.22 for repayment of the principal. At $n = 24$, the last \$235.37 payment is just sufficient to pay the interest on the unpaid principal of the loan and to repay the remaining principal. Figure 4.17 illustrates the ratios between the interest and principal payments over the life of the loan.

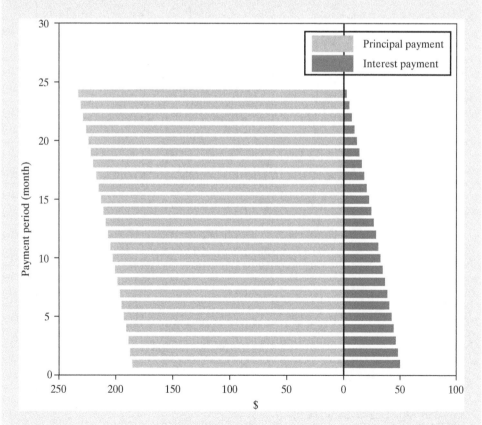

Figure 4.17 The proportions of principal and interest payments over the life of the loan (monthly payment = \$235.37) (Example 4.14).

COMMENTS: Certainly, constructing a loan repayment schedule such as that in Table 4.5 can be tedious and time consuming, unless a computer is used.

TABLE 4.5 Creating a Loan Repayment Schedule with Excel (Example 4.14)

	A	B	C	D	E	F	G
1		Example 4.14 Loan Repayment Schedule					
2							
3							
4		**Contract amount**	$ 5,000.00		**Total payment**		$ 5,648.82
5		**Contract period**	24		**Total interest**		$648.82
6		**APR (%)**	12				
7		**Monthly Payment**	($235.37)				
8							
9							
10			**Payment No.**	**Payment Size**	**Principal Payment**	**Interest payment**	**Loan Balance**
11			1	($235.37)	($185.37)	($50.00)	$4,814.63
12			2	($235.37)	($187.22)	($48.15)	$4,627.41
13			3	($235.37)	($189.09)	($46.27)	$4,438.32
14		=PMT($C$6/1200,$C$5,$C$4,0)	4	($235.37)	($190.98)	($44.38)	$4,247.33
15			5	($235.37)	($192.89)	($42.47)	$4,054.44
16			6	($235.37)	($194.82)	($40.54)	$3,859.62
17			7	($235.37)	($196.77)	($38.60)	$3,662.85
18			8	($235.37)	($198.74)	($36.63)	$3,464.11
19			9	($235.37)	($200.73)	($34.64)	$3,263.38
20			10	($235.37)	($202.73)	($32.63)	$3,060.65
21			11	($235.37)	($204.76)	($30.61)	$2,855.89
22			12	($235.37)	($206.81)	($28.56)	$2,649.08
23			13	($235.37)	($208.88)	($26.49)	$2,440.20
24			14	($235.37)	($210.97)	($24.40)	$2,229.24
25			15	($235.37)	($213.08)	($22.29)	$2,016.16
26			16	($235.37)	($215.21)	($20.16)	$1,800.96
27			17	($235.37)	($217.36)	($18.01)	$1,583.60
28			18	($235.37)	($219.53)	($15.84)	$1,364.07
29			19	($235.37)	($221.73)	($13.64)	$1,142.34
30			20	($235.37)	($223.94)	($11.42)	$918.40
31			21	($235.37)	($226.18)	($9.18)	$692.21
32			22	($235.37)	($228.45)	($6.92)	$463.77
33			23	($235.37)	($230.73)	($4.64)	$233.04
34			24	($235.37)	($233.04)	($2.33)	$0.00
35							
36							
37							
38		=PPMT($C$6/1200,C33,$C$5,$C$4,0)					
39						=G32+E33	
40		=IPMT($C$6/1200,C33,$C$5,$C$4,0)					
41							

Method 2: Remaining-Balance Method.   Alternatively, we can derive B_n by computing the equivalent payments remaining after the nth payment. Thus, the balance with $N - n$ payments remaining is

$$B_n = A(P/A, i, N - n) \qquad (4.13)$$

and the interest payment during period n is

$$I_n = (B_{n-1})i = A(P/A, i, N - n + 1)i \qquad (4.14)$$

where $A(P/A, i, N - n + 1)$ is the balance remaining at the end of period $n - 1$ and

$$P_n = A - I_n = A - A(P/A, i, N - n + 1)i$$
$$= A[1 - (P/A, i, N - n + 1)i]$$

Knowing the interest factor relationship $(P/F, i, n) = 1 - (P/A, i, n)i$ from Table 3.6, we obtain

$$P_n = A(P/F, i, N - n + 1) \qquad (4.15)$$

As we can see in Figure 4.18, this method provides more concise expressions for computing the balance of the loan.

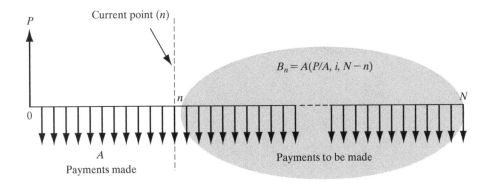

Figure 4.18   Calculating the remaining loan balance on the basis of Method 2.

Consider the home improvement loan in Example 4.14.

(a) For the sixth payment, compute both the interest and principal payments.

(b) Immediately after making the sixth monthly payment, you would like to pay off the remainder of the loan in a lump sum. What is the required amount?

SOLUTION

(a) Interest and principal payments for the sixth payment.

Given: (As for Example 4.14.)

Find: I_6 and P_6

Using Eqs. (4.14) and (4.15), we compute

$$I_6 = \$235.37(P/A, 1\%, 19)(0.01)$$

$$= (\$4,054.44)(0.01)$$

$$= \$40.54$$

$$P_6 = \$235.37(P/F, 1\%, 19) = \$194.83$$

or we simply subtract the interest payment from the monthly payment

$$P_6 = \$235.37 - \$40.54 = \$194.83$$

(b) Remaining balance after the sixth payment.

The lower half of Figure 4.19 shows the cash flow diagram that applies to this part of the problem. We can compute the amount you owe after you make the sixth payment by calculating the equivalent worth of the remaining 18 payments at the end of the sixth month, with the time scale shifted by 6.

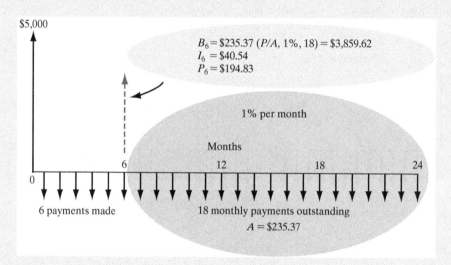

Figure 4.19 Computing the outstanding loan balance after making the sixth payment on the home improvement loan (Example 4.15).

Given: $A = \$235.37$, $i = 1\%$ per month, and $N = 18$ months.

Find: Balance remaining after six months (B_6).

$$B_6 = \$235.37(P/A, 1\%, 18) = \$3,859.62$$

If you desire to pay off the remainder of the loan at the end of the sixth payment, you must come up with $3,859.62. To verify our results, compare this answer with the value given in Table 4.5.

Add-On Interest Loans

The add-on loan is totally different from the popular amortized loan. In the add-on loan, the total interest to be paid is precalculated and added to the principal. The principal and this precalculated interest amount are then paid together in equal installments. In such a case, the interest rate quoted is not the effective interest rate, but what is known as **add-on interest**. If you borrow P for N years at an add-on rate of i with equal payments due at the end of each month, a typical financial institution might compute the monthly installment payments as

$$\text{Total add-on interest} = P(i)(N)$$

$$\text{Principal plus add-on interest} = P + P(i)(N) = P(1 + iN)$$

$$\text{Monthly installments} = \frac{P(1 + iN)}{(12 \times N)} \qquad (4.16)$$

Notice that the add-on interest is *simple interest*. Once the monthly payment is determined, the financial institution computes the APR on the basis of this payment, and you will be told what the value will be. Even though the add-on interest is specified along with the APR, many ill-informed borrowers think that they are actually paying the add-on rate quoted for this installment loan. To see how much interest you actually pay under a typical add-on loan arrangement, consider Example 4.16.

EXAMPLE 4.16 Effective Interest Rate for an Add-On Interest Loan

Consider again the home improvement loan problem in Example 4.14. Suppose that you borrow $5,000 with an add-on rate of 12% for two years. You will make 24 equal monthly payments.

(a) Determine the amount of the monthly installment.
(b) Compute the nominal and the effective annual interest rate on the loan.

SOLUTION

Given: Add-on rate = 12% per year, loan amount (P) = $5,000, and N = 2 years.
Find: (a) A, and (b) i_a and i.

(a) First we determine the amount of add-on interest:

$$iPN = (0.12)(\$5,000)(2) = \$1,200$$

Then we add this simple-interest amount to the principal and divide the total amount by 24 months to obtain A:

$$A = \frac{\$5,000 + \$1,200}{24} = \$258.33$$

(b) Putting yourself in the lender's position, compute the APR value of the loan just described. Since you are making monthly payments with monthly compounding, you need to find the effective interest rate that makes the present $5,000 sum equivalent to 24 future monthly payments of $258.33. In this situation, we are solving for i in the equation

$$258.33 = 5,000(A/P, i, 24)$$

or

$$(A/P, i, 24) = 0.0517$$

You know the value of the A/P factor, but you do not know the interest rate i. As a result, you need to look through several interest tables and determine i by interpolation. A more effective approach is to use Excel's RATE function with the following parameters:

$$= \text{RATE}(N, A, P, F, \text{type, guess})$$
$$= \text{RATE}(24, 258.33, -5000, 0, 0, 1\%) \rightarrow 1.7975\%$$

The nominal interest rate for this add-on loan is $1.7975 \times 12 = 21.57\%$, and the effective annual interest rate is $(1 + 0.017975)^{12} - 1 = 23.84\%$, rather than the 12% quoted add-on interest. When you take out a loan, you should not confuse the add-on interest rate stated by the lender with the actual interest cost of the loan.

COMMENTS: In the real world, truth-in-lending laws require that APR information always be provided in mortgage and other loan situations, so you would not have to calculate nominal interest as a prospective borrower (although you might be interested in calculating the actual or effective interest). However, in later engineering economic analyses, you will discover that solving for implicit interest rates (or rates of return on investment) is performed regularly. Our purpose in this text is to periodically give you some practice with this type of problem, even though the scenario described does not exactly model the real-world information you would be given.

4.5.2 Loan versus Lease Financing

When, for example, you choose a car, you also choose how to pay for it. If you do not have the cash on hand to buy a new car outright—and most of us don't—you can consider taking out a loan or leasing the car to spread the payments over time. Deciding whether to pay cash, take a loan, or sign a lease depends on a number of personal as well as economic factors. Leasing is an option that lets you pay for the portion of a vehicle you expect to use over a specified term, plus a charge for rent, taxes, and fees. For instance, you might want a $20,000 vehicle. Suppose that vehicle might be worth about $9,000 at the end of your three-year lease. (This is called the residual value.)

- If you have enough money to buy the car, you could purchase it in cash. If you pay cash, however, you will lose the opportunity to earn interest on the amount you spend. That could be substantial if you know of an investment that is paying a good return.
- If you purchase the vehicle using debt financing, your monthly payments will be based on the entire $20,000 value of the vehicle. You will continue to own the vehicle at the end of your financing term, but the interest you will pay on the loan will drive up the real cost of the car.
- If you lease the same vehicle, your monthly payments will be based on the amount of the vehicle you expect to "use up" over the term of the lease. This value ($11,000 in

our example) is the difference between the original cost ($20,000) and the estimated value at the end of the lease ($9,000). With leasing, the length of your lease agreement, the monthly payments, and the yearly mileage allowance can be tailored to your driving needs. The greatest financial appeal for leasing is low initial outlay costs: Usually you pay a leasing administrative fee, one month's lease payment, and a refundable security deposit. The terms of your lease will include a specific mileage allowance; if you put additional miles on your car, you will have to pay more for each extra mile.

Discount Rate to Use in Comparing Different Financing Options

In calculating the net cost of financing a car, we need to decide which interest rate to use in discounting the loan repayment series. The dealer's (bank's) interest rate is supposed to reflect the time value of money of the dealer (or the bank) and is factored into the required payments. However, the correct interest rate to use in comparing financing options is the interest rate that reflects *your* time value of money. For most individuals, this interest rate might be equivalent to the savings rate from their deposits. To illustrate, consider Example 4.17, in which we compare three auto-financing options.

EXAMPLE 4.17 Financing Your Vehicle: Paying Cash, Taking a Loan, or Leasing

Suppose you intend to own or lease a vehicle for 42 months. Consider the following three ways of financing the vehicle—say, a 2015 BMW 325 Ci 2-D Coupe.

- **Option A:** Purchase the vehicle at the normal price of $32,508, and pay for the vehicle over 42 months with equal monthly payments at 5.65% APR financing.
- **Option B:** Purchase the vehicle at a discount price of $31,020 to be paid immediately.
- **Option C:** Lease the vehicle for 42 months.

Table 4.6 lists the items of interest under each option. For each option, license, title, and registration fees, as well as taxes and insurance, are extra.

TABLE 4.6 Summary of Automobile Financing Options

Item	Debt Financing	Paying Cash	Lease Financing
Price	$32,508.00	$31,020.00	$32,508.00
Down payment	$4,500.00	$0.00	$0.00
APR (%)	5.65%		
Monthly payment	$736.53		$513.76
Length (months)	42		42
Fees			$994.00
Cash due at end of lease			$395.00
Purchase option at end of lease			$17,817.00
Cash due at signing	$4,500.00	$31,020.00	$1,507.76

For the lease option, the lessee must come up with $1,507.76 at signing. This cash due at signing consists of the first month's lease payment of $513.76 and a $994 administrative fee. The lease rate is based on 60,000 miles over the life of the contract. There will be a surcharge at the rate of 18 cents per mile for any additional miles driven over 60,000. No security deposit is required; however, a $395 disposition fee is due at the end of the lease, at which time the lessee has the option to purchase the car for $17,817. The lessee is also responsible for excessive wear and use. If the funds that would be used to purchase the vehicle are presently earning 4.5% annual interest compounded monthly, which financing option is a better choice?

DISCUSSION: With a lease payment, you pay for the portion of the vehicle you expect to use. At the end of the lease, you simply return the vehicle to the dealer and pay the agreed-upon disposal fee. With traditional financing, your monthly payment is based on the entire $32,508 value of the vehicle, and you will own the vehicle at the end of your financing terms. Since you are comparing the options over 42 months, you must explicitly consider the unused portion (resale value) of the vehicle at the end of the term. In other words, you must consider the resale value of the vehicle in order to figure out the net cost of owning it. *As the resale value, you could use the $17,817 quoted by the dealer in the lease option.* Then you have to ask yourself if you can get that kind of resale value after 42 months' ownership.

Note that the 5.65% APR represents the dealer's interest rate used in calculating the loan payments. With 5.65% interest, your monthly payments will be $A = (\$32,508 - \$4,500)(A/P, 5.65\%/12, 42) = \$736.53$. Note also, however, that the 4.5% APR represents your earning opportunity rate. In other words, if you do not buy the car, your money continues to earn 4.5% APR. Therefore, 4.5% represents your opportunity cost of purchasing the car. So which interest rate do you use in your analysis? Clearly, the 4.5% rate is the appropriate one to use.

SOLUTION

Given: Financial facts shown in Figure 4.20,

$r = 4.5\%$, payment period = monthly, and compounding period = monthly.

Find: The most economical financing option under the assumption that you will be able to sell the vehicle for $17,817 at the end of 42 months.

For each option, we will calculate the net equivalent total cost at $n = 0$. Since the loan payments occur monthly, we need to determine the effective interest rate per month, which is 4.5%/12.

- **Option A: Conventional debt financing.** The equivalent present cost of the total loan payments is

$$P_1 = \$4,500 + \$736.53(P/A, 4.5\%/12, 42)$$

$$= \$33,071.77$$

The equivalent present worth of the resale value of the car is

$$P_2 = \$17,817(P/F, 4.5\%/12, 42) = \$15,225.13$$

The equivalent net financing cost is

$$P_{\text{Option A}} = \$33,071.77 - \$15,225.13$$
$$= \$17,846.64$$

- **Option B: Cash financing.**

$$P_{\text{Option B}} = \$31,020 - \$17,817(P/F, 4.5\%/12, 42)$$
$$= \$31,020 - \$15,225.13$$
$$= \$15,794.87$$

- **Option C: Lease financing.** The equivalent present cost of the total lease payments is

$$P_{\text{Option C}} = \$1,507.76 + \$513.76(P/A, 4.5\%/12, 41)$$
$$+ \$395(P/F, 4.5\%/12, 42)$$
$$= \$1,507.76 + \$19,490.96 + \$337.54$$
$$= \$21,336.26$$

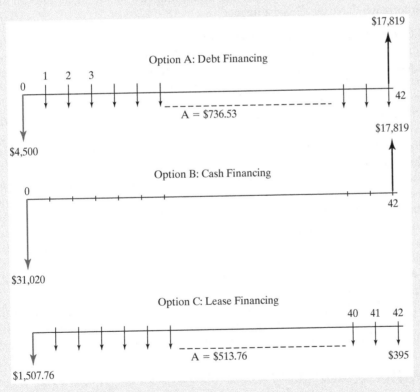

Figure 4.20   Comparing different financing options (Example 4.17).

It appears that, at 4.5% interest compounded monthly, the cash financing option is the most economical one.

4.5.3 Home Mortgage

The term **mortgage** refers to a special type of loan for buying a piece of property, such as a house. The most common mortgages are fixed-rate amortized loans, adjustable-rate loans, and graduated-payment loans. As with many of the amortized loans we have considered so far, most home mortgages involve monthly payments and monthly compounding. However, as the names of these types of mortgages suggest, a number of ways determine how monthly payments and interest rates can be structured over the life of the mortgage.

The Cost of a Mortgage

The cost of a mortgage depends on the amount you borrow, the interest you pay, and how long you take to repay the loan. Since monthly payments spread the cost of a mortgage over a long period, it is easy to forget the total expense. For example, if you borrow $100,000 for 30 years at 8.5% interest, your total repayment will be around $277,000, more than two-and-a-half times the original loan! Minor differences in the interest rate—8.5% versus 8%—can add up to a lot of money over 30 years. At 8%, the total repaid would be $264,160, almost $13,000 less than at the 8.5% rate. Other than the interest rate, any of the following factors will increase the overall cost, but a higher interest rate and longer term will have the greatest impact.

- **Loan amount.** This is the amount you actually borrow after fees and points are deducted. It is the basis for figuring the real interest or APR on the money you are borrowing.
- **Loan term.** With a shorter term, you will pay less interest overall, and your monthly payments will be somewhat larger. A 15-year mortgage, as opposed to a 30-year mortgage for the same amount, can cut your costs by more than 55%.
- **Payment frequency.** You can pay your mortgage biweekly instead of monthly, or you can make an additional payment each month. With biweekly payments, you make 26 yearly payments instead of 12. The mortgage is paid off in a little more than half the time, and you pay a little more than half the interest. With an additional payment each month, you can reduce your principal. With a fixed-rate mortgage, you pay off the loan more quickly, but regular required monthly payments remain the same.
- **Points (prepaid interest).** Points are interest that you prepay at the closing on your home. Each point is 1% of the loan amount. For example, on a $100,000 loan, three points represents a prepayment of $3,000. This is equivalent to financing $97,000, but your payments are based on a $100,000 loan.
- **Fees.** Fees include application fees, loan origination fees, and other initial costs imposed by the lender.
- **Types of mortgages.** Mortgages can have either fixed or adjustable rates (or both, in which case they are known as hybrid mortgages). As we mentioned earlier, interest rates in the financial marketplace rise and fall over time.

Calculating Monthly Payments with Fixed-Rate Mortgages

This is a mortgage on which the interest rate and monthly mortgage payment remain unchanged throughout the term of the mortgage. Calculating the required mortgage payment depends on the type of mortgage secured with the financial institution. We will review a few of them.

- **A Fully Amortizing Mortgage:** This is an installment loan in which the principal and interest will be repaid fully through regular installments by the time the loan's term ends. Typically, the loan term is either 15 years or 30 years.
- **An Interest-Only Mortgage:** A mortgage is "interest-only" if the payment the borrower is required to make consists of interest only. With this option, the payment will not include any repayment of principal. The result is that the loan balance will remain unchanged. The option to pay interest only lasts for a specified period, usually five to 10 years. One of the main advantages to go with the interest-only mortgage is for borrowers with fluctuating incomes. When their finances are tight, they can make the interest-only payment, and when they are doing well they can make a substantial payment to principal.
- **Points or No Points:** Lenders might be willing to raise the rate by a fraction (say, $\frac{1}{8}\%$ or $\frac{1}{4}\%$) and lower the points—or the reverse—as long as they make the same profit. The advantages of fewer points are lower closing costs and laying out less money when you are apt to need it most. However, if you plan to keep the house longer than five to seven years, paying more points to get a lower interest rate will reduce your long-term cost. Example 4.18 examines the effect of points on the cost of borrowing for a fixed-rate mortgage, and Example 4.19 illustrates how the points charged by the lender can affect your financing decision.

EXAMPLE 4.18 An Interest-Only versus a Fully Amortized Mortgage

Suppose that you need to finance a new home priced at $240,000 with a $40,000 down payment at 6.6% interest compounded monthly over 30 years. You are considering two options in securing a mortgage from a bank.

- **Option 1:** A fully amortized payment option.
- **Option 2:** A five-year interest-only option.

Assuming that the interest rate is fixed over the entire life of these loans, determine (a) the monthly payments and (b) the interest payments under each option.

SOLUTION

Given: $P = \$200,000$, APR $= 6.6\%$ or 0.55% per month, and $N = 30$ years.
Find: (a) the monthly payment; and (b) the interest payment during the first year of ownership of the home.

(a) Monthly payments.

- **Option 1:** $A = \$200,000\left(A/P, \dfrac{6.6\%}{12}, 360\right) = \$1,277.32$
- **Option 2:** $A = \$200,000(0.0055) = \$1,100$

(b) Interest payments are calculated using Excel as in Figure 4.21.

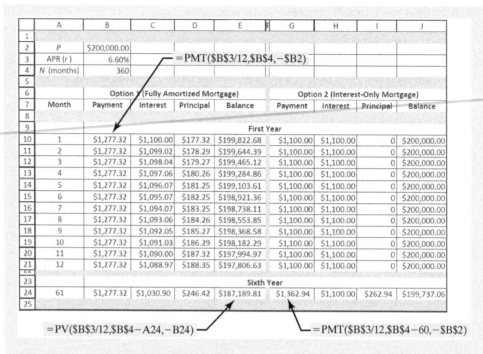

	A	B	C	D	E	F	G	H	I	J
1										
2	P	$200,000.00								
3	APR (r)	6.60%	=PMT($B$3/12,$B$4,−$B2)							
4	N (months)	360								
5										
6			Option 1 (Fully Amortized Mortgage)				Option 2 (Interest-Only Mortgage)			
7	Month	Payment	Interest	Principal	Balance		Payment	Interest	Principal	Balance
8										
9					First Year					
10	1	$1,277.32	$1,100.00	$177.32	$199,822.68		$1,100.00	$1,100.00	0	$200,000.00
11	2	$1,277.32	$1,099.02	$178.29	$199,644.39		$1,100.00	$1,100.00	0	$200,000.00
12	3	$1,277.32	$1,098.04	$179.27	$199,465.12		$1,100.00	$1,100.00	0	$200,000.00
13	4	$1,277.32	$1,097.06	$180.26	$199,284.86		$1,100.00	$1,100.00	0	$200,000.00
14	5	$1,277.32	$1,096.07	$181.25	$199,103.61		$1,100.00	$1,100.00	0	$200,000.00
15	6	$1,277.32	$1,095.07	$182.25	$198,921.36		$1,100.00	$1,100.00	0	$200,000.00
16	7	$1,277.32	$1,094.07	$183.25	$198,738.11		$1,100.00	$1,100.00	0	$200,000.00
17	8	$1,277.32	$1,093.06	$184.26	$198,553.85		$1,100.00	$1,100.00	0	$200,000.00
18	9	$1,277.32	$1,092.05	$185.27	$198,368.58		$1,100.00	$1,100.00	0	$200,000.00
19	10	$1,277.32	$1,091.03	$186.29	$198,182.29		$1,100.00	$1,100.00	0	$200,000.00
20	11	$1,277.32	$1,090.00	$187.32	$197,994.97		$1,100.00	$1,100.00	0	$200,000.00
21	12	$1,277.32	$1,088.97	$188.35	$197,806.63		$1,100.00	$1,100.00	0	$200,000.00
23					Sixth Year					
24	61	$1,277.32	$1,030.90	$246.42	$187,189.81		$1,362.94	$1,100.00	$262.94	$199,737.06
25										

=PV($B$3/12,$B$4−A24,−B24)   =PMT($B$3/12,$B$4−60,−$B$2)

Figure 4.21 Comparison of interest payments between option 1 and option 2.

COMMENTS: Note that the example assumes that the interest rate is fixed over the entire life of these loans. Most interest-only loans, however, are adjustable rate, which means that the borrower may need to renegotiate the interest rate at the end of five years.

EXAMPLE 4.19 Fixed-Rate Mortgage—Points or No Points?

When you are shopping for a home mortgage loan, you frequently encounter various types of borrowing options, including paying points up front or paying no points but accepting a slightly higher interest rate. Suppose you want to finance a home mortgage of $100,000 at a 15-year fixed-interest rate. A leading independent home lender offers the following two options with no origination fees:

- **Option 1:** Pay one point with 6.375% interest.
- **Option 2:** Pay no points with 6.75% interest.

Recall that a point is equivalent to 1% of the face value of the mortgage. In other words, with option 1, you are borrowing only $99,000, but your lender will calculate your monthly payments on the basis of $100,000. Compute the APR for option 1. Which option would you select?

DISCUSSION: Discount fees or points are a fact of life with mortgages. A point is a fee charged by a lender to increase the lender's effective yield on the money borrowed. Points are charged in lieu of interest; the more points paid, the lower is the rate of interest required on the loan to provide the same yield or return to the lender. Origination fees are the fees charged by a lender to prepare a loan document, make credit checks, inspect, and sometimes appraise a property. These fees are usually computed as a percentage of the face value of the mortgage.

SOLUTION

Given: $P = \$100,000$, $r = 6.375\%$, $N = 180$ months, and discount point $= 1$ point.
Find: APR.
We first need to find out how much the lender will calculate your monthly payments to be. Since the mortgage payments will be based on the face value of the loan, your monthly payment will be

$$A = \$100,000(A/P, 6.375\%/12, 180) = \$864.25$$

Because you have to pay $1,000 to borrow $100,000, you are actually borrowing $99,000. Therefore, you need to find out what kind of interest you are actually paying. In other words, you borrow $99,000 and you make $864.25 monthly payments over 15 years. To find the interest rate, you can set up the following equivalence equation and solve for the unknown interest rate.

$$\$99,000 = \$864.25(P/A, i, 180)$$

$$i = 0.545\% \text{ per month}$$

$$\text{APR} = 0.545\% \times 12 = 6.54\%$$

Note that with a one-point fee, the lender was able to raise its effective yield from 6.375% to 6.54%. However, the lender is still earning less than option 2 affords.

Calculating the Monthly Payments with Adjustable-Rate Mortgages (ARMs)

This is a mortgage with an interest rate that is linked to an economic index. The interest rate, and your payments, are periodically adjusted up or down as the index changes. If there is a possibility that market rates will rise above the fixed rate, some lenders may be reluctant to lock a loan into a fixed interest rate over a long term. If they did, they would have to forgo the opportunity to earn better rates, because their assets are tied up in a loan that earns a lower rate of interest. The adjustable-rate mortgage (ARM) is meant to encourage lending institutions to commit funds over long periods. Typically, the rate rises gradually each year for the first several years of the mortgage and then settles into a single rate for the remainder of the loan. Some of the main features of the ARM consist of the following.

- **Index:** The interest rate on an ARM is made up of two parts: the index and the margin. An index is a guide that lenders use to measure interest rate changes. Common indexes used by lenders include the activity of one-, three-, and five-year Constant-Maturity Treasury (CMT) securities, the Cost of Funds Index (COFI), and the London

Interbank Offered Rate (LIBOR). Figure 4.22 shows a few common indexes over a 15-year period (2000–2015). As you can see, some index rates tend to be higher than others, and some change more often. But if a lender bases interest-rate adjustments on the average value of an index over time, your interest rate would not change as dramatically.

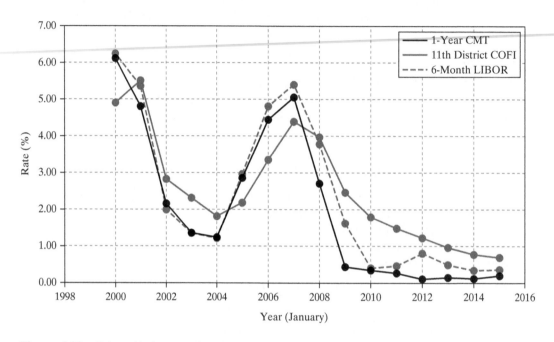

Figure 4.22 Selected index rates for ARMs over a 15-year period.

- **Margin:** To set the interest rate on an ARM, lenders add a few percentage points to the index rate, called the margin. It is basically an interest rate that represents the lender's cost of doing business plus the profit they will make on the loan. The margin is added to the index rate to determine your total interest rate. It usually stays the same during the life of your home loan.
- **Adjustment period:** The adjustment period is the period between potential interest rate adjustments. You may see an ARM described with figures such as 1/1, 3/1, and 5/1. The first figure in each set refers to the initial period of the loan, during which your interest rate will stay the same (fixed) as it was on the day you signed your loan papers. The second number is the adjustment period, showing how often adjustments can be made to the rate after the initial period has ended. The examples here are all ARMs with annual adjustments—meaning adjustments could happen every year. Typically 3/1 or 5/1 ARMs are referred to as a **hybrid-rate mortgage**.
- **Interest rate cap:** An interest rate cap places a limit on the amount your interest can change. There are two types of interest rate caps associated with ARMs.
  1. **A periodic cap** limits the amount your interest rate can increase from one adjustment period to the next. Not all ARMs have periodic rate caps.
  2. **A lifetime cap** limits how much the interest rate can increase over the life of the loan. By law, virtually all ARMs must have a lifetime cap.

- **Payment caps:**   A payment cap limits how much your monthly payment can increase at each adjustment. ARMs with payment caps often do not have periodic rate caps.

Example 4.20 illustrates how you may compute the monthly payment under a typical ARM financing.

EXAMPLE 4.20 A 5/1 Hybrid Adjustable Mortgage Plan

Suppose that you finance a home on the basis of a 5/1 hybrid mortgage (five-year fixed/adjustable) plan over 30 years. The $100,000 hybrid loan plan offers an initial rate of 6.02% fixed for 60 months. The loan rate would be adjusted thereafter every 12 months to the lowest of three options: the then-current rate on one-year Treasury bills plus 2.75 percent (margin), the previous rate plus a maximum annual cap of 2.0 percent, or a lifetime cap of 11.02 percent. There is no prepayment penalty for this type of loan. The projected interest rates by the lender after five years are as follows.

Period	Projected APR
Years 1–5	6.02% (fixed)
Year 6	6.45%
Year 7	6.60%
Year 8	6.80%
Year 9	7.15%
Year 10	7.30%

(a) Develop the payment schedule for the first 10 years.
(b) Determine the total interest paid over a 10-year ownership.

SOLUTION

Given: Varying annual mortgage rates and $N = 30$ years.

Find: (a) The monthly payment and (b) the total interest payment over the 10-year ownership of the home.

(a) Monthly payment calculation. During the first five years, you borrow $100,000 for 30 years at 6.02%. To compute the monthly payment, use $i = 6.02\%/12 = 0.5017\%$ per month and $N = 360$ months:

$$A_{1-60} = \$100,000(A/P, 0.5017\%, 360) = \$600.84$$

The balance remaining on the loan after you make the 60th payment will be

$$B_{60} = \$600.84(P/A, 0.5017\%, 300) = \$93,074$$

During the sixth year, the interest rate changes to 6.45%, or 0.5375% per month, but the remaining term is only 300 months. Therefore, the new monthly payment would be

$$A_{60-72} = \$93,074(A/P, 0.5375\%, 300) = \$625.54$$

After you make the 72nd payment, the balance remaining on the mortgage is

$$B_{72} = \$625.54(P/A, 0.5375\%, 288) = \$91,526$$

During the seventh year, the interest rate changes to 6.60%, or 0.5500% per month. The new monthly payment and the remaining balance after you make the 84th payment are then

$$A_{73-84} = \$91,526(A/P, 0.5500\%, 288) = \$634.03$$

and

$$B_{84} = \$634.03(P/A, 0.5500\%, 276) = \$89,908$$

You can compute the monthly payments in the same fashion for the remaining years. Table 4.7 gives the details over the life of the loan.

(b) To determine the total interest paid over 10 years, we first determine the total monthly mortgage payments over 10 years. Since we know the ending balance at the end of 10 years, we can easily calculate the interest payments during this home ownership period:

$$\text{Total mortgage payment} = 60 \times \$600.84$$
$$+ 12 \left(\begin{array}{c} \$625.54 + \$634.03 + \$645.09 \\ +\$664.09 + \$672.05 \end{array} \right)$$
$$= \$74,940$$

$$\text{Interest payment} = \text{Ending balance} + \text{Total mortgage payment} - \$100,000$$
$$= \$84,704 + \$74,940 - \$100,000$$
$$= \$59,644$$

TABLE 4.7 Monthly Payments Under 5/1 Hybrid Mortgage

	Year	Month	Forecast Rate	Monthly Payment	Ending Loan Balance
Fixed rate	1	1–12	6.02%	$600.84	$98,777
	2	13–24	6.02%	$600.84	$97,477
	3	25–36	6.02%	$600.84	$96,098
	4	37–48	6.02%	$600.84	$94,633
	5	49–60	6.02%	$600.84	$93,074
Variable rate	6	61–72	6.45%	$625.54	$91,526
	7	73–84	6.60%	$634.03	$89,908
	8	85–96	6.80%	$645.09	$88,229
	9	97–108	7.15%	$664.09	$86,513
	10	109–120	7.30%	$672.05	$84,704

4.6 Investing in Financial Assets

Most individual investors have three basic investment opportunities in financial assets: stocks, bonds, and cash. Cash investments include money in bank accounts, certificates of deposit (CDs), and U.S. Treasury bills. You can invest directly in any one or all three, or indirectly by buying mutual funds that pool your money with money from other people and then invest it. If you want to invest in financial assets, you have plenty of opportunities. In the United States alone, there are more than 9,000 stocks, 7,500 mutual funds, and thousands of corporate and government bonds to choose from. Even though we will discuss investment basics in the framework of financial assets in this chapter, the same economic concepts are applicable to any business assets examined in later chapters.

4.6.1 Investment Basics

Selecting the best investment for you depends on your personal circumstances as well as general market conditions. For example, a good investment strategy for a long-term retirement plan may not be a good strategy for a short-term college savings plan. In each case, the right investment is a balance of three things: liquidity, safety, and return.

- **Liquidity: How accessible is your money?** If your investment money must be available to cover financial emergencies, you will be concerned about liquidity: how easily you can convert it to cash. Money-market funds and savings accounts are very liquid, as are investments with short maturity dates, such as CDs. However, if you are investing for long-term goals, liquidity is not a critical issue. What you are after in that case is growth (building your assets). We normally consider certain stocks and stock mutual funds as growth investments.
- **Risk: How safe is your money?** Risk is the chance you take of making or losing money on your investment. For most investors, the biggest risk is losing money, so they look for investments they consider safe. Usually, that means putting money into bank accounts and U.S. Treasury bills, as these investments are either insured or default-free. The opposite (but equally important) risk is that your investments will not provide enough growth or income to offset the impact of inflation, the gradual increase in the cost of living. There are additional risks as well, including how the economy is doing. However, *the biggest risk is not investing at all*.
- **Return: How much profit will you be able to expect from your investment?** Safe investments often promise a specific, though limited, return. Those which involve more risk offer the opportunity to make—or lose—a lot of money. Both risk and reward are time dependent. On the one hand, as time progresses, low-yielding investments become more risky because of inflation. On the other hand, the returns associated with higher risk investments could become more stable and predictable over time, thereby reducing the perceived level of risk.

4.6.2 How to Determine Your Expected Return

Return is what you get back in relation to the amount you invested. Return is one way to evaluate how your investments in financial assets are doing in relation to each other and to the performance of investments in general. Let us look first at how we may derive rates of return.

Basic Concepts

Conceptually, the rate of return that we realistically expect to earn on any investment is a function of three components: (1) risk-free real return, (2) an inflation factor, and (3) a risk premium.

Suppose you want to invest in stock. First, you should expect to be rewarded in some way for not being able to use your money while you are holding the stock. Second, you would be compensated for decreases in purchasing power between the time you invest the money and the time it is returned to you. Finally, you would demand additional rewards for any chance that you would not get your money back or that it will have declined in value while invested.

For example, if you were to invest $1,000 in risk-free U.S. Treasury bills for a year, you would expect a real rate of return of about 2%. Your risk premium would be also zero. You probably think that the 2% does not sound like much. However, to that you have to add an allowance for inflation. If you expect inflation to be about 4% during the investment period, you should realistically expect to earn 6% during that interval (2% real return + 4% inflation factor + 0% for risk premium). Here is what the situation looks like in tabular form:

Real return	2%
Inflation (loss of purchasing power)	4%
Risk premium (U.S. Treasury Bills)	0%
Total expected return	6%

How would it work out for a riskier investment, say, in an Internet stock such as Google.com? Because you consider this stock to be a very volatile one, you would increase the risk premium to something like the following:

Real return	2%
Inflation (loss of purchasing power)	4%
Risk premium (Google.com)	20%
Total expected return	26%

So you will not invest your money in Google.com unless you are reasonably confident of having it grow at an annual rate of 26%. Again, the risk premium of 20% is a perceived value that can vary from one investor to another.

Return on Investment over Time

If you start out with $1,000 and end up with $2,000, your return is $1,000 on that investment (or 100%). If a similar investment grows to $1,500, your return is $500 (or 50%). However, unless you held those investments for the same period, you cannot determine which has a better performance. What you need in order to compare your return on one investment with the return on another investment is the **compound annual return**, the average percentage that you have gained on each investment over a series of one-year periods. For example, if you buy a share for $15 and sell it for $20, your profit is $5. If that happens within a year, your rate of return is an impressive 33.33%. If it takes five years, your return (compounded) will be closer to 5.92%, since the profit is spread over a five-year period. Mathematically, you are solving the following equivalence equation for i:

$$\$20 = \$15(1 + i)^5$$

$$i = 5.92\%$$

Figuring out the actual return on your portfolio investment is not always that simple for a number of reasons.

1. The amount of your investment changes. Most investment portfolios are active, with money moving in and out.
2. The method of computing the return can vary. For example, the performance of a stock can be averaged or compounded, which changes the rate of return significantly, as we will demonstrate in Example 4.21.
3. The time you hold specific investments varies. When you buy or sell can have a dramatic effect on the overall return.

EXAMPLE 4.21   Figuring Average versus Compound Return

Consider the following six different cases of the performance of a $1,000 investment over a three-year holding period. Compute the average versus compound return for each case.

Investment	Annual Investment Yield					
	Case I	Case 2	Case 3	Case 4	Case 5	Case 6
Year 1	9%	5%	0%	0%	−1%	−5%
Year 2	9%	10%	7%	0%	−1%	−8%
Year 3	9%	12%	20%	27%	29%	40%

SOLUTION

Given: Three years' worth of annual investment yield data.
Find: Compound versus average rate of return.

As an illustration, consider Case 6 for an investment of $1,000. At the end of the first year, the value of the investment decreases to $950; at the end of second year, it decreases again to $950(1 − 0.08) = \$874$; at the end of third year, it increases to $874(1 + 0.40) = \$1,223.60$. Therefore, one way you can look at the investment is to ask, "At what annual interest rate would the initial $1,000 investment grow to $1,223.60 over three years?" This is equivalent to solving the following equivalence problem.

$$\$1,223.60 = \$1,000(1 + i)^3$$

$$i = 6.96\%$$

If someone evaluates the investment on the basis of the average annual rate of return, he or she might proceed as

$$i = \frac{-5\% - 8\% + 40\%}{3} = 9\%$$

If you calculate the remaining cases, you will observe that all six cases have the same average annual rate of return, although their compound rates of return vary from 6.96% to 9%.

	Compound versus Average Rate of Return					
	Case 1	Case 2	Case 3	Case 4	Case 5	Case 6
Average return	9%	9%	9%	9%	9%	9%
Balance at end of three years	$1,295	$1,294	$1,284	$1,270	$1,264	$1,224
Compound rate of return	9.00%	8.96%	8.69%	8.29%	8.13%	6.96%

Your immediate question is "Are they indeed the same?" Certainly not: You will have the most money with Case 1, which also has the highest compound rate of return. The average rate of return is easy to calculate, but it ignores the basic principle of the time value of money. In other words, according to the average-rate-of-return concept, we may view all six cases as equivalent. However, the amount of money available at the end of year 3 would be different in each case. Although the average rate of return is popular for comparing investments in terms of their yearly performance, it is not a correct measure in comparing the performance for investments over a multiyear period.

COMMENTS: You can evaluate the performance of your portfolio by comparing it against standard indexes and averages that are widely reported in the financial press. If you own stocks, you can compare their performance with the performance of the Dow Jones Industrial Average (DJIA), perhaps one of the best-known measures of stock market performance in the world. If you own bonds, you can identify an index that tracks the type you own: corporate, government, or agency bonds. If your investments are in cash, you can follow the movement of interest rates on Treasury bills, CDs, and similar investments. In addition, total-return figures for the performance of mutual funds are reported regularly. You can compare how well your investments are doing against those numbers. Another factor to take into account in evaluating your return is the current inflation rate. Certainly, your return needs to be higher than the inflation rate if your investments are going to have real growth.

4.6.3 Investing in Bonds

Bonds are loans that investors make to corporations and governments. As shown in Figure 4.23, the borrowers get the cash they need while the lenders earn interest. Americans have more money invested in bonds than in stocks, mutual funds, or any other types of securities. One of the major appeals is that bonds pay a set amount of interest on a regular basis. That is why they are called *fixed-income securities*. Another attraction is that the issuer promises to repay the loan in full and on time.

Bond versus Loan

A bond is similar to a loan. For example, say you lend out $1,000 for 10 years in return for a yearly payment of 7% interest. Here is how that arrangement translates into bond

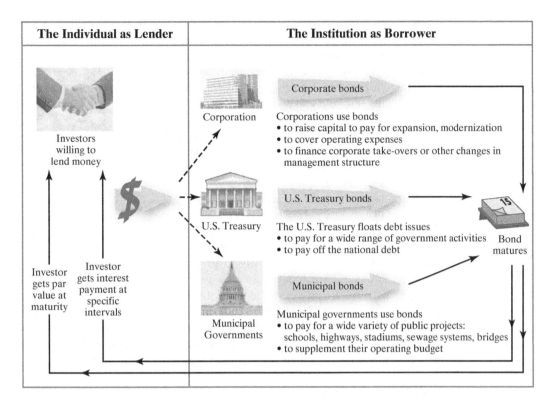

Figure 4.23   Types of bonds and how they are issued in the financial market.
Source: Virginia B. Morris and Kenneth M. Morris, *The Wall Street Guide to Understanding Money and Investing*, © 2014 by Lightbulb Press, Inc.

terminology. You did not make a loan; you bought a bond. The $1,000 of principal is the **face value** of the bond, the yearly interest payment is its **coupon**, and the length of the loan (10 years) is the bond's **maturity**. If you buy a bond at face value (or **par**) and hold it until it matures, you will earn interest at the stated (or coupon) rate. For example, if you buy a 20-year $1,000 bond paying 8%, you will earn $80 a year for 20 years. The yield (or return on your investment) will also be 8%, and you get your $1,000 back. You can also buy and sell bonds through a broker after their date of issue. This is known as the **secondary market**. Here the price fluctuates, with a bond sometimes selling at more than par value—at a premium price (premium bonds); and sometimes below par value—at a discount. Changes in price are tied directly to the interest rate of the bond. If its rate is higher than the rate being paid on similar bonds, buyers are willing to pay more to get the higher interest. If its rate is lower, the bond will sell for less in order to attract buyers. However, as the price goes up, the yield goes down, and when the price goes down, the yield goes up.

Types of Bonds

You can choose different types of bonds to fit your financial needs—be they investing for college, finding tax-free income, or operating over a range of other possibilities. That is why it is important to have a sense of how the various types work. Several types of bonds are available in the financial market.

- **Corporate bonds.**   Bonds are the major source of corporate borrowing. *Debentures* are backed by the general credit of the corporation. *Specific corporate assets, such as property or equipment, may back bonds.*
- **Municipal bonds.**   State and local governments issue bonds to finance many projects that could not be funded through tax revenues. *General-obligation* bonds are backed by the full faith and credit of the issuer, *revenue bonds* by the earnings of the particular project being financed.
- **Treasury notes and bonds.**   The U.S. Treasury floats debt issues to pay for a wide range of government activities, including financing the national debt. Intermediate (two to 10 years) and long-term (10 to 30 years) government bonds are a major source of government funding.
- **Treasury bills.**   T-bills are the largest components of the money market, where short-term (13 to 52 weeks) securities are bought and sold. Investors use T-bills for part of their cash reserve or as an interim holding place. Interest is the difference between the buying price and the amount paid at maturity.
- **Zero-coupon bonds.**   Corporations and governments sell zero-coupon bonds (zeros) at a deep discount. Investors do not collect interest; instead, the value of the bond increases to its full value when it matures. In this way, zero-coupon bonds are similar to old series savings bonds that you bought for $37.50 and could cash in for $50 after seven years. Organizations like to issue zeros because they can continue to use the loan money without paying periodic interest. Investors like zeros because they can buy more bonds for their money and then time the maturities to coincide with anticipated expenses.
- **Callable bonds.**   Callable bonds do not run their full term. In a process called redemption, the issuer may call the bond—pay off the debt—before the maturity date. Issuers will sometimes call bonds when interest rates drop so that they can reduce their debt. If they pay off their outstanding bonds, they can float another bond at the lower rate. It is the same idea as that of refinancing a mortgage to get a lower interest rate and make lower monthly payments. Callable bonds are more risky for investors than noncallable bonds, because the investors are often faced with reinvesting the money at a lower, less attractive rate. To protect bondholders who expect long-term steady income, call provisions often specify that a bond cannot be called before a certain number of years, usually five or 10.
- **Floating-rate bonds.**   These bonds promise periodic adjustments of the interest rate to persuade investors that they are not locked into what seems like an unattractively low rate.

Under the United States Constitution, any type of government bond must include a tax break. Because federal and local governments cannot interfere with each other's affairs, income from local government bonds (municipals or simply **munis**) is immune from federal taxes, and income from U.S. Treasury bonds is free from local taxes.

Understanding Bond Prices

Corporate bond prices are quoted either by a percentage of the bond's face value or in increments of points and eight fractions of a point, where a par of $1,000 is the base. The value of each point is $10 and of each fraction $1.25. For example, a bond quoted at 86 1/2 would be selling for $865 (or 86.5% of its face value), and one quoted at 100 3/4 would be selling for $1,007.50. Treasury bonds are measured in thirty-seconds, rather than hundredths, of a point. Each 1/32 equals 31.25 cents, and we normally drop the fractional part of the cent when stating a price. For example, if a bond is quoted at 100.2 (or 100 + 2/32), the price translates to $1,000.62.

Are Bonds Safe?

Just because bonds have a reputation as a conservative investment does not mean that they are always safe. There are sources of risk in holding or trading bonds.

- Not all loans are paid back. Companies, cities, and counties occasionally do go bankrupt. Only U.S. Treasury bonds are considered rock solid.
- Another source of risk for certain bonds is that your loan may be called or paid back early. While that is certainly better than it not being paid back at all, it forces you to find another, possibly less lucrative, place to put your money.
- The main danger for buy-and-hold investors, however, is a rising inflation rate. Since the dollar amount they earn on a bond investment does not change, the value of that money can be eroded by inflation. Worse yet, with your money locked away in the bond, you will not be able to take advantage of the higher interest rates that are usually available in an inflationary economy. Now you know why bond investors cringe at cheerful headlines about full employment and strong economic growth: These traditional signs of inflation hint that bondholders may soon lose their shirts.

How Do Prices and Yields Work?

You can trade bonds on the market just as you do stocks. Once a bond is purchased, you may keep it until it matures or for a variable number of interest periods before selling it. You can purchase or sell bonds at prices other than face value, depending on the economic environment. Furthermore, bond prices change over time because of the risk of nonpayment of interest or par value, supply and demand, and the economic outlook. These factors affect the **yield to maturity** (or **return on investment**) of the bond:

- The **yield to maturity** (YTM) represents the actual interest earned from a bond over the holding period. In other words, the yield to maturity on a bond is the interest rate that establishes the equivalence between all future interest and face-value receipts and the market price of the bond.
- The **current yield** of a bond is the annual interest earned as a percentage of the current market price. The current yield provides an indication of the annual return realized from investment in the bond. To illustrate, we will explain these values with numerical examples shortly.
- Bond quotes follow a few unique conventions. As an example, let's take a look at a corporate bond with a face value of $1,000 (issued by Apple Computer (AAPL) on May 3, 2013) traded on July 11, 2014:

Price:	$947.20
Coupon (%):	2.40
Maturity Date:	03 May, 2023
Yield to Maturity (%):	3.090
Current Yield (%):	2.534%

Debt Rating:	AA+
Coupon Payment Frequency:	Semiannual
First Coupon Date:	11 November, 2013
Type:	Corporate
Industry:	Information/Computer

- Prices are given as percentages of face value. The bond on the list, for instance, has just fallen to sell for $947.20 (or 94.72%) of its $1,000 face value. In other words, an investor who bought the bond when it was issued (at $1,000) could now sell it for a 5.27% discount.
- The premium over face value is explained by examining the bond's coupon rate of 2.40% and its current yield of 2.534%. The current yield will be higher than the coupon rate whenever the bond is selling for lower than its par value.
- The coupon rate is 2.40% paid semiannually, meaning that for every six-month period you will receive $24/2 = $12.
- The bond rating system helps investors determine a company's credit risk. Think of a bond rating as the report card on a company's credit rating. Blue-chip firms, which are safer investments, have a high rating, while risky companies have a low rating. Table 4.8 illustrates the different bond rating scales from Moody's, Standard and Poor's (S&P), and Fitch Ratings—the major rating agencies in the United States.

    Notice that if the company falls below a certain credit rating, its grade changes from investment quality to junk status. Junk bonds are aptly named: They are the debt of companies that are in some sort of financial difficulty. Because they are so risky, they have to offer much higher yields than any other debt. This brings up an important point: Not all bonds are inherently safer than stocks. Certain types of bonds can be just as risky, if not riskier, than stocks.
- If you buy a bond at face value, its rate of return (or yield) is just the coupon rate. However, a glance at a table of bond quotes (like the preceding one) will tell you that after they are first issued, bonds rarely sell for exactly face value. So how much is the yield then? In our example, if you purchase the bond for $947.20, you need to realize two things. First, you have effectively bought a bond with a 2.4% coupon, since the $24 coupon is 2.534% of your $947.20 purchase price. (Recall that the coupon rate, adjusted for the current price, is the current yield of the bond.) However,

TABLE 4.8 Types of Bond Rating According to Financial Risk

Bond Rating			
Moody's	S&P/Fitch	Grade	Risk
Aaa	AAA	Investment	Highest quality
Aa	AA	Investment	High quality
A	A	Investment	Strong
Baa	BBB	Investment	Medium grade
Ba, B	BB, B	Junk	Speculative
Caa/Ca/C	CCC/CC/C	Junk	Highly speculative
C	D	Junk	In default

there's more: Although you paid $947.20, you will still receive the full $1,000 face value in 2023.

Example 4.22 illustrates how you calculate the yield to maturity and the current yield, considering both the purchase price and the capital gain for a new issue.

EXAMPLE 4.22 Yield to Maturity and Current Yield

Consider buying a $1,000 corporate (Coca-Cola Enterprises Inc.) bond at the market price of $1,352.90 with a coupon rate of 8.5% on February 10, 2014. The interest will be paid semiannually, the interest rate per payment period will be simply 4.25%, and 16 interest payments over eight years are required. We show the resulting cash flow to the investor in Figure 4.24. Find (a) the yield to maturity and (b) the current yield.

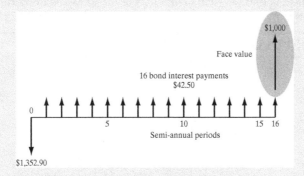

Figure 4.24 A typical cash flow transaction associated with an investment in Coca-Cola Enterprises' corporate bond.

DISCUSSION

- **Debenture:**   Coca-Cola Enterprises issued bonds totaling $750 million, and the company backed the bonds with specific pieces of property, such as buildings.
- **Par value:**   Coca-Cola's bond has a par value of $1,000.
- **Maturity date:**   Coca-Cola's bonds, which were issued on February 11, 1992, will mature on February 1, 2022; thus, they have a 30-year maturity at the time of their issue. However, if you purchased the bond on February 10, 2014, it has only an 8-year maturity.
- **Coupon rate:**   Coca-Cola's coupon rate is 2(4.25%) = 8.5%, and interest is payable semiannually. For example, Coca-Cola's bonds have a $1,000 par value, and they pay $85 in simple interest (8.5%) each year ($42.50 every six months).
- **Premium bond:**   At 135.29%, or a 35.29% premium, Coca-Cola's bonds are selling at higher than their par value. Price at offering in 1992 was 98.83 with a yield of 8.61%.

SOLUTION

Given: Purchase price = $1,352.90, coupon rate = 8.5% per year paid semiannually, and 8-year remaining maturity with a par value of $1,000.

Find: (a) Yield to maturity and (b) current yield.

(a) We find the yield to maturity by determining the interest rate that makes the present worth of the receipts equal to the market price of the bond:

$$\$1,352.90 = \$42.50(P/A, i, 16) + \$1,000(P/F, i, 16)$$

The value of i that makes the present worth of the receipts equal to $\$1,352.90$ lies between 1.5 and 2%. We could use a linear interpolation to find the yield to maturity, but it is much easier to obtain with Excel's IRR function. As shown in Table 4.9, solving for i yields $i = 1.71\%$. Note that this 1.71% is the yield to maturity *per semiannual period*. The nominal (annual) yield is $2(1.71) = 3.42\%$

TABLE 4.9 An Excel Worksheet to find the yield to maturity of the Coca-Cola Enterprises bond

	A	B	C	D	E	F	G
1			Yield to Maturity and Current Yield				
2							
3		Inputs				Output	
4							
5	Current Market Value (P)	$	1,352.90		(i) Yield to Maturity		3.4204%
6	Coupon Rate (%)		8.5000%		Current Yield		6.2828%
7	Par Value	$	1,000.00			=IRR(B14:B30,C6)×2	
8	Interest Payment (A)	$	42.50				
9	Payment Frequency		Semi-Annual			Cash Balance Over Time	
10	Maturity in Years (N)		8		$200.00		
11						=C7×C6/C5	
12	Period (n)	Deposit (P)	Cash Balance		$-		
13					0 2 4 6 8 10 12 14 16		
14	0	$ (1,352.90)	$ (1,352.90)				
15	1	$ 42.50	$ (1,333.54)		$(200.00)		
16	2	$ 42.50	$ (1,313.84)				
17	3	$ 42.50	$ (1,293.81)		$(400.00)		
18	4	$ 42.50	$ (1,273.44)				
19	5	$ 42.50	$ (1,252.72)		$(600.00)		
20	6	$ 42.50	$ (1,231.64)				
21	7	$ 42.50	$ (1,210.20)		$(800.00)		
22	8	$ 42.50	$ (1,188.40)				
23	9	$ 42.50	$ (1,166.22)		$(1,000.00)		
24	10	$ 42.50	$ (1,143.67)				
25	11	$ 42.50	$ (1,120.73)		$(1,200.00)		
26	12	$ 42.50	$ (1,097.39)				
27	13	$ 42.50	$ (1,073.66)				
28	14	$ 42.50	$ (1,049.52)		$(1,400.00)		
29	15	$ 42.50	$ (1,024.97)				
30	16	$ 1,042.50	$ 0.00		$(1,600.00)		
31							

compounded semiannually. Compared with the coupon rate of 8.5%, purchasing the bond with the premium paid at 35.29% brings the yield to maturity down significantly below that of the coupon rate of 8.5%. The effective annual interest rate is then

$$i_a = (1 + 0.0171)^2 - 1 = 3.45\%$$

The 3.45% represents the **effective annual yield** to maturity on the bond. Notice that when you purchase a bond at par value and sell at par value, the yield to maturity will be the *same* as the coupon rate of the bond.

(b) For our example of Coca-Cola Enterprises, we compute the current yield as

$$\frac{\$42.50}{\$1,352.90} = 3.14\% \text{ per semiannual period}$$

$$3.14\% \times 2 = 6.283\% \text{ per year (nominal current yield)},$$

$$i_a = (1 + 0.0314)^2 - 1 = 6.381\%$$

COMMENTS: This effective current yield is about 2.931% higher than the 3.45% yield to maturity we just computed. If the bond is selling at a discount, the current yield is smaller than the yield to maturity. If the bond is selling at a premium, the current yield is larger than the yield to maturity. A significant difference between the yield to maturity and the current yield of a bond can exist because the market price of a bond may be more or less than its face value. Moreover, both the current yield and the yield to maturity may differ considerably from the stated coupon value of the bond.

EXAMPLE 4.23 Bond Value over Time

Consider again the Coca-Cola Enterprises' bond investment introduced in Example 4.22. (a) If the nominal yield to maturity remains constant at 3.42%, what will be the value of the bond one year after it was purchased? (b) If the market interest rate drops to 3.12% a year later, what would be the market price of the bond?

SOLUTION

Given: The same data as in Example 4.22.

Find: (a) The value of the bond one year later, and (b) the market price of the bond a year later at the going rate of 3.12% interest.

(a) We can find the value of the bond one year later by using the same valuation procedure as in Example 4.22, but now the term to maturity is only seven years:

$$\$42.50\left(P/A, \frac{3.42\%}{2}, 14 \right) + \$1,000\left(P/F, \frac{3.42\%}{2}, 14 \right) = \$1,313.86$$

The value of the bond will remain at $1,313.86 as long as the yield to maturity remains constant at 3.42% over seven years.

(b) Now suppose interest rates in the economy have fallen since the Coca-Cola Enterprises' bonds were purchased, and consequently, the going rate of interest is 3.12%. Then both the coupon interest payments and the maturity value remain constant, but now 3.12% values have to be used to calculate the value of the bond. The value of the bond at the end of the first year would be

$$\$42.50\left(P/A, \frac{3.12}{2}\%, 14 \right) + \$1,000\left(P/F, \frac{3.12}{2}\%, 14 \right) = \$1,335.97$$

Thus, the bond would sell again at a premium over its par value.

COMMENTS: The arithmetic of the bond price increase should be clear, but what is the logic behind it? We can explain the reason for the increase. Because the going market interest rate for the bond has fallen to 3.12%, if we had $1,000 to invest, we could buy new bonds with a coupon rate of 3.12%. These would pay $31.20 interest each year, rather than $85. We would prefer $85 to $31.20; therefore, we would be willing to pay more than $1,000 for Coca-Cola Enterprises' bonds to obtain higher coupons. All investors would recognize these facts, and hence Coca-Cola's bonds would be bid up in price to $1,335.97. At that point, they would provide the same yield to maturity (rate of return) to a potential investor as would the new bonds, namely, 3.12%.

SUMMARY

- Interest is most frequently quoted by financial institutions as an **annual percentage rate** (**APR**). However, compounding frequently occurs more often than once annually, and the APR does not account for the effect of this more frequent compounding. The situation leads to the distinction between nominal and effective interest.

- **Nominal interest** is a stated rate of interest for a given period (usually one year).

- **Effective interest** is the actual rate of interest which accounts for the interest amount accumulated over a given period. The **effective rate** is related to the APR by the equation

$$i = \left(1 + \frac{r}{M} \right)^M - 1$$

where r is the APR, M is the number of compounding periods, and i is the effective interest rate.

In any equivalence problem, the interest rate to use is the effective interest rate per payment period:

$$i = \left[1 + \frac{r}{CK} \right]^C - 1$$

where C is the number of interest periods per payment period, K is the number of payment periods per year, and r/K is the nominal interest rate per payment period. Figure 4.10 outlines the possible relationships between compounding and payment periods and indicates which version of the effective-interest formula to use.

- The equation for determining the effective interest of continuous compounding is

$$i = e^{r/K} - 1$$

The difference in accumulated interest between continuous compounding and very frequent compounding ($M > 50$) is relatively small.

- Cash flows, as well as compounding, can be continuous. Table 4.4 shows the interest factors to use for continuous cash flows with continuous compounding.

- Nominal (and hence effective) interest rates may fluctuate over the life of a cash flow series. Some forms of home mortgages and bond yields are typical examples.

■ **Amortized loans** are paid off in equal install-ments over time, and most of these loans have interest that is compounded monthly.

■ Under a typical **add-on loan**, the lender precal-culates the total simple interest amount and adds it to the principal. The principal and this precal-culated interest amount are then paid together in equal installments.

■ The term **mortgage** refers to a special type of loan for buying a piece of property, such as a house or a commercial building. The cost of the mort-gage will depend on many factors, including the amount and term of the loan, the frequency of pay-ments, and points and fees.

■ Two types of mortgages are common: fixed-rate mortgages and variable-rate mortgages. Fixed-rate mortgages offer loans whose interest rates are fixed over the period of the contract, whereas variable-rate mortgages offer interest rates that fluctuate with market conditions. In general, the initial interest rate is lower for variable-rate mortgages, as the lenders have the flexibility to adjust the cost of the loans over the period of the contract.

■ Allocating one's assets is simply a matter of answering the following question: "Given my personal tolerance for risk and my investment objectives, what percentage of my assets should be allocated for **growth**, what percentage for **income**, and what percentage for **liquidity**?"

■ You can determine the **expected rate of return** on a portfolio by computing the weighted average of the returns on each investment.

■ You can determine the **expected risk** of a port-folio by computing the weighted average of the volatility of each investment.

■ All other things being equal, if the expected returns are approximately the same, choose the portfolio with the lowest expected risk.

■ All other things being equal, if the expected risk is about the same, choose the portfolio with the highest expected return.

■ **Asset-backed bonds:** If a company backs its bonds with specific pieces of property, such as buildings, we call these types of bonds **mortgage bonds**. These indicate the terms of repayment and the particular assets pledged to the bondholders in case of default. It is much more common, however, for a corporation simply to pledge its overall assets. A **debenture bond** represents such a promise.

■ **Par value:** Individual bonds are normally issued in even denominations of $1,000 or multiples of $1,000. The stated face value of an individual bond is termed the **par value**.

■ **Maturity date:** Bonds generally have a speci-fied **maturity** date on which the par value is to be repaid.

■ **Coupon rate:** We call the interest paid on the par value of a bond the **annual coupon rate**. The time interval between interest payments could be of any duration, but a semiannual period is the most common.

■ **Discount or premium bond:** A bond that sells below its par value is called a **discount bond**. When a bond sells above its par value, it is called a **premium bond**.

PROBLEMS

Nominal and Effective Interest Rates

4.1. Suppose you have received a credit card offer from a bank that charges interest at 1.4% per month, compounded monthly. What is the nominal interest (annual percentage) rate for this credit card? What is the effective annual interest rate?

4.2. If your credit card calculates interest based on 17.85% APR, compounded monthly:

(a) What are your monthly interest rate and annual effective interest rate?

(b) If your current outstanding balance is $2,500 and you skip payments for two months, what would be the total balance two months from now?

4.3. A local bank advertised the following informa-tion: Interest 6.89%—effective annual yield 7.128%. No mention was made of the interest period in the advertisement. Can you figure out the compounding scheme used by the bank?

4.4. Student Emergency Financial Services, Inc., which makes small loans to college students, offers to lend $550. The borrower is required to pay $42 at the end of each week for 16 weeks. Find the interest rate per week. What is the nominal interest rate per year? What is the effective interest rate per year?

4.5. What is the effective annual yield of 8.7% compounded continuously?

4.6. A financial institution is willing to lend you $600. However, $645 is repaid at the end of one week.

(a)   What is the nominal interest rate?

(b)   What is the effective annual interest rate?

4.7. If a bank advertises a savings account that pays an 8% nominal interest rate compounded continuously, what is the effective annual percentage rate?

4.8. The Toyota Motor Company is advertising a 24-month lease of a 2014 Camry for $189 payable at the beginning of each month. The lease requires a $2,399 down payment plus first month payment. No security deposit is required. Lessee pays maintenance, excess wear and tear, and $0.18 per mile over 12,000 miles per year. Lease-end purchase option is $17,784 and lease payments total $4,536. A disposition fee of $350 is due at lease-end. Assuming an interest rate of 6% compounded monthly, what is the total cost of leasing if you put 40,000 miles at the end of lease?

4.9. You have three choices in placing your money in a bank account.

■ **Bank A** pays 7.25% compounded semi-annually.
■ **Bank B** pays 7.20% compounded quarterly.
■ **Bank C** pays 7.15% compounded continuously. Which bank would you open an account with?

Which bank would you open an account with?

4.10. The APR is stated as 6% compounded monthly. What is the continuously compounded rate? (In other words, what APR at continuous compounding has the same effect?)

4.11. Bank A offers a nominal annual interest rate of 6.5% compounded daily, while Bank B offers continuous compounding at a 6.25% nominal annual rate. If you deposit $5,000 with each bank, what will be the difference in the two bank account balances after three years?

Compounding More Frequent than Annually

4.12. How much will $8,000 be worth in 8 years with 6.75% interest compounded continuously?

4.13. How much money must you deposit now to have $6,000 in ten years at 6.5% compounded continuously?

4.14. How much money will you have in seven years if you deposit $7,000 in the bank at 8.5% interest compounded daily?

4.15. How long will it take money to triple with continuous compounding at 8% interest?

4.16. Republic Finance offers money at 2.32% per month compounded monthly.

(a)   What is the nominal interest rate?

(b)   What is the effective annual interest rate?

(c)   How many years will it take an investment to triple if interest is compounded monthly?

(d)   How many years will it take an investment to quadruple if the nominal rate is compounded continuously?

4.17. Suppose your savings account pays 6% interest compounded monthly. If you deposit $18,000 for four years, how much will you have?

4.18. If you invest $25,000 today at a 7.5% interest compounded weekly, what will be your ending value after 15 years?

4.19. What will be the amount accumulated by each of these present investments?

(a) $12,250 in 15 years at 8.5% compounded quarterly.

(b) $15,750 in 12 years at 8.32% compounded weekly.

(c) $25,500 in 10 years at 7.5% compounded monthly.

4.20. How many years will it take an investment to double if the interest rate is 8.4% compounded at the given intervals?

(a)   Quarterly

(b)   Monthly

(c)   Continuously

4.21. A series of equal quarterly payments of $25,000 for 10 years is equivalent to what future worth amount at an interest rate of 8.4% compounded at the given intervals?

(a)   Quarterly

(b)   Monthly

(c)   Continuously

4.22. What is the future worth of an equal payment series of $15,000 at the end of each month for eight years if the interest rate is 7.5% compounded continuously?

4.23. Suppose that $12,000 is placed in a bank account at the end of each quarter over the next 15 years. What is the future worth at the end of 15 years when the interest rate is 8.5% compounded at the given intervals?

(a) Quarterly

(b) Monthly

(c) Continuously

4.24. A series of equal end-of-quarter deposits of $4,000 extends over a period of five years. It is desired to compute the future worth of this quarterly deposit series at 9% compounded monthly. Which of the following equations is correct?

(a) $F = \$4,000(F/A, 2.25\%, 20)$

(b) $F = 4(\$4,000)(F/A, 9\%, 5)$

(c) $F = \$4,000(F/A, 2.267\%, 20)$

(d) $F = 4(\$4,000)(F/A, 0.75\%, 20)$

4.25. If the interest rate is 9.3% compounded continuously, what is the required quarterly payment to repay a loan of $15,000 in eight years?

4.26. What is the future worth of a series of equal end-of-month payments of $2,750 if the series extends over a period of twelve years at 8.4% interest compounded at the given intervals?

(a) Quarterly

(b) Monthly

(c) Continuously

4.27. Suppose you deposit $2,000 at the end of each quarter for ten years at an interest rate of 9% compounded monthly. What equal end-of-year deposit over the ten years would accumulate the same amount at the end of the ten years under the same interest compounding? To answer the question, which of the following is correct?

(a) $A = \$2,000(F/A, 2.267\%, 4)$

(b) $A = \$2,000(F/A, \frac{9\%}{12}, 40) \times (A/F, 9\%, 10)$

(c) $A = [\$2,000(F/A, 2.25\%, 40)] \times (A/F, 9\%, 10)$

(d) None of the above.

4.28. A series of equal quarterly payments of $12,000 for 8 years is equivalent to what future lump-sum amount at the end of 14 years at an interest rate of 8.5% compounded continuously?

4.29. What will be the required monthly payment to repay a loan of $72,000 in eight years if the interest rate is 10.5% compounded continuously?

4.30. A series of equal quarterly payments of $4,200 extends over a period of twelve years. What is the present worth of this quarterly payment series at 9.42% interest compounded continuously?

4.31. What is the future worth of the following series of payments?

(a) $75,000 at the end of each six-month period for ten years at 10.2% compounded semi-annually.

(b) $25,000 at the end of each quarter for twelve years at 9.2% compounded quarterly.

(c) $22,000 at the end of each month for eight years at 8.4% compounded monthly.

4.32. You have $32,836.48 in your savings account that has been paying at 6.5% compounded continuously. If you opened the account 12 years ago and made no deposits during any other years, how much was your original deposit?

4.33. What equal series of payments must be paid into a sinking fund to accumulate the following amount?

(a) $62,000 in 12 years at 8.5% compounded semi-annually when payments are semi-annual.

(b) $35,000 in 10 years at 8.2% compounded quarterly when payments are quarterly.

(c) $40,000 in 6 years at 7.5% compounded monthly when payments are monthly.

4.34. You have a habit of drinking a cup of Starbucks coffee ($3.75 a cup) on the way to work every morning. If instead you put the money in the bank for 30 years, how much would you have at the end of that time, assuming that your account earns 5% interest compounded *daily*? Assume also that you drink a cup of coffee every day, including weekends.

4.35. Michelle Passmore is purchasing a $25,000 automobile, which is to be paid for in 60 monthly installments of $489.15. What effective annual interest is she paying for this financing arrangement?

4.36. You are planning to purchase your first home eight years from today. The required down payment will be $80,000. You currently have $25,000, but you plan to contribute $750 each quarter to a special savings account until you purchase the home. To achieve your goal, what nominal interest rate, compounded continuously, must you earn on your account?

4.37. A loan of $20,000 is to be financed to assist in buying an automobile. On the basis of monthly compounding for 48 months, the end-of-the-month equal payment is quoted as $512.55. What nominal interest rate in percentage is being charged?

4.38. Suppose a young newlywed couple is planning to buy a home three years from now. To save the down payment required at the time of purchasing a home worth $350,000 (let's assume that the down payment is 20% of the sale price, which is $70,000), the couple decides to set aside some money from each of their salaries at the end of every month. If each of them can earn 6% interest (compounded monthly) on his or her savings, determine the equal amount this couple must deposit each month until the point is reached where the couple can buy the home.

4.39. What is the present worth of the following series of payments?

(a) $5,000 at the end of each six-month period for 12 years at 9% compounded semiannually.

(b) $7,000 at the end of each quarter for 10 years at 8% compounded quarterly.

(c) $3,500 at the end of each month for five years at 6% compounded monthly.

4.40. What is the amount of the quarterly deposits A such that you will be able to withdraw the amounts shown in the cash flow diagram if the interest rate is 8% compounded quarterly?

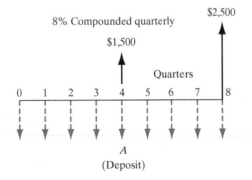

Figure P4.40

4.41. Joe Franklin deposits $22,000 in a savings account that pays 6% interest compounded monthly. Three years later, he deposits $16,000. Two years after the $16,000 deposit, he makes another deposit in the amount of $13,500. Four years after the $13,500 deposit, half of the accumulated funds are transferred to a fund that pays 8% interest compounded quarterly. How much money will be in each account six years after the transfer?

4.42. A man is planning to retire in 20 years. He wishes to deposit a regular amount every three months until he retires, so that, beginning one year following his retirement, he will receive annual payments of $95,000 for the next 18 years. How much must he deposit if the interest rate is 8% compounded quarterly?

4.43. You borrowed $35,000 to buy a new car from a bank at an interest rate of 10.2% compounded monthly. This loan will be repaid in 96 equal monthly installments over eight years. Immediately after the 36th payment, you desire to pay the remainder of the loan in a single payment. Compute this lump-sum amount.

4.44. A building is priced at $450,000. If a down payment of $50,000 is made and a payment of $8,000 every month thereafter is required, how many months will it take to pay for the building? Interest is charged at a rate of 10.2% compounded monthly.

4.45. You obtained a loan of $20,000 to finance an automobile. Based on monthly compounding over 36 months, the end-of-the-month equal payment was figured to be $650.52. What APR was used for this loan?

4.46. *The Economist*, a professional business journal, offers three types of subscriptions, payable in advance: three months auto-renewing at $45, one year at $160, and two years at $279. If money can earn 6% interest compounded monthly, which subscription should you take? Assume that you plan to subscribe to the journal over the next two years. All subscription payments should be made at the beginning of each renewal cycle.

4.47. A married couple is trying to finance their four-year-old son's college education. Money can be deposited at 8% compounded quarterly. What end-of-quarter deposit must be made from the son's 4th birthday to his 18th birthday to provide $80,000 on each birthday from the 18th to the 21st? (Note that the first deposit comes three months after his 4th birthday and the last deposit is made on the date of the first withdrawal.)

4.48. Cory Manciagli is planning to retire in 22 years. Money can be deposited at 8.2% compounded quarterly. What quarterly deposit must be made at the end of each quarter until Cory retires so that he can make a withdrawal of $65,000 semi-annually over the first 15 years of his retirement? Assume that his first withdrawal occurs at the end of six months after his retirement.

4.49. Patricia French received $500,000 from an insurance company after her husband's death. Patricia wants to deposit this amount in a savings account that earns interest at a rate of 5% compounded monthly. Then she would like to make 120 equal monthly withdrawals over the 10-year period such that, when she makes the last withdrawal, the savings account will have a balance of zero. How much can she withdraw each month?

4.50. Jenny Walters, who owns a real estate agency, bought an old house to use as her business office. She found that the ceiling was poorly insulated and that the heat loss could be cut significantly if six inches of foam insulation were installed. She estimated that with the insulation, she could cut the heating bill by $80 per month and the air-conditioning cost by $70 per month. Assuming that the summer season is three months (June, July, and August) of the year and that the winter season is another three months (December, January, and February) of the year, how much can Jenny spend on insulation if she expects to keep the property for five years? Assume that neither heating nor air-conditioning would be required during the fall and spring seasons. If she decides to install the insulation, it will be done at the beginning of May. Jenny's interest rate is 6% compounded monthly.

Continuous Payments with Continuous Compounding

4.51. A new chemical production facility that is under construction is expected to be in full commercial operation one year from now. Once in full operation, the facility will generate $95,000 cash profit daily over the plant's service life of 10 years. Determine the equivalent present worth of the future cash flows generated by the facility at the beginning of commercial operation, assuming

(a) 10% interest compounded daily, with the daily flows.

(b) 10% interest compounded continuously, with the daily flow series approximated by a uniform continuous cash flow function.

Also, compare the difference between parts (a) discrete (daily) and part (b) continuous compounding.

4.52. The productivity of a certain piece of manufacturing equipment is expected to decline at a constant rate with age from an initial value of $500,000 at time 0, to a final value of $40,000 at the end of year 3. If interest is compounded continuously at a nominal annual rate of 11%, determine the present value of this continuous cash flow.

4.53. A sum of $45,000 per year will be received uniformly over an eight-year period beginning three years from today. What is the present value of this deferred-funds flow if interest is compounded continuously at a nominal rate of 8.4%?

4.54. A small chemical company that produces an epoxy resin expects its production volume to decline exponentially according to the relationship

$$y_t = 5e^{-0.25t}$$

where y_t is the production rate at time t. Simultaneously, the unit price is expected to increase linearly over time at the rate

$$u_t = 55(1 + 0.09t)$$

What is the expression for the present worth of sales revenues from $t = 0$ to $t = 20$ at 12% interest compounded continuously?

Changing Interest Rates

4.55. How much money would be accumulated in 15 years for a deposit of $15,000 made at the end-of-year zero if the account earned interest at 8% per year for the first four years, 10% per year for the next six years, and 9% per year for the last five years?

4.56. Consider the cash flow diagram shown in Figure P4.56, which represents three different interest rates applicable over the five-year time span shown.

(a) Calculate the equivalent amount P at the present time.

(b) Calculate the single-payment equivalent to F at $n = 5$.

(c) Calculate the equivalent equal-payment series cash flow A that runs from $n = 1$ to $n = 5$.

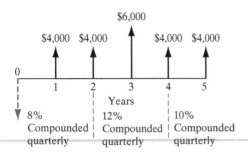

Figure P4.56

4.57. Consider the cash flow transactions depicted in Figure P4.57 with the changing interest rates specified.

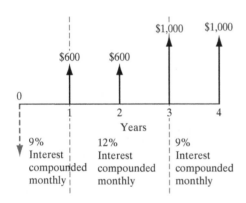

Figure P4.57

(a) What is the equivalent present worth? (In other words, how much do you have to deposit now so that you can withdraw $600 at the end of year 1, $600 at the end of year 2, $1,000 at the end of year 3, and $1,000 at the end of year 4?)

(b) What is the single, effective annual interest rate over four years?

4.58. Compute the future worth of the cash flows with the different interest rates specified in Figure P4.58. The cash flows occur at the end of each year over four years.

4.59. Suppose the continuously compounded interest rate this year is 9%, and in subsequent years, it will be 11%. What is the balance after five years of $6,500 invested today?

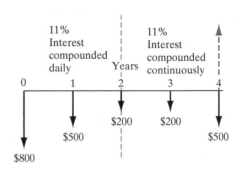

Figure P4.58

Amortized Loans

4.60. An automobile loan of $45,000 at a nominal rate of 10.5% compounded monthly for 48 months requires equal end-of-month payments of $1,152.15. Complete Table P4.60, as you would expect a bank to calculate the values.

TABLE P4.60

End of Month (n)	Interest Payment	Repayment of Principal	Remaining Loan Balance
1			$44,241.60
2		$765.04	
15	$295.37		
35			$14,099.41
48			0

4.61. Mr. Smith wants to buy a new car that will cost $35,000. He will make a down payment in the amount of $15,000. He would like to borrow the remainder from a bank at an interest rate of 12% compounded monthly. He agrees to pay off the loan monthly for a period of five years. Select the correct answer for the following questions:

(a) What is the amount of the monthly payment A?
   i. $A = \$20,000(A/F, 1\%, 60)$
   ii. $A = \$20,000(A/P, 12\%, 5)/12$
   iii. $A = \$20,000(A/P, 1\%, 60)$
   iv. $A = \$4,000(A/F, 12\%, 5)/12$

(b) Mr. Smith has made 36 payments and wants to figure out the balance remaining immediately after the 36th payment. What is that balance?

i. $B_{36} = A(P/A, 12\%, 5)/36$

ii. $B_{36} = A(P/A, 1\%, 24)$

iii. $B_{36} = \$20,000 - 36A$

iv. $B_{36} = A(P/A, 1\%, 36)$

4.62. Kelly Robins is considering purchasing a used automobile. The price including the title and taxes is \$15,455. Kelly is able to make a \$2,455 down payment. The balance of \$13,000 will be borrowed from her credit union at an interest rate of 9.45% compounded daily. The loan should be paid in 36 equal monthly payments. Compute the monthly payment. What is the total amount of interest Kelly has to pay over the life of the loan?

4.63. Suppose you are in the market for a new car worth \$28,000. You are offered a deal to make a \$4,000 down payment now and to pay the balance in equal end-of-month payments of \$527.15 over a 60-month period. Consider the following situations.

(a) Instead of going through the dealer's financing, you want to make a down payment of \$3,500 and take out an auto loan from a bank at 10.5% compounded monthly. What would be your monthly payment to pay off the loan in five years?

(b) If you were to accept the dealer's offer, what would be the effective rate of interest per month the dealer charges on your financing?

4.64. Clay Harden borrowed \$25,000 from a bank at an interest rate of 9% compounded monthly. The loan will be repaid in 36 equal monthly installments over three years. Immediately after his 20th payment, Clay desires to pay the remainder of the loan in a single payment. Compute the total amount he must pay.

4.65. You plan to buy a \$275,000 home with a 25% down payment. The bank you want to finance the loan through suggests two options: a 15-year mortgage at 6.3% APR and a 30-year mortgage at 6% APR. What is the difference in monthly payments between these two options?

4.66. Antonio Menifee borrowed money from a bank to finance a small fishing boat. The bank's terms allowed him to defer payments (including interest) on the loan for six months and to make 36 equal end-of-month payments thereafter. The original bank note was for \$16,000 with an interest rate of 9% compounded monthly. After 16 monthly payments, Antonio found himself in a financial bind and went to a loan company for assistance in lowering his monthly payments. Fortunately, the loan company offered to pay his debts in one lump sum if he would pay the company \$308.29 per month for the next 36 months. What monthly rate of interest is the loan company charging on this transaction?

4.67. Suzan Long is considering buying a home for \$350,000.

(a) If she makes a down payment of \$50,000 and takes out a mortgage on the rest of the money at 5.5% compounded monthly, what will be her monthly payment to retire the mortgage in 15 years?

(b) Consider the 60th payment. How much will the interest and principal payments be?

4.68. On a \$600,000 home mortgage loan with a 15-year term at 10.5% APR compounded monthly, compute the total payments on principal and interest over the first six years of ownership.

4.69. A lender requires that monthly mortgage payments be no more than 30% of gross monthly income with a maximum term of 25 years. If you can make only a 20% down payment, what is the minimum monthly income needed to purchase a \$450,000 house when the interest rate is 9% compounded monthly?

4.70. To buy an \$180,000 house, you take out a 10.5% (APR) mortgage for \$145,000. Ten years later, you sell the house for \$212,000 (after all other selling expenses). What equity (the amount that you can keep before tax) would you realize with a 25-year repayment term on monthly payment basis?

4.71. Consider the following three individuals. Just after their 15th payment:

■ Robert Dixon had a balance of \$90,000 on a 10.8%, 18-year mortgage on a monthly payment basis;

■ Wanda Harper had a balance of \$90,000 on a 10.8%, 22-year mortgage on a monthly payment basis; and

■ Tony Zang had a balance of \$90,000 on a 10.8%, 25-year mortgage on a monthly payment basis.

How much interest did each individual pay on the 16th payment?

4.72. Home mortgage lenders usually charge points on a loan to avoid exceeding a legal limit on interest rates or to be competitive with other lenders. As an example, for a two-point loan, the lender would lend only $98 for each $100 borrowed. The borrower would receive only $98, but would have to make payments just as if he or she had received $100. Suppose that you receive a loan of $130,000, payable at the end of each month for 30 years with an interest rate of 9% compounded monthly, but you have been charged three points. What is the effective borrowing rate on this home mortgage loan?

4.73. You are considering purchasing a lot adjacent to your laundry business to provide adequate parking space for your customers. You need to borrow $75,000 to secure the lot. You have made a deal with a local bank to pay the loan back over a five-year period with the following payment terms: 14%, 20%, 26%, 32%, and 38% of the initial loan at the end of first, second, third, fourth, and fifth years, respectively.

(a) What rate of interest is the bank earning from this loan?

(b) What would be the total interest paid over the five-year period?

4.74. Emily Dorsey's current salary is $90,000 per year, and she is planning to retire 22 years from now. She anticipates that her annual salary will increase by $4,000 each year ($90,000 the first year, $94,000 the second year, $98,000 the third year, and so forth), and she plans to deposit 6% of her yearly salary into a retirement fund that earns 9% interest compounded daily. What will be the amount of interest accumulated at the time of Emily's retirement?

4.75. Consider the following two options for financing a car:

• **Option A.** Purchase the vehicle at the normal price of $36,500 and pay for the vehicle over four years with equal monthly payments at 0% APR financing.

• **Option B.** Purchase the vehicle for a discounted price of $32,750 to be paid immediately. The funds that would be used to purchase the vehicle are presently earning 5.5% annual interest compounded monthly.

(a) What is the meaning of the APR of 0% quoted by the dealer?

(b) Under what circumstances would you prefer to go with the dealer's financing?

(c) Which interest rate (the dealer's interest rate or the savings rate) would you use in comparing the two options?

Add-On Loans

4.76. If you borrowed $32,000 for 72 months at 16.5% with an add-on loan, what would be your monthly payment and the total amount you paid on interest?

4.77. James Wilson wants to purchase a set of furniture worth $8,000. He plans to finance the furniture for four years. The furniture store tells James that the interest rate is only 1.1% per month, and his monthly payment is computed as follows.

- Installment period = 48 months.
- Interest = 48(0.011)($8,000) = $4,224.
- Loan processing fee = $40.
- Total amount owed = $8,000 + + $4,224 + $40 = $12,264.
- Monthly payment = $12,264/48 = $255.50 per month

(a) What is the annual effective interest rate that James is paying for his loan transaction? What is the nominal interest (annual percentage rate) for the loan?

(b) James bought the furniture and made 24 monthly payments. Now he wants to pay off the remaining installments in one lump sum (at the end of 24 months). How much does he owe the furniture store?

4.78. You purchase a piece of furniture worth $6,050 on credit through a local furniture store. You are told that your monthly payment will be $156.88, including an acquisition fee of $35, at an 11% add-on interest rate over 60 months. After making 24 payments, you decide to pay off the balance. Compute the remaining balance based on the conventional amortized loan.

Loans with Variable Payments

4.79. Kathy Stonewall bought a new car for $15,458. A dealer's financing was available through a local bank at an interest rate of 11.5% compounded monthly. Dealer financing required a 10% down payment and 60 equal monthly payments. Because the interest rate was rather high, Kathy checked her credit union for possible financing. The loan officer at the credit union quoted a 9.8% interest rate for a new-car loan and 10.5% for a used car. But to be eligible for the loan, Kathy has to be a member of the union for at least six months. Since she joined the union two months ago, she has to wait four more months to apply for the loan. Consequently, she decided to go ahead with the dealer's financing, and four months later she refinanced the balance through the credit union at an interest rate of 10.5%.

(a) Compute the monthly payment to the dealer.

(b) Compute the monthly payment to the union.

(c) What is the total interest payment on each loan?

4.80. A house can be purchased for $182,000, and you have $32,000 cash for a down payment. You are considering the following two financing options:

- **Option 1.** Getting a new standard mortgage with a 7.8% (APR) interest and a 28-year term.
- **Option 2.** Assuming the seller's old mortgage, which has an interest rate of 6.3% (APR), a remaining term of 24 years (the original term was 28 years), a remaining balance of $108,520, and payments of $705 per month. You can obtain a second mortgage for the remaining balance ($41,480) from your credit union at 9.3% (APR) with a 14-year repayment period.

(a) What is the effective interest rate of the combined mortgage?

(b) Compute the monthly payments for each option over the life of the mortgage.

(c) Compute the total interest payment for each option.

(d) What homeowner's interest rate makes the two financing options equivalent?

Loans with Variable Interest Rates

4.81. A loan of $25,000 is to be financed over a period of 60 months. The agency quotes a nominal rate of 8.5% for the first 36 months and a nominal rate of 9.3% for any remaining unpaid balance after 36 months, which is compounded monthly. Based on these rates, what equal end-of-the-month payment for 60 months would be required to repay the loan with interest?

4.82. Emily Wang financed her office furniture from a furniture dealer. The dealer's terms allowed her to defer payments (including interest) for six months and to make 36 equal end-of-month payments thereafter. The original note was for $15,000, with interest at 9% compounded monthly. After 26 monthly payments, Emily found herself in a financial bind and went to a loan company for assistance. The loan company offered to pay her debts in one lump sum if

she would pay the company $186 per month for the next 30 months.

(a) Determine the original monthly payment made to the furniture store.

(b) Determine the lump-sum payoff amount the loan company will make.

(c) What monthly rate of interest is the loan company charging on this loan?

4.83. You borrow $150,000 with a 20-year term at a 9.3% (APR) variable rate and the interest rate can be changed every six years.

(a) What is the initial monthly payment?

(b) If the lender's interest rate is 10.2% (APR) at the end of six years, what will the new monthly payments be?

4.84. Consider the following advertisement seeking to sell a beachfront condominium at SunDestin, Florida.

$$95\% \text{ Financing } 8\frac{1}{8}\% \text{ interest!!}$$

5% Down Payment. Own a Gulf-Front Condominium for only $100,000 with a 30-year variable-rate mortgage. We're providing incredible terms: $95,000 mortgage (30 years), year 1 at 8.125%, year 2 at 10.125%, year 3 at 12.125%, and years 4 through 30 at 13.125%.

(a) Compute the monthly payments for each year.

(b) Calculate the total interest paid over the life of the loan.

(c) Determine the equivalent single-effective annual interest rate for the loan.

Investment in Bonds

4.85. The Green Cable Corporation issued a new series of bonds on January 5, 2015. The bonds were sold at par ($1,000), have a 6% coupon rate, and mature in 30 years on December 31, 2044. Coupon interest payments are made semiannually (on June 30 and December 31).

(a) What was the yield to maturity (YTM) of the bond on January 5, 2015?

(b) Assuming that interest rates had fallen to 5%, what was the price of the bond on January 1, 2020 (five years later)?

(c) On July 1, 2017, the bonds sold for $922.38. What was the YTM at that date? What was the current yield at that date?

4.86. A $1,000, 9.50% semiannual bond is purchased for $1,010. If the bond is sold after three years and six interest payments, what should the selling price be to yield a 10% return on the investment?

4.87. Mr. Gonzalez wishes to sell a bond that has a face value of $1,000. The bond bears an interest rate of 8% with bond interest payable semiannually. Four years ago, $920 was paid for the bond. At least a 9% return (yield) on the investment is desired. What must be the minimum selling price?

4.88. Suppose you have the choice of investing in one of the following.

(a) A zero-coupon bond which costs $513.60 today, pays nothing during its life, and then pays $1,000 after five years.

(b) A bond that costs $1,000 today, pays $113 in interest semiannually, and matures at the end of five years with a par value of $1,000 to be repaid. Which bond would provide the higher yield?

4.89. Suppose you were offered a 10-year, 14% coupon, $5,000 par value bond at a price of $6,108.52. What rate of interest (yield to maturity) would you earn if you bought the bond and held it to maturity (at semiannual interest)?

4.90. The Lakeside Electronics Company has two bond issues outstanding. Both bonds pay $100 semiannual interest plus $1,000 at maturity. Bond A has a remaining maturity of 15 years; bond B has a maturity of one year. What is the value of each of these bonds now when the going rate of interest is 9%?

4.91. The Photo Film Company's bonds have four years remaining to maturity. Interest is paid annually, the bonds have a $1,000 par value, and the coupon interest rate is 8.75%.

(a) What is the yield to maturity at a current market price of $1,108?

(b) Would you pay $935 for one of these bonds if you thought that the market rate of interest was 9.5%?

4.92. Suppose Ford sold an issue of bonds with a 15-year maturity, a $1,000 par value, a 12% coupon rate, and semiannual interest payments.

(a) Two years after the bonds were issued, the going rate of interest on bonds such as these fell to 9%. At what price would the bonds sell?

(b) Suppose that, two years after the bonds' issue, the going interest rate had risen to 13%. At what price would the bonds sell?

(c) Today, the closing price of the bond is $783.58. What is the effective current yield?

4.93. Suppose you purchased a corporate bond with a 10-year maturity, a $1,000 par value, a 10% coupon rate, and semiannual interest payments. This means that you receive a $50 interest payment at the end of each six-month period for 10 years (20 times). Then, when the bond matures, you will receive the principal amount (the face value) in a lump sum. Three years after the bonds were purchased, the going rate of interest on new bonds fell to 6% (or 6% compounded semiannually). What is the current market value (P) of the bond (three years after its purchase)?

Short Case Studies

ST4.1 Anita Wu, an engineering junior, received two guaranteed line-of-credit applications from two different banks. Each bank offered a different annual fee and finance charge.

Anita expects her average monthly balance after payment to the bank to be $300 and plans to keep the credit card she chooses for only 24 months. (After graduation, she will apply for a new card.) Anita's interest rate on her savings account is 6% compounded daily. Table ST4.1 lists the terms of each bank:

(a) Compute the effective annual interest rate for each card.

(b) Which bank's credit card should Anita choose?

(c) Suppose Anita decided to go with Bank B and used the card for one year. The balance after one year is $1,500. If she makes just a minimum payment each month (say, 5% of the unpaid balance), how long will it take to pay off the card debt? Assume that she will not make any new purchases on the card until she pays off the debt.

TABLE ST4.1

Terms	Bank A	Bank B
Annual fee	$20	$30
Finance charge	1.55% Monthly interest rate	16.5% Annual percentage rate

ST4.2 Consider the following flyer that you just received.

"We have good news, Eric Moon! Your credit history allows us to offer you a guaranteed check of $2,009.06. That's right—a preapproved loan just in time to help with your vacation! It's easy. Just endorse the attached loan check and deposit it in your bank account. Your payment schedule will be 18 equal monthly payments of $150 each, beginning 30 days after your depositing the check in your bank account."

(a) Determine the annual percentage of this loan solicitation.

(b) What would be the finance charge (dollar amount the credit will cost you)?

(c) Suppose you have a credit card. Is it better by doing a cash advance in the amount of $2,009.06 on your credit card? Your credit company charges APR of 24.99% for cash advances. Determine the annual percentage rate (APR) of this loan solicitation.

(d) Suppose that you decided to go with a cash advance by making $50 monthly payment until you pay off the loan. What would be the total finance (interest) charge (dollar amount the credit will cost you) for each option?

ST4.3 You received a bill consolidation loan offer in the mail.

Dear Larry Montgomery

I'm delighted to inform you that you have been approved for a Homeowner Bill Consolidation Loan of up to $417,7000, at our lowest fixed rate of 6.00% (6.184% APR). You can use this money for anything you like; to pay off your high-interest debt, pay your urgent expenses, take a dream vacation, or whatever you want. There are no restrictions! And unlike the interest you pay on credit cards, auto loans and personal loans, the interest on your Homeowner Bill Consolidation Loan is usually fully tax-deductible.

For an example, with the following debt consolidation, a typical homeowner would save $1,224.88 per month!!

	Interest Rate	Outstanding Balance	Monthly Payment
Credit Card	21%	$12,000	$448.62
Auto Loan	8%	$15,000	$359.20
Home Improvement Loan	10%	$23,000	$488.68
Current Home Mortgage	8%	$170,000	$1,247.40
Total Current Debt		$220,000	$2,543.90
New Homeowner's Bill Consolidation Loan	6%	$220,000	$1,319.02
Monthly savings			$1,224.88

First Month's Savings = $1,224.88
First Year's Savings = $14,698.56
Potential 10-year Savings = $146,985.60
Plus, tax-deductibility actually makes your savings even greater!!

(a) Do you agree with the savings claims by the loan company?

(b) Explain any illogical claims made by the loan company.

ST4.4 Are you shopping for a mortgage to finance a home that you expect to own for no more than a few years? If so, you should know about a hybrid mortgage. Hybrid loans give prospective homebuyers the ability to buy a lot more home than they can afford—thanks to the initially lower interest rate. But with such flexibility comes greater risk. Since lenders are free to design loans to fit borrowers' needs, the terms and fees vary widely and homeowners can get burned if rates climb higher.

Hybrid mortgages allow homeowners to benefit from the best aspects of both fixed-rate and adjustable-rate mortgages (ARMs). With hybrids, borrowers choose to accept a fixed interest rate over a number of years—usually, three, five, seven, or 10 years—and afterward the loan converts to an ARM. But therein lies the danger: While you're getting an extraordinarily low rate up front for a few years, when the fixed-rate period expires you could very well end up paying more than double your current rate of interest.

At a rate of 6.16% for a 30-year mortgage, for example, a person borrowing $200,000 would pay $1,220 a month. With a seven-year hybrid, more commonly called a 7/1 loan, at the going rate of 5.61%, that monthly payment drops to $1,150. By the end of the seventh year, the homeowner would save about $7,700 in interest charges by going with a seven-year hybrid. To say that there are drawbacks is an understatement. Despite the surge in popularity, a hybrid loan can be a ticking time bomb for borrowers who plan on holding the loan for the long term. Discuss the potential drawbacks associated with the hybrid ARMs by giving a specific loan and interest rate scenario.

ST4.5 Suppose you are going to buy a home worth $110,000 and you make a down payment in the amount of $50,000. The balance will be borrowed from the Capital Savings and Loan Bank. The loan officer offers the following two financing plans for the property.

- **Option 1.** A conventional fixed-rate loan at an interest rate of 13% over 30 years with 360 equal monthly payments.
- **Option 2.** A graduated payment schedule at 11.5% interest with the monthly payment schedule given in Table ST4.5. If you go with this graduated payment loan, mortgage insurance is a must in addition to the monthly payments.

(a) Compute the monthly payment under option 1.
(b) What is the effective annual interest rate you are paying under option 2?
(c) Compute the outstanding balance at the end of five years under each option.
(d) Compute the total interest payment under each option.
(e) Assuming that your only investment alternative is a savings account that earns an interest rate of 6% compounded monthly, which option is a better deal?

TABLE ST4.5

Year (*n*)	Monthly Payment	Monthly Mortgage Insurance
1	$497.76	$25.19
2	522.65	25.56
3	548.78	25.84
4	576.22	26.01
5	605.03	26.06
6–30	635.28	25.96

PART

TWO

EVALUATION OF BUSINESS AND ENGINEERING ASSETS

5

Present-Worth Analysis

Building a New Football Stadium for Colorado State University. With ever declining state funding, Tony Frank, President of Colorado State University (CSU) at Fort Collins, had to come up with a new revenue source to make up the shortfall from state funding. One of the potential revenue sources on the table was to bring more out-of-state students as they pay $23,347 in annual tuition, three times what Colorado students pay. Currently, about 19% of the student body is classified as nonresidents.

After a series of discussions with royal alumni and local businessmen, Mr. Frank came up with a proposal to build a 40,000-seat football stadium on its Fort Collins campus. The essential idea was that a new football stadium would attract better football players, helping the Colorado State Rams win more games and make more television appearances, and increase donations to the academic programs. This in turn would raise the university profile and attract more out-of-state students. Another option hotly debated was fixing outdated Hughes Stadium, but building a new stadium on the campus was considered a better option, as it had little to no support from private funders, and would require about $60 million from the general fund.

Opponents, including some alumni, students, and faculty, did not think that the project would generate enough revenue as the CSU's team

Source: CSA Images/Printstock Collection/Getty Images

belongs to a relatively low-profile athletic conference that is hard to sell out its current 32,500-seat stadium. In 2012, CSU's average attendance per game sank to around 19,000 where only 9,300 of them were actual paying customers. The university spends about $35 million on athletics each year and football does not pay for itself—the football program generates about $4.1 million per year. Despite many doubts and reservation of the stadium project, Mr. Frank was able to convince the CSU's board of governors to accept his recommendation with a price tag of $246 million.

No matter how ambitious CSU's plan is, many financial analysts agree that one important measure of determining the long-term financial success of the **"stadium expansion"** plan is whether the $246 million investment will bring a positive Net Present Value (NPV) to the CSU.

I n Chapters 3 and 4, we presented the concept of the time value of money and developed techniques for establishing cash flow equivalence with compound-interest factors. That background provides a foundation for accepting or rejecting a capital investment: the economic evaluation of a project's desirability. The forthcoming coverage of investment worth in this chapter will allow us to go a step beyond merely accepting or rejecting an investment to making comparisons of alternative investments. We will learn how to compare alternatives on an equal basis and select the wisest alternative from an economic standpoint.

The three common measures based on cash flow equivalence are (1) equivalent present worth (PW), (2) equivalent future worth (FW), and (3) equivalent annual worth (AW). Present worth represents a measure of future cash flow relative to the time point "now," with provisions that account for earning opportunities. Future worth is a measure of cash flow at some future planning horizon and offers a consideration of the earning opportunities of intermediate cash flows. Annual worth is a measure of cash flow in terms of equivalent equal payments made on an annual basis.

Our treatment of measures of investment worth is divided into three chapters. Chapter 5 begins with a consideration of the payback period, a project-screening tool that was the first formal method used to evaluate investment projects. Then it introduces two measures based on fundamental cash-flow equivalence techniques: present-worth and future-worth analysis. Because the annual-worth approach has many useful engineering applications related to estimating the unit cost, Chapter 6 is devoted to annual cash flow analysis. Chapter 7 presents measures of investment worth based on yield—measures known as rate-of-return analysis.

We must also recognize that one of the most important parts of the capital budgeting process is the estimation of relevant cash flows. For all examples in this chapter and in Chapters 6 and 7, however, net cash flows can be viewed as before-tax values or after-tax values for which tax effects have been recalculated. Since some organizations (e.g., governments and nonprofit organizations) are not subject to tax, the before-tax situation provides a valid base for this type of economic evaluation. Taking this after-tax view will allow us to focus on our main area of concern: the economic evaluation of investment projects. The procedures for determining after-tax net cash flows in taxable situations are developed in Chapter 10.

CHAPTER LEARNING OBJECTIVES

After completing this chapter, you should understand the following concepts:

- How firms screen potential investment opportunities.
- How firms evaluate the profitability of an investment project by considering the time value of money.
- How firms compare mutually exclusive investment opportunities.

5.1 Describing Project Cash Flows

In Chapter 1, we described many engineering economic decision problems, but we did not provide suggestions on how to evaluate and solve them. What do all engineering economic decision problems have in common? The answer is that they all involve two dissimilar types of amounts. First, there is the investment, which is usually made in a lump sum at the beginning of the project. Although not literally made "today," the investment is made at a specific point in time that, for analytical purposes, is called today, or time 0. Second, there is a stream of cash benefits that are expected to result from the investment over some years in the future.

5.1.1 Loan versus Project Cash Flows

An investment made in a fixed asset is similar to an investment made by a bank when it lends money. The essential characteristic of both transactions is that funds are committed today in the expectation of their earning a return in the future. In the case of the bank loan, the future return takes the form of interest plus repayment of the principal and is known as the **loan cash flow**. In the case of the fixed asset, the future return takes the form of cash generated by productive use of the asset. As shown in Figure 5.1, the representation of these future earnings, along with the capital expenditures and annual expenses (such as wages, raw materials, operating costs, maintenance costs, and income taxes), is the **project cash flow**. This similarity between the loan cash flow and the project cash flow brings us to an important conclusion: We can use the same equivalence techniques developed in Chapter 3 to measure economic worth. Example 5.1 illustrates a typical procedure for obtaining a project's cash flows.

Loan cash flow

Project cash flow

Figure 5.1   A bank loan versus an investment project.

EXAMPLE 5.1 Identifying Project Cash Flows

XL Chemicals is thinking of installing a state-of-the-art computer process-control system in one of its process plants. The plant is used about 40% of the time (or 3,500 operating hours per year) to produce a proprietary demulsification chemical. During the remaining 60% of the time, it is used to produce other specialty chemicals. Annual production of the demulsification chemical amounts to 30,000 kilograms, and it sells for $15 per kilogram. The proposed computer process-control system will cost $650,000 and is expected to provide the following specific benefits in the production of the demulsification chemical.

- First, the selling price of the product could be increased by $2 per kilogram because the product will be of higher purity, which translates into better demulsification.
- Second, production volumes will increase by 4,000 kilograms per year as a result of higher reaction yields, without any increase in the quantities of raw material or in production time.
- Third, the number of process operators can be reduced by one per shift, which represents a savings of $25 per hour.

The new control system would result in additional maintenance costs of $53,000 per year and has an expected useful life of eight years. It is also expected to have a salvage value of $50,000 at the end of its useful life. Although the system is likely to provide similar benefits in the production of the other specialty chemicals manufactured in the process plant, these benefits have not yet been quantified. What is the net cash flow in each year?

SOLUTION
Given: The preceding cost and benefit information.
Find: Net cash flow in each year over the life of the new system.

Although we could assume that similar benefits are derivable from the production of the other specialty chemicals, let's restrict our consideration to the demulsification chemical and allocate the full initial cost of the control system and the annual maintenance costs to this chemical. (Note that you could logically argue that only 40% of these costs belong to this production activity.) The gross benefits are the additional revenues realized from the increased selling price and the extra production, as well as the cost savings resulting from having one fewer operator.

- Revenues from the price increases are 30,000 kilograms per year $\times$ $2/kilogram, or $60,000 per year. The added production volume at the new pricing adds revenues of 4,000 kilograms per year $\times$ $17 per kilogram, or $68,000 per year.
- The elimination of one operator results in an annual savings of 3,500 operating hours per year $\times$ $25 per hour, or $87,500 per year.
- The net benefits in each of the eight years that make up the useful lifetime of the new system are the gross benefits less the maintenance costs:

$$($60,000 + $68,000 + $87,500 - $53,000) = $162,500 \text{ per year}$$

- The cash flow in eight years will be adjusted by the amount of salvage of $50,000.

Now we are ready to summarize a cash flow table.

Year (n)	Cash Inflows (Benefits)	Cash Outflows (Costs)	Net Cash Flows
0	$ 0	$650,000	−$650,000
1	215,500	53,000	162,500
2	215,500	53,000	162,500
⋮	⋮	⋮	⋮
8	215,500 + $50,000	53,000	162,500 + $50,000

COMMENTS: If the company purchases the computer process-control system for $650,000 now, it can expect an annual savings of $162,500 for eight years. (Note that these savings are represented in discrete lump sums at the end of each year.) We also considered only the benefits associated with the production of the demulsification chemical. We could have quantified some benefits attributable to the production of the other chemicals from this plant. Suppose that the demulsification chemical benefits alone justify the acquisition of the new system. Then it is obvious that, had we considered the benefits associated with the other chemicals as well, the acquisition of the system would have been even more clearly justified.

We draw a cash flow diagram of this situation in Figure 5.2. Assuming that these cost savings and cash flow estimates are correct, should management give the go-ahead for installation of the system? If management decides not to purchase the computer process-control system, what should it do with the $650,000 (assuming that it has this amount in the first place)? The company could buy $650,000 of Treasury bonds, or it could invest the amount in other cost-saving projects. How would the company compare cash flows that differ both in timing and amount for the alternatives it is considering? This is an extremely important question, because virtually every engineering investment decision involves a comparison of alternatives. Indeed, these are the types of questions this chapter is designed to help you answer.

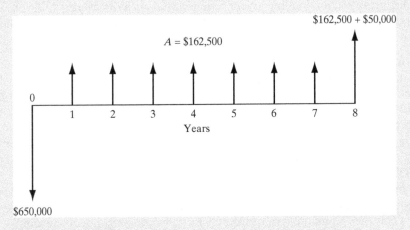

Figure 5.2   Cash flow diagram for the computer process-control system described in Example 5.1.

5.1.2 Independent versus Mutually Exclusive Investment Projects

Most firms have a number of unrelated investment opportunities available. For example, in the case of XL Chemicals, other projects being considered in addition to the computer process-control system in Example 5.1 are a new waste heat recovery boiler, a CAD system for the engineering department, and a new warehouse. The economic attractiveness of each of these projects can be measured, and a decision to accept or reject the project can be made without reference to any of the other projects. In other words, the decision regarding any one project has no effect on the decision to accept or reject another project. Such projects are said to be **independent**.

In Section 5.5, we will see that in many engineering situations, we are faced with selecting the most economically attractive project from a number of alternative projects, all of which solve the same problem or meet the same need. It is not desirable to choose more than one project in this situation, and the acceptance of one automatically entails the rejection of all of the others. Such projects are said to be **mutually exclusive**.

As long as the total cost of all the independent projects found to be economically attractive is less than the investment funds available to a firm, all of these projects could proceed. However, this is rarely the case. The selection of those projects which should proceed when investment funds are limited is the subject of capital budgeting. Apart from Chapter 15, which deals with capital budgeting, the availability of funds will not be a consideration in accepting or rejecting projects dealt with in this book.

5.2 Initial Project Screening Method

Let us suppose that you are in the market for a new punch press for your company's machine shop and you visit an equipment dealer. As you take a serious look at one of the punch press models in the display room, an observant equipment salesperson approaches you and says,

> "That press you are looking at is the state-of-the-art in its category. If you buy that top-of-the-line model, it will cost a little bit more, but *it will pay for itself* in less than two years."

Before studying the four measures of investment attractiveness, we will review a simple method commonly used to screen capital investments. One of the primary concerns of most businesspeople is whether and when the money invested in a project can be recovered. The **payback method** screens projects on the basis of how long it takes for net receipts to equal investment outlays. This calculation can take one of two forms: either ignore time-value-of-money considerations or include them. The former case is usually designated the **conventional payback method**, the latter case the **discounted payback method**.

A common standard used to determine whether to pursue a project is that the project does not merit consideration unless its payback period is shorter than some specified period. (This time limit is determined largely by management policy. For example, a high-tech firm, such as a computer chip manufacturer, would set a short time limit for any new investment, because high-tech products rapidly become obsolete.) If the payback period is within the acceptable range, a formal project evaluation (such as a present-worth analysis) may begin. It is important to remember that **payback screening** is not an *end* in itself, but rather a method of screening out certain obviously unacceptable investment alternatives before progressing to an analysis of potentially acceptable ones.

5.2.1 Payback Period: The Time It Takes to Pay Back

Determining the relative worth of new production machinery by calculating the time it will take to pay back what it costs is the single most popular method of project screening. If a company makes investment decisions solely on the basis of the payback period, it considers only those projects with a payback period *shorter* than the maximum acceptable payback period. (However, because of shortcomings of the payback screening method, which we will discuss later, it is rarely used as the only decision criterion.)

What does the payback period tell us? One consequence of insisting that each proposed investment have a short payback period is that investors can assure themselves of being restored to their initial position within a short span of time. By restoring their initial position, investors can take advantage of additional, perhaps better, investment possibilities that may come along.

In Example 5.2, dividing the initial payment by annual receipts to determine the payback period is a simplification we can make because the annual receipts are uniform. Whenever the expected cash flows vary from year to year, however, the payback period must be determined by adding the expected cash flows for each year until the sum is equal to or greater than zero. The significance of this procedure is easily explained. The cumulative cash flow equals zero at the point where cash inflows exactly match, or pay back, the cash outflows; thus, the project has reached the payback point. Similarly, if the cumulative cash flows are greater than zero, then the cash inflows exceed the cash outflows, and the project has begun to generate a profit, thus surpassing its payback point. To illustrate, consider Example 5.3.

EXAMPLE 5.2 A Conventional Payback Period for the Computer Process-Control System Project

Consider the cash flows given in Example 5.1. Determine the payback period for this computer process-control system project.

SOLUTION
Given: Initial cost = $650,000 and annual net benefits = $162,500.
Find: Conventional payback period.

Given a uniform stream of receipts, we can easily calculate the payback period by dividing the initial cash outlay by the annual receipts:

$$\text{Payback period} = \frac{\text{Initial cost}}{\text{Uniform annual benefit}} = \frac{\$650,000}{\$162,500}$$

$$= 4 \text{ years}$$

If the company's policy is to consider only projects with a payback period of five years or less, this computer process-control system project passes the initial screening.

EXAMPLE 5.3 Conventional Payback Period with Salvage Value

Autonumerics Company has just bought a new spindle machine at a cost of $105,000 to replace one that had a salvage value of $20,000. The projected annual after-tax savings via improved efficiency, which will exceed the investment cost, are as follows. What is the conventional payback period?

Period	Cash Flow	Cumulative Cash Flow
0	−$105,000 + $20,000	−$85,000
1	15,000	−70,000
2	25,000	−45,000
3	35,000	−10,000
4	45,000	35,000
5	45,000	80,000
6	35,000	115,000

SOLUTION
Given: Cash flow series as shown in Figure 5.3(a).
Find: Conventional payback period.

The salvage value of retired equipment becomes a major consideration in most justification analysis. (In this example, the salvage value of the old machine should be taken

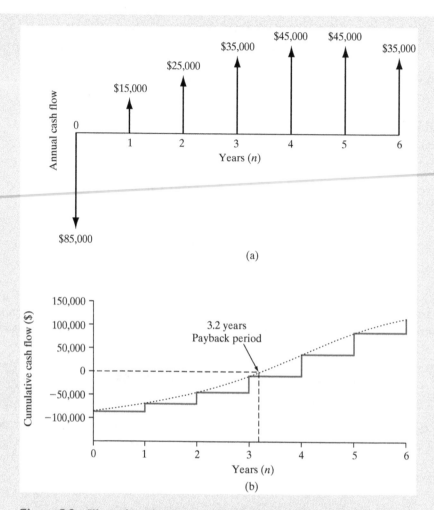

Figure 5.3   Illustration of conventional payback period (Example 5.3).

into account, as the company already had decided to replace the old machine.) When used, the salvage value of the retired equipment is subtracted from the purchase price of new equipment, revealing a closer true cost of the investment. As we see from the cumulative cash flow in Figure 5.3(b), the total investment is recovered during year 4. If the firm's stated maximum payback period is three years, the project will not pass the initial screening stage.

COMMENTS: In Example 5.2, we assumed that cash flows are represented only in discrete lumps at the ends of years. If instead cash flows occur continuously through-out the year, the payback period calculation needs adjustment. A negative balance of $10,000 remains at the start of year 4. If $45,000 is expected to be received as a more or less continuous flow during year 4, the total investment will be recovered about two-tenths ($10,000/$45,000) of the way through the fourth year. Thus, in this situa-tion, the payback period is 3.2 years.

5.2.2 Benefits and Flaws of Payback Screening

The simplicity of the payback method is one of its most appealing qualities. Initial project screening by the payback method reduces the information search by focusing on that time at which the firm expects to recover the initial investment. The method may also eliminate some alternatives, thus reducing the firm's time spent analyzing. But the much-used payback method of equipment screening has a number of serious drawbacks. The principal objection to the method is that it fails to measure profitability (i.e., no "profit" is made during the payback period). Simply measuring how long it will take to recover the initial investment outlay contributes little to gauging the earning power of a project. (In other words, you already know that the money you borrowed for the drill press is costing you 12% per year; the payback method can't tell you how much your invested money is contributing toward the interest expense.) Also, because payback period analysis ignores differences in the timing of cash flows, it fails to recognize the difference between the present and future value of money. For example, although the payback on two investments can be the same in terms of numbers of years, a front-loaded investment is better, because money available today is worth more than that to be gained later. Finally, because payback screening ignores all proceeds after the payback period, it does not allow for the possible advantages of a project with a longer economic life.

By way of illustration, consider the two investment projects listed in Table 5.1. Each requires an initial investment outlay of $90,000. Project 1, with expected annual cash proceeds of $30,000 for the first three years, has a payback period of three years. Project 2 is expected to generate annual cash proceeds of $25,000 for six years; hence, its payback period is 3.6 years. If the company's maximum payback period is set to three years, then project 1 would pass the initial project screening, whereas project 2 would fail, even though it is clearly the more profitable investment.

5.2.3 Discounted Payback Period

To remedy one of the shortcomings of the conventional payback period, we may modify the procedure so that it takes into account the time value of money—that is, the cost of funds (interest) used to support a project. This modified payback period is often referred to as the

TABLE 5.1	**Investment Cash Flows for Two Competing Projects**	
n	**Project 1**	**Project 2**
0	−$90,000	−$90,000
1	30,000	25,000
2	30,000	25,000
3	30,000	25,000
4	1,000	25,000
5	1,000	25,000
6	1,000	25,000
	$ 3,000	$60,000

discounted payback period. In other words, we may define the discounted payback period as the number of years required to recover the investment from *discounted* cash flows.

For the project in Example 5.3, suppose the company requires a rate of return of 15%. To determine the period necessary to recover both the capital investment and the cost of funds required to support the investment, we may construct Table 5.2, showing cash flows and costs of funds to be recovered over the life of the project.

To illustrate, let's consider the cost of funds during the first year: With $85,000 committed at the beginning of the year, the interest in year 1 would be $12,750 ($85,000 × 0.15). Therefore, the total commitment grows to $97,750, but the $15,000 cash flow in year 1 leaves a net commitment of $82,750. The cost of funds during the second year would be $12,413 ($82,750 × 0.15), but with the $25,000 receipt from the project, the net commitment drops to $70,163. When this process repeats for the remaining years of the project's life, we find that the net commitment to the project ends during year 5. Depending on which cash flow assumption we adopt, the project must remain in use about 4.2 years (continuous cash flows) or 5 years (year-end cash flows) in order for the company to cover its cost of capital and also recover the funds it has invested. Figure 5.4 illustrates this relationship.

The inclusion of effects stemming from time value of money has increased the payback period calculated for this example by a year. Certainly, this modified measure is an improved one, but it does not show the complete picture of the project's profitability either.

5.2.4 Where Do We Go from Here?

Should we abandon the payback method? Certainly not, but if you use payback screening exclusively to analyze capital investments, look again. You may be missing something that another method can help you spot. Therefore, it is illogical to claim that payback is

TABLE 5.2 Payback Period Calculation Taking into Account the Cost of Funds (Example 5.3)

	A	B	C	D
1				
2	Period	Cash Flow	Cost of Funds	Ending
3	(n)	A_n	15%	Cash Balance
4				
5	0	-$85,000.00		-$85,000.00
6	1	$15,000.00	-$12,750.00	-$82,750.00
7	2	$25,000.00	-$12,412.50	-$70,162.50
8	3	$35,000.00	-$10,524.38	-$45,686.88
9	4	$45,000.00	-$6,853.03	-$7,539.91
10	5	$45,000.00	-$1,130.99	$36,329.11
11	6	$35,000.00	$5,449.37	$76,778.47
12				
13	Cost of funds = Ending cash balance (prior) x 0.15			

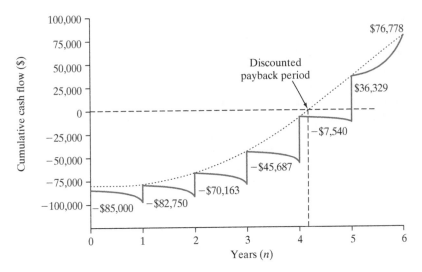

Figure 5.4   Illustration of discounted payback period.

either a good or bad method of justification. Clearly, it is not a measure of profitability. But when it is used to supplement other methods of analysis, it can provide useful information. For example, payback can be useful when a company needs a measure of the speed of cash recovery, when the company has a cash flow problem, when a product is built to last only for a short time, and when the machine the company is contemplating buying itself is known to have a short market life.

5.3 Discounted Cash Flow Analysis

Until the 1950s, the payback method was widely used as a means of making investment decisions. As flaws in this method were recognized, however, businesspeople began to search for methods to improve project evaluations. The result was the development of **discounted cash flow techniques (DCFs)**, which take into account the time value of money over the entire project life. One of the DCFs is the net-present-worth (or net-present-value) method. A capital investment problem is essentially a problem of determining whether the anticipated cash inflows from a proposed project are sufficient to attract investors to invest funds in the project. In developing the decision criterion, we will use the concept of cash flow equivalence discussed in Chapter 3.

5.3.1 Net-Present-Worth Criterion

As we observed, the most convenient point at which to calculate the equivalent values is often at time 0. Under the DCF analysis, the present worth of all cash inflows is compared against the present worth of all cash outflows associated with an investment project. The difference between the present worth of these cash flows is referred to as the **net present worth (NPW)**. This value determines whether the project is an acceptable investment. When two or more projects are under consideration, NPW analysis further allows us to select the best project by comparing their NPW figures. We will first summarize the basic procedure for applying the net-present-worth criterion to a typical investment project.

- Determine the interest rate that the firm wishes to earn on its investments. The interest rate you determine represents the rate at which the firm can always invest the money in its **investment pool**. This interest rate is often referred to as either a **required rate of return** or a **minimum attractive rate of return (MARR)**. Usually, selection of the MARR is a policy decision made by top management. It is possible for the MARR to change over the life of a project, as we saw in Section 4.4, but for now, we will use a single rate of interest in calculating the NPW.
- Estimate the service life of the project.
- Estimate the cash inflow for each period over the service life.
- Estimate the cash outflow over each service period.
- Determine the net cash flows (net cash flow = cash inflow − cash outflow).
- Find the present worth of each net cash flow at the MARR. Add up all the present-worth figures; their sum is defined as the project's NPW, given by

$$
\begin{aligned}
\text{PW}(i) = \text{NPW calculated at } i &= \frac{A_0}{(1+i)^0} + \frac{A_1}{(1+i)^1} + \cdots \frac{A_N}{(1+i)^N} \\
&= \sum_{n=0}^{N} \frac{A_n}{(1+i)^n} \\
&= \sum_{n=0}^{N} A_n(P/F, i, n)
\end{aligned}
\tag{5.1}
$$

where

A_n = Net cash flow at end of period n

i = MARR (or cost of capital)

N = Service life of the project

A_n will be positive if the corresponding period has a net cash inflow, and negative if there is a net cash outflow.

Single Project Evaluation

In this context, a positive NPW means that the equivalent worth of the inflows is greater than the equivalent worth of outflows, so the project makes a profit. Therefore, if the PW(i) is positive for a single project, the project should be accepted; if the PW(i) is negative, the project should be rejected.[1] The decision rule is

If PW(i) > 0, accept the investment.

If PW(i) = 0, remain indifferent.

If PW(i) < 0, reject the investment.

---

[1] Some projects (e.g., the installation of pollution control equipment) cannot be avoided. In a case such as this, the project would be accepted, even though its NPW ≤ 0.

Comparing Multiple Alternatives

Compute the PW(i) for each alternative and select the one with the largest PW(i). As you will learn in Section 5.5, when you compare mutually exclusive alternatives with the *same revenues*, they could be compared on a *cost-only basis*. In this situation (because you are minimizing costs, rather than maximizing profits), you should accept the project that results in the *smallest*, or *least negative*, NPW.

EXAMPLE 5.4 **Net Present Worth: Uniform Flows**

Consider the investment cash flows associated with the computer process-control system project discussed in Example 5.1. If the firm's MARR is 15%, compute the NPW of this project. Is the project acceptable?

SOLUTION
Given: Cash flows in Figure 5.2 and MARR = 15% per year.
Find: NPW.

Since the computer process-control system project requires an initial investment of $650,000 at $n = 0$ followed by the eight equal annual savings of $162,000, plus $50,000 salvage value at the end of eight years, we can easily determine the NPW as follows.

$$PW(15\%)_{\text{Outflow}} = \$650,000$$

$$PW(15\%)_{\text{Inflow}} = \$162,500(P/A, 15\%, 8) + \$50,000(P/F, 15\%, 8)$$

$$= \$745,535$$

Then the NPW of the project is

$$PW(15\%) = PW(15\%)_{\text{Inflow}} - PW(15\%)_{\text{Outflow}}$$

$$= \$745,535 - \$650,000$$

$$= \$95,535$$

or, from Eq. (5.1),

$$PW(15\%) = -\$650,000 + \$162,500(P/A, 15\%, 8) + \$50,000(P/F, 15\%, 8)$$

$$= \$95,535$$

Since PW(15%) > 0, the project is acceptable.

Now let's consider another example in which the investment cash flows are not uniform over the service life of the project.

In Example 5.5, we compute the NPW of the project at a fixed interest rate of 12%. If we compute the NPW at varying interest rates, we obtain the data in Table 5.3. Plotting

EXAMPLE 5.5 Machine Tool Investment Justification

Tiger Machine Tool Company is considering acquiring a new metal-cutting machine. The required initial investment of $76,000 and the projected cash benefits[2] over the project's four-year life are as follows.

Period (n)	Net Cash Flow (A_n)
0	−$76,000
1	$35,560
2	$37,360
3	$31,850
4	$34,400

You have been asked by the president of the company to evaluate the economic merit of the acquisition. The firm's MARR is known to be 12%.

SOLUTION

Given: Cash flows as tabulated and MARR $=$ 12% per year.
Find: NPW.

If we bring each flow to its equivalent at time 0, we find that

$$PW(12\%) = -\$76,000 + \$35,560(P/F, 12\%, 1) + \$37,360(P/F, 12\%, 2)$$
$$+ \$31,850(P/F, 12\%, 3) + \$34,400(P/F, 12\%, 4)$$
$$= \$30,065$$

Since the project results in a surplus of $30,065, the project is acceptable.

the NPW as a function of interest rate gives Figure 5.5, which is the present-worth profile. (You may use a spreadsheet program such as Excel to generate Table 5.3 or Figure 5.5.)

Figure 5.5 indicates that the investment project has a positive NPW if the interest rate is below 30% and a negative NPW if the interest rate is above 30%. As we will see in Chapter 7, this **break-even interest rate** is known as the **internal rate of return**. If the firm's MARR is 12%, then the project has an NPW of $30,065 and so may be accepted. The $30,065 figure measures the equivalent immediate gain in present worth to the firm following acceptance of the project. By contrast, at $i = 35\%$, PW(35%) $= -\$5,858$, and the firm should reject the project. (Note that either accepting or rejecting an investment is influenced by the choice of a MARR, so it is crucial to estimate the MARR correctly. We will defer this important issue until Section 5.3.3. For now, we will assume that the firm has an accurate MARR estimate available for use in investment analysis.)

---

[2]As we stated at the beginning of this chapter, we treat net cash flows as before-tax values or as having their tax effects precalculated. Explaining the process of obtaining cash flows requires an understanding of income taxes and the role of depreciation, which are discussed in Chapter 9.

TABLE 5.3 Present-Worth Amounts at Varying Interest Rates (Example 5.5)

	A	B	C	D	E	F	G	H	I	J
1										
2			Input					Output		
3										
4	(*i*) MARR (%)			12%		MARR(%)	PW(*i*)		MARR(%)	PW(*i*)
5	(*N*) Project life			4						
6						0	$63,170		26	$5,325
7		Cash	Interest	Project		2	$56,565		28	$2,586
8	Period	Flow	Earned (Paid)	Balance		4	$50,454		30	$2
9						6	$44,787		32	($2,440)
10	0	$ (76,000)		$ (76,000)		8	$39,525		34	($4,750)
11	1	$ 35,560	$ (9,120)	$ (49,560)		10	$34,628		36	($6,937)
12	2	$ 37,360	$ (5,947)	$ (18,147)		12	$30,065		38	($9,010)
13	3	$ 31,850	$ (2,178)	$ 11,525		14	$25,806		40	($10,977)
14	4	$ 34,400	$ 1,383	$ 47,308		16	$21,823		42	($12,845)
15						18	$18,095		44	($14,622)
16						20	$14,599		46	($16,312)
17						22	$11,316		48	($17,922)
18						24	$8,230		50	($19,457)
19										

=NPV(I18%,$B$11:$B$14)+$B$10

5.3.2 Meaning of Net Present Worth

In present-worth analysis, we assume that all the funds in a firm's treasury can be placed in investments that yield a return equal to the MARR. We may view these funds as an **investment pool**. Alternatively, if no funds are available for investment, we assume that the firm can borrow them at the MARR (or cost of capital) from the capital market. In

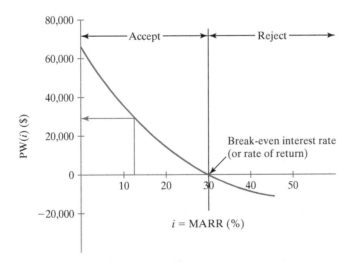

Figure 5.5 Present-worth profile described in Example 5.5— the machine tool project is acceptable as long as the firm's MARR is less than 30%.

this section, we will examine these two views as we explain the meaning of the MARR in NPW calculations.

Investment Pool Concept

An investment pool is equivalent to a firm's treasury. All fund transactions are administered and managed by the firm's comptroller. The firm may withdraw funds from this investment pool for other investment purposes, but if left in the pool, these funds will earn at the MARR. Thus, in investment analysis, net cash flows will be net cash flows relative to an investment pool. To illustrate the investment pool concept, we consider again the project in Example 5.5 that required an investment of $76,000.

- **Option 1: Do not invest in the project.** If the firm did not invest in the project and left $76,000 in the investment pool for four years, these funds would grow as

$$76,000(F/P, 12\%, 4) = \$119,587.$$

- **Option 2: Invest in the project.** Suppose the company decided instead to invest $76,000 in the project described in Example 5.5. Then the firm would receive a stream of cash inflows during the project's life of three years in the following amounts.

Period (n)	Net Cash Flow (A_n)
1	$35,560
2	$37,360
3	$31,850
4	$34,400

Since the funds that return to the investment pool earn interest at a rate of 12%, it is worthwhile to see how much the firm would benefit from its $75,000 investment. For this alternative, the returns after reinvestment are

$$\$35,560(F/P, 12\%, 3) = \$49,959$$

$$\$37,360(F/P, 12\%, 2) = \$46,864$$

$$\$31,850(F/P, 12\%, 1) = \$35,672$$

$$\$34,400(F/P, 12\%, 0) = \$34,400$$

These returns total $166,896. At the end of three years, the additional cash accumulation from investing in the project is

$$\$166,896 - \$119,587 = \$47,309$$

If we compute the equivalent present worth of this net cash surplus at time 0, we obtain

$$\$47,309(P/F, 12\%, 4) = \$30,065$$

which is exactly what we get when we compute the NPW of the project with Eq. (5.1). Clearly, on the basis of its positive NPW, the alternative of purchasing a new machine should be preferred to that of simply leaving the funds in the investment pool at the MARR. Thus, in PW analysis, any investment is assumed to be returned at the MARR.

Option A

If $76,000 were left in the investment pool for 4 years ⟶ $76,000(F/P,12%,4) ⟶ $119,587

Option B

If $76,000 withdrawal from the investment pool were invested in the project

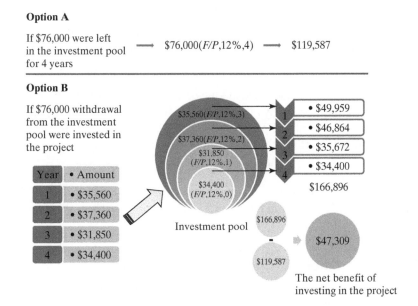

Year	• Amount
1	• $35,560
2	• $37,360
3	• $31,850
4	• $34,400

$35,560(F/P,12%,3) ⟶ 1 • $49,959
$37,360(F/P,12%,2) ⟶ 2 • $46,864
$31,850 (F/P,12%,1) ⟶ 3 • $35,672
$34,400 (F/P,12%,0) ⟶ 4 • $34,400

$166,896

Investment pool

$166,896 − $119,587 ⟹ $47,309

The net benefit of investing in the project

Figure 5.6 The concept of an investment pool with the company as a lender and the project as a borrower.

If a surplus exists at the end of the project, then PW(MARR) > 0. Figure 5.6 illustrates the reinvestment concept as it relates to the firm's investment pool.

Borrowed-Funds Concept

Suppose that the firm does not have $76,000 at the outset. In fact, the firm doesn't have to have an investment pool at all. Suppose further that the firm borrows all of its capital from a bank at an interest rate of 12%, invests in the project, and uses the proceeds from the investment to pay off the principal and interest on the bank loan. How much is left over for the firm at the end of the project's life?

- $n = 1$: At the end of the first year, the interest on the project's use of the bank loan would be $76,000(0.12) = $9,120. Therefore, the total loan balance grows to $76,000(1.12) = $85,120. Then, the firm receives $35,560 from the project and applies the entire amount to repay the loan portion, leaving a balance due of

$$\text{PB}(12\%)_1 = -\$76,000(1 + 0.12) + \$35,560 = -\$49,560$$

- $n = 2$: This amount becomes the net amount the project is borrowing at the beginning of year 2, which is also known as the **project balance**. At the end of year 2, the bank debt grows to $49,560(1.12) = $55,507, but with the receipt of $37,360, the project balance is reduced to

$$\text{PB}(12\%)_2 = -\$49,560(1 + 0.12) + \$37,360 = -\$18,147$$

- $n = 3$: With the receipt of $31,850 from the project, the firm should be able to pay off the remaining balance and come out with a surplus in the amount of $11,525. Or simply, at the end of year 3, the project balance becomes

$$\text{PB}(12\%)_3 = -\$18,147(1 + 0.12) + \$31,850 = \$11,525$$

- $n = 4$: Now we need to ask the company what to do with the surplus at the end of year 3. If the company can invest this surplus at the same rate as the borrowing rate, say 12%, then the project balance at the end of year 4 would be

$$PB(12\%)_4 = \$11,525(1 + 0.12) + \$34,400 = \$47,308$$

This terminal project balance is also known as the **net future worth** of the project. In other words, the firm repays its initial bank loan and interest at the end of year 3 with a resulting profit of $11,525 reinvested at 12% for one year. If we compute the equivalent present worth of this net profit at time 0, we obtain

$$PW(12\%) = PB(12\%)_4(P/F, 12\%, 4) = \$30,065$$

The result is identical to the case in which we directly computed the NPW of the project at $i = 12\%$, shown in Example 5.5. Figure 5.7 illustrates the project balance as a function of time.[3]

5.3.3 Basis for Selecting the MARR

The basic principle used to determine the discount rate in project evaluations is similar to the concept of the required return on investment for financial assets discussed in Section 4.6.2. The first element to cover is the **cost of capital**, which is the required return necessary to make an investment project worthwhile. The cost of capital would include both the *cost of debt* (the interest rate associated with borrowing) and the *cost of equity* (the return that stockholders require for a company). Both the cost of debt and the cost

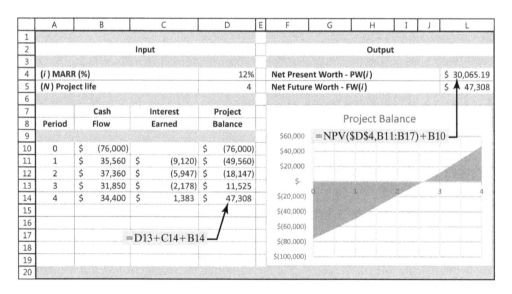

Figure 5.7   Project balance diagram as a function of time. (A negative project balance indicates the amount of investment to be recovered or the degree of risk exposure if the project terminates prematurely.)

---

[3]Note that the sign of the project balance changes from negative to positive during year 3. The time at which the project balance becomes zero is known as the **discounted payback period**.

Figure 5.8 Elements of minimum attractive rate of return.

of equity reflect the presence of inflation in the economy. *The cost of capital determines how a company can raise money (through issuing a stock, borrowing, or a mix of the two). Therefore, this is normally considered as the rate of return that a firm would receive if it invested its money someplace else with a similar risk.*

The second element is a consideration of any additional risk associated with the project. If the project belongs to the normal risk category, the cost of capital may already reflect such a risk premium. However, if you are dealing with a project with higher risk, the additional risk premium may be added onto the cost of capital.

As shown in Figure 5.8, the discount rate (or **MARR**) to use for project evaluation would be equivalent to the firm's cost of capital for a project of normal risk, but could be much higher if you are dealing with a risky project. Chapter 15 will detail the analytical process of determining this discount rate. For now, we assume that such a rate is already known to us, and we will focus on the evaluation of the investment project.

5.4 Variations of Present-Worth Analysis

As variations of present-worth analysis, we will consider two additional measures of investment worth: **future-worth analysis** and **capitalized equivalent-worth analysis**. (The equivalent annual-worth measure is another variation of the present-worth measure, but we present it in Chapter 6.) Future-worth analysis calculates the future worth of an investment undertaken. Capitalized equivalent-worth analysis calculates the present worth of a project with a perpetual life span.

5.4.1 Future-Worth Analysis

Net present worth measures the surplus in an investment project at time 0. **Net-future-worth (NFW)** analysis is particularly useful in an investment situation in which we need to compute the equivalent worth of a project at other points in the future, rather than at its beginning. For example, it may take seven to 10 years to build a nuclear power plant because of the complexities of engineering design and the many time-consuming regulatory procedures that must be followed to ensure public safety. In this situation, it is more common to measure the worth of the investment at the time of the project's commercialization (i.e., we conduct an economic analysis at the end of the construction period). That way, we get to know exactly how much money has been invested in the project, and this will be the basis for utility rate settings in many states in the U.S.

To illustrate the process of determining the net future worth, let A_n represent the cash flow at time n for $n = 0, 1, 2, \ldots, N$ for a typical investment project that extends over N periods. Then the net-future-worth (NFW) expression at the end of period N is

$$FW(i) = A_0(1 + i)^N + A_1(1 + i)^{N-1} + A_2(1 + i)^{N-2} + \cdots + A_N$$

$$= \sum_{n=0}^{N} A_n(1 + i)^{N-n}$$

$$= \sum_{n=0}^{N} A_n(F/P, i, N - n) \tag{5.2}$$

As you might expect, the decision rule for the NFW criterion is the exactly same as that for the NPW criterion for a single project evaluation.

> If $FW(i) > 0$, accept the investment.
>
> If $FW(i) = 0$, remain indifferent to the investment.
>
> If $FW(i) < 0$, reject the investment.

EXAMPLE 5.6 A Robot Development Project and Its Evaluation

Higgins Corporation (HC), a Detroit-based robot-manufacturing company, has developed a new robot called Helpmate, which incorporates advanced technology, such as vision systems, tactile sensing, and voice recognition. These features allow the robot to roam the corridors of a hospital or office building without following a predetermined track or bumping into objects. HC's marketing department plans to target sales of the robot toward major hospitals. The robots will ease nurses' workloads by performing low-level duties, such as delivering medications and meals to patients.

- The firm would need a new plant to manufacture the Helpmates; this plant could be built and made ready for production in two years. It would require a 30-acre site, which can be purchased for $1.5 million in year 0. Building construction would begin early in year 1 and continue throughout year 2. The building would cost an estimated $10 million, with a $4 million payment due to the contractor at the end of year 1, and with another $6 million payable at the end of year 2.

- The necessary manufacturing equipment would be installed late in year 2 and would be paid for at the end of year 2. The equipment would cost $13 million, including transportation and installation. When the project terminates, the land is expected to have an after-tax market value of $2 million, the building an after-tax value of $3 million, and the equipment an after-tax value of $3 million.

For capital budgeting purposes, assume that the cash flows occur at the end of each year. Because the plant would begin operations at the beginning of year 3, the first operating cash flows would occur at the end of year 3. The Helpmate plant's estimated economic life is six years after completion, with the following expected after-tax operating cash flows in millions given in Table 5.4.

TABLE 5.4 **Projected Cash Flows Associated with the Helpmates Project**

Calendar Year	2015	2016	2017	2018	2019	2020	2021	2022	2023
End of Year	0	1	2	3	4	5	6	7	8
After-tax cash flows									
A. Operating revenue				$6	$8	$13	$18	$14	$8
B. Investment									
Land	−$1.5								+2
Building		−$4	−$6						+3
Equipment			−13						+3
Net cash flow	−$1.5	−$4	−$19	$6	$8	$13	$18	$14	$16

Compute the equivalent worth of this investment at the start of operations. Assume that HC's MARR is 15%.

SOLUTION

Given: Preceding cash flows and MARR $= 15\%$ per year.
Find: Equivalent worth of project at the end of calendar year 2.

One easily understood method involves calculating the present worth and then transforming it to the equivalent worth at the end of year 2. First, we can compute PW(15%) at time 0 of the project:

$$\begin{aligned}
\text{PW}(15\%) = &-\$1.5 - 4(P/F, 15\%, 1) - \$19(P/F, 15\%, 2) \\
&+ \$6(P/F, 15\%, 3) + \$8(P/F, 15\%, 4) + \$13(P/F, 15\%, 5) \\
&+ \$18(P/F, 15\%, 6) + \$14(P/F, 15\%, 7) + \$16(P/F, 15\%, 8) \\
= &\$13.91 \text{ million}
\end{aligned}$$

Then the equivalent future project worth at the start of operation is

$$\text{FW}(15\%) = \text{PW}(15\%)(F/P, 15\%, 2) = \$18.40 \text{ million}$$

A second method brings all flows prior to year 2 up to that point and discounts future flows back to year 2. The equivalent worth of the earlier investment (when the plant begins full operation) is

$$-\$1.5(F/P, 15\%, 2) - \$4(F/P, 15\%, 1) - \$19 = -\$25.58 \text{ million}$$

which produces an equivalent flow as shown in Figure 5.9. If we discount the future flows to the start of operation, we obtain

$$\begin{aligned}
\text{FW}(15\%) = &-\$25.58 + \$6(P/F, 15\%, 1) + \$8(P/F, 15\%, 2) + \cdots \\
&+ \$16(P/F, 15\%, 6) \\
= &\$18.40 \text{ million}
\end{aligned}$$

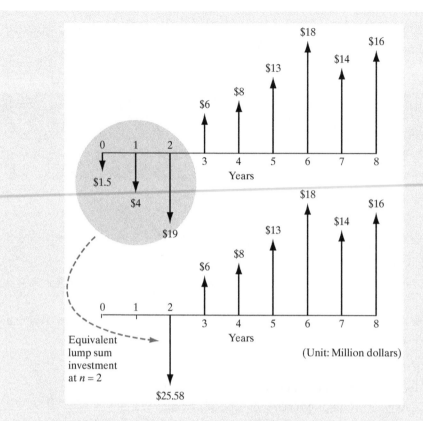

Figure 5.9 Cash flow diagram for the Helpmate project (Example 5.6).

COMMENTS: If another company were willing to purchase the plant and the right to manufacture the robots immediately after completion of the plant (year 2), HC would set the price of the plant at $43.98 million ($18.40 + $25.58) at a minimum.

5.4.2 Capitalized Equivalent Method

Another special case of the PW criterion is useful when the life of a proposed project is **perpetual** or the planning horizon is extremely long (say, 40 years or more). Many public projects (such as bridges, waterway structures, irrigation systems, and hydroelectric dams) are expected to generate benefits over an extended period (or forever). In this section, we will examine the **capitalized equivalent (CE(i))** method for evaluating such projects.

Perpetual Service Life

Consider the cash flow series shown in Figure 5.10. How do we determine the PW for an infinite (or almost infinite) uniform series of cash flows or a repeated cycle of cash flows? The process of computing the PW cost for this infinite series is referred to as the **capitalization** of the project cost. The cost, known as the **capitalized cost**, represents the

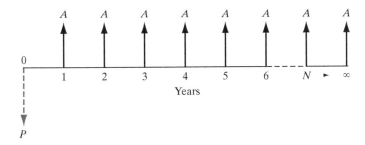

Figure 5.10 Equivalent present worth of an infinite cash flow series.

amount of money that must be invested today to yield a certain return A at the end of each and every period forever, assuming an interest rate of i. Observe the limit of the uniform series present-worth factor as N approaches infinity:

$$\lim_{N \to \infty} (P/A, i, N) = \lim_{N \to \infty} \left[\frac{(1 + i)^N - 1}{i(1 + i)^N} \right] = \frac{1}{i}$$

Thus,

$$PW(i) = A(P/A, i, N \to \infty) = \frac{A}{i} \tag{5.3}$$

Another way of looking at this problem is to ask what constant income stream could be generated by $PW(i)$ dollars today in perpetuity. Clearly, the answer is $A = i\,PW(i)$. If withdrawals were greater than A, you would be eating into the principal, which would eventually reduce it to 0.

EXAMPLE 5.7 Capitalized Equivalent Cost

An engineering school has just initiated a campaign targeting alumni to raise funds to create student scholarships to attract high-quality students. The amount of annual scholarships to be awarded would be $5 million. Assuming that the school can create a trust fund that earns 4% interest annually, how much has to be raised now to cover the perpetual string of $5 million in annual costs?

SOLUTION
Given: $A = \$5$ million, $i = 4\%$ per year, and $N = \infty$.
Find: CE(4%).

The capitalized cost equation is

$$CE(i) = \frac{A}{i}$$

so

$$CE(4\%) = \frac{\$5,000,000}{0.04}$$

$$= \$125,000,000$$

COMMENTS: It is easy to see that this lump-sum amount should be sufficient to pay scholarship for the school forever. Suppose the school deposited $125 million in a bank that paid 4% interest annually. Then at the end of the first year, the $125 million would earn 4%(125 million) = $5 million interest. If this interest were withdrawn, the $125 million would remain in the account. At the end of the second year, the $125 million balance would again earn 4%($125 million) = $5 million. This annual withdrawal could be continued forever, and the scholarship fund would always remain at $125 million.

Project's Service Life Is Extremely Long

The benefits of typical civil engineering projects, such as bridge and highway construction, although not perpetual, can last for many years. In this section, we will examine the use of the CE(i) criterion to approximate the NPW of engineering projects with long lives.

EXAMPLE 5.8 Comparison of Present Worth for Long Life and Infinite Life

Mr. Gaynor L. Bracewell amassed a small fortune developing real estate in Florida and Georgia over the past 30 years. He sold more than 700 acres of timber and farmland to raise $800,000 with which he built a small hydroelectric plant, known as High Shoals Hydro. The plant was a decade in the making. The design for Mr. Bracewell's plant, which he developed using his Army training as a civil engineer, is relatively simple. A 22-foot-deep canal, blasted out of solid rock just above the higher of two dams on his property, carries water 1,000 feet along the river to a "trash rack" where leaves and other debris are caught. A 6-foot-wide pipeline capable of holding three million pounds of liquid then funnels the water into the powerhouse at 7.5 feet per second, thereby creating 33,000 pounds of thrust against the turbines. Under a 1978 Federal law designed to encourage alternative power sources, Georgia Power Company is required to purchase any electricity Mr. Bracewell can supply. Mr. Bracewell estimates that his plant can generate 6 million kilowatt-hours per year.

Suppose that, after paying income taxes and operating expenses, Mr. Bracewell's annual income from the hydroelectric plant will be $120,000. With normal maintenance, the plant is expected to provide service for at least 50 years. Figure 5.11 illustrates when and in what quantities Mr. Bracewell spent his $800,000 (not taking into account the time value of money) during the last 10 years. Was Mr. Bracewell's $800,000 investment a wise one? How long will he have to wait to recover his initial investment, and will he ever make a profit? Examine the situation by computing the project's worth at varying interest rates.

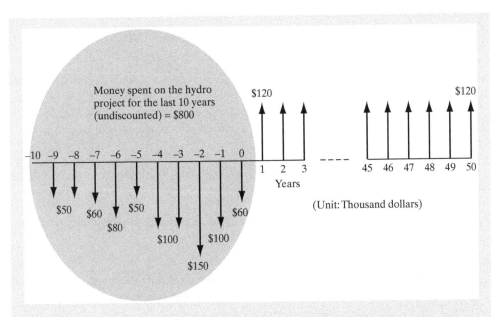

Figure 5.11 Net cash flow diagram for Mr. Bracewell's hydroelectric project (Example 5.8).

(a) If Mr. Bracewell's interest rate is 8%, compute the NPW (at time 0 in Figure 5.11) of this project with a 50-year service life and infinite service, respectively.

(b) Repeat part (a), assuming an interest rate of 12%.

SOLUTION
Given: Cash flow in Figure 5.11 (to 50 years or ∞) and $i = 8\%$ or 12%, respectively.
Find: NPW at time 0.

One of the main questions is whether Mr. Bracewell's plant will be profitable. Now we will compute the equivalent total investment and the equivalent worth of receiving future revenues at the start of power generation (i.e., at time 0).

(a) Let $i = 8\%$.
 • **With a plant service life of 50 years:** We can make use of single-payment compound-amount factors in the invested cash flow to help us find the equivalent total investment at the start of power generation. Using K to indicate thousands, we obtain

$$V_1 = -\$50K(F/P, 8\%, 9) - \$50K(F/P, 8\%, 8)$$
$$-\$60K(F/P, 8\%, 7) \cdots -\$100K(F/P, 8\%, 1) - \$60K$$
$$= -\$1,101K$$

The equivalent total benefit at the start of generation is

$$V_2 = \$120K(P/A, 8\%, 50) = \$1,468K$$

Summing, we find the net equivalent worth at the start of power generation:

$$V_1 + V_2 = -\$1{,}101K + \$1{,}468K$$
$$= \$367K$$

- **With an infinite service life:** The net equivalent worth is called the capitalized equivalent worth. The investment portion prior to time 0 is identical, so the capitalized equivalent worth is

$$CE(8\%) = -\$1{,}101K + \$120K/(0.08)$$
$$= \$399K$$

Note that the difference between the infinite situation and the planning horizon of 50 years is only $32,000.

(b) Let $i = 12\%$.

- **With a service life of 50 years:** Proceeding as we did in part (a), we find that the equivalent total investment at the start of power generation is

$$V_1 = -\$50K(F/P, 12\%, 9) - \$50K(F/P, 12\%, 8)$$
$$-\$60K(F/P, 12\%, 7) \cdots -\$100K(F/P, 12\%, 1) - \$60K$$
$$= -\$1{,}299K$$

Equivalent total benefits at the start of power generation are

$$V_2 = \$120K(P/A, 12\%, 50) = \$997K$$

The net equivalent worth at the start of power generation is

$$V_1 + V_2 = -\$1{,}299K + \$997K$$
$$= -\$302K$$

- **With infinite cash flows:** The capitalized equivalent worth at the current time is

$$CE(12\%) = -\$1{,}299K + \$120K/(0.12)$$
$$= -\$299K$$

Note that the difference between the infinite situation and a planning horizon of 50 years is merely $3,000, which demonstrates that we may approximate the present worth of long cash flows (i.e., 50 years or more) by using the capitalized equivalent value. The accuracy of the approximation improves as the interest rate increases (or the number of years is greater).

COMMENTS: At $i = 12\%$, Mr. Bracewell's investment is not a profitable one, but at 8%, it is. This outcome indicates the importance of using the appropriate i in investment analysis. The issue of selecting an appropriate i will be presented again in Chapter 15.

5.5 Comparing Mutually Exclusive Alternatives

Until now, we have considered situations involving either a single project alone or projects that were independent of each other. In both cases, we made the decision to reject or accept each project individually according to whether it met the MARR requirements evaluated with either the PW or FW criterion.

In the real world of engineering practice, however, it is typical for us to have two or more choices of projects that are not independent of one another in seeking to accomplish a business objective. (As we will see, even when it appears that we have only one project to consider, the implicit "do-nothing" alternative must be factored into the decision-making process.) In this section, we extend our evaluation techniques to multiple projects that are mutually exclusive. Other dependencies between projects will be considered in Chapter 15.

Up to now in this chapter, all available options in a decision problem were also assumed to have equal lifetimes. Often, however, various projects or investments under consideration do not have the same duration or do not match the desired study period. Adjustments must then be made to account for the differences. In this section, we explain the concept of an analysis period and the process of accommodating for different lifetimes, two important considerations that apply in selecting among several alternatives.

5.5.1 Meaning of Mutually Exclusive and "Do Nothing"

As we briefly mentioned in Section 5.1.2, several alternatives are **mutually exclusive** when any one of them will fulfill the same need and the selection of one of them implies that the others will be excluded. Take, for example, buying versus leasing an automobile for business use; when one alternative is accepted, the other is excluded. We use the terms **alternative** and **project** interchangeably to mean basically "decision option."

"Do Nothing" Is a Decision Option

When considering an investment, we are in one of two situations: Either the project is aimed at replacing an existing asset or system, or it is a new endeavor. In either case, a do-nothing alternative may exist. On the one hand, if a process or system already in place to accomplish our business objectives is still adequate, then we must determine which, if any, new proposals are economical replacements. If none are feasible, then we do nothing. On the other hand, if the existing system has failed, then the choice among proposed alternatives is mandatory (i.e., "do nothing" is not an option).

For most new endeavors, "do nothing" is generally an alternative (zero revenues ad zero costs; i.e., nothing currently exists), as we won't proceed unless at least one of the proposed alternatives is economically sound. In fact, undertaking even a single project entails making a decision between two alternatives, because the do-nothing alternative is implicitly included. Occasionally, a new initiative must be undertaken, cost notwithstanding, and in this case, the goal is to choose the most economical alternative, since "do nothing" is not an option.

When the option of retaining an existing asset or system is available, there are two ways to incorporate it into the evaluation of the new proposals.

- One way is to treat the do-nothing option as a distinct alternative; we cover this approach primarily in Chapter 14, where methodologies specific to replacement analysis are presented.
- The second approach, used mostly in this chapter, is to generate the cash flows of the new proposals relative to that of the do-nothing alternative. That is, for each new

alternative, the **incremental costs** (and incremental savings or revenues, if applicable) relative to "do nothing" are used in the economic evaluation. For a replacement-type problem, these costs are calculated by subtracting the do-nothing cash flows from those of each new alternative. For new endeavors, the incremental cash flows are the same as the absolute amounts associated with each alternative, since the do-nothing values are all zero.

Because the main purpose of this chapter is to illustrate how to choose among mutually exclusive alternatives, most of the problems are structured so that one of the options presented must be selected. Therefore, unless otherwise stated, it is assumed that "do nothing" is not an option, and costs and revenues can be viewed as incremental to "do nothing."

5.5.2 Service Projects versus Revenue Projects

When comparing mutually exclusive alternatives, we need to classify investment projects into either service or revenue projects. This classification is based on the amount of revenue expected from each alternative.

Service Projects

The revenues of service projects do not depend on the choice of project; rather, such projects *must produce the same amount of output (revenue).* In this situation, we certainly want to choose an alternative with the least input (or cost). For example, suppose an electric utility is considering building a new power plant to meet the peak-load demand during either hot summer or cold winter days. Two alternative service projects could meet this demand: a combustion turbine plant and a fuel-cell power plant. No matter which type of plant is selected, the firm will collect the same amount of revenue from its customers. The only difference is how much it will cost to generate electricity from each plant. If we were to compare these service projects, we would be interested in knowing which plant could provide the cheaper power (lower production cost). Further, if we were to use the NPW criterion to compare alternatives so as to minimize expenditures, we would choose the alternative with the **lower present-value** production cost over the service life of the plant.

Revenue Projects

With revenue projects, revenues depend on the choice of alternative. We are not limiting the amount of input going into the project or the amount of output that the project would generate; our decision is to select the alternative with the largest net gains (output–input). For example, a TV manufacturer is considering marketing two types of high-definition screen monitors. With its present production capacity, the firm can market only one of them. Distinct production processes for the two models could incur very different manufacturing costs, so the revenues from each model would be expected to differ due to divergent market prices and potentially different sales volumes. In this situation, if we were to use the NPW criterion, we would select the model that promises to bring in the higher net present worth.

5.5.3 Application of Investment Criteria

In comparing the mutually exclusive alternatives, we may adopt one of two approaches— (1) total-investment approach and (2) incremental-investment approach.

Total-Investment Approach

Applying an evaluation criterion to each mutually exclusive alternative individually and then comparing the results to make a decision is referred to as the **total-investment approach**. We compute the PW for each individual alternative, as we did in Section 5.3, and select the one with the highest PW. Note that this approach guarantees valid results only for absolute measures such as PW, FW, or AE which we will discuss in Chapter 6.

Incremental-Investment Approach

As we discussed in Section 1.6, in principle, an economic decision should be based on the differences among the alternatives considered. As we will see in Chapters 7 and 15, the total-investment approach does not work for any decision criterion based on either a percentage (rate of return) or a ratio (e.g., a benefit–cost ratio). With percentages or ratios, you need to use the incremental-investment approach, which also works with any decision criterion, including PW, FW, and AE. The incremental-investment approach will be discussed in detail in Chapter 7. Figure 5.12 illustrates a road map to follow in comparing mutually exclusive alternatives.

5.5.4 Scale of Investment

Frequently, mutually exclusive investment projects may require different levels of investments. At first, it seems unfair to compare a project requiring a smaller investment with one requiring a larger investment. However, the disparity in scale of investment should not be of concern in comparing mutually exclusive alternatives, as long as you understand the basic assumption: *Funds not invested in the project will continue to earn interest at the MARR.* We will look at mutually exclusive alternatives that require different levels of investments for both service projects and revenue projects.

- **Service projects.** Typically, what you are asking yourself to do here is to decide whether the higher initial investment can be justified by additional savings that will occur in the future. More efficient machines are usually more expensive to acquire initially, but they will reduce future operating costs, thereby generating more savings.
- **Revenue projects.** If two mutually exclusive revenue projects require different levels of investments with varying future revenue streams, then your question is what to

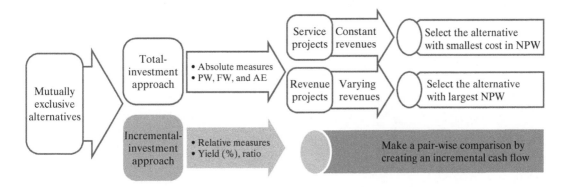

Figure 5.12 Diagram illustrating the process of making a choice among mutually exclusive alternatives.

do with the difference in investment funds if you decide to go with the project that requires the smaller investment.

To illustrate, consider the two mutually exclusive revenue projects illustrated in Figure 5.13. Our objective is to compare these two projects at a MARR of 10%.

Suppose you have exactly $4,000 to invest. If you choose Project B, you do not have any leftover funds. However, if you go with Project A, you will have $3,000 in unused funds. Our assumption is that these unused funds will continue to earn an interest rate that is the MARR. Therefore, the unused funds will grow at 10%, or $3,933, at the end of the project term (three years from now). Consequently, selecting Project A is equivalent to having a modified project cash flow, as shown in Figure 5.13.

Let's calculate the net present worth for each option at 10%.

- **Project A:**

$$PW(10\%)_A = -\$1,000 + \$450(P/F, 10\%, 1) + \$600(P/F, 10\%, 2)$$
$$+ \$500(P/F, 10\%, 3)$$
$$= \$281$$

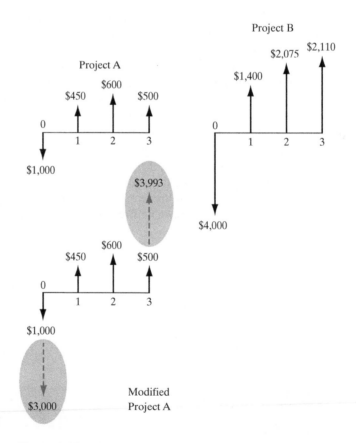

Figure 5.13 Comparing mutually exclusive revenue projects requiring different levels of investment.

- **Project B:**

$$\text{PW}(10\%)_B = -\$4{,}000 + \$1{,}400(P/F, 10\%, 1)$$

$$+ \$2{,}075(P/F, 10\%, 2) + \$2{,}110(P/F, 10\%, 3)$$

$$= \$573$$

Clearly, Project B is the better choice. But how about the modified Project A? If we calculate the present worth for the modified Project A, we have the following.

- **Modified Project A:**

$$\text{PW}(10\%)_A = -\$4{,}000 + \$450(P/F, 10\%, 1) + \$600(P/F, 10\%, 2)$$

$$+ \$4{,}493(P/F, 10\%, 3)$$

$$= \$281$$

This is exactly the same as the net present worth without including the investment consequence of the unused funds. It is not a surprising result, as the return on investment in the unused funds will be exactly 10%. If we also discount the funds at 10%, there will be no surplus. So, what is the conclusion? It is this: If there is any disparity in investment scale for mutually exclusive revenue projects, go ahead and calculate the net present worth for each option without worrying about the investment differentials.

5.5.5 Analysis Period

The **analysis period** is the time span over which the economic effects of an investment will be evaluated. The analysis period also may be called the **study period** or **planning horizon**. The length of the analysis period may be determined in several ways: It may be a predetermined amount of time set by company policy, or it may be either implied or explicit in the need the company is trying to fulfill. (For example, a diaper manufacturer sees the need to dramatically increase production over a 10-year period in response to an anticipated "baby boom.") In either of these situations, we consider the analysis period to be the **required service period**.

- *When the required service period is not stated at the outset*, the analyst must choose an appropriate analysis period over which to study the alternative investment projects. In such a case, one convenient choice of analysis period is the period of the useful life of the investment project.
- *When the useful life of an investment project does not match the analysis or required service period,* we need to introduce different techniques in our analysis. A further complication in a consideration of two or more mutually exclusive projects is that the investments themselves may have different useful lives. Accordingly, we must compare projects with different useful lives over an **equal time span**, which may require further adjustments in our analysis.

Figure 5.14 is a flowchart showing the possible combinations of the analysis period and the useful life of an investment. In the sections that follow, we will explore in more detail how to handle situations in which project lives differ from the analysis period and from each other. But we begin with the most straightforward situation: when the project lives and the analysis period coincide.

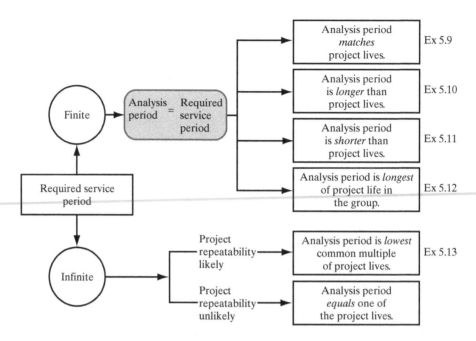

Figure 5.14 Analysis period implied in comparing mutually exclusive alternatives.

5.5.6 Analysis Period Matches Project Lives

When the project lives equal the analysis period, we compute the NPW for each project and select the one with the highest NPW. Example 5.9 illustrates this point.

EXAMPLE 5.9 Revenue Projects with Equal Lives: Three Alternatives

Bullard Company (BC) is considering expanding its range of industrial machinery products by manufacturing machine tables, saddles, machine bases, and other similar parts. Several combinations of new equipment and personnel could serve to fulfill this new function.

- **Option 1 (M1):** New machining center with three operators.
- **Option 2 (M2):** New machining center with an automatic pallet changer and three operators.
- **Option 3 (M3):** New machining center with an automatic pallet changer and two task-sharing operators.

Each of these arrangements incurs different costs and revenues. The time taken to load and unload parts is reduced in the pallet-changer cases. Certainly, it costs more to acquire, install, and tool-fit a pallet changer, but because the device is more efficient and versatile, it can generate larger annual revenues. Although saving on labor costs, task-sharing operators take longer to train and are more inefficient initially.

As the operators become more experienced at their tasks and get used to collaborating with each other, it is expected that the annual benefits will increase by 13% per year over the five-year study period. BC has estimated the investment costs and additional revenues as in Table 5.5.

TABLE 5.5 Summary of Financial Data for Three Machining Center Design Options

	Machining Center Design Options		
	M1	**M2**	**M3**
Investment:			
Machine tool purchase	$121,000	$121,000	$121,000
Automatic pallet changer	—	$ 66,600	$ 66,600
Installation	$ 30,000	$ 42,000	$ 42,000
Tooling expense	$ 58,000	$ 65,000	$ 65,000
Total investment	$209,000	$294,600	$294,600
Annual Benefits: Year 1			
Additional revenues	$ 55,000	$ 69,300	$ 36,000
Direct labor savings	—	—	$ 17,300
Setup savings	—	$ 4,700	$ 4,700
Year 1: Net revenues	$ 55,000	$ 74,000	$ 58,000
Years 2–5: Net revenues	Constant	Constant	$g = 13\%$/year
Salvage value in year 5	$ 80,000	$120,000	$120,000

All cash flows include all tax effects. "Do nothing" is obviously an option, since BC will not undertake this expansion if none of the proposed methods is economically viable. If a method is chosen, BC expects to operate the machining center over the next five years. On the basis of the use of the PW measure at $i = 12\%$, which option would be selected?

SOLUTION

Given: Cash flows for three revenue projects and $i = 12\%$ per year.

Find: NPW for each project and which project to select.

For these revenue projects, the net-present-worth figures at $i = 12\%$ would be as follows.

- For option M1,

$$PW(12\%)_{M1} = -\$209,000 + \$55,000(P/A, 12\%, 5)$$
$$+ \$80,000(P/F, 12\%, 5)$$
$$= \$34,657$$

TABLE 5.6 **An Excel Worksheet to Calculate the NPW for Each Option**

	A	B	C	D	E	F	G	H
1								
2	Option 1		0	1	2	3	4	5
3								
4		Revenues:						
5		Additional revenues		$55,000	$55,000	$55,000	$55,000	$55,000
6		Investment:						
7		Machine tool	-$121,000					
8		Automatic pallet changer						
9		Installation	-$30,000					
10		Tooling	-$58,000					
11		Salvage value						$80,000
12								
13		Net cash flows	-$209,000	$55,000	$55,000	$55,000	$55,000	$135,000
14		NPW	$34,657					
15	Option 2							
16				=NPV(12%,D13:H13)+C13				
17		Revenues:						
18		Additional revenues		$69,300	$69,300	$69,300	$69,300	$69,300
19		Setup savings		$4,700	$4,700	$4,700	$4,700	$4,700
20		Investment:						
21		Machine tool	-$121,000					
22		Automatic pallet changer	-$66,600					
23		Installation	-$42,000					
24		Tooling	-$65,000					
25		Salvage value						$120,000
26								
27		Net cash flows	-$294,600	$74,000	$74,000	$74,000	$74,000	$194,000
28		NPW	$40,245					
29	Option 3							
30								
31		Revenues:						
32		Additional revenues		$36,000	$40,680	$45,968	$51,944	$58,697
33		Direct labor savings		$17,300	$19,549	$22,090	$24,962	$28,207
34		Setup savings		$4,700	$5,311	$6,001	$6,782	$7,663
35		Investment:						
36		Machine tool	-$121,000					
37		Automatic pallet changer	-$66,600					
38		Installation	-$42,000					
39		Tooling	-$65,000					
40		Salvage value						$120,000
41								
42		Net cash flows	-$294,600	$58,000	$65,540	$74,060	$83,688	$214,567
43		NPW	$37,085					
44								

- For option M2,

$$PW(12\%)_{M2} = -\$294,600 + \$74,000(P/A, 12\%, 5)$$
$$+ \$120,000(P/F, 12\%, 5)$$
$$= \$40,245$$

- For option M3,

$$PW(12\%)_{M3} = -\$294,600 + \$58,000(P/A_1 \ 13\%, 12\%, 5)$$
$$+ \$120,000(P/F, 12\%, 5)$$
$$= \$37,085$$

Clearly, option M2 is the most profitable. Given the nature of BC parts and shop orders, management decides that the best way to expand would be with an automatic pallet changer, but without task sharing. Table 5.6 demonstrates how an Excel worksheet might be created to compare these mutually exclusive alternatives.

5.5.7 Analysis Period Differs from Project Lives

In Example 5.9, we assumed the simplest scenario possible when analyzing mutually exclusive projects: The projects had useful lives equal to each other and to the required service period. In practice, this is seldom the case. Often, project lives do not match the required analysis period or do not match each other (or both). For example, two machines may perform exactly the same function, but one lasts longer than the other, and both of them last longer than the analysis period over which they are being considered. In the sections and examples that follow, we will develop some techniques for dealing with these complications.

Case I: Project's Life Is Longer Than Analysis Period

Project lives rarely conveniently coincide with a firm's predetermined required analysis period; they are often too long or too short. The case of project lives that are too long is the easier one to address.

Consider the case of a firm that undertakes a five-year production project when all of the alternative equipment choices have useful lives of seven years. In such a case, we analyze each project for only as long as the required service period (in this case, five years). We are then left with some unused portion of the equipment (in this case, two years' worth), which we include as salvage value in our analysis. **Salvage value** is the amount of money for which the equipment could be sold after its service to the project has been rendered. Alternatively, salvage value is the dollar measure of the remaining usefulness of the equipment.

A common instance of project lives that are longer than the analysis period occurs in the construction industry: A building project may have a relatively short completion time, but the equipment that is used in construction has a much longer useful life.

EXAMPLE 5.10 Service Projects: Project Lives Longer Than the Analysis Period

Waste Management Company (WMC) has won a contract that requires the firm to remove radioactive material from government-owned property and transport it to a designated dumping site. This task requires a specially made ripper–bulldozer to dig and load the material onto a transportation vehicle. Approximately 400,000 tons of waste must be moved in a period of two years.

- Model A costs $150,000 and has a life of 6,000 hours before it will require any major overhaul. Two units of model A would be required to remove the material within two years, and the operating cost for each unit would run to $40,000/year for 2,000 hours of operation. At this operational rate, the model would be operable for three years, at the end of which time it is estimated that the salvage value will be $25,000 for each machine.

- A more efficient model B costs $240,000 each, has a life of 12,000 hours without any major overhaul, and costs $22,500 to operate for 2,000 hours per year to complete the job within two years. The estimated salvage value of model B at the end of six years is $30,000. Once again, two units of model B would be required to remove the material within two years.

Since the lifetime of either model exceeds the required service period of two years (Figure 5.15), WMC has to assume some things about the used equipment at the end of that time. Therefore, the engineers at WMC estimate that, after two years, the model A units could be sold for $45,000 each and the model B units for $125,000 each. After considering all tax effects, WMC summarizes the resulting cash flows (in thousands of dollars) for each project as follows.

Period	Model A	Model B
0	−$300	−$480
1	−80	−45
2	−80 +90	−45 +250
3	−80 +50	−45
4		−45
5		−45
6		−45 +60

Here, the figures in the boxes represent the estimated salvage values at the end of the analysis period (the end of year 2). Assuming that the firm's MARR is 15%, which option would be acceptable?

SOLUTION

Given: Cash flows for two alternatives, as shown in Figure 5.15, and $i = 15\%$ per year.
Find: NPW for each alternative and which alternative is preferred.

First, note that these are service projects, so we can assume the same revenues for both configurations. Since the firm explicitly estimated the market values of the assets at the end of the analysis period (two years), we can compare the two models directly. Because the benefits (removal of the waste) are equal, we can concentrate on the costs:

$$PW(15\%)_A = -\$300 - \$80(P/A, 15\%, 2) + \$90(P/F, 15\%, 2)$$
$$= -\$362$$
$$PW(15\%)_B = -\$480 - \$45(P/A, 15\%, 2) + \$250(P/F, 15\%, 2)$$
$$= -\$364$$

Model A is cheaper to operate than Model B, and thus would be preferred.

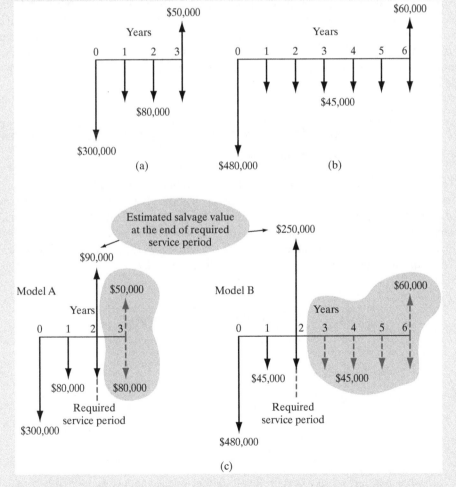

Figure 5.15 (a) Cash flow for model A; (b) cash flow for model B; (c) comparison of service projects with unequal lives when the required service period is shorter than the individual project life (Example 5.10).

Case II: Project's Life is Shorter Than Analysis Period

When project lives are shorter than the required service period, we must consider how, at the end of the project lives, we will satisfy the rest of the required service period. Replacement projects—additional projects to be implemented when the initial project has reached the limits of its useful life—are needed in such a case. A sufficient number of replacement projects that match or exceed the required service period must be analyzed.

To simplify our analysis, we could assume that the replacement project will be exactly the same as the initial project, with the same costs and benefits. However, this assumption is not necessary. For example, depending on our forecasting skills, we may decide that a different kind of technology—in the form of equipment, materials, or processes—is a preferable replacement. Whether we select exactly the same alternative or a new technology as the replacement project, we are ultimately likely to have some unused portion of the equipment to consider as salvage value, just as in the case when the project lives are longer than the analysis period. Of course, we may instead decide to lease the necessary equipment or subcontract the remaining work for the duration of the analysis period. In this case, we can probably match our analysis period and not worry about salvage values.

In any event, at the outset of the analysis period, we must make some initial guess concerning the method of completing the analysis. Later, when the initial project life is closer to its expiration, we may revise our analysis with a different replacement project. This is only reasonable, since economic analysis is an ongoing activity in the life of a company and an investment project, and we should always use the most reliable, up-to-date data we can reasonably acquire.

EXAMPLE 5.11 Service Projects: Project Lives Shorter Than the Analysis Period

The Smith Novelty Company, a mail-order firm, wants to install an automatic packaging system to fill the mail orders. The firm has a choice between two different types of packaging machines. The two machines are designed differently, but have identical capacities and do exactly the same job. The $12,500 semiautomatic model A will last three years, while the fully automatic model B will cost $15,000 and last four years. The expected operating cash flows for the two machines, including maintenance, salvage value, and tax effects, are as follows.

n	Model A	Model B
0	−$12,500	−$15,000
1	−5,000	−4,000
2	−5,000	−4,000
3	−5,000 + 2,000	−4,000
4		−4,000 + 1,500
5 Analysis period		

As business grows to a certain level, neither of the models may be able to handle the expanded volume at the end of year 5. If that happens, a fully computerized packaging system will need to be installed to handle the increased business volume. In the scenario just presented, which model should the firm select at MARR = 15%?

SOLUTION

Given: Cash flows for two alternatives, analysis period of five years, and $i = 15\%$.

Find: NPW of each alternative and which alternative to select.

STRATEGY: Since both models have a shorter life than the required service period of five years, we need to make an explicit assumption of how the service requirement is to be met. Suppose that the company considers leasing equipment comparable to model A at an annual payment of $6,000 (after taxes) and with an annual operating cost of $5,000 for the remaining required service period. In this case, the cash flow would look like that shown in Figure 5.16.

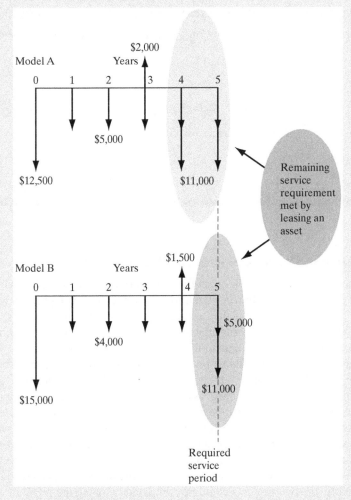

Figure 5.16   Comparison for service projects with unequal lives when the required service period is longer than the individual project life (Example 5.11).

n	Model A	Model B
0	−$12,500	−$15,000
1	−5,000	−4,000
2	−5,000	−4,000
3	−5,000 + 2,000	−4,000
4	−5,000 − ⌷6,000⌷	−4,000 + 1,500
5	−5,000 − ⌷6,000⌷	−5,000 − ⌷6,000⌷

Here, the boxed figures in the table represent the annual lease payments. (It costs $6,000 to lease the equipment and $5,000 to operate it annually. Other maintenance costs will be paid by the leasing company.) Note that both alternatives now have the same required service period of five years. Therefore, we can use NPW analysis:

$$PW(15\%)_A = -\$12,500 - \$5,000(P/A, 15\%, 2) - \$3,000(P/F, 15\%, 3)$$

$$-\$11,000(P/A, 15\%, 2)(P/F, 15\%, 3)$$

$$= -\$34,359$$

$$PW(15\%)_B = -\$15,000 - \$4,000(P/A, 15\%, 3) - \$2,500(P/F, 15\%, 4)$$

$$-\$11,000(P/F, 15\%, 5)$$

$$= -\$31,031$$

Since these are service projects, model B is the better choice.

Case III: Analysis Period Coincides with Longest Project Life

As seen in the preceding examples, equal future periods are generally necessary to achieve comparability of alternatives. In some situations, however, revenue projects with different lives can be compared directly if they require only a one-time investment, because the task or need within the firm is a one-time task or need. An example of this situation is the extraction of a fixed quantity of a natural resource, such as oil or coal.

Consider two mutually exclusive processes: One requires ten years to recover some coal, and the other can accomplish the task in only eight years. There is no need to continue the project if the short-lived process is used and all the coal has been retrieved. In this example, the two processes can be compared over an analysis period of ten years (the longest project life of the two being considered), assuming that no cash flows after eight years for the shorter-lived project. Because of the time value of money, the revenues must be included in the analysis, even if the price of coal is constant. Even if the total (undiscounted) revenue is equal for either process, that for the faster process has a larger present worth. Therefore, the two projects could be compared by using the NPW of each over its own life. Note that in this case, the analysis period is determined by, and coincides with, the longest project life in the group. (Here we are still, in effect, assuming an analysis period of ten years.)

EXAMPLE 5.12 Revenue Projects: Analysis Period Coincides with the Project with the Longest Life in the Mutually Exclusive Group

The family-operated Foothills Ranching Company (FRC) owns the mineral rights to land used for growing grain and grazing cattle. Recently, oil was discovered on this property. The family has decided to extract the oil, sell the land, and retire. The company can lease the necessary equipment and extract and sell the oil itself, or it can lease the land to an oil-drilling company.

- **Drill option.** If the company chooses to drill, it will require $300,000 leasing expenses up front, but the net annual cash flow after taxes from drilling operations will be $600,000 at the end of each year for the next five years. The company can sell the land for a net cash flow of $1,000,000 in five years, when the oil is depleted.
- **Lease option.** If the company chooses to lease, the drilling company can extract all the oil in only three years, and FRC can sell the land for a net cash flow of $800,000 at that time. (The difference in resale value of the land is due to the increasing rate of land appreciation anticipated for this property.) The net cash flow from the lease payments to FRC will be $630,000 at the *beginning* of each of the next three years.

All benefits and costs associated with the two alternatives are shown in Figure 5.17. Which option should the firm select at $i = 15\%$?

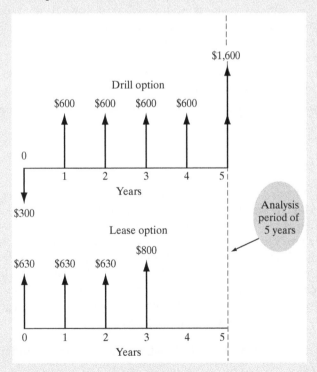

Figure 5.17 Comparison of revenue projects with unequal lives when the analysis period coincides with the project with the longest life in the mutually exclusive group (Example 5.12). In our example, the analysis period is five years, assuming no cash flow in years 4 and 5 for the lease option.

SOLUTION

Given: Cash flows shown in Figure 5.17 and $i = 15\%$ per year.
Find: NPW of each alternative and which alternative to select.

As illustrated in Figure 5.17, after depletion of the oil, the project will terminate. The present worth of each of the two options is as follows:

$$PW(15\%)_{Drill} = -\$300,000 + \$600,000(P/A, 15\%, 4)$$
$$+ \$1,600,000(P/F, 15\%, 5)$$
$$= \$2,208,470$$

$$PW(15\%)_{Lease} = \$630,000 + \$630,000(P/A, 15\%, 2)$$
$$+ \$800,000(P/F, 15\%, 3)$$
$$= \$2,180,210$$

Note that these are revenue projects; therefore, the drill option appears to be the marginally better option.

COMMENTS: The relatively small difference between the two NPW amounts ($28,260) suggests that the actual decision between drilling and leasing might be based on noneconomic issues. Even if the drilling option is slightly better, the company might prefer to forgo the small amount of additional income and select the lease option, rather than undertake an entirely new business venture and do its own drilling. A variable that might also have a critical effect on this decision is the sales value of the land in each alternative. The value of land is often difficult to forecast over any long period, and the firm may feel some uncertainty about the accuracy of its guesses. In Chapter 12, we will discuss sensitivity analysis, a method by which we can factor uncertainty about the accuracy of project cash flows into our analysis.

5.5.8 Analysis Period is Not Specified

Our coverage so far has focused on situations in which an analysis period is known. When an analysis period is not specified, either explicitly by company policy or practice or implicitly by the projected life of the investment project, it is up to the analyst to choose an appropriate one. In such a case, the most convenient procedure is to choose an analysis on the basis of the useful lives of the alternatives. When the alternatives have equal lives, this is an easy selection. When the lives of at least some of the alternatives differ, we must select an analysis period that allows us to compare projects with different lifetimes on an equal time basis—that is, a **common service period**.

Case I: Project Repeatability Likely—Lowest Common Multiple of Project Lives

A required service period of infinity may be assumed if we anticipate that an investment project will be proceeding at roughly the same level of production for some indefinite period. It is certainly possible to make this assumption mathematically, although the analysis is likely to be complicated and tedious. Therefore, in the case of an indefinitely ongoing investment project, we typically select a finite analysis period by using the **lowest common multiple** of project lives. For example, if alternative A has a three-year useful life and alternative B has a four-year life, we may select 12 years as the analysis or common service period. We would consider alternative A through four life-cycles and alternative B through three life-cycles; in each case, we would use the alternatives completely. We then accept the finite model's results as a good prediction of what will be the economically wisest course of action for the foreseeable future. Example 5.13 is a case in which we conveniently use the lowest common multiple of project lives as our analysis period.

EXAMPLE 5.13 Unequal Lives, Lowest-Common-Multiple Method

Consider Example 5.11. Suppose that models A and B can each handle the increased future volume and that the system is not going to be phased out at the end of five years. Instead, the current mode of operation is expected to continue indefinitely. Suppose also that the two models will be available in the future without significant changes in price and operating costs. At MARR $= 15\%$, which model should the firm select?

SOLUTION
Given: Cash flows for two alternatives, as shown in Figure 5.18, $i = 15\%$ per year, and an indefinite period of need.
Find: NPW of each alternative and which alternative to select.

Recall that the two mutually exclusive alternatives have different lives, but provide identical annual benefits. In such a case, we ignore the common benefits and can make the decision solely on the basis of costs, as long as a common analysis period is used for both alternatives.

 To make the two projects comparable, let's assume that, after either the three- or four-year period, the system would be reinstalled repeatedly, using the same model, and that the same costs would apply. The lowest common multiple of three and four is 12, so we will use 12 years as the common analysis period. Note that any cash flow difference between the alternatives will be revealed during the first 12 years. After that, the same cash flow pattern repeats every 12 years for an indefinite period. The replacement cycles and cash flows are shown in Figure 5.18. Here is our analysis.

- **Model A.**   Four replacements occur in a 12-year period. The PW for the first investment cycle is

$$PW(15\%) = -\$12,500 - \$5,000(P/A, 15\%, 3)$$
$$+ \$2,000(P/F, 15\%, 3)$$
$$= -\$22,601$$

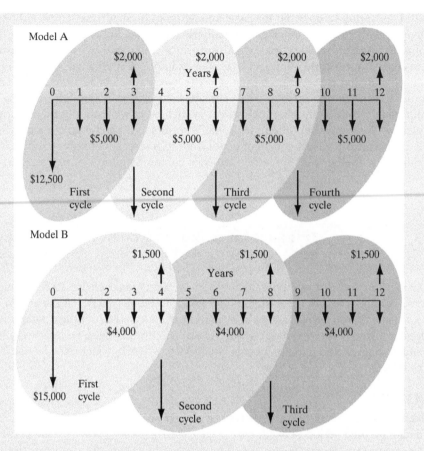

Figure 5.18 Comparison of projects with unequal lives when the required service period is infinite and the project is likely to be repeatable with the same investment and operations and maintenance costs in all future replacement cycles (Example 5.13).

With four replacement cycles, the total PW is

$$PW(15\%) = -\$22,601[1 + (P/F, 15\%, 3)$$

$$+ (P/F, 15\%, 6) + (P/F, 15\%, 9)]$$

$$= -\$53,657$$

- **Model B.** Three replacements occur in a 12-year period. The PW for the first investment cycle is

$$PW(15\%) = -\$15,000 - \$4,000(P/A, 15\%, 4)$$

$$+ \$1,500(P/F, 15\%, 4)$$

$$= -\$25,562$$

With three replacement cycles in 12 years, the total PW is

$$PW(15\%) = -\$25,562[1 + (P/F, 15\%, 4) + (P/F, 15\%, 8)]$$

$$= -\$48,534$$

Clearly, model B is a better choice, as before.

COMMENTS: In Example 5.13, an analysis period of 12 years seems reasonable. The number of actual reinvestment cycles needed with each type of system will depend on the technology of the future system, so we may or may not actually need the four (model A) or three (model B) reinvestment cycles we used in our analysis. The validity of the analysis also depends on the costs of the system and labor remaining constant. If we assume **constant-dollar prices** (see Chapter 11), this analysis would provide us with a reasonable result. (As you will see in Example 6.3, the annual equivalent worth approach makes it mathematically easier to solve this type of comparison.) If we cannot assume constant-dollar prices in future replacements, we need to estimate the costs for each replacement over the analysis period. This will certainly complicate the problem significantly.

Case II: Project Repeatability Unlikely—Other Common Analysis Periods

In some cases, the lowest common multiple of project lives is an unwieldy analysis period to consider. Suppose, for example, that you were considering two alternatives with lives of seven and 12 years, respectively. Besides making for tedious calculations, an 84-year analysis period may lead to inaccuracies, since, over such a long time, we can be less and less confident about the ability to install identical replacement projects with identical costs and benefits. In a case like this, it would be reasonable to use the useful life of one of the alternatives by either factoring in a replacement project or salvaging the remaining useful life, as the case may be. The important idea is that we must compare both projects on the same time basis.

SUMMARY

In this chapter, we presented the concept of present-worth analysis, based on cash flow equivalence along with the payback period. We observed the following important results.

- Present worth is an equivalence method of analysis in which a project's cash flows are discounted to a single present value. It is perhaps the most efficient analysis method we can use in determining the acceptability of a project on an economic basis. Other analysis methods, which we will study in Chapters 6 and 7, are built on a sound understanding of present worth.

- The minimum attractive rate of return (MARR) is the interest rate at which a firm can always earn or borrow money under a normal operating environment. It is generally dictated by management

and is the rate at which NPW analysis should be conducted.

- Revenue projects are those for which the income generated depends on the choice of project. Service projects are those for which the income remains the same, regardless which project is selected.

- Several alternatives that meet the same need are **mutually exclusive**; if one of them is selected, the others will be rejected.

- When not specified by management or company policy, the analysis period to use in a comparison of mutually exclusive projects may be chosen by an individual analyst. Several efficiencies can be applied when an analysis period is selected. In general, the analysis period should be chosen to cover the required service period, as highlighted in Figure 5.14.

PROBLEMS

Note: *Unless otherwise stated, all cash flows are after-tax cash flows. The interest rate (MARR) is also given on an after-tax basis.*

Identifying Cash Inflows and Outflows

5.1 Business at your design engineering firm has been brisk. To keep up with the increasing workload, you are considering the purchase of a new state-of-the-art CAD/CAM system costing $500,000, which would provide 5,000 hours of productive time per year. Your firm puts a lot of effort into drawing new product designs. At present, this is all done by design engineers on an old CAD/CAM system installed five years ago. If you purchase the system, 40% of the productive time will be devoted to drawing (CAD) and the remainder to CAM. While drawing, the system is expected to out-produce the old CAD/CAM system by a factor of 3:1. You estimate that the additional annual out-of-pocket cost of maintaining the new CAD/CAM system will be $175,000, including any tax effects. The expected useful life of the system is eight years, after which the equipment will have no residual value. As an alternative, you could hire more design engineers. Each normally works 2,000 hours per year, and 60% of this time is productive. The total cost for a design engineer is $45 per hour. There are five design engineers.) Identify the net cash flows (benefits and costs) associated with the drawing activities if the CAD/CAM system is purchased instead of hiring more design engineers.

5.2 Camptown Togs, Inc., a children's clothing manufacturer, has always found payroll processing to be costly because it must be done by a clerk. The number of piece–goods coupons received by each employee is collected and the types of tasks performed by each employee are calculated. Not long ago, an industrial engineer designed a system that partially automates the process by means of a scanner that reads the piece–goods coupons. Management is enthusiastic about this system, because it utilizes some personal computer systems that were purchased recently. It is expected that this new automated system will save $45,000 per year in labor. The new system will cost about $30,000 to build and test prior to operation. It is expected that operating costs, including income taxes, will be about $5,000 per year. The system will have a five-year useful life. The expected net salvage value of the system is estimated to be $3,000.

(a) Identify the cash inflows over the life of the project.
(b) Identify the cash outflows over the life of the project.
(c) Determine the net cash flows over the life of the project.

Payback Period

Unless otherwise mentioned, problems use "payback period" to mean "conventional payback period."

5.3 If a project costs $120,000 and is expected to return $40,000 annually, how long does it take to recover the initial investment? What would be the discounted payback period at $i = 15\%$?

5.4 Refer to Problem 5.2, and answer the following questions:

(a) How long does it take to recover the investment?
(b) If the firm's interest rate is 15% after taxes, what would be the discounted payback period for this project?

5.5 Adidas will put on sale what it bills as the world's first computerized "smart shoe." But consumers will decide whether to accept the bionic running shoe's $250 price tag—four times the average shoe price at stores such as Foot Locker. Adidas uses a sensor, a microprocessor, and a motorized cable system to automatically adjust the shoe's cushioning. The sensor under the heel measures compression and decides whether the shoe is too soft or firm. That information is sent to the microprocessor, and while the shoe is in the air, the cable adjusts the heel cushion. The whole system weighs less than 40 grams. Adidas's computer-driven shoe—three years in the making—is the latest innovation in the $16.4 billion U.S. sneaker industry. The top-end running shoe from New Balance lists for $199.99. With runners typically replacing shoes every 500 miles, the $250 Adidas could push costs to 50 cents per mile. Adidas is spending an estimated $20 million on the rollout.

The investment required to develop a full-scale commercial rollout cost Adidas $70 million (including the $20 million ad campaign), which will be financed at an interest rate of 10%. With a price tag of $250, Adidas will have about $100 net cash profit from each sale. The product will have a five-year market life. Assuming that the annual demand for the product remains constant over the market life, how many units does Adidas have to sell each year to pay off the initial investment and interest?

5.6 Consider the investment projects in Table P5.6, all of which have a four-year investment life.

TABLE P5.6

	Project's Cash Flow ($)			
n	**A**	**B**	**C**	**D**
0	−$3,800	−$5,200	−$3,800	−$2,800
1	0	$2,200	−$1,600	−$800
2	0	$3,800	$6,700	−$,600
3	0	$4,400	$6,800	$9,400
4	$12,500	$3,600	$6,900	$8,800

(a) What is the payback period of each project?

(b) What is the discounted payback period at an interest rate of 15% for each project?

NPW Criterion

5.7 What is the present worth of the project, which requires $130,000 investment now and receives $40,000 every year for six years at an interest rate of 14%?

5.8 Larson Manufacturing is considering purchasing a new injection-molding machine for $250,000 to expand its production capacity. It will cost an additional $20,000 to do the site preparation. With the new injection-molding machine installed, Larson Manufacturing expects to increase its revenue by $90,000. The machine will be used for five years, with an expected salvage value of $75,000. At an interest rate of 12%, would the purchase of the injection-molding machine be justified?

5.9 Consider the cash flows from an investment project.

Year	Cash Flows
0	−$35,600
1	29,220
2	32,540
3	38,350
4	44,580

(a) Compute the net present worth of the project at $i = 12\%$.

(b) Plot the present worth as a function of the interest.

5.10 You need to know whether the building of a new warehouse is justified under the following conditions.

- The proposal is for a warehouse costing $250,000.
- The warehouse has an expected useful life of 35 years and a net salvage value (net proceeds from sale after tax adjustments) of $50,000.
- Annual receipts of $67,000 are expected, annual maintenance and administrative costs will be $12,000/year, and annual income taxes are $15,000.

Given the foregoing data, which of the following statements are correct?

(a) The proposal is justified for a MARR of 15%.

(b) The proposal has a net present worth of −$50,254 when 20% is used as the interest rate.

(c) The proposal is acceptable, as long as MARR ≤ 15.93%.

(d) All of the preceding are correct.

5.11 Your firm is considering purchasing an old office building with an estimated remaining service life of 25 years. Recently, the tenants signed a long-term lease, which leads you to believe that the current rental income of $250,000 per year will remain constant for the first five years. Then the rental income will increase by 10% for every five-year interval over the remaining life of the asset. That is, the annual rental income would be $275,000 for years 6 through 10, $302,500 for years 11 through 15, $332,750 for years 16 through 20, and $366,025 for years 21 through 25. You estimate that operating expenses, including income taxes, will be $85,000 for the first year and that they will increase by $5,000 each year thereafter. You also estimate that razing the building and selling the lot on which it stands will realize a net amount of $50,000 at the end of the 25-year period. If you had the opportunity to invest your money elsewhere and thereby earn interest at the rate of 12% per annum, what would be the maximum amount you would be willing to pay for the building and lot at the present time?

5.12 Beloit Co. is a manufacturer of mini-doughnut machine makers. Early in 2015 a customer asked Beloit to quote a price for a custom-designed doughnut machine to be delivered by the end of 2015. Once purchased, the customer intends to place the machine in service in January 2016 and will use it for four years. The expected annual operating net cash flow is estimated to be $120,000. The expected salvage value of the equipment at the end of four years is about 10% of the initial purchase price. To expect a 15% required rate of return on investment, what would be the maximum amount that should be spent on purchasing the doughnut machine?

5.13 FootballComm LLC is a manufacturer of devices for football coaches and their coaching staffs that enables them to quickly and effectively communicate with one another during a game through a wireless sideline headset system. Currently, the company purchases all of the headsets included in their systems from a China-based supplier, but they have experienced a significant decrease in overall headset quality. To correct the problem, the company is considering producing their own headsets in-house. The company estimates that the transition will take place over 12 months. The system is expected to have an eight-year service life and produce savings and expenditures given in Table P5.13.

If the firm's MARR is 15%, use the NPW method to calculate the economic worth of producing the headsets in house.

TABLE P5.13

In-House Headset Manufacturing	
Investment	
Now (building)	$500,000
First year (equipment and facilities)	$2,200,000
Annual savings in materials and quality inspection	$5,000,000
Incremental annual expenses	$1,500,000
Incremental annual income taxes	$800,000
Economic service life	8 years
Net salvage value	$1,500,000

5.14 A large food-processing corporation is considering using laser technology to speed up and eliminate waste in the potato-peeling process. To implement the system, the company anticipates needing $3.5 million to purchase the industrial-strength lasers. The system will save $1,550,000 per year in labor and materials. However, it will require an additional operating and maintenance cost of $350,000. Annual income taxes will also increase by $150,000. The system is expected to have a 10-year service life and will have a salvage value of about $200,000. If the company's MARR is 18%, use the NPW method to justify the project.

5.15 You are considering purchasing a new injection-molding machine. This machine will have an estimated service life of 12 years with a negligible after-tax salvage value. Its annual net after-tax operating cash flows are estimated to be $75,000. At a 15% required rate of return on investment, what would be the maximum amount that should be spent on purchasing the injection-molding machine?

Future Worth and Project Balance

5.16 Consider the following investment project.

n	A_n
0	−$55,000
1	42,500
2	44,600
3	42,800
4	42,400
5	42,200

Calculate the net future worth of this investment at 14% and determine the acceptability of the investment.

5.17 Consider the sets of investment projects in Table P5.17, all of which have a three-year investment life.

TABLE P5.17

Period	Project's Cash Flow			
(n)	A	B	C	D
0	−$12,500	−12,500	22,500	−14,000
1	6,400	−3,000	−7,000	7,500
2	14,400	23,000	−2,500	4,500
3	7,200	13,000	4,000	8,500

(a) Compute the net future worth of each project at $i = 15\%$. Which project or projects are acceptable?

(b) Compute the terminal project balance for each project at $i = 15\%$, and compare them with the net future worth of the projects in (a). What conclusions can you reach?

5.18 Which of the following investment options will maximize your future wealth at the end of 18 years? Assume any funds that remain invested will earn a nominal rate of 12% compounded monthly.

(a) Deposit $8,000 now.

(b) Deposit $120 at the end of each month for the first 12 years.

(c) Deposit $105 at the end of each month for 18 years.

(d) Deposit a lump sum in the amount of $35,000 at the end of year 12.

5.19 Consider the project balances in Table P5.19 for a typical investment project with a service life of four years.

TABLE P5.19

n	A_n	Project Balance
0	−$10,000	−$10,000
1	—	−9,000
2	—	−7,000
3	1,840	−6,000
4	—	100

(a) Determine the interest rate used in computing the project balance.

(b) Reconstruct the original cash flows of the project.

(c) Would the project be acceptable at $i = 15\%$?

5.20 Your R&D group has developed and tested a computer software package that helps engineers control the proper chemical mix for various process-manufacturing industries. If you decide to market the software, your first-year operating net cash flow is estimated to be $1,200,000. Because of market competition, product life will be about four years, and the product's market share will decrease by 25% each year over the previous year's share.

(a) To protect the erosion of the market share, you need to upgrade the software every year at the expense of $200,000. Is it worth upgrading the software? Your interest rate is 15%.

(b) You are approached by a big software house which wants to purchase the right to produce and distribute the software. For what minimum price would you be willing to sell the software?

5.21 Consider the accompanying project balance diagram for a typical investment project with a service life of five years (Figure P5.21). The numbers in the figure indicate the beginning project balances.

(a) From the project balance diagram, reconstruct the project's original cash flows.

(b) What is the project's conventional payback period (without interest)?

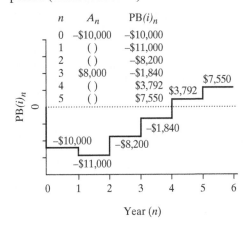

n	A_n	$PB(i)_n$
0	−$10,000	−$10,000
1	( )	−$11,000
2	( )	−$8,200
3	$8,000	−$1,840
4	( )	$3,792
5	( )	$7,550

Figure P5.21

5.22 Consider the following cash flows and present-worth profile.

(a) Determine the values of X and Y.

(b) Calculate the terminal project balance of project 1 at MARR = 24%.

(c) Find the values of a, b, and c in the NPW plot.

TABLE P5.22

Year	Net Cash Flows ($)	
	Project 1	Project 2
0	−$1,000	−$1,000
1	400	300
2	800	Y
3	X	800

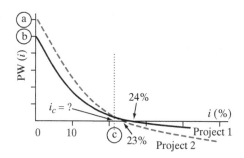

Figure P5.22

TABLE P5.25

| n | Project's Cash Flow | | | |
	A	B	C	D
0	−$1,000	−$1,000	−$1,000	−$1,000
1	0	594	687	500
2	0	594	588	600
3	1,965	594	487	700

5.23 Consider the project balances for a typical investment project with a service life of five years, as shown in Table P5.23.

(a) Determine the interest rate used in the project balance calculation, and compute the present worth of this project at the computed interest rate.

(b) Construct the original cash flows of the project and the terminal balance, and fill in the blanks in Table P5.23.

TABLE P5.23 Investment Project Balances

n	A_n	Project Balance
0	−$3,000	−$3,000
1	ⓐ	−2,700
2	1,470	−1,500
3	ⓑ	0
4	ⓒ	−300
5	600	ⓓ

5.24 Refer to Problem 5.6 and answer the following questions.

(a) Graph the project balances (at $i = 12\%$) of each project as a function of n.

(b) By examining the graphical results in part (a), determine which project appears to be the safest to undertake if there is some possibility of premature termination of the projects at the end of year 2.

5.25 Consider the following investment projects using the information in Table P5.25.

(a) Compute the future worth at the end of life for each project at $i = 10\%$.

(b) Determine the discounted payback period for each project.

(c) Compare the area of negative project balance for each project.

(d) Which project would you prefer and why?

5.26 Suppose that you undertook an investment project with the following estimated cash flows:

n	A_n
0	−$152,000
1	42,400
2	53,400
3	52,500
4	62,500
5	33,000

At the end of year 3, you have received an offer to buy the project from another firm; what minimum price would be the fair price to ask? Here your required return on investment is 15%.

5.27 Covington Corporation purchased a vibratory finishing machine for $20,000 in year 0. The useful life of the machine is 10 years, at the end of which the machine is estimated to have a salvage value of zero. The machine generates net annual revenues of $6,000. The annual operating and maintenance expenses are estimated to be $1,000. If Covington's MARR is 15%, how many years will it take before this machine becomes profitable?

5.28 Gene Research, Inc., just finished a four-year program of R&D and clinical trials. It expects a quick approval from the Food and Drug Administration. If Gene markets the product on its own, the company will require $30 million immediately ($n = 0$) to build a new manufacturing facility, and it is expected to have a 10-year product life. The R&D expenditure in the previous years and the anticipated revenues that the company can generate over the next 10 years are summarized in Table P5.28.

TABLE P5.28

Period (n)	Cash Flow (Unit: $ million)
−4	−$10
−3	−10
−2	−10
−1	−10
0	−10 − 30
1–10	100

Merck, a large drug company, is interested in purchasing the R&D project and the right to commercialize the product from Gene Research, Inc.; it wants to do so immediately ($n = 0$). What would be a starting negotiating price for the project from Merck? Assume that Gene's MARR = 20%.

5.29 Consider the independent investment projects in Table P5.29.

Compute the project worth of each project at the end of six years with variable MARRs as follows: 10% for $n = 0$ to $n = 3$ and 15% for $n = 4$ to $n = 6$.

TABLE P5.29

	Project Cash Flows		
n	A	B	C
0	−$800	−$500	$200
1	150	150	−40
2	150	150	−60
3	350	150	−140
4	−200	100	—
5	500	100	—
6	400	—	—

5.30 Consider the project balance profiles shown in Table P5.30 for proposed investment projects. Project balance figures are rounded to nearest dollars.

(a) Compute the net present worth of projects A and C.
(b) Determine the cash flows for project A.
(c) Identify the net future worth of project C.
(d) What interest rate would be used in the project balance calculations for project B?

TABLE P5.30 Profiles for Proposed Investment Projects

	Project Balances		
n	A	B	C
0	−$1,000	−$1,000	−$1,000
1	−1,000	−650	−1,200
2	−900	−348	−1,440
3	−690	−100	−1,328
4	−359	85	−1,194
5	105	198	−1,000
Interest rate used	10%	?	20%
NPW	?	$79.57	?

5.31 Consider the project balance profiles for proposed investment projects in Table P5.31.

Project balance figures are rounded to the nearest dollar.

(a) Compute the net present worth of each investment.

(b) Determine the project balance for project C at the end of period 2 if $A_2 = \$500$.

(c) Determine the cash flows for each project.

(d) Identify the net future worth of each project.

TABLE P5.31 **Profiles for Proposed Investment Projects**

	Project Balances		
n	A	B	C
0	−$1,000	−$1,000	−$1,000
1	−800	−680	−530
2	−600	−302	X
3	−400	−57	−211
4	−200	233	−89
5	0	575	0
Interest rate used	0%	18%	12%

Capitalized Equivalent Worth

5.32 Maintenance money for an athletic complex has been sought. Mr. Kendall, Athletic Director, would like to solicit a donation to cover all future expected maintenance costs for the building. These maintenance costs are expected to be $1million each year for the first five years, $1.3 million each year for years 6 through 10, and $1.5 million each year after that. (The building has an indefinite service life.)

If the money is placed in an account that will pay 5% interest compounded annually, how large should the gift be?

5.33 Geo-Star Manufacturing Company is considering a new investment in a punch-press machine that will cost $100,000 and has an annual maintenance cost of $10,000. There is also an additional overhauling cost of $20,000 for the equipment once every four years. Assuming that this equipment will last

infinitely under these conditions, what is the capitalized equivalent cost of this investment at an interest rate of 10%?

5.34 Consider an investment project, the cash flow pattern of which repeats itself every five years forever, as shown in the accompanying diagram (Figure P5.34). At an interest rate of 12%, compute the capitalized equivalent amount for this project.

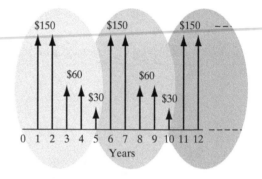

Figure P5.34

5.35 A group of concerned citizens has established a trust fund that pays 7.5% interest, compounded monthly, to preserve a historical building by providing annual maintenance funds of $140,000 forever. Compute the capitalized equivalent amount for these building maintenance expenses.

5.36 A newly constructed bridge costs $15,000,000. The same bridge is estimated to need renovation every 15 years at a cost of $3,000,000. Annual repairs and maintenance are estimated to be $1,000,000 per year.

(a) If the interest rate is 5%, determine the capitalized cost of the bridge.

(b) Suppose that in (a), the bridge must be renovated every 20 years, not every 15 years. What is the capitalized cost of the bridge?

(c) Repeat (a) and (b) with an interest rate of 10%. What can you say about the effect of interest on the results?

5.37 To decrease the costs of operating a lock in a large river, a new system of operation is proposed. The system will cost $830,000 to design and build. It is estimated that it will have to be reworked every 10 years at a cost of $120,000. In addition, an expenditure of $80,000 will have to be made at the end of the fifth year for a new type of gear that will not be available until then. Annual operating costs are expected to be $70,000 for the first 15 years and $100,000 a year thereafter. Compute the capitalized cost of perpetual service at $i = 7\%$.

Mutually Exclusive Alternatives

5.38 You are considering two investment options. In option A, you have to invest $5,000 now and $1,000 three years from now. In option B, you have to invest $3,500 now, $1,500 a year from now, and $1,000 three years from now. In both options, you will receive four annual payments of $2,000 each. (You will get the first payment a year from now.) Which of these options would you choose based on (a) the conventional payback criterion, and (b) the present worth criterion, assuming 10% interest?

5.39 Consider the cash flow data in Table P5.39 for two competing investment projects. At $i = 14\%$, which of the two projects would be a better choice?

5.40 Consider the cash flows in Table P5.40 for the following investment projects.

(a) Suppose projects A and B are mutually exclusive. On the basis of the NPW criterion, which project would be selected? Assume that MARR = 15%.

TABLE P5.39

	Cash Flow Data (Unit: $ thousand)	
n	Project A	Project B
0	−$1,600	−$3,300
1	−1,450	−870
2	−680	1,250
3	1,275	1,180
4	1,735	1,460
5	1,825	1,980
6	2,075	1,750
7	845	960
8	935	760
9	625	450
10	840	1,080

TABLE P5.40

	Project's Cash Flow				
n	A	B	C	D	E
0	−$1,800	−$1,500	−$4,000	$1,400	−$1,500
1	1,150	1,200	1,500	−450	500
2	700	700	X	−450	500
3	200	700	1,800	−450	500
4	200	100	X	−450	500

(b) Find the minimum value of *X* that makes project C acceptable, still using MARR = 15%.

(c) Would you accept project D at *i* = 18%?

(d) Assume that projects D and E are mutually exclusive. On the basis of the NPW criterion, which project would you select?

5.41 Consider two mutually exclusive investment projects, each with MARR = 12%, as shown in Table P5.41. On the basis of the NPW criterion, which alternative would be selected?

TABLE P5.41 **Two Mutually Exclusive Investment Projects**

	Project's Cash Flow	
n	A	B
0	−$22,500	−$16,900
1	18,610	15,210
2	15,930	16,720
3	16,300	12,500

5.42 Consider the two mutually exclusive investment projects in Table P5.42. For what range of MARR would you prefer project B?

TABLE P5.42

	Project's Cash Flow	
n	A	B
0	−$5,300	−$7,300
1	900	2.200
2	14,000	15,500

5.43 Two methods of carrying away surface runoff water from a new subdivision are being evaluated.

■ **Method A.**   Dig a ditch. The first cost would be $60,000, and $25,000 of re-digging and shaping would be required at five-year intervals forever.

■ **Method B.**   Lay concrete pipe. The first cost would be $150,000, and a replacement would be required at 50-year intervals at a net cost of $180,000 forever.

At *i* = 12%, which method is the better one?

5.44 A local car dealer is advertising a standard 24-month lease of $1,150 per month for its new XT 3000 series sports car. The standard lease requires a down payment of $4,500, plus a $1,000 refundable initial deposit *now*. The first lease payment is due at the beginning of month 1. In addition, the company offers a 24-month lease plan that has a single up-front payment of $29,500, plus a refundable initial deposit of $1,000. Under both options, the initial deposit will be refunded at the end of month 24. Assume an interest rate of 6% compounded monthly. With the present-worth criterion, which option is preferred?

5.45 You are considering two types of machines for a manufacturing process.

■ **Machine A** has a first cost of $75,200, and its salvage value at the end of six years of estimated service life is $21,000. The operating costs of this machine are estimated to be $6,800 per year. Extra income taxes are estimated at $2,400 per year.

■ **Machine B** has a first cost of $44,000, and its salvage value at the end of six years' service is estimated to be negligible. The annual operating costs will be $11,500.

Compare these two mutually exclusive alternatives by the present-worth method at *i* = 13%.

5.46 An electric motor is rated at 10 horsepower (HP) and costs $1,200. Its full-load efficiency is specified to be 85%. A newly designed high-efficiency motor of the same size has an efficiency of 90%, but it costs $1,600. It is estimated that the motors will operate at a rated 10 HP output for 2,000 hours a year, and the cost of energy will be $0.09 per kilowatt-hour. Each motor is expected to have a 15-year life. At the end of 15 years, the first motor will have a salvage value of $50 and the second motor will have a salvage value of $100. Consider the MARR to be 8%. (*Note*: 1 HP = 0.7457 kW.)

(a) Use the NPW criterion to determine which motor should be installed.

(b) In part (a), what if the motors operated 1,000 hours a year instead of 2,000 hours a year? Would the motor you chose in part (a) still be the best choice?

5.47 Consider the two mutually exclusive service projects in Table P5.47.

TABLE P5.47

	Project's Cash Flow	
n	A	B
0	−$22,000	−$27,000
1	17,500	20,500
2	17,000	28,000
3	15,000	—

Assuming that you need service of either machine for an indefinite period, which project would be selected at $i = 16\%$? Assume that both types of machine are available with the same costs and benefits during the planning horizon.

5.48 Consider the two mutually exclusive investment projects in Table P5.48, which have unequal service lives.

TABLE P5.48

	Project's Cash Flow	
n	A1	A2
0	−$900	−$1,800
1	−400	−300
2	−400	−300
3	−400 + 200	−300
4	—	−300
5	—	−300
6	—	−300 + 500

(a) What assumption(s) do you need in order to compare a set of mutually exclusive service projects with unequal service lives?

(b) With the assumption(s) defined in part (a) and using 10%, determine which project should be selected.

(c) If your analysis period (study period) is just three years, what should be the salvage value of project A2 at the end of year 3 to make the two alternatives economically indifferent?

5.49 Consider the two mutually exclusive projects in Table P5.49.

Salvage values represent the net proceeds (after tax) from disposal of the assets if they are sold at the end of each year. Both projects B1 and B2 will be available (or can be repeated) with the same costs and salvage values for an indefinite period.

(a) Assuming an infinite planning horizon, which project is a better choice at MARR = 12%?

(b) With a 10-year planning horizon, which project is a better choice at MARR = 12%?

5.50 You need a lathe for your machine shop for 10 years. You narrowed down to two models: Kendall and Toyota. You have also collected the financial data given in Table P5.50. (*Note*: The salvage values represent the values at the end of useful life.)

If your interest rate is 12%, what should be the salvage value of the Toyota model at the end of 10 years so that you would be indifferent between the two models?

TABLE P5.49

	B1		B2	
n	Cash Flow	Salvage Value	Cash Flow	Salvage Value
0	−$20,000	—	−$17,000	—
1	−2,000	10,000	−2,500	9,000
2	−2,000	8,000	−2,500	6,000
3	−2,000	5,000	−2,500	3,000
4	−2,000	3,000	—	—
5	−2,000	2,000	—	—

TABLE P5.50

	Kendall	Toyota
First Cost	$25,000	$32,000
Annual O&M Costs	$11,000	$9,700
Useful Life	10 years	14 years
Salvage Value	$3,000	$2,000

5.51 Consider the cash flows for two types of models given in Table P5.51.

Both models will have no salvage value upon their disposal (at the end of their respective service lives). The firm's MARR is known to be 12%.

TABLE P5.51

	Project's Cash Flow	
n	**Model A**	**Model B**
0	−$8,000	−$15,000
1	3,500	10,000
2	3,500	10,000
3	3,500	—

(a) Notice that the models have different service lives. However, model A will be available in the future with the same cash flows, whereas Model B is available at one time only. If you select model B now, you will have to replace it with model A at the end of year 2. If your firm uses the present worth as a decision criterion, which model should be selected, assuming that the firm will need either model for an indefinite period?

(b) Suppose that your firm will need either model for only two years. Determine the salvage value of model A at the end of year 2 that makes both models indifferent (equally likely).

5.52 An electric utility is taking bids on the purchase, installation, and operation of microwave towers. Table P5.52 has some details associated with the two bids that were received.

(a) Which is the most economical bid if the interest rate is considered to be 11%? Neither tower will have any salvage value after 20 years of use.

(b) Use the NPW method to compare these two mutually exclusive plans.

TABLE P5.52

	Cost per Tower	
	Bid A	**Bid B**
Equipment cost	$112,000	$98,000
Installation cost	$25,000	$30,000
Annual maintenance and inspection fee	$2,000	$2,500
Annual extra income taxes	—	$800
Life	40 years	35 years
Salvage value	$0	$0

5.53 Consider the following two investment alternatives given in Table P5.53. The firm's MARR is known to be 15%.

(a) Compute PW (15%) for project A1.

(b) Compute the unknown cash flow X in years 2 and 3 for project A2.

(c) Compute the project balance (at 15%) of project A1 at the end of period 3.

(d) If these two projects are mutually exclusive alternatives, which one would you select?

TABLE P5.53

	Project's Cash Flow	
n	**A1**	**A2**
0	−$20,000	−$35,000
1	14,750	-
2	16,250	X
3	12,500	X
PW(15%)	?	$12,50

5.54 Consider each of the after-tax cash flows shown in Table P5.54.

Suppose that projects B and C are mutually exclusive. Suppose also that the required service period is eight years and that the company is considering leasing comparable equipment with an annual lease expense of $4,000, payable at the end of each year for the remaining years of the required service period. Which project is a better choice at 14%?

TABLE P5.54 After-Tax Cash Flows

n	B	C
0	−$12,000	−$10,500
1	−$4,200	−$3,800
2	−$3,500	−$3,800
3	−$3,200	−$3,800
4	−$2,500	−$3,800
5	−$2,500	−$3,800
6	−$2,500	−$3,800
7	—	−$3,800
8	—	—

5.55 A mall with two levels is under construction. The plan is to install only 9 escalators at the start, although the ultimate design calls for 16. The question arises as to whether to provide necessary facilities (stair supports, wiring conduits, motor foundations, etc.) that would permit the installation of the additional escalators at the mere cost of their purchase and installation or to defer investment in these facilities until the escalators need to be installed.

■ **Option 1:** Provide these facilities now for all seven future escalators at $300,000.

■ **Option 2:** Defer the investment in the facility as needed. Install two more escalators in two years, three more in five years, and the last two in eight years. The installation of these facilities at the time they are required is estimated to cost $140,000 in year 2, $160,000 in year 5, and $180,000 in year 8.

Additional annual expenses are estimated at $7,000 for each escalator facility installed. Assume that these costs begin one year subsequent to the actual addition. At an interest rate of 12%, compare the net present worth of each option over eight years.

5.56 An electrical utility is experiencing a sharp power demand that continues to grow at a high rate in a certain local area.

Two alternatives are under consideration. Each is designed to provide enough capacity during the next 25 years, and both will consume the same amount of fuel, so fuel cost is not considered in the analysis.

• **Alternative A.** Increase the generating capacity now so that the ultimate demand can be met without additional expenditures later. An investment of $25 million would be required, and it is estimated that this plant facility would be in service for 25 years and have a salvage value of $0.85 million. The annual operating and maintenance costs (including income taxes) would be $0.4 million.

• **Alternative B.** Spend $15 million now and follow this expenditure with future additions during the 10th year and the 15th year. These additions would cost $18 million and $12 million, respectively. The facility would be sold 25 years from now with a salvage value of $1.5 million. The annual operating and maintenance costs (including income taxes) will be $250,000 initially and will increase to $0.35 million after the second addition (from the 11th year to the 15th year) and to $0.45 million during the final 10 years. (Assume that these costs begin one year subsequent to the actual addition.)

On the basis of the present-worth criterion, if the firm uses 15% as a MARR, which alternative should be undertaken?

5.57 A large refinery–petrochemical complex is to manufacture caustic soda, which will use feedwater of 10,000 gallons per day. Two types of feedwater storage installation are being considered over the 40 years of their useful life.

• **Option 1:** Build a 20,000-gallon tank on a tower. The cost of installing the tank and tower is estimated to be $164,000. The salvage value is estimated to be negligible.

• **Option 2:** Place a tank of 20,000-gallon capacity on a hill, which is 150 yards away from the refinery. The cost of installing the tank on the hill, including the extra length of service lines, is estimated to be $120,000 with negligible salvage value. Because of the tank's location on the hill, an additional investment of $12,000 in pumping equipment is required. The pumping equipment is expected to have a service life of 20 years with a salvage value of $1,000 at the end of that time. The annual operating and maintenance cost (including any income tax effects) for the pumping operation is estimated at $1,000.

If the firm's MARR is known to be 12%, which option is better on the basis of the present-worth criterion?

5.58 A professional football player has a contract on hand at the beginning of 2015 season. He will earn either $11,920,000 (deferred plan) over 12 years or $10,450,000 (non-deferred plan) over 7 years. He must declare which plan he prefers. The $11.92 million package is deferred through the year 2026, while the non-deferred arrangement ends after the 2021 season. Regardless of which plan is chosen, he will be playing through the 2021 season.

(a) Details of the two plans are given in Table P5.58. If his interest rate is 7.5%, which plan is more attractive?

(b) At what interest rate would the two plans be economically equivalent?

5.59 Currently, you have $10,000 to invest over five years. The interest rate in the United States is 4% for invested dollars. The interest rate in Europe is 5% for invested euros. The interest rate in Japan is 6% for invested yen. Assume that these interest rates are expected to remain unchanged over the next five years. The current and expected exchange rates are given in Table P5.59.

Which of the following options (if any) will maximize your wealth in U.S. dollars at the end of five years?

(a) Investing in the United States.
(b) Investing in Europe.
(c) Investing in Japan.

TABLE P5.59

Current Exchange Rates	Expected Exchange Rates Five Years from Now
$1 = 0.74 euro	$1 = 0.880 euro
$1 = 101.67 yen	$1 = 110 yen

TABLE P5.58

Deferred Plan		Non-deferred Plan	
2015	$2,500,000	2015	$2,500,000
2016	625,000	2016	1,050,000
2017	730,000	2017	1,150,000
2018	850,000	2018	1,250,000
2019	880,000	2019	1,350,000
2020	940,000	2020	1,550,000
2021	940,000	2021	1,600,000
2022	720,000		
2023	640,000		
2024	560,000		
2025	1,255,000		
2026	1,280,000		
Total	$11,920,000	Total	$10,450,000

Short Case Studies

ST5.1 Apex Corporation requires a chemical finishing process for a product under contract for a period of six years. Three options are available. Neither option 1 nor option 2 can be repeated after its process life. However, option 3 will always be available from H&H Chemical Corporation at the same cost during the period that the contract is operative. Here are the options:

- **Option 1:** Process device A, which costs $100,000, has annual operating and labor costs of $60,000 and a useful service life of four years with an estimated salvage value of $10,000.

- **Option 2:** Process device B, which costs $150,000, has annual operating and labor costs of $50,000 and a useful service life of six years with an estimated salvage value of $30,000.

- **Option 3:** Subcontract out the process at a cost of $100,000 per year.

According to the present-worth criterion, which option would you recommend at $i = 12\%$?

ST5.2 A utility company is in the process of considering two alternative methods of providing transformer capacity.

- **Option 1:** Purchase a 16 MVA transformer now and add a similar-sized unit at a later date when load growth is warranted. The price of a 16 MVA unit is $680,000.

- **Option 2:** Purchase a 25 MVA transformer at the outset which will meet the future load growth for at least the next 10 years. The price of a 25 MVA unit is $920,000.

Technical specs for the transformers are as given in Table ST5.2.

TABLE ST5.2

Descriptions	Option 1	Option 2
No-load losses	16.1 kW	21.8 kW
Full-load losses	88 kW	121 kW
Cost of the no-load losses	$680/kW	$680/kW
Cost of the full-load series losses	$280/kW	$280/kW

Suppose the expected load growth over the next 20 years is 11 MVA during the first year and 1 MVA each following year over the previous year. Each transformer has a useful life of 20 years. Which transformer unit would be selected using an interest rate of 8% with the study period of 10 years? What about for 20 years? Assume the future replacement cost would be the same as the initial purchase price.

ST5.3 Technology is taking much of the fun out of finding a place to park the car. Now, in cities from New York to Seattle, the door is open to a host of wireless technologies seeking to improve the parking meter even further. Chicago and Sacramento, California, among others, are equipping enforcement vehicles with infrared cameras capable of scanning license plates even at 30 miles an hour. Using a global positioning system, the cameras can tell which individual cars have parked too long in a two-hour parking zone. At a cost of $75,000 a camera, the system is an expensive upgrade of the old method of chalking tires and then coming back two hours later to see if the car has moved.

Parking czars in municipalities across the country are starting to realize parking meters' original goals: generating revenue and creating a continuous turnover of parking spaces on city streets. Clearly, their main questions are "Would there be enough new revenues from installing the expensive parking monitoring devices?" and "How many devices should be installed to maximize the revenue streams?" From the device manufacturer's point of view, the question is "Would there be enough demand for their products to justify the investment required in new facilities and marketing?" If the manufacturer decides to go ahead and market the products, but the actual demand is far less than its forecast or the adoption of the technology is too slow, what would be the potential financial risk?

6

Annual Equivalent-Worth Analysis

Robots May Revolutionize China's Electronics Manufacturing[1]

Can you imagine the entire assembly of iPhones done by robots someday? One iPhone supplier, Foxconn, continues to rely on city-sized factories where more than 1.1 million workers do the bulk of the assembly of iPhones and other devices by hands. This is not a far stretch of the imagination.

With soaring wages and an aging population, electronics factory managers say the day is approaching when robotic workers will replace people on the Chinese factory floor. Industrial robots can give factories a helping hand—working 24 hours a day, 7 days a week for 7 to 10 years. A new wave of industrial robots is in development, ranging from high-end humanoid machines with vision, touch, and even learning capabilities, to low-cost robots vying to undercut China's minimum wage.

In response to this need, Delta is testing a one-armed, four-jointed robot that can move objects, join components, and complete other similar tasks. By 2016, Delta hopes to sell a version for as little as $10,000, which would be less than half the cost of current mainstream robots. That price is also cheaper than the salary of a Chinese worker, and the robot can work around the clock. Foxconn originally planned to install one million robotic arms in its factories by 2014,

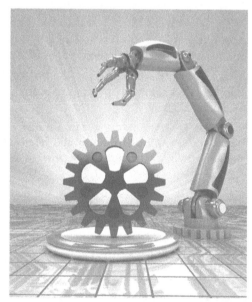

Source: Adam Vilimek/Shutterstock

---

[1] *Source*: "Robots May Revolutionize China's Electronics Manufacturing," *The Wall Street Journal*, September 23, 2013.

but executives said it would take much longer to reach that target. It's clear that automation is the future trend in China, but the big question is how to bring down the costs for robots. Replacing people with robots would reduce the *operating cost* at the expense of increasing *capital cost.*

Now put yourself in a factory as a manager. Over the next five years these technologies will transform China's factories and also fill a growing labor shortage as the country's youth become increasingly unwilling to perform manual labor. At what annual equivalent cost of owning and operating the robots could you replace the human assemblers?

Suppose the robot assembly plant is fully operational by 2016 and you expect to operate it continuously for 20 years. How would you calculate the operating cost per hour? Suppose you are considering buying a new car. If you expect to drive 12,000 miles per year, can you figure out how much the car costs per mile? You would have good reason to want to know the cost if you were being reimbursed by your employer on a per-mile basis for the business use of your car. Or consider a real-estate developer who is planning to build a 500,000-square-foot shopping center. What would be the minimum annual rental fee per square foot required to recover the initial investment?

Annual equivalence analysis is the method by which these and other unit costs (or profits) are calculated. Along with present-worth analysis, annual equivalence analysis is the second major equivalence technique for putting alternatives on a common basis of comparison. In this chapter, we develop the annual equivalent-worth criterion and demonstrate a number of situations in which annual equivalence analysis is preferable to other methods of comparison.

CHAPTER LEARNING OBJECTIVES

After completing this chapter, you should understand the following concepts:

- How to determine the equivalent annual worth (cost) for a given project.
- Why the annual equivalent approach facilitates the comparison of unequal service-life problems.
- How to determine the capital cost (or ownership cost) when you purchase an asset.
- How to determine the unit cost or unit profit.
- How to conduct a life-cycle cost analysis.
- How to optimize design parameters in engineering design.

6.1 Annual Equivalent-Worth Criterion

In this section, we will describe a fundamental decision rule based on annual equivalent worth by considering both revenue and cost streams of a project. If revenue streams are irrelevant, then we make a decision solely on the basis of cost. This leads to a popular decision tool known as "life-cycle cost analysis," which we will discuss in Section 6.4.

6.1.1 Fundamental Decision Rule

The **annual equivalent worth (AE)** criterion provides a basis for measuring the worth of an investment by determining equal payments on an annual basis. Knowing that any lump-sum cash amount can be converted into a series of equal annual payments, we may first find the net present worth (NPW) of the original series and then multiply this amount by the capital recovery factor:

$$AE(i) = PW(i)(A/P, i, N) \tag{6.1}$$

- **Single-project evaluation:** The accept–reject selection rule for a single *revenue* project is as follows:

  If $AE(i) > 0$, accept the investment.

  If $AE(i) = 0$, remain indifferent to the investment.

  If $AE(i) < 0$, reject the investment.

  Notice that the factor $(A/P, i, N)$ in Eq. (6.1) is positive for $-100\% < i < \infty$, which indicates that the value of $AE(i)$ will be positive if, and only if, $PW(i)$ is positive. In other words, accepting a project that has a positive $AE(i)$ is equivalent to accepting a project that has a positive $PW(i)$. Therefore, the AE criterion for evaluating a project is consistent with the NPW criterion.
- **Comparing mutually exclusive alternatives:** As with present-worth analysis, when you compare mutually exclusive *service* projects whose revenues are the same, you may compare them on a *cost*-only basis. In this situation, the alternative with the minimum *annual equivalent cost* (AEC) or least negative annual equivalent worth is selected.

  Example 6.1 illustrates how to find the equivalent annual worth for a proposed energy-savings project. As you will see, you first calculate the net present worth of the project and then convert this present worth into an equivalent annual basis.

EXAMPLE 6.1 Annual Equivalent Worth: A Single-Project Evaluation

A utility company is considering adding a second feedwater heater to its existing system to increase the efficiency of the system and thereby reduce fuel costs. The second feedwater heater to go with the 150 MW system will cost $1,650,000 and has a service life of 25 years. The expected salvage value of the unit is considered negligible. With the second unit installed, the efficiency of the dual system will improve from 55% to 56%. The fuel cost to run the feedwater is estimated at $0.05 kWh. The system unit will have a load factor of 85%, meaning that the system will run 85% of the year. Is it worth having a mere 1% improvement in efficiency at the expense of $1,650,000?

(a) Determine the equivalent annual worth of adding the second unit with an interest rate of 12%.

(b) If the fuel cost increases at the annual rate of 4% after first year, what is the equivalent annual worth of having the second feedwater heater at $i = 12\%$?

DISCUSSION: Whenever we compare machines with different efficiency ratings, we need to determine the input powers required to operate the machines. Since the percent efficiency is equal to the ratio of the output power to the input power, we can determine the input power by dividing the output power by the motor's percent efficiency:

$$\text{Input power} = \frac{\text{output power}}{\text{percent efficiency}}$$

For example, a 30-HP motor with 90% efficiency will require an input power of

$$\text{Input power} = \frac{(30 \text{ HP} \times 0.746 \text{ kW/HP})}{0.90}$$

$$= 24.87 \text{ kW}$$

Therefore, energy consumption with and without the second unit can be calculated as

- Before adding the second unit, $\dfrac{150{,}000 \text{ kW}}{0.55} = 272{,}727 \text{ kW}.$

- After adding the second unit, $\dfrac{150{,}000 \text{ kW}}{0.56} = 267{,}857 \text{ kW}.$

So the reduction in energy consumption is 4,870 kW. Since the system unit will operate only 85% of the year, the total annual operating hours are calculated as

$$\text{Annual operating hours} = (365)(24)(0.85) = 7{,}446 \text{ hours/year}$$

SOLUTION

Given: $I = \$1{,}650{,}000$, $N = 25$ years, $S = 0$, annual fuel savings, and $i = 12\%$.
Find: AE of fuel savings due to improved efficiency.

(a) With the assumption of constant fuel cost over the service life of the second heater,

$$A_{\text{fuel savings}} = (\text{reduction in fuel requirement}) \times (\text{fuel cost})$$

$$\times (\text{operating hours per year})$$

$$= (4{,}870 \text{ kW}) \times (\$0.05/\text{kWh})$$

$$\times (7{,}446 \text{ hours/year})$$

$$= \$1{,}813{,}101 \text{ per year}$$

$$\text{PW}(12\%) = -\$1{,}650{,}000 + \$1{,}813{,}101(P/A, 12\%, 25)$$

$$= \$12{,}570{,}403$$

$$\text{AE}(12\%) = \$12{,}570{,}403(A/P, 12\%, 25)$$

$$= \$1{,}602{,}726$$

(b) With the assumption of escalating energy cost at the annual rate of 4%, since the first year's fuel savings is already calculated in part (a), we use it as A_1 in the geometric-gradient series present-worth factor $(P/A_1, g, i, N)$:

$$A_1 = \$1,813,101$$

$$\text{PW}(12\%) = -\$1,650,000 + \$1,813,101(P/A_1, 4\%, 12\%, 25)$$

$$= \$17,459,783$$

$$\text{AE}(12\%) = \$17,459,783(A/P, 12\%, 25)$$

$$= \$2,226,122$$

Clearly, either situation generates enough fuel savings to justify adding the second unit of the feedwater heater. Figure 6.1 illustrates the cash flow series associated with the required investment and fuel savings over the heater's service life of 25 years.

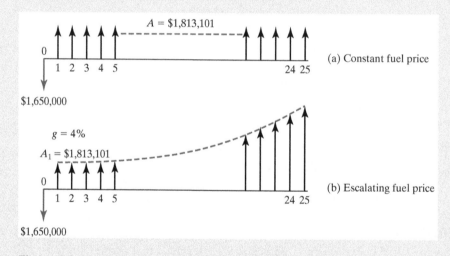

Figure 6.1 Cash flow diagram (Example 6.1).

6.1.2 Annual-Worth Calculation with Repeating Cash Flow Cycles

In some situations, a **cyclic cash flow pattern** may be observed over the life of the project. Unlike the situation in Example 6.1, where we first computed the NPW of the entire cash flow and then calculated the AE, we can compute the AE by examining the first cash flow cycle. Then we can calculate the NPW for the first cash flow cycle and derive the AE over that cycle. This shortcut method gives the same solution when the NPW of the entire project is calculated, and then the AE can be computed from this NPW.

EXAMPLE 6.2 Annual Equivalent Worth: Repeating Cash Flow Cycles

SOLEX Company is producing electricity directly from a solar source by using a large array of solar cells and selling the power to the local utility company. SOLEX decided to use amorphous silicon cells because of their low initial cost, but these cells degrade over time, thereby resulting in lower conversion efficiency and power output. The cells must be replaced every four years, which results in a particular cash flow pattern that repeats itself, as shown in Figure 6.2. Determine the annual equivalent cash flows at $i = 12\%$.

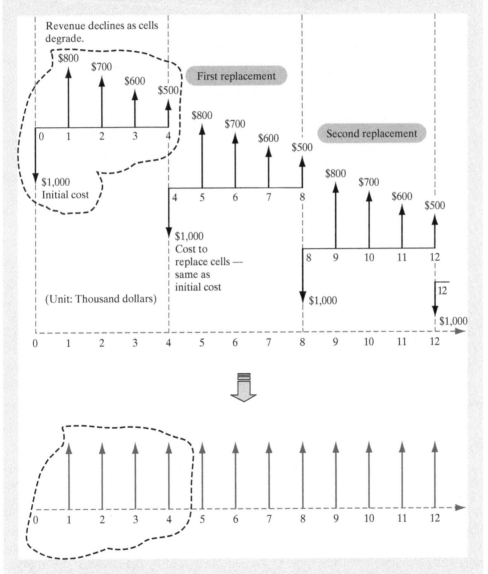

Figure 6.2 Conversion of repeating cash flow cycles into an equivalent annual payment (Example 6.2).

SOLUTION

Given: Cash flows in Figure 6.2 and $i = 12\%$.
Find: Annual equivalent benefit.

To calculate the AE, we need only consider one cycle over the four-year replacement period of the cells. For $i = 12\%$, we first obtain the NPW for the first cycle as

$$PW(12\%) = -\$1,000,000$$
$$+ [(\$800,000 - \$100,000(A/G, 12\%, 4)](P/A, 12\%, 4)$$
$$= -\$1,000,000 + \$2,017,150$$
$$= \$1,017,150$$

Then we calculate the AE over the four-year life cycle:

$$AE(12\%) = \$1,017,150(A/P, 12\%, 4)$$
$$= \$334,880$$

We can now say that the two cash-flow series are equivalent:

Original Cash Flows		Annual Equivalent Flows	
n	A_n	n	A_n
0	−$1,000,000	0	0
1	800,000	1	$334,880
2	700,000	2	334,880
3	600,000	3	334,880
4	500,000	4	334,880

We can extend this equivalency over the remaining cycles of the cash flow. The reasoning is that each similar set of five values (one disbursement and four receipts) is equivalent to four annual receipts of $334,880 each. In other words, the $1 million investment in the solar project will recover the entire investment and generate equivalent annual savings of $334,880 over a four-year life cycle.

6.1.3 Comparing Mutually Exclusive Alternatives

In this section, we consider a situation in which two or more mutually exclusive alternatives need to be compared on the basis of annual equivalent worth. In Section 5.5, we discussed the general principles that should be applied when mutually exclusive alternatives with unequal service lives were compared. The same general principles should be applied in comparing mutually exclusive alternatives on the basis of annual equivalent worth: Mutually exclusive alternatives in equal time spans must be compared. Therefore, we must give careful consideration to the period covered by the analysis: the **analysis period**. We will consider two situations: (1) The analysis period equals project lives, and (2) the analysis period differs from project lives.

With situation (1), we compute AE (for revenue projects) or AEC (for service projects) for each project and select the project that has the largest AE for revenue projects (or the least AEC for service projects). With situation (2), we need to consider the issue of unequal project lives. As we saw in Chapter 5, comparing projects with unequal service lives is complicated by the need to determine the lowest common multiple life. For the special situation of an indefinite service period and replacement with identical projects, we can avoid this complication by the use of AE analysis, provided that the following criteria are met.

1. The service of the selected alternative is required on a continuous basis.
2. Each alternative will be replaced by an *identical* asset that has the same costs and performance.

When these two criteria are met, we may solve for the AE of each project on the basis of its initial life span, rather than on that of the lowest common multiple of the projects' lives. Example 6.3 illustrates the process of comparing unequal service projects.

EXAMPLE 6.3 Annual Equivalent Cost Comparison for Service Projects: Unequal Project Lives

Consider again Example 5.13, in which we compared two types of equipment with unequal service lives. Apply the AEC approach to select the most economical equipment.

SOLUTION
Given: Cost cash flows shown in Figure 6.3 and $i = 15\%$ per year.
Find: AEC and which alternative is the preferred one.

An alternative procedure for solving Example 5.13 is to compute the annual equivalent cost of an outlay of $12,500 for model A every three years and the annual equivalent cost of an outlay of $15,000 for model B every four years. Notice that the AEC of each 12-year cash flow is the same as that of the corresponding three- or four-year cash flow (Figure 6.3). From Example 5.13, we calculate the following.

- **Model A:** For a three-year life,

$$PW(15\%) = -\$22,601$$
$$AEC(15\%) = \$22,601(A/P, 15\%, 3)$$
$$= \$9,899$$

For a 12-year period (the entire analysis period),

$$PW(15\%) = -\$53,657$$
$$AEC(15\%) = \$53,657(A/P, 15\%, 12)$$
$$= \$9,899$$

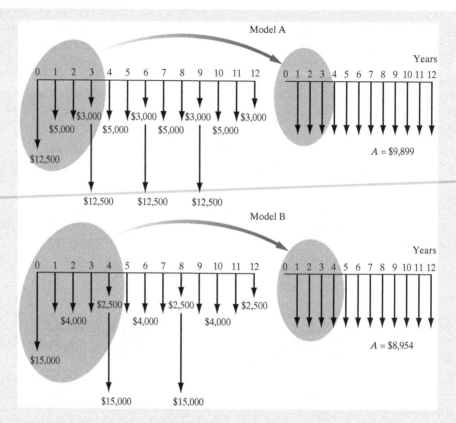

Figure 6.3 Comparison of projects with unequal lives and an indefinite analysis period using the annual equivalent-worth criterion (Example 6.3).

- **Model B:** For a four-year life,

$$PW(15\%) = -\$25,562$$

$$AEC(15\%) = \$25,562(A/P, 15\%, 4)$$

$$= \$8,954$$

For a 12-year period (the entire analysis period),

$$PW(15\%) = -\$48,534$$

$$AEC(15\%) = \$48,534(A/P, 15\%, 12)$$

$$= \$8,954$$

Notice that the annual equivalent costs that were calculated on the basis of the common service period are the same as those which were obtained over their initial life-spans. Thus, for alternatives with unequal lives, we will obtain the same selection by comparing the NPW over a common service period using repeated projects or by comparing the AE for initial lives.

6.2 Capital Costs Versus Operating Costs

As mentioned previously, when only costs are involved, the AE method is sometimes called the **annual equivalent cost (AEC)** method. In this case, revenues must cover two kinds of costs: operating costs and capital costs.

- **Operating costs** are incurred through the operation of physical plant or equipment needed to provide service; examples include items such as labor and raw materials.
- **Capital costs (capital recovery cost)** are incurred by purchasing assets to be used in production and service. Normally, capital costs are nonrecurring (i.e., one-time) costs, whereas operating costs recur for as long as an asset is owned.

Because operating costs recur over the life of a project, they tend to be estimated commonly on an annual basis, so for the purposes of annual equivalent cost analysis, no special calculation is required. However, because capital costs tend to be one-time costs, we must translate this one-time cost into its annual equivalent over the life of the project. The annual equivalent of a capital cost is given a special name: **capital recovery cost with return**, which is designated $CR(i)$.

As shown in Figure 6.4, two general monetary transactions are associated with the purchase and eventual retirement of a capital asset: its initial cost (I) and its salvage value (S). Taking into account these sums, we calculate the capital recovery factor as

$$CR(i) = [I - S(P/F, i, N)](A/P, i, N)$$

$$= I(A/P, i, N) - S(A/F, i, N) \qquad (6.2)$$

Now recall algebraic relationships between factors in Table 3.6, and notice that the factor $(A/F, i, N)$ can be expressed as

$$(A/F, i, N) = (A/P, i, N) - i$$

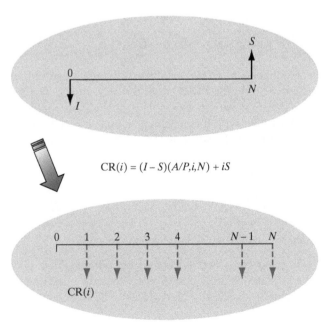

$$CR(i) = (I - S)(A/P, i, N) + iS$$

Figure 6.4 Capital recovery (ownership) cost calculation.

Then we may rewrite CR(i) as

$$CR(i) = I(A/P, i, N) - S[(A/P, i, N) - i]$$
$$= (I - S)(A/P, i, N) + iS \qquad (6.3)$$

Since we are calculating the equivalent annual costs, we treat cost items with a positive sign. The salvage value is treated as having a negative sign in Eq. (6.3). We may interpret this situation thus: To obtain the machine, one borrows a total of I dollars, S dollars of which are returned at the end of the Nth year. The first term, $(I - S)(A/P, i, N)$, implies that the balance $(I - S)$ will be paid back in equal installments over the N-year period at a rate of i. The second term, iS implies that simple interest in the amount iS is paid on S until it is repaid. Thus, the amount to be financed is $I - S$ $(P/F, i, N)$, and the installments of this loan over the N-year period are

$$AEC(i) = [I - S(P/F, i, N)](A/P, i, N)$$
$$= I(A/P, i, N) - S(P/F, i, N)(A/P, i, N)$$
$$= [I(A/P, i, N) - S(A/F, i, N)]$$
$$= CR(i) \qquad (6.4)$$

Therefore, CR(i) tells us what the bank would charge each year. Many auto leases are based on this arrangement, because most banks require a guarantee of S dollars in salvage. From an industry viewpoint, CR(i) is the annual cost to the firm of owning the asset.

With this information, the amount of annual savings required to recover the capital and operating costs associated with a project can be determined, as Example 6.4 illustrates.

EXAMPLE 6.4 Capital Recovery Cost

Consider a machine that costs $20,000 and has a five-year useful life. At the end of the five years, it can be sold for $4,000 after tax adjustment. The annual operating and maintenance (O&M) costs are about $500. If the firm could earn an after-tax revenue of $5,000 per year with this machine, should it be purchased at an interest rate of 10%? (All benefits and costs associated with the machine are accounted for in these figures.)

SOLUTION
Given: $I = \$20,000, S = \$4,000, O\&M = \$500, A = \$5,000, N = 5$ years, and $i = 10\%$ per year.
Find: AE, and determine whether to purchase the machine.

The first task is to separate cash flows associated with acquisition and disposal of the asset from the normal operating cash flows. Since the operating cash flows—the $5,000 yearly revenue—are already given in equivalent annual flows, we need to

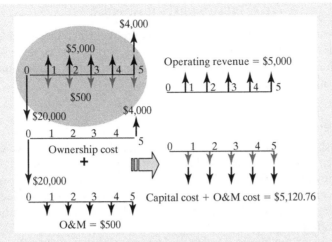

Figure 6.5 Separating ownership cost (capital cost) and operating cost from operating revenue, which must exceed the annual equivalent cost to make the project acceptable.

convert only the cash flows associated with acquisition and disposal of the asset into equivalent annual flows (Figure 6.5). Using Eq. (6.3), we obtain

$$CR(i) = (I - S)(A/P, i, N) + iS$$
$$= (\$20{,}000 - \$4{,}000)(A/P, 10\%, 5) + (0.10)\$4{,}000$$
$$= \$4{,}620.76$$

$$O\&M(i) = \$500$$

$$AEC(10\%) = CR(10\%) + O\&M(10\%)$$
$$= \$4{,}620.76 + \$500$$
$$= \$5{,}120.76$$

$$AE(10\%) = \$5{,}000 - \$5{,}120.76$$
$$= -\$120.76$$

This negative AE value indicates that the machine does not generate sufficient revenue to recover the original investment, so we must reject the project. In fact, there will be an equivalent loss of $120.76 per year over the life of the machine.

COMMENTS: We may interpret the value found for the annual equivalent cost by asserting that the annual operating revenues must be at least $5,120.76 in order to recover the cost of owning and operating the asset. Unlike the NPW approach, the AE approach provides us with a specific target to achieve to make the project worthwhile.

6.3 Applying Annual-Worth Analysis

In general, most engineering economic analysis problems can be solved by the present-worth methods that were introduced in Chapter 5. However, we can solve some economic analysis problems more efficiently by annual-worth analysis. In this section, we introduce several applications that call for such an analysis.

6.3.1 Benefits of AE Analysis

Example 6.1 should look familiar to you: It is exactly the situation we encountered in Chapter 3 when we converted a mixed cash flow into a single present value and then into a series of equivalent cash flows. In the case of Example 6.1, you may wonder why we bother to convert NPW to AE at all, since we already know that the project is acceptable from NPW analysis. In fact, the example was mainly an exercise to familiarize you with the AE calculation.

However, in the real world, a number of situations can occur in which AE analysis is preferred (or even demanded) over NPW analysis. For example, corporations issue annual reports and develop yearly budgets. For these purposes, a company may find it more useful to present the annual cost or benefit of an ongoing project rather than its overall cost or benefit. The following are some additional situations in which AE analysis is preferred.

1. **Consistency of report formats.** Financial managers commonly work with annual rather than overall costs in any number of internal and external reports. Engineering managers may be required to submit project analyses on an annual basis for consistency and ease of use by other members of the corporation and stockholders.
2. **Need for unit costs or profits.** In many situations, projects must be broken into unit costs (or profits) for ease of comparison with alternatives. Make-or-buy and reimbursement analyses are key examples, and these will be discussed in this chapter.
3. **Life-cycle cost analysis.** When there is no need for estimating the revenue stream for a proposed project, we can consider only the cost streams of the project. In that case, it is common to convert this life-cycle cost (LCC) into an equivalent annual cost for purposes of comparison. Industry has used the LCC to help determine which design option will cost less over the life of a product. LCC analysis has had a long tradition in the Department of Defense, having been applied to virtually every new weapon system proposed or under development.

6.3.2 Unit Profit or Cost Calculation

In many situations, we need to know the unit profit (or cost) of operating an asset. To obtain this quantity, we may proceed as follows.

Step 1. Determine the number of units to be produced (or serviced) each year over the life of the asset.

Step 2. Identify the cash flow series associated with production or service over the life of the asset.

Step 3. Calculate the net present worth of the project cash-flow series at a given interest rate, and then determine the equivalent annual worth.

Step 4. Divide the equivalent annual worth by the number of units to be produced or serviced during each year. When the number of units varies each year, you may need to convert them into equivalent annual units.

To illustrate the procedure, Example 6.5 uses the annual equivalent concept in estimating the savings per machine hour for the proposed acquisition of a machine.

EXAMPLE 6.5 Unit Profit per Machine Hour When Annual Operating Hours Remain Constant

Consider the investment in the metal-cutting machine of Example 5.5. Recall that this four-year investment was expected to generate an NPW of $30,065. Suppose that the machine will be operated for 3,000 hours per year. Compute the equivalent savings per machine hour at $i = 12\%$.

SOLUTION

Given: NPW $= \$30,065$, $N = 4$ years, $i = 12\%$ per year, and 3,000 machine hours per year.
Find: Equivalent savings per machine hour.

We first compute the annual equivalent savings from the use of the machine. Since we already know the NPW of the project, we obtain the AE by the formula

$$AE(12\%) = \$30,065(A/P, 12\%, 4) = \$9,898$$

With an annual usage of 3,000 hours, the equivalent savings per machine hour would be

$$\text{Savings per machine hour} = \$9,898/3,000 \text{ hours} = \$3.30/\text{hour}$$

COMMENTS: Note that we cannot simply divide the NPW ($30,065) by the total number of machine hours over the four-year period (12,000 hours) to obtain $2.51/hour. This $2.51 figure represents the instant savings in present worth for each hourly use of the equipment, but it does not consider the time over which the savings occur. Once we have the annual equivalent worth, we can divide by the desired time unit if the compounding period is one year. If the compounding period is shorter, then the equivalent worth should be calculated for the compounding period.

EXAMPLE 6.6 Unit Profit per Machine Hour When Annual Operating Hours Fluctuate

Consider again Example 6.5, and suppose that the metal-cutting machine will be operated according to varying hours: 3,500 hours the first year, 4,000 hours the second year, 1,700 hours the third year, and 2,800 hours the fourth year. The total operating hours still remain at 12,000 over four years. Compute the equivalent savings per machine hour at $i = 12\%$.

SOLUTION

Given: NPW $= \$30,065$, $N = 4$ years, $i = 12\%$ per year, and operating hours of 3,500 the first year, 4,000 the second year, 1,700 the third year, and 2,800 hours the fourth year.
Find: Equivalent savings per machine hour.

As calculated in Example 6.5, the annual equivalent savings is $9,898. Let C denote the equivalent annual savings per machine hour, which we need to determine. Now, with varying annual usages of the machine, we can set up the equivalent annual savings as a function of C:

$$\text{Equivalent annual savings} = [C(3,500)(P/F, 12\%, 1)$$
$$+ \ C(4,000)(P/F, 12\%, 2)$$
$$+ \ C(1,700)(P/F, 12\%, 3)$$
$$+ \ C(2,800)(P/F, 12\%, 4)](A/P, 12\%, 4)$$
$$= 3,062.95C$$

We can equate this amount to the $9,898 and solve for C. This gives us

$$C = \$9,898/3,062.05 = \$3.23/\text{hour}$$

which is about 7 cents less than the $3.30 we found in Example 6.5.

6.3.3 Make-or-Buy Decision—Outsourcing Decisions

Make-or-buy problems are among the most common of business decisions. At any given time, a firm may have the option of either buying an item or producing it. Unlike the make-or-buy situation we will consider in Chapter 8, *if either the "make" or the "buy" alternative requires the acquisition of machinery or equipment, then it becomes an investment decision.* Since the cost of an outside service (the "buy" alternative) is usually quoted in terms of dollars per unit, it is easier to compare the two alternatives if the differential costs of the "make" alternative are also given in dollars per unit. This unit cost comparison requires the use of annual-worth analysis. The specific procedure is as follows.

Step 1. Determine the time-span (planning horizon) for which the part (or product) will be needed.

Step 2. Determine the annual quantity of the part (or product).

Step 3. Obtain the unit cost of purchasing the part (or product) from the outside firm.

Step 4. Determine the equipment, manpower, and all other resources required to make the part (or product) in-house.

Step 5. Estimate the net cash flows associated with the "make" option over the planning horizon.

Step 6. Compute the annual equivalent cost of producing the part (or product).

Step 7. Compute the unit cost of making the part (or product) by dividing the annual equivalent cost by the required annual volume.

Step 8. Choose the option with the minimum unit cost.

EXAMPLE 6.7 Outsourcing Production of Electric Compressors

JK Engineering Company manufactures a variety of air-conditioning units for auto-mobiles. The company has just received an offer from an outside supplier to pro-vide the electric compressor for the company's heat pump system line, at a price of $42 per unit, over the next 7 years. The company is interested in this offer, since its own production of electric compressor strains many parts of heat pump production systems. The company estimates that if the supplier's offer were accepted, the direct labor and variable overhead costs of the heat pump system line would be reduced by 20% and the direct materials cost by 25%. Under present operations, the company manufactures all of its own heat pump system from start to finish. Fixed overhead costs charged to the heat pump system line total $500,000 each year. The cost of producing 50,000 units of heat pump system per year with and without outsourcing is given Table 6.1.

TABLE 6.1 Unit Cost Data before and after Outsourcing

	Unit Cost	
Item	Current	With Outsourcing
Direct materials	$150	$112.50
Direct labor	$100	$80
Variable overhead	$20	$16
Fixed manufacturing overhead (based on production of 50,000 units each year)	$10	$10
Total	$270	$218.50 + $42

So, if JK outsourced the electric compressor, it would cut down the manufacturing cost by $9.50 per unit. However, the company must also buy specialized inspection equipment to test the quality of the electric compressor units, since its current inspec-tion is done during each process of manufacturing of the electric compressor, rather than on the completed part. The inspection machine costs $325,000 with an estimated salvage value of $60,000 at the end of the 7-year useful life. The operating and main-tenance cost of the inspection machine is expected to be $120,000 per year. The accounting department has itemized the annual costs associated with each option as follows. Assuming that JK's MARR is 18%, calculate the unit cost under each option.

SOLUTION

Given: Cash flows for two options and $i = 18\%$.
Find: Unit cost for each option and the option preferred.

We may itemize the annual costs associated with each option as follows:

- **Make Option:** $270 per unit
- **Buy option:** The two additional cost components are capital cost and operating cost.

Capital cost:

$$CR(18\%) = (\$325,000 - \$60,000)(A/P, 18\%, 7)$$
$$+ \$60,000(0.18)$$
$$= \$80,326$$

Operating cost:

$$OC(18\%) = \$120,000$$

Total annual equivalent cost related to the owning and operating the inspection machine:

$$AEC(18\%)_{\text{Inspection}} = CR(18\%) + OC(18\%) = \$200,326$$

$$\text{Unit cost increment due to inspection} = \frac{\$200,326}{50,000} = \$4.01$$

Total unit cost:

$$\text{Unit cost} = \$218.50 \text{ (with outsourcing)} + \$42 + \$4.01 = \$264.51$$

Buying the electric compressor units from the outside vendor will save JK $5.49 per unit before any consideration of taxes. Figure 6.6 summarizes the make-or-buy analysis for this decision problem.

Figure 6.6 Unit cost comparison between make-or-buy decision.

COMMENTS: Certainly, knowing how much it costs to produce will provide important information in setting the price of the product. Two important noneconomic factors should also be considered. The first is the question of whether the quality of the supplier's component is better than, equal to, or worse than the component the firm is presently manufacturing. The second is the reliability of the supplier in terms of providing the needed quantities of the product on a timely basis. A reduction in quality or reliability should virtually always rule out buying.

6.3.4 Pricing the Use of an Asset

Companies often need to calculate the cost of equipment that corresponds to a **unit of use** of that equipment. For example, if you own an asset, such as a building, you would be interested in knowing the cost per square foot of owning and operating the asset. This information will be the basis for determining the rental fee for the asset. A familiar example is an employer's reimbursement of costs for the use of an employee's personal car for business purposes. If an employee's job depends on obtaining and using a personal vehicle on the employer's behalf, reimbursement on the basis of the employee's overall costs per mile seems fair.

EXAMPLE 6.8 **Pricing an Apartment Rental Fee**

Sunbelt Corporation, an investment company, is considering building a 50-unit apartment complex in a growing area near Tucson, Arizona. Since the long-term growth potential of the town is excellent, it is believed that the company could average 85% full occupancy for the complex each year. If the following financial data are reasonably accurate estimates, determine the minimum monthly rent that should be charged if a 15% rate of return is desired:

- Land investment cost = $1,000,000
- Building investment cost = $2,500,000
- Annual upkeep cost = $150,000
- Property taxes and insurance = 5% of total initial investment
- Study period = 25 years
- Salvage value = only land cost can be recovered in full

SOLUTION

Given: Preceding financial data, study period = 25 years, and $i = 15\%$.
Find: Minimum monthly rental charge.

First we need to determine the capital cost associated with ownership of the property:

$$\text{Total investment required} = \text{land cost} + \text{building cost} = \$3,500,000$$

$$\text{Salvage value} = \$1,000,000 \text{ at the end of 25 years}$$

$$\text{CR}(15\%) = (\$3,500,000 - \$1,000,000)(A/P, 15\%, 25)$$
$$+ (\$1,000,000)(0.15)$$
$$= \$536,749$$

Second, the annual operating cost has two elements: (1) property taxes and insurance; and (2) annual upkeep cost. Thus,

$$\text{O\&M cost} = (0.05)(\$3,500,000) + \$150,000$$
$$= \$325,000$$

So the total annual equivalent cost is

$$\text{AEC}(15\%) = \$536,749 + \$325,000$$
$$= \$861,749$$

which is the minimum annual rental required to achieve a 15% rate of return. Therefore, with annual compounding, the monthly rental amount is

$$\text{Required monthly charge} = \frac{\$861,749}{(12 \times 50)(0.85)}$$
$$= \$1,690.$$

COMMENTS: The rental charge that is exactly equal to the cost of owning and operating the building is known as the **break-even point**, or a sure 15% return on investment.

6.4 Life-Cycle Cost Analysis

Life-cycle cost analysis (LCCA) is a process of evaluating the economic performance of various design options that have differing initial costs, operating costs, maintenance costs, and possibly different life cycles. Actually all analyses we have done so far are basically LCCAs. However, LCCA is based upon the further assumptions that multiple design options can meet the stated objectives and achieve acceptable performance, and it enables the analyst to make sure that the selection of a design alternative is not based solely on the lowest initial costs, but also takes into account all the future costs over the project's usable life. This is because about 80% of the total life-cycle cost of a system occurs after the system has entered service. Therefore, the best long-term system acquisition and support decisions should be based on a full understanding of the total cost of acquiring, operating, and supporting the system (Figure 6.7). Some of the unique features of LCCA are as follows:

- LCCA is used appropriately only to select from among design alternatives that would yield the same level of performance or benefits to the project's users during normal operations. If benefits vary among the design alternatives, then the alternatives cannot be compared solely on the basis of cost. Rather, the analyst would need to employ

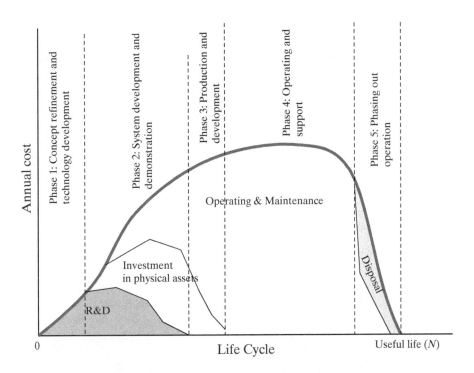

Figure 6.7 Stages of life-cycle cost. These include the concept refinement and technology development phase, the system development and demonstration phase, the production and deployment phase, the operating phase, and the disposal phase.

present-worth analysis or **benefit–cost analysis (BCA)**, which measures the monetary value of life-cycle benefits as well as costs. BCA is discussed in Economic Analysis in the Service Sectors.

- LCCA is a way to predict the most cost-effective solution; it does not guarantee a particular result, but it allows the plant designer or manager to make a reasonable comparison among alternative solutions within the limits of the available data.
- To make a fair comparison, the plant designer or manager might need to consider the measure used. For example, the same process output volume should be considered, and if the two items being examined cannot give the same output volume, it may be appropriate to express the figures in cost per unit of output (e.g., $/ton, or euro/kg), which requires a calculation of annual equivalent dollars generated. This calculation is typically based on the annual output.

In many situations, we need to compare a set of different design alternatives—each of which would produce the same number of units (constant revenues), but would require different amounts of investment and operating costs (because of different degrees of mechanization). Example 6.9 illustrates the use of the annual equivalent-cost concept to compare the cost of operating an existing pumping system with that of an improved system.

EXAMPLE 6.9 HVAC Retrofit of an Existing Laboratory Building

A university wants to evaluate a number of HVAC retrofit approaches to one of its laboratory buildings on campus and determine which, if any, are worth implanting. The building contains 71,000 gross square feet of laboratory and office space. Of that area, roughly 38,000 square feet are assignable. Three design options are considered:

- **Option A:** Variable-air-volume (VAV) distribution system for non-laboratory areas
- **Option B:** VAV distribution system for laboratory areas
- **Option C:** Combination of Option A and Option B

Costs for each design option are summarized in Table 6.2.

TABLE 6.2 HVAC Retrofit Cost Summary for Alternatives

		Annual Utility Cost				
Alternatives	First Cost	Electricity	Chilled Water	Steam	Total Utility Costs	Annual Energy Savings Over Base
Base Case		$460,847	$209,485	$235,004	$905,336	
Option A	500,000	389,535	166,583	104,032	660,150	$245,186
Option B	650,000	436,363	199,588	194,015	829,966	75,370
Option C	875,000	324,271	138,892	63,809	526,972	378,364

The university uses a 30-year study period for a building project of this nature with a discount rate of 7%. Here, the university also assumes that the salvage value of these retrofit projects would be virtually "nil" after considering the removal of the system at the end of its service life.

SOLUTION:

Given: Financial data as summarized in Table 6.2, MARR = 7%, and $N = 30$ years.
Find: Which design option to choose.

For each option, we calculate both the capital cost and equivalent annual energy savings over the base case. Since typical operating and maintenance costs are already included in the annual utility costs, we are basically looking at whether or not each option provides sufficient energy savings to cover the capital costs.

- **Option A:**

$$CR(7\%)_A = (\$500,000 - 0\$)(A/P, 7\%, 30) + (0.07)(\$0)$$
$$= \$40,293$$

$$\text{Energy savings} = \$245,186 \text{ per year}$$

$$AE(7\%) = \$245,186 - \$40,293 = \$204,893$$

- **Option B:**

$$CR(7\%)_B = (\$650,000 - 0\$)(A/P, 7\%, 30) + (0.07)(\$0)$$
$$= \$52,381$$

$$\text{Energy savings} = \$75,370 \text{ per year}$$

$$AE(7\%) = \$75,370 - \$52,381 = \$22,989$$

- **Option C:**

$$CR(7\%)_C = (\$875,000 - 0\$)(A/P, 7\%, 30) + (0.07)(\$0)$$
$$= \$70,513$$

$$\text{Energy savings} = \$378,364 \text{ per year}$$

$$AE(7\%) = \$378,364 - \$70,513 = \$307,851$$

The analysis shows that Option C has the highest first cost but also the most energy savings, resulting in the maximum annual net savings. Therefore, over the long run, Option C comes out the most cost effective.

COMMENTS: Many practitioners may be tempted to pick Option A with the notion that it has the shortest payback period of 2.04 years, whereas Option C has a 2.31 per year payback period. Therefore, Option A could be seen as the best choice, balancing first-cost impacts with energy savings. But clearly, this violates the very principle of LCCA—do not just compare different design options based on the speed of recovery of the initial investment.

EXAMPLE 6.10 Pumping System with a Problem Valve[2]

Consider a single-pump circuit that transports a process fluid containing some solids from a storage tank to a pressurized tank. A heat exchanger heats the fluid, and a control valve regulates the rate of flow into the pressurized tank to 80 cubic meters per hour (m^3/h), or 350 gallons per minute (gpm). The process is depicted in Figure 6.8.

The plant engineer is experiencing problems with a fluid control valve (FCV) that fails due to erosion caused by cavitations. The valve fails every 10 to 12 months at a cost of $4,000 per repair. A change in the control valve is being considered: Replace

---

[2]*Pump Life Cycle Costs: A Guide to LCC Analysis for Pumping Systems.* DOE/GO-102001-1190 December 2000, Office of Industrial Technologies, U.S. Department of Energy and Hydraulic Institute.

Figure 6.8 Sketch of a pumping system in which the control valve fails.

the existing valve with one that can resist cavitations. Before the control valve is repaired again, the project engineer wants to look at other options and perform an LCCA on alternative solutions.

ENGINEERING SOLUTION ALTERNATIVES: The first step is to determine how the system is currently operating and why the control valve fails. Then the engineer can see what can be done to correct the problem. The control valve currently operates between 15% and 20% open and with considerable cavitation noise from the valve. It appears that the valve was not sized properly for the application. After reviewing the original design calculations, it was discovered that the pump was oversized: 110 m³/h (485 gpm) instead of 80 m³/h (350 gpm). This resulted in a larger pressure drop across the control valve than was originally intended.

As a result of the large differential pressure at the operating rate of flow, and because the valve is showing cavitation damage at regular intervals, the engineer determines that the control valve is not suitable for this process. The following four options are suggested.

- **Option A:** A new control valve can be installed to accommodate the high pressure differential.
- **Option B:** The pump impeller can be trimmed so that the pump does not develop as much head, resulting in a lower pressure drop across the current valve.
- **Option C:** A variable-frequency drive (VFD) can be installed and the flow control valve removed. The VFD can vary the pump speed and thus achieve the desired process flow.
- **Option D:** The system can be left as it is with a yearly repair of the flow control valve to be expected.

COST SUMMARY: Pumping systems often have a lifespan of 15 to 20 years. Some cost elements will be incurred at the outset, and others will be incurred at different times throughout the lives of the different solutions evaluated. Therefore, you need to calculate a present value of the LCC to accurately assess the different solutions.

Some of the major LCC elements related to a typical pumping system are summarized as follows.

$$\text{LCC} = C_{ic} + C_{in} + C_e + C_o + C_m + C_s + C_{env} + C_d$$

where

   LCC = life-cycle cost;
   C_{ic} = initial costs, purchase price (costs of pump, system, pipe, auxiliary services);
   C_{in} = installation and commissioning cost (including cost of training);
   C_e = energy costs (predicted cost for system operation, including costs of pump driver, controls, and any auxiliary services);
   C_o = operation cost (labor cost of normal system supervision);
   C_m = maintenance and repair costs (costs of routine and predicted repairs);
   C_s = downtime costs (cost of loss of production);
   C_{env} = environmental costs (costs due to contamination from pumped liquid and auxiliary equipment);
   C_d = decommissioning and disposal costs (including the cost of restoration of the local environment and disposal auxiliary services).

For each option, the major cost elements identified are as follows.

- **Option A:** The cost of a new control valve that is properly sized is $5,000. The process operates at 80 m³/h for 6,000 h/year. The energy cost is $0.08 per kWh and the motor efficiency is 90%.

- **Option B:** By trimming the impeller to 375 mm, the pump's total head is reduced to 42.0 m (138 ft) at 80 m³/h. This drop in pressure reduces the differential pressure across the control valve to less than 10 m (33 ft), which better matches the valve's original design intent. The resulting annual energy cost with the smaller impeller is $6,720 per year. It costs $2,250 to trim the impeller. This cost includes the machining cost as well as the cost to disassemble and reassemble the pump.

- **Option C:** A 30-kW VFD costs $20,000 and an additional $1,500 to install. The VFD will cost $500 to maintain each year. It is assumed that it will not need any repairs over the project's eight-year life.

- **Option D:** The option to leave the system unchanged will result in a yearly cost of $4,000 for repairs to the cavitating flow control value.

Table 6.3 summarizes financial as well as technical data related to the various options.

TABLE 6.3 Cost Comparison for Options A Through D in the System with a Failing Control Valve

Cost	Change Control Valve (A)	Trim Impeller (B)	VFD (C)	Repair Control Valve (D)
Pump Cost Data				
Impeller diameter	430 mm	375 mm	430 mm	430 mm
Pump head	71.7 m (235 ft)	42.0 m (138 ft)	34.5 m (113 ft)	71.7 m (235 ft)
Pump efficiency	75.1%	72.1%	77%	75.1%
Rate of flow	80 m³/h	80 m³/h	80 m³/h	80 m³/h
	(350 gpm)	(350 gpm)	(350 gpm)	(350 gpm)
Power consumed	23.1 kW	14.0 kW	11.6 kW	23.1 kW
Energy cost/year	$11,088	$ 6,720	$ 5,568	$11,088
New valve	$ 5,000	0	0	0
Modify impeller	0	$ 2,250	0	0
VFD	0	0	$20,000	0
Installation of VFD	0	0	$ 1,500	
Valve repair/year	0	0	0	$ 4,000
Maintenance cost for VFD	**0**	**0**	$ 500	**0**

SOLUTION

Given: Financial data as summarized in Table 6.3.

Find: Which design option to choose.

Assumptions:

- The current energy price is $0.08/kWh.
- The process is operated for 6,000 hours/year.
- The company has a cost of $500 per year for routine maintenance of pumps of this size, with a repair cost of $2,500 every second year.
- There is no decommissioning cost or environmental disposal cost associated with this project.
- The project has an eight-year life.
- The interest rate for new capital projects is 8%, and an inflation rate of 4% is expected.

Table 6.4 summarizes the life-cycle cost data for four design options.

TABLE 6.4 Comparison of LCC for Options A Through D

	Option A Change Control Valve	Option B Trim Impeller	Option C VFD and Remove Control Valve	Option D Repair Control Valve
Input				
Initial investment cost:	$ 5,000	$ 2,250	$21,500	0
Energy price (present) per kWh:	0.080	0.080	0.080	0.080
Weighted average power of equipment in kW:	23.1	14.0	11.6	23.1
Average operating hours/year:	6,000	6,000	6,000	6,000
Energy cost/year = energy price × weighted average power × average operating hours/year:	11,088	6,720	5,568	11,088
Maintenance cost (routine maintenance/year):	500	500	1,000	500
Repair every second year:	2,500	2,500	2,500	2,500
Other yearly costs:	0	0	0	4,000
Downtime cost/year:	0	0	0	0
Environmental cost:	0	0	0	0
Decommissioning/disposal (salvage) cost:	0	0	0	0
Lifetime in years:	8	8	8	8
Interest rate (%):	8.0%	8.0%	8.0%	8.0%
Inflation rate (%):	4.0%	4.0%	4.0%	4.0%
Output				
Present LCC value:	$ 91,827	$ 59,481	$ 74,313	$113,930
Cost per operating hour	**$2.66**	**$1.73**	**$2.16**	**$3.30**

A sample LCC calculation for option A is shown in Table 6.5. Note that the energy cost and other cost data are escalated at the annual rate of 4%. For example, the current estimate of energy cost is $11,088. To find the cost at the end of year 1, we multiply $11,088 by (1 + 0.04), yielding $11,532. For year 2, we multiply $11,532 by (1 + 0.04) to obtain $11,993. Once we calculate the LCC in present worth ($91,827), we find the equivalent annual value ($15,979) at 8% interest. Finally, to calculate the cost per hour, we divide $15,979 by 6,000 hours, resulting in $2.66. We can calculate the unit costs for other options in a similar fashion.

Option B, trimming the impeller, has the lowest life-cycle cost ($1.73 per hour) and is the preferred option for this example.

TABLE 6.5 LCC Calculation for Option A

	A	B	C	D	E	F	G	H	I	J
1										
2	**Sample Calculation Chart for Life-Cycle-Cost: Option A**									
3										
4	End of Year	0	1	2	3	4	5	6	7	8
5										
6	Initial investment cost:	$5,000								
7	Installation and commissioning cost:	$0								
8										
9	Energy cost/year (calculated)		$11,532	$11,993	$12,472	$12,971	$13,490	$14,030	$14,591	$15,175
10	Routine maintenance		$520	$541	$562	$585	$608	$633	$658	$684
11	Repair cost every 2nd year			$2,704		$2,925		$3,163		$3,421
12	Operating costs		$0	$0	$0	$0	$0	$0	$0	$0
13	Other costs		$0	$0	$0	$0	$0	$0	$0	$0
14										
15	Downtime costs		$0	$0	$0	$0	$0	$0	$0	$0
16										
17	Decommissioning costs									$0
18	Environmental & Disposal costs									$0
19										
20	**Sum costs**	$5,000	$12,052	$15,238	$13,035	$16,481	$14,099	$17,826	$15,249	$19,280
21										
22	**Costs in Present Dollars**	$5,000	$11,159	$13,064	$10,348	$12,114	$9,595	$11,233	$8,898	$10,417
23	Present costs of energy		$10,677	$10,282	$9,901	$9,534	$9,181	$8,841	$8,514	$8,198
24	Present costs of routine maintenance		$481	$464	$446	$430	$414	$399	$384	$370
25										
26	**Sum of Present costs**	$91,826.93	←——— =NPV(8%,C20:J20)+B20							
27	Sum of energy costs	$75,129								
28	Sum of routine maintenance costs	$3,388								
29	**Annual equivalent cost (8%)**	$15,979	←——— =PMT(8%,8,-B26)							
30	**Cost per operating hour**	$2.66	←——— =B29/6000							
31										

6.5 Design Economics

Engineers are frequently involved in making design decisions that provide the required functional quality at the lowest cost. Another valuable extension of AE analysis is minimum-cost analysis, which is used to determine optimal engineering designs. The AE analysis method is useful when two or more cost components are affected differently by the same design element (i.e., for a single design variable, some costs may increase while others decrease). When the equivalent annual total cost of a design variable is a function of increasing and decreasing cost components, we can usually find the optimal value that will minimize the cost of the design with the formula

$$\text{AEC}(i) = a + bx + \frac{c}{x} \tag{6.5}$$

where x is a common design variable and a, b, and c are constants.

To find the value of the common design variable that minimizes AE(i), we need to take the first derivative, equate the result to zero, and solve for x:

$$\frac{d\mathrm{AEC}(i)}{dx} = b - \frac{c}{x^2}$$

$$= 0$$

$$x = x^* = \sqrt{\frac{c}{b}} \tag{6.6}$$

(It is common engineering practice to denote the optimal solution with an asterisk.)

The logic of the first-order requirement is that an activity should be carried to a point, if possible, where its **marginal yield** $d\mathrm{AEC}(i)/dx$ is zero. However, to be sure whether we have found a maximum or a minimum when we have located a point whose marginal yield is zero, we must examine it in terms of what are called the **second-order conditions**. Second-order conditions are equivalent to the usual requirements that the second derivative be negative in the case of a maximum and positive for a minimum. In our situation,

$$\frac{d^2\mathrm{AEC}(i)}{dx^2} = \frac{2c}{x^3}$$

As long as $c > 0$, the second derivative will be positive, indicating that the value x^* is the minimum-cost point for the design alternative. To illustrate the optimization concept, two examples are given, the first having to do with designing the optimal cross-sectional area of a conductor and the second dealing with selecting an optimal size for a pipe.

EXAMPLE 6.11 Optimal Cross-Sectional Area

A constant electric current of 5,000 amps is to be transmitted a distance of 1,000 feet from a power plant to a substation for 24 hours a day, 365 days a year. A copper conductor can be installed for $8.25 per pound. The conductor will have an estimated life of 25 years and a salvage value of $0.75 per pound. The power loss from a conductor is inversely proportional to the cross-sectional area A of the conductor.

It is known that the resistance of the conductor sought is 0.8145×10^{-5} ohm per foot for 1 square inch of cross section. The cost of energy is $0.05 per kilowatt-hour, the interest rate is 9%, and the density of copper is 555 lb/ft^3. For the given data, calculate the optimum cross-sectional area A of the conductor.

DISCUSSION: The resistance of the conductor is the most important cause of power loss in a transmission line. The resistance of an electrical conductor varies directly with its length and inversely with its cross-sectional area according to the equation

$$R = \rho\left(\frac{L}{A}\right) \tag{6.7}$$

where R is the resistance of the conductor, L is the length of the conductor, A is the cross-sectional area of the conductor, and ρ is the resistivity of the conductor material.

Any consistent set of units may be used. In power work in the United States, L is usually given in feet, A in circular mils (cmil), and ρ in ohm-circular mils per foot.

A circular mil is the area of a circle that has a diameter of 1 mil equal to 1×10^{-3} in. The cross-sectional area of a wire in square inches equals its area in circular mils multiplied by 0.7854×10^{-6}. More specifically, one unit relates to another as follows.

$$1 \text{ linear mil} = 0.001 \text{ in.}$$
$$= 0.0254 \text{ millimeter}$$
$$1 \text{ circular mil} = \text{area of circle 1 linear mil in diameter}$$
$$= (0.5)^2 \pi \text{ mil}^2$$
$$= 0.7854 \times 10^{-6} \text{ in.}^2$$
$$1 \text{ in.}^2 = \frac{1}{0.7854 \times 10^{-6}}$$
$$= 1.27324 \times 10^6 \text{ cmil}$$

In SI units (the official designation for the Système International d'Unités), L is in meters, A in square meters, and ρ in ohm-meters. In terms of SI units, the copper conductor has a ρ value of 1.7241×10^{-8} Ω-meter. We can easily convert this value into units of Ω-in.^{2} per foot. From Eq. (6.7), solving for ρ yields

$$\rho = \frac{RA}{L} = 1.7241 \times 10^{-8} \ \Omega (1 \text{ meter})^2/1 \text{ meter}$$

$$= 1.7241 \times 10^{-8} \ \Omega (39.37 \text{ in.})^2/3.2808 \text{ ft}$$

$$= 0.8145421 \times 10^{-5} \ \Omega \text{ in.}^2/\text{ft}$$

When current flows through a circuit, power is used to overcome resistance. The unit of electrical work is the kilowatt-hour (kWh), which is equal to the power in kilowatts, multiplied by the hours during which work is performed. If the current I is steady, then the charge passing through the wire in time T is equivalent to the power that is converted to heat (known as energy loss) and is equal to

$$\text{Power} = I^2 \frac{RT}{1,000} \tag{6.8}$$

where I is the current in amperes, R is the resistance in ohms, and T is the duration in hours during which work is performed.

SOLUTION

Given: Cost components as a function of cross-sectional area (A), $N = 25$ years, and $i = 9\%$.

Find: Optimal value of A.

Step 1. This classic minimum-cost example—the design of the cross-sectional area of an electrical conductor—involves increasing and decreasing cost components. Since the resistance of a conductor is inversely proportional to the size of the conductor, energy loss will decrease as the size of the conductor increases. More specifically, the annual energy loss in kilowatt-hours (kWh) in a conductor due to resistance is equal to

$$\text{Energy loss in kilowatt-hours} = \frac{I^2R}{1{,}000A}T$$

$$= \frac{(5{,}000^2)(0.008145)}{1{,}000A}(24 \times 365)$$

$$= \frac{1{,}783{,}755}{A}\text{ kWh}$$

Step 2. Again, since the electrical resistance is inversely proportional to the cross-sectional area A of the conductor, the total energy loss in dollars per year for a specified conductor material is

$$\text{Cost of energy loss} = \frac{1{,}783{,}755}{A}(\phi)$$

$$= \frac{1{,}783{,}755}{A}(\$0.05)$$

$$= \frac{\$89{,}188}{A}$$

where ϕ is the cost of energy in dollars per kWh.

Step 3. As we increase the size of the conductor, however, it costs more to build. First, we need to calculate the total amount of conductor material in pounds. Since the cross-sectional area is given in square inches, we need to convert the total length in feet to inches before finding the weight of the material:

$$\text{Weight of material in pounds} = \frac{(1{,}000)(12)(555)A}{12^3}$$

$$= 3{,}854(A)$$

$$\text{Total material cost} = 3{,}854(A)(\$8.25)$$

$$= \$31{,}797(A)$$

Here, we are looking for the trade-off between the cost of installation and the cost of energy loss.

Step 4. Since, at the end of 25 years, the copper material will be salvaged at the rate of $0.75 per pound, we can compute the capital recovery cost as

$$\text{CR}(9\%) = [31{,}797A - 0.75(3{,}854A)](A/P, 9\%, 25) + 0.75(3{,}854A)(0.09)$$

$$= 2{,}943A + 260A$$

$$= 3{,}203A$$

Step 5. Using Eq. (6.5), we express the total annual equivalent cost as a function of the design variable A:

$$\text{AEC}(9\%) = \overbrace{3{,}203A}^{\text{Capital cost}} + \underbrace{\frac{89{,}188.}{A}}_{\text{Operating cost}}$$

To find the minimum annual equivalent cost, we use Eq. (6.6):

$$\frac{d\text{AEC}(9\%)}{dA} = 3{,}203 - \frac{89{,}188}{A^2} = 0$$

$$A^* = \sqrt{\frac{89{,}188}{3{,}203}}$$

$$= 5.277 \text{ in.}^2$$

The minimum annual equivalent total cost is

$$\text{AEC}(9\%) = 3{,}203(5.277) + \frac{89{,}188}{5.277}$$

$$= \$33{,}804$$

Figure 6.9 illustrates the nature of this design trade-off problem.

Figure 6.9 Optimal cross-sectional areas for a copper conductor. Note that the minimum point almost coincides with the crossing point of the capital-cost and operating-cost lines. This is not always true. Since the cost components can have a variety of cost patterns, the minimum point does not in general occur at the crossing point (Example 6.11).

COMMENTS: We may attempt to solve this optimization problem using Excel. Table 6.6 illustrates how this task may be carried out. The total AEC is calculated as a function of cross-sectional area by varying the parameter from 0 to 6 in^2 in the increment of 0.5 in^2.

TABLE 6.6 Excel Output: Determining the Optimal Cross-Sectional Area (Example 6.11)

	A	B	C	D	E	F	G	H	I
1									
2		Example 6.11 Optimal Cross-Sectional Area							
3									
4			Input Parameter			Input Data			
5									
6		Electric current (amperes)				5000			
7		Transmission line length (feet)				1000			
8		Annual operating hours				8760			
9		Service life (years)				25			
10		Material cost ($/lb)				8.25			
11		Scrap value ($/lb)				0.75			
12		Resistivity of the conductor (ohms-in^2/ft)				8.15E-06			
13		Cost of energy ($/kWh)				0.05			
14		Material Density (lb/ft^3)				555			
15		MARR(%)				9			
16									
17		Cross-					Capital	Annual	Total
18		Sectional	Energy	Material	Material	Salvage	Recovery	Energy	AE
19		Area (A)	Loss	Weight	Cost	Value	Cost	Loss	Cost
20		(in^2)	(kWh)	(lbs)	($)	($)	($)	($)	($)
21									
22		1.000	1,783,755	3,854	31,797	2,891	3,203	89,188	92,391
23		1.500	1,189,170	5,781	47,695	4,336	4,804	59,459	64,263
24		2.000	891,878	7,708	63,594	5,781	6,406	44,594	51,000
25		2.500	713,502	9,635	79,492	7,227	8,007	35,675	43,683
26		3.000	594,585	11,563	95,391	8,672	9,609	29,729	39,338
27		3.500	509,644	13,490	111,289	10,117	11,210	25,482	36,693
28		4.000	445,939	15,417	127,188	11,563	12,812	22,297	35,109
29		4.500	396,390	17,344	143,086	13,008	14,413	19,820	34,233
30		5.000	356,751	19,271	158,984	14,453	16,015	17,838	33,853
31		5.275	338,153	20,331	167,729	15,248	16,896	16,908	33,803
32		5.500	324,319	21,198	174,883	15,898	17,616	16,216	33,832
33		6.000	297,293	23,125	190,781	17,344	19,218	14,865	34,083
34	Optimal Cross-								
35	Sectional Area								
36									
37		Cell Formulas:							
38									
39		C22:=($F$6)^2*($F$12*$F$7)*($F$8)/(1000*B22)							
40		D22:=($F$7)*(12)*($F$14)*(B22)/(12^3)							
41		E22:=D22*$F$10							
42		F22:=$F$11*D22							
43		G22:=-PMT($F$15%,$F$9,E22-F22)+($F$15/100)*F22							
44		H22:=C20*$F$11							
45		I22:=G22+H22							
46									
47									

EXAMPLE 6.12 Economical Pipe Size

To meet the ever-increasing demand for oil, Kuwait is studying the feasibility of running a steel pipeline across the Arabian Peninsula to the Red Sea. The pipeline will be designed to handle 3 million barrels of crude oil per day under optimum conditions. The length of the line will be 600 miles. Calculate the optimum pipeline diameter that will be used for 20 years for the following data at $i = 10\%$.

$$\text{Pumping power} = \frac{1.333Q\Delta P}{1,980,000}\text{HP}$$

$$Q = \text{volume flow rate, ft}^3/\text{hour}$$

$$\Delta P = \frac{128Q\mu L}{g\pi D^4}, \text{ pressure drop, lb/ft}^2$$

$$L = \text{pipe length, ft}$$

$$D = \text{inside pipe diameter, ft}$$

$$t = 0.01D, \text{ pipeline wall thickness, ft}$$

$$\mu = 8,500 \text{ lb/hour-ft, oil viscosity}$$

$$g = 32.2 \times 12,960,000 \text{ ft/hour}^2$$

Power cost, $0.015 per HP-hour

Pipeline cost, $1.00 per pound of steel

Pump and motor costs, $195 per HP

The salvage value of the steel after 20 years is assumed to be zero, because removal costs exhaust scrap profits from steel. (See Figure 6.10 for the relationship between D and t.)

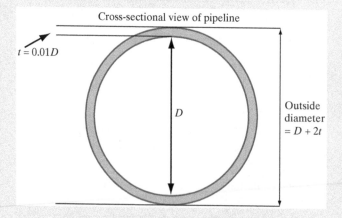

Figure 6.10 Designing economical pipe size to handle 3 million barrels of crude oil per day (Example 6.11).

DISCUSSION: In general, when a progressively larger size of pipe is used to carry a given fluid at a given volume flow rate, the energy required to move the fluid will progressively decrease. However, as we increase the pipe size, the cost of its construction will increase. In practice, to obtain the best pipe size for a particular situation, you may choose a reasonable, but small, starting size. Compute the energy cost of pumping fluid through this size of pipe and the total construction cost, and then compare the difference in energy cost with the difference in construction cost. When the savings in energy cost exceeds the added construction cost, the process may be repeated with progressively larger pipe sizes until the added construction cost exceeds the savings in energy cost. As soon as this happens, the best pipe size to use in the particular application is identified. However, this search process can be simplified by using the minimum-cost concept encompassed by Eqs. (6.5) and (6.6).

SOLUTION

Given: Cost components as a function of pipe diameter (D), $N = 20$ years, and $i = 10\%$.

Find: Optimal pipe size (D).

Step 1. Several different units are introduced; however, we need to work with common units. We will assume the following conversions.

$$1 \text{ mile} = 5{,}280 \text{ ft}$$

$$600 \text{ miles} = 600 \times 5{,}280 = 3{,}168{,}000 \text{ ft}$$

$$1 \text{ barrel} = 42 \text{ U. S. gallons}$$

$$1 \text{ barrel} = 42 \text{ gallons} \times 231 \text{ in.}^3/\text{gallon} = 9{,}702 \text{ in.}^3$$

$$1 \text{ barrel} = 9{,}702 \text{ in.}^3/12^3 = 5.6146 \text{ ft}^3$$

$$\text{Density of steel} = 490.75 \text{ lb/ft}^3$$

Step 2. Power required to pump oil: It is well known that, for any given set of operating conditions involving the flow of a noncompressible fluid, such as oil, through a pipe of constant diameter, a small-diameter pipe will have a high fluid velocity and a high fluid pressure drop through the pipe. This will require a pump that will deliver a high discharge pressure and a motor with large energy consumption. To determine the pumping power required to boost the pressure drop, we need to determine both the volume flow rate and the amount of pressure drop. Then we can calculate the cost of the power required to pump oil.

- **Volume flow rate per hour:**

$$Q = 3{,}000{,}000 \text{ barrels/day} \times 5.6146 \text{ ft}^3/\text{barrel}$$

$$= 16{,}843{,}800 \text{ ft}^3/\text{day}$$

$$= 701{,}825 \text{ ft}^3/\text{hour}$$

- **Pressure drop:**

$$\Delta P = \frac{128 Q \mu L}{g \pi D^4}$$

$$= \frac{128 \times 701{,}825 \times 8{,}500 \times 3{,}168{,}000}{32.2 \times 12{,}960{,}000 \times 3.14159 D^4}$$

$$= \frac{1{,}845{,}153{,}595}{D^4} \text{ lb/ft}^2$$

- **Pumping power required to boost the pressure drop:**

$$\text{Power} = \frac{1.333 Q \Delta P}{1{,}980{,}000}$$

$$= \frac{1.333 \times 701{,}825 \times \dfrac{1{,}845{,}153{,}595}{D^4}}{1{,}980{,}000}$$

$$= \frac{871{,}818{,}975}{D^4} \text{ HP}$$

- **Power cost to pump oil:**

$$\text{Power cost} = \frac{871{,}818{,}975}{D^4} \text{ HP} \times \$0.015/\text{HP-hour}$$

$$\times 24 \text{ hours/day} \times 365 \text{ days/year}$$

$$= \frac{\$114{,}557{,}013{,}315}{D^4}/\text{year}$$

Step 3. Pump and motor cost calculation: Once we determine the required pumping power, we can find the size of the pump and the motor costs. This is because the capacity of the pump and motor is proportional to the required pumping power. Thus,

$$\text{Pump and motor cost} = \frac{871{,}818{,}975}{D^4} \times \$195/\text{HP}$$

$$= \frac{\$170{,}004{,}700{,}125}{D^4}$$

Step 4. Required amount and cost of steel: The pumping cost will be counterbalanced by the lower costs for the smaller pipe, valves, and fittings. If the pipe diameter is made larger, the fluid velocity drops markedly, and the pumping costs become substantially lower. Conversely, the capital cost for the larger pipe, fittings, and valves becomes greater. For a given cross-sectional area of the pipe, we can determine the total volume of the pipe as well as the weight. Once the total weight of the pipe is determined, we can easily convert it into the required investment cost. The calculations are as follows.

$$\text{Cross-sectional area} = 3.14159[(0.51D^2 - (0.50D^2)]$$

$$= 0.032D^2 \text{ ft}^2$$

$$\text{Total volume of pipe} = 0.032D^2 \text{ ft}^2 \times 3{,}168{,}000 \text{ ft}$$

$$= 101{,}376D^2 \text{ ft}^3$$

$$\text{Total weight of steel} = 101{,}376D^2 \text{ ft}^3 \times 490.75 \text{ lb/ft}^3$$

$$= 49{,}750{,}272D^2 \text{ lb}$$

$$\text{Total pipeline cost} = \$1.00/\text{lb} \times 49{,}750{,}272D^2 \text{ lb}$$

$$= \$49{,}750{,}272D^2$$

Step 5. Annual equivalent cost calculation: For a given total pipeline cost and its salvage value at the end of 20 years of service life, we can find the equivalent annual capital cost by using the formula for the capital recovery factor with the return:

$$\text{Capital cost} = \left(\$49{,}750{,}272D^2 + \frac{\$170{,}004{,}700{,}125}{D^4} \right)(A/P, 10\%, 20)$$

$$= 5{,}843{,}648D^2 + \frac{19{,}968{,}752{,}076}{D^4}$$

$$\text{Annual power cost} = \frac{\$114{,}557{,}013{,}315}{D^4}$$

Step 6. Economical pipe size: Now that we have determined the annual pumping and motor costs and the equivalent annual capital cost, we can express the total equivalent annual cost as a function of the pipe diameter (D):

$$\text{AEC}(10\%) = 5{,}843{,}648D^2 + \frac{19{,}968{,}752{,}076}{D^4} + \frac{114{,}557{,}013{,}315}{D^4}$$

To find the optimal pipe size (D) that results in the minimum annual equivalent cost, we take the first derivative of AEC(10%) with respect to D, equate the result to zero, and solve for D:

$$\frac{d\text{AEC}(10\%)}{dD} = 11{,}687{,}297D - \frac{538{,}103{,}061{,}567}{D^5}$$

$$= 0;$$

$$11{,}687{,}297D^6 = 538{,}103{,}061{,}567$$

$$D^6 = 46{,}041.70$$

$$D^* = 5.9868 \text{ ft}$$

Note that, ideally, the velocity in a pipe should be no more than approximately 10 ft/sec, because friction wears the pipe. To check whether the preceding answer is reasonable, we may compute

$$Q = \text{velocity} \times \text{pipe inner area}$$

$$701{,}825 \text{ ft}^3/\text{hr} \times \frac{1}{3{,}600} \text{ hr/sec} = V\frac{3.14159(5.9868)^2}{4}$$

$$V = 6.93 \text{ ft/sec}$$

which is less than 10 ft/sec. Therefore, the optimal answer as calculated is practical.

Step 7. Equivalent annual cost at optimal pipe size:

$$\text{Capital cost} = \left[\$49{,}750{,}272(5.9868)^2 + \frac{\$170{,}004{,}700{,}125}{5.9868^4} \right] (A/P, 10\%, 20)$$

$$= 5{,}843{,}648(5.9868)^2 + \frac{19{,}968{,}752{,}076}{5.9868^4}$$

$$= \$224{,}991{,}039$$

$$\text{Annual power cost} = \frac{114{,}557{,}013{,}315}{5.9868^4}$$

$$= \$89{,}174{,}911$$

$$\text{Total annual cost} = \$224{,}991{,}039 + \$89{,}174{,}911$$

$$= \$314{,}165{,}950$$

Step 8. Total annual oil revenue:

$$\text{Annual oil revenue} = (\$100/\text{bbl}) \times (3{,}000{,}000 \text{ bbls/day})$$

$$\times 365 \text{ day/year}$$

$$= \$109{,}500{,}000{,}000/\text{year}$$

Enough revenues are available to offset the capital cost as well as the operating cost.

COMMENTS: A number of other criteria exist for choosing the pipe size for a particular fluid transfer application. For example, low velocity may be required when erosion or corrosion concerns must be considered. Alternatively, higher velocities may be desirable for slurries when settling is a concern. Ease of construction may also weigh significantly in choosing a pipe size. On the one hand, a small pipe size may not accommodate the head and flow requirements efficiently; on the other, space limitations may prohibit the selection of large pipe sizes.

SUMMARY

- Annual equivalent worth analysis (AE) is—along with present-worth analysis—one of the two main analysis techniques based on the concept of equivalence. The equation for AE is

$$AE(i) = PW(i)(A/P, i, N)$$

AE analysis yields the same decision result as PW analysis.

- The capital recovery cost factor (CR(i)) is one of the most important applications of AE analysis, because it allows managers to calculate an annual equivalent cost of capital for ease of itemization with annual operating costs. The equation for CR(i) is

$$CR(i) = (I - S)(A/P, i, N) + iS$$

where I = initial cost and S = salvage value.

- AE analysis is recommended over NPW analysis in many key real-world situations for the following reasons.

1. In many financial reports, an annual equivalent value is preferred to a present-worth value.

2. Unit costs often must be calculated to determine reasonable pricing for items that are on sale.

3. The cost per unit of use of an item must be calculated in order to reimburse employees for the business use of personal cars.

4. Make-or-buy decisions usually require the development of unit costs for the various alternatives.

5. Minimum-cost analysis is easy to do when it is based on annual equivalent cost.

- LCCA is a way to predict the most cost-effective solution by allowing engineers to make a reasonable comparison among alternatives within the limit of the available data.

PROBLEMS

Note: *Unless otherwise stated, all cash flows given in the problems that follow represent after-tax cash flows.*

Annual Equivalent-Worth Calculation

6.1 Consider the cash flows in Figure P6.1 and compute the equivalent annual worth at $i = 12\%$.

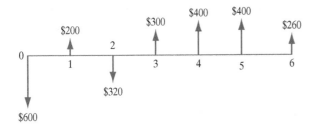

Figure P6.1

6.2 Consider the accompanying cash flow diagram. Compute the equivalent annual worth at $i = 14\%$.

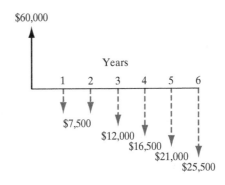

Figure P6.2

6.3 Calculate the equivalent annual worth of the following scheduled payments at an interest rate of 13%.

Figure P6.3

6.4 Consider the accompanying cash flow diagram. Compute the equivalent annual worth at $i = 14\%$.

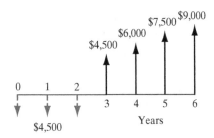

Figure P6.4

6.5 Consider the accompanying cash flow diagram. Compute the equivalent annual worth at $i = 13\%$.

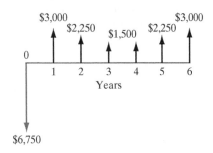

Figure P6.5

6.6 Consider the accompanying cash flow diagram. Compute the equivalent annual worth at $i = 11\%$.

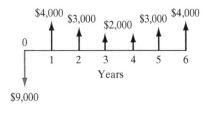

Figure P6.6

6.7 Consider the cash flows in Table P6.7 for the following investment projects (MARR $= 14\%$).

TABLE P6.7

	Project's Cash Flow		
n	A	B	C
0	−$8,500	−$8,500	−$8,500
1	$3,000	$3,000	$2,500
2	$3,200	$3,000	$3,200
3	$3,400	$3,000	$3,200
4	$1,600	$2,400	$3,200

Determine the annual equivalent worth for each project at $i = 15\%$ and determine the acceptability of each project.

6.8 The Solar Energy Company is producing electricity directly from a solar source by using a large array of solar cells and selling the power to the local utility company. Because these cells degrade over time, thereby resulting in lower conversion efficiency and power output, the cells must be replaced every four years, which results in a particular cash flow pattern that repeats itself as follows: $n = 0$, −$600,000; $n = 1$, $400,000; $n = 2$, $300,000; $n = 3$, $200,000; and $n = 4$, $100,000. Determine the annual equivalent cash flows at $i = 10\%$.

6.9 Consider the sets of investment projects from Table P6.9. Compute the equivalent annual worth of each project at $i = 15\%$, and determine the acceptability of each project.

TABLE P6.9

	Project's Cash Flow			
n	A	B	C	D
0	−$8,500	−$6,800	−$6,500	−$8,200
1	$4,600	$2,200	$4,400	$0
2	$4,600	$3,000	$3,200	$0
3	$4,600	$3,800	$2,000	$12,200

6.10 The repeating cash flows for a certain project are as given in Table P6.10. Find the equivalent annual worth for this project at $i = 12\%$, and determine the acceptability of the project.

TABLE P6.10

	Net Cash Flow		
n	Investment	Operating Income	
0	−$5,400	$0	1st cycle
1	$0	$3,500	
2	$0	$2,200	
3	−$5,400	$1,600	2nd cycle
4	$0	$3,500	
5	$0	$2,200	
6	−$5,400	$1,600	3rd cycle
7	$0	$3,500	
8	$0	$2,200	
9	$0	$1,600	

6.11 Beginning next year, a foundation will support an annual event on campus with the earnings of a $420,000 gift it received this year. It is felt that 8.4% interest will be realized for the first 12 years, but that plans should be made to anticipate an interest rate of 6% after that time. What amount should be added to the foundation now to fund the event at the $45,000 level into infinity?

Capital (Recovery) Cost/Annual Equivalent Cost

6.12 K.T. Labs bought a Gene gun for $25,000. The accounting department has estimated that the machine would have an annualized capital cost of $3,880 over its 10-year service life. What salvage value was assumed in calculating the capital cost? The firm's interest rate is known to be 10%.

6.13 You are considering investing $65,000 in new equipment. You estimate that the net cash flows will be $18,000 during the first year, but will increase by $2,500 per year the next year and each year thereafter. The equipment is estimated to have a 10-year service life and a net salvage value of $5,000 at that time. Assume an interest rate of 9%.

(a) Determine the annual capital cost (ownership cost) for the equipment.

(b) Determine the equivalent annual savings (revenues).

(c) Determine whether this is a wise investment.

6.14 What is the annual equivalent cost of purchasing a lift truck that has an initial cost of $65,000, an annual operating cost of $13,500, and an estimated salvage value of $23,000 after six years of use at an annual interest rate of 6%?

6.15 You are considering purchasing a dump truck. The truck will cost $75,000 and have operating and maintenance costs that start at $18,000 the first year and increases by $2,000 per year. Assume that the salvage value at the end of five years is $22,000 and interest rate is 12%. What is the equivalent annual cost of owning and operating the truck?

6.16 The Emerson Electronics Company just purchased a soldering machine to be used in its assembly cell for flexible disk drives. The soldering machine cost $350,000. Because of the specialized function it performs, its useful life is estimated to be five years. It is also estimated that at that time its salvage value will be $60,000. What is the capital recovery cost for this investment if the firm's interest rate is 15%?

6.17 The present price (year 0) of kerosene is $4.30 per gallon, and its cost is expected to increase by 10% per year. (At the end of year 1, kerosene will cost $4.73 per gallon.) Mr. Garcia uses about 800 gallons of kerosene for space heating during a winter season. He has an opportunity to buy a storage tank for $600, and at the end of four years, he can sell the storage tank for $100. The tank has a capacity to supply four years of Mr. Garcia's heating needs. So, he can buy four years' worth of kerosene at its present price ($4.30), or he can invest his money elsewhere at 6% interest. Should he purchase the storage tank? Assume that kerosene purchased on a pay-as-you-go basis is paid for at the end of the year. (However, kerosene purchased for the storage tank is purchased now.)

6.18 The city of Carlsbad in California is considering building a $300 million water-desalination plant. The facility would be the largest in the Western Hemisphere, producing 50 million gallons of drinking water a day—enough to supply about 100,000 homes. This plant, which uses improved membranes and pumping system to bring down the energy cost to $1.10 in electricity to produce 1,000 gallons of water. Suppose the desalination plant is fully operational and you expect to operate it continuously for 20 years, with an estimated salvage value of $20 million. The operating and maintenance costs (excluding the energy cost) amount to $15 million the first year and will increase at a rate of 3% each year. If the city plans to sell the water for about $950 per acre-foot. (an acre-foot is 325,851 gallons—enough water for four people per year.), can it recover the capital cost and all other operating costs? Assume the city's interest rate for this water project is known to be 6%.

6.19 The Geo-Star Manufacturing Company is considering a new investment in a punch-press machine that will cost $100,000 and has an annual maintenance cost of $10,000. There is also an additional overhauling cost of $20,000 for the equipment once every four years. Assuming that this equipment will last infinitely under these conditions, what is the capitalized equivalent cost of this investment at an interest rate of 10%?

6.20 The cash flows for two investment projects are as given in Table P6.20.

(a) For project A, find the value of X that makes the equivalent annual receipts equal the equivalent annual disbursement at $i = 15\%$.

(b) For A to be preferred over project B, determine the minimum acceptable value of X in year 2 at $i = 12\%$ based on an AE criterion.

TABLE P6.20

	Project's Cash Flow	
n	A	B
0	−$4,500	$6,500
1	$1,000	−$1,400
2	X	−$1,400
3	$1,000	−$1,400
4	$1,000	−$1,400

6.21 An industrial firm can purchase a special machine for $70,000. A down payment of $5,000 is required, and the unpaid balance can be paid off in five equal year-end installments at 9% interest. As an alternative, the machine can be purchased for

$66,000 in cash. If the firm's MARR is 10%, use the annual equivalent method to determine which alternative should be accepted.

6.22 An industrial firm is considering purchasing several programmable controllers and automating the company's manufacturing operations. It is estimated that the equipment will initially cost $120,000, and the labor to install it will cost $25,000. A service contract to maintain the equipment will cost $5,000 per year. Trained service personnel will have to be hired at an annual salary of $50,000. Also estimated is an approximate $10,000 annual income-tax savings (cash inflow). How much will this investment in equipment and services have to increase the annual revenues after taxes in order to break even? The equipment is estimated to have an operating life of 10 years with no salvage value because of obsolescence. The firm's MARR is 12%.

6.23 A construction firm is considering establishing an engineering computing center. The center will be equipped with three engineering workstations that cost $45,000 each, and each has a service life of five years. The expected salvage value of each workstation is $2,000. The annual operating and maintenance cost would be $25,000 for each workstation. At a MARR of 15%, determine the equivalent annual cost for operating the engineering center.

Unit-Cost Profit Calculation

6.24 You have purchased a machine costing $30,000. The machine will be used for two years, and at the end of this time, its salvage value is expected to be $18,000. The machine will be used 6,000 hours during the first year and 8,000 hours during the second year. The expected annual net savings will be $35,000 during the first year and $42,000 during the second year. If your interest rate is 12%, what would be the equivalent net savings per machine hour?

6.25 The General Mills Company (GMC) purchased a milling machine for $100,000, which it intends to use for the next five years. This machine is expected to save GMC $35,000 during the first operating year. Then the annual savings are expected to decrease by *3% each subsequent year over the previous year* due to increased maintenance costs.

Assuming that GMC would operate the machine for an average of 3,000 hours per year and that it would have no appreciable salvage value at the end of the five-year period, determine the equivalent dollar savings per operating hour at 14% interest compounded annually.

6.26 The City of Washington is considering adding new buses for its current mass-transit system that links from the Tacoma International Airport to clustering city destinations on non-stop basis. The total investment package is worth $10 million and expected to last 10 years with a salvage value of $850,000. The annual operating and maintenance costs for buses would be $5 million during the first year and will grow by 5% each year over the previous year's O&M costs thereafter. If the system is used for 800,000 trips per year, what would be the fair price to charge per trip? Assume that the City of

Washington uses 5% interest rate for any city-sponsored project. Assume the fare will be the same for all destinations.

6.27 Jennifer Ozman is a sales engineer at Montana Chemical Engineering Company. Jennifer owns two vehicles, and one of them is entirely dedicated to business use. Her business car is a small, used pickup truck, which she purchased with $11,000 of personal savings. On the basis of her own records and with data compiled by the U.S. Department of Transportation, Jennifer has estimated the costs of owning and operating her business vehicle for the first three years as given in Table P6.27. If her interest rate is 7%, what should be Jennifer's reimbursement rate per mile so that she can break even?

TABLE P6.27

	First Year	**Second Year**	**Third Year**
Depreciation	$2,879	$1,776	$1,545
Scheduled maintenance	100	153	220
Insurance	635	635	635
Registration and taxes	78	57	50
Total ownership cost	$3,692	$2,621	$2,450
Nonscheduled repairs	35	85	200
Replacement tires	35	30	27
Accessories	15	13	12
Gasoline and taxes	688	650	522
Oil	80	100	100
Parking and tolls	135	125	110
Total operating costs	$988	$1,003	$971
Total of all costs	$4,680	$3,624	$3,421
Expected miles driven	14,500	13,000	11,500

6.28 Two 180-horsepower (HP) motors are being considered for installation at a municipal sewage-treatment plant. The first costs $5,800 and has an operating efficiency of 83%. The second costs $4,600 and has an efficiency of 80%. Both motors are projected to have zero salvage value after a life of 10 years. If all the annual charges, such as insurance,

maintenance, etc., amount to a total of 15% of the original cost of each motor, and if power costs are a flat five cents per kilowatt-hour, how many minimum hours of full-load operation per year are necessary to justify purchasing the more expensive motor at $i = 6\%$? (A conversion factor you might find useful is 1 HP = 746 watts = 0.746 kilowatts.)

6.29 The Novelty Company, a manufacturer of farm equipment, currently produces 20,000 units of gas filters per year for use in its lawn-mower production. The costs, based on the previous year's production, are reported in Table P6.29.

TABLE P6.29 Production Costs

Item	Expense ($)
Direct materials	$60,000
Direct labor	$180,000
Variable overhead (power and water)	$135,000
Fixed overhead (light and heat)	$70,000
Total cost	**$445,000**

It is anticipated that gas-filter production will last five years. If the company continues to produce the product in-house, annual direct material costs will increase at the rate of 5%. (For example, annual material costs during the first production year will be $63,000.) Direct labor will increase at the rate of 6% per year, and variable overhead costs will increase at the rate of 3%. However, the fixed overhead will remain at its current level over the next five years. The John Holland Company has offered to sell Novelty 20,000 units of gas filters for $25 per unit. If Novelty accepts the offer, some of the manufacturing facilities currently used to manufacture the filter could be rented to a third party for $35,000 per year. In addition, $3.5 per unit of the fixed overheard applied to the production of gas filters would be eliminated. The firm's interest rate is known to be 15%. What is the unit cost of buying the gas filters from the outside source? Should Novelty accept John Holland's offer, and if so, why?

6.30 Southern Environmental Consulting (SEC), Inc. designs plans and specifications for asbestos abatement (removal) projects in public, private, and government buildings. Currently, SEC must conduct an air test before allowing the occupancy of a building after the asbestos has been removed. SEC subcontracts air-test samples to a laboratory for analysis by transmission electron microscopy (TEM). To offset the cost of TEM analysis, SEC charges its clients $100 more than the subcontractor's fee. The only expenses in this system are the costs of shipping the air-test samples to the subcontractor and the labor involved in shipping the samples. With the growth of the business, SEC must consider either continuing to subcontract the TEM analysis to outside companies or developing its own TEM laboratory. Because of the passage of the Asbestos Hazard Emergency Response Act (AHERA) by the U.S. Congress, SEC expects about 1,000 air-sample tests per year over eight years. The firm's MARR is known to be 15%.

- **Subcontract option.** The client is charged $400 per sample, which is $100 above the subcontracting fee of $300. Labor expenses are $1,500 per year, and shipping expenses are estimated to be $0.50 per sample.

- **TEM purchase option.** The purchase and installation cost for the TEM is $415,000. The equipment would last for eight years, at which time it would have no salvage value. The design and renovation cost is estimated to be $9,500. The client is charged $300 per sample based on the current market price. One full-time manager and two part-time technicians are needed to operate the laboratory. Their combined annual salaries will be $50,000. Material required to operate the lab includes carbon rods, copper grids, filter equipment, and acetone. The costs of these materials are estimated at $6,000 per year. Utility costs, operating and maintenance costs, and the indirect labor needed to maintain the lab are estimated at $18,000 per year. The extra income-tax expenses would be $20,000.

(a) Determine the cost of an air-sample test by the TEM laboratory (in-house).

(b) What is the required number of air samples per year to make the two options equivalent?

6.31 A company is currently paying its employees $0.56 per mile to drive their own cars on company business. The company is considering supplying employees with cars, which would involve purchasing at $25,000 with an estimated three-year life, a net salvage value of $8,000, taxes and insurance at a cost of $1,200 per year, and operating and maintenance expenses of $0.30 per mile. If the interest rate is 10% and the company anticipates an employee's annual travel to be 30,000 miles, what is the equivalent cost per mile (neglecting income taxes)?

6.32 An automobile that runs on electricity can be purchased for $25,000. The automobile is estimated to have a life of 12 years with annual travel of 20,000 miles. Every three years, a new set of batteries will have to be purchased at a cost of $3,000. Annual maintenance of the vehicle is estimated to cost $700. The cost of recharging the batteries is estimated at $0.015 per mile. The salvage value of the batteries and the vehicle at the end of 12 years is estimated to be $2,000. Suppose the MARR is 7%. What is the cost per mile to own and operate this vehicle based on the preceding estimates? The $3,000 cost of the batteries is a net value with the old batteries traded in for the new ones.

6.33 The estimated cost of a completely installed and ready-to-operate 40-kilowatt generator is $32,000. Its annual maintenance costs are estimated at $800. The energy that can be generated annually at full load is estimated to be 100,000 kilowatt-hours. If the value of the energy generated is $0.09 per kilowatt-hour, how long will it take before this machine becomes profitable? Take the MARR to be 10% and the salvage value of the machine to be $2,000 at the end of its estimated life of 15 years.

6.34 A large land-grant university that is currently facing severe parking problems on its campus is considering constructing parking decks off campus. A shuttle service could pick up students at the off-campus parking deck and transport them to various locations on campus. The university would charge a small fee for each shuttle ride, and the students could be quickly and economically transported to their classes. The funds raised by the shuttle would be used to pay for trolleys, which cost about $170,000 each. Each trolley has a 12-year service life, with an estimated salvage value of $12,000. To operate each trolley, additional expenses will be incurred, as given in Table P6.34.

TABLE P6.34

Item	Annual Expenses ($)
Driver	$70,000
Maintenance	$15,000
Insurance	$5,000

If students pay 10 cents for each ride, determine the annual ridership per trolley (number of shuttle rides per year) required to justify the shuttle project, assuming an interest rate of 6%.

6.35 Eradicator Food Prep, Inc. has invested $10 million to construct a food irradiation plant. This technology destroys organisms that cause spoilage and disease, thus extending the shelf life of fresh foods and the distances over which they can be shipped. The plant can handle about 300,000 pounds of produce in an hour, and it will be operated for 4,000 hours a year. The net expected operating and maintenance costs (taking into account income-tax effects) would be $4 million per year. The plant is expected to have a useful life of 15 years with a net salvage value of $800,000. The firm's interest rate is 15%.

(a) If investors in the company want to recover the plant investment within six years of operation (rather than 15 years), what would be the equivalent annual revenues that must be generated?

(b) To generate annual revenues determined in part (a), what minimum processing fee per pound should the company charge to its producers?

6.36 The local government of Santa Catalina Island, off the coast of Long Beach, California, is completing plans to build a desalination plant to help ease a critical drought on the island. The drought has combined with new construction on Catalina to leave the island with an urgent need for a new water source. On Catalina, the cost of acquiring water from natural sources is about the same as that for desalting. The $5 million plant, with a daily desalting capacity of 0.4 acre-foot, can produce 132,000 gallons of fresh water a day (enough to supply 295 households daily), more than a quarter of the island's total needs. The desalination plant has an estimated service life of 20 years with no appreciable salvage value. The annual operating and maintenance costs would be about $250,000. Assuming an interest rate of 10%, what should be the minimum monthly water bill for each household?

6.37 A California utility firm is considering building a 50-megawatt geothermal plant that generates electricity from naturally occurring underground heat. The binary geothermal system will cost $85 million to build and $6 million (including any income-tax effect) to operate per year. (Virtually no fuel costs will accrue compared with fuel costs related to a conventional fossil-fuel plant.) The geothermal plant is to last for 25 years. At that time, its expected salvage value will be about the same as the cost to remove the plant. The plant will be in operation for 70% (plant utilization factor) of the year (or 70% of 8,760 hours per year). If the firm's MARR is 14% per year, determine the cost per kilowatt-hour of generating electricity.

6.38 A corporate executive jet with a seating capacity of 20 has the cost factors given in Table P6.38.

The company flies three round trips from Boston to London per week, which is a distance of 3,280 miles one way. How many passengers must be carried on an average trip in order to justify the use of the jet if the first-class round-trip fare is $3,400? The firm's MARR is 15%. (Ignore income-tax consequences.)

TABLE P6.38

Item	Cost
Initial cost	$12,000,000
Service life	15 years
Salvage value	$2,000,000
Crew costs per year	$225,000
Fuel cost per mile	$1.10
Landing fee	$250
Maintenance per year	$237,500
Insurance cost per year	$166,000
Catering per passenger trip	$75

Comparing Mutually Exclusive Alternatives by the AE Method

6.39 The cash flows in Table P6.39 represent the potential annual savings associated with two different types of production processes, each of which requires an investment of $40,000.

Assume an interest rate of 12%.

(a) Determine the equivalent annual savings for each process.

(b) Determine the hourly savings for each process if it will be in operation of 3,000 hours per year.

(c) Which process should be selected?

TABLE P6.39

n	Process A	Process B
0	−$40,000	−$40,000
1	$19,120	$17,350
2	$17,840	$17,350
3	$16,560	$17,350
4	$15,280	$17,350

6.40 Washington Air Company is considering the purchase of a helicopter for connecting services between the company's base airport and the new intercounty airport being built about 30 miles away. It is believed that the chopper will be needed only for six years until the Rapid Transit Connection is phased in. The estimates on two types of helicopters under consideration, the Whirl 2B and the ROT 8, are given in Table P6.40.

TABLE P6.40

	The Whirl 2B	The ROT 8
First cost	$95,000	$120,000
Annual maintenance	$3,000	$9,000
Salvage value	$12,000	$25,000
Useful life in years	3	6

Assuming that the Whirl 2B will be available in the future with identical costs, what is the annual cost advantage of selecting the ROT8? (Use an interest rate of 10%.)

6.41 Birmingham Steel, Inc. is considering whether to buy conventional 25-HP, 230-V, 60-Hz, 1800-rpm induction motors in its plant or modern premium efficiency (PE) motors with the same specifications. Both types of motor have a power output of 18.650 kW per motor (25 HP × 0.746 kW/HP). Conventional motors have a published efficiency of 89.5%, while the PE motors are 93% efficient. The initial cost of the conventional motors is $14,000, whereas the initial cost of the proposed PE motors is $16,600. The motors are operated 12 hours per day, 5 days per week, 52 weeks per year, with a local utility cost of $0.08 per kilowatt-hour (kWh). The motors are to be operated at a 75% load, and the lifecycle of both the conventional and PE motors is 20 years with no appreciable salvage value.

(a) At an interest rate of 12%, what are the savings per kWh achieved by switching from the conventional motors to the PE motors?

(b) At what operating hours are the two motors equally economical?

6.42 A certain factory building has an old lighting system. Lighting the building currently costs, on average, $20,000 a year. A lighting consultant tells the factory supervisor that the lighting bill can be reduced to $8,000 a year if $50,000 is invested in new lighting in the building. If the new lighting system is installed, an incremental maintenance cost of $3,000 per year must be taken into account. The new lighting system has zero salvage value at the end of its life. If the old lighting system also has zero salvage value, and the new lighting system is estimated to have a life of 20 years, what is the net annual benefit for this investment in new lighting? Take the MARR to be 12%. Assume the old lighting system will last 20 years.

6.43 Your company needs a machine for the next seven years, and you have two choices (assume an annual interest rate of 15%).

- **Machine A** costs $100,000 and has an annual operating cost of $47,000. Machine A has a useful life of seven years and a salvage value of $15,000.
- **Machine B** costs $150,000 and has an annual operating cost of $30,000. Machine B has a useful life of five years and no salvage value. However, the life of Machine B can be extended by two years with a certain amount of investment. If Machine B's life is extended, it will still cost $30,000 annually to operate and still have no salvage value.

What would you pay at the end of year 5 to extend the life of Machine B by two years?

6.44 Travis Wenzel has $2,000 to invest. Usually, he would deposit the money in his savings account, which earns 6% interest compounded monthly. However, he is considering three alternative investment opportunities.

- **Option 1.** Purchasing a bond for $2,000. The bond has a face value of $2,000 and pays $100 every six months for three years, after which time the bond matures.
- **Option 2.** Buying and holding a stock that grows 11% per year for three years.
- **Option 3.** Making a personal loan of $2,000 to a friend and receiving $150 per year for three years.

Determine the equivalent annual cash flows for each option, and select the best option.

6.45 A chemical company is considering two types of incinerators to burn solid waste generated by a chemical operation. Both incinerators have a burning capacity of 20 tons per day. The data in Table P6.45 have been compiled for comparison.

TABLE P6.45

	Incinerator A	Incinerator B
Installed cost	$1,300,000	$850,000
Annual O&M costs	$70,000	$100,000
Service life	20 years	10 years
Salvage value	$60,000	$30,000
Income taxes	$40,000	$30,000

If the firm's MARR is known to be 13%, determine the processing cost per ton of solid waste incurred by each incinerator. Assume that incinerator B will be available in the future at the same cost.

6.46 Company X has been contracting its overhauling work to Company Y for $40,000 per machine per year. Company X estimates that by building a $500,000 maintenance facility with a life of 15 years and a salvage value of $100,000 at the end of its life, it could handle its own overhauling at a cost of only $30,000 per machine per year. What is the minimum annual number of machines that Company X must operate to make it economically feasible to build its own facility? (Assume an interest rate of 10%.)

Lifecycle Cost Analysis

6.47 Two types of heating systems are considered to provide heat for a 35,000 ft² office building. Cost data for each system is given in Table P6.47.

TABLE P6.47

	Heat Pump	Geothermal
First cost	$194,000	$256,500
Life	15	15
Salvage value	0	0
Annual electricity requirement	145,000 kWh	25,590 kWh
Cost of electricity	$0.06 per kWh	$0.06 per kWh
Annual operating and maintenance cost	$3,860	$3,650
Compressor replacement cycle and cost	$2,500 every 10 years	$0

At 12% interest, which system would be more economical?

6.48 An airline is considering two types of engine systems for use in its planes. Each has the same life and the same maintenance and repair record.

- **System A** costs $100,000 and uses 40,000 gallons per 1,000 hours of operation at the average load encountered in passenger service.
- **System B** costs $200,000 and uses 32,000 gallons per 1,000 hours of operation at the same level.

Both engine systems have three-year lives before any major overhaul is required. On the basis of the initial investment, the systems have 10% salvage values. If jet fuel currently costs $2.10 a gallon and fuel consumption is expected to increase at the rate of 6% per year because of degrading engine efficiency, which engine system should the firm install? Assume 2,000 hours of operation per year and an MARR of 10%. Use the AE criterion. What is the equivalent operating cost per hour for each engine?

6.49 Mustang Auto Parts, Inc. is considering one of two forklift trucks for its assembly plant.

- **Truck A** costs $15,000 and requires $3,000 annually in operating expenses. It will have a $5,000 salvage value at the end of its three-year service life.
- **Truck B** costs $20,000, but requires only $2,000 annually in operating expenses; its service life is four years, at which time its expected salvage value will be $8,000.

The firm's MARR is 12%. Assuming that the trucks are needed for 12 years and that no significant changes are expected in the future price and functional capacity of each truck, select the most economical truck on the basis of AE analysis.

6.50 A small manufacturing firm is considering purchasing a new machine to modernize one of its current production lines. Two types of machines are available on the market. The lives of machine A and machine B are four years and six years, respectively, but the firm does not expect to need the service of either machine for more than five years. The machines have the expected receipts and disbursements given in Table P6.50.

TABLE P6.50

Item	Machine A	Machine B
First cost	$6,500	$8,500
Service life	4 years	6 years
Estimated salvage value	$600	$1,000
Annual O&M costs	$800	$520
Change oil filter every other year	$100	None
Engine overhaul	$200 (every 3 years)	$280 (every 4 years)

The firm also has another option: leasing a machine at $3,000 per year, which is fully maintained by the leasing company. After four years of use, the salvage value for machine B will remain at $1,000.

(a) How many decision alternatives are there?

(b) Which decision appears to be the best at $i = 10\%$?

6.51 A plastic-manufacturing company owns and operates a polypropylene production facility that converts the propylene from one of its cracking facilities to polypropylene plastics for outside sale. The polypropylene production facility is currently forced to operate at less than capacity due to an insufficiency of propylene production capacity in its hydrocarbon cracking facility. The chemical engineers are considering alternatives for supplying additional propylene to the polypropylene production facility. Two feasible alternatives are to build a pipeline to the nearest outside supply source and to provide additional propylene by truck from an outside source. The engineers also gathered the following projected cost estimates.

- Future costs for purchased propylene excluding delivery: $0.215 per lb.
- Cost of pipeline construction: $200,000 per pipeline mile.
- Estimated length of pipeline: 180 miles.
- Transportation costs by tank truck: $0.05 per lb, utilizing a common carrier.
- Pipeline operating costs: $0.005 per lb, excluding capital costs.
- Projected additional propylene needs: 180 million lb per year.
- Projected project life: 20 years.
- Estimated salvage value of the pipeline: 8% of the installed costs.

Determine the propylene cost per pound under each option if the firm's MARR is 18%. Which option is more economical?

6.52 The City of Prescott is comparing two plans for supplying water to a newly developed subdivision:

- **Plan A** will take care of requirements for the next 15 years, and at the end of this time, the initial cost of $400,000 will have to be duplicated to meet the requirements of subsequent years. The facilities

installed at dates 0 and 15 may be considered permanent. However, certain supporting equipment will have to be replaced every 30 years from the installation dates at a cost of $75,000. Operating costs are $31,000 a year for the first 15 years and $62,000 thereafter, although they are expected to increase by $1,000 a year beginning in the 21st year.

- **Plan B** will supply all requirements for water indefinitely into the future, although it will be operated only at half capacity for the first 15 years. Annual costs over this period will be $35,000 and will increase to $55,000 beginning in the 16th year. The initial cost of Plan B is $550,000. The facilities can be considered permanent, although it will be necessary to replace $150,000 worth of equipment every 30 years after the initial installation.

The city will charge the subdivision for the use of water on the basis of the equivalent annual cost. At an interest rate of 10%, determine the equivalent annual cost for each plan, and make a recommendation to the city.

Minimum-Cost Analysis

6.53 A continuous electric current of 2,000 amps is to be transmitted from a generator to a transformer located 200 feet away. A copper conductor can be installed for $6 per pound, will have an estimated life of 25 years, and can be salvaged for $1 per pound. Power loss from the conductor will be inversely proportional to the cross-sectional area of the conductor and may be expressed as $6.516/A$ kilowatt, where A is in square inches. The cost of energy is $0.0825 per kilowatt-hour, the interest rate is 11%, and the density of copper is 555 pounds per cubic foot.

(a) Calculate the optimum cross-sectional area of the conductor.

(b) Calculate the annual equivalent total cost for the value you obtained in part (a).

(c) Graph the two individual cost factors (capital cost and power-loss cost) and the total cost as a function of the cross-sectional area A, and discuss the impact of increasing energy cost on the optimum obtained in part (a).

6.54 All industrial facilities have a network of piping that carries water or other liquids. According to a U.S. Department of Energy study[3], 16% of a typical

---

[3] "Energy Tips: Reduce Pumping Costs Through Optimum Pipe Size," Office of Industrial Technologies, Energy Efficiency, and Renewable Energy, U.S. Department of Energy.

facility's electricity costs are for its pumping systems. Suppose that you are considering a pumping facility having 10,000 ft to carry 600 gpm of water continuously to a storage tank. The general equation for estimating fractional pumping costs is

$$\text{Cost(\$)} = \frac{1}{1705}(\text{Friction Factor})$$

$$\frac{(\text{Flow in gpm})^3(\text{Pipe length in ft})}{(\text{Pipe inner diameter in inches})^5}$$

$$\times \frac{(\text{Number of hours})(\$/\text{kWh})}{(\text{Combined pump and motor efficiency as a percent})}$$

where the friction factor (based on the pipe roughness), pipe diameter, and the Reynolds number can be obtained from an engineering handbook. For most applications, the value of this friction factor will be between 0.015 and 0.0225.

(a) Using clean iron and 6-inch steel pipe (schedule 40) for pumping 70°F water, an electricity cost of $0.05/kWh, 8,760 operating hours annually, and combined pump and motor efficiency of 70%, determine the pumping cost.

(b) With the same condition, determine the pumping cost with a 10-inch pipe.

(c) Find the current price of steel and determine the optimal size of pipe, assuming that the pumping operation would last 15 years and the salvage value of the pipe at 10% of the initial investment. Assume also a discount rate of 12%.

Short Case Studies

ST6.1 Absorption chillers provide cooling to buildings by using heat. This seemingly paradoxical (but highly efficient) technology is most cost-effective in large facilities with significant heat loads. Not only do absorption chillers use less energy than conventional equipment does, but they also cool buildings without the use of ozone-depleting chlorofluorocarbons (CFCs). Unlike conventional electric chillers, which use mechanical energy in a vapor compression process to provide refrigeration, absorption chillers primarily use heat energy with limited mechanical energy for pumping. Absorption chillers can be powered by natural gas, steam, or waste heat.

- The most promising markets for absorption chillers are in commercial buildings, government facilities, college campuses, hospital complexes, industrial parks, and municipalities.

- Absorption chillers generally become economically attractive when there is a source of inexpensive thermal energy at temperatures between 212°F and 392°F.

- An absorption chiller transfers thermal energy from the heat source to the heat sink through an absorbent fluid and a refrigerant. The absorption chiller accomplishes its refrigerative effect by absorbing and then releasing water vapor into and out of a lithium–bromide solution. Absorption chiller systems are classified by single-, double-, or triple-stage effects, which indicate the number of generators in the given system.

- The greater the number of stages, the higher is the overall efficiency of the system. Double-effect absorption chillers typically have a higher first cost (but a significantly lower energy cost) than single-effect chillers, resulting in a lower net present worth.

Some of the known economic benefits of the absorption chiller over the conventional mechanical chiller are as follows.[4]

[4]*Source*: EcoGeneration Solutions™ LLC. Companies, 12615 Jones Rd., Suite 209, Houston, Texas, 77070 (http://www.cogeneration.net/Absorption_Chillers.htm).

- In a plant where low-pressure steam is currently being vented to the atmosphere, a mechanical chiller with a coefficient of performance (COP) of 4.0 is used 4,000 hours per year to produce an average of 300 tons of refrigeration.
- The plant's cost of electricity is $0.05 per kilowatt-hour. An absorption unit requiring 5,400 lb/hr of 15-psig steam could replace the mechanical chiller, providing the following annual electrical cost savings:

$$\text{Annual Savings} = 300 \text{ tons} \times (12{,}000 \text{ Btu/ton/}4.0)$$
$$\times 4.000 \text{ hrs/yr}$$
$$\times \$0.05/\text{kWh}$$
$$\times 1 \text{ kWh/}3{,}413 \text{ Btu} = \$52{,}740$$

Suppose you plan to install the chiller and expect to operate it continuously for 10 years. What would be the maximum amount that can be invested in installing the chiller with the capacity as described above? If the energy cost were expected to increase at an annual rate of 10%, how would your answer change? Assume $i = 10\%$.

ST6.2 Automotive engineers at H. Motor Co. are considering the laser blank welding (LBW) technique to produce a windshield frame rail blank. The engineers believe that, compared with the conventional sheet metal blanks, LBW would result in a significant savings as follows.

1. Scrap reduction through more efficient blank nesting on coil.
2. Scrap reclamation (weld scrap offal into a larger, usable blank).

The use of a laser-welded blank provides a reduction in engineered scrap for the production of a window-frame rail blank. On the basis of an annual volume of 3,000 blanks, the engineers have estimated the financial data given in Table ST6.2.

The LBW technique appears to achieve significant savings, so the engineers are leaning toward adopting it. Since the engineers have had no previous experience with LBW, they are not sure whether producing the windshield frames in-house at this time is a good strategy. For this windshield frame, it may be cheaper to use the services of a supplier that

TABLE ST6.2 **Blanking Method**

Description	Conventional	Laser Blank Welding
Weight per blank (lb/part)	63.764	34.870
Steel cost/part	$14.98	$8.19
Transportation/part	$0.67	$0.42
Blanking/part	$0.50	$0.40
Die investment	$106,480	$83,000

has both the experience with, and the machinery for, laser blanking. The make-or-buy decision depends on two factors: the amount of new investment that is required in laser welding and whether additional machinery will be required for future products. Assume a lifetime of 10 years and an interest rate of 16% to recommend the best course of action. Assume also that the salvage value at the end of 10 years is estimated to be insignificant for either system. If H. Motor Co. considers the subcontracting option, what would be the acceptable range of contract bid (unit cost per part)?

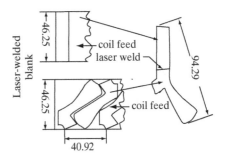

Figure ST6.2

ST6.3 A Veterans Administration (VA) hospital is to decide which type of boiler fuel system will most efficiently provide the required steam energy output for heating, laundry, and sterilization purposes. The current boilers were installed in the early 1950s and are now obsolete. Much of the auxiliary equipment is also old and in need of repair. Because of these general conditions, an engineering recommendation was made to replace the entire plant with a new boiler plant building that would house modern equipment. The cost of demolishing the old boiler plant would be almost a complete loss, as the salvage value of the scrap steel and used brick was estimated to be only about $1,000. The VA hospital's engineer finally selected two alternative proposals as being worthy of more intensive analysis. The hospital's annual energy requirement, measured in terms of steam output, is approximately 145,000,000 pounds of steam. As a rule of thumb for analysis, 1 pound of steam is approximately 1,000 Btu, and 1 cubic foot of natural gas is also approximately 1,000 Btu. The two alternatives are as follows.

- **Proposal 1.** Replace the old plant with a new coal-fired boiler plant that costs $1,770,300. To meet the requirements for particulate emission as set by the Environmental Protection Agency, this coal-fired boiler, even if it burned low-sulfur coal, would need an electrostatic precipitator, which would cost approximately $100,000. The plant would last for 20 years. One pound of dry coal yields about 14,300 Btu. To convert the 145,000,000 pounds of steam energy to the common denominator of Btu, it is necessary to multiply by 1,000. To find the Btu input requirements, it is necessary to divide by the relative boiler efficiency for the type of fuel. The boiler efficiency for coal is 0.75. The price of coal is estimated to be $65.50 per ton.

- **Proposal 2.** Build a gas-fired boiler plant with No. 2 fuel oil and use the new plant as a standby. This system would cost $889,200 and have an expected service life of 20 years. Since small

household or commercial gas users that are entirely dependent on gas have priority, large plants must have an oil switch-over capability. It has been estimated that 6% of 145,000,000 pounds of steam energy (or 8,700,000 pounds) would come about as a result of the switch to oil. The boiler efficiency with each fuel would be 0.78 for gas and 0.81 for oil, respectively. The heat value of natural gas is approximately 1,000,000 Btu/MCF (thousand cubic feet), and for No. 2 fuel oil, it is 139,400 Btu/gal. The estimated gas price is $6.50/MCF, and the price of No. 2 fuel oil is $2.85 per gallon.

(a) Calculate the annual fuel costs for each proposal.

(b) Determine the unit cost per steam pound for each proposal. Assume that $i = 10\%$.

(c) Which proposal is the more economical?

ST6.4 The following is a letter that I received from a local city engineer:

> Dear Professor:
>
> Thank you for taking the time to assist with this problem. I'm really embarrassed at not being able to solve it myself, since it seems rather straightforward. The situation is as follows:
>
> A citizen of Opelika paid for concrete drainage pipe to be installed on his property approximately 20 years ago. (We have a policy that if drainage trouble exists on private property and the owner agrees to pay for the material, city crews will install it.) That was the case in this problem. Therefore, we are dealing with only material costs, disregarding labor.
>
> However, this past year, we removed the pipe purchased by the citizen, due to a larger area drainage project. He thinks, and we agree, that he is due some refund for salvage value of the pipe due to its remaining life.

Problem:

- Known: 80′ of 48″ pipe purchased 20 years ago. Current quoted price of 48″pipe = $52.60/foot, times 80 feet = $4,208 total cost in today's dollars.

- Unknown: Original purchase price.

- Assumptions: 50-year life; therefore, assume 30 years of life remaining at removal after 20 years. A 4% price increase per year, average, over 20 years.

Thus, we wish to calculate the cost of the pipe 20 years ago. Then we will calculate, in today's dollars, the present salvage value after 20 years' use, with 30 years of life remaining. Thank you again for your help.

After reviewing this letter, recommend a reasonable amount of compensation to the citizen for the replaced drainage pipe.

7

Rate-of-Return Analysis

Cost $1,000, Sold for $21 Million[1] Recently, the *New York Times* printed a story that stated "A 1958 Ferrari 335S that had been bailed out of a United States Customs warehouse in New York for $1,000 in 1963 was sold for $21.5 million in a private transaction to an Austrian collector, Andreas Mohringer."

Clearly, it was an impressive **return on investment** of 2,155%. No sale of an automobile in recent memory has realized a percentage gain that compares. The story goes on, "This car was sold originally to Alan Connell, a Texan active in Sports Car Club of America racing. Connell drove his 335S in 1958 and part of 1959, but the engine was troublesome, and the car went back to Italy for repairs. The bill is reported

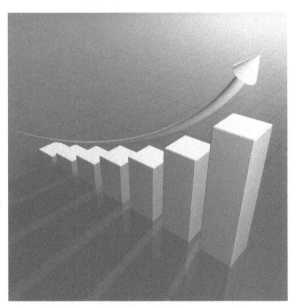

Source: Johan Swanepoel/Shutterstock

by Ferrari historians to have been on the order of $70,000. When the car arrived back in New York, Connell pointed out that a new car cost about $12,000, so there was little sense in retrieving this one. The car languished in a Customs warehouse. That's where Gordon Tatum, a Maryland auto dealer, found it. He paid the storage fee of $1,000 and took it home."

If Gordon Tatum kept the car for all those years, his return on investment would be 2,155%. What does the 2,155% rate of return for the investment really represent?

[1] John Lamm, "Cost $1,000, Sold for $21 Million," Collectible Cars, *New York Times*, February 21, 2014.

The story of a huge return on investment from a collectible car is certainly intriguing. We could hear similar stories in the investment of art collection, or investments in stock or real estate. Let's take a few of more examples related to stock investment:

- Wal-Mart Corporation went public in 1971 at $16.50 per share. If you purchased 100 shares (say $1,650), and kept the stock until today, it was worth $15,767,500 in July 2014. The rate of return on this investment is 23.76%—this is a phenomenal return as you are earning a compound interest of 23.76% per year over 43 years!

- You don't have to go that far to find the impressive stock return. If you purchased 100 shares of Apple Computer stock at $12.42 in February 23, 2009, it was worth $131.95 or 48.26% return on investment as of February 23, 2015.

How do we compute these rate of return figures? And once computed, how do we use this figure when evaluating an investment project? Another question is, if we make an investment decision based on this rate of return, can we can we reach the same conclusion from the NPW criterion? Our consideration of the concept of "rate of return" in this chapter will answer these and other questions.

Along with the NPW and the AE criteria, the third primary measure of investment worth is **rate of return**. As shown in Chapter 5, the NPW measure is easy to calculate and apply. Nevertheless, many engineers and financial managers prefer rate-of-return analysis to the NPW method because they find it intuitively more appealing to analyze investments in terms of percentage rates of return rather than dollars of NPW. Consider the following statements regarding an investment's profitability.

- This project will bring in a 15% rate of return on the investment.
- This project will result in a net surplus of $10,000 in the NPW.

Neither statement describes the nature of the investment project in any complete sense. However, the rate of return is somewhat easier to understand, because many of us are so familiar with savings-and-loan interest rates, which are, in fact, rates of return on your deposits.

In this chapter, we will examine four aspects of rate-of-return analysis: (1) the concept of return on investment; (2) the calculation of a rate of return; (3) the development of a set of decision rules, known an internal-rate-of-return criterion, on how to use this rate of return in making an accept or reject decision of an investment; and (4) the comparison of mutually exclusive alternatives based on a rate of return.

CHAPTER LEARNING OBJECTIVES

After completing this chapter, you should understand the following concepts:

- The meaning of the rate of return.
- The various methods to compute the rate of return.
- How you make an accept or reject decision with the rate of return.
- How to resolve the multiple rates of return problem.
- How you conduct an incremental analysis with the rate of return.

7.1 Rate of Return

Many different terms are used to refer to **rate of return**, including **yield** (i.e., the yield to maturity, commonly used in bond valuation), **internal rate of return**, and **marginal efficiency of capital**. First we will review three common definitions of rate of return. Then we will use the definition of internal rate of return as a measure of profitability for a single investment project throughout the text.

7.1.1 Return on Investment

There are several ways of defining the concept of a rate of return on investment. The first is based on a typical loan transaction, the second on the mathematical expression of the present-worth function, and the third on the project cash-flow series.

Definition 1

The rate of return is the interest rate earned on the unpaid balance of an amortized loan.

Suppose that you borrow $10,000 from a bank with a repayment schedule of $4,021 each year for three years. How would you determine the interest rate that the bank charges on this transaction? As we learned in Chapter 3, you would set up the equivalence equation

$$\$10,000 = \$4,021(P/A, i, 3)$$

and solve for i. It turns out that $i = 10\%$. In this situation, the bank will earn a return of 10% on its investment of $10,000 made at time 0. The bank calculates the balances over the life of the loan as shown in Table 7.1.

A negative balance indicates an unpaid balance. In other words, you still owe money to the bank.

TABLE 7.1 Unpaid Loan Balance at End of Each Year

	A	B	C	D	E
1					
2		**Unpaid Loan**	**Return on**	**Payment**	**Unpaid Loan**
3		**Balance at**	**Unpaid**	**Received**	**Balance at**
4	*n*	**Beginning of Year**	**Balance (10%)**	**from Borrower**	**End of Year**
5					
6	0	$0	$0	-$10,000	-$10,000
7	1	-$10,000	-$1,000	$4,021	-$6,979
8	2	-$6,979	-$698	$4,021	-$3,656
9	3	-$3,656	-$365	$4,021	$0
10					

Observe that, for the repayment schedule shown, the 10% interest is calculated only on each year's outstanding balance. In this situation, only part of the $4,021 annual payment represents interest; the remainder goes toward repaying the principal. Thus, the three annual payments repay the loan itself and additionally provide a return of 10% on the *amount still outstanding each year.*

Note that when the last payment is made, the outstanding principal is eventually reduced to zero.[2] If we calculate the NPW of the loan transaction at its rate of return (10%), we see that

$$PW(10\%) = -\$10,000 + \$4,021(P/A, 10\%, 3) = 0$$

which indicates that the bank can break even at a 10% rate of interest. If the customer pays back more than $4,021, the bank will earn more than 10% return. If the customer pays less than $4,021, the bank earns less than 10% return on its investment. In other words, the rate of return becomes the rate of interest that equates the present value of future cash repayments to the amount of the loan. This observation prompts the second definition of rate of return.

Definition 2

The rate of return is the break-even interest rate i^* that equates the present worth of a project's cash outflows to the present worth of its cash inflows, or

$$PW(i^*) = PW_{\text{Cash inflows}} - PW_{\text{Cash outflows}}$$
$$= 0$$

Note that the expression for the NPW is equivalent to

$$PW(i^*) = \frac{A_0}{(1 + i^*)^0} + \frac{A_1}{(1 + i^*)^1} + \cdots + \frac{A_N}{(1 + i^*)^N} = 0 \qquad (7.1)$$

---

[2] As we learned in Section 5.3.2, this terminal balance is equivalent to the net future worth of the investment. If the net future worth of the investment is zero, its NPW also should be zero.

Here we know the value of A_n for each period, but not the value of i^*. Since it is the only unknown, however, we can solve for i^*. (Inevitably, there will be N values of i^* that satisfy this equation. In most project cash flows, you would be able to find a unique positive i^* that satisfies Eq. (7.1). However, you may encounter some cash flows that cannot be solved for a single rate of return greater than -100%. By the nature of the NPW function in Eq. (7.1), it is possible to have more than one rate of return for certain types of cash flows. For some cash flows, we may not find a specific rate of return at all.[3]

Note that the formula in Eq. (7.1) is simply the NPW formula solved for the particular interest rate (i^*) at which PW(i) is equal to zero. By multiplying both sides of Eq. (7.1) by $(1 + i^*)^N$, we obtain

$$\mathrm{PW}(i^*)(1 + i^*)^N = \mathrm{FW}(i^*) = 0$$

If we multiply both sides of Eq. (7.1) by the capital recovery factor ($A/P, i^*, N$), we obtain the relationship $\mathrm{AE}(i^*) = 0$. Therefore, the i^* of a project may be defined as the rate of interest that equates the present worth, future worth, and annual equivalent worth of the entire series of cash flows to zero.

$$\mathrm{PW}(i^*) = \mathrm{FW}(i^*) = \mathrm{PB}(i^*)_N = \mathrm{AE}(i^*) = 0 \tag{7.2}$$

7.1.2 Return on Invested Capital

Investment projects can be viewed as analogous to bank loans. We will now introduce the concept of rate of return based on the return on invested capital in terms of a project investment. A project's return is referred to as the internal rate of return (IRR) or the **yield** promised by an **investment project** over its **useful life**.

Definition 3

The internal rate of return is the interest rate earned on the unrecovered project balance of the investment by the firm such that, when the project terminates, the unrecovered project balance will be exactly zero.

Suppose a company invests $10,000 in an equipment with a three-year useful life and equivalent annual labor savings of $4,021. Here, we may view the investing firm as the lender and the project as the borrower. The cash flow transaction between them would be identical to the amortized loan transaction described under Definition 1.

In our project balance calculation (Table 7.2), we see that 10% is earned (or charged) on $10,000 during year 1, 10% is earned on $6,979 during year 2, and 10% is earned on $3,656 during year 3. This indicates that the firm earns a 10% rate of return on funds that remain *internally* invested in the project. Since it is a return that is *internal* to the project, we refer to it as the **internal rate of return (IRR)**. This means that the equipment project under consideration brings in enough cash to pay for itself in three years and also will provide the firm with a return of 10% on its invested capital. Put differently, if the computer is financed with funds costing 10% annually, the cash generated by the investment will be exactly sufficient to repay the principal and the annual interest charge on the fund in three years.

---

[3] You will always have N rates of return. The issue is whether they are real or imaginary. If they are real, the question, "Are they in the $(-100\%, \infty)$ interval?" should be asked. A **negative rate of return** implies that you never recover your initial investment in the full amount.

TABLE 7.2 Unrecovered Project Balance at End of Each Year

	A	B	C	D	E
1					
2		Beginning	Return on		Ending
3		Project	Invested	Cash	Project
4	*n*	Balance	Capital	Payment	Balance
5					
6	0	$0	$0	-$10,000	-$10,000
7	1	-$10,000	-$1,000	$4,021	-$6,979
8	2	-$6,979	-$698	$4,021	-$3,656
9	3	-$3,656	-$365	$4,021	$0
10					

Notice that only one cash outflow occurs at time 0, and the present worth of this outflow is simply $10,000. There are three equal receipts, and the present worth of these inflows is $4,021(*P/A*, 10%, 3) = $10,000. Since

$$\text{NPW} = \text{PW}_{\text{Inflow}} - \text{PW}_{\text{Outflow}} = \$10,000 - \$10,000 = 0$$

10% also satisfies Definition 2 of the rate of return. Even though the preceding simple example implies that i^* coincides with IRR, only Definitions 1 and 3 correctly describe the true meaning of the internal rate of return. As we will see later, if the cash expenditures of an investment are not restricted to the initial period, several break-even interest rates (i^*s) may exist that satisfy Eq. (7.1). However, there may not be a rate of return that is *internal* to the project.

7.2 Methods for Finding the Rate of Return

We may find i^* by several procedures, each of which has its advantages and disadvantages. To facilitate the process of finding the rate of return for an investment project, we will first classify various types of investment cash flow.

7.2.1 Simple versus Nonsimple Investments

We can classify an investment project by counting the number of sign changes in its net cash-flow sequence. A change from either "+" to "−", or "−" to "+" is counted as one sign change. (We ignore a zero cash flow.) Then

- A **simple investment** is an investment in which the initial cash flows are negative and only one sign change occurs in the remaining net cash-flow series. If the initial flows are positive and only one sign change occurs in the subsequent net cash flows, they are referred to as **simple borrowing** cash flows.
- A **nonsimple investment** is an investment in which more than one sign change occurs in the cash flow series.

Multiple i^*s, as we will see later, occur only in nonsimple investments. Three different types of investment possibilities are illustrated in Example 7.1.

EXAMPLE 7.1 Investment Classification

Consider the following three cash flow series and classify them into either simple or nonsimple investments:

Period	Net Cash Flow		
n	Project A	Project B	Project C
0	−$1,000	−$1,000	$1,000
1	−500	3,900	−450
2	800	−5,030	−450
3	1,500	2,145	−450
4	2,000	0	0

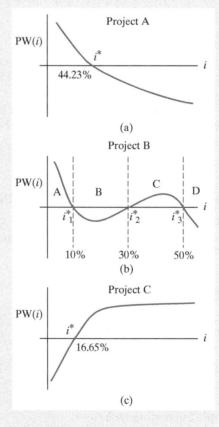

Figure 7.1 Present-worth profiles:
(a) Simple investment, (b) nonsimple
investment with multiple rates of return,
and (c) simple borrowing cash flows.

SOLUTION

Given: Preceding cash flow sequences.

Find: Classify the investments shown as either simple or nonsimple investments.

- **Project A** represents many common simple investments. That is, many investment projects involve a series of cash outflows followed by a series of cash inflows. This situation reveals the NPW profile shown in Figure 7.1(a). The curve crosses the *i*-axis only once.

- **Project B** represents a nonsimple investment. The NPW profile for this investment has the shape shown in Figure 7.1(b). The *i*-axis is crossed at 10%, 30%, and 50%.

- **Project C** represents neither a simple nor a nonsimple investment, even though only one sign change occurs in the cash flow sequence. Since the first cash flow is positive, this is a **simple borrowing** cash flow, not an investment flow. Figure 7.1(c) depicts the NPW profile for this type of investment.

COMMENTS: Not all NPW profiles for nonsimple investments have multiple crossings of the *i*-axis. Clearly then, we should place a high priority on discovering this situation early in our analysis of a project's cash flows. The quickest way to predict multiple *i**s is to generate an NPW profile on a computer and check whether it crosses the horizontal axis more than once. In the next section, we illustrate when to expect such multiple crossings by examining types of cash flows.

7.2.2 Predicting Multiple *i**s

As hinted at in Example 7.1, for certain series of project cash flows, we may uncover the complication of multiple *i** values that satisfy Eq. (7.1). By analyzing and classifying cash flows, we may anticipate this difficulty and adjust our analysis approach later. Here we will focus on the initial problem of whether we can predict a unique *i** for a project by examining its cash flow pattern. Two useful rules allow us to focus on sign changes: (1) in net cash flows; and (2) in accounting net profit (accumulated net cash flows).

Net Cash Flow Rule of Signs

One useful method for predicting an upper limit on the number of positive *i**s of a cash flow stream is to apply the rule of signs:

*The number of real positive i*s for a project with N periods is never greater than the number of sign changes in the sequence of the A_n's. A zero cash flow is ignored.*

An example is given in Figure 7.2. Three sign changes occur in the cash flow sequence, so three or fewer real positive *i**s exist.

It must be emphasized that the rule of signs provides an indication only of the *possibility* of multiple rates of return: The rule predicts only the *maximum* number of possible *i**s. Many projects have multiple sign changes in their cash flow sequence, but still possess a unique real positive *i**.

Accumulated Cash-Flow Sign Test

The accumulated cash flow is the sum of the net cash flows up to and including a given time. If the rule of cash flow signs indicates multiple *i**s, we should proceed to the **accumulated cash-flow sign test** to eliminate some possibility of multiple rates of return.

Period	A_n	Sign Change
0	−$100	
1	−20	
2	+50	1
3	0	
4	+60	
5	−30	1
6	+100	1

Figure 7.2 How to count the number of sign changes.

If we let A_n represent the net cash flow in period n and S_n represent the accumulated cash flow (the accounting sum) up to period n, we have the data in Table 7.3. We then examine the sequence of accumulated cash flows ($S_0, S_1, S_2, S_3, \ldots, S_N$) to determine the number of sign changes.

If the series S_n starts negatively and changes sign only once, then a unique positive i^* exists.

This cumulative cash flow sign rule is a more discriminating test for identifying the uniqueness of positive i^* than the previously described method.

TABLE 7.3 Process of Performing Accumulated Cash-Flow Sign Test

Period (n)	Cash Flow (A_n)	Accumulated Cash Flow (S_n)
0	A_0	$S_0 = A_0$
1	A_1	$S_1 = S_0 + A_1$
2	A_2	$S_2 = S_1 + A_2$
⋮	⋮	⋮
N	A_N	$S_N = S_{N-1} + A_N$

EXAMPLE 7.2 Predicting the Number of i's

Predict the number of real positive rates of return for each of the following cash flow series.

Period	A	B	C	D
0	−$100	−$100	$ 0	−$100
1	−200	+50	−50	+50
2	+200	−100	+115	0
3	+200	+60	−66	+200
4	+200	−100	0	−50

SOLUTION

Given: Four cash flow series.

Find: The upper limit on number of positive i^*s for each series.

The cash flow rule of signs indicates the following possibilities for the positive values of i^*:

Project	Number of Sign Changes in Net Cash Flows	Possible Number of Positive Values of i^*
A	1	1 or 0
B	4	4, 3, 2, 1, or 0
C	2	2, 1, or 0
D	2	2, 1, or 0

For cash flows B, C, and D, we would like to apply the more discriminating cumulative cash-flow test to see if we can specify a smaller number of possible values of i^*. Accordingly, we write

	Project B		Project C		Project D	
n	A_n	S_n	A_n	S_n	A_n	S_n
0	−$100	−$100	$0	$0	−$100	−$100
1	+50	−50	−50	−50	+50	−50
2	−100	−150	+115	+65	0	−50
3	+60	−90	−66	−1	+200	+150
4	−100	−190	0	0	−50	+100

Recall the test: If the series starts *negatively* and changes sign only once, a unique positive i^* exists.

- Only project D begins negatively and passes the test; therefore, we may predict a unique positive i^* value, rather than 2, 1, or 0 as predicted by the cash flow rule of signs.

$$(i_1^* = -75.16\% \text{ and } i_2^* = 35.05\%)$$

- Project B, with no sign change in the cumulative cash flow series, has no rate of return.
- Project C fails the test, and we cannot eliminate the possibility of multiple positive i^*s.

$$(i_1^* = 10\% \text{ and } i_2^* = 20\%)$$

is continued until PW(i) is approximately equal to zero. Whenever we reach the point where PW(i) is bounded by one negative and one positive value, we use **linear interpolation** to approximate i^*. This process is somewhat tedious and inefficient. (The trial-and-error method does not work for nonsimple investments in which the NPW function is not, in general, a monotonically decreasing function of the interest rate.)

EXAMPLE 7.4 Finding i^* by Trial and Error

The Imperial Chemical Company is considering purchasing a chemical analysis machine worth $13,000. Although the purchase of this machine will not produce any increase in sales revenues, it will result in a reduction of labor costs. In order to operate the machine properly, it must be calibrated each year. The machine has an expected life of six years, after which it will have no salvage value. The following table summarizes the annual savings in labor cost and the annual maintenance cost in calibration over six years.

Year (n)	Costs ($)	Savings ($)	Net Cash Flow ($)
0	13,000	0	−$13,000
1	2,300	6,000	3,700
2	2,300	7,000	4,700
3	2,300	9,000	6,700
4	2,300	9,000	6,700
5	2,300	9,000	6,700
6	2,300	9,000	6,700

Find the rate of return for this project.

SOLUTION

Given: Cash flows over six years as shown in Figure 7.3.
Find: i^*.

We start with a guessed interest rate of 25%. The present worth of the cash flows is

$$PW(25\%) = -\$13,000 + \$3,700(P/F, 25\%, 1) + \$4,700(P/F, 25\%, 2)$$
$$+ \$6,700(P/A, 25\%, 4)(P/F, 25\%, 2)$$
$$= \$3,095$$

Since this present worth is positive, we must raise the interest rate to bring PW toward zero. When we use an interest rate of 35%, we find that

$$PW(35\%) = -\$13,000 + \$3,700(P/F, 35\%, 1) + \$4,700(P/F, 35\%, 2)$$
$$+ \$6,700(P/A, 35\%, 4)(P/F, 35\%, 2)$$
$$= -\$339$$

Figure 7.3 Cash flow diagram for a simple investment (Example 7.4).

We have now bracketed the solution: PW(i) will be zero at i somewhere between 25% and 35%. Using straight-line interpolation, we approximate

$$i^* \approx 25\% + (35\% - 25\%)\left[\frac{3{,}095 - 0}{3{,}095 - (-339)}\right]$$

$$= 25\% + 10\%(0.9013)$$

$$= 34.01\%$$

Now we will check to see how close this value is to the precise value of i^*. If we compute the present worth at this interpolated value, we obtain

$$PW(34\%) = -\$13{,}000 + \$3{,}700(P/F, 34\%, 1) + \$4{,}700(P/F, 34\%, 2)$$

$$+ \$6{,}700(P/A, 34\%, 4)(P/F, 34\%, 2)$$

$$= -\$50.58$$

As this is not zero, we may recompute i^* at a lower interest rate, say, 33%:

$$PW(33\%) = -\$13{,}000 + \$3{,}700(P/F, 33\%, 1) + \$4{,}700(P/F, 33\%, 2)$$

$$+ \$6{,}700(P/A, 33\%, 4)(P/F, 33\%, 2)$$

$$= \$248.56$$

With another round of linear interpolation, we approximate

$$i^* \approx 33\% + (34\% - 33\%)\left[\frac{248.56 - 0}{248.56 - (-50.58)}\right]$$

$$= 33\% + 1\%(0.8309)$$

$$= 33.83\%$$

At this interest rate,

$$PW(33.83\%) = -\$13,000 + \$3,700(P/F, 33.83\%, 1)$$
$$+ \$4,700(P/F, 33.83\%, 2)$$
$$+ \$6,700(P/A, 33.83\%, 4)(P/F, 33.83\%, 2)$$
$$= -\$0.49$$

which is practically zero, so we may stop here. In fact, there is no need to be more precise about these interpolations, because the final result can be no more accurate than the basic data, which ordinarily are only rough estimates.

COMMENT: With Excel, you can evaluate the IRR for the project as = IRR (range, guess), where you specify the cell range for the cash flow (e.g., B3:B9) and the initial guess, such as 10%. Computing i^* for this problem in Excel, incidentally, gives us 33.8283%. Instead of using the factor notations, you may attempt to use a tabular approach as given in Table 7.4.

TABLE 7.4 A Tabular Approach to Locating an IRR

		Internal Rate of Return: What It Looks Like			
		Discount Rate: 25%		Discount Rate: 35%	
Year	Cash Flow	Factor	Amount	Factor	Amount
0	−$13,000	1.0000	−$13,000	1.0000	−$13,000
1	3,700	0.8000	2,960	0.7407	2,741
2	4,700	0.6400	3,008	0.5487	2,579
3	6,700	0.5120	3,430	0.4064	2,723
4	6,700	0.4096	2,744	0.3011	2,017
5	6,700	0.3277	2,196	0.2230	1,494
6	6,700	0.2621	1,756	0.1652	1,107
Total	+$22,200	NPW =	+$3,094	NPW =	−$339
				IRR = close to 34%	

Graphical Method

We don't need to do laborious manual calculations to find i^*. Many financial calculators have built-in functions for calculating i^*. It is worth noting that many online financial calculators or spreadsheet packages have i^* functions, which solve Eq. (7.1) very rapidly,[6]

[6] An alternative method of solving for i^* is to use a computer-aided economic analysis program. Cash Flow Analyzer (CFA) finds i^* visually by specifying the lower and upper bounds of the interest search limit and generates NPW profiles when given a cash flow series. In addition to the savings in calculation time, the advantage of computer-generated profiles is their precision.

usually with the user entering the cash flows via a computer keyboard or by reading a cash flow data file. (For example, Microsoft Excel has an IRR financial function that analyzes investment cash flows, as illustrated in Figure 7.3.)

The most easily generated and understandable graphic method of solving for $i*$ is to create the **NPW profile** on a computer. On the graph, the horizontal axis indicates the interest rate and the vertical axis indicates the NPW. For a given project's cash flows, the NPW is calculated at an interest rate of zero (which gives the vertical-axis intercept) and several other interest rates. Points are plotted and a curve is sketched. Since $i*$ is defined as the interest rate at which $PW(i*) = 0$, the point at which the curve crosses the horizontal axis closely approximates $i*$. The graphical approach works for both simple and nonsimple investments.

EXAMPLE 7.5 Graphical Approach to Estimate $i*$

Consider the cash flow series shown in Figure 7.4. Estimate the rate of return by generating the NPW profile on a computer.

SOLUTION

Given: Cash flow series in Figure 7.4.

Find: (a) $i*$ by plotting the NPW profile and (b) $i*$ by using Excel.

(a) The present-worth function for the project cash-flow series is

$$PW(i) = -\$10,000 + \$20,000(P/A, i, 2) - \$25,000(P/F, i, 3)$$

First we use $i = 0$ in this equation to obtain NPW $= \$5,000$, which is the vertical-axis intercept. Then we substitute several other interest rates—10%, 20%, ..., 140%—and plot these values of $PW(i)$ as well. The result is Figure 7.4, which shows the curve crossing the horizontal axis at roughly 140%. This value can be verified by other methods if we desire to do so. Note that, in addition to establishing the interest rate that makes NPW $= 0$, the NPW profile indicates where positive and negative NPW values fall, thus giving us a broad picture of those interest rates for which the project is acceptable or unacceptable. (Note also that a trial-and-error method would lead to some confusion: As you increase the interest rate from 0% to 20%, the NPW value also keeps increasing, instead of decreasing.) Even though the project is a nonsimple investment, the curve crosses the horizontal axis only once. As mentioned in the previous section, however, most nonsimple projects have more than one value of $i*$ that makes NPW $= 0$ (i.e., more than one $i*$ per project). In such a case, the NPW profile would cross the horizontal axis more than once.[7]

---

[7] In Section 7.2.2, we discuss methods of predicting the number of $i*$ values by looking at cash flows. However, generating an NPW profile to discover multiple $i*$s is as practical and informative as any other method.

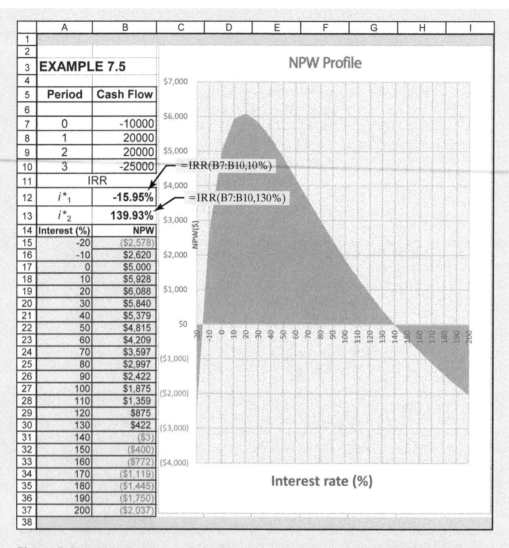

	A	B
3	**EXAMPLE 7.5**	
5	**Period**	**Cash Flow**
7	0	-10000
8	1	20000
9	2	20000
10	3	-25000
11	IRR	
12	i^*_1	**-15.95%**
13	i^*_2	**139.93%**
14	**Interest (%)**	**NPW**
15	-20	($2,578)
16	-10	$2,620
17	0	$5,000
18	10	$5,928
19	20	$6,088
20	30	$5,840
21	40	$5,379
22	50	$4,815
23	60	$4,209
24	70	$3,597
25	80	$2,997
26	90	$2,422
27	100	$1,875
28	110	$1,359
29	120	$875
30	130	$422
31	140	($3)
32	150	($400)
33	160	($772)
34	170	($1,119)
35	180	($1,445)
36	190	($1,750)
37	200	($2,037)

=IRR(B7:B10,10%)

=IRR(B7:B10,130%)

Figure 7.4 Graphical solution to rate-of-return problem for a typical nonsimple investment (Example 7.5). In fact, the project has two i^*s—($i^*_1 = -15.95\%$, and $i^*_2 = 140\%$).

(b) With Excel, you can evaluate the IRR for the project with the function

$$=\text{IRR(range,guess)}$$

in which you specify the cell range for the cash flow and the initial guess, such as 10%. One drawback with the IRR function provided by Excel is that when the project has multiple rates of return the function only finds the rate of return closest to the guess value. Therefore, use the graphical method to screen the number of potential i^*s and adjust the guess value close to each intersection point, with the interest rate axis to read the precise value of each rate of **return**.

7.3 Internal-Rate-of-Return Criterion

Now that we have classified investment projects and learned methods for determining the $i*$ value for a given project's cash flows, our objective is to develop an accept-reject decision rule that gives results consistent with those obtained from NPW analysis.

7.3.1 Relationship to PW Analysis

As we already observed in Chapter 5, NPW analysis depends on the rate of interest used for the computation of NPW. A different rate may change a project from being considered acceptable to being unacceptable, or it may change the ranking of several projects.

- Consider again the NPW profile as drawn for the simple project in Figure 7.1(a). For interest rates below $i*$, this project should be accepted because NPW > 0; for interest rates above $i*$, it should be rejected.
- By contrast, for certain nonsimple projects, the NPW may look like the one shown in Figure 7.1(b). NPW analysis would lead you to accept the projects in regions **A** and **C**, but reject those in regions **B** and **D**. Of course, this result goes against intuition: A higher interest rate would change an unacceptable project into an acceptable one. The situation graphed in Figure 7.1(b) is one of the cases of multiple $i*$s mentioned in Definition 2.

Therefore, for the simple investment situation in Figure 7.1(a), $i*$ can serve as an appropriate index for either accepting or rejecting the investment. However, for the non-simple investment of Figure 7.1(b), it is not clear which $i*$ to use to make an accept-reject decision. Therefore, the $i*$ value fails to provide an appropriate measure of profitability for an investment project with multiple rates of return.

7.3.2 Net-Investment Test: Pure versus Mixed Investments

To develop a consistent accept-reject decision rule with the NPW, we need to further classify a project into either a pure or a mixed investment:

- A project is said to be a **net investment** when the project balances computed at the project's $i*$ values, $PB(i*)_n$ are either less than or equal to zero throughout the life of the investment, with the first cash flow being negative ($A_0 < 0$). The investment is *net* in the sense that the firm does not overdraw on its return at any point, and hence, investment is *not indebted* to the project. This type of project is called a **pure investment**. In contrast, **pure borrowing** is defined as the situation in which $PB(i*)_n$ values are positive or zero throughout the life of the loan, with $A_0 > 0$. *Simple investments will always be pure investments.*
- If any of the project balances calculated at the project's $i*$ is positive, the project is not a pure investment. A positive project balance indicates that, at some time during the project life, the firm acts as a borrower [$PB(i*)_n > 0$] rather than an investor in the project [$PB(i*)_n < 0$]. This type of investment is called a **mixed investment**.

EXAMPLE 7.6 Pure versus Mixed Investments

Consider the following four investment projects with known i^* values.

	Project Cash Flows			
n	A	B	C	D
0	-$1,000	-$1,000	-$1,000	-$1,000
1	-1,000	1,600	500	3,900
2	2,000	-300	-500	-5,030
3	1,500	-200	2,000	2,145
i^*	33.64%	21.95%	29.95%	10%, 30%, 50%

Determine which projects are pure investments.

SOLUTION

Given: Four projects with cash flows and i^*s as shown.
Find: Which projects are pure investments?

We will first compute the project balances at the projects' respective i^*s. If multiple rates of return exist, we may use the largest value of i^* greater than zero.[8]

- **Project A:**

    $PB(33.64\%)_0 = -\$1,000$

    $PB(33.64\%)_1 = -\$1,000(1 + 0.3364) + (-\$1,000) = -\$2,336.40$

    $PB(33.64\%)_2 = -\$2,336.40(1 + 0.3364) + \$2,000 = -\$1,122.36$

    $PB(33.64\%)_3 = -\$1,122.36(1 + 0.3364) + \$1,500 = 0$

    $(-, -, -, 0)$ passes the net-investment test (pure investment).

- **Project B:**

    $PB(21.95\%)_0 = -\$1,000$

    $PB(21.95\%)_1 = -\$1,000(1 + 0.2195) + \$1,600 = \$380.50$

    $PB(21.95\%)_2 = +\$380.50(1 + 0.2195) - \$300 = \$164.02$

    $PB(21.95\%)_3 = +\$164.02(1 + 0.2195) - \$200 = 0$

    $(-, +, +, 0)$ fails the net-investment test (mixed investment).

- **Project C:**

    $PB(29.95\%)_0 = -\$1,000$

    $PB(29.95\%)_1 = -\$1,000(1 + 0.2995) + \$500 = -\$799.50$

    $PB(29.95\%)_2 = -\$799.50(1 + 0.2995) - \$500 = -\$1,538.95$

    $PB(29.95\%)_3 = -\$1,538.95(1 + 0.2995) + \$2,000 = 0$

    $(-, -, -, 0)$ passes the net-investment test (pure investment).

[8] In fact, it does not matter which rate we use in applying the net-investment test. If one value passes the test, they will all pass. If one value fails, they will all fail.

- **Project D:** There are three rates of return. We can use any of them for the net investment test. Thus,

$$PB(50\%)_0 = -\$1,000$$
$$PB(50\%)_1 = -\$1,000(1 + 0.50) + \$3,900 = \$2,400$$
$$PB(50\%)_2 = +\$2,400(1 + 0.50) - \$5,030 = -\$1,430$$
$$PB(50\%)_3 = -\$1,430(1 + 0.50 + \$2,145 = 0$$

$(-, +, -, 0)$ fails the net-investment test (mixed investment).

COMMENTS: As shown in Figure 7.5, projects A and C are the only pure investments. Project B demonstrates that the existence of a unique i^* is a necessary but not sufficient condition for a pure investment.

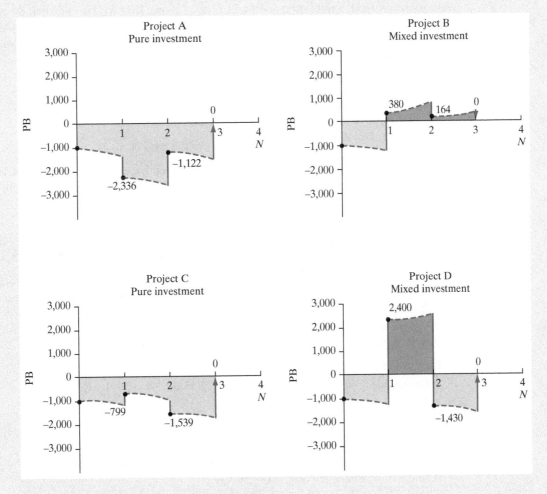

Figure 7.5 Net-investment test (Example 7.6).

7.3.3 Decision Rule for Pure Investments

Suppose we have a pure investment. (Recall that all simple investments are pure investments as well.) Why are we interested in finding the particular interest rate that equates a project's cost with the present worth of its receipts? Again, we may easily answer this question by examining Figure 7.1(a). In this figure, we notice two important characteristics of the NPW profile. First, as we compute the project's PW(i) at a varying interest rate i, we see that the NPW is positive for $i < i^*$, indicating that the project would be acceptable under PW analysis for those values of i. Second, the NPW is negative for $i > i^*$, indicating that the project is unacceptable for those values of i. Therefore, i^* serves as a **benchmark (break-even)** interest rate, knowledge of which will enable us to make an accept-reject decision consistent with NPW analysis.

Note that, for a pure investment, i^* is indeed the IRR of the investment. (See Section 7.1.2.) Merely knowing i^*, however, is not enough to apply this method. Because firms typically wish to do better than break even (recall that at NPW = 0, we were indifferent to the project), a minimum acceptable rate of return (MARR) is indicated by company policy, management, or the project decision maker. If the IRR exceeds this MARR, we are assured that the company will more than break even. Thus, the IRR becomes a useful gauge against which to judge a project's acceptability, and the decision rule for a pure project is as follows.

> If IRR > MARR, accept the project.
>
> If IRR = MARR, remain indifferent.
>
> If IRR < MARR, reject the project.

Note that this decision rule is designed to be applied for a single project evaluation. When we have to compare mutually exclusive investment projects, we need to apply the **incremental analysis approach**, as we will see in Section 7.4.2.

EXAMPLE 7.7 Investment Decision for a Pure Investment

Merco, Inc., a machinery builder in Louisville, Kentucky, is considering investing $2,500,000 in a complete structural beam-fabrication system. The increased productivity resulting from the installation of the drilling system is central to the project's justification. Merco estimates the following figures as a basis for calculating productivity:

- Increased fabricated steel production: 2,000 tons/year
- Average sales price/ton fabricated steel: $1,355/ton
- Labor rate: $18.50/hour
- Tons of steel produced in a year: 15,000 tons
- Cost of steel per ton (2,205 lb): $750/ton
- Number of workers on layout, hole making, sawing, and material handling: 17
- Additional maintenance cost: $188,500/year

With the cost of steel at $750 per ton and the direct labor cost of fabricating one lb at 10 cents, the cost of producing a ton of fabricated steel is about $970.50. With a selling price of $1,355 per ton, the resulting contribution to overhead and profit becomes

$384.50 per ton. Assuming that Merco will be able to sustain an increased production of 2,000 tons per year by purchasing the system, engineers have estimated the projected additional contribution to be 2,000 tons $\times$ \$384.5 = \$769,000.

Since the drilling system has the capacity to fabricate the full range of structural steel, two workers can run the system: one on the saw and the other on the drill. A third operator is required to operate a crane for loading and unloading materials. Merco estimates that to do the equivalent work of these three workers with conventional manufacturing techniques would require, on the average, an additional 14 people for center punching, hole making with a radial or magnetic drill, and material handling. This translates into a labor savings in the amount of \$518,000 per year ($14 \times \$18.50 \times 40$ hours/week $\times$ 50 weeks/year). The system can last for 15 years with an estimated after-tax salvage value of \$180,000. However, after subtracting \$326,000 in annual corporate income taxes, the net investment costs, as well as savings, are as follows:

- Project investment cost: \$2,500,000
- Projected annual net savings:

$$(\$769,000 + \$518,000) - \$188,500 - \$326,000 = \$772,500$$

- Projected after-tax salvage value at the end of year 15: \$180,000

(a) What is the projected IRR on this fabrication investment?
(b) If Merco's MARR is known to be 18%, is this investment justifiable?

SOLUTION

Given: Projected cash flows as shown in Figure 7.6 and MARR = 18%.
Find: (a) The IRR, and (b) whether to accept or reject the investment.

(a) Since only one sign change occurs in the net cash flow series, the fabrication project is a simple investment. This indicates that there will be a unique rate of return that is internal to the project:

$$PW(i) = -\$2,500,000 + \$772,500(P/A, i, 15)$$
$$+ \$180,000(P/F, i, 15)$$
$$= 0$$
$$i^* = 30.36\%$$

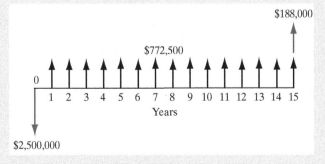

Figure 7.6 Cash flow diagram (Example 7.7).

> With Excel, you will also find that the IRR is about 30.36% for the net investment of $2,500,000.
>
> (b) The IRR figure far exceeds Merco's MARR, indicating that the fabrication system project is an economically attractive one. Merco's management believes that, over a broad base of structural products, there is no doubt that the installation of the fabricating system would result in a significant savings, even after considering some potential deviations from the estimates used in the analysis.

7.3.4 Decision Rule for Mixed Investments

Applied to pure projects, $i*$ provides an unambiguous criterion for measuring profitability. However, when multiple rates of return occur, none of them is an accurate portrayal of a project's acceptability or profitability. However, there is a correct method, which uses an **external interest rate**, for refining our analysis when we do discover multiple $i*$s. An external rate of return allows us to calculate a single, accurate rate of return; if you choose to avoid these more complicated applications of rate-of-return techniques, you must be able to predict multiple $i*$s via the NPW profile and, when they occur, select an alternative method, such as modified internal rate of return (MIRR), which we will discuss in Section 7.3.5 for determining the project's acceptability.

Need for an External Interest Rate for Mixed Investments

In the case of a mixed investment, we can extend the economic interpretation of the IRR to the return-on-invested-capital measure if we are willing to make an assumption about what happens to the extra cash that the investor gets (borrowed) from the project during the intermediate years.

First, the **project balance (PB)** (or investment balance) can also be interpreted from the viewpoint of a financial institution that borrows money from an investor and then pays interest on the PB. Thus, a negative PB means that the investor has money in a bank account; a positive PB means that the investor has borrowed money from the bank. Negative PBs represent interest paid by the bank to the investor; positive PBs represent interest paid by the investor to the bank.

Now, can we assume that the interest paid by the bank and the interest received from the investor are the same for the same amount of balance? In our banking experience, we know that is not the case. Normally, the borrowing rate (interest charged by the financial institutions) is higher than the interest rate earned on your deposit (interest paid by the bank).

However, when we calculate the project balance at an $i*$ for mixed investments, we notice an important point: Cash borrowed (released) from the project is assumed to earn the same interest rate through external investment as money that remains internally invested. In other words, in solving a cash flow for an unknown interest rate, it is assumed that money released from a project can be reinvested to yield a rate of return equal to that received from the project. In fact, we have been making this assumption regardless of whether a cash flow does or does not produce a unique positive $i*$. Note that money is borrowed only when $\text{PB}(i*) > 0$, and the magnitude of the borrowed amount is the project balance. When $\text{PB}(i*) < 0$, no money is borrowed, even though the cash flow may be positive at that time.

In reality, it is not always possible for cash borrowed (released) from a project to be reinvested to yield a rate of return equal to that received from the project. Instead, it is likely that the rate of return available on a capital investment in the business is much different—usually higher—from the rate of return available on other external investments. Thus, it may be necessary to compute the project balances for a project's cash flow at two rates of interest—one on the internal investment and one on the external investments. As we will see later, by separating the interest rates, we can measure the **true rate of return** of any internal portion of an investment project.

Calculation of Return on Invested Capital for Mixed Investments

For a mixed investment, we must calculate a rate of return on the portion of capital that remains invested internally. This rate is defined as the **true IRR** for the mixed investment and is commonly known as the **return on invested capital (RIC)**. Then what interest rate should we assume for the portion of external investment? Insofar as a project is not a net investment, one or more periods when the project has a net outflow of money (a positive project balance) must later be returned to the project. This money can be put into the firm's investment pool until such time as it is needed in the project. The interest rate of this investment pool is the interest rate at which the money can, in fact, be invested outside the project.

Recall that the NPW method assumed that the interest rate charged to any funds withdrawn from a firm's investment pool would be equal to the MARR. In this text, *we will use the MARR as an established external interest rate* (i.e., the rate earned by money invested outside of the project). We can then compute the RIC as a function of the MARR by finding the value of the RIC that will make the terminal project balance equal to zero. (This implies that the firm wants to fully recover any investment made in the project and pay off any borrowed funds at the end of the project life.) This way of computing the rate of return is an accurate measure of the profitability of the project as represented by the cash flow. The following procedure outlines the steps for determining the IRR for a mixed investment:

Step 1. Identify the MARR (or external interest rate).

Step 2. Calculate $PB(i, MARR)_n$ (or simply PB_n) according to the rule

$$PB(i, MARR)_0 = A_0$$

$$PB(i, MARR)_1 = \begin{cases} PB_0(1 + i) + A_1, \text{ if } PB_0 < 0 \\ PB_0(1 + MARR) + A_1, \text{ if } PB_0 > 0 \end{cases}$$

$$\vdots$$

$$PB(i, MARR)_n = \begin{cases} PB_{n-1}(1 + i) + A_n, \text{ if } PB_{n-1} < 0 \\ PB_{n-1}(1 + MARR) + A_n, \text{ if } PB_{n-1} > 0 \end{cases}$$

(As defined in the text, A_n stands for the net cash flow at the end of period n. Note that the terminal project balance must be zero.)

Step 3. Determine the value of i by solving the terminal project balance equation

$$PB(i, MARR)_N = 0.$$

The interest rate i is the RIC (or IRR) for the mixed investment.

Using the MARR as an external interest rate, we may accept a project if the IRR exceeds the MARR, and we should reject the project otherwise. Figure 7.7 summarizes the IRR computation for a mixed investment.

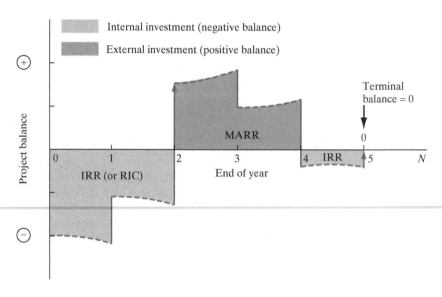

Figure 7.7 Computational logic for IRR (mixed investment).

EXAMPLE 7.8 IRR for a Mixed Investment

A defense contractor has received a government contract from the U.S. Department of Defense worth $460 million, to build an advanced vehicle for future ground operations. The DOD will pay $50 million when the contract is signed, another $360 million at the end of the first year, and the $50 million balance at the end of second year. The expected cash outflows required to produce these weapon systems are estimated to be $150 million now, $100 million during the first year, and $218 million during the second year. The firm's MARR is 25% for this type of technically challenging project. The cash flow is as follows.

n	Outflow ($ million)	Inflow ($ million)	Net Cash Flow ($ million)
0	$150	$50	−$100
1	100	360	260
2	218	50	−168

In normal situations, the defense contractor would not even consider a marginal project such as this. However, hoping that the company could continue to maintain its position as a technology leader in space exploration, management felt that it was worth outbidding its competitors. Financially, what is the economic worth of outbidding the competitors for this project? That is,

(a) Compute the NPW at 25%.
(b) Compute the values of $i*$s for this project.
(c) Make an accept-reject decision based on the results in part (b).

SOLUTION

Given: Cash flow shown and MARR $= 25\%$.

Find: (a) Compute the NPW, (b) i^*, and (c) RIC at MARR $= 25\%$. Also, determine whether to accept the project.

(a) $\text{PW}(25\%) = -\$100 + \$260(P/F, 25\%, 1) - \$168(P/F, 25\%, 2)$

$\qquad\qquad = \$0.48(\text{or } \$480{,}000) > 0$

(b) Since the project has a two-year life, we may solve the net-present-worth equation directly via the quadratic formula:

$$-\$100 + \frac{\$260}{(1 + i^*)} - \frac{\$168}{(1 + i^*)^2} = 0$$

If we let $X = \dfrac{1}{(1 + i^*)}$, we can rewrite the preceding expression as

$$-100 + 260X - 168X^2 = 0$$

Solving for X gives $X = 5/6$ and $5/7$, or $i^* = 20\%$ and 40%. As shown in Figure 7.8, the NPW profile intersects the horizontal axis twice: once at 20% and again at 40%. The investment is obviously not a simple one; thus, neither 20% nor 40% represents the true internal rate of return of this government project.

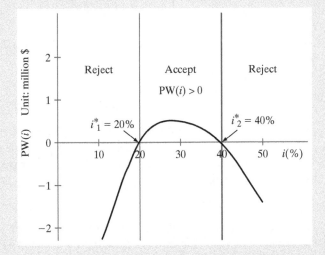

Figure 7.8 NPW plot for a nonsimple investment with multiple rates of return (Example 7.8).

(c) As calculated in (b), the project has multiple rates of return. The project fails the net investment test, as shown in Table 7.5.

At $n = 0$, there is a net investment to the firm, so the project balance expression becomes

$$\text{PB}(i, 25\%)_0 = -\$100.$$

TABLE 7.5 Net Investment Test Using Either 20% or 40%

	A	B	C	D	E
1	Net investment test using $i^* = 20\%$				
2					
3		Project	Return on		Project
4		Balance at	Unrecovered	Payment	Balance at
5	n	Beginning of Year	Balance (20%)	Received	End of Year
6					
7	0	0	0	-100	-100
8	1	-100	-20	260	140
9	2	140	28	-168	0
10					
11	Net investment test using $i^* = 40\%$				
12					
13		Project	Return on		Project
14		Balance at	Unrecovered	Payment	Balance at
15	n	Beginning of Year	Balance (40%)	Received	End of Year
16					
17	0	0	0	-100	-100
18	1	-100	-40	260	120
19	2	120	48	-168	0
20					

The net investment of $100 million that remains invested internally grows at the interest rate i for the next period. With the receipt of $260 in year 1, the project balance becomes

$$\text{PB}(i, 25\%)_1 = -\$100(1 + i) + \$260$$

$$= 160 - 100i$$

At this point, we do not know whether $\text{PB}(i, 25\%)_1$ is positive or negative; we want to know this in order to test for net investment. It depends on the value of i, which we want to determine. Therefore, we need to consider two situations: (1) $i < 1.6$, and (2) $i > 1.6$.

Case 1: $i < 1.6 \rightarrow \text{PB}(i, 25\%)_1 > 0$.

Since this indicates a positive balance, the cash released from the project would be returned to the firm's investment pool to grow at the MARR until it is required back in the project. By the end of year 2, the cash placed in the investment pool would have grown at the rate of 25% [to $(160 - 100i)(1 + 0.25)$] and should be available to meet the required investment in the amount of $168 in period 2. This also happens to be the last period and then the terminal balance must be zero

$$\text{PB}(i, 25\%)_2 = (160 - 100i)(1 + 0.25) - 168$$

$$= -125i + 32$$

$$= 0$$

Solving for i yields

$$\text{RIC} = \text{IRR} = 0.2560, \text{ or } 25.60\% > 25\%$$

Case 2: $i > 1.6 \rightarrow \text{PB}(i, 25\%)_1 < 0.$

The firm is still in an investment mode. Therefore, the balance at the end of year 1 that remains invested will grow at the rate i for the next period. With the investment of \$168 required in year 2 and the fact that the net investment must be zero at the end of the project life, the balance at the end of year 2 should be

$$\text{PB}(i, 25\%)_2 = (160 - 100i)(1 + i) - 168$$
$$= -8 + 60i - 100i^2$$
$$= 0$$

Solving for i gives

$$\text{IRR} = 0.20 \text{ or } 0.40 < 1.6$$

which violates the initial assumption that $i > 1.6$. Therefore, case 1 is the only correct situation. Since it indicates that $\text{IRR}(= \text{RIC}) > \text{MARR}$, the project is acceptable, resulting in the same decision as obtained in part (a) by applying the NPW criterion.

COMMENTS: In this example, we have seen that the RIC is dependent on the choice of a MARR of 25%. We could easily develop a functional relationship between RIC and MARR. Knowing that case 1 is the correct situation, we simply substitute 25% with MARR to obtain the equation

$$\text{PB}(i, 25\%)_2 = (160 - 100i)(1 + \text{MARR}) - 168$$
$$= 0$$
$$i = \text{RIC} = \frac{160\,\text{MARR} - 8}{100(1 + \text{MARR})}$$

As you can see in Figure 7.9, as MARR increases, the investment return (RIC) increases. If we set $\text{MARR} = i^*$ (either 20% or 40%), then we see $\text{RIC} = i^*$, indicating that the interest rates that make the project's NPW equal to zero become the true IRR. Graphically, if we plot the function between RIC and MARR, the true IRR is the point at which the function crosses the 45° line. It is the point at which the firm just breaks even on the investment. It becomes clear that the project just breaks even if MARR is equal to 20% or 40%, and then it is a profitable investment if MARR is greater than 20% but less than 40%. You will see that this is the only range in which the project's NPW becomes positive, giving the same accept signal.

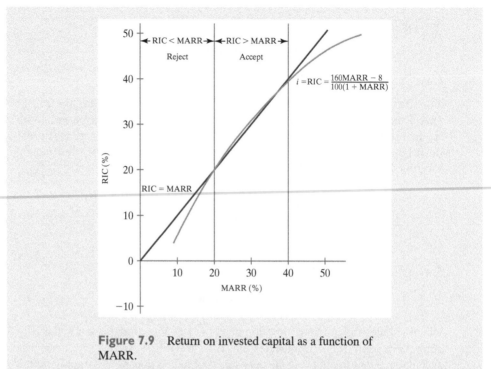

Figure 7.9 Return on invested capital as a function of MARR.

Trial-and-Error Method for Computing IRR for Mixed Investments

The trial-and-error approach to finding the IRR (RIC) for a mixed investment is similar to the trial-and-error approach to finding i^*. We begin with a given MARR and a guess for IRR and solve for the project balance. (A value of IRR close to the MARR is a good starting point for most problems.) Since we desire the project balance to approach zero, we can adjust the value of IRR as needed after seeing the result of the initial guess. For example, for a given pair of interest rates (IRR guess, MARR), if the terminal project balance is positive, the IRR guess value is too low, so we raise it and recalculate. We can continue adjusting our IRR guesses in this way until we obtain a project balance equal or close to zero.

EXAMPLE 7.9 IRR for a Mixed Investment with Excel

Consider project D in Example 7.6. The project has the following cash flow.

n	A_n
0	−$1,000
1	3,900
2	−5,030
3	2,145

We know from Example 7.6 that this is a mixed investment. Compute the IRR for this project. Assume that MARR = 6%.

SOLUTION

Given: Cash flow as stated for mixed investment and MARR = 6%.
Find: IRR.

For MARR = 6%, we must compute i by trial and error. Suppose we guess i = 8%:

$$PB(8\%, 6\%)_0 = -\$1,000$$

$$PB(8\%, 6\%)_1 = -\$1,000(1 + 0.08) + \$3,900 = \$2,820$$

$$PB(8\%, 6\%)_2 = +\$2,820(1 + 0.06) - \$5,030 = -\$2,040.80$$

$$PB(8\%, 6\%)_3 = -\$2,040.80(1 + 0.08) + \$2,145 = -\$59.06$$

The net investment is negative at the end of the project life, indicating that our trial i = 8% is in error. After several trials, we conclude that for MARR = 6% the IRR is approximately 6.13%. To verify the results, we write

$$PB(6.13\%, 6\%)_0 = -\$1,000$$

$$PB(6.13\%, 6\%)_1 = -\$1,000.00(1 + 0.0613) + \$3,900 = \$2,838.70$$

$$PB(6.13\%, 6\%)_2 = +\$2,838.70(1 + 0.0600) - \$5,030 = -\$2,020.98$$

$$PB(6.13\%, 6\%)_3 = -\$2,020.98(1 + 0.0613) + \$2,145 \approx 0$$

The positive balance at the end of year 1 indicates the need to borrow from the project during year 2. However, note that the net investment becomes zero at the end of the project life, confirming that 6.13% is the IRR for the cash flow. Since IRR > MARR, the investment is acceptable.

With Excel, this trial-and-error process can be easily streamlined using the Goal Seek function, as shown in Figure 7.10. Again, our goal is to find the RIC, and the condition to find the RIC is to observe the terminal project balance (cell C12) to be zero by changing the guess value at cell B5.

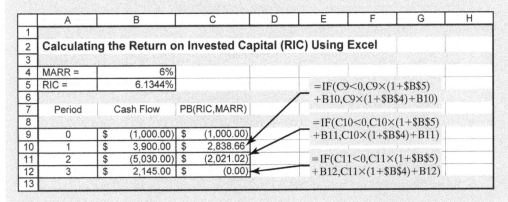

Figure 7.10 Calculating the return on invested capital (or true IRR) using Excel.

COMMENTS: On the basis of the NPW criterion, the investment would be acceptable if the MARR was between zero and 10% or between 30% and 50%. The rejection region is $10\% < i < 30\%$ and $i > 50\%$. This can be verified in Figure 7.1(b). Note that the project also would be marginally accepted under the NPW analysis at MARR $= i = 6\%$:

$$PW(6\%) = -\$1,000 + 3,900(P/F, 6\%, 1)$$

$$-\$5,030(P/F, 6\%, 2) + 2,145(P/F, 6\%, 3)$$

$$= \$3.55 > 0.$$

The flowchart in Figure 7.11 summarizes how you should proceed to apply the net cash-flow sign rule, accumulated cash-flow sign test, and net-investment test to calculate an IRR and make an accept-reject decision for a single project.

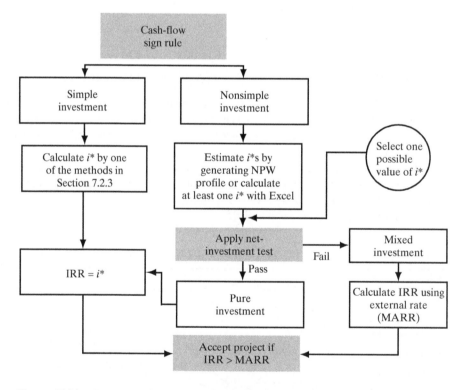

Figure 7.11 Summary of IRR criterion: A flowchart that summarizes how you may proceed to apply the net cash-flow sign rule and net-investment test to calculate IRR for a pure as well as a mixed investment.

7.3.5 Modified Internal Rate of Return (MIRR)

Given the complications involved in finding the RIC, an alternative measure, commonly known as **modified internal rate of return** (MIRR) approach, is proposed. Just like the RIC calculation, the MIRR approach requires two specific interest rates—positive cash flows (cash inflows) are reinvested at the firm's MARR and the negative cash flows (cash outflows) are financed at the firm's cost of capital. Then, we find the equivalent worth of all positive cash flows at the end of project life and the equivalent worth of all cash outlays at $n = 0$. In this way, a unique rate of return can be computed. This MIRR is defined mathematically as follows:

$$\frac{\left(\begin{array}{c}\text{Future value of all cash inflows}\\\text{at the end of project life}\end{array}\right)}{\left(\begin{array}{c}\text{Present value of all cash outflows}\\\text{at the beginning of project life}\end{array}\right)} = (1 + \text{MIRR})^N$$

Or,

$$\frac{\sum_{n=0}^{N} \max(A_n, 0)(1 + i)^{N-n}}{\sum_{n=0}^{N} \min(A_n, 0)(1 + k)^{-n}} = (1 + \text{MIRR})^N \tag{7.4}$$

Where $i = $ MARR,

$\quad k = $ cost of capital cost (or borrowing interest rate),

and then solve for MIRR.

The MIRR will always give a unique solution as the modified cash flow series is basically a simple investment. If we further assume the reinvestment rate is the same as the financing cost ($i = k$), the MIRR criterion will be consistent with the NPW criterion as both criteria assume the same reinvestment assumptions. Therefore, the same accept-reject decision rule that is developed for the pure investment can be applied. In summary, the MIRR will be different from the IRR, as we use different reinvestment assumptions:

- **IRR:** all available money at the end of each period (project balance, not cash flow) will be reinvested at its solving rate of return.
- **MIRR:** all future money received at the end of each period (cash flow, not project balance) will be reinvested at the MARR, and all future expenditures will be discounted back at the cost of financing (or MARR).

The MIRR approach technically eliminated the complexity of dealing with multiple IRR problems, but its economic interpretation is not clear as we manipulated the cash flows. For IRR, recall that we are reinvesting the surplus funds (project balance) at the end of each period. When we receive the money from the project, typically we pay off the required debt obligations first, and any remaining money will be reinvested for the next period. For MIRR, all future cash flows are reinvested or discounted (regardless of debt positions) at MARR, this is exactly the same assumption used in NPW criterion. If that is the case, we just use the NPW criterion.

EXAMPLE 7.10 Calculation of MIRR

Reconsider the project cash flows given in Example 7.9. At MARR of 6%, determine the MIRR and make an accept-reject decision. Assume that the financing cost is the same as MARR of 6%.

n	A_n	Outflow	Inflow
0	−$1,000	$1,000	
1	3,900		$3,900
2	−5,030	5,030	
3	2,145		2,145

SOLUTION:

Given: Cash flow series as stated for mixed investment and $i = k = 6\%$.
Find: MIRR.

Using Eq. (7.4),

$$\frac{\$3,900(F/P, 6\%, 2) + \$2,145}{\$1,000 + \$5,030(P/F, 6\%, 2)} = (1 + \text{MIRR})^3$$

$$= \frac{\$6,527.04}{\$5,476.38}$$

$$= 1.1919$$

$$1.1919 = (1 + \text{MIRR})^3$$

$$\text{MIRR} = 6.026\% > 6\%$$

So, we accept the investment. Note that this MIRR figure is different from the RIC of 6.13% in Example 7.9. Figure 7.12 illustrates the process of obtaining the MIRR.

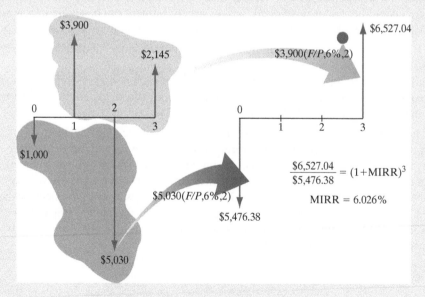

Figure 7.12 Process of computing the MIRR.

COMMENTS: Since it is simpler than the RIC calculation, the method has been advocated in many engineering economy textbooks or by the practitioners. However, that does not mean that the MIRR more accurately reflect the cost and profitability of a project. Once again, if all we want is to modify the cash flow series to produce a unique rate of return, we just use the NPW criterion.

7.4 Mutually Exclusive Alternatives

In this section, we present the decision procedures that should be used in comparing two or more mutually exclusive projects on the basis of the rate-of-return measure. We will consider two situations: (1) alternatives that have the same economic service life, and (2) alternatives that have unequal service lives.

7.4.1 Flaws in Project Ranking by IRR

Under NPW or AE analysis, the mutually exclusive project with the highest worth was preferred. (This is known as the "total investment approach.") Unfortunately, the analogy does not carry over to IRR (or MIRR) analysis: The project with the highest IRR (or MIRR) may *not* be the preferred alternative.

To illustrate the flaws inherent in comparing IRRs in order to choose from mutually exclusive projects, suppose you have two mutually exclusive alternatives, each with a one-year service life: One requires an investment of $1,000 with a return of $2,000, and the other requires $5,000 with a return of $7,000. You already obtained the IRRs and NPWs at MARR $= 10\%$ as given in Table 7.6.

Assuming that you have enough money in your investment pool to select either alternative, would you prefer the first project simply because you expect a higher rate of return?

We can see that A2 is preferred over A1 by the NPW measure. On the other hand, the IRR measure gives a numerically higher rating for A1. This inconsistency in ranking occurs because the NPW, NFW, and AE are **absolute (dollar)** measures of investment worth, whereas the IRR is a **relative (percentage)** measure and cannot be applied in the same way. That is, the IRR measure ignores the **scale** of the investment. Therefore, the answer to our question in the previous paragraph is no; instead, you would prefer the second project with the lower rate of return but higher NPW. Either the NPW or the AE measure would lead to that choice, but a comparison of IRRs would rank the smaller project higher. How do we resolve this inconsistency in ranking the projects? Another approach, referred to as **incremental analysis**, is needed.

TABLE 7.6 **Flaws in Project Ranking by IRR**

n	A1	A2
0	−$1,000	−$5,000
1	$2,000	$7,000
IRR	100%	40%
PW(10%)	$818	$1,364

7.4.2 Incremental Investment Analysis

In the previous example, the more costly option requires an incremental investment of $4,000 at an incremental return of $5,000. Let's assume that you have exactly $5,000 in your investment pool.

- If you decide to invest in option A1, you will need to withdraw only $1,000 from your investment pool. The remaining $4,000 will continue to earn 10% interest. One year later, you will have $2,000 from the outside investment and $4,400 from the investment pool. With an investment of $5,000, in one year you will have $6,400. The equivalent present worth of this change in wealth is

$$\text{PW}(10\%) = -\$5,000 + \$6,400(P/F, 10\%, 1) = \$818.$$

- If you decide to invest in option A2, you will need to withdraw $5,000 from your investment pool, leaving no money in the pool, but you will have $7,000 from your outside investment. Your total wealth changes from $5,000 to $7,000 in a year. The equivalent present worth of this change in wealth is

$$\text{PW}(10\%) = -\$5,000 + \$7,000(P/F, 10\%, 1) = \$1,364.$$

In other words, if you decide to take the more costly option, certainly you would be interested in knowing that this additional investment can be justified at the MARR. The 10%-of-MARR value implies that you can always earn that rate from other investment sources (i.e., $4,400 at the end of 1 year for a $4,000 investment). However, by investing the additional $4,000 in the second option, you would make an additional $5,000, which is equivalent to earning at the rate of 25%. Therefore, the incremental investment can be justified.

Decision Rules

Now we can generalize the decision rule for comparing mutually exclusive projects. For a pair of mutually exclusive projects (A and B, with B defined as the more costly option), we may rewrite B as

$$
\begin{cases}
A = A \\
B = A + \overbrace{(B - A)}^{\text{incremental investment}}
\end{cases}
$$

In other words, B has two cash flow components: (1) the same cash flow as A and (2) the incremental component $(B - A)$. Therefore, the situation in which B is preferred to A is only when the rate of return on the incremental component $(B - A)$ exceeds the MARR. Therefore, for two mutually exclusive projects, rate-of-return analysis is done by computing the *internal rate of return on the incremental investment* (IRR_Δ) between the projects. Since we want to consider increments of investment, we compute the cash flow for the difference between the projects by subtracting the cash flow for the lower investment-cost project (A) from that of the higher investment-cost project (B). Then the decision rule is with $\Delta = B - A$

- If $IRR_{B-A} > $ MARR, select B
- If $IRR_{B-A} = $ MARR, select either project.
- If $IRR_{B-A} < $ MARR, select A

where $B - A$ is an investment increment (negative cash flow).

If a "do-nothing" alternative is allowed, the smaller cost option must be profitable (its IRR must be greater than the MARR) at first. This means that you compute the rate of return for each alternative in the mutually exclusive group and then eliminate the alternatives whose IRRs are less than the MARR before applying the incremental analysis.

It may seem odd to you how this simple rule always allows us to select the right project. Example 7.11 illustrates the incremental investment decision rule.

EXAMPLE 7.11 IRR on Incremental Investment: Two Alternatives

John Covington, a college student, wants to start a small-scale painting business during his off-school hours. To limit start-up costs, he decides to purchase some used painting equipment. He has two mutually exclusive options: Do most of the painting by himself by limiting his business to residential painting jobs (B1) or purchase more painting equipment and hire some helpers to do both residential and commercial painting jobs that he expects will have a higher equipment cost but will provide higher revenues as well (B2). In either case, John expects to fold up the business in three years when he graduates from college.

The cash flows for the two mutually exclusive alternatives are as follows.

n	B1	B2	B2 − B1
0	−$3,000	−$12,000	−$9,000
1	1,350	4,200	2,850
2	1,800	6,225	4,425
3	1,500	6,330	4,830
IRR	25%	17.43%	15%

Knowing that both alternatives are revenue projects, which project would John select at MARR = 10%? (Note that both projects are also profitable at 10%.)

SOLUTION

Given: Incremental cash flow between two alternatives and MARR = 10%.

Find: (a) IRR on the increment, and (b) which alternative is preferable.

(a) To choose the best project, we compute the incremental cash flow B2 − B1. Then we compute the IRR on this increment of investment by solving the equation

$$-\$9,000 + \$2,850(P/F, i, 1) + \$4,425(P/F, i, 2)$$
$$+ \$4,830(P/F, i, 3) = 0$$

(b) We obtain $i^*_{B2-B1} = 15\%$, as plotted in Figure 7.13. By inspection of the incremental cash flow, we know that it is a simple investment, so $\text{IRR}_{B2-B1} = i^*_{B2-B1}$. Since $\text{IRR}_{B2-B1} > \text{MARR}$, we select B2, which is consistent with the NPW analysis. Note that, at MARR > 25%, neither project would be acceptable.

Figure 7.13 NPW profiles for B1 and B2 (Example 7.11).

COMMENTS: Why did we choose to look at the increment B2 − B1 instead of B1 − B2? Because we want the first flow of the incremental cash flow series to be negative (an investment flow), so that we can calculate an IRR. By subtracting the lower initial investment project from the higher, we guarantee that the first increment will be an investment flow. If we ignore the investment ranking, we might end up with an increment that involves *borrowing cash flow* and has no rate of return on investment. This is indeed the case for B1 − B2. (i^*_{B1-B2} is also 15%, not −15%, but it has a different meaning: It is a borrowing rate, not a rate of return on your investment.) If, erroneously, we had compared this i^* with the MARR, we might have accepted project B1 over B2. This undoubtedly would have damaged our credibility with John Covington!!

The next example indicates that the inconsistency in ranking between NPW and IRR can also occur when differences in the timing of a project's future cash flows exist, even if their initial investments are the same.

EXAMPLE 7.12 IRR on Incremental Investment When Initial Flows are Equal

Consider the following two mutually exclusive investment projects that require the same amount of investment.

n	C1	C2
0	−$9,000	−$9,000
1	480	5,800
2	3,700	3,250
3	6,550	2,000
4	3,780	1,561
IRR	18%	20%

Which project would you select on the basis of the rate of return on incremental investment, assuming that MARR = 12%? (Once again, both projects are profitable at 12%.)

SOLUTION

Given: Cash flows for two mutually exclusive alternatives (as shown) and MARR = 12%.

Find: (a) IRR on incremental investment, and (b) which alternative is preferable.

(a) First check to see if IRRs for both projects exceed the MARR. Since the initial investments are equal, we progress through the cash flows until we find the first difference and then set up the increment so that this first nonzero flow is negative (i.e., an investment). Thus, we set up the incremental investment by taking (C1 − C2):

n	C1 − C2
0	$0
1	−5,320
2	450
3	4,550
4	2,219

We next set the PW equation equal to zero:

$$-\$5,320 + \$450(P/F, i, 1) + \$4,550(P/F, i, 2)$$
$$+ \$2,219(P/F, i, 3) = 0$$

(b) Solving for i yields $i^* = 14.71\%$, which is also an IRR, since the increment is a simple investment. Since $\text{IRR}_{C1-C2} = 14.71\% > \text{MARR}$, we would select C1. If we used NPW analysis, we would obtain $\text{PW}(12\%)_{C1} = \$1,443$ and $\text{PW}(12\%)_{C2} = \$1,185$, indicating that C1 is preferred over C2.

When you have more than two mutually exclusive alternatives, they can be compared in pairs by successive examination. Example 7.13 illustrates how to compare three mutually exclusive alternatives. (In Chapter 15, we will examine some multiple-alternative problems in the context of capital budgeting.)

EXAMPLE 7.13 IRR on Incremental Investment: Three Alternatives

Consider the following three sets of mutually exclusive alternatives. Which project would you select on the basis of the rate of return on incremental investment, assuming that MARR = 15%?

SOLUTION

Given: Cash flows for three alternatives and MARR = 15%.

Find: IRR on incremental investment and which alternative is preferable.

n	D1	D2	D3
0	−$2,000	−$1,000	−$3,000
1	1,500	800	1,500
2	1,000	500	2,000
3	800	500	1,000
IRR	34.37%	40.76%	24.81%

Step 1. Examine the IRR of each alternative. At this point, we can eliminate any alternative that fails to meet the MARR. In this example, all three alternatives exceed the MARR.

Step 2. Compare D1 and D2 in pairs.[9] Because D2 has a lower initial cost, compute the rate of return on the increment (D1–D2), which represents an increment of investment.

n	D1 − D2
0	−$1,000
1	700
2	500
3	300

The incremental cash flow represents a simple investment. To find the incremental rate of return, we write

$$-\$1,000 + \$700(P/F, i, 1) + \$500(P/F, i, 2) + \$300(P/F, i, 3) = 0$$

[9] When faced with many alternatives, you may arrange them in order of increasing initial cost. This is not a required step, but it makes the comparison more tractable.

Solving for i^*_{D1-D2} yields 27.61%, which exceeds the MARR; therefore, D1 is preferred over D2. Now you eliminate D2 from further consideration.

Step 3. Compare D1 and D3. Once again, D1 has a lower initial cost. Examine the increment (D3 − D1):

n	D3 − D1
0	−$1,000
1	0
2	1,000
3	200

Here, the incremental cash flow represents another simple investment. The increment (D3 − D1) has an unsatisfactory 8.8% rate of return; therefore, D1 is preferred over D3. Accordingly, we conclude that D1 is the best alternative.

EXAMPLE 7.14 Incremental Analysis for Cost-Only Projects

Falk Corporation is considering two types of manufacturing systems to produce its shaft couplings over six years: (1) a cellular manufacturing system (CMS), and (2) a flexible manufacturing system (FMS). The average number of pieces to be produced with either system would be 544,000 per year. The operating cost, initial investment, and salvage value for each alternative are estimated as given in Table 7.7.

TABLE 7.7 A Summary of Financial Data on CMS and FMS Options

Items	CMS Option	FMS Option
Annual O&M costs:		
Annual labor cost	$1,169,600	$707,200
Annual material cost	832,320	598,400
Annual overhead cost	3,150,000	1,950,000
Annual tooling cost	470,000	300,000
Annual inventory cost	141,000	31,500
Annual income taxes	1,650,000	1,917,000
Total annual costs	$7,412,920	$5,504,100
Investment	$4,500,000	$12,500,000
Net salvage value	$500,000	$1,000,000

Figure 7.14 illustrates the cash flows associated with each alternative. The firm's MARR is 15%. On the basis of the IRR criterion, which alternative would be a better choice?

DISCUSSION: Since we can assume that both manufacturing systems would provide the same level of revenues over the analysis period, we can compare the two alternatives on the basis of cost only. (These are service projects.) Although we cannot compute the IRR for each option without knowing the revenue figures, we can still calculate the IRR on incremental cash flows (Table 7.8). Since the FMS option requires a higher initial investment than that for the CMS, the incremental cash flow is the difference (FMS − CMS).

SOLUTION

Given: Cash flows shown in Figure 7.13 and $i = 15\%$ per year.

Find: Incremental cash flows, and select the better alternative on the basis of the IRR criterion.

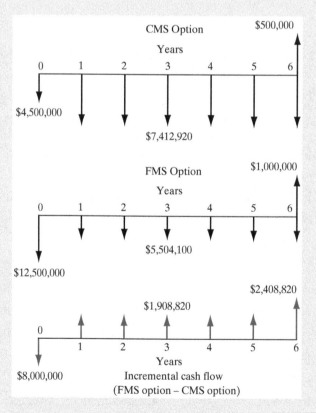

Figure 7.14 Comparison of mutually exclusive alternatives with equal revenues (cost only) (Example 7.14).

TABLE 7.8 Incremental Cash Flows (FMS − CMS)

n	CMS Option	FMS Option	Incremental (FMS − CMS)
0	−$4,500,000	−$12,500,000	−$8,000,000
1	−7,412,920	−5,504,100	1,908,820
2	−7,412,920	−5,504,100	1,908,820
3	−7,412,920	−5,504,100	1,908,820
4	−7,412,920	−5,504,100	1,908,820
5	−7,412,920	−5,504,100	1,908,820
6	−7,412,920	−5,504,100	1,908,820
	+ $500,000	+ $1,000,000	$2,408,820

First, we have

$$PW(i)_{FMS-CMS} = -\$8,000,000 + \$1,908,820(P/A, i, 5)$$
$$+ \$2,408,820(P/F, i, 6)$$
$$= 0$$

Solving for i by trial and error yields 12.43%. Since $IRR_{FMS-CMS} = 12.43\% < 15\%$, we would select CMS. Although the FMS would provide an incremental annual savings of $1,908,820 in operating costs, the savings are not large enough to justify the incremental investment of $8,000,000.

COMMENTS: Note that the CMS option is marginally preferred to the FMS option. However, there are risks in relying solely on the easily quantified savings in input factors—such as labor, energy, and materials—from the FMS and in not considering gains from improved manufacturing performance that are more difficult and subjective to quantify. Factors such as improved product quality, increased manufacturing flexibility (rapid response to customer demand), reduced inventory levels, and increased capacity for product innovation are frequently ignored, because we have inadequate means for quantifying their benefits. If these intangible benefits were considered, the FMS option could come out better than the CMS option.

7.4.3 Handling Unequal Service Lives

In Chapters 5 and 6, we discussed the use of the NPW and AE criteria as bases for comparing projects with unequal lives. The IRR measure also can be used to compare projects with unequal lives, as long as we can establish a common analysis period. The decision procedure is then exactly the same as for projects with equal lives. It is likely, however, that we will have a multiple-root problem, which creates a substantial computational burden. For example, suppose we apply the IRR measure to a case in which one project has a five-year life and the other project has an eight-year life, resulting in a least common multiple of 40 years. Then when we determine the incremental cash flows over the analysis period, we are bound to observe many sign changes. This leads to the possibility of having many $i*$s. Example 7.14 uses $i*$ to compare mutually exclusive projects in which the incremental cash flows reveal several sign changes. (Our purpose is not to encourage you to use the IRR approach to compare projects with unequal lives; rather, it is to show the correct way to compare them if the IRR approach *must* be used.)

EXAMPLE 7.15 IRR Analysis for Projects with Different Lives in Which the Increment Is a Nonsimple Investment

Consider Example 5.13, in which a mail-order firm wants to install an automatic mailing system to handle product announcements and invoices. The firm has a choice between two different types of machines. Using the IRR as a decision criterion, select the best machine. Assume a MARR of 15%, as before.

SOLUTION

Given: Cash flows for two projects with unequal lives, as shown in Example 5.13, and MARR $= 15\%$.

Find: The alternative that is preferable.

Since the analysis period is equal to the least common multiple of 12 years, we may compute the incremental cash flow over this 12-year period. As shown in Figure 7.15, we subtract the cash flows of model A from those of model B to form the increment of investment. (Recall that we want the first cash-flow difference to be a negative value.) We can then compute the IRR on this incremental cash flow.

- Five sign changes occur in the incremental cash flows, indicating a nonsimple incremental investment.

  - Even though there are five sign changes in the cash flow, there is only one positive $i*$ for this problem: 63.32%.

  - Unfortunately, however, the net investment test shows that it is a mixed investment, so we need to calculate either RIC or use the MIRR on incremental investment.

  RIC: We need to employ an external rate to compute the RIC in order to make a proper accept-reject decision. Assuming that the firm's MARR is 15%, we find that the RIC is 50.68%. Since $\text{IRR}_{B-A} > \text{MARR}$, model B would be selected, which is consistent with the NPW analysis. In fact, we can use the Goal Seek function in Excel to find the RIC, as shown in Table 7.9.

  MIRR: We can also calculate the MIRR on this incremental investment at $i = 15\%$.

$$\frac{\$1{,}000(F/P, 15\%, 11) + \$1{,}000(F/P, 15\%, 10) + \cdots \$500}{\$2{,}500 + \$12{,}500(P/F, 15\%, 4) + \$12{,}500(P/F, 15\%, 8)}$$

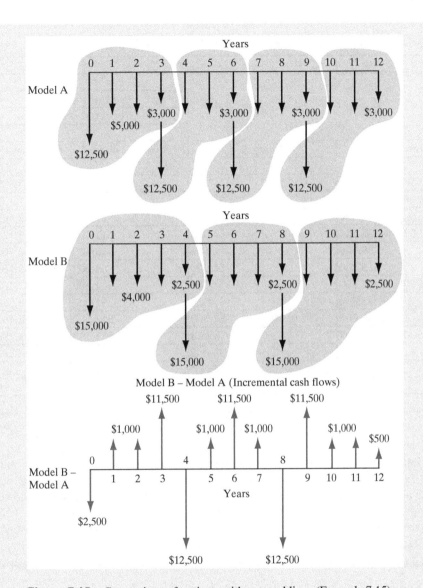

Figure 7.15 Comparison of projects with unequal lives (Example 7.15).

$$= \frac{\$100,887.62}{\$13,733.18}$$

$$= 7.3463$$

$$= (1 + \text{MIRR})^{12}$$

$$\text{MIRR} = 18.08\% > 15\%$$

which indicates that model B be preferred.

TABLE 7.9 Comparison of Mutually Exclusive Alternatives Based on RIC

	A	B	C	D	E	F	G
1							
2						**MARR =**	15%
3						**Guess RIC =**	51%
4							
5	*n*	Model A		Model B		Model B - Model A	PB(RIC,MARR)
6							
7	0	-$12,500		-$15,000		-$2,500	-$2,500.00
8	1		-$5,000		-$4,000	$1,000	-$2,767.00
9	2		-$5,000		-$4,000	$1,000	-$3,169.32
10	3	-$12,500	-$3,000		-$4,000	$11,500	$6,724.47
11	4		-$5,000	-$15,000	-$2,500	-$12,500	-$4,766.86
12	5		-$5,000		-$4,000	$1,000	-$6,182.71
13	6	-$12,500	-$3,000		-$4,000	$11,500	$2,183.89
14	7		-$5,000		-$4,000	$1,000	$3,511.47
15	8		-$5,000	-$15,000	-$2,500	-$12,500	-$8,461.80
16	9	-$12,500	-$3,000		-$4,000	$11,500	-$1,250.25
17	10		-$5,000		-$4,000	$1,000	-$883.88
18	11		-$5,000		-$4,000	$1,000	-$331.83
19	12		-$3,000		-$2,500	$500	$0.00
20							

$$=IF(G17<0, G17\times(1+\$G\$3)+F18, G17\times(1+\$G\$2)+F18)$$

COMMENTS: Given the complications inherent in IRR analysis in comparing mutually exclusive alternatives, it is usually more desirable to employ one of the other equivalence techniques for this purpose. As an engineering manager, you should keep in mind the intuitive appeal of the rate-of-return measure. Once you have selected a project on the basis of NPW or AE analysis, you may also wish to express its worth as a rate of return for the benefit of your associates.

SUMMARY

- The **rate of return (ROR)** is the interest rate earned on unrecovered project balances such that an investment's cash receipts make the terminal project balance equal to zero. The rate of return is an intuitively familiar and understandable measure of project profitability that many managers prefer to NPW or other equivalence measures.

- Mathematically, we can determine the rate of return for a given project cash-flow series by locating an interest rate that equates the net present worth of the project's cash flows to zero. This break-even interest rate is denoted by the symbol i^*.

- The **internal rate of return (IRR)** is another term for ROR, which stresses the fact that we are concerned with the interest earned on the portion of the project that is internally invested, not those portions released by (borrowed from) the project.

■ To apply rate-of-return analysis correctly, we need to classify an investment as either simple or nonsimple. A **simple investment** is defined as an investment in which the initial cash flows are negative and only one sign change in the net cash flow occurs, whereas a **nonsimple investment** is an investment for which more than one sign change in the cash flow series occurs. Multiple $i*$s occur only in nonsimple investments. However, not all nonsimple investments will have multiple $i*$ (s) In this regard,

1. The possible presence of multiple $i*$s (rates of return) can be predicted by

   - The net cash-flow sign test
   - The accumulated cash-flow sign test

     When multiple rates of return cannot be ruled out by the two methods, it is useful to generate an NPW profile to approximate the value of -300.

2. All $i*$ values should be exposed to the **net-investment test**. Passing this test indicates that $i*$ is an internal rate of return and is therefore a suitable measure of project profitability. Failing to pass the test indicates project borrowing, a situation that requires further analysis with the use of an **external interest rate**.

3. **Return-on-invested-capital** analysis uses one rate (the firm's MARR) on externally invested balances and solves for another rate ($i*$) on internally invested balances.

4. The IRR assumes the cash flows from a project are reinvested at the IRR.

5. The modified IRR (MIRR) assumes that positive cash flows are reinvested at the firm's MARR, and the initial outlays are financed at the firm's financing cost. If the outlays are financed at the same rate of MARR, then the MIRR appraoch will be exactly the same as the NPW criterion.

■ For a pure investment, $i*$ is the rate of return that is internal to the project. For a mixed investment, the RIC calculated with the use of the external interest rate (or MARR) is the true IRR; so the decision rule is as follows.

   If IRR > MARR, accept the project.
   If IRR = MARR, remain indifferent.
   If IRR < MARR, reject the project.

IRR analysis yields results consistent with NPW and other equivalence methods.

■ In properly selecting among alternative projects by IRR analysis, **incremental investment** must be used. In creating an incremental investment, we always subtract the lower-cost investment from the higher-cost one. Basically, you want to know that the extra investment required can be justified on the basis of the extra benefits generated in the future.

PROBLEMS

Note: *The symbol $i*$ represents the interest rate that makes the net present value of the project in question equal to zero. The symbol IRR represents the internal rate of return of the investment. For a simple investment, IRR = $i*$. For a nonsimple investment, $i*$ is generally not equal to IRR.*

Concept of Rate of Return

7.1 Suppose that you invest $3,000 in stock. Five years later, your investment yields $8,568. What is the rate of return of your investment?

7.2 You are going to buy new machinery worth $36,200. The dealer computes your monthly payment to be $735.25 for 72 months of financing. What is the dealer's effective rate of return on this loan transaction?

7.3 You wish to sell a bond that has a face value of $5,000. The bond bears an interest rate of 7.5%, which is payable quarterly. Six years ago, the bond was purchased at $4,800. At least a 9% annual return on the investment is desired. What must be the minimum selling price of the bond now in order to make the desired return on the investment?

7.4 In 1970, Wal-Mart offered 300,000 shares of its common stock to the public at a price of $16.50 per share. Since that time, Wal-Mart has had 11 two-for-one stock splits. On a purchase of 100 shares at $16.50 per share on the company's first offering, the number of shares has grown to 204,800 shares worth $15,767,500 in July 2014. What is the return on investment for investors who purchased the stock in 1970 (over a 44-year ownership period)? Assume that no dividends were received during that period.

7.5 Johnson Controls spent more than $2.5 million retrofitting a government complex and installing a computerized energy-management system for the State of Massachusetts. As a result, the state's energy bill dropped from an average of $6 million a year to $3.5 million. Moreover, both parties will benefit from the 10-year life of the contract. Johnson recovers half the money it saved in reduced utility costs (about $1.2 million a year over 10 years); Massachusetts has more money to spend on other things. What is the rate of return realized by Johnson Controls in this energy-control system?

7.6 Suppose that you invest $74,000 in a restaurant business. Two years later, you sell half of this business to a partner for $165,000. Then a year later, the business is in the red, and you have to pay $36,000 to close the business. What is the rate of return on your investment from this restaurant business?

7.7 As a rule of thumb, 8% annualized net return over a minimum holding period of five years is a realistic figure. Pablo Picasso's 1905 portrait *Boy with a Pipe* sold for $104.2 million in an auction at Sotheby's Holdings, Inc., on June 24, 2004, shattering the existing record for art and ushering in a new era in pricing for 20th-century paintings. The Picasso, sold by the philanthropic Greentree Foundation, cost Mr. Whitney about $30,000 in 1950. Determine the annual rate of appreciation of the artwork over 54 years. Is this a good investment?

Investment Classification and Calculation of i^*

7.8 Consider four investments with the sequences of cash flows given in Table P7.8.
(a) Identify all the simple investments.
(b) Identify all the nonsimple investments.
(c) Compute i^* for each investment.
(d) Which project has no rate of return?

TABLE P7.8

	Net Cash Flow			
n	Project A	Project B	Project C	Project D
0	$46,500	−$52,200	−$58,000	−$42,000
1	−19,400	18,250	$42,000	$38,000
2	−19,400	25,759	$27,000	$38,000
3	−19,400	32,216	$12,000	−$26,000

7.9 Consider the infinite cash-flow series with repeated cash-flow patterns given in Table P7.9.

TABLE P7.9

	Net Cash Flow
n	A_n
0	−$6,200
1	2,600
2	3,400
3	−6,200 + 4,200
4	2,600
5	3,400
6	−$6,200 + 4,200
7	2,600
8	3,400
⋮	⋮

Determine i^* for this infinite cash-flow series.

7.10 Consider the investment projects given in Table P7.10.
(a) Classify each project as either simple or nonsimple.
(b) Use the quadratic equation to compute i^* for project C.
(c) Obtain the rate(s) of return for each project by plotting the NPW as a function of the interest rate.

7.11 Consider the projects given in Table P7.11.
(a) Classify each project as either simple or nonsimple.
(b) Identify all positive i^*s for each project.
(c) For each project, plot the present worth as a function of the interest rate (i).

TABLE P7.10

n	A	B	C	D	E
			Net Cash Flow		
0	−$650	−$250	−$560	−$500	−$300
1	225	−140	860	220	140
2	40	−80	−215	150	90
3	95	0	—	225	−280
4	180	430	—	90	110
5	110	280	—	60	85
6	−120	150	—	—	50
7	450	120	—	—	—

TABLE P7.11

n	A	B	C	D
		Net Cash Flow		
0	−$5,800	−$8,500	−$6,200	−$5,400
1	−1,280	4,650	1,800	2,450
2	−850	4,200	1,800	3,200
3	4,816	3,050	1,200	2,850
4	5,255	−900	1,200	—
5	2,615	1,800	—	—
6	—	−1,250	—	—
7	—	1,800	—	—

7.12 Consider the financial data for a project given in Table P7.12.

(a) What is i^* for this project?

(b) If the annual expense increases at an 8% rate over the previous year's expenses, but the annual income is unchanged, what is the new i^*?

(c) (c) In part (b), at what annual rate will the annual income have to increase to maintain the same i^* obtained in part (a)?

TABLE P7.12

Initial investment	$120,000
Project life	8 years
Salvage value	$15,000
Annual revenue	$38,000
Annual expenses	$12,000

7.13 Consider two investments, A and B, with the sequences of cash flows given in Table P7.13.

TABLE P7.13

	Net Cash Flow	
n	Project A	Project B
0	−$56,000	−$56,000
1	8,500	20,500
2	12,500	16,500
3	16,400	14,500
4	25,200	12,500
5	30,500	9,500

(a) Compute $i*$ for each investment.
(b) Plot the present-worth curve for each project on the same chart, and find the interest rate that makes the two projects equivalent.

7.14 Consider the investment projects given in Table P7.14.

(a) For each project, apply the sign rule to predict the number of possible $i*$s.
(b) For each project, plot the NPW profile as a function of i between 0% and 200%.
(c) For each project, compute the value(s) of $i*$.

7.15 Consider an investment project with the cash flows given in Table P7.15.

(a) Find the IRR for this investment.
(b) Plot the present worth of the cash flow as a function of i.
(c) On the basis of the IRR criterion, should the project be accepted at MARR = 16%?

TABLE P7.15

n	Net Cash Flow
0	−$325,000
1	92,000
2	220,000
3	105,000

Mixed Investments, RIC and MIRR

7.16 Consider the investment projects given in Table P7.16.

Assume that MARR = 15% in the following questions.

(a) Compute $i*$ for each investment. If the problem has more than one $i*$, identify all of them.
(b) Compute IRR(true) for each project.
(c) Compute the MIRR at MARR = 15%.
(d) Determine the acceptability of each investment.

TABLE P7.16

	Net Cash Flow		
n	Project 1	Project 2	Project 3
0	−$4,200	−$4,200	−$4,200
1	800	6,550	7,930
2	3,240	−2,090	−640

7.17 Consider the investment projects given in Table P7.17.

(a) Use the quadratic formula to compute $i*$ for B.
(b) Classify each project as either simple or nonsimple.

TABLE P7.14

	Project Cash Flows					
n	A	B	C	D	E	F
0	−$290	−$250	−$360	−$420	−$450	−$250
1	220	−180	270	350	0	280
2	280	340	50	−120	−50	230
3	−60	420	350	650	940	−210

TABLE P7.17

			Project Cash Flow		
n	A	B	C	D	E
0	−$245	−$255	−$265	$275	−$285
1	90	220	120	120	250
2	100	125	140	−$600	100
3	125	—	0	−$640	−$180
4	135	—	100	310	—
5	—	—	80	650	—

(c) Apply the cash-flow sign rules to each project, and determine the number of possible positive $i*$s. Identify all projects having a unique $i*$.

(d) Compute the IRRs for all the projects, excepting B. Assume MARR = 15%.

(e) Compute the MIRR for project D at MARR = 15%.

(f) Apply the net-investment test to project D.

7.18 You established an investment fund with an initial deposit of $2,500. You further contributed $2,500 at the end of the each year up to the fourth year. The fund value, including interest, is $16,850 at the end of fifth year. Find the internal rate of return from your investment fund.

7.19 Consider the investment projects given in Table P7.19.

Assume that MARR = 14% in the following questions:

(a) Identify the $i*('s)$ for each investment. If the project has more than one $i*$, identify all of them.

(b) Which project(s) is (are) a mixed investment?

(c) Compute the IRR for each project.

(d) Compute the MIRR for each project at MARR = 14%.

(e) Determine the acceptability of each project.

7.20 Consider the investment projects given in Table P7.20.

Assume that MARR = 14% for the following questions.

(a) Identify the pure investment(s).

(b) Identify the mixed investment(s).

(c) Determine the IRR for each investment.

(d) Which project would be acceptable?

7.21 By outbidding its competitors, Trane Image Processing (TIP), a defense contractor, received a contract worth $7,300,000 to build navy flight simulators for U.S. Navy pilot training over two years. With some defense contracts, the U.S. government makes an advance payment when the contract is signed, but in this case, the government will make two progressive payments: $4,300,000 at the end of the first year and the $3,000,000 balance at the end of the second year. The expected cash outflows required to produce the simulators are estimated to be $1,000,000 now,

TABLE P7.19

		Net Cash Flow	
n	Project 1	Project 2	Project 3
0	−$10,500	−$9,250	−$10,200
1	20,100	18,400	6,500
2	60,200	−8,500	8,250
3	−80,500	—	—

TABLE P7.20

| | Net Cash Flow | | |
n	Project 1	Project 2	Project 3
0	−$2,500	−$720	−$1,850
1	1,050	2,952	7,600
2	1,050	−4,018	0
3	1,100	1,814	−3,250
4	1,100	—	—
IRR	(25.57%)	(20%, 40%, 50%)	(−27.99%, 299.82%)

$2,000,000 during the first year, and $4,320,000 during the second year. The expected net cash flows from this project are summarized in Table P7.21.

In normal situations, TIP would not even consider a marginal project such as this one. However, hoping that the company can establish itself as a technology leader in the field, management felt that it was worth outbidding its competitors. Financially, what is the economic worth of outbidding the competitors for this project?

(a) Determine whether this project is or is not a mixed investment.

(b) Compute the IRR for this investment. Assume MARR = 12%.

(c) Should Tip accept the project?

7.22 Consider the investment projects given in Table P7.22.

(a) Compute i^* for project C. If there is more than one i^*, identify all of them.

(b) Identify the mixed investment(s).

(c) Assuming that MARR = 15%, determine the acceptability of each project on the basis of the IRR criterion.

TABLE P7.21

Year	Cash Inflow	Cash Outflow	Net Cash Flow
0	—	$1,000,000	−$1,000,000
1	$4,300,000	2,000,000	2,300,000
2	3,000,000	4,320,000	−1,320,000

TABLE P7.22

| | Net Cash Flow | | |
n	Project A	Project B	Project C
0	—	−$1,850	−$2,240
1	−1,020	1,420	4,420
2	840	−460	−2,130
3	480	1,280	—
4	260	—	—
i^*	32.42%	10.73%	?

TABLE P7.23

	Net Cash Flow				
n	A	B	C	D	E
0	−$2,550	−$2,800	−$3,150	−$3,400	$3,860
1	3,280	1,450	4,655	−400	8,200
2	−720	2,022	−1,324	6,240	4,350
3	—	0	—	−650	−4000
i*	−77.89%,	14.73	−61.73%,	23.90	−47.35

7.23 Consider the investment projects given in Table P7.23.

Assume that MARR = 14% in the following questions.

(a) Classify each project as either a pure or a mixed investment.

(b) Compute the true IRR for each investment.

(c) Compute the MIRR for each project at MARR = 14%.

(d) Determine the acceptability of each project.

7.24 Consider an investment project whose cash flows are as in Table P7.24.

(a) Apply the net investment test.

(b) Is this a mixed investment? Then, calculate the RIC at MARR = 16%.

(c) Calculate the MIRR, assuming that all inflows and outflows are compounded and discounted at MARR = 16%. Is this a good investment?

TABLE P7.24

n	Net Cash Flow
0	−$14,500
1	13,250
2	42,500
3	−43,800

7.25 Consider the cash flows of a certain project given in Table P7.25.

The project's i*s are computed as 5.36% and 106%, respectively. The firm's MARR is 15%.

(a) Show why this investment project fails the net–investment test.

(b) Compute the IRR, and determine the acceptability of this project.

(c) Compute the MIRR at MARR = 15%. and determine the acceptability of the project.

TABLE P7.25

n	Net Cash Flow
0	−$176,000
1	548,000
2	−382,000

7.26 Consider the investment projects given in Table P7.26.

TABLE P7.26

	Net Cash Flow		
n	Project 1	Project 2	Project 3
0	−$4,000	$8,000	−$12,000
1	2,220	−2,500	16,240
2	5,160	−400	−750
3	−1,850	15,600	−420

Which of the following statements is *not* correct?
(a) Project 1 will have two real rates of return.
(b) Project 2 should have a unique real rate of return.
(c) Project 3 is a nonsimple investment.
(d) None of the above

IRR Analysis

7.27 You are considering an investment that costs $4,500. It is expected to have a useful life of three years. You are very confident about the revenues during the first two years but you are unsure about the revenue in year 3. If you hope to make at least a 15% rate of return on your investment ($4,500), what should be the minimum revenue in year 3?

Year	Net Cash Flow
0	−$4,500
1	$2,240
2	$2,450
3	X

7.28 Agdist Corporation distributes agricultural equipment. The board of directors is considering a proposal to establish a facility to manufacture an electronically controlled "intelligent" crop sprayer invented by a professor at a local university. This crop sprayer project would require an investment of $10 million in assets and would produce an annual after-tax net benefit of $1.8 million over a service life of eight years. All costs and benefits are included in these figures. When the project terminates, the net proceeds from the sale of the assets will be $1 million. Compute the rate of return of this project. Is this a good project at MARR = 10%?

7.29 Consider an investment project with the cash flows given in Table P7.29.

Compute the IRR for this investment. Is the project acceptable at MARR = 12%?

7.30 Consider the cash flow of a certain project given in Table P7.30. If the project's IRR is 12%:

If the project's IRR is 10%:
(a) Find the value of X.
(b) Is this project acceptable at MARR = 14%?

TABLE P7.29

Year	Cash Flow
0	−$48,500
1	24,250
2	22,495
3	20,285

TABLE P7.30

Year	Net Cash Flow
0	−$34,400
1	$14,200
2	$16,600
3	X

7.31 A company invests $6,250 at the beginning of a seven-year project. At the end of every year for the first five years, the project generates $1,550. At the end of the sixth year, the project generates no money. At the end of the seventh year, the project is terminated. How much must the project generate at the end of the seventh year to realize 14% return on the initial investment?

7.32 You are considering a luxury apartment building project that requires an investment of $18,250,000. The building has 65 units. You expect the maintenance cost for the apartment building to be $420,000 the first year and $495,000 the second year. The maintenance cost will continue to increase by $75,000 in subsequent years. The cost to hire a Manager for the building is estimated to be $90,000 per year. After five years of operation, the apartment building can be sold for $19,200,000. What is the annual rent per apartment unit that will provide a return on investment of 16%? Assume that the building will remain fully occupied during its five years of operation.

7.33 A machine costing $35,000 to buy and $4,000 per year to operate will mainly save labor expenses in packaging over six years. The anticipated salvage value of the machine at the end of the six years is

$5,000. To receive a 12% return on investment (rate of return), what is the minimum required annual savings in labor from this machine?

7.34 Champion Chemical Corporation is planning to expand one of its propylene manufacturing facilities. At $n = 0$, a piece of property costing $1.5 million must be purchased to build a plant. The building, which needs to be expanded during the first year, costs $3 million. At the end of the first year, the company needs to spend about $4 million on equipment and other start-up costs. Once the building becomes operational, it will generate revenue in the amount of $3.5 million during the first operating year. This will increase at the annual rate of 5% over the previous year's revenue for the next nine years. After 10 years, the sales revenue will stay constant for another three years before the operation is phased out. (It will have a project life of 13 years after construction.) The expected salvage value of the land at the end of the project's life would be about $2 million, the building about $1.4 million, and the equipment about $500,000. The annual operating and maintenance costs are estimated to be approximately 40% of the sales revenue each year. What is the IRR for this investment? If the company's MARR is 15%, determine whether the investment is a good one. (Assume that all figures represent the effect of the income tax.)

7.35 You are considering purchasing a CNC machine which costs $150,000. This machine will have an estimated service life of 10 years with a net after-tax salvage value of $15,000. Its annual after-tax operating and maintenance costs are estimated to be $50,000. To expect an 18% rate of return on investment, what would be the required minimum annual after-tax revenues?

7.36 Recent technology has made possible a computerized vending machine that can grind coffee beans and brew fresh coffee on demand. The computer also makes possible such complicated functions as changing $5 and $10 bills, tracking the age of an item, and moving the oldest stock to the front of the line, thus cutting down on spoilage. With a price tag of $4,500 for each unit, Easy Snack has estimated the cash flows in millions of dollars over the product's six-year useful life, including the initial investment, as given in Table P7.36.

TABLE P7.36

n	Net Cash Flow
0	−$30
1	9
2	18
3	20
4	18
5	10
6	5

(a) On the basis of the IRR criterion, if the firm's MARR is 18%, is this product worth marketing?

(b) If the required investment remains unchanged, but the future cash flows are expected to be 10% higher than the original estimates, how much of an increase in IRR do you expect?

(c) If the required investment has increased from $30 million to $35 million, but the expected future cash flows are projected to be 10% smaller than the original estimates, how much of a decrease in IRR do you expect?

Comparing Alternatives

7.37 Consider two investments A and B with the sequences of cash flows given in Table P7.37.

(a) Compute the IRR for each investment.

(b) At MARR $= 13\%$, determine the acceptability of each project.

(c) If A and B are mutually exclusive projects, which project would you select based on the rate of return on incremental investment?

TABLE P7.37

	Net Cash Flow	
n	Project A	Project B
0	−$156,000	−$184,000
1	54,000	72,000
2	76,000	98,000
3	82,000	66,000

7.38 With $15,000 available, you have two investment options. The first is to purchase a bond for $15,000 and invest the bond's interest in the bank at an interest rate of 8%. The bond pays 6.3% interest annually and will mature to its face value of $15,000 in five years. The second choice is to buy a certificate of deposit from a bank at an interest rate of 8.5% annually for five years. Which option is better? Assume that your MARR is 8% per year.

7.39 A pharmaceutical firm is considering the mutually exclusive alternatives given in Table P7.39. Determine which project is a better choice at a MARR = 14% based on the IRR criterion.

TABLE P7.39

	Net Cash Flow	
n	Project A$_1$	Project A$_2$
0	−$9,200	−$12,500
1	7,700	9,600
2	6,600	8,600

7.40 Consider the two mutually exclusive alternatives given in Table P7.40.

TABLE P7.40

	Net Cash Flow	
n	Project A$_1$	Project A$_2$
0	−$18,750	−$22,250
1	8,400	9,800
2	8,400	9,800
3	8,400	9,800

(a) Determine the IRR on the incremental investment in the amount of $3,500.
(b) If the firm's MARR is 12%, which alternative is the better choice?

7.41 Consider the two mutually exclusive investment alternatives given in Table P7.41.

(a) Which project would be selected based on the rate of return decision criterion? (Assume that MARR = 15%.)

TABLE P7.41

	Net Cash Flow	
n	Project A$_1$	Project A$_2$
0	−$29,900	−$34,400
1	14,200	21,100
2	16,600	17,700
3	18,600	14,800
IRR	28.43%	27.88%

(b) Determine the MIRR on the incremental investment. Which project would be chosen at MARR = 15%?

7.42 Which of the capital investment projects in Figure P7.42 would you choose, and why?

(a) Select either one because both have the same NPW.
(b) Select Project A because it has a larger IRR.
(c) Select Project B because it has a larger cash flow.
(d) None of the above.

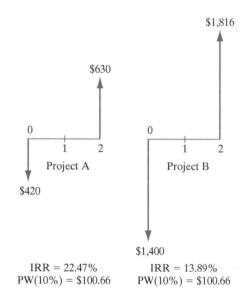

IRR = 22.47%   IRR = 13.89%
PW(10%) = $100.66   PW(10%) = $100.66

Figure P7.42

7.43 Consider the two mutually exclusive investment projects given in Table P7.43.

TABLE P7.43

	Net Cash Flow	
n	**Project A**	**Project B**
0	−$300	−$800
1	0	1,150
2	690	40
i*	51.66%	47.14%

Assume that MARR = 15%.

(a) According to the IRR criterion, which project would be selected?

(b) Sketch the PW(i) function on the incremental investment (B − A).

7.44 You are considering two types of motor cars. Model A costs $182,250 and model B costs $195,000. Although the two models are essentially the same, after four years of use, model A can be sold for $78,400, while model B can be sold for $95,500. Model B commands a better resale value because its styling is popular among young executives. Determine the rate of return on the incremental investment of $12,750. For what range of values of your MARR is model B preferable?

7.45 A plant engineer is considering the two types of solar water heating systems given in Table P7.45.

TABLE P7.45

	Model A	Model B
Initial cost	$22,500	$34,500
Annual savings	$2,750	$4,240
Annual maintenance	$410	$445
Expected life	25 years	25 years
Salvage value	$1,600	$1,850

The firm's MARR is 9%. On the basis of the IRR criterion, which system is the better choice? "Do nothing" is not an option.

7.46 Consider the investment projects given in Table P7.46.

Assume that MARR = 16%.

(a) Projects A and B are mutually exclusive. Assuming that both projects can be repeated for an indefinite period, which one would you select on the basis of the IRR criterion?

(b) Suppose projects C and D are mutually exclusive. According to the IRR criterion, which project would be selected?

(c) Suppose projects E and F are mutually exclusive. Which project is better according to the MIRR criterion?

7.47 Fulton National Hospital is reviewing ways of cutting the cost of stocking medical supplies. Two new systems are being considered to lower the hospital's holding and handling costs. The hospital's industrial engineer has compiled the

TABLE P7.46

	Net Cash Flow					
n	**A**	**B**	**C**	**D**	**E**	**F**
0	−$475	−$350	−$2,800	−$3,850	−$3,600	−$4,800
1	225	235	2,015	2,650	3,700	3,260
2	245	250	2,254	2,860	1,640	3,630
3	265	—	—	—	—	—
i*	24.65%	24.51%	32.65%	27.22%	36.22%	27.32%

relevant financial data for each system as given in Table P7.47 (dollar values are in millions).

The system life of eight years represents the period that the contract with the medical suppliers is in force. If the hospital's MARR is 10%, which system is the most economical?

7.48 Consider the cash flows for the investment projects given in Table P7.48.

Assume that the MARR = 10%.

(a) Suppose A, B, and C are mutually exclusive projects. Which project would be selected on the basis of the IRR criterion?

(b) What is the borrowing rate of return (BRR) for project D?

(c) Would you accept project D at MARR = 12%?

(d) Assume that projects C and E are mutually exclusive. Using the IRR criterion, which project would you select?

7.49 Consider the investment projects given in Table P7.49.

Assume that MARR = 16%.

(a) Compute the IRR for each project.

(b) On the basis of the MIRR criterion, if the three projects are mutually exclusive investments, which project should be selected?

TABLE P7.49

	Net Cash Flow		
n	**Project 1**	**Project 2**	**Project 3**
0	−$6,250	−$7,750	−$6,150
1	1,400	4,800	1,850
2	7,650	6,200	7,550

7.50 Consider the two investment alternatives given in Table P7.50.

TABLE P7.50

	Net Cash Flow	
n	**Project A**	**Project B**
0	−$45,500	−$32,400
1	0	17,200
2	0	17,200
3	82,300	17,200
IRR	?	27.43%
PW(14%)	10,050	?

TABLE P7.47

	Current Practice	Just-in-Time System	Stockless Supply System
Start-up cost	$0	$3	$6
Annual stock holding cost	$6	$1.8	$0.4
Annual operating cost	$3	$2	$1.5
System life	8 years	8 years	8 years

TABLE P7.48

	Net Cash Flow				
n	**A**	**B**	**C**	**D**	**E**
0	−$4,250	−$4,250	−$4,850	$2,500	−$4,250
1	3,200	2,850	2,100	−835	1,500
2	2,850	2,900	2,100	−835	3,250
3	800	1,050	2,100	−835	1,600
4	300	500	2,100	−835	1,200

The firm's MARR is known to be 14%.

(a) Compute the IRR of project A and NPW of project B.

(b) Compute the MIRR of project B.

(c) Suppose that projects A and B are mutually exclusive. Using the IRR, which project would you select?

7.51 The following information on four mutually exclusive projects is given here.

All four projects have the same service life and require investment in year 0 only once. Suppose that you are provided with the following additional information about incremental rates of return between projects.

$$IRR(A - B) = 36\%$$
$$IRR(C - B) = 18\%$$
$$IRR(D - C) = 22\%$$
$$IRR(A - D) = 44\%$$

Which project would you choose based on the rate-of-return criterion at a MARR of 29%?

TABLE P7.51

Project	Required Investment at Year 0	IRR
A	1000	56%
B	1200	67%
C	1500	43%
D	2500	49%

7.52 The E. F. Fedele Company is considering acquiring an automatic screwing machine for its assembly operation of a personal computer. Three different models with varying automatic features are under consideration. The required investments are $360,000 for model A, $380,000 for model B, and $405,000 for model C. All three models are expected to have the same service life of eight years. The financial information from Table P7.52, in which model (B − A) represents the incremental cash flow determined by subtracting model A's cash flow from model B's, is available. If the firm's MARR is known to be 12%, which model should be selected?

TABLE P7.52

Model	IRR (%)
A	30
B	15
C	25

Model	Incremental IRR (%)
(B − A)	5
(C − B)	40
(C − A)	15

7.53 The GeoStar Company, a leading manufacturer of wireless communication devices, is considering three cost-reduction proposals in its batch job-shop manufacturing operations. The company has already calculated rates of return for the three projects, along with some incremental rates of return, as given in Table P7.53.

TABLE P7.53

Incremental Investment	Incremental Rate of Return (%)
$A_1 - A_0$	18
$A_2 - A_0$	20
$A_3 - A_0$	25
$A_2 - A_1$	10
$A_3 - A_1$	18
$A_3 - A_2$	23

A_0 denotes the do-nothing alternative. The required investments are $420,000 for A_1 $550,000 for A_2 and $720,000 for A_3. If the MARR is 15%, what system should be selected?

7.54 A manufacturer of electronic circuit boards is considering six mutually exclusive cost-reduction projects for its PC-board manufacturing plant. All have lives of 10 years and zero salvage values. The required investment and the estimated after-tax reduction in annual disbursements for each alternative are given in Table P7.54a, along with computed rates of return on incremental investments in Table P7.54b.

TABLE P7.54A

Proposal A_j	Required After-Tax Investment	Savings	Rate of Return (%)
A_1	$60,000	$22,000	34.8
A_2	100,000	28,200	25.2
A_3	110,000	32,600	26.9
A_4	120,000	33,600	25.0
A_5	140,000	38,400	24.3
A_6	150,000	42,200	25.1

TABLE P7.54B

Incremental Investment	Incremental Rate of Return (%)
$A_2 - A_1$	8.9
$A_3 - A_2$	42.7
$A_4 - A_3$	0.0
$A_5 - A_4$	20.2
$A_6 - A_5$	36.3

If the MARR is 15%, which project would you select based on the rate of return on incremental investment?

7.55 Baby Doll Shop manufactures wooden parts for dollhouses. A worker is paid $8.10 an hour and using a handsaw can produce a year's required production (1,600 parts) in just eight 40-hour weeks. That is, the worker averages five parts per hour when working by hand. The shop is considering purchasing a power band saw with associated fixtures to improve the productivity of this operation. Three models of power saw could be purchased: Model A (the economy version), model B (the high-powered version), and model C (the deluxe high-end version). The major operating difference between these models is their speed of operation. The investment costs, including the required fixtures and other operating characteristics, are summarized in Table P7.55.

TABLE P7.55

Category	By Hand	Model A	Model B	Model C
Production rate (parts/hour)	5	10	15	20
Labor hours required (hours/year)	320	160	107	80
Annual labor cost (@ $8.10/hour)	$2,592	$1,296	$867	$648
Annual power cost ($)	—	400	420	480
Initial investment ($)	—	4,000	6,000	7,000
Salvage value ($)	—	400	600	700
Service life (years)	—	20	20	20

Assume that MARR $= 10\%$. Are there enough savings to purchase any of the power band saws? Which model is most economical, based on the rate-of-return principle? (Assume that any effect of income tax has been already considered in the dollar estimates.) (*Source*: This problem is adapted with the permission of Professor Peter Jackson of Cornell University.)

Unequal Service Lives

7.56 Consider the two mutually exclusive investment projects given in Table P7.56 for which MARR $= 15\%$. On the basis of the IRR criterion, which project would be selected under an infinite planning horizon with project repeatability likely?

TABLE P7.56

n	Net Cash Flow	
	Project A	Project B
0	−$5,000	−$10,000
1	3,000	8,000
2	4,000	8,000
3	4,000	—
IRR	49.49%	39.97%

7.57 Consider the two mutually exclusive investment projects given in Table P7.57.

(a) To use the IRR criterion, what assumption must be made in comparing a set of mutually exclusive investments with unequal service lives?

(b) With the assumption made in part (a), determine the range of MARRs which will indicate that project A1 should be selected.

TABLE P7.57

n	Net Cash Flow	
	Project A1	Project A2
0	−$10,000	−$15,000
1	5,000	20,000
2	5,000	—
3	5,000	—

Short Case Studies

ST7.1 *Will That Be Cash, Credit—or Fingertip?*[10] Have you ever found yourself short of cash or without a wallet when you want to buy something? Consider the following two types of technologies available in retail stores to speed up checkouts.

- **Pay By Touch**™ takes fingerprints when customers enroll in the program. The image is then converted to about 40 unique points of the finger. Those points are stored in a computer system with "military-level encryption." They want this to be your cash replacement because of the time savings, and a lot of customers who are paying cash will find it more convenient now to use these cards.

- **A contactless card** allows the shopper to pay in seconds by waving his or her contactless card in

front of a reader, which lights up and beeps to tell the shopper the transaction is done. A contactless payment is twice as fast as a no-signature credit card purchase and three times as fast as using cash. That's why it's catching on at fast-food restaurants and convenience stores.

Profits depend, in part, on how quickly customers— typically with small purchases—get through the line. These new technologies being rolled out at convenience stores, supermarkets, and gas stations could someday make it passé to carry bulky wallets. Without the need to dig for cash and checks at the register, the quick stop-and-go payments promise speedier transactions for consumers—and perhaps fatter profits for retailers.

---

[10] Will That be Cash, Credit—or Fingertip? Kathy Chu, *USA Today*, Section B1, Friday, December 2, 2005.

The appeal is that there is no need to run anything through a machine. A contactless-card transaction is usually more expensive for a retailer to process than a cash payment. But retailers that adopt contactless payments hope they'll bring in more customers, off-setting higher costs. If that turns out to be false, then some could turn their backs on the new technology.

One retailer who just installed a Pay By Touch system hopes to increase its customer traffic so that a 10% return on investment can be attained. The Pay By Touch scanners cost about $50 each, the monthly service fee ranges between $38 and $45, and each transaction fee costs 10 cents. In a society driven by convenience, anything that speeds up the payment process attracts consumers. But technology providers will need to convince consumers of the safety of their information before the technologies can become a staple in the checkout line.

Suppose that you are a national franchise operator with 28,000 stores. As a part of growth strategy, you are thinking about installing the Pay-By-Touch systems (say, two systems for each store location). What kinds of additional revenue per store would justify the investment, if you are looking for at least 20% return on your investment?

ST7.2 Critics have charged that, in carrying out an economic analysis, the commercial nuclear power industry does not consider the cost of decommissioning, or "mothballing," a nuclear power plant and that the analysis is therefore unduly optimistic. As an example, consider the Tennessee Valley Authority's Bellefont twin nuclear generating facility under construction at Scottsboro, in northern Alabama. The initial cost is $1.5 billion and the estimated life is 40 years. The annual operating and maintenance costs are assumed to be 4.6% of the initial cost during the first year and are expected to increase at an annual rate of 0.05% thereafter. Annual revenues are estimated to be three times the annual operating and maintenance costs throughout the life of the plant.

(a) The criticism that the economic analysis is over-optimistic because it omits "mothballing" costs is not justified, since the addition of a cost of 50% of the initial cost to "mothball" the plant decreases the 8.95% rate of return only to approximately 8.77%.

(b) If the estimated life of the plants is more realistically taken to be 25 years instead of 40 years, then the criticism is justified. By reducing the life to 25 years, the rate of return of approximately 9% without a "mothballing" cost drops to

approximately 7.7% when a cost to "mothball" the plant equal to 50% of the initial cost is added to the analysis.

Comment on these statements.

ST7.3 The B&E Cooling Technology Company, a maker of automobile air-conditioners, faces a three-year deadline to phase out the traditional chilling technique, which uses chlorofluorocarbons (CFCs)—a family of refrigerant chemicals believed to attack the earth's protective ozone layer. B&E has been pursuing other means of cooling and refrigeration. As a near-term solution, the engineers recommend a cold technology known as an absorption chiller, which uses plain water as a refrigerant and semiconductors that cool down when charged with electricity. B&E is considering two options.

- **Option 1.** Retrofit the plant now to adapt the absorption chiller and continue to be a market leader in cooling technology. Because of untested technology on a large scale, it may cost more to operate the new facility while personnel are learning the new system.

- **Option 2.** Defer the retrofitting until the federal deadline, which is three years away. With expected improvement in cooling technology and technical know-how, the retrofitting cost will be cheaper. However, there will be tough market competition, and the revenue would be less than that of option 1.

The financial data for the two options are as given in Table ST7.3.

(a) What assumptions must be made to compare these two options?

(b) If B&E's MARR is 15%, which option is the better choice based on the IRR criterion?

ST7.4 An oil company is considering changing the size of a small pump that is currently operational in wells in an oil field. If this pump is kept, it will extract 50% of the known crude-oil reserve in the first year of its operation and the remaining 50% in the second year. A pump larger than the current pump will cost $1.6 million, but it will extract 100% of the known reserve in the first year. The total oil revenues over the two years are the same for both pumps, namely, $20 million. The advantage of the large pump is that it allows 50% of the revenues to be realized a year earlier than with the small pump.

If the firm's MARR is known to be 20%, what do you recommend based on the IRR criterion?

TABLE ST7.3

	Option 1	Option 2
Investment timing	Now	3 years from now
Initial investment	$6 million	$5 million
System life	8 years	8 years
Salvage value	$1 million	$2 million
Annual revenue	$15 million	$11 million
Annual O&M costs	$6 million	$7 million

TABLE ST7.4

	Current Pump	Larger Pump
Investment, year 0	0	$1.6 million
Revenue, year 1	$10 million	$20 million
Revenue, year 2	$10 million	0

ST7.5 You have been asked by the president of the company you work for to evaluate the proposed acquisition of a new injection-molding machine for the firm's manufacturing plant. Two types of injection-molding machines have been identified, with the estimated cash flows given in Table ST7.5.

TABLE ST7.5

	Net Cash Flow	
n	Project 1	Project 2
0	−$30,000	−$40,000
1	20,000	43,000
2	18,200	5,000
IRR	18.1%	18.1%

You return to your office, quickly retrieve your old engineering economics text, and then begin to smile: Aha—this is a classic rate-of-return problem! Now, using a calculator, you find out that both projects have about the same rate of return: 18.1%. This figure seems to be high enough to justify accepting the project, but you recall that the ultimate justification should be done with reference to the firm's MARR. You call the accounting department to find out the current MARR the firm should use in justifying a project. "Oh boy, I wish I could tell you, but my boss will be back next week, and he can tell you what to use," says the accounting clerk.

A fellow engineer approaches you and says, "I couldn't help overhearing you talking to the clerk. I think I can help you. You see, both projects have the same IRR, and on top of that, project 1 requires less investment, but returns more cash flows

$$(-\$30,000 + \$20,000 + \$18,200 = \$8,200$$

and

$$-\$40,000 + \$43,000 + \$5,000 = \$8,000)$$

Thus, project 1 dominates project 2. For this type of decision problem, you don't need to know a MARR!"

(a) Comment on your fellow engineer's statement.

(b) At what range of MARRs would you recommend the selection of project 2?

PART
THREE

ANALYSIS
OF PROJECT
CASH FLOWS

CHAPTER 8

Cost Concepts Relevant to Decision Making

What Does It Cost to Make an iPhone? Have you thought about the cost of building a smartphone? According to an IHS report,[1] the 16GB iPhone 6 costs Apple at least $200.10 to build, whereas Apple's 6 Plus costs just $15.50 more for components and manufacturing.

The 16GB iPhone 6 sells for $649 and iPhone 6 Plus for $749 in the U.S. without a contract with a wireless carrier. This will provide Apple a healthy gross margin of about 69% for the iPhone 6 and 71% for the iPhone Plus. The devices are similar in production cost, but the biggest expense for both devices is the touch screen, which costs $45 on the 4.7-in. iPhone 6 and $52.50 on the 5.5-in. iPhone 6 Plus. The iPhone 6 Plus comes with a fancier camera featuring optical image stabilization. That costs $12.50, compared to $11 for the iPhone 6's camera. Apple also spends money on research and development, software, shipping, marketing, licensing, and other costs. But even with all expenses included, Apple's margins are huge. The company had gross margins of 39.4% in the most recent fiscal quarter.

Source: Mauro Fabbro/Shutterstock

---

[1] "iPhone 6 Plus: $100 Costlier for Consumers to Buy—Just $15.50 More Expensive for Apple to Make," IHS Inc. Press Release, September 23, 2014.

Preliminary Apple iPhone 6 & 6 Plus Teardown Estimates

Components/ Hardware Elements	Apple		Apple	
	iPhone 6 (Sprint Version)		iPhone 6 Plus (Sprint Version)	
Retail Price		$649.00		$749.00
Total BOM Cost		$196.10		$211.10
Manufacturing Cost		**$4.00**		**$4.50**
BOM + Manufacturing		$200.10		$215.60
Major Cost Drivers				
Memory				
NAND Flash + DRAM	16GB NAND + 1GB LPDDR3	$15.00	16GB NAND + 1GB LPDDR3	$15.00
Display & Touch-Screen	4.7" 1334x750 IPS LCD, w/In-Cell Touch	$45.00	5.5" 1920x1080 IPS LCD, w/ In-Cell Touch	$52.50
Processor - AP	Apple 64-bit A8 Processor	$20.00	Apple 64-bit A8 Processor	$20.00
Camera(s)	8MP w/'Focus Pixels'/1.2MP	$11.00	8MP w/ 'Focus Pixels' and OIS/1.2MP	$12.50
Wireless Section - BB/ RF/PA	Contains Qualcomm MDM9625M + WTR1625L + WFR1620 + Front End	$33.00	Contains Qualcomm MDM9625M + WTR1625L + WFR1620 + Front End	$33.00
User Interface & Sensors	Contains Audio Codec, NFC Controller, & Sensors	$22.00	Contains Audio Codec, NFC Controller, & Sensors	$22.00
BT / WLAN	Contains Broadcom BCM4345	$4.50	Contains Broadcom BCM4345	$4.50
Power Management	Dialog + Qualcomm	$7.00	Dialog + Qualcomm	$7 00
Battery	3.8V, 1810mAh Li-Polymer	$3.60	3.8V, 2915mAh Li-Polymer	$4.60
Mechanical/ Electro-Mechanical		$30.00		$35.00
Box Contents		$5.00		$5.00

© 2014 IHS, Inc.

Source: Groundbreaking iPhone 5s Carries $199 BOM and Manufacturing Cost, IHS Teardown Reveals, IHS Technology, Press Release, September 25, 2013.

You may be curious about how all of these cost data were estimated in manufacturing iPhones. To price iPhones, Apple first needs to come up with an accurate estimate of these component costs. Knowing the right price of the iPhones allows Apple to estimate the potential demand of the products and their revenues.

For any investment decision, we need to understand the concept of various costs. For any pricing decision, we need to know how much it costs to manufacture the product. At the level of plant operations, engineers must make decisions involving materials, plant facilities, and the in-house capabilities of company personnel. Consider, for example, the manufacture of iPhones. In terms of selecting materials, several of the parts could be made of plastic, whereas others must be made of metal. Once materials have been chosen, engineers must consider the production methods, the shipping weight, and the method of packaging necessary to protect the different types of materials. In terms of actual production, parts may be made in-house or purchased from an outside vendor, depending on the availability of machinery and labor.

All of these operational decisions (commonly known as **present economic studies** in traditional engineering economic texts) require estimating the costs associated with various production or manufacturing activities. Because these costs also provide the basis for estimating project cash flows, developing successful business strategies, and planning future operations, it is important to understand how various costs respond to changes in levels of business activity. In this chapter, we discuss many of the possible uses of cost data. We also discuss how to define, classify, and estimate costs for each use. Our ultimate task in doing so is to explain how costs are classified in making various engineering economic decisions. This will be the first step toward estimating the project cash flows in later chapters.

CHAPTER LEARNING OBJECTIVES

After completing this chapter, you should understand the following concepts:

- Various cost terminologies that are common in cost accounting and engineering economic studies.

- How a cost item reacts or responds to changes in the level of production or business activities.

- The types of cost data that management needs in making a choice between alternative courses of action.

- The types of present economic studies frequently performed by engineers in a manufacturing and business environment.

- How to develop a production budget related to operating activities.

8.1 General Cost Terms

In engineering economics, the term **cost** is used in many different ways. Because there are many types of costs, each is classified differently according to the immediate needs of management. For example, engineers may want cost data to prepare planning budgets or to make operational or investment decisions. Also, each different use of cost data demands a different classification and definition of *cost*. For example, the preparation of external financial reports requires the use of historical cost data, whereas decision-making may require current cost data or estimated future cost data.

The most common cost objective of interest is a product. A product can be either a tangible good, such as a smartphone, or a service, such as a repair job on a heating and air-conditioning system. Our initial focus in this chapter is on manufacturing goods, because their basic activities (acquiring raw materials, producing finished goods, marketing, etc.) are commonly found in most other businesses. Therefore, the understanding of costs in a manufacturing company can be helpful in understanding costs in other types of business organizations.

8.1.1 Manufacturing Costs

Several types of manufacturing costs incurred by a typical manufacturer are illustrated in Figure 8.1. In converting raw materials into finished goods, a manufacturer incurs various costs associated with operating a factory. First a firm purchases raw materials, which are then processed by the production workers using machinery. Once a final product is manufactured, it is transferred to the warehouse to be sold to the customers. In terms of identifying the manufacturing costs, most firms divide them into three broad categories: direct raw-material costs, direct labor costs, and manufacturing overhead.

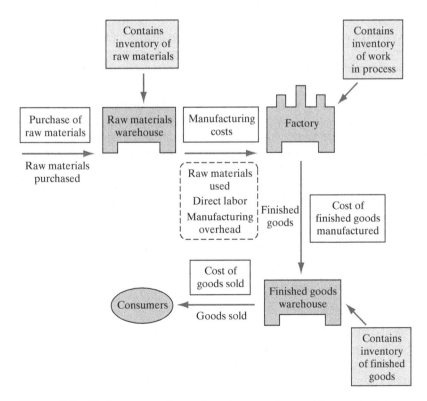

Figure 8.1   Various types of manufacturing costs incurred by a manufacturer.

Direct Raw Material Cost

Direct raw materials are any materials that are used in the final product and that can be easily traced to it. Some examples are wood in furniture, steel in bridge construction, paper in printing firms, and fabric for clothing manufacturers. It is important to note that the finished product of one company can become the raw materials of another company. For example, the computer chips produced by Intel are a raw material used by Dell in its personal computers.

Direct Labor Cost

Like direct raw materials, direct labor incurs costs that go into the production of a product. The labor costs of assembly-line workers, for example, would be direct labor costs, as would the labor costs of welders in metal-fabricating industries, carpenters or bricklayers in home building, and machine operators in various manufacturing operations.

Manufacturing Overhead Cost

Manufacturing overhead, the third element of manufacturing costs, includes all costs of manufacturing except those of direct materials and direct labor. In particular, manufacturing overhead includes such items as the costs of indirect materials; indirect labor; maintenance and repairs on production equipment; heat and light, property taxes, depreciation, and insurance on manufacturing facilities; and overtime premiums. The most important thing to note about manufacturing overhead is the fact that, unlike direct materials and direct labor, it is not easily traceable to specific units of output. In addition, many manufacturing overhead costs do not change as output changes, as long as the production volume stays within the capacity of the plant. For example, depreciation of factory buildings is unaffected by the amount of production during any particular period. If, however, a new building is required to meet any increased production, manufacturing overhead will certainly increase.

- Sometimes it may not be worth the effort to trace the costs of materials that are relatively insignificant in the finished products. Such minor items include the solder used to make electrical connections in a computer circuit board and the glue used to bind this textbook. Materials such as solder and glue are called *indirect materials* and are included as part of manufacturing overhead.
- Sometimes we may not be able to trace some of the labor costs to the creation of a product. We treat this type of labor cost as a part of manufacturing overhead, along with indirect materials. *Indirect labor* includes the wages of janitors, supervisors, material handlers, and night security guards. Although the efforts of these workers are essential to production, it would be either impractical or impossible to trace their costs to specific units of product. Therefore, we treat such labor costs as indirect labor costs.

8.1.2 Nonmanufacturing Costs

Two additional costs incurred in supporting any manufacturing operations are (1) marketing or selling costs and (2) administrative costs. Marketing or selling costs include all costs necessary to secure customer orders and get the finished product or service into the hands of the customer. Breakdowns of these types of costs provide data for control over selling and administrative functions in the same way that manufacturing cost breakdowns provide data for control over manufacturing functions. For example, a company incurs costs for

- **Overhead:** Heat and light, property taxes, and depreciation or similar items associated with the company's selling and administrative functions.
- **Marketing:** Advertising, shipping, sales travel, sales commissions, and sales salaries. Marketing costs include all executive, organizational, and clerical costs associated with sales activities.
- **Administrative functions:** Executive compensation, general accounting, public relations, and secretarial support associated with the general management of an organization.

8.2 Classifying Costs for Financial Statements

For purposes of preparing financial statements, we often classify costs as either period costs or product costs. To understand the difference between them, we must introduce the matching concept essential to any accounting studies. In financial accounting, the **matching principle** states that *the costs incurred in generating a certain amount of revenue should be recognized as expenses in the same period that the revenue is recognized.* This matching principle is the key to distinguishing between period costs and product costs. Some costs are matched against periods and become expenses immediately. Other costs are matched against products and do not become expenses until the products are sold, which may be in the next accounting period.

8.2.1 Period Costs

Period costs are costs charged to expenses in the period in which they are incurred. The underlying assumption is that the associated benefits are received in the same period the cost is incurred. Some specific examples are all general and administrative expenses, selling expenses, and insurance and income tax expenses. Therefore, advertising costs, executives' salaries, sales commissions, public-relations costs, and other nonmanufacturing costs discussed earlier all would be period costs. Such costs are not related to the production and flow of manufactured goods but are deducted from revenue in the income statement. In other words, period costs will appear on the income statement as expenses during the time in which they occur.

8.2.2 Product Costs

Some costs are better matched against products than they are against periods. Costs of this type—called **product costs**—consist of the costs involved in the purchase or manufacture of goods. In the case of manufactured goods, product costs are the costs of direct materials, direct labor costs, and manufacturing overhead. Product costs are not viewed as expenses immediately; rather, they are the cost of creating inventory. Thus, product costs are considered an asset until the associated goods are sold. At the time they are sold, the costs are released from inventory as expenses (typically called cost of goods sold) and matched against sales revenue. Since product costs are assigned to inventories, they are also known as *inventory costs*. In theory, product costs include all manufacturing costs— that is, all costs relating to the manufacturing process. As shown in Figure 8.2, product costs appear on financial statements when the inventory (or final goods) is sold, not when the product is manufactured.

To understand product costs more fully, let us look briefly at the flow of costs in a manufacturing company. By doing so, we will be able to see how product costs move

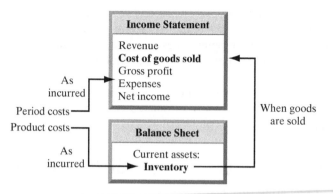

Figure 8.2 How the period costs and product costs flow through financial statements from the manufacturing floor to sales.

through the various accounts and affect the balance sheet and the income statement in the course of the manufacture and sale of goods. The flows of period costs and product costs through the financial statements are illustrated in Figure 8.3. All product costs filter through the balance-sheet statement as "inventory costs." If the product gets sold, the inventory costs in the balance-sheet statement are transferred to the income statement under the heading "cost of goods sold."

- **Raw-materials inventory.** This account in the balance-sheet statement represents the unused portion of the raw materials on hand at the end of the fiscal year.
- **Work-in-process inventory.** This balance-sheet entry consists of the partially completed goods on hand in the factory at year-end. When raw materials are used in production, their costs are transferred to the work-in-process inventory account as direct materials. Note that direct labor cost and manufacturing overhead cost are

Figure 8.3 Cost flows and classifications in a manufacturing company.

also added directly to work-in-process, which can be viewed as the assembly line in a manufacturing plant, where workers are stationed and where products slowly take shape as they move from one end of the line to the other.

- **Finished-goods inventory.**   This account shows the cost of finished goods that are on hand and awaiting sale to customers at year-end. As the goods are completed, accountants transfer the cost from the work-in-process account into the finished-goods account. At this stage, the goods await sale to a customer. As the goods are sold, their cost is transferred from the finished-goods account into the cost-of-goods-sold (or cost-of-revenue) account. At this point, we finally treat the various material, labor, and overhead costs that were involved in the manufacture of the units being sold as *expenses* in the income statement.

Example 8.1 serves to explain the classification scheme for financial statements.

EXAMPLE 8.1 Classifying Costs for Uptown Ice Cream Shop

Here is a look at why it costs $2.50 for a single-dip ice cream cone at a typical store in Washington, D.C. The annual sales volume (the number of ice cream cones sold) averages around 185,000 cones, bringing in revenue of $462,500. This is equivalent to selling more than 500 cones a day, assuming a seven-day operation. Table 8.1 shows the unit price of an ice cream cone and the costs that go into producing the product.

TABLE 8.1   Breakdowns of Unit Cost Items

Items	Total Cost	Unit Price*	% of Total Cost
Ice cream (cream, sugar, milk, and milk solids)	$120,250	$0.65	26%
Cone	9,250	0.05	2
Rent	112,850	0.61	24
Wages	46,250	0.25	10
Payroll taxes	9,250	0.05	2
Sales taxes	42,550	0.23	9
Business taxes	14,800	0.08	3
Debt service	42,550	0.23	9
Supplies	16,650	0.09	4
Utilities	14,800	0.08	3
Other expenses (insurance, advertising, fees, and heating and lighting for shop)	9,250	0.05	2
Profit	24,050	0.13	5
Total	**$462,500**	**$2.50**	**100%**

*Based on an annual volume of 185,000 cones.

If you were to classify the operating costs into either product costs or period costs, how would you do it?

SOLUTION

Given: Financial data just described.

Find: Classify the cost elements into product costs and period costs.

The following is a breakdown of the two kinds of costs.

- **Product costs:**   Costs matched with the sales revenues in the period in which the goods are actually sold to arrive at gross margin.

Raw materials:	
Ice cream @ $0.65	$120,250
Cone @ $0.05	9,250
Labor:	
Wages @ $0.25	46,250
Overhead:	
Supplies @ $0.09	16,650
Utilities @ $0.08	14,800
Total product cost	**$207,200**

- **Period costs:**   Costs matched with revenue in a given accounting period.

Business taxes:	
Payroll taxes @ $0.05	$ 9,250
Sales taxes @ $0.23	42,550
Business taxes @ $0.08	14,800
Operating expenses:	
Rent @ $0.61	112,850
Debt service @ $0.23	42,550
Other @ $0.05	9,250
Total period cost	**$231,250**

8.3 Cost Classification for Predicting Cost Behavior

In engineering economic analysis, we need to predict how a certain cost will behave in response to a change in activity. For example, a manager will want to estimate the impact a 5% increase in production will have on the company's total wages before he or she decides whether to alter production. **Cost behavior** describes how a cost item will react or respond to changes in the level of business activity.

8.3.1 Volume Index

In general, the operating costs of any company are likely to respond in some way to changes in the company's operating volume. In studying cost behavior, we need to determine some measurable volume or activity that has a strong influence on the amount of cost incurred. The unit of measure used to define volume is called a **volume index**. A volume index may be based on production inputs, such as tons of coal processed, direct labor hours used, or machine-hours worked; or it may be based on production outputs, such as the number of kilowatt-hours generated. For a vehicle, the number of miles driven per year may be used as a volume index. Once we identify a volume index, we try to find out how costs change in response to changes in the index.

8.3.2 Cost Behaviors

Accounting systems typically record the cost of resources acquired and track their subsequent usage. Fixed costs and variable costs are the two most common cost behavior patterns. An additional category known as "mixed costs" contains two parts, the first of which is fixed and the other of which varies with the volume of output.

Fixed Costs

The costs of providing a company's basic operating capacity are known as the company's **fixed cost** or **capacity cost**. For a cost item to be classified as fixed, it must have a relatively wide span of output over which costs are expected to remain constant. This span is called the **relevant range**. In other words, fixed costs do not change within a given period, although volume may change. In the case of an automobile, for example, the annual insurance premium, property tax, and license fee are fixed costs, since they are independent of the number of miles driven per year. Some other typical examples are building rents; depreciation of buildings, machinery, and equipment; and salaries of administrative and production personnel. In our Uptown Scoop Ice Cream Store example, we may classify expenses such as rent, business taxes, debt service, and other (insurance, advertising, professional fees) as fixed costs (costs that are fixed in total for a given period of time and for given production levels).

Variable Costs

In contrast to fixed operating costs, **variable operating costs** have a close relationship to the level of volume of a business. If, for example, volume increases 10%, a total variable cost will also increase by approximately 10%. Gasoline is a good example of a variable automobile cost, because fuel consumption is directly related to miles driven. Similarly, the cost of replacing tires will increase as a vehicle is driven more.

In a typical manufacturing environment, direct labor and material costs are major variable costs. In our Uptown Scoop example, the variable costs would include the cost of the ice cream and cone (direct materials), wages, payroll taxes, sales taxes, and supplies. Both payroll and sales taxes are related to sales volume. In other words, if the store becomes busy, more servers are needed, which will increase the payroll as well as taxes.

Mixed Costs

Some costs do not fall precisely into either the fixed or the variable category, but contain elements of both. We refer to these as **mixed costs** (or **semi-variable costs**). In our automobile example, **depreciation** (loss of value) is a mixed cost. On the one hand, some depreciation occurs simply from the passage of time, regardless of how many miles a car is driven, and this represents the fixed portion of depreciation. On the other hand, the more miles an automobile is driven per year, the faster it loses its market value, and this represents the variable portion of depreciation. A typical example of a mixed cost in manufacturing is the cost of electric power. Some components of power consumption, such as lighting, are independent of the operating volume, while other components (e.g., the number of machine-hours equipment is operated) may vary directly with volume. In our Uptown Scoop example, the utility cost can be a mixed-cost item: some lighting and heating requirements might stay the same, but the use of power mixers will be in proportion to sales volume.

Average Unit Cost—Cost–Volume Chart

The foregoing description of fixed, variable, and mixed costs was expressed in terms of total volume over a given period, commonly known as **cost-volume chart**. We often use the term **average cost** to express activity cost on a per unit basis. In terms of unit costs, the description of cost is quite different:

- The variable cost per unit of volume is a constant.
- Fixed cost per unit varies with changes in volume: As the volume increases, the fixed cost per unit decreases.
- The mixed cost per unit also changes as volume changes, but the amount of change is smaller than that for fixed costs.

To explain the behavior of the fixed, variable, mixed, and average costs in relation to volume, let's consider a four-wheel drive (4WD) sport-utility vehicle. In calculating the vehicle's operating costs, we may use the procedure outlined by the American Automobile Association (AAA).

The AAA report[2] factors many elements in for its report, including gas prices, routine car maintenance, tire replacement, depreciation, financing and licensing, registration, and taxes. It also adds in the typical insurance paid by a married 47-year-old man with a good driving record who drives three to 10 miles daily to work. Specific assumptions used in establishing the cost base for key items are

1. Fuel–use of unleaded gasoline weighted 60% city and 40% highway driving.
2. Depreciation–based on the difference between the new-vehicle purchase price and the estimated trade-in value at the end of five years.
3. Finance–a five-year loan at 6% APR with a 10% down payment.

On the basis of these assumptions, the operating and ownership costs of a typical 4WD sport-utility vehicle (2014) based on 15,000 miles are estimated as follows:

1. Operating cost per mile for gas, maintenance and tires: 23.75 cents
2. Ownership costs per year including insurance, license, registration, taxes, depreciation, and finance charges): $7,476

[2] *Source:* Your Driving Costs, 2014 Edition, American Automobile Association (AAA).

3. Total annual cost:

   Operating cost: $0.2375/mile x 15,000 miles/year = $3,563 per year

   Ownership costs: $7,476 per year

   Total cost: $11,039

4. Total cost per mile: $11,039/15,000 miles = $0.7359 per mile

Even though estimating the cost of owning a car isn't an exact science.

   Now, if you drive the same vehicle for 20,000 miles or 12,000 miles instead of 15,000 miles, you may be interested in knowing how the average cost per mile would change. Example 8.2 illustrates how you determine the average cost as a function of mileage.

EXAMPLE 8.2 Calculating Average Driving Cost per Mile as a Function of Mileage

Table 8.2 itemizes the operating and ownership costs associated with driving a passenger car by fixed, variable, and mixed classes. Note that the depreciation expense is divided into two categories—fixed and variable expenses. Using the given data, develop a cost–volume chart and calculate the average cost per mile as a function of the annual mileage.

TABLE 8.2 Cost Classifications of Owning and Operating a 4WD Sport-Utility Vehicle

Classification	Cost
Variable costs	
Fuel and oil per mile	$0.1694
Maintenance per mile	$0.0550
Tires per mile	$0.0131
Total variable	$0.2375
Fixed costs	
Insurance (comprehensive)	$1,025
License, registration, taxes	$797
Finance charge	$1,076
Total fixed	$2,898
Mixed costs	
Depreciation:	
Fixed portion per year (15,000 miles)	$4,578
Variable portion per mile (above 15,000 miles)	$0.2130

DISCUSSION: First we may examine the effect of driving an additional 1,000 miles over the 15,000 allotted miles. Since the loss in the car's value due to driving an additional 1,000 miles over 15,000 miles is estimated to be $213, the total cost per year and the average cost per mile, based on 16,000 miles, can be recalculated as follows.

- Added depreciation cost: $213
- Added operating cost: 1,000 miles × 23.75 cents = $237.5
- Total cost per year: $11,039 + $213 + $237.5 = $11,489.50
- Average cost per mile ($11,489.50/16,000 miles): 71.81 cents

Note that the average cost comes down as you drive more, as the ownership cost per mile is further reduced.

SOLUTION

Given: Financial data.

Find: The average cost per mile at an annual operating volume between 15,000 and 20,000 miles.

In Table 8.3, we summarize the costs of owning and operating the automobile at various annual operating volumes from 15,000 to 20,000 miles. Once the total cost figures are available at specific volumes, we can calculate the effect of volume on unit (per mile) costs by converting the total cost figures to average unit costs.

TABLE 8.3 Operating Costs as a Function of Mileage Driven

	A	B	C	D	E	F	G
1							
2	Volume Index (miles)	15,000	16,000	17,000	18,000	19,000	20,000
3							
4	Variable costs:	$3,563	$3,800	$4,038	$4,275	$4,513	$4,750
5	($0.2375 per mile)						
6	Fixed costs:	$2,898	$2,898	$2,898	$2,898	$2,898	$2,898
7							
8	Mixed costs:						
9	Depreciation:						
10	Variable portion	$0	$213	$426	$639	$852	$1,065
11	Fixed portion	$4,578	$4,578	$4,578	$4,578	$4,578	$4,578
12							
13	Total variable cost	$3,563	$4,013	$4,464	$4,914	$5,365	$5,815
14							
15	Total fixed cost	$7,476	$7,476	$7,476	$7,476	$7,476	$7,476
16							
17	Total costs	$11,039	$11,489	$11,940	$12,390	$12,841	$13,291
18							
19	Cost per mile	$0.7359	$0.7181	$0.7023	$0.6883	$0.6758	$0.6646
20							

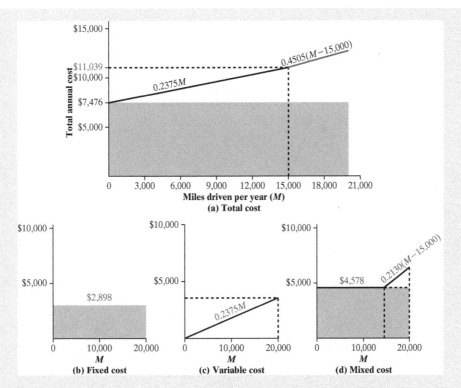

Figure 8.4 Cost–volume relationships pertaining to annual automobile costs (Example 8.2).

To estimate annual costs for any assumed mileage, we construct a **cost–volume chart**, as shown in Figure 8.4(a). Furthermore, we can show the relation between volume (miles driven per year) and the three cost classes separately, as in Figures 8.4(b) through (d). We can use these cost–volume graphs to estimate both the separate and combined costs of operating the car at other possible volumes. For example, an owner who expects to drive 16,500 miles in a given year may estimate the total cost at $11,714 (or 71 cents per mile). In Figure 8.4, all costs more than those necessary to operate at the zero level are known as variable costs. Since the fixed cost is $7,476 a year, the remaining $4,238 is variable cost. By combining all the fixed and variable elements of cost, we can state simply that the cost of owning and operating an automobile is $7,476 per year plus 25.68 cents per mile ($4,238/16,500 miles) driven during the year.

Figure 8.5 graphically illustrates the average unit cost of operating the automobile. The average fixed unit cost, represented by the height of the middle curve in the figure, declines steadily as the volume increases. The average unit cost is high when volume is low, because the total fixed cost is spread over a relatively few units of volume. In other words, the total fixed costs remain the same regardless of the number of miles driven, but the average fixed cost decreases on a per mile basis as the number of miles driven increases.

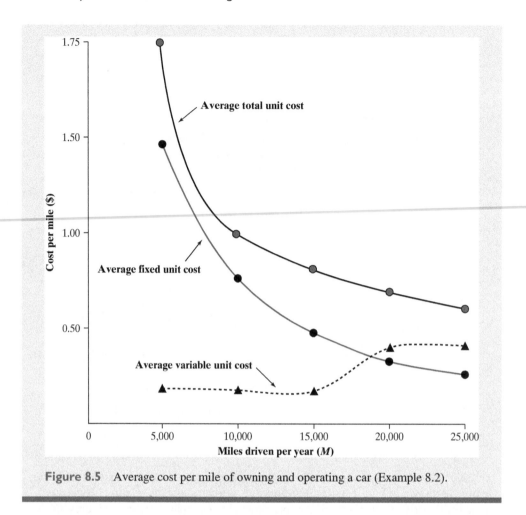

Figure 8.5   Average cost per mile of owning and operating a car (Example 8.2).

8.3.3 Cost–Volume–Profit Analysis

We can extend our cost–volume analysis to include changes in profit as a function of volume. The fundamental assumption in all types of cost–volume–profit decisions is that the firm commits a fixed set of resources for at least a short-run period. The decision problem facing a manager is to determine the most efficient and productive use of these resources relative to output levels and output mixes.

In principle, to maximize a firm's profits in the short run, management should expand its output as long as the marginal revenue from additional units of output exceeds the marginal costs of producing and selling them. The output that maximizes profit can be found by differentiating the firm's profit function with respect to output and setting this derivative equal to zero.

For example, suppose that the firm's **total revenue (TR)** and **total cost (TC)** as non-linear functions of volume (x) in multiples of 1,000 units are as follows:

$$TR(x) = 330x - 6x^2$$

$$TC(x) = 300 + 20x - 5x^2 + \frac{1}{2}x^3$$

The profit function (Φ) is then

$$\Phi = TR(x) - TC(x)$$

$$= 330x - 6x^2 - 300 - 20x + 5x^2 - \frac{1}{2}x^3$$

$$= -300 + 310x - x^2 - \frac{1}{2}x^3$$

To maximize Φ, we take a first derivative of Φ with respect to x:

$$\frac{d\Phi}{dx} = 310 - 2x - \frac{3}{2}x^2 = 0$$

$$x^* = 13.72$$

which implies an optimal output of $x = 13,720$ units. Referring to Figure 8.6, we note that total profit increases to a maximum near the point $x = 14$ (or 14,000 units) and then decreases as output is expanded beyond this point. This is a very simple model of profit maximizing behavior in the short run when a firm's facilities are fixed.

The Use of Linear Functions

One critical assumption in constructing Figure 8.6 is that management has a fairly good idea about the behavior of the firm's revenue and cost functions for ranges of output at which the firm typically operates, and this becomes the basis for making production decisions. We may add another simplifying but practical assumption in estimating the cost and revenue curves in linear function.

A straight-line approximation to the total cost curve in Figure 8.7 can be reasonably accurate, as long as we focus on the behavior of these curves in short-run increments. In other words, the change in total cost in relation to varying output is likely approximately linear. The same thing can be said of the total revenue curve. For example, in Figure 8.6, the change in total costs as output increases from 2,000 to 4,000 units is $332,000 - $324,000 = $8,000. The average change of $4 per unit is a good measure

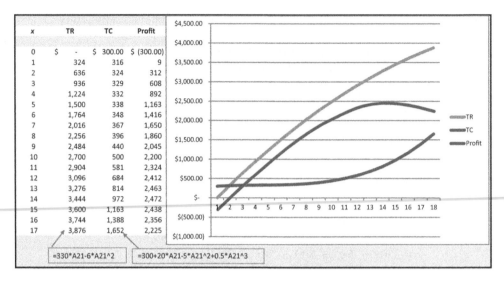

x	TR	TC	Profit
0	$ -	$ 300.00	$ (300.00)
1	324	316	9
2	636	324	312
3	936	329	608
4	1,224	332	892
5	1,500	338	1,163
6	1,764	348	1,416
7	2,016	367	1,650
8	2,256	396	1,860
9	2,484	440	2,045
10	2,700	500	2,200
11	2,904	581	2,324
12	3,096	684	2,412
13	3,276	814	2,463
14	3,444	972	2,472
15	3,600	1,163	2,438
16	3,744	1,388	2,356
17	3,876	1,652	2,225

=330*A21-6*A21^2   =300+20*A21-5*A21^2+0.5*A21^3

Figure 8.6 Cost–Volume–Profit Curve—A profit maximization (unit: 1,000).

of the marginal change in costs as output increases. For the same range of output (x), the change in total revenue is $1,224,000 − $636,000 = $588,000, which is equivalent to a slope of $294 per unit. However, it is important to remember that these linear approximations given in Figure 8.7 are usually accurate only over restricted ranges of output in short-run increments to the output level.

Contribution Margin and Break-Even Sales

The linear break-even chart may be used to evaluate the effects of alternative policies on management's original output decisions, with the assumption that these policies do not result in a significant change in the shapes of the revenue or cost functions. With the total revenue and total cost function expressed in terms of sales volume, the expected profit may be calculated as

$$\Phi = TR(x) − TC(x)$$
$$= px − (vx + F) \tag{8.1}$$

where Φ = profit
$\quad p$ = selling price per unit
$\quad x$ = sales volume
$\quad v$ = variable cost per unit
$\quad F$ = fixed costs

By definition, the firm will break even when $\Phi = 0$, or when its total sales equal its total costs, both fixed and variable. The difference between selling price and variable cost ($p − v$) is referred to as the **contribution margin** or **(marginal contribution)**. Basically,

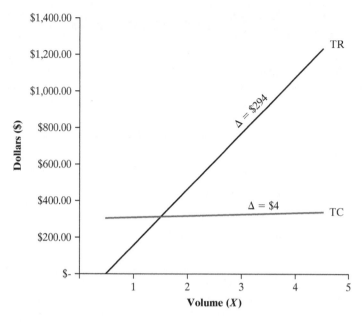

Figure 8.7 Linear approximations of total revenue and total cost curves.

this is the contribution that each unit makes toward absorbing fixed costs and showing a profit. Then the break-even volume for a single product will occur when the contribution is exactly equal to fixed costs, or

$$\text{Break-even volume} = x_b = \frac{F}{p - v} \tag{8.2}$$

The calculation of a break-even point in units for each individual product line does not have much meaning for a multiproduct firm. It might make more sense to talk about the break-even volume in terms of aggregated dollar sales. This figure can be easily calculated by multiplying both sides of Eq. (8.2) by the selling price (p), and then dividing the numerator and denominator of the right-hand component by the selling price.

$$\begin{aligned} \text{Break-even sales} = p(x_b) &= \frac{p(F)}{p - v} \\ &= \frac{F}{1 - \dfrac{v}{p}} \end{aligned} \tag{8.3}$$

where the term $(1 - \frac{v}{p})$ is defined as the **marginal contribution rate**. Therefore, the marginal contribution rate is viewed as the percentage of every sales dollar that is available to cover the fixed costs and provide for profits.

EXAMPLE 8.3 Contribution Margin and Break-Even Volume

The Bragg & Stratton Company manufactures a specialized motor for chain saws. The company expects to manufacture and sell 30,000 motors in the year 2015. It can manufacture an additional 10,000 motors without adding new machinery and equipment. Bragg & Stratton's projected total costs for the 30,000 units are given as follows:

Direct materials	$150,000
Direct labor	300,000
Manufacturing overhead:	
Variable portion	100,000
Fixed portion	80,000
Selling and administrative costs:	
Variable portion	180,000
Fixed portion	70,000

The selling price for the motor is $80.

(a) What is the total manufacturing cost per unit if 30,000 motors are produced?
(b) What is the total manufacturing cost per unit if 40,000 motors are produced?
(c) What is the contribution margin for the motor?
(d) What is the break-even sales volume?
(e) What is the break-even price on the motors if the firm is looking for a $300,000 before-tax profit?

SOLUTION

Given: Financial data.
Find: (a) & (b) total manufacturing cost per unit, (c) unit contribution margin, (d) break-even volume, and (e) break-even price.

(a) Total unit manufacturing cost if 30,000 units are produced:

$$\frac{(\$150,000 + \$300,000 + \$100,000 + \$80,000)}{30,000} = \$21$$

(b) Average total manufacturing costs per unit if 40,000 units are produced:

$$\frac{\left(\dfrac{\$150,000 + \$300,000 + \$100,000}{30,000}\right)40,000 + \$80,000}{40,000} = \$20.33$$

Total variable cost for producing 40,000 units
Average variable unit cost
Fixed cost

(c) Unit contribution margin:

$$p = \$80$$

$$v = \frac{\$150,000 + \$300,000 + \$100,000 + \$180,000}{30,000}$$

$$= \$24.33$$

$$p - v = \$80 - \$24.33$$

$$= \$55.67$$

(d) Break-even sales volume:

$$x_b = \frac{F}{p - v} = \frac{\$80,000 + \$70,000}{\$55.67}$$

$$= 2,695$$

(e) Break-even price with 30,000 units produced:

$$\$21 + \frac{\$250,000 + \$300,000}{30,000} = \$39.33$$

8.4 Future Costs for Business Decisions

In the previous sections, our focus has been on classifying cost data that serve management's need to control and evaluate the operations of a firm. However, these data are historical in nature; they may not be suitable for management's planning of future business operations. We are not saying that historical cost data are of no use in making decisions about the future. In fact, they serve primarily as the first step in predicting the uncertain future. However, the types of cost data that management needs in making choices between alternative courses of action are different from historical cost data.

8.4.1 Differential Cost and Revenue

As we have seen throughout the text, decisions involve choosing among alternatives. In business decisions, each alternative has certain costs and benefits that must be compared with the costs and benefits of the other available alternatives. A difference in cost between any two alternatives is known as a **differential cost**. Similarly, a difference in revenue between any two alternatives is known as **differential revenue**. A differential cost is also known as an incremental cost, although technically, an incremental cost should refer only to an increase in cost from one alternative to another.

Cost–volume relationships based on differential costs find many engineering applications. In particular, they are useful in making a variety of short-term operational decisions. Many short-run problems have the following characteristics.

- The base case is the status quo (the current operation or existing method), and we propose an alternative to the base case. If we find the alternative to have lower costs than the base case, we accept the alternative, assuming that nonquantitative factors do not offset the cost advantage. The **differential (incremental) cost** is the difference in total cost that results from selecting one alternative instead of another. If several alternatives are possible, we select the one with the maximum savings from the base. Decision problems of this type are often called trade-off problems, because one type of cost is traded off for another.
- New investments in physical assets are not required.
- The planning horizon is relatively short (a week or a month—certainly less than a year).
- Relatively few cost items are subject to change by management decision.

Some common examples of short-run problems are method changes, operations planning, and make-or-buy decisions.

Method Changes

Often, we may derive the best information about future costs from an analysis of historical costs. Suppose the proposed alternative is to consider some new method of performing an activity. Then, as Example 8.4 shows, if the differential costs of the proposed method are significantly lower than the current method, we adopt the new method.

EXAMPLE 8.4 Differential Cost Associated with Adopting a New Production Method

The engineering department at an auto-parts manufacturer recommends that the current dies (the base case) be replaced with higher quality dies (the alternative), which would result in substantial savings in manufacturing one of the company's products. The higher cost of materials would be more than offset by the savings in machining time and electricity. If estimated monthly costs of the two alternatives are as shown in Table 8.4, what is the differential cost of going with better dies?

SOLUTION

Given: Financial data.

Find: The preferred production method.

In this problem, the differential cost is $5,000 a month. The differential cost being negative indicates a savings, rather than an addition to total cost. An important point to remember is that differential costs usually include variable costs, but fixed costs are affected only if the decision involves going outside of the relevant range. Although the production volume remains unchanged in our example, the slight increase in depreciation expense is due to the acquisition of new machine tools (dies). All other items of fixed cost remained within their relevant ranges.

TABLE 8.4 Differential Cost of Going with Better Dies

	Current Dies	Better Dies	Differential Cost
Variable costs:			
Materials	$150,000	$170,000	+$20,000
Machining labor	85,000	64,000	−21,000
Electricity	73,000	66,000	−7,000
Fixed costs:			
Supervision	25,000	25,000	0
Taxes	16,000	16,000	0
Depreciation	40,000	43,000	+3,000
Total	**$389,000**	**$384,000**	**−$5,000**

Operations Planning

In a typical manufacturing environment, when demand is high, managers are interested in whether to use a one-shift-plus-overtime operation or to add a second shift. When demand is low, it is equally possible to explore whether to operate temporarily at very low volume or to shut down until operations at normal volume become economical. In a chemical plant, several routes exist for scheduling products through the plant. The problem is which route provides the lowest cost. Example 8.5 illustrates how engineers may use cost–volume relationships in a typical operational analysis.

EXAMPLE 8.5 Break-Even Volume Analysis

Sandstone Corporation has one of its manufacturing plants operating on a single-shift five-day work week. The plant is operating at its full capacity (24,000 units of output per week) without the use of overtime or extra shifts. Fixed costs for single-shift operation amount to $90,000 per week. The average variable cost is a constant $30 per unit at all output rates—up to 24,000 units per week. The company has received an order to produce an extra 4,000 units per week beyond the current single-shift maximum capacity. Two options are being considered to fill the new order:

- **Option 1.** Increase the plant's output to 36,000 units a week by adding overtime, Saturday operations, or both. No increase in fixed costs is entailed, but the variable cost is $36 per unit for any output in excess of 24,000 units per week—up to a 36,000-unit capacity.
- **Option 2.** Operate a second shift. The maximum capacity of the second shift is 21,000 units per week. The variable cost of the second shift is $31.50 per unit, and the operation of a second shift entails additional fixed costs of $13,500 per week.

Determine the range of operating volume that will make option 2 profitable.

SOLUTION

Given: Financial data.

Find: The break-even volume that will make both options indifferent.

In this example, the operating costs related to the first-shift operation will remain unchanged if one alternative is chosen instead of another. Therefore, those costs are irrelevant to the current decision and can safely be left out of the analysis. Consequently, we need to examine only the increased total cost due to the additional operating volume under each option. (This kind of study is known as **incremental analysis.**) Let Q denote the additional operating volume. Then we have

- **Option 1.** Overtime and Saturday operation: $36Q$.
- **Option 2.** Second-shift operation: $13,500 + \$31.50Q$.

We can find the break-even volume (Q_b) by equating the incremental cost functions and solving for Q:

$$36Q = 13,500 + 31.5Q$$
$$4.5Q = 13,500$$
$$Q_b = 3,000 \text{ units}$$

If the additional volume exceeds 3,000 units, the second-shift operation becomes more efficient than overtime or Saturday operation. A break-even (or cost-volume) graph based on the foregoing data is shown in Figure 8.8. The horizontal scale

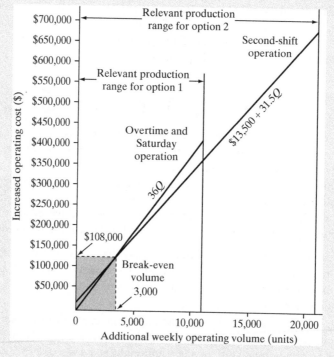

Figure 8.8 Cost-volume relationships of operating overtime, and a Saturday operation versus second-shift operation beyond 24,000 units (Example 8.5).

represents additional volume per week. The upper limit of the relevant volume range for option 1 is 12,000 units, whereas the upper limit for option 2 is 21,000 units. The vertical scale is in dollars of cost. The operating savings expected at any volume may be read from the cost–volume graph. For example, the break-even point (zero savings) is 3,000 units per week. In this scenario, option 2 is a better choice, since the additional weekly volume from the new order exceeds 3,000 units.

Make-or-Buy (Outsourcing) Decision

In business, the make-or-buy decision arises on a fairly frequent basis. Many firms perform certain activities using their own resources and pay outside firms to perform certain other activities. It is a good policy to constantly seek to improve the balance between these two types of activities by asking whether we should outsource some function that we are now performing ourselves or vice versa. Make-or-buy decisions are often grounded in the concept of **opportunity cost**.

8.4.2 Opportunity Cost

Opportunity cost may be defined as the potential benefit that is given up as you seek an alternative course of action. In fact, virtually every alternative has some opportunity cost associated with it. For example, suppose you have a part-time job while attending college that pays you $200 per week. You would like to spend a week at the beach during spring break, and your employer has agreed to give you the week off. What would be the opportunity cost of taking the time off to be at the beach? The $200 in lost wages would be an opportunity cost.

In an economic sense, opportunity cost could mean the contribution to income that is forgone by not using a limited resource in the best way possible. Or we may view opportunity costs as cash flows that could be generated from an asset the firm already owns, provided that such flows are not used for the alternative in question. In general, *accountants do not post opportunity cost in the accounting records of an organization. However, this cost must be explicitly considered in every economic decision.*

- An opportunity cost arises when a project uses a resource that already may have been paid for by the firm.
- When a resource that is already owned by a firm is being considered for use in a project, that resource has to be priced on its next-best alternative use, which may be one of the following options.
  1. A sale of the asset, in which case the opportunity cost is the expected proceeds from the sale, net of any taxes on gains.
  2. Renting or leasing the asset out, in which case the opportunity cost is the expected present value of the after-tax revenue from the rental or lease.
  3. Some use elsewhere in the business, in which case the opportunity cost is the cost of replacing the resource.
  4. That the asset has been abandoned or is of no use. Then the opportunity cost is zero.

EXAMPLE 8.6 Opportunity Cost: Lost Rental Income

Benson Company is a farm-equipment manufacturer that currently produces 20,000 units of gas filters annually for use in its lawn-mower production. The expected annual production cost of the gas filters is summarized as follows.

Variable costs:	
Direct materials	$100,000
Direct labor	190,000
Power and water	35,000
Fixed costs:	
Heating and light	20,000
Depreciation	100,000
Total cost	**$445,000**

Tompkins Company has offered to sell Benson 20,000 units of gas filters for $17.00 per unit. If Benson accepts the offer, some of the manufacturing facilities currently used to manufacture the filters could be rented to a third party at an annual rent of $35,000. Should Benson accept Tompkins's offer, and if so, why?

SOLUTION

Given: Financial data; production volume = 20,000 units.
Find: Whether Benson should outsource the gas-filter operation.

This problem is unusual in the sense that the buy option would generate a rental fee of $35,000. In other words, Benson could rent out the current manufacturing facilities if it were to purchase the gas filters from Tompkins. To compare the two options, we need to examine the cost of each option, as given in Table 8.5.

TABLE 8.5 **Breakdowns of Cost Items**

	Make Option	Buy Option	Differential Cost (Make-Buy)
Variable costs:			
Direct materials	$100,000		$100,000
Direct labor	190,000		190,000
Power and water	35,000		35,000
Gas filters		340,000	−340,000
Fixed costs:			
Heating and light	20,000	20,000	0
Depreciation	100,000	100,000	0
Rental income lost	35,000		35,000
Total cost	**$480,000**	**$460,000**	**$20,000**
Unit cost	**$24.00**	**$23.00**	**$1.00**

The buy option has a lower unit cost and saves $1 for each use of a gas filter. If the lost rental income (opportunity cost) were not considered, however, the decision would favor the make option.

8.4.3 Sunk Costs

A sunk cost is a cost that already has been incurred by past actions. **Sunk costs** are not relevant to decisions, because they cannot be changed regardless of what decision is made now or in the future. The only costs relevant to a decision are costs that vary among the alternative courses of action being considered. To illustrate a sunk cost, suppose you have a very old car that requires frequent repairs. You want to sell the car, and you figure that the current market value would be about $1,200 at best. While you are in the process of advertising the car, you find that the car's water pump is leaking. You decided to have the pump repaired, which cost you $200. A friend of yours is interested in buying your car and has offered $1,300 for it. Would you take the offer, or would you decline it simply because you cannot recoup the repair cost with that offer? In this example, the $200 repair cost is a sunk cost. You cannot change this repair cost, regardless of whether you keep or sell the car. Since your friend's offer is $100 more than the best market value, it would be better to accept the offer.

8.4.4 Marginal Cost

We make decisions every day, as entrepreneurs, professionals, executives, investors, and consumers, with little thought as to where our motivations come from or how our assumptions of logic fit a particular academic regimen. As you have seen, the engineering economic decisions that we describe throughout this text owe much to English economist Alfred Marshall (1842–1924) and his concepts of **marginalism**. In our daily quest for material gain, rarely do we recall the precepts of microeconomics or marginal utility, yet the ideas articulated by Marshall remain some of the most useful core principles guiding economic decision making.

Marginal Cost and Marginal Revenues

Another cost term useful in cost–volume analysis is marginal cost. We define **marginal cost** as the added cost that would result from increasing the rate of output by a single unit. The accountant's differential-cost concept can be compared to the economist's marginal-cost concept. In speaking of changes in cost and revenue, the economist employs the terms *marginal cost* and *marginal revenue*. The revenue that can be obtained from selling one more unit of product is called **marginal revenue**. The cost involved in producing one more unit of product is called **marginal cost**. Example 8.7 illustrates how we calculate the marginal cost associated with a typical consumption of electricity.

EXAMPLE 8.7 Marginal Costs versus Average Costs

Consider a company that has an available electric load of 37 horsepower and that purchases its electricity at the following rates:

kWh/Month	@$/kWh	Average Cost($/kWh)
First 1,500	$0.050	$0.050
Next 1,250	0.035	$\dfrac{\$75 + 0.0350(X - 1,500)}{X}$
Next 3,000	0.020	$\dfrac{\$118.75 + 0.020(X - 2,750)}{X}$
All over 5,750	0.010	$\dfrac{\$178.75 + 0.010(X - 5,750)}{X}$

According to this rate schedule, the unit variable cost in each rate class represents the marginal cost per kilowatt-hours (kWh). Alternatively, we may determine the average costs in the third column by finding the cumulative total cost and dividing it by the total number of kWh (X). Suppose that the current monthly consumption of electric power averages 3,200 kWh. On the basis of this rate schedule, determine the marginal cost of adding one more kWh and, for a given operating volume (3,200 kWh), the average cost per kWh.

SOLUTION

Given: Marginal cost schedule for electricity; operating volume = 3,200 kWh.
Find: Marginal and average cost per kWh.

The marginal cost of adding one more kWh is $0.020. The average variable cost per kWh is calculated as follows.

kWh	Cumulative kWh	Rate ($/kWh)	Cost
First 1,500	1,500	0.050	$75.00
Next 1,250	2,750	0.035	43.75
Remaining 450	3,200	0.020	9.00
Total			**$127.75**

The average variable cost per kWh is $\dfrac{\$127.75}{3,200 \text{ kWh}} = \$0.0399 \text{ kWh}$. Or we can find the value by using the formulas in the third column of the rate schedule:

$$\frac{\$118.75 + \$0.020(3,200 - 2,750)}{3,200} = \frac{\$127.75}{3,200 \text{ kWh}} = \$0.0399 \text{ kWh}$$

Changes in the average variable cost per unit are the result of changes in the marginal cost. As shown in Figure 8.9, the average variable cost continues to fall because the marginal cost is lower than the average variable cost over the entire volume.

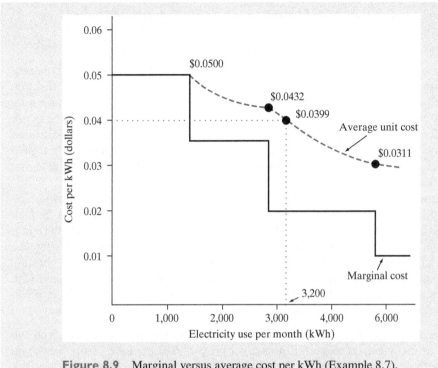

Figure 8.9 Marginal versus average cost per kWh (Example 8.7).

Marginal Analysis

The fundamental concept of marginal thinking is that you are where you are, and what is past is irrelevant. (This means that we should ignore the sunk cost.) The point is whether you move forward, and you will do so if the benefits outweigh the costs—even if they do so by a smaller margin than before. You will continue to produce a product or service until the cost of doing so equals the revenue derived.

Microeconomics suggests that businesspersons and consumers measure progress not in great leaps, but in small incremental steps. Rational persons will reevaluate strategies and take new actions if benefits exceed costs on a marginal basis. *If there is more than one alternative to choose, always choose the alternative with the greatest marginal benefit.*

As we have mentioned, the economist's marginal-cost concept is the same as the accountant's differential-cost concept. For a business problem on maximization of profit, economists place a great deal of emphasis on the importance of marginal analysis: Rational individuals and institutions should perform the most cost-effective activities first. Specifically,

- When you examine marginal costs, you want to use the least expensive methods first and the most expensive last (if at all).
- Similarly, when you examine marginal revenue (or demand), the most revenue-enhancing methods should be used first and the least revenue-enhancing one last.
- If we need to consider the revenue along with the cost, the volume at which the total revenue and total cost are the same is known as the *break-even point*, which can be calculated using Eq (8.2). Example 8.8 illustrates these key issues.

EXAMPLE 8.8 Profit-Maximization Problem: Marginal Analysis[3]

Suppose you are a chief executive officer (CEO) of a small pharmaceutical company that manufactures generic aspirin. You want the company to maximize its profits. You can sell as much aspirin as you make at the prevailing market price. You have only one manufacturing plant, which is the constraint. You have the plant working at full capacity Monday through Saturday, but you close the plant on Sunday, because on Sundays you have to pay workers overtime rates, and it is not worth it. The marginal costs of production are constant Monday through Saturday. Marginal costs are higher on Sundays only because labor costs are higher.

Now you obtain a long-term contract to manufacture a brand-name aspirin. The costs of making the generic aspirin or the brand-name aspirin are identical. In fact, there is no cost or time involved in switching from the manufacture of one to the other. You can manufacture both the brand-name and the generic aspirin, and you will make much larger profits from the brand-name aspirin, but the demand is limited. One day of manufacturing each week will permit you to fulfill the contract. Compared with the situation before you obtained the contract, your profits will be much higher if you now begin to manufacture on Sundays—even if you have to pay overtime wages.

- *Generic aspirin.* Each day, you can make 1,000 cases of generic aspirin. You can sell as many as you make for the market price of $10 per case. Every week you have fixed costs of $5,000 (land tax and insurance). No matter how many cases you manufacture, the cost of materials and supplies is $2 per case; the cost of labor is $5 per case—except on Sundays when it is $10 per case.

- *Brand-name aspirin.* Your order for the brand-name aspirin requires that you manufacture 1,000 cases per week, which you sell for $30 per case. The cost for the brand-name aspirin is identical to the cost of the generic aspirin.

What do you do?

SOLUTION

Given: Sales price: $10 per case for generic aspirin, $30 per case for brand-name aspirin; weekly fixed cost, $5,000; variable cost, $7 per case during weekdays, $12 per case during Sunday operation; weekly production, 6,000 cases of generic aspirin, 1,000 cases of brand-name aspirin.

Find: (a) Optimal production mix and (b) break-even volume.

(a) Optimal production mix:
- The marginal revenue from manufacturing on Sunday is $10,000 (1,000 cases times $10 per case).
- The marginal cost from manufacturing on Sunday is $12,000 (1,000 cases times $12 per case—$10 labor + $2 materials).
- Total profits will be shrinking by $2,000 if the plant operates on Sunday.

The marginal costs of manufacturing brand-name aspirin are constant Monday through Saturday; they rise substantially on Sunday and are above the marginal revenue from

---

[3]*Source:* "Profit Maximization Problem," by David Hemenway and Elon Kohlberg, *Economic Inquiry*, October 1, 1997, p. 862. Copyright 1997, Western Economic Association International.

manufacturing generic aspirin. In other words, your company should manufacture the brand-name aspirin first. Your marginal revenue is the highest the "first" day, when you manufacture the brand-name aspirin. It then falls and remains constant for the rest of the week. On the seventh day (Sunday), the marginal revenue from manufacturing the generic aspirin falls below the marginal cost. You should manufacture brand-name aspirin one day a week and generic aspirin five days a week. On Sundays, the plant should close.

(b) Break-even volume: As shown in Figure 8.10, the total revenue and cost functions can be represented as

$$\text{Total revenue function:} \begin{cases} 30Q & \text{for } 0 \le Q \le 1{,}000 \\ 30{,}000 + 10(Q - 1{,}000) & \text{for } 1{,}000 \le Q \le 7{,}000 \end{cases}$$

$$\text{Total cost function:} \begin{cases} 5{,}000 + 7Q & \text{for } 0 \le Q \le 6{,}000 \\ 47{,}000 + 12(Q - 6{,}000) & \text{for } 6{,}000 \le Q \le 7{,}000 \end{cases}$$

Table 8.6 shows the various factors involved in the production of the brand-name and the generic aspirin.

- If you produce the brand-name aspirin first, the break-even volume is

$$30Q - 7Q - 5{,}000 = 0$$
$$Q_b = 217.39$$

- If you produce the generic aspirin first, the break-even volume is

$$10Q - 7Q - 5{,}000 = 0$$
$$Q_b = 1{,}666.67$$

Clearly, scheduling the production of the brand-name aspirin first is the better strategy, as you can recover the fixed cost ($5,000) much faster by selling just 217.39 cases of brand-name aspirin. Also, Sunday operation is not economical, as the marginal cost exceeds the marginal revenue by $2,000.

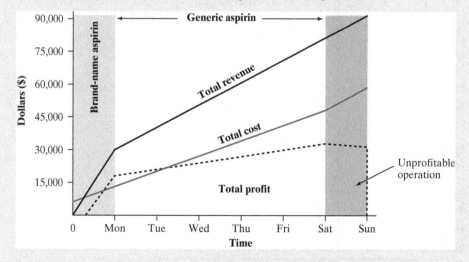

Figure 8.10 Weekly profits as a function of time. Sunday operation becomes unprofitable, because the marginal revenue stays at $10 per case, whereas the marginal cost increases to $12 per case (Example 8.8).

TABLE 8.6 Net Profit Calculation as a Function of Production Volume (Example 8.8)

	A	B	C	D	E	F	G	H	I	J
1										
2		Daily	Cumulative							
3		Production	Production	Product		Total	Variable	Fixed	Total	Net
4	Day	Capacity	Volume (Q)	Mix	Revenue	Revenue	Cost	Cost	Cost	Profit
5										
6		-	-					$5,000	$5,000	-$5,000
7	Monday	1,000	1,000	Brand	$30,000	$30,000	$7,000	$0	$12,000	$18,000
8	Tuesday	1,000	2,000	Generic	$10,000	$40,000	$7,000	$0	$19,000	$21,000
9	Wednesday	1,000	3,000	Generic	$10,000	$50,000	$7,000	$0	$26,000	$24,000
10	Thursday	1,000	4,000	Generic	$10,000	$60,000	$7,000	$0	$33,000	$27,000
11	Friday	1,000	5,000	Generic	$10,000	$70,000	$7,000	$0	$40,000	$30,000
12	Saturday	1,000	6,000	Generic	$10,000	$80,000	$7,000	$0	$47,000	$33,000
13	Sunday	1,000	7,000	Generic	$10,000	$90,000	$12,000	$0	$59,000	$31,000
14										

8.5 Estimating Profit from Operation

Up to this point, we have defined various cost elements and their behaviors in a manufacturing environment. Our ultimate objective is to develop a project's cash flows; we do so in Chapter 10, where the first step is to formulate the budget associated with the sales and production of the product. As we will explain in that chapter, the income statement developed in the current chapter will be the focal point for laying out the project's cash flows on yearly basis.

8.5.1 Calculation of Operating Income

Accountants measure the net income of a specified operating period by subtracting expenses from revenues for that period. These terms can be defined as follows:

1. The **project revenue** is the income earned[4] by a business as a result of providing products or services to customers. Revenue comes from sales of merchandise to customers and from fees earned by services performed for clients or others.

2. The **project expenses** that are incurred[5] are the cost of doing business to generate the revenues of the specified operating period. Some common expenses are the cost of the goods sold (labor, material, inventory, and supplies), depreciation, the cost of employees' salaries, the business's operating costs (such as the cost of renting buildings and the cost of insurance coverage), and income taxes.

The business expenses just listed are accounted for in a straightforward fashion on a company's income statement and balance sheet: The amount paid by the organization for each item would translate (dollar for dollar) into expenses in financial reports for the period. One additional category of expenses—the purchase of new assets—is treated by depreciating the total cost gradually over time. Because capital goods are given this unique accounting treatment, depreciation is accounted for as a separate expense in financial reports. Because of its significance in engineering economic analysis, we will treat depreciation accounting in a separate chapter, namely Chapter 9.

Figure 8.11 will be used as a road map to construct the budgeted income statement related to the proposed manufacturing activities.

8.5.2 Annual Sales Budget for a Manufacturing Business

The first step in laying out an annual sales budget for a manufacturing business is to predict the dollar sales, which is the key to the entire process. The estimates are typically expressed in both dollars and units of the product. Using the expected selling price, we can easily extend unit sales to revenues. Often, the process is initiated by individual salespersons or managers predicting sales in their own area for the budget period. Of course, an aggregate sales budget is influenced by economic conditions, pricing decisions, competition, marketing programs, etc.

To illustrate, suppose that you want to create a sales budget schedule on a quarterly basis. Projected sales units are 1,000 units during the first quarter, 1,200 units during the second quarter, 1,300 units during the third quarter, and 1,500 units during the fourth quarter. The projected sales price is $15 per unit. Then an estimated sales schedule looks like Table 8.7.

[4] Note that the cash may be received in a different accounting period.
[5] Note that the cash may be paid in a different accounting period.

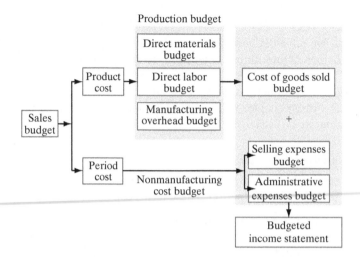

Figure 8.11 Process of creating a master production budget.

TABLE 8.7 Sales Budget Schedule (Year 2016)—Product X

	1Q	2Q	3Q	4Q	Annual Total
Budgeted units	1,000	1,200	1,300	1,500	5,000
Sales price	$15	$15	$15	$15	$15
Estimated sales	$15,000	$18,000	$19,500	$22,500	$75,000

8.5.3 Preparing the Annual Production Budget

Once the sales budget is known, we can prepare the production activities. Using the sales budget for the number of projected units needed, we prepare an estimate of the number of units to be produced in the budget period. Note that the production budget is the basis for projecting the cost-of-goods-manufactured budget, which we will discuss subsequently. To illustrate, we may use the following steps.

Step 1. From the sales budget, record the projected number of units to be sold.

Step 2. Determine the desired number of units to carry in the ending inventory; usually, that number is a percentage of next quarter's needs. In our example, we will assume that 20% of the budgeted units will be the desirable ending inventory position in each quarter.

Step 3. *Add* to determine the total number of units needed each period.

Step 4. Determine the projected number of units in the beginning inventory. Note that the beginning inventory is last quarter's ending inventory. In our example, we will assume that the first quarter's beginning inventory is 100 units.

Step 5. *Subtract* to determine the projected production budget.

Then a typical production budget may look like Table 8.8.

TABLE 8.8 Production Budget (Year 2016)—Product X

	1Q	2Q	3Q	4Q	Annual Total
Budgeted units to be produced	1,000	1,200	1,300	1,500	5,000
Desired ending inventory	200	240	260	300	1,000
Total units needed	1,200	1,440	1,560	1,800	6,000
Less beginning inventory	100	200	240	260	800
Units to produce	1,100	1,240	1,320	1,540	5,200

Materials Budgets

Once we know how many units we need to produce, we are ready to develop direct materials budgets. We may use the following steps.

Step 1. From the production budget, copy the projected number of units to be sold.

Step 2. *Multiply* by the amount of raw materials needed per unit to calculate the amount of materials needed. In our example, we will assume that each production unit consumes $4 of materials.

Step 3. Calculate the desired ending inventory (the number of units required in the ending inventory $\times$ $4).

Step 4. *Add* to calculate the total amount of materials needed.

Step 5. *Subtract* the beginning inventory, which is last quarter's ending inventory, to calculate the amount of raw materials needed to be purchased.

Step 6. Calculate the net cost of raw materials.

Then a typical direct materials budget may look like Table 8.9.

TABLE 8.9 Direct Materials Budget (Year 2016)—Product X

	1Q	2Q	3Q	4Q	Annual Total
Units to be sold	1,000	1,200	1,300	1,500	5,000
Unit cost of materials	$4	$4	$4	$4	
Cost of materials for units to be sold	$4,000	$4,800	$5,200	$6,000	$20,000
Cost of materials in ending inventory	$800	$960	$1,040	$1,200	$4,000
Cost of materials in beginning inventory	$400	$800	$960	$1,040	$3,200
Cost of materials to purchase	$4,400	$4,960	$5,280	$6,160	$20,800

Direct Labor Budget for a Manufacturing Business

As with the materials budgets, once the production budget has been completed, we can easily prepare the direct labor budget. This budget allows the firm to estimate labor requirements—both labor hours and dollars—in advance. To illustrate, use the following steps.

Step 1. From the production budget, copy the projected number of units to be produced.

Step 2. *Multiply* by the direct labor cost per unit to calculate the total direct labor cost. In our example, we will assume that the unit direct labor cost is $1.27 in 1Q, $1.30 in 2Q, $1.32 in 3Q, and $1.35 in 4Q.

Then a typical direct labor budget may look like Table 8.10.

TABLE 8.10 **Direct Labor Budget (Year 2016)—Product X**

	1Q	2Q	3Q	4Q	Annual Total
Units to produce	1,100	1,240	1,320	1,540	5,200
× Direct labor cost per unit	$1.27	$1.30	$1.32	$1.35	
Total direct labor cost ($)	$1,379	$1,612	$1,742	$2,079	$6,812

Overhead Budget for a Manufacturing Business

The overhead budget should provide a schedule of all costs of production other than direct materials and direct labor. Typically, the overhead budget is expressed in dollars based on a predetermined overhead rate. In preparing a manufacturing overhead budget, we may take the following steps.

Step 1. From the production budget, copy the projected number of units to be produced.

Step 2. *Multiply* by the variable overhead rate to calculate the budgeted variable overhead. In our example, we will assume the variable overhead rate to be $1.50 per unit. There are several ways to determine this overhead rate. One common approach (known as traditional standard costing) is to divide the expected total overhead cost by the budgeted number of direct labor hours (or units). Another approach is to adopt an **activity-based costing** concept, allocating indirect costs against the activities that caused them. We will not review this accounting method here, but it can more accurately reflect indirect cost improvement than **traditional standard costing** can.

Step 3. *Add* any budgeted fixed overhead to calculate the total budgeted overhead. In our example, we will assume the fixed overhead to be $230 each quarter.

Then a typical manufacturing overhead budget may look like Table 8.11.

8.5.4 Preparing the Cost-of-Goods-Sold Budget

The production budget developed in the previous section shows how much it would cost to produce the required production volume. Note that the number of units to be produced includes both the anticipated number of sales units and the number of units in inventory.

TABLE 8.11 Manufacturing Overhead Budget (Year 2016)—Product X

	1Q	2Q	3Q	4Q	Annual Total
Units to produce	1,100	1,240	1,320	1,540	5,200
Variable manufacturing over-head rate per unit ($1.50)	$1,650	$1,860	$1,980	$2,310	$7,800
Fixed manufacturing overhead	$230	$230	$230	$230	$920
Total overhead	$1,880	$2,090	$2,210	$2,540	$8,720

The cost-of-goods-sold budget is different from the production budget, because we do not count the costs incurred to carry the inventory. Therefore, we need to prepare a budget that details the costs related to the sales, not the inventory. Typical steps in preparing a cost-of-goods-sold budget are:

Step 1. From the sales budget, and not the production budget, copy the budgeted number of sales units.

Step 2. *Multiply* by the direct material cost per unit to estimate the amount of direct materials.

Step 3. *Multiply* by the direct labor cost per unit to estimate the direct labor.

Step 4. *Multiply* by the manufacturing overhead per unit to estimate the overhead.

Then a typical cost-of-goods-sold budget may look like Table 8.12

TABLE 8.12 Cost of Goods Sold (Year 2016)—Product X

	1Q	2Q	3Q	4Q	Annual Total
Budgeted sales units	1,000	1,200	1,300	1,500	5,000
Direct materials ($4/unit)	$4,000	$4,800	$5,200	$6,000	$20,000
Direct labor	$1,270	$1,560	$1,716	$2,025	$6,571
Manufacturing overhead:					
Variable ($1.50/unit)	$1,500	$1,800	$1,950	$2,250	$7,500
Fixed	$230	$230	$230	$230	$920
Cost of goods sold	$8,730	$10,420	$11,276	$12,985	$43,411

8.5.5 Preparing the Nonmanufacturing Cost Budget

To complete the entire production budget, we need to add two more items related to production: the selling expenses and the administrative expenses.

Selling Expenses Budget for a Manufacturing Business

Since we know the sales volume, we can develop a selling expenses budget by considering the budgets of various individuals involved in marketing the products. The budget includes both variable cost items (shipping, handling, and sales commission) and fixed items, such as advertising and salaries for marketing personnel. To prepare the selling expenses budget, we may take the following steps.

Step 1. From the sales budget, copy the projected number of unit sales.

Step 2. List the variable selling expenses, such as the sales commission. In our example, we will assume that the sales commission is calculated at 5% of unit sales.

Step 3. List the fixed selling expenses, typically rent, depreciation expenses, advertising, and other office expenses. In our example, we will assume the following: rent, $500 per quarter; advertising, $300 per quarter; office expense, $200 per quarter.

Step 4. *Add* to calculate the total budgeted selling expenses.

Then a typical selling expenses budget may look like Table 8.13.

TABLE 8.13 Selling Expenses (Year 2016)—Product X

	1Q	2Q	3Q	4Q	Annual Total
Budgeted unit sales ($)	$ 15,000	$ 18,000	$ 19,500	$ 22,500	$ 75,000
Variable expenses:					
Commission	$750	$900	$975	$1,125	$3,750
Fixed expenses:					
Rent	$500	$500	$500	$500	$2,000
Advertising	$300	$300	$300	$300	$1,200
Office expenses	$200	$200	$200	$200	$800
Total selling expenses	$1,750	$1,900	$1,975	$2,125	$7,750

Administrative Expenses Budget for a Manufacturing Business

Another nonmanufacturing expense category to consider is administrative expenses, a category that contains mostly fixed-cost items, such as executive salaries and the depreciation of administrative buildings and office furniture. To prepare an administrative expense budget, we may take the following steps.

Step 1. List the variable administrative expenses, if any.

Step 2. List the fixed administrative expenses, which include salaries, insurance, office supplies, and utilities. We will assume the following quarterly expenses: salaries, $1,400; insurance, $135; office supplies, $300; other office expenses, $150.

Step 3. *Add* to calculate the total administrative expenses.

Then a typical administrative expenses budget may look like Table 8.14.

TABLE 8.14 Administrative Expenses (Year 2016)—Product X

	1Q	2Q	3Q	4Q	Annual Total
Variable expenses:					
Fixed expenses:					
Salaries	$1,400	$1,400	$1,400	$1,400	$5,600
Insurance	$135	$135	$135	$135	$540
Office supplies	$300	$300	$300	$300	$1,200
Utilities (phone, power, water, etc.)	$500	$500	$500	$500	$2,000
Other office expenses	$150	$150	$150	$150	$600
Total administrative expenses	$2,485	$2,485	$2,485	$2,485	$9,940

8.5.6 Putting It All Together: The Budgeted Income Statement

Now we are ready to develop a consolidated budget that details a company's projected sales, costs, expenses, and net income. To determine the net income, we need to include any other operating expenses, such as interest expenses and renting or leasing expenses. Once we calculate the taxable income, we can determine the federal taxes to pay. The procedure to follow is

Step 1. Copy the net sales from the sales budget.
Step 2. Copy the cost of goods sold.
Step 3. Copy both the selling and administrative expenses.
Step 4. Gross income = Sales − Cost of goods sold.
Step 5. Operating income = Sales − (Cost of goods sold + Operating expenses).
Step 6. Determine any interest expense incurred during the budget period.
Step 7. Calculate the federal income tax at a percentage of taxable income. In our example, let's assume a 35% tax rate.
Step 8. Net income = Income from operations − Federal income tax, which summarizes the projected profit from the budgeted sales units.

Then the budgeted income statement would look like the format shown in Table 8.15. Once you have developed a budgeted income statement, you may be able to examine the profitability of the manufacturing operation. Three margin figures are commonly used to quickly assess the profitability of the operation: gross margin, operating margin, and net profit margin.

Gross Margin

The gross margin reveals how much a company earns, taking into consideration the costs that it incurs for producing its products or services. In other words, gross margin is equal to gross income divided by net sales and is expressed as a percentage:

$$\text{Gross margin} = \frac{\text{Gross income}}{\text{Net sales}} \tag{8.2}$$

TABLE 8.15 Budgeted Income Statement (Year 2016)—Product X

	1Q	2Q	3Q	4Q	Annual Total
Sales	$15,000	$18,000	$19,500	$22,500	$75,000
Cost of goods sold	$8,730	$10,420	$11,276	$12,985	$43,411
Gross income	$6,270	$7,580	$8,224	$9,515	$31,589
Operating expenses:					
Selling expenses	$1,750	$1,900	$1,975	$2,125	$7,750
Administrative expenses	$2,485	$2,485	$2,485	$2,485	$9,940
Operating income	$2,035	$3,195	$3,764	$4,905	$13,899
Interest expenses	—	—	—	—	—
Net income before taxes	$2,035	$3,195	$3,764	$4,905	$13,899
Income taxes (35%)	$712	$1,118	$1,317	$1,717	$4,865
Net income	**$1,323**	**$2,077**	**$2,447**	**$3,188**	**$9,034**

In our example, gross margin = $31,589/$75,000 = 42.12%. Gross margin indicates how profitable a company is at the most fundamental level. Companies with higher gross margins will have more money left over to spend on other business operations, such as research and development or marketing.

Operating Margin

The operating margin is defined as the operating profit for a certain period, divided by revenues for that period:

$$\text{Operating margin} = \frac{\text{Operating income}}{\text{Net sales}} \tag{8.3}$$

In our example, operating margin = $13,899/$75,000 = 18.53%. Operating profit margin indicates how effective a company is at controlling the costs and expenses associated with its normal business operations.

Net Profit Margin

As we defined the term in Chapter 2, the net profit margin is obtained by dividing net profit by net revenues, often expressed as a percentage:

$$\text{Net profit margin} = \frac{\text{Net income}}{\text{Net sales}} \tag{8.4}$$

This number is an indication of how effective a company is at cost control. The higher the net profit margin, the more effective the company is at converting revenue into actual profit. In our example, the net profit margin = $9,034/$75,000 = 12.05%. The net profit margin is a good way of comparing companies in the same industry, since such companies are generally subject to similar business conditions.

8.5.7 Looking Ahead

The operational decision problem described in this chapter tends to have a relatively short-term horizon (weekly or monthly). That is, such decisions do not commit a firm to a certain course of action over a relatively long period. If operational decision problems significantly affect the amount of funds that must be invested in a firm, fixed costs will have to increase. In Example 8.7, if the daily aspirin production increases to 10,000 cases per day, exceeding the current production capacity, the firm must make new investments in plant and equipment. Since the benefits of the expansion will continue to occur over an extended period, we need to consider the economic effects of the fixed costs over the life of the assets. Also, we have not discussed how we actually calculate the depreciation amount related to production or the amount of income taxes to be paid from profits generated from the operation. Doing so requires an understanding of the concepts of **capital investment**, **depreciation**, and **income taxes**, which we will discuss in the next chapter.

SUMMARY

In this chapter, we examined several ways in which managers classify costs. How the costs will be used dictates how they will be classified.

- Most manufacturing companies divide **manufacturing costs** into three broad categories: *direct materials, direct labor*, and *manufacturing overhead*. **Nonmanufacturing costs** are classified into two categories: *marketing* or *selling costs* and *administrative costs.*

- For the purpose of valuing inventories and determining expenses for the balance sheet and income statement, costs are classified as either **product costs** or **period costs**.

- For the purpose of predicting **cost behavior**—how costs will react to changes in activity—managers commonly classify costs into two categories: variable costs and fixed costs.

- An understanding of the following **cost–volume–profit relationships** is essential to developing successful business strategies and to planning future operations.

  Fixed operating costs: Costs that do not vary with production volume.

  Variable operating costs: Costs that vary with the level of production or sales.

  Average costs: Costs expressed in terms of units obtained by dividing total costs by total volumes.

  Differential (incremental) costs: Costs that represent differences in total costs, which result from selecting one alternative instead of another.

  Opportunity costs: Benefits that could have been obtained by taking an alternative action.

  Sunk costs: Past costs not relevant to decisions because they cannot be changed no matter what actions are taken.

  Marginal costs: Added costs that result from increasing rates of outputs, usually by single units.

  Marginal analysis: In economic analysis, we need to answer the apparently trivial question, "Is it worthwhile?"—whether the action in question will add sufficiently to the benefits enjoyed by the decision maker to make performing the action worth the cost. This is the heart of marginal decision-making—the statement that an action merits performance if, and only if, as a result, we can expect to be better off than we were before.

- At the level of plant operations, engineers must make decisions involving materials, production processes, and the in-house capabilities of company personnel. Most of these operating decisions do not require any investments in physical assets; therefore, they depend solely on the cost and volume of business activity without any consideration of the time value of money.

■ Engineers are often asked to prepare the production budgets related to their operating division. Doing this requires a knowledge of budgeting scarce resources, such as labor and materials, and an understanding of the overhead cost. The same budgeting practice will be needed in preparing the estimates of costs and revenues associated with undertaking a new project.

PROBLEMS

Classifying Costs

8.1 Identify which of the following transactions and events are *product costs* and which are *period costs*:
- Office fire insurance.
- Cost of electric power.
- Rental charges of warehouses.
- Costs of basic raw materials used in production.
- Depreciation of electrical equipment in the plant.
- Depreciation of the furniture in the office.
- Sales incentives paid to the sales personnel.
- Inspection costs of raw materials.
- Direct overhead costs.
- Costs of packing materials.
- Overtime premiums.

Cost Behavior

8.2 Identify which of the following costs are *fixed*, which are *variable*, and which are *mixed*.
- Maintenance costs of vehicles used in supplying goods.
- Direct labor cost.
- Wages paid to machine operators.
- Depreciation of factory buildings.
- Salaries paid to quality control inspectors.
- Advertising expenses.
- Costs of direct materials.
- Employee fidelity insurance.
- Cost of electric power.
- Sales taxes.

8.3 Intelligent Inc. specializes in the manufacture of components for personal computers. Management has determined that the daily cumulative cost of producing these components (in dollars) is given by:

$$C(x) = 0.0001x^3 - 0.08x^2 + 40x + 5000$$

where x is the number of components produced.

(a) What is the actual cost incurred for manufacturing the 301st component?

(b) What is the marginal cost of the 200th component? What can you say about the marginal cost as a function of x (increasing or decreasing)?

(c) What is the average unit cost if you produce just 300 units?

8.4 Harris Company manufactures a single product. Costs for the year 2015 for output levels of 1,000 and 2,000 units are given in Table P8.4.

At each level of output, compute the following:

(a) Total manufacturing costs.
(b) Manufacturing costs per unit.
(c) Total variable costs.
(d) Total variable costs per unit.
(e) Total costs that have to be recovered if the firm is to make a profit.

TABLE P8.4

Units Produced	1,000	2,000
Direct labor	$30,000	$60,000
Direct materials	20,000	40,000
Overhead:		
Variable portion	12,000	24,000
Fixed portion	36,000	36,000
Selling and administrative costs:		
Variable portion	5,000	10,000
Fixed portion	22,000	22,000

Cost–Volume–Profit Relationships

8.5 Hyundai Automobile U.S.A. is sponsoring a charity golf tournament to raise monies for a children's hospital in Birmingham. There are two possible venues:

- **RTJ Golf Resort at Prattville,** which has a fixed rental cost of $6,810. The hotel provides for meals and waiters and waitresses to serve drinks and finger foods at $60 per person. The green fees and cart for each person will be $35.
- **Wynlakes Golf & Country Club,** which has a fixed rental cost of $2,400 plus a charge of $100 per person for its own catering of meals and serving of drinks and finger foods. The green fees and cart for each person will be $40.

Hyundai budgets $5,000 in costs for administration and marketing for the event. The band will cost a fixed amount of $3,000. Tickets to this local community event will be $300 per person. All proceeds from the event will be donated to the children's hospital.

(a) Compute the break-even point for each venue in terms of tickets sold.

(b) At what level of tickets sold will the two venues have the same operating income?

8.6 The accompanying chart (Figure P8.6) shows the expected monthly profit or loss of Cypress Manufacturing Company within the range of its monthly practical operating capacity. Using the information provided in the chart, answer the following questions.

(a) What is the company's break-even sales volume?

(b) What is the company's marginal contribution rate?

(c) What effect would a 5% decrease in selling price have on the break-even point in part (a)?

(d) What effect would a 10% increase in fixed costs have on the marginal contribution rate in part (b)?

(e) What effect would a 6% increase in variable costs have on the break-even point in part (a)?

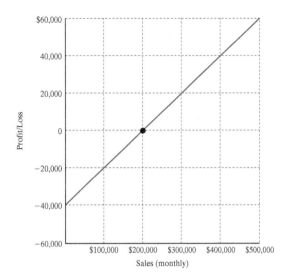

Figure P8.6

(f) If the chart also reflects $20,000 monthly depreciation expenses, compute the sales volume at the break-even point for cash costs.

8.7 The J&J Company has three product lines of belts—A, B, and C—having contribution margins of $3, $2, and $1, respectively. The president foresees sales up to 200,000 units in the coming period, consisting of 20,000 units of A, 100,000 units of B, and 80,000 units of C. The company's fixed costs for the period are $255,000.

(a) What is the company's break-even point in units, assuming that the given sales mix is maintained?

(b) If the mix is maintained, what is the total contribution margin when 200,000 units are sold? What is the operating income?

(c) What would the operating income become if 20,000 units of A, 80,000 units of B, and 100,000 units of C were sold? What is the new break-even point in units if these relationships persist in the next period?

8.8 The accompanying graph (Figure P8.8) is a cost–volume–profit graph. In the graph, identify the following line segments or points.

(a) Line EF represents _____.

(b) The horizontal axis AB represents _____, and the vertical axis AD represents _____.

(c) Point V represents _____.

(d) The distance CB divided by the distance AB is _____.

(e) The point V_b is a break-even _____.

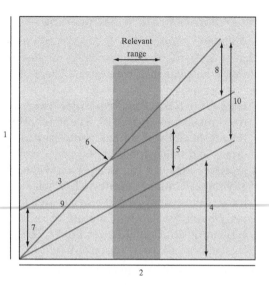

Figure P8.9

(a) Identify the numbered components in the accompanying CVP graph.

No.	Description	No.	Description
1		6	
2		7	
3		8	
4		9	
5		10	

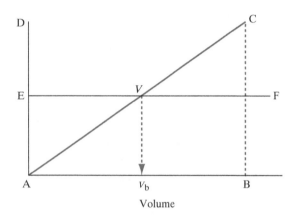

Figure P8.8

8.9 A cost–volume–profit (CVP) graph (Figure P8.9) is a useful technique for showing relationships between costs, volume, and profits in an organization.

(b) Using the typical CVP relationship shown, fill in the missing amounts in each of the following four situations (each case is independent of the others):

Case	Units Sold	Sales	Variable Expenses	Contribution Margin per Unit	Fixed Expenses	Net Income (Loss)
A	9,000	$270,000	$162,000	—	$90,000	—
B	—	$350,000	—	$15	$170,000	$40,000
C	20,000	—	$280,000	$6	—	$35,000
D	5,000	$100,000	—	—	$82,000	($12,000)

Cost Concepts Relevant to Decision Making

8.10 The procurement manager from a large merchandising firm has called your vice president for production to get a price quote for an additional 250 units of a given product. The vice president has asked you to prepare a cost estimate. The number of hours required to produce a unit is 4. The average labor rate is $16 per hour. The materials cost $22 per unit. Overhead for an additional 250 units is estimated at 40% of the direct labor cost. If the company wants to have a 25% profit margin, what should be the total price to quote?

8.11 The Morton Company produces and sells two products: A and B. Financial data related to producing these two products are summarized as in the table below.

(a) If these products are sold in the ratio of four A's for every three B's, what is the break-even point?

(b) If the product mix has changed to five A's to five B's, what would happen to the break-even point?

(c) In order to maximize the profit, which product mix should be pushed?

(d) If both products must go through the same manufacturing machine and there are only 30,000 machine hours available per period, which product should be pushed? Assume that product A requires 0.5 hour per unit and B requires 0.25 hour per unit.

	Product A	**Product B**
Selling price	$10.00	$12.00
Variable costs	$5.00	$10.00
Fixed costs	$2,000	$600

8.12 The Holtz Company manufactures a variety of electronic printed circuit boards (PCBs) that go into cellular phones. The company has just received an offer from an outside supplier to provide the electrical soldering for Holtz's Motorola product line (Z-7 PCB, slimline). The quoted price is $4.80 per unit. Holtz is interested in this offer, since its own soldering operation of the PCB is at its peak capacity.

- **Outsourcing option:** The company estimates that if the supplier's offer were accepted, the direct labor and variable overhead costs of the Z-7 slimline would be reduced by 15%, and the direct material cost would be reduced by 20%.

- **In-house production option:** Under the present operations, Holtz manufactures all of its own PCBs from start to finish. The Z-7 slimlines are sold through Motorola at $20 per unit. Fixed overhead charges to the Z-7 slimline total $20,000 each year. The further breakdown of producing one unit is

Direct materials	$7.50
Direct labor	$5.00
Manufacturing overhead	$4.00
Total cost	$16.50

The manufacturing overhead of $4.00 per unit includes both variable ($3.80) and fixed ($0.20) manufacturing overhead based on a production of 100,000 units each year.

(a) Should Holtz Company accept the outside supplier's offer?

(b) What is the maximum unit price that Holtz Company should be willing to pay the outside supplier?

8.13 Given the following information, answer the questions:

- The ratio of variable cost per unit divided by selling price per unit equals 0.3.

- Fixed costs amount to $60,000.

(a) Draw the cost-volume-profit chart.

(b) What is the break-even point?

(c) What effect would an 8% decrease in selling price have on the break-even point from part (b)?

Short Case Studies

ST8.1 The Hamilton Flour Company is currently operating its mill six days per week, 24 hours per day, on three shifts. At current prices, the company could easily obtain a sufficient volume of sales to take the entire output of a seventh day of operation each week. The mill's practical capacity is 6,000 hundredweight of flour per day. Note that:

- Flour sells for $20.24 per hundredweight (cwt) and the price of wheat is $8.35 per bushel. About 2.35 bushels of wheat are required per cwt of flour.

Fixed costs now average $8,520 per day (or $1.42 per cwt). The average variable cost of mill operation, almost entirely wages, is $0.68 per cwt.

- With Sunday operation, wages would be doubled for Sunday work, which would bring the variable cost of Sunday operation to $1.36 per cwt. If the mill were to operate on Sunday, the daily fixed cost would increase by $850, such that the total fixed cost for a 7-day operation would be $51,970.

  (a) Using the information provided, compute the break-even volumes for six-day and seven-day operation.

  (b) What are the marginal contribution rates for six-day and seven-day operation?

  (c) Compute the average total cost per cwt for six-day operation and the net profit margin per cwt before taxes.

  (d) Would it be economical for the mill to operate on Sundays? (Justify your answer numerically.)

ST8.2 Florida Citrus Inc. (FCI) produces and sells a highly popular sports drink in the North American market. In the current year, it will sell 20 million barrels of drinks in the Asian market. For many years, it has sold in the Asian market through a Tokyo-based importer. The contract with the importer is up for renewal, and FCI decides to reconsider its Asian strategy. After much analysis, it decides that three alternatives warrant further consideration.

- **Option 1. Stay with the importer:** Sell through the current importer who manages all the marketing and distribution of FCI's sports drink in the Asian market. FCI receives a net payment of $5 per barrel (net after transportation costs and import duties) sold in the Asian market.

- **Option 2. Move to production licensing:** License production of FCI drinks to a Japanese beverage firm who also will manage its marketing and distribution. This firm will charge FCI a fixed fee of $5 million each year to cover its costs of maintaining the quality of FCI products. It will pay FCI $10 per barrel of FCI products it sells in Asia.

- **Option 3. Turn to self-production:** Purchase a fully operational beverage plant from a Japanese company with excess capacity. The annual fixed

costs of operating the plant are $30 million, and the variable costs are $60 per barrel. FCI will sell to independent wholesalers in Asia at $100 a barrel.

  (a) Compute the break-even point for each of the three options that FCI is considering.

  (b) At which unit sales level (barrels) of FCI products in Asia would FCI report the same operating income under options 2 and 3?

  (c) If FCI expects the sports drink sales volume in Asia to be somewhere between 800,000 barrels and 950,000 barrels, which option would you recommend?

ST8.3 The Ace Bicycle Company[6] produces bicycles. This year's expected production is 10,000 units. Currently Ace also makes the chains for its bicycles. Ace's accountant reports the following costs for making the 10,000 chains in Table ST8.3.

Ace has received an offer from an outside vendor to supply any number of chains Ace requires, at $8.20 per chain. The following additional information is available.

- Inspection, setup, and material handling costs vary with the number of batches in which the chains are produced. Ace produces chains in batch sizes of 1,000 units. Ace estimates that it will produce the 10,000 units in 10 batches.

TABLE ST8.3

Description	Per Unit Costs	Costs for 10,000 Units
Direct materials	$4.00	$40,000
Direct manufacturing labor	2.00	20,000
Variable manufacturing overhead (power & utilities)	1.50	15,000
Inspection, setup, material handling		2,000
Machine lease		3,000
Allocated fixed plant administration, taxes, and insurance		30,000
Total		**$110,000**

[6] *Source*: HORNGREN, CHARLES T.; FOSTER, GEORGE; DATAR, SRIKANT M.; RAJAN, MADHAV V.; ITTNER, CHRIS M., COST ACCOUNTING: A MANAGERIAL EMPHASIS, 13th Ed., © 2009, pp.423–424. Reprinted and Electronically reproduced by permission of Pearson Education, Inc., New York, NY.

- The costs for the machine lease are the payments Ace makes for renting the equipment used in making the chains. If Ace buys all its chains from the outside vendor, it does not need this machine.

(a) Assume that if Ace purchases the chains from the outside supplier, the facility where the chains are currently made will remain idle. Should Ace accept the outside supplier's offer at the anticipated production (and sales) volume of 10,000 units?

(b) For this question, assume that if the chains are purchased outside, the facilities where the chains are currently made will be used to upgrade the bicycles by adding mud flaps and reflector bars. As a consequence, the selling price on bicycles will be raised by $20. The variable per unit cost of the upgrade would be $18, and additional totaling costs of $16,000 would be incurred. Should Ace make or buy the chains, assuming that 10,000 units are produced (and sold)?

(c) The sales manager at Ace is concerned that the estimate of 10,000 units may be high and believes that only 6,200 units will be sold. With production cut back, which opens up more work space, now there is room to add the mud flaps and reflectors whether Ace goes outside for the chains or makes them in-house. At this lower output, Ace will produce the chains in eight batches of 775 units each. Should Ace purchase the chains from the outside vendor?

ST8.4 J.K Mfg. Company produces golf carts. The company expects to sell 10,000 units yearly. Currently the company produces the engines to power the golf carts. To produce the 10,000 engines, J.K needs to budget the following amount for the current fiscal year:

Description	Per unit Costs	Costs for 10,000 Units
Direct materials	$20.00	$200,000
Direct manufacturing labor	10.00	100,000
Variable manufacturing overhead (power & utilities)	6.00	60,000
Inspection, setup, material handling		10,000
Testing machine lease		15,000

Description	Per unit Costs	Costs for 10,000 Units
Allocated fixed plant administration, taxes, and insurance		150,000
Total		$535,000

J.K has received an offer from an outside vendor to supply any number of engines to go with the golf carts. The outside vendor quotes $40 per engine. The following additional information is available:

- Inspection, setup, and material handling costs vary with the number of batches in which the engines are produced. J.K produces engines in batch sizes of 1,000 units. J.K estimates that it will produce the 10,000 units in 10 batches.

- J.K rents a special testing machine for $15,000 a year to produce the engines in-house. If J.K buys all its engines from the outside vendor, it does not need this testing machine.

(a) Assume that if J.K purchases the engines from the outside supplier, the facility where the engines are currently made will remain idle. Should J.K accept the outside supplier's offer at the anticipated production (and sales) volume of 10,000 units?

(b) If the engines are purchased outside, the facilities where the engines are currently made will be used to upgrade the golf carts by adding GPS monitors to display the accurate distance information to the players. As a consequence, the selling price on golf carts will be raised by $100. The variable per unit cost of the upgrade would be $90, and additional tooling costs of $80,000 would be incurred. Should J.K make or buy the engines, assuming that 10,000 units are produced (and sold)?

(c) The sales manager at J.K is concerned that the estimate of 10,000 units may be too optimistic and believes that only 6,000 units would be sold. At this lower estimate, J.K will produce the engines in 8 batches of 750 units each. Then, the inspection, setup, and material handling costs will be reduced to $6,000. Should J.K purchase the engines from the outside vendor?

CHAPTER 9

Depreciation and Corporate Taxes

As bag fees rise, are airlines improving baggage handling?[1]
Most of us have experienced delayed or even lost baggage during air travel. Most airlines try to reduce the number of delayed and lost bags and charge

hefty baggage fees to do so, and this also improves their profitability. About 70% of the 100,000 bags Delta handles in Atlanta each day connect between flights,[2] and connecting baggage most often goes astray, airlines say. Delta used to have a conveyor-belt system that moved luggage under its passenger check-in terminal. Once sorted, luggage had to be carted to five different terminals, a process that contributed to missed connections. Carriers have invested in better baggage handling technology and equipment; Delta Airlines has invested $100 million to fix its baggage operation in Atlanta since

Source: Lorelyn Medina/Shutterstock

2008, including hand-held scanners that beep when bags are being loaded onto the wrong flight. Delta also launched a new smartphone app this year that lets passengers track their bags at every stage of the journey. The rate at which airlines lose luggage has improved by a whopping 29% compared with the old system. It costs a carrier an average of $100 to reunite each delayed bag with its owner, and the airlines must also refund the fees they charge for checked bags.

Now ask yourself, how does the cost of this baggage handling system ($100 million) affect the financial position of the airline? In the long run, the

---

[1] "As Bag Fees Rise, Are Airline Improving Baggage Handling?" Charisse Jones, *USA Today*.
[2] "Why Do 26 Million Checked Bags Go Missing Every Year?" *Conde Nast Traveler*, July 2012.

system promises to create greater cost savings for the airline by enhancing productivity, improving customer goodwill, and cutting down the cost of delayed or lost bags. In the short run, however, the high initial cost of the baggage handling system will adversely affect the organization's "bottom line," because that cost is only gradually rewarded by the benefits the improved system offers.

Another consideration should come to mind as well: This state-of-the-art baggage handling system must inevitably wear out over time, and even if its productive service extends over many years, the cost of maintaining its high level of functioning will increase as the individual pieces of hardware wear out and need to be replaced. Of even greater concern is the question of how long this baggage handling system will be the "state of the art." When will the competitive advantage the airline has just acquired become a competitive disadvantage through obsolescence?

One of the facts of life that organizations must deal with and account for is that fixed assets lose their value—even as they continue to function and contribute to the engineering projects that use them. This loss of value, called **depreciation**, can involve deterioration and obsolescence.

The main function of **depreciation accounting** is to account for the cost of fixed assets in a pattern that matches their decline in value over time. The cost of the robot we have just described, for example, will be allocated over several years in the airline's financial statements, so that the pattern of the airline's costs roughly matches its pattern of service.

457

In this way, as we will see, depreciation accounting enables the firm to stabilize the statements about its financial position that it distributes to stockholders and the outside world.

On a project level, engineers must be able to assess how the practice of depreciating fixed assets influences the investment value of a given project. To do this, the engineers need to estimate the allocation of capital costs over the life of the project, which requires an understanding of the conventions and techniques that accountants use to depreciate assets. In this chapter, we will review the conventions and techniques of asset depreciation and income taxes.

We begin by discussing the nature and significance of depreciation, distinguishing its general economic definition from the related, but different, accounting view of depreciation. We then focus our attention almost exclusively on the rules and laws that govern asset depreciation and the methods that accountants use to allocate depreciation expenses. Knowledge of these rules will prepare you for applying them in assessing the depreciation of assets acquired in engineering projects. Then we turn our attention to the subject of depletion, which utilizes similar ideas, but specialized techniques, to allocate the cost of the depletion of natural-resource assets.

Once we understand the effect of depreciation at the project level, we need to address the effect of corporate taxes on project cash flows. There are many forms of government taxation, including sales taxes, property taxes, user taxes, and state and federal income taxes. In this chapter, we will focus on federal income taxes. When you are operating a business, any profits or losses you incur are subject to income-tax consequences. Therefore, we cannot ignore the impact of income taxes in project evaluation. The chapter will give you a good idea of how the U.S. tax system operates and how federal income taxes affect economic analysis. Although tax law is subject to frequent changes, the analytical procedures presented here provide a basis for tax analysis that can be adapted to reflect future changes in tax law. Thus, while we present many examples based on current tax rates, in a larger context, we present a general approach to the analysis of *any* tax law.

CHAPTER LEARNING OBJECTIVES

After completing this chapter, you should understand the following concepts:

- How to account for the loss of value of an asset in business.
- The meaning and types of depreciation.
- The difference between book depreciation and tax depreciation.
- The effects of depreciation on net income calculation.
- The general scheme of U.S. corporate taxes.
- How to determine ordinary gains and capital gains.
- How to determine the appropriate tax rate to use in project analysis.
- The relationship between net income and net cash flow.

9.1 Asset Depreciation

Fixed assets, such as equipment and real estate, are economic resources that are acquired to provide future cash flows. Generally, **depreciation** can be defined as a gradual decrease in the utility of fixed assets with use and time. While this general definition does not capture the subtleties inherent in a more specific definition of depreciation, it does provide

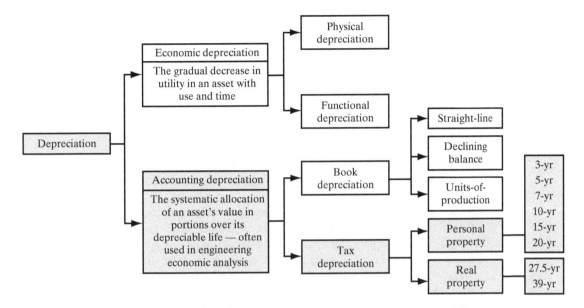

Figure 9.1 Classification of types of depreciation.

us with a starting point for examining the variety of underlying ideas and practices that are discussed in this chapter. Figure 9.1 will serve as a road map for understanding the different types of depreciation that we will explore here.

9.1.1 Economic Depreciation

This chapter is concerned primarily with accounting depreciation, which is the form of depreciation that provides an organization with the information it uses to assess its financial position. It also would be useful, however, to discuss briefly the economic ideas upon which accounting depreciation is based. In the course of the discussion, we will develop a precise definition of economic depreciation that will help us distinguish between various concepts of depreciation. We can classify economic depreciation into the categories of physical or functional depreciation.

Physical depreciation can be defined as a reduction in an asset's capacity to perform its intended service due to physical impairment. Physical depreciation can occur in any fixed asset in the form of (1) deterioration from interaction with the environment, including such agents as corrosion, rotting, and other chemical changes, and (2) wear and tear from use. Physical depreciation leads to a decline in performance and high maintenance costs.

Functional depreciation occurs as a result of changes in the organization or in technology that decrease or eliminate the need for an asset. Examples of functional depreciation include obsolescence attributed to advances in technology, a declining need for the services performed by an asset, and the inability to meet increased quantity or quality demands.

If you have ever owned a car, you are probably familiar with the term *depreciation* as it is used to describe the decreasing value of your vehicle. Because a car's reliability and appearance usually decline with age, the vehicle is worth less with each passing year. You can calculate the economic depreciation accumulated for your car by subtracting the current market value, or "blue book" value, of the car from the price you originally paid for it. We can define **economic depreciation** as

$$\text{Economic depreciation} = \text{Purchase price} - \text{Market value}$$

Physical and functional depreciation are categories of economic depreciation. The measurement of economic depreciation does not require that an asset be sold: The market value of the asset can be closely estimated without actually testing it in the marketplace. The need to have a precise scheme for recording the ongoing decline in the value of an asset as a part of the accounting process leads us to an exploration of how organizations account for depreciation.

9.1.2 Accounting Depreciation

The acquisition of fixed assets is an important activity for a business organization, whether the organization is starting up or acquiring new assets to remain competitive. Like other disbursements, the cost of these fixed assets must be recorded as expenses on a firm's balance sheet and income statement. However, unlike costs such as maintenance, material, and labor, the costs of fixed assets are not treated simply as expenses to be accounted for in the year that they are acquired. Rather, these assets are **capitalized**; that is, their costs are distributed by subtracting them as expenses from gross income—one part at a time over a number of periods. The systematic allocation of the initial cost of an asset in parts over a time, known as the asset's depreciable life, is what we mean by **accounting depreciation**. Because accounting depreciation is the standard of the business world, we sometimes refer to it more generally as **asset depreciation**.

Accounting depreciation is based on the **matching concept**: A fraction of the cost of the asset is chargeable as an expense in each of the accounting periods in which the asset provides service to the firm, and each charge is meant to be a percentage of the whole cost that "matches" the percentage of the value utilized in the given period. The matching concept suggests that the accounting depreciation allowance generally reflects, at least to some extent, the actual economic depreciation of the asset. *In engineering economic analysis, we use the concept of accounting depreciation exclusively.* This is because accounting depreciation provides a basis for determining the income taxes associated with any project undertaken.

9.2 Factors Inherent in Asset Depreciation

The process of depreciating an asset requires that we make several preliminary determinations: (1) What is the cost of the asset?; (2) What is the asset's value at the end of its useful life?; (3) What is the depreciable life of the asset?; and finally, (4) What method of depreciation do we choose? In this section, we will examine each of these questions.

9.2.1 Depreciable Property

As a starting point, it is important to recognize what constitutes a **depreciable asset**—that is, a property for which a firm may take depreciation deductions against income. For the purposes of U.S. tax law, any depreciable property has the following characteristics.

1. It must be used in business or must be held for the production of income.
2. It must have a definite service life, and that life must be longer than one year.
3. It must be something that wears out, decays, gets used up, becomes obsolete, or loses value from natural causes.

Depreciable property includes buildings, machinery, equipment, and vehicles. Inventories are not depreciable property, because they are held primarily for sale to customers in the ordinary course of business. If an asset has no definite service life, the asset cannot be depreciated. For example, *you can never depreciate land*.[3]

As a side note, we mention the fact that, while we have been focusing on depreciation within firms, individuals may also depreciate assets as long as they meet the conditions we have listed. For example, an individual may depreciate an automobile if the vehicle is used exclusively for business purposes.

9.2.2 Cost Basis

The **cost basis** of an asset represents the total cost that is claimed as an expense over the asset's life (i.e., the sum of the annual depreciation expenses). The cost basis generally includes the actual cost of the asset and all other incidental expenses, such as freight, site preparation, and installation. This total cost, rather than the cost of the asset only, must be the depreciation basis charged as an expense over the asset's life.

Besides being used in figuring depreciation deductions, an asset's cost basis is used in calculating the gain or loss to the firm if the asset is ever sold or salvaged. (We will discuss these topics in Section 9.8.)

EXAMPLE 9.1   Cost Basis without Trade-in Allowance

Lanier Corporation purchased an automatic hole-punching machine priced at $62,500. The vendor's invoice included a sales tax of $3,263. Lanier also paid the inbound transportation charges of $725 on the new machine, as well as the labor cost of $2,150 to install the machine in the factory. In addition, Lanier had to prepare the site at a cost of $3,500 before installation. Determine the cost basis for the new machine for depreciation purposes.

SOLUTION

Given: Invoice price = $62,500, freight = $725, installation cost = $2,150, and site preparation = $3,500.
Find: The cost basis.

The cost of machine that is applicable for depreciation is computed as follows.

Cost of new hole-punching machine	$62,500
Freight	725
Installation labor	2,150
Site preparation	3,500
Cost of machine (cost basis)	**$68,875**

---

[3] This also means that you cannot depreciate the cost of clearing, grading, planting, and landscaping. All four expenses are considered part of the cost of the land.

COMMENTS: Why do we include all the incidental charges relating to the acquisition of a machine in its cost? Why not treat these incidental charges as expenses incurred during the period in which the machine is acquired? The matching of costs and revenue is the basic accounting principle. Consequently, the total cost of the machine should be viewed as an asset and allocated against the future revenue that the machine will generate. All costs incurred in acquiring the machine are costs of the services to be received from using the machine.

If the asset is purchased by trading in a similar asset, the difference between the book value (the cost basis minus the total accumulated depreciation) and the trade-in allowance must be considered in determining the cost basis for the new asset. If the trade-in allowance exceeds the book value, the difference (known as **unrecognized gain**) needs to be subtracted from the cost basis of the new asset. If the opposite is true (**unrecognized loss**), the difference should be added to the cost basis for the new asset.

EXAMPLE 9.2 Cost Basis with Trade-in Allowance

In Example 9.1, suppose Lanier purchased the hole-punching press by trading in a similar machine and paying cash for the remainder. The trade-in allowance is $5,000, and the book value of the hole-punching machine that was traded in is $4,000. Determine the cost basis for this hole-punching press.

SOLUTION

Given: Accounting data from Example 9.1. Trade allowance = $5,000.
Find: The revised cost basis.

Old hole-punching machine (**book value**)	$4,000
Less: Trade-in allowance	5,000
Unrecognized gains	**$1,000**
Cost of new hole-punching machine	$62,500
Less: Unrecognized gains	(1,000)
Freight	725
Installation labor	2,150
Site preparation	3,500
Cost of machine (cost basis)	**$67,875**

COMMENTS: If the trade-in allowance were $3,000 instead of $5,000, you will have an **unrecognized loss** in the amount of $1,000. Then the cost basis of the new machine would be $69,875.

9.2.3 Useful Life and Salvage Value

Over how many periods will an asset be useful to a company? What do published statutes allow you to choose as the life of an asset? These are the central questions to be answered in determining an asset's depreciable life (i.e., the number of years over which the asset is to be depreciated).

Historically, depreciation included choosing a depreciable life that was based on the service life of an asset. Determining the service life of the asset, however, was often very difficult, and the uncertainty of these estimates often led to disputes between taxpayers and the Internal Revenue Service (IRS). To alleviate the problems, the IRS published guidelines on lives for categories of assets. The guidelines, known as **asset depreciation ranges (ADRs)**, specified a range of lives for classes of assets based on historical data, and taxpayers were free to choose a depreciable life within the specified range for a given asset. An example of ADRs for some assets is given in Table 9.1.

The **salvage value** is an asset's estimated value at the end of its life—the amount eventually recovered through sale, trade-in, or salvage. The eventual salvage value of an

TABLE 9.1 Asset Depreciation: Some Selected Asset Guideline Classes

	Asset Depreciation Range (Years)		
Assets Used	Lower Limit	Midpoint Life	Upper Limit
Office furniture, fixtures, and equipment	8	10	12
Information systems (computers)	5	6	7
Airplanes	5	6	7
Automobiles, taxis	2.5	3	3.5
Buses	7	9	11
Light trucks	3	4	5
Heavy trucks (concrete ready-mixer)	5	6	7
Railroad cars and locomotives	12	15	18
Tractor units	5	6	7
Vessels, barges, tugs, and water transportation systems	14.5	18	21.5
Industrial steam and electrical generation and/or distribution systems	17.5	22	26.5
Manufacturer of electrical and nonelectrical machinery	8	10	12
Manufacturer of electronic components, products, and systems	5	6	7
Manufacturer of motor vehicles	9.5	12	14.5
Telephone distribution plant	28	35	42

Source: IRS Publication 534, *Depreciation*, Washington, D.C.: U.S. Government Printing Office, 2009.

asset must be estimated when the depreciation schedule for the asset is established. If this estimate subsequently proves to be inaccurate, then an adjustment must be made. We will discuss these specific issues in Section 9.6.

9.2.4 Depreciation Methods: Book and Tax Depreciation

Most firms calculate depreciation in two different ways, depending on whether the calculation is intended for

1. Financial reports (the **book depreciation method**), such as for the balance sheet or income statement.
2. The Internal Revenue Service (IRS), for the purpose of determining taxes (the **tax depreciation method**).

In the United States, this distinction is totally legitimate under IRS regulations, as it is in many other countries. Calculating depreciation differently for financial reports and for tax purposes allows the following benefits.

- It enables firms to report depreciation to stockholders and other significant outsiders on the basis of the matching concept. Therefore, the actual loss in value of the assets is generally reflected.
- It allows firms to benefit from the tax advantages of depreciating assets more quickly than would be possible with the matching concept. In many cases, tax depreciation allows firms to defer paying income taxes. *This does not mean that they pay less tax overall, because the total depreciation expense accounted for over time is the same in either case.* However, because tax depreciation methods usually permit a higher depreciation in earlier years than do book depreciation methods, the tax benefit of depreciation is enjoyed earlier, and firms generally pay lower taxes in the initial years of an investment project. Typically, this leads to a better cash position in early years, and the added cash leads to greater future wealth because of the time value of the funds.

As we proceed through the chapter, we will make increasing use of the distinction between depreciation accounting for financial reporting and depreciation accounting for income tax calculation. Now that we have established the context for our interest in both tax and book depreciation, we can survey the different methods with an accurate perspective.

9.3 Book Depreciation Methods

Three different methods can be used to calculate the periodic depreciation allowances: (1) the straight-line method, (2) accelerated methods, and (3) the unit-of-production method. In engineering economic analysis, we are interested primarily in depreciation in the context of income tax computation. Nonetheless, a number of reasons make the study of book depreciation methods useful.

- First, tax depreciation methods are based largely on the same principles that are used in book depreciation methods.
- Second, firms continue to use book depreciation methods for financial reporting to stockholders and outside parties.

- Third, book depreciation methods are still used for state income tax purposes in many states, and even for federal income tax purposes for assets that were put into service before 1981.
- Finally, our discussion of depletion in Section 9.5 is based largely on one of these three book depreciation methods.

9.3.1 Straight-Line Method

The **straight-line (SL) method** of depreciation interprets a fixed asset as an asset that offers its services in a uniform fashion. The asset provides an equal amount of service in each year of its useful life.

The straight-line method charges (as an expense) an equal fraction of the net cost of the asset each year, as expressed by the relation

$$D_n = \frac{(I - S)}{N} \tag{9.1}$$

where

D_n = depreciation charge during year n

I = cost of the asset, including installation expenses

S = salvage value at the end of the asset's useful life

N = useful life

The book value of the asset at the end of n years is then defined as

Book value in a given year = Cost basis − total depreciation charges made to date

or

$$B_n = I - (D_1 + D_2 + D_3 + \ldots + D_n) \tag{9.2}$$

EXAMPLE 9.3 Straight-Line Depreciation

Consider the following data on an automobile:

Cost basis of the asset, I = $10,000

Useful life, N = 5 years

Estimated salvage value, S = $2,000

Use the straight-line depreciation method to compute the annual depreciation allowances and the resulting book values.

SOLUTION

Given: I = $10,000, S = $2,000, and N = 5 years.
Find: D_n and B_n for n = 1 to 5.

The straight-line depreciation rate is $\frac{1}{5}$, or 20%. Therefore, the annual depreciation charge is

$$D_n = (0.20)(\$10,000 - \$2,000) = \$1,600$$

The asset then would have the following book values during its useful life.

n	B_{n-1}	D_n	B_n
1	$10,000	1,600	$8,400
2	8,400	1,600	6,800
3	6,800	1,600	5,200
4	5,200	1,600	3,600
5	3,600	1,600	2,000

Here, B_{n-1} represents the book value before the depreciation charge for year n. This situation is illustrated in Table 9.2.

TABLE 9.2 An Excel Worksheet for Straight-Line Depreciation Method

	A	B	C	D	E	F	G	H	I	J
1	Depreciation Method			SL						
2	Cost Basis ($)			10,000						
3	Depreciable Life			5						
4	Salvage Value ($)			2,000						
5										
6	Period	Beginning	Depreciation	Ending						
7	(n)	Book Value	Amount	Book Value						
8										
9	1	$10,000	$1,600	$8,400						
10	2	$8,400	$1,600	$6,800						
11	3	$6,800	$1,600	$5,200						
12	4	$5,200	$1,600	$3,600						
13	5	$3,600	$1,600	$2,000						
14										

9.3.2 Declining Balance Method

The second concept of depreciation recognizes that the stream of services provided by a fixed asset may decrease over time; in other words, the stream may be greatest in the first year of an asset's service life and least in its last year. This pattern may occur because the mechanical efficiency of an asset tends to decline with age, because maintenance costs tend to increase with age, or because of the increasing likelihood that better equipment will become available and make the original asset obsolete. This kind of reasoning leads to a method that charges a larger fraction of the cost as an expense of the early years than of the later years. Any such method is called an **accelerated method**. The most widely used accelerated method is the **double-declining-balance method**.

The **declining-balance (DB) method** of calculating depreciation allocates a fixed fraction of the beginning book balance each year. The fraction, α, is obtained as

$$\alpha = \left(\frac{1}{N}\right)(\text{multiplier}) \tag{9.3}$$

The most commonly used multipliers in the United States are 1.5 (called 150% DB) and 2.0 (called 200% or **double-declining balance, DDB**). As N increases, α decreases, resulting in a situation in which depreciation is highest in the first year and then decreases over the asset's depreciable life.

The fractional factor can be utilized to determine depreciation charges for a given year, D_n as follows:

$$D_1 = \alpha I,$$
$$D_2 = \alpha(I - D_1) = \alpha I(1 - \alpha)$$
$$D_3 = \alpha(I - D_1 - D_2) = \alpha I(1 - \alpha)^2$$

Thus, for any year n, we have a depreciation charge

$$D_n = \alpha I(1 - \alpha)^{n-1} \tag{9.4}$$

We can also compute the **total DB depreciation (TDB)** at the end of n years as follows:

$$\begin{aligned}
\text{TDB} &= D_1 + D_2 + \ldots + D_n \\
&= \alpha I + \alpha I(1 - \alpha) + \alpha I(1 - \alpha)^2 + \ldots + \alpha I(1 - \alpha)^{n-1} \\
&= \alpha I[1 + (1 - \alpha) + (1 - \alpha)^2 + \ldots + (1 - \alpha)^{n-1}] \\
&= I[1 - (1 - \alpha)^n]
\end{aligned} \tag{9.5}$$

The book value B_n at the end of n years will be the cost I of the asset minus the total depreciation at the end of n years:

$$\begin{aligned}
B_n &= I - \text{TDB} \\
&= I - I[1 - (1 - \alpha)^n] \\
B_n &= I(1 - \alpha)^n
\end{aligned} \tag{9.6}$$

The salvage value (S) of the asset must be estimated at the outset of depreciation analysis. We will consider three cases in relation to the final book value (B_N): When $B_N = S$, we simply use Eq. (9.5) to calculate the yearly depreciation amount. When $B_N \neq S$, we would want to make adjustments in our depreciation methods. We will consider each case.

Case 1: $B_N = S$

In Example 9.4, the final book value (B_N) conveniently equals the estimated salvage value, which is a coincidence that is rather unusual in the real world.

EXAMPLE 9.4 Declining-Balance Depreciation ($B_N = S$)

Consider the following accounting information for a computer system:

$$\text{Cost basis of the asset, } I = \$10,000$$

$$\text{Useful life, } N = 5 \text{ years}$$

$$\text{Estimated salvage value, } S = \$778$$

Use the double-declining-depreciation method to compute the annual depreciation allowances and the resulting book values.

SOLUTION

Given: $I = \$10,000$, $S = \$778$, and $N = 5$ years.
Find: D_n and B_n for $n = 1$ to 5.

The declining balance is illustrated in terms of the book value of time in Figure 9.2. The book value at the beginning of the first year is $10,000, and the declining-balance rate (α) is $(\frac{1}{5})(2) = 40\%$. Then the depreciation deduction for the first year will be $4,000$($40\% \times \$10,000 = \$4,000$). To figure the depreciation deduction in the second year, we must first adjust the book value for the amount of depreciation we deducted in the first year. The first year's depreciation from the beginning book value is subtracted ($\$10,000 - \$4,000 = \$6,000$), and the resulting amount is multiplied

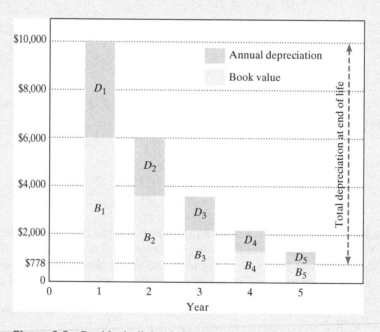

Figure 9.2 Double-declining-balance method (Example 9.4).

by the rate of depreciation ($6,000 × 40% = $2,400). By continuing the process, we obtain the following table.

n	B_{n-1}	D_n	B_n
1	10,000	4,000	6,000
2	6,000	2,400	3,600
3	3,600	1,440	2,160
4	2,160	864	1,296
5	1,296	518	778

Case 2: $B_N > S$

When $B_N > S$, we are faced with a situation in which we have not depreciated the entire cost of the asset and thus have not taken full advantage of depreciation's tax-deferring benefits. If you would prefer to reduce the book value of an asset to its salvage value as quickly as possible, it can be done by switching from DB to SL whenever SL depreciation results in larger depreciation charges, and therefore is a more rapid reduction in the book value of the asset. The switch from DB to SL depreciation can take place in any of the n years, the objective being to identify the optimal year to switch. The switching rule is as follows: If depreciation by DB in any year is less than (or equal to) what it would be by SL, we should switch to and remain with the SL method for the duration of the project's depreciable life. The straight-line depreciation in any year n is calculated by

$$D_n = \frac{\text{Book value at beginning of year } n - \text{salvage value}}{\text{Remaining useful life at beginning of year } n} \tag{9.7}$$

EXAMPLE 9.5 Declining Balance with Conversion to Straight-Line Depreciation ($B_N > S$)

Suppose the asset given in Example 9.4 has a zero salvage value instead of $778; that is,

$$\text{Cost basis of the asset, } I = \$10,000$$

$$\text{Useful life, } N = 5 \text{ years}$$

$$\text{Salvage value, } S = \$0$$

$$\alpha = (1/5)(2) = 40\%$$

Determine the optimal time to switch from DB to SL depreciation and the resulting depreciation schedule.

SOLUTION

Given: $I = \$10,000$, $S = 0$, $N = 5$ years, and $\alpha = 40\%$.
Find: Optimal conversion time, D_n and B_n for $n = 1$ to 5.

We will first proceed by computing the DDB depreciation for each year, as in Example 9.4.

Year n	D_n	B_n
1	$4,000	$6,000
2	2,400	3,600
3	1,440	2,160
4	864	1,296
5	518	778

Then using Eq. (9.7), we compute the SL depreciation for each year. We compare SL with DDB depreciation for each year and use the aforementioned decision rule for when to change.

If Switch to SL at Beginning of Year	SL Depreciation	DDB Depreciation	Switching Decision
2	($6,000 − 0)/4 = $1,500	<$2,400	Do not switch
3	(3,600 − 0)/3 = 1,200	<1,440	Do not switch
4	(2,160 − 0)/2 = 1,080	>864	Switch to SL

The optimal time (year 4) in this situation corresponds to n' in Figure 9.3(a). The resulting depreciation schedule is as follows:

Year n	DDB with Switch to SL	End-of-Year Book Value
1	$ 4,000	$6,000
2	2,400	3,600
3	1,440	2,160
4	1,080	1,080
5	<u>1,080</u>	0
	$10,000	

Table 9.3 illustrates how an Excel worksheet may be developed to calculate the depreciation amount based on the DDB with switch to SL.

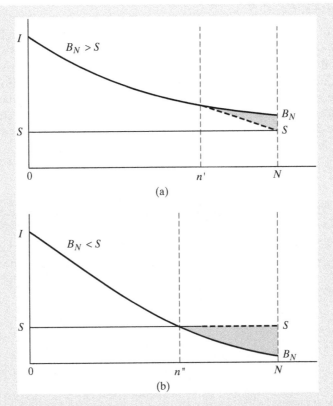

Figure 9.3   Adjustments to the declining-balance method:
(a) Switch from declining balance to straight line after n';
(b) no further depreciation allowances are available after n''
(Examples 9.5 and 9.6).

TABLE 9.3   **An Excel Worksheet for Declining Balance Depreciation When ($B_N > S$)**

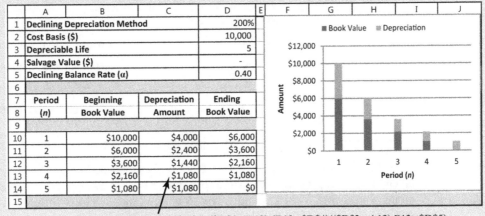

	A	B	C	D	E	F	G	H	I	J
1	Declining Depreciation Method			200%						
2	Cost Basis ($)			10,000						
3	Depreciable Life			5						
4	Salvage Value ($)			-						
5	Declining Balance Rate (α)			0.40						
6										
7	Period	Beginning	Depreciation	Ending						
8	(n)	Book Value	Amount	Book Value						
9										
10	1	$10,000	$4,000	$6,000						
11	2	$6,000	$2,400	$3,600						
12	3	$3,600	$1,440	$2,160						
13	4	$2,160	$1,080	$1,080						
14	5	$1,080	$1,080	$0						
15										

$$=\text{IF}(B13 \times \$D\$5 < (B13 - \$D\$4)/(\$D\$3 - A12), (B13 - \$D\$4)/(\$D\$3 - A12), B13 \times \$D\$5)$$

Case 3: $B_N < S$

With a relatively high salvage value, it is possible that the book value of the asset could decline below the estimated salvage value. When $B_N < S$, we must readjust our analysis, because tax law does not permit us to depreciate assets below their salvage value. To avoid deducting depreciation charges that would drop the book value below the salvage value, you simply stop depreciating the asset whenever you get down to $B_N = S$. In other words, if at any period the implied book value is lower than S, then the depreciation amounts are adjusted so that $B_N = S$. Example 9.6 illustrates this situation.

EXAMPLE 9.6 Declining Balance, $(B_N < S)$

Compute the double-declining-balance (DDB) depreciation schedule for the data used in Example 9.5, this time with the asset having a salvage value of $2,000. Then

$$\text{Cost basis of the asset, } I = \$10,000$$
$$\text{Useful life, } N = 5 \text{ years}$$
$$\text{Salvage value, } S = \$2,000$$
$$\alpha = (\tfrac{1}{5})(2) = 40\%$$

SOLUTION
Given: $I = \$10,000$, $S = \$2,000$, $N = 5$ years, and $\alpha = 40\%$.
Find: D_n and B_n for $n = 1$ to 5.

The given data result in the accompanying table.

End of Year	D_n	B_n
1	0.4($10,000) = $4,000	$10,000 − $4,000 = $6,000
2	0.4(6,000) = 2,400	6,000 − 2,400 = 3,600
3	0.4(3,600) = 1,440	3,600 − 1,440 = 2,160
4	0.4(2,160) = 864 > $\boxed{160}$	2,160 − 160 = 2,000
5	$\underline{0}$	2,000 − 0 = 2,000
	Total = $8,000	

Note that B_4 would be less than $S = \$2,000$ if the full deduction ($864) had been taken. Therefore, we adjust D_4 to $160, making $B_4 = \$2,000$. D_5 is zero, and B_5 remains at $2,000. Year 4 is equivalent to n'' in Figure 9.3(b). An Excel solution to this problem can be found in Table 9.4.

TABLE 9.4 **An Excel Worksheet to Calculate the Depreciation Amounts When ($B_N < S$)**

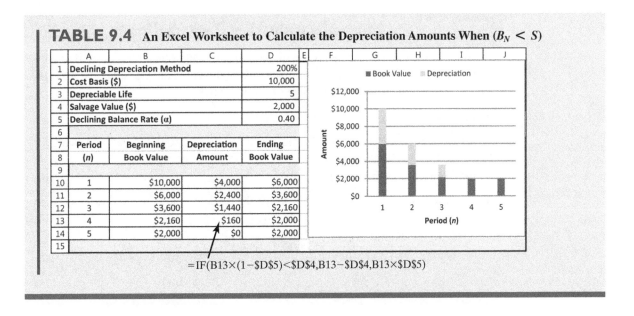

	A	B	C	D	E	F	G	H	I	J
1	Declining Depreciation Method			200%						
2	Cost Basis ($)			10,000						
3	Depreciable Life			5						
4	Salvage Value ($)			2,000						
5	Declining Balance Rate (α)			0.40						
6										
7	Period	Beginning	Depreciation	Ending						
8	(n)	Book Value	Amount	Book Value						
9										
10	1	$10,000	$4,000	$6,000						
11	2	$6,000	$2,400	$3,600						
12	3	$3,600	$1,440	$2,160						
13	4	$2,160	$160	$2,000						
14	5	$2,000	$0	$2,000						
15										

$$=\text{IF}(B13\times(1-\$D\$5)<\$D\$4,B13-\$D\$4,B13\times\$D\$5)$$

9.3.3 Units-of-Production Method

Straight-line depreciation can be defended only if the machine is used for exactly the same amount of time each year. What happens when a punch press machine runs 1,670 hours one year and 780 the next or when some of its output is shifted to a new machining center? This leads us to a consideration of another depreciation method that views the asset as consisting of a bundle of service units. Unlike the SL and accelerated methods, however, this one does not assume that the service units will be consumed in a time-phased pattern. Rather, the cost of each service unit is the net cost of the asset divided by the total number of such units. The depreciation charge for a period is then related to the number of service units consumed in that period. The result is the **units-of-production method**, according to which the depreciation in any year is given by

$$D_n = \frac{\text{Service units consumed during year } n}{\text{Total service units}}(I - S) \tag{9.8}$$

When the units-of-production method is used, depreciation charges are made proportional to the ratio of the actual output to the total expected output. Usually, this ratio is figured in machine hours. The advantages of using the units-of-production method include the fact that depreciation varies with production volume, so the method gives a more accurate picture of machine usage. A disadvantage of the method is that collecting data on machine usage is somewhat tedious, as are the accounting methods.

This method can be useful in depreciating equipment used to exploit natural resources if the resources will be depleted before the equipment wears out. It is not, however, considered a practical method for general use in depreciating industrial equipment.

EXAMPLE 9.7 Units-of-Production Depreciation

A truck for hauling coal has an estimated net cost of $55,000 and is expected to give service for 250,000 miles, resulting in a $5,000 salvage value. Compute the allowed depreciation amount for a truck usage of 30,000 miles.

SOLUTION

Given: $I = \$55,000$, $S = \$5,000$, total service units $= 250,000$ miles, and usage for this year $= 30,000$ miles.

Find: Depreciation amount in this year.

The depreciation expense in a year in which the truck traveled 30,000 miles would be

$$\frac{30,000 \text{ miles}}{250,000 \text{ miles}}(\$55,000 - \$5,000) = \left(\frac{3}{25}\right)(\$50,000)$$

$$= \$6,000$$

9.4 Tax Depreciation Methods

Prior to the Economic Recovery Act of 1981, taxpayers could choose among several methods when depreciating assets for tax purposes. The most widely used methods were the straight-line method and the declining-balance method. The subsequent imposition of the Accelerated Cost Recovery System (ACRS) and the Modified Accelerated Cost Recovery System (MACRS) superseded these methods for use in tax purposes.

9.4.1 MACRS Depreciation

From 1954 to 1981, congressional changes in tax law evolved fairly consistently toward simpler, more rapid depreciation methods. Prior to 1954, the straight-line method was required for tax purposes, but that year, accelerated methods, such as double-declining balance and sum-of-years digits, were permitted. In 1981, these conventional accelerated methods were replaced by the simpler ACRS. In 1986, Congress modified the ACRS and introduced the MACRS—sharply reducing depreciation allowances that were enacted in the Economic Recovery Tax Act of 1981. This section will present some of the primary features of MACRS tax depreciation.

MACRS Recovery Periods

Historically, for both tax and accounting purposes, an asset's depreciable life was determined by its estimated useful life; it was intended that the asset be fully depreciated at approximately the end of its useful life. The MACRS scheme, however, totally abandoned this practice, and simpler guidelines were established that created several classes of assets—each with a more or less arbitrary life called a **recovery period**. (Note: *Recovery periods do not necessarily bear any relationship to expected useful lives.*)

A major effect of the original ACRS method of 1981 was to shorten the depreciable lives of assets, thus giving businesses larger depreciation deductions that (in early years) resulted in lower taxes and increased cash flows available for reinvestment. As shown in Table 9.5, the MACRS method of 1986 reclassified certain assets on the basis of the midpoint of their lives under the ADR system. The MACRS scheme includes eight categories of assets, with lives of 3, 5, 7, 10, 15, 20, 27.5, and 39 years.

- Investments in some short-lived assets are depreciated over three years by using 200% DB and then switching to SL depreciation.
- Computers, automobiles, and light trucks are written off over five years by using 200% DB and then switching to SL depreciation.
- Most types of manufacturing equipment are depreciated over seven years, but some long-lived assets are written off over 10 years. Most equipment write-offs are calculated first by the 200% DB method and are followed by a switch to SL depreciation, which allows faster write-offs in the first few years after an investment is made.
- Sewage-treatment plants and telephone-distribution plants are written off over 15 years by using 150% DB and then switching to SL depreciation.
- Sewer pipes and certain other very long-lived equipment are written off over 20 years by using 150% DB and then switching to SL depreciation.
- Investments in residential rental property are written off in straight-line fashion over 27.5 years. Nonresidential real estate (commercial buildings), by contrast, is written off by the SL method over 39 years.

Under the MACRS, *the salvage value of property is always treated as zero.*

TABLE 9.5   MACRS Property Classifications

Recovery Period	ADR* Midpoint Class	Applicable Property
3 years	ADR ≤ 4	Special tools for the manufacture of plastic products, fabricated metal products, and motor vehicles
5 years	4 < ADR ≤ 10	Automobiles,[+] light trucks, high-tech equipment, equipment used for research and development, computerized telephone switching systems
7 years	10 < ADR ≤ 16	Manufacturing equipment, office furniture, fixtures
10 years	16 < ADR ≤ 20	Vessels, barges, tugs, railroad cars
15 years	20 < ADR ≤ 25	Wastewater plants, telephone-distribution plants, similar utility property
20 years	25 < ADR	Municipal sewers, electrical power plant
27.5 years		Residential rental property
39 years		Nonresidential real property, including elevators and escalators

*ADR = Asset depreciation range; guidelines are published by the IRS.

[+]Automobiles have a midpoint life of three years in the ADR guidelines, but are classified into a five-year property class.

9.4.2 MACRS Depreciation Rules

Under earlier depreciation methods, the rate at which the value of an asset declined was estimated and then used as the basis for tax depreciation. Thus, different assets were depreciated along different paths over time. The MACRS method, however, established prescribed depreciation rates, called **recovery allowance percentages**, for all assets within each class. These rates, as set forth in 1986 and 1993, are shown in Table 9.6. The yearly recovery (or depreciation expense) is determined by multiplying the asset's depreciation base by the applicable recovery allowance percentage.

TABLE 9.6 MACRS Depreciation Schedules for Personal Properties with Half-Year Convention, Declining-Balance Method

Year n	Depreciation Rate	3	5	7	10	15	20
		200%	200%	200%	200%	150%	150%
1		33.33	20.00	14.29	10.00	5.00	3.750
2		44.45	32.00	24.49	18.00	9.50	7.219
3		14.81*	19.20	17.49	14.40	8.55	6.677
4		7.41	11.52*	12.49	11.52	7.70	6.177
5			11.52	8.93*	9.22	6.93	5.713
6			5.76	8.92	7.37	6.23	5.285
7				8.93	6.55*	5.90*	4.888
8				4.46	6.55	5.90	4.522
9					6.56	5.91	4.462*
10					6.55	5.90	4.461
11					3.28	5.91	4.462
12						5.90	4.461
13						5.91	4.462
14						5.90	4.461
15						5.91	4.462
16						2.95	4.461
17							4.462
18							4.461
19							4.462
20							4.461
21							2.231

*Year to switch from declining balance to straight line.

Source: IRS Publication 534, *Depreciation*, Washington, D.C.: U.S. Government Printing Office, 2009.

EXAMPLE 9.8 MACRS Depreciation: Personal Property

A taxpayer wants to place in service a $10,000 asset that is assigned to the five-year class. Compute the MACRS percentages and the depreciation amounts for the asset.

SOLUTION

Given: Five-year asset, half-year convention, $\alpha = 40\%$, and $S = 0$.
Find: MACRS depreciation percentages D_n for $10,000 asset.

For this problem, we use the following equations:

$$\text{Straight-line rate} = \tfrac{1}{5} = 0.20$$

$$200\% \text{ declining balance rate} = 2(0.20) = 40\%$$

$$\text{Under MACRS, salvage } (S) = 0$$

Then beginning with the first taxable year and ending with the sixth year, MACRS deduction percentages are computed in Table 9.7.

TABLE 9.7 **Computational Process of Determining MACRS Percentages**

Year n	Calculation (%)	MACRS Percentage
1	$\tfrac{1}{2}$-year DDB	= 20.00%
	depreciation $= \left(\dfrac{1}{2}\right)(0.40)(100\%)$	
2	DDB depreciation $= (0.40)(100\% - 20\%)$	= 32.00%
	SL depreciation $= \left(\dfrac{1}{4.5}\right)(100\% - 20\%)$	= 17.78%
3	DDB depreciation $= (0.40)(100\% - 52\%)$	= 19.20%
	SL depreciation $= \left(\dfrac{1}{3.5}\right)(100\% - 52\%)$	= 13.71%
4	DDB depreciation $= (0.40)(100\% - 71.20\%)$	= 11.52%
	SL depreciation $= \left(\dfrac{1}{2.5}\right)(100\% - 71.20\%)$	= 11.52%
5	SL depreciation $= \left(\dfrac{1}{1.5}\right)(100\% - 82.72\%)$	= 11.52%
6	$\tfrac{1}{2}$-year SL depreciation $= \left(\dfrac{1}{2}\right)(11.52\%)$	= 5.76%

In year 2, we check to see what the SL depreciation would be. Since 4.5 years are left to depreciate, SL depreciation $= (1/4.5)(100\% - 20\%) = 17.78\%$. The DDB depreciation is greater than the SL depreciation, so DDB still applies. Note that SL depreciation $\geq$ DDB depreciation in year 4, so we switch to SL then. Figure 9.4 summarizes the difference between the conventional DDB and the DDB under MACRS.

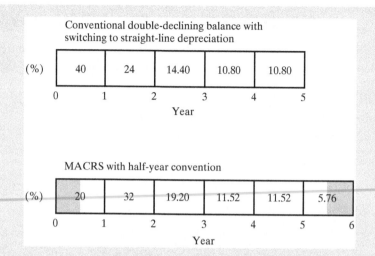

Figure 9.4 MACRS with a five-year recovery period (Example 9.8).

We can calculate the depreciation amounts from the percentages we just determined. In practice, the percentages are taken directly from Table 9.6, as supplied by the IRS. Or we can develop an Excel worksheet to automate the process as shown in Table 9.8.

Note that when an asset is disposed of before the end of a recovery period, only half of the normal depreciation is allowed. For example, if the $10,000 asset were to be disposed of in year 2, the MACRS deduction for that year would be $1,600.

TABLE 9.8 **An Excel Worksheet to Calculate the Depreciation Amounts by MACRS**

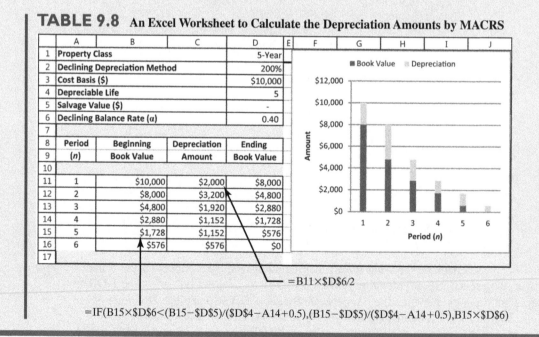

Half-Year Convention

The MACRS recovery percentages shown in Table 9.6 use the **half-year convention**; that is, it is assumed that all assets are placed in service at midyear and that they will have *zero* salvage value. As a result, only a half year of depreciation is allowed for the first year that property is placed in service. With half of one year's depreciation being taken in the first year, a full year's depreciation is allowed in each of the remaining years of the asset's recovery period, and the remaining half-year's depreciation is taken in the year following the end of the recovery period. A half year of depreciation is also allowed for the year in which property is disposed of (or is otherwise retired from service) any time before the end of the recovery period.

Switching from DB to the Straight-Line Method

The MACRS asset is depreciated initially by the DB method and then by SL depreciation. Consequently, the MACRS adopts the switching convention illustrated in Example 9.5. To demonstrate how the MACRS depreciation percentages were calculated by the IRS with the use of the half-year convention, consider Example 9.8.

Midmonth Convention for Real Property

Table 9.9 shows the useful lives and depreciation percentages for real property. In depreciating such property, the straight-line method and the midmonth convention are used. For example, a property placed in service in March would be allowed $9\frac{1}{2}$ months of depreciation for year 1. If it is disposed of before the end of the recovery period, the depreciation percentage must take into account the number of months the property was in service during the year of its disposal.

Table 9.9 is developed under these conditions.

EXAMPLE 9.9 MACRS for Real Property

On May 1st, Jack Costanza paid $100,000 for a residential rental property. This purchase price represents $80,000 for the cost of the building and $20,000 for the cost of the land. Three years and five months later, on October 1st, he sold the property for $130,000. Compute the MACRS depreciation for each of the four calendar years during which Jack owned the property.

SOLUTION

Given: Residential real property, cost basis = $80,000. The building was put into service on May 1st.

Find: The depreciation in each of four tax years the property was in service.

In this example, the midmonth convention assumes that the property is placed in service on May 15th, which gives 7.5 months of depreciation in the first year. Remembering that only the building (not the land) may be depreciated, we use the SL method to compute the depreciation over a 27.5-year recovery period:

Year	Calculation		D_n	Recovery Percentages
1	$\left(\dfrac{7.5}{12}\right)\dfrac{80,000-0}{27.5}$	=	$1,818	2.273%
2	$\dfrac{80,000-0}{27.5}$	=	$2,909	3.636%

TABLE 9.9 MACRS Percentages for Real Property

Year n	Month Property Is Placed in Service											
	1	2	3	4	5	6	7	8	9	10	11	12
(a) Residential Rental Property: Straight-Line over 27.5 Years with Midmonth Convention												
1	3.485%	3.182%	2.879%	2.576%	2.273%	1.970%	1.667%	1.364%	1.061%	0.758%	0.455%	0.152%
2–9	3.636	3.636	3.636	3.636	3.636	3.636	3.636	3.636	3.636	3.636	3.636	3.636
10	3.637	3.637	3.637	3.637	3.637	3.637	3.636	3.636	3.636	3.636	3.636	3.636
11	3.636	3.636	3.636	3.636	3.636	3.636	3.637	3.637	3.637	3.637	3.637	3.637
12	3.637	3.637	3.637	3.637	3.637	3.637	3.636	3.636	3.636	3.636	3.636	3.636
13	3.636	3.636	3.636	3.636	3.636	3.636	3.637	3.637	3.637	3.637	3.637	3.637
14–25	3.637	3.637	3.637	3.637	3.637	3.637	3.636	3.636	3.636	3.636	3.636	3.636
26	3.637	3.637	3.637	3.637	3.637	3.637	3.636	3.636	3.636	3.636	3.636	3.636
27	3.636	3.636	3.636	3.636	3.636	3.636	3.637	3.637	3.637	3.637	3.637	3.637
28	1.97	2.273	2.576	2.879	3.182	3.485	3.636	3.636	3.636	3.636	3.636	3.636
29							0.152	0.455	0.758	1.061	1.364	1.667
(b) Nonresidential Real Property: Straight-Line Over 39 Years with Midmonth Convention												
1	2.4573	2.2436	2.0299	1.8162	1.6026	1.3889	1.1752	0.9615	0.7479	0.5342	0.3205	0.1068
2–39	2.5641	2.5641	2.5641	2.5641	2.5641	2.5641	2.5641	2.5641	2.5641	2.5641	2.5641	2.5641
40	0.1068	0.3205	0.5342	0.7479	0.9615	1.1752	1.3889	1.6026	1.8162	2.0299	2.2436	2.457

Source: IRS Publication 534, *Depreciation*, Washington, D.C.: U.S. Government Printing Office, 2009.

Year	Calculation		D_n	Recovery Percentages
3	$\dfrac{80,000 - 0}{27.5}$	=	$2,909	3.636%
4	$\left(\dfrac{9.5}{12}\right)\dfrac{80,000 - 0}{27.5}$	=	$2,303	2.879%

Notice that the midmonth convention also applies to the year of disposal. Now compare the percentages with those in Table 9.9. As with personal property, calculations for real property generally use the precalculated percentages in the table. Table 9.10 shows how an Excel worksheet might be developed to generate a depreciation schedule for a real property.

TABLE 9.10 An Excel Worksheet to Generate a Depreciation Schedule for a Real Property

$=D8 \times (12\text{-}MONTH(D5)+0.5)/12$

$=D8 \times (MONTH(D6)-0.5)/12$

9.5 Depletion

If you own mineral property (distinguished from personal and real properties), such as oil, gas, a geothermal well, or standing timber, you may be able to claim a deduction as you deplete the resource. A capital investment in natural resources needs to be recovered as those resources are being removed and sold. The process of amortizing the cost of natural resources in the accounting periods is called depletion. The objective of depletion is the same as that of depreciation: to amortize the cost in a systematic manner over the asset's useful life.

There are two ways of figuring depletion: **cost depletion** and **percentage depletion**. These depletion methods are used for book as well as tax purposes. In most instances, depletion is calculated by both methods, and the larger value is taken as the depletion allowance for the year. For standing timber and for most oil and gas wells, only cost depletion is permissible.

9.5.1 Cost Depletion

The cost depletion method is based on the same concept as the units-of-production method. To determine the amount of cost depletion, the adjusted basis of the mineral property is divided by the total number of recoverable units in the deposit, and the resulting rate is multiplied by the number of units sold:

$$\text{Cost depletion} = \frac{\text{Adjusted basis of mineral property}}{\text{Total number of recoverable units}} \times \text{Number of units sold} \quad (9.9)$$

The **adjusted basis** represents all the depletion allowed (or the allowable cost on the property) less the salvage value of the property. Estimating the number of recoverable units in a natural deposit is largely an engineering problem.

9.5.2 Percentage Depletion

Percentage depletion is an alternative method of calculating the depletion allowance for certain mineral properties. For a given mineral property, the depletion allowance calculation is based on a prescribed percentage of the gross income from the property during the tax year. Notice the distinction between depreciation and depletion: **Depreciation** is the allocation of cost over a useful life, whereas **percentage depletion** is an annual allowance of a percentage of the gross income from the property.

EXAMPLE 9.10 Cost Depletion for a Timber Tract

Suppose you bought a timber tract for $200,000 and the land was worth $80,000. The basis for the timber is therefore $120,000. The tract has an estimated 1.5 million board feet (1.5 MBF) of standing timber. If you cut 0.5 MBF of timber, determine your depletion allowance.

SOLUTION

Given: Basis = $120,000, total recoverable volume = 1.5 MBF, and amount sold this year = 0.5 MBF.

Find: The depletion allowance this year.

Timber depletion may be figured only by the cost method. Percentage depletion does not apply to timber, nor does your depletion basis include any part of the cost of the land. Because depletion takes place when standing timber is cut, you may figure your depletion deduction only after the timber is cut and the quantity is first accurately measured. We have

$$\text{Depletion allowance per MBF} = \$120,000/1.5 \text{ MBF}$$

$$= \$80,000/\text{MBF}$$

$$\text{Depletion allowance for the year} = 0.5 \text{ MBF} \times \$80,000/\text{MBF}$$

$$= \$40,000$$

TABLE 9.11 Percentage Depletion Allowances for Mineral Properties

Deposits	Percentage Allowed
Oil and gas wells (only for certain domestic and gas production)	15
Sulfur and uranium; if from deposits in the United States, asbestos, lead, zinc, nickel, mica, and certain other ores and minerals	22
Gold, silver, copper, iron ore, and oil shale, if from deposits in the United States	15
Coal, lignite, and sodium chloride	10
Clay and shale (used in making sewer pipe or bricks)	7.5
Clay (used for roofing tile), gravel, sand, and stone	5
Most other minerals; includes carbon dioxide produced from a well and metallic ores	14

Since percentage depletion is computed on the basis of the income from (rather than the cost of) the property, the total depletion on a property may exceed the cost of the property. To prevent this from happening, the annual allowance under the percentage method cannot be more than 50% of the taxable income from the property (figured without the deduction for depletion). Table 9.11 shows the allowed percentages for selected mining properties.

EXAMPLE 9.11 Percentage Depletion versus Cost Depletion

A gold mine with an estimated deposit of 120,000 ounces of gold has a basis of $30 million (cost minus land value). The mine has a gross income of $16,425,000 for the year from selling 18,000 ounces of gold (at a unit price of $912.50 per ounce). Mining expenses before depletion equal $12,250,000. Compute the percentage depletion allowance. Would it be advantageous to apply cost depletion rather than percentage depletion?

SOLUTION

Given: Basis = $30 million, total recoverable volume = 120,000 ounces of gold, amount sold this year = 18,000 ounces, gross income = $16,425,000, and this year's expenses before depletion = $12,250,000.

Find: Maximum depletion allowance (cost or percentage).

Percentage depletion: Table 9.11 indicates that gold has a 15% depletion allowance. The percentage depletion allowance is computed from the gross income:

Gross income from sale of 18,000 ounces	$16,425,000
Depletion percentage	× 15%
Computed percentage depletion	$2,463,750

Next, we need to compute the taxable income. The percentage depletion allowance is limited to the computed percentage depletion or 50% of the taxable income, whichever is smaller.

Gross income from sale of 18,000 ounces	$16,425,000
Less mining expenses	12,250,000
Taxable income from mine	4,175,000
Deduction limitation	× 50%
Maximum depletion deduction	$2,087,500

Since the maximum depletion deduction ($2,087,500) is less than the computed percentage depletion ($2,463,750), the allowable percentage deduction is $2,087,500.

Cost depletion: It is worth computing the depletion allowance with the cost depletion method as

$$\text{Cost depletion} = \left(\frac{\$30,000,000}{120,000}\right)(18,000)$$
$$= \$4,500,000$$

Note that percentage depletion is less than the cost depletion. Since the law allows the taxpayer to take whichever deduction is larger in any one year, in this situation, it would be advantageous to apply the cost depletion method. Figure 9.5 illustrates the steps for finding optimal depletion between cost and percentage methods.

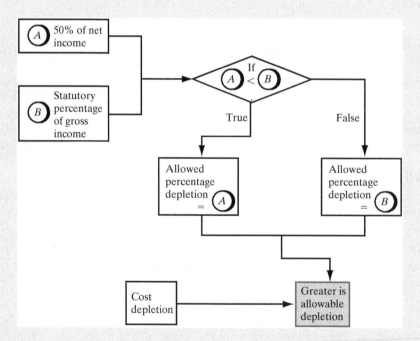

Figure 9.5 Calculating the allowable depletion deduction for federal tax purposes.

9.6 Repairs or Improvements Made to Depreciable Assets

If any major repairs (e.g., an engine overhaul) or improvements (say, an addition) are made during the life of the asset, we need to determine whether these actions will extend the life of the asset or will increase the originally estimated salvage value. When either of these situations arises, a revised estimate of the useful life of the asset should be made, and the periodic depreciation expense should be updated accordingly. We will examine how repairs or improvements affect both book and tax depreciations.

9.6.1 Revision of Book Depreciation

Recall that book depreciation rates are based on estimates of the useful lives of assets. Such estimates are seldom precise. Therefore, after a few years of use, you may find that the asset could last for a considerably longer or shorter period than was originally estimated. If this happens, the annual depreciation expense based on the estimated useful life may be either excessive or inadequate. (If repairs or improvements do not extend the life or increase the salvage value of the asset, these costs may be treated as maintenance expenses during the year in which they were incurred.) The procedure for correcting the book depreciation schedule is to revise the current book value and to allocate this cost over the remaining years of the asset's useful life.

9.6.2 Revision of Tax Depreciation

For tax depreciation purposes, repairs or improvements made to any property are treated as *separate* property items. The recovery period for a repair or improvement to the initial property normally begins on the date the repaired or improved property is placed in service. The recovery class of the repair or improvement is the recovery class that would apply to the property if it were placed in service at the same time as the repair or improvement. Example 9.12 illustrates the procedure for correcting the depreciation schedule for an asset that had repairs or improvements made to it during its depreciable life.

EXAMPLE 9.12 Depreciation Adjustment for an Overhauled Asset

In January 2013, Ogletree Manufacturing Company purchased a new numerical control machine at a cost of $60,000. The machine had an expected life of 10 years at the time of purchase and a zero expected salvage value at the end of the 10 years.

- For book depreciation purposes, no major overhauls had been planned over the 10-year period, and the machine was being depreciated toward a zero salvage value ($6,000 per year) with the straight-line method.
- For tax purposes, the machine was classified as a seven-year MACRS property.

In December 2015, however, the machine was thoroughly overhauled and rebuilt at a cost of $15,000. It was estimated that the overhaul would extend the machine's useful life by five years.

(a) Calculate the book depreciation for the year 2018 on a straight-line basis.

(b) Calculate the tax depreciation for the year 2018 for this machine.

SOLUTION

Given: $I = \$60,000$, $S = \$0$, $N = 10$ years, machine overhaul $= \$15,000$, and extended life $= 15$ years from the original purchase.

Find: D_6 for book depreciation and D_6 for tax depreciation.

(a) Since an improvement was made at the end of the year 2015, the book value of the asset at that time consisted of the original book value plus the cost added to the asset. First, the original book value at the end of 2015 is calculated as

$$B_3 \text{ (before improvement)} = \$60,000 - 3(\$6,000) = \$42,000$$

After the improvement cost of $15,000 is added, the revised book value is

$$B_3 \text{ (after improvement)} = \$42,000 + \$15,000 = \$57,000$$

To calculate the book depreciation in the year 2018, which is three years after the improvement, we need to calculate the annual straight-line depreciation amount with the extended useful life. The remaining useful life before the improvement was made was seven years. Therefore, the revised remaining useful life should be 12 years. The revised annual depreciation is then $57,000/12 = \$4,750$. Using the straight-line depreciation method, we compute the depreciation amount for 2018 as

$$D_6 = \$4,750$$

(b) For tax depreciation purposes, the improvement made is viewed as a separate property with the same recovery period as that of the initial asset. Thus, we need to calculate both the tax depreciation under the original asset and that of the new asset. For the seven-year MACRS property, the sixth-year depreciation allowance is 8.92% of $60,000 ($5,352). The third-year depreciation for the improved asset is 17.49% of $15,000 ($2,624). Therefore, the total tax depreciation in 2018 is

$$D_6 = \$5,352 + \$2,624 = \$7,976$$

Figure 9.6 illustrates how additions or improvements are treated in revising depreciation amounts for book and tax purposes.

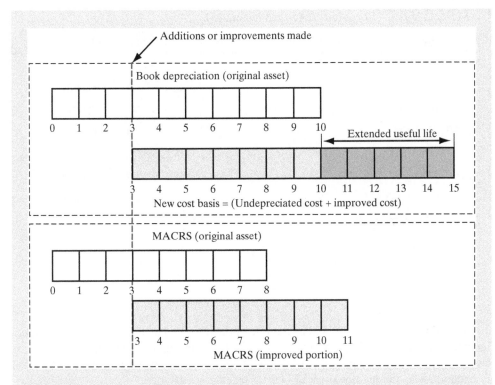

Figure 9.6 Revision of depreciation amount as additions or improvements are made as described in Example 9.12.

9.7 Corporate Taxes

Now that we have learned what elements constitute taxable income, we turn our attention to the process of computing income taxes. The **corporate tax rate** is applied to the taxable income of a corporation. As we briefly discussed in Section 8.5, the allowable deductions include the cost of goods sold, salaries and wages, rent, interest, advertising, depreciation, amortization,[4] depletion, and various tax payments other than federal income tax.

9.7.1 Income Taxes on Operating Income

As an example, the corporate tax rate structure for 2015 is relatively simple. As shown in Table 9.12, there are four basic rate brackets (15%, 25%, 34%, and 35%) plus two surtax rates (5% and 3%) based on taxable incomes. U.S. tax rates are progressive; that is, businesses with lower taxable incomes are taxed at lower rates than those with higher taxable incomes.

---

[4] The **amortization expense** is a special form of depreciation for an intangible asset, such as patents, goodwill, and franchises. More precisely, the amortization expense is the systematic write-off of expenses of the cost of an intangible asset over the period of its economic usefulness. Normally a straight-line method is used to calculate the amortization expense.

TABLE 9.12 Corporate Tax Schedule for 2015

Taxable Income (X)	Tax Rate	Tax Computation Formula
$0–$50,000	15%	$0 + 0.15X$
50,001–75,000	25%	$7,500 + 0.25(X - $50,000)$
75,001–100,000	34%	$13,750 + 0.34(X - 75,000)$
100,001–335,000	34% + 5%	$22,250 + 0.39(X - 100,000)$
335,001–10,000,000	34%	$113,900 + 0.34(X - 335,000)$
10,000,001–15,000,000	35%	$3,400,000 + 0.35(X - 10,000,000)$
15,000,001–18,333,333	35% + 3%	$5,150,000 + 0.38(X - 15,000,000)$
18,333,334 and up	35%	$6,416,666 + 0.35(X - 18,333,333)$

Marginal Tax Rate

The **marginal tax rate** is defined as the rate applied to the last dollar of income earned. Income of up to $50,000 is taxed at a 15% rate (meaning that if your taxable income is less than $50,000, your marginal tax rate is 15%); income between $50,000 and $75,000 is taxed at 25%; and income over $75,000 is taxed at a 34% rate.

An additional 5% surtax (resulting in 39%) is imposed on a corporation's taxable income in excess of $100,000 with the maximum additional tax limited to $11,750(235,000 × 0.05). This surtax provision phases out the benefit of graduated rates for corporations with taxable incomes between $100,000 and $335,000. Another 3% surtax is imposed on corporate taxable income in the range from $15,000,001 to $18,333,333.

Corporations with incomes in excess of $18,333,333 in effect pay a flat tax of 35%. As shown in Table 9.12, the corporate tax is progressive up to $18,333,333 in taxable income, but essentially is constant thereafter.

Effective (Average) Tax Rate

Effective tax rates can be calculated from the data in Table 9.12. For example, if your corporation had a taxable income of $16,000,000 in 2015, then the income tax owed by the corporation would be as given in Table 9.13.

Alternatively, using the tax formulas in Table 9.11, we obtain

$$\$5,150,000 + 0.38(\$16,000,000 - \$15,000,000) = \$5,530,000$$

The effective (average) tax rate would then be

$$\frac{\$5,530,000}{\$16,000,000} = 0.3456, \text{ or } 34.56\%$$

as opposed to the marginal rate of 38%. In other words, on the average, the company paid 34.56 cents for each taxable dollar it generated during the accounting period.

TABLE 9.13 Income Tax Calculation

Taxable Income	Tax Rate	Taxes	Cumulative Taxes
First $50,000	15%	$7,500	$7,500
Next $25,000	25%	6,250	13,750
Next $25,000	34%	8,500	22,250
Next $235,000	39%	91,650	113,900
Next $9,665,000	34%	3,286,100	3,400,000
Next $5,000,000	35%	1,750,000	5,150,000
Remaining $1,000,000	38%	380,000	$5,530,000

EXAMPLE 9.13 Corporate Taxes

A mail-order computer company sells personal computers and peripherals. The company leased showroom space and a warehouse for $20,000 a year and installed $290,000 worth of inventory-checking and packaging equipment. The allowed depreciation expense for this capital expenditure ($290,000) amounted to $58,000 using the category of a five-year MACRS. The store was completed and operations began on January 1st. The company had a gross income of $1,250,000 for the calendar year. Supplies and all operating expenses (other than the lease expense) were itemized as

Merchandise sold in the year	$600,000
Employee salaries and benefits	150,000
Other supplies and expenses	90,000
	$840,000

Compute the taxable income for this company. How much will the company pay in federal income taxes for the year?

SOLUTION
Given: Income, preceding cost information, and depreciation.
Find: Taxable income and federal income taxes.

First we compute the taxable income as follows.

Gross revenues	$1,250,000
Expenses	−840,000
Lease expense	−20,000
Depreciation	−58,000
Taxable income	**$332,000**

Note that capital expenditures are not deductible expenses. Since the company is in the 39% marginal tax bracket, its income tax can be calculated by using the formula given in Table 9.12, namely, $22{,}250 + 0.39(X - 100{,}000)$. Therefore,

$$\text{Income tax} = \$22{,}250 + 0.39(\$332{,}000 - \$100{,}000)$$

$$= \$112{,}730$$

The firm's current marginal tax rate is 39%, but its average corporate tax rate is

$$\frac{\$112{,}730}{\$332{,}000} = 33.95\%$$

COMMENTS: Instead of using the tax formula in Table 9.12, we can compute the federal income tax in the following fashion:

First $50,000 at 15%	$ 7,500
Next $25,000 at 25%	6,250
Next $25,000 at 34%	8,500
Next $232,000 at 39%	90,480
Income tax	**$112,730**

9.8 Tax Treatment of Gains or Losses on Depreciable Assets

As in the disposal of capital assets, gains or losses generally are associated with the sale (or exchange) of depreciable assets. To calculate a gain or loss, we first need to determine the book value of the depreciable asset at the time of its disposal.

9.8.1 Disposal of a MACRS Property

For a MACRS property, one important consideration at the time of disposal is whether the property is disposed of *during* or *before* its specified recovery period. Moreover, with the half-year convention, which is now mandated by all MACRS depreciation methods, the year of disposal is charged one-half of that year's annual depreciation amount if it should occur during the recovery period.

9.8.2 Calculations of Gains and Losses on MACRS Property

When a depreciable asset used in business is sold for an amount that differs from its book value, the gain or loss has an important effect on income taxes. The gain or loss is found using the following examples.

Case 1: Salvage value < Cost basis

If the salvage value of the asset is less than its cost basis, then

$$\text{Gains (losses)} = \text{Salvage value} - \text{book value}$$

EXAMPLE 9.14 Book Value in the Year of Disposal

Consider a five-year MACRS asset purchased for $10,000. Note that property belonging to the five-year MACRS class is depreciated over six years due to the half-year convention. The applicable depreciation percentages shown in Table 9.6 are 20%, 32%, 19.20%, 11.52%, 11.52%, and 5.76%. Compute the allowed depreciation amounts and the book value when the asset is disposed of (a) in year 3, (b) in year 5, and (c) in year 6.

SOLUTION

Given: Five-year MACRS asset and cost basis = $10,000.

Find: Total depreciation and book value at disposal if the asset is sold in year 3, 5, or 6.

(a) If the asset is disposed of in year 3 (or at the end of year 3), the total accumulated depreciation amount and the book value would be

$$\text{Total depreciation} = \$10,000\left(0.20 + 0.32 + \frac{0.192}{2}\right)$$

$$= \$6,160$$

$$\text{Book value} = \$10,000 - \$6,160$$

$$= \$3,840$$

(b) If the asset is disposed of during[5] year 5, the depreciation and book values will be

$$\text{Total depreciation} = \$10,000\left(0.20 + 0.32 + 0.192 + 0.1152 + \frac{0.1152}{2}\right)$$

$$= \$8,848$$

$$\text{Book value} = \$10,000 - \$8,848$$

$$= \$1,152$$

(c) If the asset is disposed of after the recovery period, there will be no penalty due to early disposal. Since the asset is depreciated fully, we have

$$\text{Total depreciation} = \$10,000$$

$$\text{Book value} = \$0$$

Figure 9.7 illustrates the amount depreciation allowances associated with disposal of a MACRS property.

---

[5] Note that even though you dispose of the asset at the *end* of the recovery period, the half-year convention still applies.

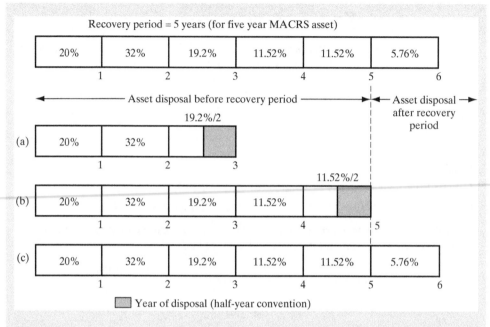

Figure 9.7 Disposal of a MACRS property and its effect on depreciation allowances (Example 9.14).

where the salvage value represents the proceeds from the sale (i.e., the selling price) less any selling expense or removal cost. These gains, commonly known as **depreciation recapture**, are taxed as ordinary income under current tax law.

Case 2: Salvage value > Cost basis

In the unlikely event that an asset is sold for an amount greater than its cost basis, the gains (salvage value − book value) are divided into two parts for tax purposes:

$$\text{Gains} = \text{Salvage value} - \text{book value}$$

$$= \underbrace{(\text{Salvage value} - \text{cost basis})}_{\text{Capital gains}}$$

$$+ \underbrace{(\text{Cost basis} - \text{book value})}_{\text{Ordinary gains}} \qquad (9.10)$$

Recall from Section 9.2.2 that the cost basis is the cost of purchasing an asset plus any incidental costs, such as freight and installation costs. As illustrated in Figure 9.8,

$$\text{Capital gains} = \text{Salvage value} - \text{cost basis}$$
$$\text{Ordinary gains} = \text{Cost basis} - \text{book value}$$

This distinction is necessary only when capital gains are taxed at the capital gains tax rate and ordinary gains (or depreciation recapture) at the ordinary income tax rate. Current tax law does not provide a special low rate of taxation for capital gains. Instead, capital

Figure 9.8 Capital gains and ordinary gains (or depreciation recapture) when the salvage value exceeds the cost basis.

gains are treated as ordinary income, but the maximum tax rate is set at the U.S. statutory rate of 35%. Nevertheless, the statutory structure for capital gains has been retained in the tax code. This provision could allow Congress to restore capital gains' preferential treatment at some future time.

EXAMPLE 9.15 Gains and Losses on Depreciable Assets

A company purchased a drill press costing $230,000 in year 0. The drill press, classified as seven-year recovery property, is being depreciated by the MACRS method. If it is sold at the end of three years, compute the gains (losses) for the following four salvage values:

(a) $150,000, (b) $120,693, (c) $100,000, and (d) $250,000. Assume that both capital gains and ordinary income are taxed at 34%.

SOLUTION

Given: Seven-year MACRS asset, cost basis = $230,000, and the asset is sold three years after its purchase.

Find: Gains or losses, tax effects, and net proceeds from the sale if the asset is sold for $150,000, $120,693, $100,000, or $250,000.

In this example, we first compute the current book value of the machine. From the MACRS depreciation schedule in Table 9.6, the allowed annual depreciation percentages for the first three years of a seven-year MACRS property are 14.29%, 24.49%, and 17.49%. Since the asset is disposed of before the end of its recovery period, the depreciation amount in year 3 will be reduced by half. The total depreciation and the final book value will be

$$\text{Total allowed depreciation} = \$230{,}000\left(0.1429 + 0.2449 + \frac{0.1749}{2}\right)$$
$$= \$109{,}308$$
$$\text{Book value} = \$230{,}000 - \$109{,}308$$
$$= \$120{,}692$$

(a) **Case 1: Book value < Salvage value < Cost basis:** In this case, there are no capital gains to consider. All gains are ordinary gains. Thus, we have

$$\text{Ordinary gains} = \text{Salvage value} - \text{book value}$$
$$= \$150{,}000 - \$120{,}692 = \$29{,}308$$
$$\text{Gains tax (34\%)} = 0.34(\$29{,}308) = \$9{,}965$$
$$\text{Net proceeds from sale} = \text{Salvage value} - \text{gains tax}$$
$$= \$150{,}000 - \$9{,}965 = \$140{,}035$$

This situation (in which the asset's salvage value exceeds its book value) is denoted as Case 1 in Figure 9.9.

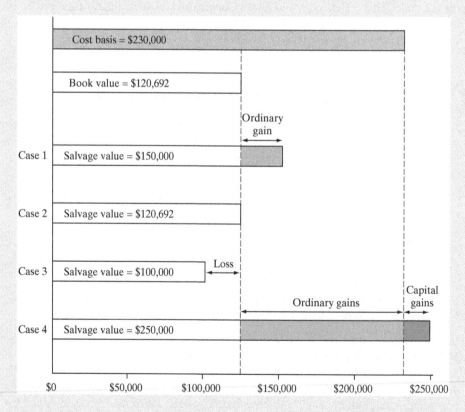

Figure 9.9 Calculations of gains or losses on MACRS property (Example 9.15).

(b) **Case 2: Salvage value = Book value:**   The book value is again $120,692. Thus, if the drill press's salvage value equals $120,692—its book value—no taxes are levied on that salvage value. Therefore, the net proceeds equal the salvage value.

(c) **Case 3: Salvage value < Book value:**   This case illustrates a loss when the salvage value (say, $100,000) is less than the book value. We compute the net salvage value after tax as

$$\text{Gain (loss)} = \text{Salvage value} - \text{book value}$$

$$= \$100,000 - \$120,692$$

$$= (\$20,692)$$

$$\text{Tax savings} = 0.34(\$20,692)$$

$$= \$7,035$$

$$\text{Net proceeds from sale} = \$100,000 + \$7,035$$

$$= \$107,035$$

(d) **Case 4: Salvage value > Cost basis:**   This situation is not likely for most depreciable assets (except real property). But it is the only situation in which both capital gains and ordinary gains can be observed. The tax treatment on this gain is as follows:

$$\text{Capital gains} = \text{Salvage value} - \text{cost basis}$$

$$= \$250,000 - \$230,000$$

$$= \$20,000$$

$$\text{Capital gains tax} = \$20,000(0.34)$$

$$= \$6,800$$

$$\text{Ordinary gains} = \$230,000 - \$120,692$$

$$= \$109,308$$

$$\text{Gains tax} = \$109,308(0.34)$$

$$= \$37,165$$

$$\text{Net proceeds from sale} = \$250,000 - (\$6,800 + \$37,165)$$

$$= \$206,035$$

COMMENTS: Note that in (c), the reduction in tax, due to the loss, actually increases the net proceeds. This is realistic when the incremental tax rate (34% in this case) is positive, indicating the corporation is still paying tax, but less than if the asset had not been sold at a loss. The incremental tax rate will be discussed in Section 9.9.

9.9 Income Tax Rate to Be Used in Economic Analysis

As we have seen in earlier sections, average income tax rates for corporations vary with the level of taxable income from 0 to 35%. Suppose that a company now paying a tax rate of 25% on its current operating income is considering a profitable investment. What tax rate should be used in calculating the taxes on the investment's projected income?

9.9.1 Incremental Income Tax Rate

The choice of a corporation's rate depends on the incremental effect of an investment on the company's taxable income. In other words, the tax rate to use is the rate that applies to the additional taxable income projected in the economic analysis.

To illustrate, consider ABC Corporation, whose taxable income from operations is expected to be $70,000 for the current tax year. ABC management wishes to evaluate the incremental tax impact of undertaking a project during the same tax year. The revenues, expenses, and taxable incomes before and after the project are estimated as

	Before	**After**	**Incremental**
Gross revenue	$200,000	$240,000	$40,000
Salaries	100,000	110,000	10,000
Wages	30,000	40,000	10,000
Taxable income	$70,000	$90,000	$20,000

Because the income tax rate is progressive, the tax effect of the project cannot be isolated from the company's overall tax obligations. The base operations of ABC without the project are expected to yield a taxable income of $70,000. With the new project, the taxable income increases to $90,000. From the tax computation formula in Table 9.12, the corporate income taxes with and without the project are

$$\text{Income tax without the project} = \$7,500 + 0.25(\$70,000 - \$50,000)$$
$$= \$12,500$$
$$\text{Income tax with the project} = \$13,750 + 0.34(\$90,000 - \$75,000)$$
$$= \$18,850$$

The additional income tax is then $18,850 - $12,500 = $6,350. This amount (on the additional $20,000 of taxable income) is based on a rate of 31.75%, which is an incremental rate. This is the rate we should use in evaluating the project in isolation from the rest of ABC's operations. As shown in Figure 9.10, the 31.75% is not an arbitrary figure, but it is a weighted average of two distinct marginal rates. Because the new project pushes ABC into a higher tax bracket, the first $5,000 it generates is taxed at 25%; the remaining $15,000 it generates is taxed in the higher bracket at 34%. Thus, we could have calculated the incremental tax rate with the formula

$$0.25\left(\frac{5,000}{20,000}\right) + 0.34\left(\frac{15,000}{20,000}\right) = 31.75\%$$

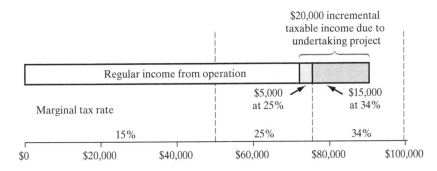

Figure 9.10 Illustration of incremental tax rate.

The average tax rates before and after the new project being considered is undertaken are as follows.

	Before	**After**	**Incremental**
Taxable income	$70,000	$90,000	$20,000
Income taxes	12,500	18,850	6,350
Average tax rate	17.86%	20.94%	
Incremental tax rate			31.75%

Note that neither 17.86% nor 20.94% is a correct tax rate to use in evaluating the new project.

A corporation with continuing base operations that place it consistently in the highest tax bracket will have both marginal and average federal tax rates of 35%. For such firms, the tax rate on an additional investment project is, naturally, 35%. But for corporations in lower tax brackets and for those which fluctuate between losses and profits, the marginal and average tax rates are likely to vary. For such corporations, estimating a prospective incremental tax rate for a new investment project may be difficult. The only solution may be to perform scenario analysis, in which we examine how much the income tax fluctuates due to undertaking the project. (In other words, we calculate the total taxes and the incremental taxes for each scenario.) A typical scenario is presented in Example 9.16.

EXAMPLE 9.16 Scenario Analysis for a Small Company

EverGreen Grass Company expects to have an annual taxable income of $320,000 from its regular grass-sodding business over the next two years. EverGreen has just won a contract to sod grasses for a new golf complex for just those years. This two-year project requires a purchase of new equipment costing $50,000. The equipment falls into the MACRS five-year class, with depreciation allowances of 20%, 32%, 19.2%, 11.52%, 11.52%, and 5.76% in each of the six years, respectively, during which the equipment will be depreciated. After the contract is terminated, the equipment will be retained for future use (instead of being sold), indicating no salvage cash flow, gain, or loss on this asset. The project will bring in an additional annual revenue of $150,000, but it is expected to incur additional annual operating costs of $90,000. Compute the incremental (marginal) tax rates applicable to the project's operating profits for years 1 and 2.

SOLUTION

Given: Base taxable income = $320,000 per year, incremental income, expenses, and depreciation amounts as stated.

Find: Incremental tax rate for this new project in years 1 and 2.

First, we compute the additional taxable income from the golf course project over the next two years:

Year	1	2
Gross Revenue	$150,000	$150,000
Expenses	90,000	90,000
Depreciation	10,000	16,000
Taxable income	$50,000	$44,000

Next, we compute the income taxes. To do this, we need to determine the applicable marginal tax rate, but because of the progressive income tax rate on corporations, the project cannot be isolated from the other operations of EverGreen.

We can solve this problem much more efficiently by using the incremental tax rate concept discussed at the beginning of this section. Because the golf course project pushes EverGreen into the 34% tax bracket from the 39% bracket, we want to know what proportion of the incremental taxable income of $50,000 in year 1 is taxed at 39% and what proportion is taxable at 34%. Without the project, the firm's taxable income is $320,000, and its marginal tax rate is 39%. With the additional taxable income of $50,000, EverGreen's tax bracket reverts to 34%, as its combined taxable income changes from $320,000 to $370,000. Since the rate changes at $335,000, the first $15,000 of the $50,000 taxable income will still be in the 39% bracket, and the remaining $35,000 will be in the 34% bracket. In year 2, we can divide the additional taxable income of $44,000 in a similar fashion. Then we can calculate the incremental tax rates for the first two years as

$$0.39\left(\frac{\$15,000}{\$50,000}\right) + 0.34\left(\frac{\$35,000}{\$50,000}\right) = 0.3550$$

$$0.39\left(\frac{\$15,000}{\$44,000}\right) + 0.34\left(\frac{\$29,000}{\$44,000}\right) = 0.3570$$

Note that these incremental tax rates vary slightly from year to year. Much larger changes can occur if a company's taxable income fluctuates drastically from its continuing base operation.

9.9.2 Consideration of State Income Taxes

For large corporations, the top federal marginal tax rate is 35%. In addition to federal income taxes, state income taxes are levied on corporations in most states. State income taxes are an allowable deduction in computing federal taxable income, and two ways are available to consider explicitly the effects of state income taxes in an economic analysis.

The first approach is to estimate explicitly the amount of state income taxes before calculating the federal taxable income. We then reduce the federal taxable income by the amount of the state taxes and apply the marginal tax rate to the resulting federal taxes. The total taxes would be the sum of the state taxes and the federal taxes.

The second approach is to calculate a single tax rate that reflects both state and federal income taxes. This single rate is then applied to the federal taxable income without subtracting state income taxes. Taxes computed in this fashion represent total taxes. If state income taxes are considered, the combined state and federal marginal tax rate may be higher than 35%. Since state income taxes are deductible as expenses in determining federal taxes, the marginal rate for combined federal and state taxes can be calculated with the expression

$$t_m = t_f + t_s - (t_f)(t_s) \tag{9.11}$$

where

t_m = combined marginal tax rate

t_f = federal marginal tax rate

t_s = state marginal tax rate

This second approach provides a more convenient and efficient way to handle taxes in an economic analysis in which the incremental tax rates are known. Therefore, incremental tax rates will be stated as combined marginal tax rates, unless indicated otherwise. (For large corporations, these would be about 40%, but they vary from state to state.)

EAMPLE 9.17 Combined State and Federal Income Taxes

Consider a corporation whose revenues and expenses before income taxes are

Gross revenue	$1,000,000
All expenses	400,000

If the marginal federal tax rate is 35% and the marginal state rate is 7%, compute the combined state and federal taxes using the two methods just described.

SOLUTION

Given: Gross income = $1,000,000, deductible expenses = $400,000, t_f = 35%, and t_s = 7%.

Find: Combined income taxes t_m.

(a) Explicit calculation of state income taxes:

Let's define FT as federal taxes and ST as state taxes. Then

$$\text{State taxable income} = \$1,000,000 - \$400,000$$
$$= \$600,000$$

and

$$\text{ST} = (0.07)(\$600,000)$$
$$= \$42,000$$

Also,

$$\text{Federal taxable income} = \$1,000,000 - \$400,000 - \text{ST}$$
$$= \$558,000$$

so that

$$\text{FT} = (0.35)(\$558,000)$$
$$= \$195,300$$

Thus,

$$\text{Combined taxes} = \text{FT} + \text{ST}$$
$$= \$237,300$$

(b) Tax calculation based on the combined tax rate:
Compute the combined tax rate directly from the formula

$$\text{Combined tax rate } (t_m) = 0.35 + 0.07 - (0.35)(0.07)$$
$$= 39.55\%$$

Hence,

$$\text{Combined taxes} = \$600,000(0.3955)$$
$$= \$237,300$$

(c) As expected, these two methods always produce exactly the same results.

9.10 The Need for Cash Flow in Engineering Economic Analysis

Traditional accounting stresses net income as a means of measuring a firm's profitability, but it is desirable to discuss why cash flows are relevant in project evaluation. As seen in Section 8.2, net income is an accounting measure based in part on the **matching concept**. Costs become expenses as they are matched against revenue. The actual timing of cash inflows and outflows is ignored.

9.10.1 Net Income versus Net Cash Flow

Over the life of a firm, net incomes and net cash inflows will usually be the same. However, the timing of incomes and cash inflows can differ substantially. Given the time value of money, it is better to receive cash now rather than later, because cash can be invested to earn more cash. (You cannot invest net income.) For example, consider the two firms in Table 9.14 showing their income and cash flow schedules over two years.

Both companies have the same amount of net income and cash over two years, but Company A returns $1 million cash yearly, while Company B returns $2 million at the end of the second year. If you received $1 million at the end of the first year from Company A, you could, for example, invest it at 10%. While you would receive only $2 million in total from Company B at the end of the second year, you would receive $2.1 million in total from Company A.

TABLE 9.14 **Timing of Cash Flows and Income Recognition**

		Company A	Company B
Year 1	Net income	$1,000,000	$1,000,000
	Cash flow	1,000,000	0
Year 2	Net income	1,000,000	1,000,000
	Cash flow	1,000,000	2,000,000

9.10.2 Treatment of Noncash Expenses

Apart from the concept of the time value of money, certain expenses do not even require a cash outflow. Depreciation and amortization are the best examples of this type of expense. Even though depreciation (or amortization expense) is deducted from revenue on a daily basis, no cash is paid to anyone.

In Example 9.13, we learned that the annual depreciation allowance has an important impact on both taxable and net income. However, although depreciation has a direct impact on net income, it is *not* a cash outlay; hence, it is important to distinguish between annual income in the presence of depreciation and annual cash flow.

The situation described in Example 9.13 serves as a good vehicle to demonstrate the difference between depreciation costs as expenses and the cash flow generated by the purchase of a fixed asset. In that example, cash in the amount of $290,000 was expended in year 0, but the $58,000 depreciation charged against the income in year 1 was not a cash outlay. Figure 9.11 summarizes the difference.

Net income (**accounting profit**) is important for accounting purposes, but **cash flows** are more important for project evaluation purposes. However, as we will now demonstrate, net income can provide us with a starting point to estimate the cash flow of a project.

Table 9.15 shows the procedure for calculating net cash flow (after tax), which is identical to that used to obtain net income from operations with the exception of depreciation, which is excluded from the net cash flow computation. (Depreciation is needed only for computing income taxes.) Assuming that revenues are received and expenses are paid in cash, we can obtain the net cash flow by adding the **noncash expense** (depreciation) to net income, which cancels the operation of subtracting it from revenues.

Example 9.18 illustrates this relationship.

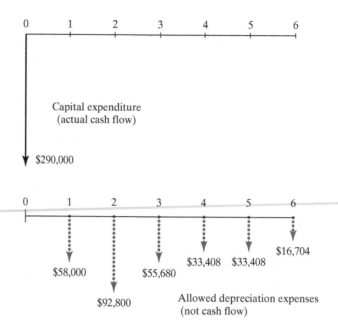

Figure 9.11 Cash flows versus depreciation expenses for an asset with a cost basis of $290,000, which was placed in service in year 0.

TABLE 9.15 **Estimating Net Cash Flow from Operations**

Item
Gross income
Expenses:
Cost of goods sold
Depreciation
Operating expenses
Taxable operating income
Income taxes
Net income
Net cash flow = Net income + Depreciation

EXAMPLE 9.18 Cash Flow versus Net Income

A company buys a numerically controlled (NC) machine for $28,000 (year 0) and uses it for five years, after which it is scrapped. The allowed depreciation deduction during the first year is $4,000, as the equipment falls into the category of seven-year MACRS property. (The first-year depreciation rate is 14.29%.) The cost of the goods produced by this NC machine should include a charge for the depreciation of the machine. Suppose the company estimates the following revenues and expenses, including depreciation, for the first operating year:

$$\text{Gross income} = \$50,000$$

$$\text{Cost of goods sold} = \$20,000$$

$$\text{Depreciation on NC machine} = \$4,000$$

$$\text{Operating expenses} = \$6,000$$

(a) If the company pays taxes at the rate of 40% on its taxable income, what is its net income from the project during the first year?
(b) Assume that (1) all sales are cash sales, and (2) all expenses except depreciation were paid during year 1. How much cash would be generated from operations?

SOLUTION

Given: Net-income components.
Find: Cash flow.

We can generate a cash flow statement simply by examining each item in the income statement and determining which items actually represent receipts or disbursements (some of the assumptions listed in the statement of the problem make this process simpler).

Item	Income	Cash Flow
Gross income (revenues)	$50,000	$50,000
Expenses:		
Cost of goods sold	20,000	−20,000
Depreciation	4,000	
Operating expenses	6,000	−6,000
Taxable income	20,000	
Taxes (40%)	8,000	−8,000
Net income	$12,000	
Net cash flow		$16,000

Column 2 shows the income statement, while column 3 shows the statement on a cash flow basis. The sales of $50,000 are all cash sales. Costs other than depreciation were $26,000; these were paid in cash, leaving $24,000. Depreciation is not a cash flow: The firm did not pay out $4,000 in depreciation expenses. Taxes, however, are paid in cash, so the $8,000 for taxes must be deducted from the $24,000, leaving a net cash flow from operations of $16,000. As shown in Figure 9.12, this $16,000 is exactly equal to net income plus depreciation: $12,000 + $4,000 = $16,000.

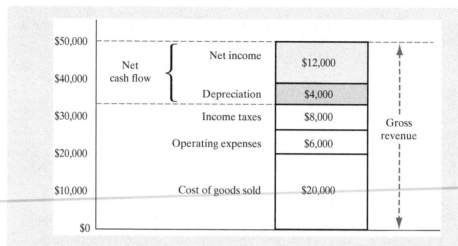

Figure 9.12 Net income versus net cash flow (Example 9.18). Cash flow versus depreciation expenses for an asset with a cost basis of $28,000, which was placed in service in year 0.

As we've just seen, depreciation has an important impact on annual cash flow in its role as an accounting expense that reduces taxable income and thus taxes. (Although depreciation expenses are not actual cash flows, depreciation has a positive impact on the after-tax cash flow of the firm.) Of course, during the year in which an asset is actually acquired, the cash disbursed to purchase it creates a significant negative cash flow, and during the depreciable life of the asset, the depreciation charges will affect the taxes paid and, therefore, cash flows.

As shown in Example 9.18, through its influence on taxes, depreciation plays a critical role in project cash-flow analysis, which we will explore further in Chapter 10.

SUMMARY

■ Machine tools, other manufacturing equipment, and even the factory buildings themselves are subject to wear over time. However, it is not always obvious how to account for the cost of their replacement. Clearly, the choice of estimated service life for a machine and the method used to calculate the cost of operating it can have significant effects on an asset's management.

■ The entire cost of replacing a machine cannot be properly charged to any one year's production; rather, the cost should be spread (or capitalized) over the years in which the machine is in service. The cost charged to operations during a particular year is called **depreciation**. Several different meanings and applications of depreciation have been presented in this chapter. From an engineering economics point of view, our primary concern is with **accounting depreciation**: the systematic allocation of an asset's value over its depreciable life.

■ Accounting depreciation can be broken into two categories:
   1. Book depreciation—the method of depreciation used in financial reports and for pricing products.

2. Tax depreciation—the method of depreciation used for calculating taxable income and income taxes; this method is governed by tax legislation.

■ The four components of information required to calculate depreciation are as follows.

1. The cost basis of the asset.

2. The salvage value of the asset.

3. The depreciable life of the asset.

4. The method of its depreciation.

Table 9.16 summarizes the differences in the way these components are treated for purposes of book and tax depreciation.

■ Because it employs accelerated methods of depreciation and shorter-than-actual depreciable lives, the **Modified Accelerated Cost Recovery System (MACRS)** gives taxpayers a break, allowing them to take earlier and faster advantage of the tax-deferring benefits of depreciation.

■ Many firms select straight-line depreciation for book depreciation because of its relative ease of calculation.

■ Depletion is a cost allocation method used particularly for natural resources. **Cost depletion** is based on the units-of-production method of depreciation. **Percentage depletion** is based on a prescribed percentage of the gross income of a property during a tax year.

■ Given the frequently changing nature of depreciation and tax law, we must use whatever percentages, depreciable lives, and salvage values are in effect *at the time an asset is acquired*.

■ Explicit consideration of taxes is a necessary aspect of any complete economic study of an investment project.

TABLE 9.16 **Summary of Book versus Tax Depreciation**

Component of Depreciation	Book Depreciation	Tax Depreciation (MACRS)
Cost basis	Based on the actual cost of the asset, plus all incidental costs, such as the cost of freight, site preparation, installation, etc.	Same as for book depreciation.
Salvage value	Estimated at the outset of depreciation analysis. If the final book value does not equal the estimated salvage value, we may need to make adjustments in our depreciation calculations.	Salvage value is zero for all depreciable assets.
Depreciable Life	Firms may select their own estimated useful lives or follow government guidelines for asset depreciation ranges (ADRs).	Eight recovery periods—3, 5, 7, 10, 15, 20, 27.5, and 39 years—have been established; all depreciable assets fall into one of these eight categories.

Method of Depreciation	Firms may select from the following:
	
	• straight line
	• accelerated methods:
	  declining balance
	  double-declining balance
	• units of production
		Exact depreciation percentages are mandated by tax legislation, but are based largely on DDB and straight-line methods.

- For corporations, the United States tax system has the following characteristics.

  1. Tax rates are progressive: The more you earn, the more you pay.
  2. Tax rates increase in stair-step fashion: four brackets for corporations and two additional surtax brackets, giving a total of six brackets.
  3. Allowable exemptions and deductions may reduce the overall tax assessment.

- Three distinct terms to describe taxes were used in this chapter.

  1. **Marginal tax rate**, which is the rate applied to the last dollar of income earned.
  2. **Effective (average) tax rate**, which is the ratio of income tax paid to net income.
  3. **Incremental tax rate**, which is the average rate applied to the incremental income generated by a new investment project.

- **Capital gains** are currently taxed as ordinary income, and the maximum rate is capped at 35%. **Capital losses** are deducted from capital gains; net remaining losses may be carried backward and forward for consideration in years other than the current tax year.

- Since we are interested primarily in the measurable financial aspects of depreciation, we consider its effects on two important measures of an organization's financial position: **net income** and **cash flow**. Once we understand that depreciation has a significant influence on the income and cash position of a firm, we will be able to appreciate fully the importance of utilizing depreciation as a means of maximizing the value both of engineering projects and of the organization as a whole.

PROBLEMS

Note: *Unless otherwise specified, use current tax rates for corporate taxes.*

Economic Depreciation

9.1 A machine now in use was purchased four years ago at a cost of $30,000. It has a book value of $9,369. It can be sold for $12,000, but it could be used for three more years, at the end of which time, it would have no salvage value. What is the current amount of economic depreciation for this asset?

9.2 You sold an automobile at a price of $14,000. The automobile was purchased three years ago for $20,000. The car had been depreciated according to a five-year MACRS property class, and the book value was $5,760. What is the amount of economic depreciation?

Cost Basis

9.3 The General Service Contractor Company paid $400,000 for a house and lot. The value of the land was appraised at $155,000 and the value of the house at $245,000. The house was then torn down at an additional cost of $15,000 so that a warehouse could be built on the combined lots at a cost of $1,250,000. What is the value of the property with the warehouse? For depreciation purposes, what is the cost basis for the warehouse?

9.4 To automate one of its production processes, the Milwaukee Corporation bought three flexible manufacturing cells at a price of $400,000 each. When they were delivered, Milwaukee paid freight charges of $20,000 and handling fees of $15,000. Site preparation for these cells cost $45,000. Six foremen, each earning $20 an hour, worked five 40-hour weeks to set up and test the manufacturing cells. Special wiring and other materials applicable to the new manufacturing cells cost $3,500. Determine the cost basis (amount to be capitalized) for these cells.

9.5 A new drill press was purchased for $148,000 by trading in a similar machine that had a book value of $46,220. Assuming that the trade-in allowance is $40,000 and that $108,000 cash is to be paid for the new asset, what is the cost basis of the new asset for depreciation purposes?

9.6 A lift truck priced at $38,000 is acquired by trading in a similar lift truck and paying cash for the remaining balance. Assuming that the trade-in allowance is $9,000 and the book value of the asset traded in is $7,808, what is the cost basis of the new asset for the computation of depreciation for tax purposes?

Book Depreciation Methods

9.7 Consider the following data on an asset:

Cost of the asset, I	$235,000
Useful life, N	5 years
Salvage value, S	$ 60,000

Compute the annual depreciation allowances and the resulting book values, using

(a) The straight-line depreciation method.

(b) The double-declining-balance method.

9.8 A firm is trying to decide whether to keep an item of construction equipment for another year. The firm has been using DDB for book purposes, and this is the fourth year of ownership of the equipment, which cost $125,000 new. The useful life of the asset was eight years. What was the depreciation in year 3?

9.9 Consider the following data on an asset:

Cost of the asset, I	$123,000
Useful life, N	7 years
Salvage value, S	$13,000

Compute the annual depreciation allowances and the resulting book values, initially using the DDB and then switching to SL.

9.10 The double-declining-balance method is to be used for an asset with a cost of $88,000, an estimated salvage value of $13,000, and an estimated useful life of six years.

(a) What is the depreciation for the first three fiscal years, assuming that the asset was placed in service at the beginning of the year?

(b) If switching to the straight-line method is allowed, when is the optimal time to switch?

9.11 Compute the double-declining-balance (DDB) depreciation schedule for the following asset:

Cost of the asset, I	$200,000
Useful life, N	8 years
Salvage value, S	$ 32,000

9.12 Compute the DDB depreciation schedule for the following asset:

Cost of the asset, I	$42,000
Useful life, N	6 years
Salvage value, S	$6000

(a) What is the value of α?

(b) What is the amount of depreciation for the second full year of use of the asset?

(c) What is the book value of the asset at the end of the fourth year?

9.13 The Upjohn Company purchased new packaging equipment with an estimated useful life of five years. The cost of the equipment was $55,000, and the salvage value was estimated to be $5,000 at the end of year 5. Compute the annual depreciation expenses over the five-year life of the equipment under each of the following methods of book depreciation:

(a) Straight-line method

(b) Double-declining-balance method

9.14 A secondhand bulldozer acquired at the beginning of the fiscal year at a cost of $56,000 has an estimated salvage value of $6,500 and an estimated useful life of 12 years. Determine the following.

(a) The amount of annual depreciation by the straight-line method.

(b) The amount of depreciation for the third year, computed by the double-declining-balance method.

Units-of-Production Method

9.15 If a dump truck for hauling coal has an estimated net cost of $90,000 and is expected to give service for 250,000 miles, resulting in a salvage value of $5,000, what would be the depreciation per mile? Compute the allowed depreciation amount for the same truck's usage amounting to 30,000 miles.

9.16 A diesel-powered generator with a cost of $68,000 is expected to have a useful operating life of 50,000 hours. The expected salvage value of this generator is $7,500. In its first operating year, the generator was operated for 5,500 hours. Determine the depreciation for the year.

9.17 The Collins Metal Shop purchased a stamping machine for $257,000 on March 1, 2015.

The machine is expected to have a useful life of 10 years, a salvage value of $32,000, a production of 250,000 units, and working hours of 30,000. During 2016, Denver used the stamping machine for 2,450 hours to produce 23,450 units. From the information given, compute the book depreciation expense for 2016 under each of the following methods.

(a) Straight-line

(b) Units-of-production

(c) Working hours

(d) Double-declining-balance (without conversion to straight-line)

(e) Double-declining-balance (with conversion to straight-line)

Tax Depreciation

9.18 The M&M Company purchased a new punch press in 2005 at a cost of $265,000. M&M also paid $46,000 to have the punch press delivered and installed. The punch press has an estimated useful life of 12 years, but it will be depreciated using MACRS over its seven-year property class life.

(a) What is the cost basis of the punch press?

(b) What will be the depreciation allowance in each year over seven years?

9.19 The Vermont Construction Company purchased a hauling truck on January 1, 2009 at a cost of $35,000. The truck has a useful life of eight years with an estimated salvage value of $6,000. The straight-line method is used for book purposes. For tax purposes, the truck would be depreciated with the MACRS method over its five-year class life. Determine the annual depreciation amount to be taken over the useful life of the hauling truck for both book and tax purposes.

9.20 The Harris Foundry Company purchased new casting equipment in 2010 at a cost of $190,000. Harris also paid $25,000 to have the equipment delivered and installed. The casting machine has an estimated useful life of 10 years, but it will be depreciated with MACRS over its seven-year class life.

(a) What is the cost basis of the casting equipment?

(b) What will be the depreciation allowance in each year of the seven-year class life of the casting equipment?

9.21 The ABC Corporation placed an asset in service three years ago. The company uses the MACRS method (seven-year life) for tax purposes and the straight-line method (seven-year useful life) for financial reporting purposes. The cost of the asset is $100,000, and the salvage value used for depreciating purposes is $20,000. What is the difference in the current book value obtained using both methods?

9.22 A machine is classified as seven-year MACRS property. Compute the book value for tax purposes at the end of three years. The cost basis is $185,000.

9.23 A piece of machinery purchased at a cost of $92,000 has an estimated salvage value of $12,000 and an estimated useful life of five years. It was placed in service on May 1st of the current fiscal year, which ends on December 31st. The asset falls into a seven-year MACRS property category. Determine the depreciation amounts over the useful life.

9.24 Suppose that a taxpayer places in service a $50,000 asset that is assigned to the six-year class (say, a new property class) with a half-year convention. Develop the MACRS deductions (assuming a 200% declining-balance rate), followed by switching to straight-line with the half-year convention.

9.25 On April 1st, Leo Smith paid $310,000 for a residential rental property. This purchase price represents $250,000 for the building and $60,000 for the land. Five years later, on November 1st, he sold the property for $400,000. Compute the MACRS depreciation for each of the five calendar years during which he had the property.

9.26 In 2016, you purchased a spindle machine (seven-year MACRS property) for $53,000, which you placed in service in January. Using the calendar year as your tax year, compute the depreciation allowances over seven years.

9.27 On October 1st, you purchased a residential home in which to locate your professional office for $350,000. The appraisal is divided into $80,000 for the land and $270,000 for the building.

(a) In your first year of ownership, how much can you deduct for depreciation for tax purposes?

(b) Suppose that the property was sold at $375,000 at the end of fourth year of ownership. What is the book value of the property?

9.28 For each of four assets in the Table P9.28, determine the missing amounts (for asset type III, the annual usage is 15,000 miles).

TABLE P9.28

Types of Asset	I	II	III	IV
Depreciating Methods	SL	DDB	UP	MACRS
End of year	6	3	3	4
Initial cost ($)	30,000	25,000	41,000	20,000
Salvage value ($)	6,000	5,000	5,000	2,000
Book value ($)	12,000	5,400		3,456
Depreciable life	8 yr	5 yr	90,000 mi	
Depreciable Amount ($)				
Accumulated Depreciable ($)		19,600		

9.29 At the beginning of the fiscal year, the Borland Company acquired new equipment at a cost of $89,000. The equipment has an estimated life of five years and an estimated salvage value of $10,000.

(a) Determine the annual depreciation (for financial reporting) for each of the five years of the estimated useful life of the equipment, the accumulated depreciation at the end of each year, and the book value of the equipment at the end of each year. Use (1) the straight-line method and (2) the double-declining-balance method for each.

(b) Determine the annual depreciation for tax purposes, assuming that the equipment falls into a seven-year MACRS property class.

(c) Assume that the equipment was depreciated under seven-year MACRS. In the first month of the fourth year, the equipment was traded in for similar equipment priced at $92,000. The trade-in allowance on the old equipment was $20,000, and cash was paid for the balance. What is the cost basis of the new equipment for computing the amount of depreciation for income tax purposes?

Depletion

9.30 Early in 2015, the Atlantic Mining Company began operation at its West Virginia mine. The mine had been acquired at a cost of $8,900,000 in 2013, and is expected to contain 4 million tons of silver and to have a residual value of $1,500,000 (once the silver is depleted). Before beginning mining operations, the company installed equipment at a cost of $2,500,000. This equipment will have no economic usefulness once the silver is depleted. Therefore, depreciation of the equipment is based upon the estimated number of tons of ore produced each year. Ore removed from the West Virginia mine amounted to 550,000 tons in 2015 and 688,000 tons in 2016.

(a) Compute the per ton depletion rate of the mine and the per ton depreciation rate of the mining equipment.

(b) Determine the depletion expense for the mine and the depreciation expense for the mining equipment.

9.31 Suppose you bought a timber tract for $850,000, and the land was worth as much as $250,000. An estimated 6.8 million board feet (MBF) of standing timber was in the tract. If you cut 1.5 MBF of timber for this fiscal year, determine your depletion allowance for the year.

9.32 A gold mine with an estimated deposit of 500,000 ounces of gold is purchased for $80 million. The mine has a gross income of $48,365,000 for the year, obtained from selling 52,000 ounces of gold. Mining expenses before depletion equal $22,250,000.

Compute the percentage depletion allowance. Would it be advantageous to apply cost depletion rather than percentage depletion?

9.33 The Oklahoma Oil Company incurred acquisition, exploration, and development costs during 2016 as given in Table P9.33.

The market price of oil during 2016 was $60 per barrel.

(a) Determine the cost basis for depletion on each parcel.

(b) During 2016, 1,200,000 barrels were extracted from parcel A at a production cost of $3,600,000. Determine the depletion charge allowed for parcel A.

(c) In part (b), if Oklahoma Oil Company sold 1,000,000 of the 1,200,000 barrels extracted during 2016 at a price of $75 per barrel, the sales revenue would be $75,000,000. If it qualified for the use of percentage depletion (15%), what would be the allowed depletion amount for 2016?

(d) During 2016, 800,000 barrels were extracted (and sold) from parcel B at a production cost of $3,000,000. Assume that, during 2017, it is ascertained that the remaining proven reserves on parcel B total only 4,000,000 barrels (instead of the expected 4,200,000). This revision in proven reserves is considered a change in an accounting estimate that must be corrected during the current and future years. (A correction of previous years' depletion amounts is not permitted.) If 1,000,000 barrels are extracted during 2017, what is the total depletion charge allowed, according to the unit cost method?

9.34 A coal mine expected to contain 6.5 million tons of coal was purchased at a cost of $30 million. One million tons of coal for steel making are produced this year. The gross income for this coal is $15,000,000 and operating costs (excluding depletion expenses) are $1,850,000. If you know that coal has a 10% depletion allowance, what will be the depletion allowance?

(a) Cost depletion.

(b) Percentage depletion.

Revision of Depreciation Rates

9.35 The Perkins Construction Company bought a building for $800,000 to be used as a warehouse. A number of major structural repairs completed at the beginning of the current year at a cost of $125,000 are expected to extend the life of the building 10 years beyond the original estimate. For book purposes, the building has been depreciated by the straight-line method for 25 years. The salvage value is expected to be negligible and has been ignored. The book value of the building before the structural repairs is $400,000.

(a) What has the amount of annual depreciation been in past years?

(b) What is the book value of the building after the repairs have been recorded?

(c) What is the amount of depreciation for the current year, according to the straight-line method? (Assume that the repairs were completed at the very beginning of the year.)

(d) For tax purposes, what is the amount of depreciation for the current year, assuming that the building belongs to a 39-year real property class?

TABLE P9.33

Items*	Site		
	Parcel A	Parcel B	Total
Acquisition costs	6	4	10
Exploration costs	13	9	22
Development costs	20	11	31
Recoverable oil (millions of barrels)	9	5	14

*Units are in millions of dollars, except recoverable oil.

9.36 The Dow Ceramic Company purchased a glass-molding machine in January 2010 for $180,000. The company has been depreciating the machine over an estimated useful life of 10 years, assuming no salvage value, by the straight-line method of depreciation. For tax purposes, the machine has been depreciated as seven-year MACRS property. At the beginning of 2013, Dow overhauled the machine at a cost of $45,000. As a result of the overhaul, Dow estimated that the useful life of the machine would extend five years beyond the original estimate.

(a) Calculate the book depreciation for the year 2016.

(b) Calculate the tax depreciation for the year 2016.

9.37 On January 2, 2013, the Hines Food Processing Company purchased a machine that dispenses a premeasured amount of tomato juice into a can. The machine cost $85,000, and its useful life was estimated at eight years, with a salvage value of $4,000. At the time it purchased the machine, Hines incurred the following additional expenses:

Freight-in	$800
Installation cost	2,500
Testing costs prior to regular operation	1,200

Book depreciation was calculated by the straight-line method, but for tax purposes, the machine was classified as a seven-year MACRS property. In January 2015, accessories costing $5,000 were added to the machine to reduce its operating costs. These accessories neither prolonged the machine's life nor provided any additional salvage value.

(a) Calculate the book depreciation expense for 2016.

(b) Calculate the tax depreciation expense for 2016.

Corporate Tax Systems

9.38 In tax year 1, an electronics-packaging firm had a gross income of $35,000,000, $6,000,000 in salaries, $7,000,000 in wages, $800,000 in depreciation expenses, a loan principal payment of $200,000, and a loan interest payment of $150,000. Determine the net income of the company in tax year 1.

9.39 A consumer electronics company was formed to develop cell phones that run on or are recharged by fuel cells. The company purchased a warehouse and converted it into a manufacturing plant for $6,000,000. It completed installation of assembly equipment worth $1,500,000 on December 31st. The plant began operation on January 1st. The company had a gross income of $8,500,000 for the calendar year. Manufacturing costs and all operating expenses, excluding the capital expenditures, were $2,280,000. The depreciation expenses for capital expenditures amounted to $456,000.

(a) Compute the taxable income of this company.

(b) How much will the company pay in federal income taxes for the year?

9.40 The Huron Roofing Company had gross revenues of $3,500,000 from operations. Financial transactions were posted during the year as

Manufacturing expenses (including depreciation)	$650,000
Operating expenses (excluding interest expenses)	320,000
A new short-term loan from a bank	90,000
Interest expenses on borrowed funds (old and new)	40,000
Dividends paid to common stockholders	80,000
Old equipment sold	60,000

The old equipment had a book value of $75,000 at the time of its sale.

(a) What is Huron's income tax liability?

(b) What is Huron's operating income?

Gains or Losses

9.41 Company J bought a piece of equipment for $200,000. This equipment has a useful life of 10 years and a salvage value of $40,000. The company has been using the seven-year MACRS property class to depreciate the asset for tax purposes. At the end of year 4, the company sold the equipment for $120,000. The tax rate is 40%. What are the net proceeds (after tax) from the sale of the equipment?

9.42 Consider a five-year MACRS asset purchased at $80,000. (Note that a five-year MACRS property class is depreciated over six years due to the half-year convention. The applicable salvage values would be $40,000 in year 3, $30,000 in year 5, and $10,000 in year 6.) Compute the gain or loss amounts when the asset is disposed of in

(a) Year 3.

(b) Year 5.

(c) Year 6.

9.43 In year 0, an electrical appliance company purchased an industrial robot costing $350,000. The robot is to be used for welding operations, classified as seven-year recovery property, and has been depreciated by the MACRS method. If the robot is to be sold after five years, compute the amounts of gains (losses) for the following two salvage values (assume that both capital gains and ordinary incomes are taxed at 34%):

(a) $20,000

(b) $99,000

9.44 You purchased a lift truck that cost $50,000 four years ago. At that time, the truck was estimated to have a service life of eight years with a salvage value of $5,000. These estimates are still good. The property has been depreciated according to a five-year MACRS property class. Now (at the end of year 4 from purchase), you are considering selling the truck for $20,000. What book value should be used in determining the taxable gains?

9.45 Sunrise, Inc., bought a machine for $80,000 on January 2, 2014. Management expects to use the machine for eight years, at the end of which time, it will have a $20,000 salvage value. Consider the following questions independently.

(a) If Sunrise uses straight-line depreciation, what will be the book value of the machine on December 31, 2017?

(b) If Sunrise uses double-declining-balance depreciation, what will be the depreciation expense for 2017?

(c) If Sunrise uses double-declining-balance depreciation, followed by switching to straight-line depreciation, when will be the optimal time to switch?

(d) If Sunrise uses a seven-year MACRS and sells the machine on April 1, 2018 at a price of $38,000, what will be the taxable gains?

Marginal Tax Rate in Project Evaluation

9.46 The Hillside Machine Shop expects to have an annual taxable income of $425,000 from its regular business over the next six years. The company is considering acquiring a new milling machine during year 0. The machine's price is $200,000 installed. The machine falls into the MACRS seven-year class, and it will have an estimated salvage value of $50,000 at the end of six years. The machine is expected to generate additional before-tax revenue of $80,000 per year.

(a) What is the total amount of economic depreciation for the milling machine if the asset is sold at $50,000 at the end of six years?

(b) Determine the company's marginal tax rates over the next six years with the machine.

(c) Determine the company's average tax rates over the next six years with the machine.

9.47 The Columbus Electrical Company expects to have an annual taxable income of $650,000 from its residential accounts over the next two years. The company is bidding on a two-year wiring service for a large apartment complex. This commercial service requires the purchase of a new truck equipped with wire-pulling tools at a cost of $60,000. The equipment falls into the MACRS five-year class and will be retained for future use (instead of being sold) after two years, indicating no gain or loss on the property. The project will bring in additional annual revenue of $220,000, but it is expected to incur additional annual operating costs of $150,000. Compute the marginal tax rate applicable to the project's operating profits for the next two years.

9.48 The Florida Citrus Corporation estimates its taxable income for next year at $2,500,000. The company is considering expanding its product line by introducing pineapple–orange juice for the next year. The market responses could be good, fair, or poor. Depending on the market response, the expected additional taxable incomes are (1) $2,000,000 for a good response, (2) $500,000 for a fair response, and (3) a $100,000 loss for a poor response.

(a) Determine the marginal tax rate applicable to each situation.

(b) Determine the average tax rate that results from each situation.

9.49 The Simon Machine Tools Company is considering purchasing a new set of machine tools to process special orders. The following financial information is available.

- Without the project, the company expects to have a taxable income of $350,000 each year from its regular business over the next three years.

- With the three-year project, the purchase of a new set of machine tools at a cost of $50,000 is required. The equipment falls into the MACRS three-year class. The tools will be sold for $10,000 at the end of project life. The project will be bringing in additional annual revenue of $80,000, but it is expected to incur additional annual operating costs of $20,000.

  (a) What are the additional taxable incomes (due to undertaking the project) during each of years 1 through 3?

  (b) What are the additional income taxes (due to undertaking the new orders) during each of years 1 through 3?

  (c) Compute the gain taxes when the asset is disposed of at the end of year 3.

Combined Marginal Income Tax Rate

9.50 Consider a corporation whose taxable income without state income tax is

Gross revenue	$3,500,000
All expenses	1,800,000

If the marginal federal tax rate is 35% and the marginal state rate is 5%, compute the combined state

and federal taxes using the two methods described in the text.

9.51 A corporation has the following financial information for a typical operating year.

Gross revenue	$6,500,000
Cost of goods sold	3,450,000
Operating costs	650,000
Federal taxes	332,000
State taxes	193,120

(a) On the basis of this financial information, determine both federal and state marginal tax rates.

(b) Determine the combined marginal tax rate for this corporation.

9.52 Van-Line Company, a small electronics repair firm, expects an annual income of $170,000 from its regular business. The company is considering expanding its repair business to include flat-panel HDTV installation service. The expansion would bring in an additional annual income of $50,000, but it also will require an additional expense of $20,000 each year over the next three years. Using applicable current tax rates, answer the following.

(a) What is the marginal tax rate in tax year 1?

(b) What is the average tax rate in tax year 1?

(c) Suppose that the new business expansion requires a capital investment of $20,000 worth of special tools (a three-year MACRS property). At $i = 10\%$, what is the PW of the total income taxes to be paid over the project life?

9.53 A company purchased a new forging machine to manufacture disks for airplane turbine engines. The new press cost $3,500,000, and it falls into a seven-year MACRS property class. The company has to pay property taxes to the local township for ownership of this forging machine at a rate of 1.2% on the beginning book value of each year.

(a) Determine the book value of the asset at the beginning of each tax year.

(b) Determine the amount of property taxes over the machine's depreciable life.

Short Case Studies

ST9.1 On January 2, 2010, the Allen Flour Company purchased a new machine at a cost of $82,000. Installation costs for the machine were $3,000. The machine was expected to have a useful life of 10 years, with a salvage value of $3,000. The company uses straight-line depreciation for financial reporting. On January 3, 2013, the machine broke down, and an extraordinary repair had to be made to the machine at a cost of $8,000. The repair extended the machine's life to 13 years, but left the salvage value unchanged. On January 2, 2016, an improvement was made to the machine in the amount of $5,000 that increased the machine's productivity and increased the salvage value (to $6,000), but did not affect the remaining useful life. Determine depreciation expenses every December 31st for the years 2010, 2013, and 2016.

ST9.2 On March 17, 2013, the Wildcat Oil Company began operations at its Louisiana oil field. The oil field had been acquired several years earlier at a cost of $32.5 million. The field is estimated to contain 6.5 million barrels of oil and to have a salvage value of $3 million both before and after all of the oil is pumped out. Equipment costing $480,000 was purchased for use at the oil field. The equipment will have no economic usefulness once the Louisiana field is depleted; therefore, it is depreciated on a units-of-production method. In addition, Wildcat Oil built a pipeline at a cost of $2,880,000 to serve the Louisiana field. Although this pipeline is physically capable of being used for many years, its economic usefulness is limited to the productive life of the Louisiana field; therefore, the pipeline has no salvage value. Depreciation of the pipeline is based on the estimated number of barrels of oil to be produced. Production at the Louisiana oil field amounted to 420,000 barrels in 2015 and 510,000 barrels in 2016.

(a) Compute the per barrel depletion rate of the oil field during the years 2015 and 2016.

(b) Compute the per barrel depreciation rates of the equipment and the pipeline during the years 2015 and 2016.

ST9.3 Electronic Measurement and Control Company (EMCC) has developed a laser speed detector that emits infrared light, which is invisible to humans and radar detectors alike. For full-scale commercial marketing, EMCC needs to invest $5 million in new manufacturing facilities. The system is priced at $3,000 per unit. The company expects to sell 5,000 units annually over the next five years. The new manufacturing facilities will be depreciated according to a seven-year MACRS property class. The expected salvage value of the manufacturing facilities at the end of five years is $1.6 million. The manufacturing cost for the detector is $1,200 per unit, excluding depreciation expenses. The operating and maintenance costs are expected to run to $1.2 million per year. EMCC has a combined federal and state income tax rate of 35%, and undertaking this project will not change this current marginal tax rate.

(a) Determine, for the next five years, the incremental taxable income, income taxes, and net income due to undertaking this new product.

(b) Determine the gains or losses associated with the disposal of the manufacturing facilities at the end of five years.

ST9.4 Diamonid is a start-up diamond-coating company that is planning to manufacture a microwave plasma reactor that synthesizes diamonds. Diamonid anticipates that the industry demand for diamonds will skyrocket over the next decade for use in industrial drills, high-performance microchips, and artificial human joints, among other things. Diamonid has decided to raise $50 million through issuing common stocks for investment in plant ($10 million) and equipment ($40 million) improvements. Each reactor can be sold at a price of $100,000 per unit. Diamonid can expect to sell 300 units per year during the next 8 years. The unit manufacturing cost is estimated at $30,000, excluding depreciation. The operating and maintenance cost for the plant is estimated at $12 million per year. Diamonid expects to phase out the operation at the end of eight years, revamp the plant and equipment, and adopt a new diamond-manufacturing technology. At that time, Diamonid estimates that the salvage values for the plant and equipment will be about 60% and 10% of the original investments, respectively. The plant and equipment will be depreciated according to 39-year real property (placed in service in January) and seven-year MACRS, respectively. Diamonid pays 5% of state and local income taxes on its taxable income.

(a) If the 2015 corporate tax system continues over the project life, determine the combined state and federal income tax rate each year.

(b) Determine the gains or losses at the time the plant is revamped.

(c) Determine the net income each year over the plant life.

CHAPTER 10

Developing Project Cash Flows

Tiny Jet Plans a Big Comeback[1] Eclipse Aviation Inc. spent $1.4 billion over 10 years to design and build six-passenger jets powered by tiny engines. On March 12, 2014, the first customer delivery of an Eclipse 550 Twin Engine Jet was made in Albuquerque, New Mexico. With the Eclipse 550, the company has built a high performance light jet that includes advanced technologies and safety features that have traditionally been found only on much larger cabin class business jets. It also has teamed up with what it considers a more stable group of suppliers, hoping to win back the loyalty of the jets' pilots and owners with the following two premises:

OWN IT:[2] The Eclipse 550 twin-engine jet is currently delivering with a base price of $2,895,000. It is the only twin-engine jet in production under $3M.

FLY IT: Because the Eclipse 550 is the most fuel-efficient twin-engine jet on the planet, it burns only 59 gallons of jet fuel per hour while flying 430 miles per hour. The operating costs are comparable to flying a high performance piston aircraft, about $648 per hour.

Source: Absemetov/Fotolia

[1] *Tiny Jet Plans a Big Comeback* by Jon Ostrower, The Wall Street Journal, July 11, 2012.

[2] Eclipse Corporation, http://www.eclipseaerospace.net/550.php

While pursuing charter operators and individual owners, Eclipse hopes to sell its six-seat 550s to the Pentagon and other governments for use in special missions or as replacements for training aircraft.

The company already delivered 47 jets in 2014 and plans to deliver 50 to 100 in 2016, with revenue estimates of more than $125 million in 2014. Eclipse still faces an uphill battle to deliver the jet at a profitable volume in an unpredictable market, as this is still a very difficult market segment with a lot of competitors.

Clearly, Eclipse's first step would be to estimate the magnitude of the revenue stream over the life of the aircraft. In doing so, the final number of aircraft units sold would be the basis for possible revenue projections. The revenue is correlated with two important factors: small aircraft market trends and how well their competitors are doing in terms of bringing more technically advanced products to market. Although Eclipse is the market leader in the twin-engine jet market, there is no way of knowing how long it would take to recover the investment made in the project. Under this circumstance, any project's justification depends upon the ability to estimate potential cash flows.

Developing project cash flows is the most important—and the most difficult—step in the analysis of a capital project. Typically, a capital project initially requires investment outlays and only later produces annual net cash inflows. A great many variables are involved in forecasting cash flows, and many individuals, ranging from engineers to cost accountants and marketing people, participate in the process. This chapter provides the general principles on which the determination of a project's cash flows are based.

To help us imagine the range of activities that are typically initiated by project proposals, we begin the chapter with an overview in Section 10.1 of how firms classify projects. A variety of projects exists, each having its own characteristic set of economic concerns. Section 10.2 then defines the typical cash-flow elements of engineering projects. Once we have defined these elements, we will examine how to develop cash flow statements that are used to analyze the economic value of projects. In Section 10.3, we use several examples to demonstrate the development of after-tax project cash flow statements. Then Section 10.4 presents some alternative techniques for developing a cash flow statement based on a generalized cash-flow approach. By the time you have finished this chapter, not only will you understand the format and significance of after-tax cash flow statements, but also, you will know how to develop them yourself.

CHAPTER LEARNING OBJECTIVES

After completing this chapter, you should understand the following concepts:

- What constitute project cash-flow elements.
- The use of the income statement approach in developing a project cash flow.
- How to treat the gains and losses related to disposal of an asset in the project cash flow statement.
- How to determine the working capital requirement and its impact on project cash flows.
- How to incorporate the cost associated with financing a project in developing the project's cash flow statement.
- How to develop a generalized cash-flow model.
- The analysis of a lease-or-buy decision on an after-tax basis.

10.1 Cost–Benefit Estimation for Engineering Projects

Before we evaluate the economics of an engineering project, it is necessary to estimate the various cost and revenue components that describe the project. Engineering projects may range from something as simple as the purchase of a new milling machine to the design and construction of a multibillion-dollar process or resource recovery complex.

The engineering projects appearing in this text as examples and problems already include the necessary cost and revenue estimates. Developing adequate estimates for these quantities is extremely important and can be clearly a time-consuming activity. Although cost-estimating techniques are not the focus of this text, it is worthwhile to mention some of the approaches that are taken. Some are used in the context of simple projects that are straightforward and involve little or no engineering design; others are employed on complex projects that tend to be large and may involve many thousands of hours of engineering design. Obviously, there are also projects that fall between these extremes, and some combination of approaches may be appropriate in these cases.

10.1.1 Simple Projects

Projects in this category usually involve a single "off-the-shelf" component or a series of such components that are integrated together in a simple manner. The acquisition of a new milling machine is an example.

The installed cost is the price of the equipment as determined from catalogs or supplier quotations, shipping and handling charges, and the cost of building modifications and changes to utility requirements. The latter may call for some design effort to define the scope of the work, which is the basis for contractor quotations.

Project benefits are in the form of new revenue and/or cost reduction. Estimating new revenue requires agreement on the total number of units produced and the selling price per unit. These quantities are related through supply-and-demand considerations in the marketplace. In highly competitive product markets, sophisticated marketing studies are often required to establish price–volume relationships as we discussed in Chapter 8. These studies are undertaken as one of the first steps in an effort to define the appropriate scale for the project.

The ongoing costs required to operate and maintain equipment can be estimated at various levels of detail and accuracy. Familiarity with the cost–volume relationships for similar facilities allows the engineer to establish a "ballpark" cost. Some of this information may be used in conjunction with more detailed estimates in other areas. For example, maintenance costs may be estimated as a percentage of the installed cost. Such percentages are derived from historical data and are frequently available from equipment suppliers. Other costs, such as the cost of manpower and energy, may be estimated in detail to reflect specific local considerations. The most comprehensive and time-consuming type of estimate is a set of detailed estimates around each type of cost associated with the project.

10.1.2 Complex Projects

The estimates developed for complex projects involve the same general considerations as those which apply to simple projects. However, complex projects usually include specialized equipment that is not "off the shelf" and must be fabricated from detailed engineering drawings. For certain projects, such drawings are not even available until after a commitment has been made to proceed with the project. The typical phases of a project are:

- Development
- Conceptual design
- Preliminary design
- Detailed design

Depending upon the project, some phases may be combined. Project economics is performed during each of these phases to confirm the attractiveness of the project and the incentive to continue working on it. The types of estimates and estimating techniques are a function of the stage of the project.

When the project involves the total or partial replacement of an existing facility, the benefits of the project are usually well known at the outset in terms of pricing–volume relationships or cost-reduction potential. In the case of natural-resource projects, however, oil, gas, and mineral price forecasts are subject to considerable uncertainty.

At the development phase, work is undertaken to identify technologies that may be needed and to confirm their technical viability. Installed cost estimates and operating estimates are based on evaluations of the cost of similar existing facilities or their parts.

Conceptual design examines the issues of the scale of the project and alternative technologies. Again, estimates tend to be based on large pieces of (or processes within) the overall project that correspond to similar facilities already in operation elsewhere.

Preliminary design takes the most attractive alternative from the conceptual design phase to a level of detail that yields specific sizing and layout estimates for the actual equipment and associated infrastructure. Estimates at this stage tend to be based on individual pieces of equipment. Once more, the estimating basis is similar pieces of equipment in use elsewhere.

For very large projects, the cost of undertaking detailed engineering is prohibitive unless the project is going forward. At this stage, detailed fabrication and construction drawings that provide the basis for an actual vendor quotation become available.

The accuracy of the estimates improves with each phase as the project becomes defined in greater detail. Normally, the installed cost estimate from each phase includes a contingency that is some fraction of the actual estimate calculated. The contingency at the development phase can be 50 to 100%; after preliminary design, it decreases to something on the order of 10%.

More information on cost-estimating techniques can be found in reference books on project management and cost engineering. Industry-specific data books are also available for some sectors, for which costs are summarized on some normalized basis, such as dollars per square foot of building space or dollars per ton of material moved in operating mines. In these cases, the data are categorized by type of building and type of mine. Engineering design companies maintain extensive databases of such cost information.

10.2 Incremental Cash Flows

When a company purchases a fixed asset, such as equipment, it makes an investment. The company commits funds today in the expectation of earning a return on those funds in the future. Such investments are similar to those made by a bank when it lends money. In the case of a bank loan, the future cash flow consists of interest plus repayment of the principal. For a fixed asset, the future return is in the form of cash flows generated by the profitable use of the asset. In evaluating a capital investment, we are concerned only with those cash flows which result directly from the investment. These cash flows, called **differential** or **incremental cash flows**, represent the change in the firm's total cash flow that occurs as a direct result of the investment. Figure 10.1 sums up the various types of cash flows that are common in engineering investment projects. In this section, we will look into some of these cash flow elements.

10.2.1 Elements of Cash Outflows

We first consider the potential uses of cash in undertaking an investment project.

- **Purchase of New Equipment.** A typical project usually involves a cash outflow in the form of an initial investment in equipment. The relevant investment costs are incremental ones, such as the cost of the asset, shipping and installation costs, and the cost of training employees to use the new asset. If the purchase of a new asset results in the sale of an existing asset, the net proceeds from this sale reduce the amount of the incremental investment. In other words, the incremental investment represents the total amount of additional funds that must be committed to the investment project. When existing equipment is sold, the transaction results in an accounting gain or loss, which is dependent on whether the amount realized from the sale is greater or less than the equipment's book value. In either case, when existing assets are disposed of,

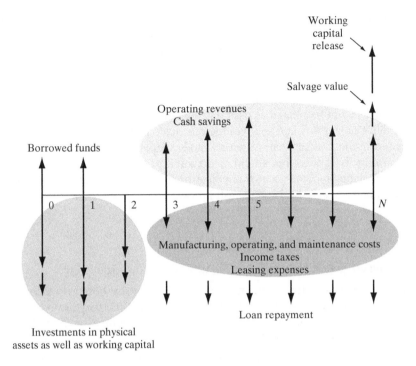

Figure 10.1   Types of cash flow elements used in project analysis.

the relevant amount by which the new investment is reduced consists of the proceeds of the sale, adjusted for tax effects.

- **Investments in Working Capital.**   Some projects require an investment in non-depreciable assets. If, for example, a project increases a firm's revenues, then more funds will be needed to support the higher level of operation. Investment in non-depreciable assets is often called **investment in working capital**. In accounting, working capital is the amount carried in cash, accounts receivable, and inventory that is available to meet day-to-day operating needs. For example, additional working capital may be needed to meet the greater volume of business that will be generated by a project. Part of this increase in current assets may be supplied from increased accounts payable, but the remainder must come from permanent capital. This additional working capital is as much a part of the initial investment as the equipment itself. (We explain the amount of working capital required for a typical investment project in Section 10.3.2.)

- **Manufacturing, Operating, and Maintenance Costs.**   The costs associated with manufacturing a new product need to be determined. Typical manufacturing costs include labor, materials, and overhead costs, the last of which cover items such as power, water, and indirect labor. Investments in fixed assets normally require periodic outlays for repairs and maintenance and for additional operating costs, all of which must be considered in investment analysis.

- **Leasing Expenses.**   When a piece of equipment or a building is leased (instead of purchased) for business use, leasing expenses become cash outflows. Many firms lease computers, automobiles, industrial equipment, and other items that are subject to technological obsolescence.

- **Interest and Repayment of Borrowed Funds.** When we borrow money to finance a project, we need to make interest payments as well as payments on the principal. Proceeds from both short-term borrowing (bank loans) and long-term borrowing (bonds) are treated as cash inflows, but repayments of debts (both principal and interest) are classified as cash outflows.
- **Income Taxes and Tax Credits.** Any income tax payments following profitable operations should be treated as cash outflows for capital budgeting purposes. As we learned in Chapter 9, when an investment is made in depreciable assets, depreciation is an expense that offsets part of what would otherwise be additional taxable income. This strategy is called a **tax shield** or **tax savings**, and we must take it into account when calculating income taxes. If any investment **tax credit** is allowed, it will directly reduce income taxes, resulting in cash inflows.

10.2.2 Elements of Cash Inflows

The following are potential sources of cash inflow over the project life.

- **Borrowed Funds.** If you finance your investment by borrowing, the borrowed funds will appear as cash inflow to the project at the time they are borrowed. From these funds, the purchase of new equipment or any other investment will be paid out.
- **Operating Revenues.** If the primary purpose of undertaking the project is to increase production or service capacity to meet increased demand, the new project will be bringing in additional revenues.
- **Cost Savings (or Cost Reduction).** If the primary purpose of undertaking a new project is to reduce operating costs, the amount involved should be treated as a cash inflow for capital budgeting purposes. A reduction in costs is equivalent to an increase in revenues—even though the actual sales revenues may remain unchanged.
- **Salvage Value.** In many cases, the estimated salvage value of a proposed asset is so small and occurs so far in the future that it may have no significant effect on the decision to undertake a project. Furthermore, any salvage value that is realized may be offset by removal and dismantling costs. In situations where the estimated salvage value is significant, the net salvage value is viewed as a cash inflow at the time of disposal. The **net salvage value** of the existing asset is the selling price of the asset, minus any costs incurred in selling, dismantling, and removing it, and this value is subject to taxable gain or loss.
- **Working-capital Release.** As a project approaches termination, inventories are sold off and receivables are collected. That is, at the end of the project, these items can be liquidated at their cost. As this occurs, the company experiences an end-of-project cash flow that is equal to the net working-capital investment that was made when the project began. This recovery of working capital is not taxable income, since it merely represents a return of investment funds to the company.

10.2.3 Classification of Cash Flow Elements

Once the cash flow elements (both inflows and outflows) are determined, we may group them into three categories: (1) cash flow elements associated with operations, (2) cash flow elements associated with investment activities (capital expenditures), and (3) cash flow elements associated with project financing (such as borrowing). The main purpose of grouping cash flows this way is to provide information about the operating, investing, and financing activities of a project.

Operating Activities

In general, cash flows from operations include current sales revenues, the cost of goods sold, operating expenses, and income taxes. Cash flows from operations should generally reflect the cash effects of transactions entering into the determination of net income. The interest portion of a loan repayment is a deductible expense that is allowed in determining net income, and it is included in operating activities. Since we usually look only at yearly flows, it is logical to express all cash flows on a yearly basis.

As mentioned in Chapter 9, we can determine the net cash flow from operations on the basis of either (1) net income, or (2) cash flow as determined by computing income taxes in a separate step. When we use net income as the starting point for cash-flow determination, we should add any noncash expenses (mainly depreciation and amortization expenses) to net income to estimate the net cash flow from the operation. Recall that depreciation (or amortization) is not a cash flow, but is deducted along with operating expenses and lease costs from gross income to find taxable income and, therefore, taxes. Accountants calculate net income by subtracting taxes from taxable income, but depreciation—which is not a cash flow—was already subtracted to find taxable income, so it must be added back to net income if we wish to use the net-income figure as an intermediate step along the path to after-tax cash flow. Mathematically, it is easy to show that the two approaches are identical. Thus,

$$\text{Cash flow from operation} = \text{Net income} + (\text{Depreciation or amortization})$$

Investing Activities

In general, three types of investment flows are associated with buying a piece of equipment: the original investment, the salvage value at the end of the useful life of the equipment, and the working-capital investment or recovery. We will assume that our outflow for both capital investment and working-capital investment take place in year 0. It is possible, however, that both investments will occur, not instantaneously, but rather over a few months as the project gets into gear; we could then use year 1 as an investment year. (Capital expenditures may occur over several years before a large investment project becomes fully operational. In this case, we should enter all expenditures as they occur.) For a small project, either method of timing these flows is satisfactory, because the numerical differences are likely to be insignificant.

Financing Activities

Cash flows classified as financing activities include (1) the amount of borrowing, and (2) the repayment of principal. Recall that interest payments are tax-deductible expenses, so they are classified as operating, not financing, activities.

The net cash flow for a given year is simply the sum of the net cash flows from operating, investing, and financing activities. Table 10.1 can be used as a checklist when you set up a cash flow statement, because it classifies each type of cash flow element as an operating, investing, or financing activity.

TABLE 10.1 **Classifying Cash Flow Elements and Their Equivalent Terms as Practiced in Business**

Activities	Cash Flow Element	Other Terms Used in Business
Operating activities	Gross income	Gross revenue, sales revenue, gross profit, operating revenue
	Cost savings	Cost reduction
	Manufacturing expenses	Cost of goods sold, cost of revenue
	Operations and maintenance cost	Operating expenses
	Operating income	Operating profit, operating margin
	Interest expenses	Interest payments, debt cost, cost of financing
	Income taxes	Income taxes owed
Investing activities	Capital investment	Purchase of new equipment, capital expenditure
	Salvage value	Net selling price, disposal value, resale value
	Investment in working capital	Working-capital requirement
	Release of working capital	Working-capital recovery
	Gains taxes	Capital gains taxes, ordinary gains taxes
Financing activities	Borrowed funds	Borrowed amounts, loan amount
	Repayment of principal	Loan repayment, debt repayment

10.3 Developing Cash Flow Statements

In this section, we will illustrate through a series of numerical examples how we actually prepare a project's cash flow statement; a generic version is shown in Figure 10.2, in which we first determine the net income from operations and then adjust the net income by adding any noncash expenses, mainly depreciation (or amortization). Then, we consider the cash flows associated with the investing and financing activities.

10.3.1 When Projects Require Only Operating and Investing Activities

We will start with the simple case of generating after-tax cash flows for an investment project with only operating and investment activities. In the sections ahead, we will add complexities to this problem by including working-capital investments (Section 10.3.2) and borrowing activities (Section 10.3.3). We will also consider a case in which a project generates a negative taxable income (operating loss) for an operating year.

Figure 10.2 A popular format used for presenting a cash flow statement.

EXAMPLE 10.1 **Cash Flow Statement: Operating and Investing Activities for an Expansion Project**

A computerized machining center has been proposed for a tool-manufacturing company. If the new system (which costs $125,000) is installed, it will generate annual revenues of $100,000 and will require $20,000 in annual labor, $12,000 in annual material expenses, and another $8,000 in annual overhead (power and utility) expenses. The automation facility would be classified as a seven-year MACRS property.

The company expects to phase out the facility at the end of five years, at which time it will be sold for $50,000. Find the year-by-year after-tax net cash flow for the project at a 40% marginal tax rate, and determine the after-tax net present worth of the project at the company's MARR of 15%.

DISCUSSION: We can approach the problem in two steps by using the format shown in Figure 10.2 to generate an income statement and then a cash flow statement. We will follow this form in our subsequent listing of givens and unknowns. In year 0 (that is, at present) we have an investment cost of $125,000 for the equipment.[3] This cost will be depreciated in years 1 to 5. The revenues and costs are uniform annual flows in years 1 to 5. We can see that once we find depreciation allowances for each year, we can easily compute the results for years 1 to 4, which have fixed revenue and expense entries along with the variable depreciation charges. In year 5, we will need to incorporate the salvage value of the asset and any gains tax from its disposal.

---

[3] We will assume that the asset is purchased and placed in service at the beginning of year 1 (or the end of year 0), and the first year's depreciation will be claimed at the end of year 1.

We will use the business convention that no signs (positive or negative) be used in preparing the income statement, except in the situation where we have a negative taxable income or tax savings. In that situation, we will use parentheses to denote a negative entry. However, in preparing the cash flow statement, we will explicitly observe the sign convention: A positive sign indicates a cash inflow; a negative sign or parentheses indicate a cash outflow.

SOLUTION

Given: Preceding cash flow information.

Find: After-tax cash flow.

Before presenting the cash flow table, we need to do some preliminary calculations. The following notes explain the essential items in Table 10.2.

TABLE 10.2 Cash Flow Statement for the Automated Machining Center Project (Example 10.1)

	A	B	C	D	E	F	G
1							
2	**Income Statement**						
3	End of Year	0	1	2	3	4	5
4							
5	Revenue		$ 100,000	$ 100,000	$ 100,000	$ 100,000	$ 100,000
6	Expenses:						
7	Labor		$ 20,000	$ 20,000	$ 20,000	$ 20,000	$ 20,000
8	Materials		$ 12,000	$ 12,000	$ 12,000	$ 12,000	$ 12,000
9	Overhead		$ 8,000	$ 8,000	$ 8,000	$ 8,000	$ 8,000
10	Depreciation		$ 17,863	$ 30,613	$ 21,863	$ 15,613	$ 5,581
11							
12	Taxable income		$ 42,137	$ 29,387	$ 38,137	$ 44,387	$ 54,419
13	Income taxes (40%)		$ 16,855	$ 11,755	$ 15,255	$ 17,755	$ 21,768
14							
15	Net income		$ 25,282	$ 17,632	$ 22,882	$ 26,632	$ 32,651
16							
17	**Cash Flow Statement**						
18	**Operating Activities:**						
19	Net income		$ 25,282	$ 17,632	$ 22,882	$ 26,632	$ 32,651
20	Depreciation		$ 17,863	$ 30,613	$ 21,863	$ 15,613	$ 5,581
21	**Investment Activities:**						
22	Milling machine	$(125,000)					
23	Salvage value		=-(G23-(-B22-SUM(C10:G10)))*0.4				$ 50,000
24	Gains tax						$ (6,613)
25	Working capital						
26							
27	**Net Cash Flow**	$(125,000)	$ 43,145	$ 48,245	$ 44,745	$ 42,245	$ 81,619
28							
29							
30		PW(15%)=	$ 43,152	=NPV(15%,C27:G27)+B27			
31		IRR =	27.62%				
32			=IRR(B27:G27,15%)				
33							

- **Calculation of depreciation:**
  1. The depreciation allowances for a seven-year property are in respective percentages of 14.29%, 24.49%, 17.49%, 12.49%, 8.93%, 8.92%, 8.93%, and 4.46%. (See Table 9.6.)
  2. Since the asset is disposed of in the fifth tax year (the last year's depreciation), the applicable depreciation amounts would be $17,863, $30,613, $21,863, $15,613, and $5,581. Note that the fifth year depreciation allowance, which would ordinarily be $11,163, is halved due to the half-year convention.
- **Salvage value and gain taxes:** In year 5, we must deal with two aspects of the asset's disposal: salvage value and gains[4] (both ordinary as well as capital). Taxable gains are calculated as follows:
  1. The total depreciation in years 1 to 5 is

  $$\$17,863 + \$30,613 + \$21,863 + \$15,613 + \$5,581 = \$91,533$$

  2. The book value at the end of period 5 is the cost basis minus the total depreciation:

  $$\$125,000 - \$91,533 = \$33,467$$

  3. The taxable gains on the sale are the salvage value minus the book value:

  $$\$50,000 - \$33,467 = \$16,533$$

  (The salvage value is less than the cost basis, so the gains are ordinary gains.)
  4. The tax on the ordinary gains is $16,533 \times 40\% = \$6,613$. This is the amount placed in the table under "Gains Tax."
- **Investment analysis:** Once we obtain the project's after-tax net cash flows, we can determine their equivalent present worth at the firm's interest rate. The after-tax cash flow series from the cash flow statement is shown in Figure 10.3. Since this series does not contain any patterns to simplify our calculations, we must find the net present worth of each payment. Using $i = 15\%$, we have

$$
\begin{aligned}
PW(15\%) = {}& -\$125,000 + \$43,145(P/F, 15\%, 1) \\
& + \$48,245(P/F, 15\%, 2) + \$44,745(P/F, 15\%, 3) \\
& + \$42,245(P/F, 15\%, 4) + \$81,619(P/F, 15\%, 5) \\
= {}& \$43,152
\end{aligned}
$$

---

[4] Even though gains from equipment disposal have an effect on income tax calculations, they should not be viewed as ordinary operating income. Therefore, in preparing the income statement, capital expenditures and related items such as gains tax and salvage value, are not included. However, these items represent actual cash flows in the year they occur and must be shown in the cash flow.

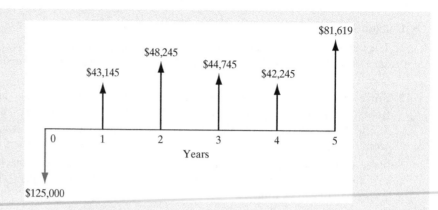

Figure 10.3 Cash flow diagram (Example 10.1).

This means that investing $125,000 in this automated facility would bring in enough revenue to recover the initial investment and the cost of funds with a surplus of $43,152.

10.3.2 When Projects Require Working-Capital Investments

In many cases, changing a production process by replacing old equipment or by adding a new product line will have an impact on cash balances, accounts receivable, inventory, and accounts payable. For example, if a company is going to market a new product, inventories of the product and larger inventories of raw materials will be needed. Accounts receivable from sales will increase, and management might also decide to carry more cash because of the higher volume of activities. These investments in working capital are investments just like those in depreciable assets, except that they have no tax effects: The flows always sum to zero over the life of a project, but the inflows and outflows are shifted in time, so they do affect net present worth.

Consider the case of a firm that is planning a new product line. The new product will require a two-month's supply of raw materials at a cost of $40,000. The firm could provide $40,000 in cash on hand to pay for them. Alternatively, the firm could finance these raw materials via a $30,000 increase in accounts payable (60-day purchases) by buying on credit. The balance of $10,000 represents the amount of net working capital that must be financed (or invested).

Working-capital requirements differ according to the nature of the investment project. For example, larger projects may require greater average investments in inventories and accounts receivable than would smaller ones. Projects involving the acquisition of improved equipment entail different considerations. If the new equipment produces more rapidly than the old equipment, the firm may be able to decrease its average inventory holdings, because new orders can be filled faster as a result of using the new equipment. One of the main advantages cited in installing advanced manufacturing systems (such as flexible manufacturing systems) is the reduction in inventory made possible by the ability to respond to market demand more quickly. Therefore, it is also possible for working-capital needs to *decrease* because of an investment. If inventory levels were to decrease at the start of a project, the decrease would be considered a *cash inflow*, since the cash freed up from inventory could be put to use in other places. (See Example 10.5.)

Two examples illustrate the effects of working capital on a project's cash flows. Example 10.2 shows how the net working-capital requirement is computed, and Example 10.3 examines the effects of working capital on the automated machining-center project discussed in Example 10.1.

EXAMPLE 10.2 Working-Capital Requirement

Suppose that in Example 10.1 the tool-manufacturing company's annual revenue projection of $100,000 is based on an annual volume of 10,000 units (or 833 units per month). Assume the accounting information in Table 10.3.

TABLE 10.3 Elements of Working Capital

Price (revenue) per unit	$10.00
Unit variable manufacturing costs:	
Labor	$2.00
Material	$1.20
Overhead	$0.80
Monthly volume	833 units
Finished-goods inventory to maintain	2-month supply
Raw-materials inventory to maintain	1-month supply
Accounts payable	30 days
Accounts receivable	60 days

The accounts receivable period of 60 days means that revenues from the current month's sales will be collected two months later. Similarly, accounts payable of 30 days indicates that payment for materials will be made approximately one month after the materials are received. Determine the working-capital requirement for this operation.

SOLUTION

Given: Preceding information.
Find: Working-capital requirement.

Figure 10.4 illustrates the working-capital requirement for the first 12-month period. Accounts receivable of $16,667 (two months' sales) means that in year 1, the company will have cash inflows of $83,333—less than the projected sales of $100,000 ($8,333 × 12). In years 2 to 5, collections will be $100,000—equal to sales, because the beginning and ending accounts receivable will be $16,667, with sales of $100,000. Collection of the final accounts receivable of $16,667 would occur in the first two months of year 6, but can be added to the year 5 revenue to simplify the calculations. The important point is that cash inflow lags sales by $16,667 in the first year.

During year 1	Income/Expense reported	Actual cash received/paid	Difference
Sales	$100,000 (10,000 units)	$83,333	−$16,667
Expenses	$40,000 (10,000 units)	$46,667 (11,667 units)	+$6,667
Income taxes	$16,855	$16,855	0
Net amount	$43,145	$19,812	−$23,333

This differential amount must be invested at the beginning of the year.

Figure 10.4 Illustration of working-capital requirement (Example 10.2).

Assuming that the company wishes to build up two months' inventory during the first year, it must produce $833 \times 2 = 1,667$ more units than are sold the first year. The extra cost of these goods in the first year is 1,667 units ($4 variable cost per unit) or $6,667. The finished-goods inventory of $6,667 represents the variable cost incurred to produce 1,667 more units than are sold in the first year. In years 2 to 4, the company will produce and sell 10,000 units per year while maintaining its 1,667-unit supply of finished goods. In the final year of operations, the company will produce only 8,333 units (for ten months) and will use up the finished-goods inventory. As 1,667 units of the finished-goods inventory get liquidated during the last year (exactly in the first two months of year 6), working capital in the amount of $6,667 will be released. Along with the collection of the final accounts receivable of $16,667, a total working-capital release of $23,333 will remain when the project terminates. Now we can calculate the working-capital requirements as given in Table 10.4.

TABLE 10.4 Calculation of Net Change in Working Capital

Current assets	Accounts receivable (833 units/month × 2 months × $10)	$16,667
	Finished-goods inventory (833 units/month × 2 months × $4)	6,667
	Raw-materials inventory (833 units/month × 1 month × $1.20)	1,000
Current liabilities	Accounts payable (purchase of raw materials) (833 units/month × 1 month × $1.20)	(1,000)
	Net change in working capital	$23,333

COMMENTS: During the first year, the company produces 11,667 units to maintain two months' finished-goods inventory, but it sells only 10,000 units. On what basis should the company calculate the net income during the first year (i.e., use 10,000 or 11,667 units)? *Any* increases in inventory expenses will reduce the taxable income; therefore, this calculation is based on 10,000 units. The reason is that the accounting measure of net income is based on the **matching concept**. If we report revenue when it is *earned* (whether it is actually received or not), and we report expenses when they are *incurred* (whether they are paid or not), we are using the *accrual method* of accounting. By tax law, this method *must* be used for purchases and sales whenever business transactions involve an inventory. Therefore, most manufacturing and merchandising businesses use the accrual basis in recording revenues and expenses. Any cash inventory expenses not accounted for in the net-income calculation will be reflected in changes in working capital.

EXAMPLE 10.3 Cash Flow Statement, Including Working Capital

Update the after-tax cash flows for the automated machining center project of Example 10.1 by including a working-capital requirement of $23,333 in year 0 and full recovery of the working capital at the end of year 5.

SOLUTION

Given: Flows as in Example 10.1, with the addition of a $23,333 working-capital requirement.

Find: Net after-tax cash flows with working capital and present worth.

Using the procedure outlined in Example 10.2, we group the net after-tax cash flows for this machining-center project, as shown in Table 10.5. As the table indicates, investments in working capital are cash outflows when they are expected to occur, and recoveries are treated as cash inflows at the times they are expected to materialize. In this example, we assume that the investment in working capital made at period 0 will be recovered at the end of the project's life. (See Figure 10.5.)[5] Moreover, we also assume a full recovery of the initial working capital. However, in many situations, the investment in working capital may not be fully recovered (e.g., inventories may deteriorate in value or become obsolete over time).

Period	0	1	2	3	4	5
Investment	−$23,333	−$23,333	−$23,333	−$23,333	−$23,333	0
Recovery	0	$23,333	$23,333	$23,333	$23,333	$23,333
Net flow	−$23,333	0	0	0	0	$23,333

[5] In fact, we could assume that the investment in working capital would be recovered at the end of the first operating cycle (say, year 1). However, the same amount of investment in working capital has to be made again at the beginning of year 2 for the second operating cycle, and the process repeats until the project terminates. Therefore, the net cash flow transaction looks as though the initial working capital will be recovered at the end of the project's life.

TABLE 10.5   Cash Flow Statement for the Automated Machining-Center Project with Working-Capital Requirement

	A	B	C	D	E	F	G
1							
2	**Income Statement**						
3	End of Year	0	1	2	3	4	5
4							
5	Revenue		$ 100,000	$ 100,000	$ 100,000	$ 100,000	$ 100,000
6	Expenses:						
7	Labor		$ 20,000	$ 20,000	$ 20,000	$ 20,000	$ 20,000
8	Materials		$ 12,000	$ 12,000	$ 12,000	$ 12,000	$ 12,000
9	Overhead		$ 8,000	$ 8,000	$ 8,000	$ 8,000	$ 8,000
10	Depreciation		$ 17,863	$ 30,613	$ 21,863	$ 15,613	$ 5,581
11							
12	Taxable income		$ 42,137	$ 29,387	$ 38,137	$ 44,387	$ 54,419
13	Income taxes (40%)		$ 16,855	$ 11,755	$ 15,255	$ 17,755	$ 21,768
14							
15	Net Income		$ 25,282	$ 17,632	$ 22,882	$ 26,632	$ 32,651
16							
17	**Cash Flow Statement**						
18	**Operating Activities:**						
19	Net income		$ 25,282	$ 17,632	$ 22,882	$ 26,632	$ 32,651
20	Depreciation		$ 17,863	$ 30,613	$ 21,863	$ 15,613	$ 5,581
21	**Investment Activities:**						
22	Milling machine	$(125,000)					
23	Salvage value			=-(G23-(-B22-SUM(C10:G10)))*0.4			$ 50,000
24	Gains tax						$ (6,613)
25	Working capital	$ (23,331)					$ 23,331
26							
27	**Net Cash Flow**	$(148,331)	$ 43,145	$ 48,245	$ 44,745	$ 42,245	$ 104,950
28							
29							
30		PW(15%)=	$ 31,420	=NPV(15%,C27:G27)+B27			
31		IRR =	22.55%				
32			=IRR(B27:G27,15%)				
33							
34							

The equivalent net present worth of the after-tax cash flows, including the effects of working capital, is calculated as

$$PW(15\%) = -\$148,333 + \$43,145(P/F, 15\%, 1) + \cdots$$
$$+ \$104,952(P/F, 15\%, 5)$$
$$= \$31,420$$

This present-worth value is $11,732 less than that in the situation with no working-capital requirement (Example 10.1), demonstrating that working-capital requirements play a critical role in assessing a project's worth.

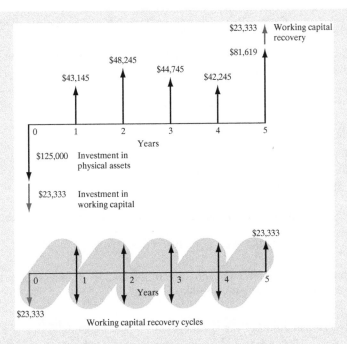

Figure 10.5 Cash flow diagram (Example 10.3).

COMMENTS: The $11,732 reduction in present worth is just the present worth of an annual series of 15% interest payments on the working capital, which is borrowed by the project at time 0 and repaid at the end of year 5:

$$\underbrace{23,333(15\%)}_{\text{annual interest lost}}(P/A, 15\%, 5) = 11,732$$

The investment tied up in working capital results in lost earnings.

10.3.3 When Projects are Financed with Borrowed Funds

Many companies use a mixture of debt and equity to finance their physical plant and equipment. The ratio of total debt to total investment, generally called the **debt ratio**, represents the percentage of the total initial investment provided by borrowed funds. For example, a debt ratio of 0.3 indicates that 30% of the initial investment is borrowed and the rest is provided from the company's earnings (also known as **equity**). Since interest is a tax-deductible expense, companies in high tax brackets may incur lower after-tax financing costs by raising money through debt. (Along with the effect of debt on taxes, the method of repaying the loan can have a significant impact on taxes. We will discuss the issue of project financing in Chapter 15.)

EXAMPLE 10.4 Cash Flow Statement with Financing (Borrowing)

Rework Example 10.3, assuming that $62,500 of the $125,000 paid for the investment is obtained through debt financing (debt ratio = 0.5). The loan is to be repaid in equal annual installments at 10% interest over five years. The remaining $62,500 will be provided by equity (e.g., from retained earnings).

SOLUTION

Given: Same as in Example 10.3, but $62,500 is borrowed and then repaid in equal installments over five years at 10% interest.

Find: Net after-tax cash flows in each year.

We first need to compute the size of the annual loan repayment installments as

$$\$62,500(A/P, 10\%, 5) = \$16,487$$

Next, we determine the repayment schedule of the loan by itemizing both the interest and principal represented in each annual repayment.

Year	Beginning Balance	Interest Payment	Principal Payment	Ending Balance
1	$62,500	$6,250	$10,237	$52,263
2	52,263	5,226	11,261	41,002
3	41,002	4,100	12,387	28,615
4	28,615	2,861	13,626	14,989
5	14,989	1,499	14,988	0

The resulting after-tax cash flow is detailed in Table 10.6. The present-value equivalent of the after-tax cash flow series is

$$PW(15\%) = -\$85,333 + \$29,158(P/F, 15\%, 1) + \cdots$$
$$+ \$89,065(P/F, 15\%, 5)$$
$$= \$44,439$$

When this amount is compared with the amount found in the case when there was no borrowing ($31,420), we see that debt financing actually increases the present worth by $13,018. This surprising result is largely caused by the firm's being able to borrow the funds at a cheaper rate (10%) than its MARR (opportunity cost rate) of 15%. We should be careful in interpreting the result, however: It is true, to some extent, that firms can usually borrow money at lower rates than their MARR, but if the firm can borrow money at a significantly lower rate, that also affects its MARR, because the borrowing rate is one of the elements determining the MARR. Therefore, a significant difference in present values between "with borrowings" and "without borrowings" is not expected in practice. We will also address this important issue in Chapter 15.

TABLE 10.6 Cash Flow Statement for the Automated Machining-Center Project with Debt Financing (Example 10.4)

	A	B	C	D	E	F	G
1							
2	**Income Statement**						
3	End of Year	0	1	2	3	4	5
4							
5	Revenue		$ 100,000	$ 100,000	$ 100,000	$ 100,000	$ 100,000
6	Expenses:	=IPMT(12%,5,5,-64800)					
7	Labor		$ 20,000	$ 20,000	$ 20,000	$ 20,000	$ 20,000
8	Materials		$ 12,000	$ 12,000	$ 12,000	$ 12,000	$ 12,000
9	Overhead		$ 8,000	$ 8,000	$ 8,000	$ 8,000	$ 8,000
10	Debt interest		$ 6,250	$ 5,226	$ 4,100	$ 2,861	$ 1,499
11	Depreciation		$ 17,863	$ 30,613	$ 21,863	$ 15,613	$ 5,581
12							
13	Taxable income		$ 35,887	$ 24,161	$ 34,037	$ 41,526	$ 52,920
14	Income taxes (40%)		$ 14,355	$ 9,664	$ 13,615	$ 16,610	$ 21,168
15							
16	Net income		$ 21,532	$ 14,497	$ 20,422	$ 24,916	$ 31,752
17			=C5-SUM(C7:C11)				=G13-G14
18	**Cash Flow Statement**						
19	Operating Activities:						
20	Net income		$ 21,532	$ 14,497	$ 20,422	$ 24,916	$ 31,752
21	Depreciation		$ 17,863	$ 30,613	$ 21,863	$ 15,613	$ 5,581
22	Investment Activities:						
23	Milling machine	$ (125,000)					
24	Salvage value		=-(G24-(-B23-SUM(C11:G11)))*0.4				$ 50,000
25	Gains tax						$ (6,613)
26	Working capital	$ (23,331)					$ 23,331
27	Financing Activities:				=PPMT(12%,5,5,-64800)		
28	Borrowed funds	$ 62,500					
29	Principal repayment		$ (10,237)	$ (11,261)	$ (12,387)	$ (13,626)	$ (14,988)
30							
31	Net Cash Flow	$ (85,831)	$ 29,158	$ 33,849	$ 29,898	$ 26,903	$ 89,063
32							
33						=SUM(G20:G29)	
34	PW(15%)=	$ 44,439	=NPV(15%,C31:G31)+B31				
35	IRR =	31.88%				=IRR(B31:G31,15%)	
36							

10.3.4 When Projects Result in Negative Taxable Income

In a typical project year, revenues may not be large enough to offset expenses, thereby resulting in an operating loss. Now, a negative taxable income does *not* mean that a firm does not need to pay income tax; rather, the loss figure can be used to reduce the taxable incomes generated by other business operations.[6] Therefore, a negative taxable income usually results in a **tax savings**. When we evaluate an investment project with the use of an incremental tax rate, we also assume that the firm has sufficient taxable income from other activities, so that changes due to the project under consideration will not change the incremental tax rate.

---

[6] Even if the firm does not have any other taxable income to offset in the current tax year, the operating loss can be carried back to each of the preceding three years and forward for the next 15 years to offset taxable income in those years.

When we compare **cost-only** mutually exclusive projects (service projects), we have no revenues to consider in their cash flow analysis. In this situation, we typically assume no revenue (zero), but proceed as before in constructing the after-tax cash flow statement. With no revenue to match expenses, we have a negative taxable income, again resulting in tax savings. Example 10.5 illustrates how we may develop an after-tax cash flow statement for this type of project.

EXAMPLE 10.5 Project Cash Flows for a Cost-Only Project

Alcoa Aluminum's McCook plant produces aluminum coils, sheets, and plates. Its annual production runs at 400 million pounds. In an effort to improve McCook's current production system, an engineering team led by the divisional vice president went on a fact-finding tour of Japanese aluminum and steel companies to observe their production systems and methods. Cited among the team's observations were large fans, which the Japanese companies used to reduce the time that coils need to cool down after various processing operations. Cooling the hot process coils with the fans was estimated to significantly reduce the queue or work-in-process (WIP) inventory buildup allowed for cooling. The approach also reduced production lead-time and improved delivery performance. The possibility of reducing production time and, as a consequence, the WIP inventory, excited the team members, particularly the vice president. After the trip, Neal Donaldson, the plant engineer, was asked to investigate the economic feasibility of installing cooling fans at the McCook plant. Neal's job was to justify the purchase of cooling fans for his plant. He was given one week to prove that the idea was a good one. Essentially, all he knew was the number of fans, their locations, and the project cost. Everything else was left to Neal's devices. Suppose that he compiled the following financial data.

- The project will require an investment of $536,000 in cooling fans now.
- The cooling fans would provide 16 years of service with no appreciable salvage values, considering the removal costs.
- It is expected that the amount of time required between hot rolling and the next operation would be reduced from five days to two days. Cold-rolling queue time also would be reduced from two days to one day for each cold-roll pass. The net effect of these changes would be a reduction in the WIP inventory at a value of $2,121,000. Because of the lead-time involved in installing the fans, as well as the consumption of stockpiled WIP inventory, this working-capital release would be realized one year after the fans are installed.
- The cooling fans will be depreciated according to seven-year MACRS.
- Annual electricity costs are estimated to rise by $86,000.
- The firm's after-tax required rate of return is known to be 20% for this type of cost-reduction project.

Develop the project cash flows over the service period of the fans, and determine whether the investment is a wise one, based on 20% interest and 40% marginal tax rate.

SOLUTION

Given: Required investment = \$536,000, service period = 16 years, salvage value = \$0, depreciation method for cooling fans = 7-year MACRS, working-capital release = \$2,121,000 one year later, annual operating cost (electricity) = \$86,000, t_m = 40%, and MARR = 20%.

Find: (a) Annual after-tax cash flows, (b) the investment decision on the basis of the NPW, and (c) the investment decision on the basis of the IRR.

(a) Because we can assume that the annual revenues would stay the same as they were before and after the installation of the cooling fans, we can treat these unknown revenue figures as zero. Table 10.7 summarizes the cash flow statement for the cooling-fan project. With no revenue to offset the expenses, the taxable income will be negative, resulting in tax savings. Note that the working-capital recovery (as opposed to working-capital investment for a typical investment project) is shown in year 1. Note also that there is no gains tax, because the cooling fans are fully depreciated with a zero salvage value.

(b) At i = 20%, the NPW of this investment would be

$$PW(20\%) = -\$536,000 + \$2,100,038(P/F, 20\%, 1)$$
$$+\$906(P/F, 20\%, 2) + \cdots$$
$$-\$2,172,600(P/F, 20\%, 16)$$
$$= \$991,008$$

Even with only a one-time savings in WIP, this cooling-fan project is economically justifiable.

(c) If we are to justify the investment on the basis of the internal rate of return, we first need to determine whether the investment is a simple or complex one. Since there are several sign changes in the cash flow series, this is not a simple investment, indicating the possibility of multiple rates of return. In fact, as shown in Figure 10.6, the project has two rates of return, one at 4.24% and the other at 291.56%. Neither is a true rate of return, as we learned in Section 7.3.2, so we may proceed to abandon the rate-of-return approach and use the NPW criterion. If you desire to find the true IRR for this project, you need to follow the procedures outlined in Section 7.3.4. At a MARR of 20%, the true IRR (or return on invested capital) is 241.87%, which is significantly larger than the MARR and indicates acceptance of the investment. Note that this is the same conclusion that we reached in part (b).

COMMENTS: As the cooling fans reach the end of their service lives, we need to add the working-capital investment (\$2,120,000) at the end of year 16, working under the assumption that the plant will return to the former system without the cooling fans and thus will require the additional investment in working capital.

 If the new system has proven to be effective and the plant would remain in service, we need to make another investment to purchase a new set of cooling fans at the end of year 16. As a consequence of this new investment, there will be a working-capital release in the amount of \$2,120,000. However, this investment should bring benefits to the second cycle of the operation, so that it should be charged against the cash flows for the second cycle, not the first cycle.

TABLE 10.7 Cash Flow Statement for the Cooling-Fan Cost-Only Project (Example 10.5)

	A	B	C	D	E	F	G	H	I	J	K	R
1	**Income Statement**											
2	End of year	0	1	2	3	4	5	6	7	8	9 - 15	16
3												
4	Revenues											
5	Expenses:											
6	Depreciation		$ 76,594	$ 131,266	$ 93,746	$ 66,946	$ 47,865	$ 47,811	$ 47,865	$ 23,906		
7	Electricity cost		86,000	86,000	86,000	86,000	86,000	86,000	86,000	86,000	86,000	86,000
8	Taxable income		(162,594)	(217,266)	(179,746)	(152,946)	(133,865)	(133,811)	(133,865)	(109,906)	(86,000)	(86,000)
9	Income taxes		(65,038)	(86,906)	(71,898)	(61,178)	(53,546)	(53,524)	(53,546)	(43,962)	(34,400)	(34,400)
10												
11	Net income		$ (97,556)	$ (130,360)	$ (107,848)	$ (91,768)	$ (80,319)	$ (80,287)	$ (80,319)	$ (65,944)	$ (51,600)	$ (51,600)
12												
13	**Cash Flow Statement**											
14	Operating activities:											
15	Net income		(97,556)	(130,360)	(107,848)	(91,768)	(80,319)	(80,287)	(80,319)	(65,944)	(51,600)	(51,600)
16	Depreciation		76,594	131,266	93,746	66,946	47,865	47,811	47,865	23,906	-	
17	Investment activities:											
18	Cooling fans	(536,000)										
19	Salvage value											
20	Gains tax											
21	Working capital		2,121,000									(2,121,000)
22												
23	Net cash flow	$ (536,000)	$ 2,100,038	$ 906	$ (14,102)	$ (24,822)	$ (32,454)	$ (32,476)	$ (32,454)	$ (42,038)	$ (51,600)	$ (2,172,600)
24	PW(20%) = $ 991,009		→	=NPV(20%,C23:R23)+B23								
25												

Note: The working-capital release attributable to a reduction in work-in-process inventories will be realized at the end of year 1.

Figure 10.6 NPW plot for the cooling-fan project (Example 10.5).

10.3.5 When Projects Require Multiple Assets

Up to this point, our examples have been limited to situations in which only one asset was employed in a project. In many situations, however, a project may require the purchase of multiple assets with different property classes. For example, a typical engineering project may involve more than just the purchase of equipment: It may need a building in which to carry out manufacturing operations. The various assets may even be placed in service at different points in time. What we then have to do is itemize the timing of the investment requirements and the depreciation allowances according to the placement of the assets. Example 10.6 illustrates the development of project cash flows that require multiple assets.

EXAMPLE 10.6 A Project Requiring Multiple Assets

Langley Manufacturing Company (LMC), a manufacturer of fabricated metal products, is considering purchasing a new computer-controlled milling machine to produce a custom-ordered metal product. The following summarizes the relevant financial data related to the project.

- The machine costs $90,000. The costs for its installation, site preparation, and wiring are expected to be $10,000. The machine also needs special jigs and dies, which will cost $12,000. The milling machine is expected to last ten years, the jigs and dies five years. The machine will have a $10,000 salvage value at the end of its life. The special jigs and dies are worth only $1,000 as scrap metal at any time in their lives. The milling machine is classified as a seven-year MACRS property and the jigs and dies as a three-year MACRS property.

- LMC needs to either purchase or build an 8,000-ft^2 warehouse in which to store the product before it is shipped to the customer. LMC has decided to purchase a building near the plant at a cost of $160,000. For depreciation purposes, the warehouse cost of $160,000 is divided into $120,000 for the building (39-year real property) and $40,000 for land. At the end of ten years, the building will have a salvage value of $80,000, but the value of the land will have appreciated to $110,000.

- The revenue from increased production is expected to be $150,000 per year. The additional annual production costs are estimated as follows: materials, $22,000; labor, $32,000; energy $3,500; and other miscellaneous costs, $2,500.

- For the analysis, a ten-year life will be used. LMC has a marginal tax rate of 40% and a MARR of 18%. No money is borrowed to finance the project. Capital gains also will be taxed at 40%.[7]

DISCUSSION: Three types of assets are to be considered in this problem: the milling machine, the jigs and dies, and the warehouse. The first two assets are personal properties and the last is a real property. The cost basis for each asset has to be determined separately. For the milling machine, we need to add the site-preparation expense to the cost basis, whereas we need to subtract the land cost from the warehouse cost to establish the correct cost basis for the real property. The various cost bases are:

- The milling machine: $90,000 + $10,000 = $100,000
- Jigs and dies: $12,000
- Warehouse (building): $120,000
- Warehouse (land): $40,000

Since the jigs and dies last only five years, we need to make a specific assumption regarding the replacement cost at the end of that time. In this problem, we will assume that the replacement cost would be approximately equal to the cost of the initial purchase. We will also assume that the warehouse property will be placed in service in January, which indicates that the first year's depreciation percentage will be 2.4573%. (See Table 9.9.)

SOLUTION

Given: Preceding cash flow elements, $t_m = 40\%$, and MARR = 18%.
Find: Net after-tax cash flow and NPW.

Table 10.8 and Figure 10.7 summarize the net after-tax cash flows associated with the multiple-asset investment. In the table, we see that the milling machine and the jigs and dies are fully depreciated during the project life, whereas the building is not. We need to calculate the book value of the building at the end of the project life.

---

[7] Capital gains for corporations are taxed at a maximum rate of 35%. However, capital gains are also subject to state taxes, so the combined tax rate will be approximately 40%.

TABLE 10.8 Cash Flow Statement for LMC's Machining-Center Project with Multiple Assets (Example 10.6)

	A	B — 0	C — 1	D — 2	E — 3	F — 4	G — 5	H — 6	I — 7	J — 8	K — 9	L — 10
1	**Income Statement**											
2	End of Year	0	1	2	3	4	5	6	7	8	9	10
3												
4	Revenue		$150,000	$150,000	$150,000	$150,000	$150,000	$150,000	$150,000	$150,000	$150,000	$150,000
5	Expenses:											
6	Materials		22,000	22,000	22,000	22,000	22,000	22,000	22,000	22,000	22,000	22,000
7	Labor		32,000	32,000	32,000	32,000	32,000	32,000	32,000	32,000	32,000	32,000
8	Energy		3,500	3,500	3,500	3,500	3,500	3,500	3,500	3,500	3,500	3,500
9	Others		2,500	2,500	2,500	2,500	2,500	2,500	2,500	2,500	2,500	2,500
10	Depreciation:											
11	Building		2,949	3,077	3,077	3,077	3,077	3,077	3,077	3,077	3,077	2,949
12	Machines		14,290	24,490	17,490	12,490	8,930	8,920	8,930	4,460		
13	Tools		4,000	5,333	1,778	889		4,000	5,333	1,778	889	
14	Taxable income		68,761	57,100	67,655	73,544	77,993	74,003	72,660	80,685	86,034	87,051
15	Income taxes		27,504	22,840	27,062	29,418	31,197	29,601	29,064	32,274	34,414	34,820
16												
17	Net Income		$41,257	34,260	40,593	44,126	46,796	44,402	43,596	48,411	51,620	52,231
18	**Cash Flow Statement**											
19	Operating Activities:											
20	Net income		41,257	34,260	40,593	44,126	46,796	44,402	43,596	48,411	51,620	52,231
21	Depreciation		21,239	32,900	22,345	16,456	12,007	15,997	17,340	9,315	3,966	2,949
22	Investment Activities:											
23	Land	(40,000)										110,000
24	Building	(120,000)										80,000
25	Machines	(100,000)										10,000
26	Tools (1st cycle)	(12,000)					1,000					
27	Tools (2nd cycle)						(12,000)					1,000
28	Gains tax:											
29	Land											(28,000)
30	Building											3,794
31	Machines											(4,000)
32	Tools						(400)					(400)
33												
34	Net Cash Flow	$(272,000)	$62,496	$67,160	$62,938	$60,582	$47,403	$60,399	$60,936	$57,726	$55,586	$227,574
35												
36												

=B34+NPV(18%,C34:L34)

=-(L24-(-B24-SUM(C11:L11)))*0.4

PW(18%) = $32,343   IRR = 21%   =IRR(B34:L34,18%)

Note: Investment in tools (jigs and dies) will be repeated at the end of year 5, at the cost of the initial purchase.

541

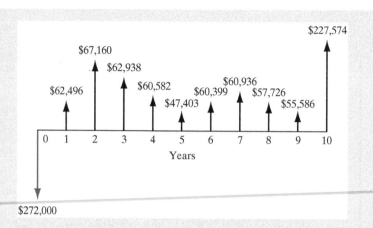

Figure 10.7 Cash flow diagram (Example 10.6).

We assume that the building will be disposed of December 31 in the 10th year, so that the midmonth convention also applies to the book-value calculation:

$$B_{10} = \$120,000 - (\$2,949 + \$3,077 \times 8 + \$2,949) = \$89,486$$

Then the gains (losses) are

$$\text{Salvage value} - \text{book value} = \$80,000 - \$89,486 = (\$9,486)$$

We can now calculate the gains or losses associated with the disposal of each asset as

Property (Asset)	Cost Basis	Salvage Value	Book Value	Gains (Losses)	Gains Taxes
Land	$40,000	$110,000	$40,000	$70,000	$28,000
Building	120,000	80,000	89,486	(9,486)	(3,794)
Milling machine	100,000	10,000	0	10,000	4,000
Jigs and dies	12,000	1,000	0	1,000	400

The NPW of the project is

$$\text{PW}(18\%) = -\$272,000 + \$62,496(P/F, 18\%, 1) + \$67,160(P/F, 18\%, 2)$$
$$+ \cdots + \$227,574(P/F, 18\%, 10)$$
$$= \$32,343 > 0$$

and the IRR for this investment is about 21%, which exceeds the MARR. Therefore, the project is acceptable.

COMMENTS: Note that the gains (losses) posted can be classified into two types: ordinary gains (losses) and capital gains. Only the $70,000 for land represents true capital gains, whereas the other figures represent ordinary gains (losses).

10.4 Generalized Cash-Flow Approach

If we analyze project cash flows for a corporation that consistently operates in the highest tax bracket, we can assume that the firm's marginal tax rate will remain the same, whether the project is accepted or rejected. In this situation, we may apply the top marginal tax rate (currently 35%) to each taxable item in the cash profile, thus obtaining the after-tax cash flows. By aggregating individual items, we obtain the project's net cash flows. This approach is referred to as the **generalized cash-flow approach**. As we will see in later chapters, it affords several analytical advantages. In particular, when we compare service projects, the generalized cash-flow method is computationally efficient, as we will demonstrate through a buy-versus-lease decision in Section 10.4.3.

10.4.1 Setting up Net Cash-Flow Equations

To construct a generalized cash-flow table, we first examine each cash-flow element. We can do this as follows, using the scheme for classifying cash flows that we have just developed:

$$
\begin{aligned}
A_n = \ &+\text{Revenues at time } n, (R_n) \\
&-\text{Expenses (excluding depreciation and} \\
&\quad \text{debt interest) at time } n, (E_n) \\
&-\text{Debt interest at time } n \ (IP_n) \\
&-\text{Income taxes at time } n, (T_n)
\end{aligned}
\left.\vphantom{\begin{aligned}1\\2\\3\\4\end{aligned}}\right\} \text{Operating activities}
$$

$$
\begin{aligned}
&-\text{Investment at time } n, (I_n) \\
&+\text{Net proceeds from sale at time } n, (S_n - G_n) \\
&-\text{Working capital investment at time } n, (W_n)
\end{aligned}
\left.\vphantom{\begin{aligned}1\\2\\3\end{aligned}}\right\} \text{Investing activities}
$$

$$
\begin{aligned}
&+\text{Proceeds from loan } n, (B_n) \\
&-\text{Repayment of principal } n, (PP_n)
\end{aligned}
\left.\vphantom{\begin{aligned}1\\2\end{aligned}}\right\} \text{Financing activities}
$$

Here, A_n is the net after-tax cash flow at the end of period n.

Depreciation (D_n) is *not* a cash flow and is therefore excluded from E_n (although it must be considered in calculating income taxes). Note that ($S_n - G_n$) represents the net proceeds from sale after adjustments for gains tax or loss credits (G_n). Not all terms are relevant in calculating a cash flow in every year; for example, the term ($S_n - G_n$) appears only when the asset is disposed of.

In terms of symbols, we can express A_n as

$$
\begin{aligned}
A_n = \ &R_n - E_n - IP_n - T_n && \leftarrow \text{Operating activities} \\
&-I_n + (S_n - G_n) - W_n && \leftarrow \text{Investing activities} \\
&+ B_n - PP_n && \leftarrow \text{Financing activities}
\end{aligned}
\tag{10.1}
$$

If we designate T_n as the total income taxes paid at time n and t_m as the marginal tax rate, income taxes on this project are

$$T_n = (\text{Taxable income})(\text{marginal tax rate})$$

$$= (R_n - E_n - \text{IP}_n - D_n)t_m$$

$$= (R_n - E_n)t_m - \underbrace{(\text{IP}_n + D_n)t_m}_{\text{tax shield (tax savings)}} \tag{10.2}$$

The term $(\text{IP}_n + D_n)t_m$ is known as the tax shield (or tax savings) from financing and asset depreciation. Now substituting the result of Eq. (10.2) into Eq. (10.1), we obtain

$$A_n = (R_n - E_n - \text{IP}_n)(1 - t_m) + \overbrace{t_m D_n}^{\text{Depreciation tax shield}}$$

$$-I_n + (S_n - G_n) - W_n$$

$$+B_n - \text{PP}_n \tag{10.3}$$

10.4.2 Presenting Cash Flows in Compact Tabular Formats

After-tax cash flow components over the project life can be grouped by type of activity in a compact tabular format as given in Table 10.9. However, in preparing their after-tax

TABLE 10.9 **Generalized Cash Flow Approach in Tabular Format**

Cash Flow Elements	End of Period			
	0	1	2	...N
Investment activities:				
$-I_n$				
$+S_n - G_n$				
$-W_n$				
Operating activities:				
$+(1 - t_m)(R_n)$				
$-(1 - t_m)(E_n)$				
$-(1 - t_m)(\text{IP}_n)$				
$+t_m D_n$				
Financing activities:				
$+B_n$				
$-\text{PP}_n$				
Net cash flow:				
A_n				

cash flow, most business firms adopt the income statement approach presented in previous sections, because they want to know the accounting income along with the cash flow statement.

EXAMPLE 10.7 Generalized Cash-Flow Approach

Consider again Example 10.4. Use the generalized cash-flow approach to obtain the after-tax cash flows.

SOLUTION
Given:

- Investment cost (I_0) = $125,000
- Investment in working capital (W_n) = $23,333
- Annual revenues (R_n) = $100,000
- Annual expenses other than depreciation and debt interest (E_n) = $40,000
- Debt interest (IP_n), years 1 to 5 = $6,250, $5,226, $4,100, $2,861, $1,499, respectively
- Principal repayment (PP_n), years 1 to 5 = $10,237, $11,261, $12,387, $13,626, $14,988, respectively
- Depreciation (D_n), years 1 to 5 = $17,863, $30,613, $21,863, $15,613, $5,581, respectively
- Marginal tax rate (t_m) = 40%
- Salvage value (S_n) = $50,000

Find: Annual after-tax cash flows (A_n).

Step 1. Find the cash flow at year 0:

(a) Investment in depreciable asset (I_0) = $125,000
(b) Investment in working capital (W_0) = $23,333
(c) Borrowed funds (B_0) = $62,500
(d) Net cash flow (A_0) = $-\$125,000 - \$23,333 + \$62,500 = -\$85,833$

Step 2. Find the cash flow in years 1 to 4:

$$A_n = (R_n - E_n - \mathrm{IP}_n)(1 - t_m) + D_n t_m - \mathrm{PP}_n$$

n	Net Operating Cash Flow ($)
1	$(100,000 - 40,000 - 6,250)(0.60) + 17,863(0.40) - 10,237 = \$29,158$
2	$(100,000 - 40,000 - 5,226)(0.60) + 30,613(0.40) - 11,261 = \$33,849$
3	$(100,000 - 40,000 - 4,100)(0.60) + 21,863(0.40) - 12,387 = \$29,898$
4	$(100,000 - 40,000 - 2,861)(0.60) + 15,613(0.40) - 13,626 = \$26,903$

Step 3. Find the cash flow for year 5:

(a) Operating cash flow:

$$(100,000 - 40,000 - 1,499)(0.60) + 5,581(0.40) - 14,988 = \$22,345$$

(b) Net salvage value

(as calculated in Example 10.1) $= \$50,000 - \$6,613 = \$43,387$

(c) Recovery of working capital, $W_5 = \$23,333$.

(d) Net cash flow in year 5, $A_5 = \$22,345 + \$43,387 + \$23,333 = \$89,065$.

Our results and overall calculations are summarized in Table 10.10. Checking them against the results we obtained in Table 10.6 confirms them.

TABLE 10.10 Cash Flow Statement for Example 10.4 Using the Generalized Cash-Flow Approach (Example 10.7)

	A	B	C	D	E	F	G
1							
2	**Generalized Cash Flow Approach**						
3	End of year	**0**	**1**	**2**	**3**	**4**	**5**
4							
5	**Operating Activities:**						
6	+ (1 - 0.40)(Revenue)		$ 60,000	$ 60,000	$ 60,000	$ 60,000	$ 60,000
7	- (1 - 0.40)(Expenses)		(24,000)	(24,000)	(24,000)	(24,000)	(24,000)
8	- (1 - 0.40)(Debt interest)		(3,750)	(3,136)	(2,460)	(1,717)	(899)
9	+ (0.40)(Depreciation)		7,145	12,246	8,745	6,246	2,232
10	**Investing Activities:**						
11	- Capital expenditure	(125,000)					
12	+ Net proceeds from sale						43,387
13	- Investment in working -capital	(23,331)					
14	+ Release of working-capital						23,331
15	**Financing Activities:**						
16	+ Borrowed funds	62,500					
17	- Repayment of principal		(10,237)	(11,261)	(12,387)	(13,626)	(14,988)
18							
19	Net cash flow	$ (85,831)	$ 29,158	$ 33,849	$ 29,898	$ 26,903	$ 89,063
20							

10.4.3 Lease-or-Buy Decision

A lease-or-buy decision begins only after a company has decided that the acquisition of a piece of equipment is necessary to carry out an investment project. Having made this decision, the company may be faced with several alternative methods of financing the acquisition: cash purchase, debt purchase, or acquisition via a lease. Here we will examine the lease financing option.

Advantages and Disadvantages of Leasing Equipment

Some of the issues you need to consider when you're looking at leasing are

- **Ownership.** Who does own the equipment? The most obvious downside to leasing is that when the lease runs out, you don't own the piece of equipment. Of course, this may also be an advantage, particularly for equipment like computers where your technology needs may change very quickly.
- **Cash Flow.** Do you need to conserve your cash for other business operations? This is the primary advantage to leasing. It eliminates a large, single expense that may drain your cash flow, freeing funds for other day-to-day needs.
- **Taxes.** What are the tax consequences in either leasing or owning the equipment? Leasing almost always allows you to expense your equipment costs, meaning that your lease payments can be deducted as business expenses. On the other hand, buying may allow you to deduct up to $500,000 worth of equipment in the year it is purchased;[8] anything above that amount gets depreciated over several years. With the first-year expense deduction, the "real cost" of a $10,000 machine may be only $6,000 if you are in the 40% tax bracket.
- **Technology Needs.** Do you need to worry about the technological obsolescence of the equipment if you do purchase it? Technology advances at a rapid rate for most high-tech equipment. If you buy a computer or other high-tech equipment outright, you may find yourself with outdated equipment in a few years with no discernible resale value. Leasing may allow you to try out new equipment configurations and update your system regularly to stay on top of the technology curve. On the other hand, if you have a "pass-down" policy in your company (where older technology gets used by certain departments), buying may be more effective.

Economic Analysis of Buy versus Lease Decision

With a debt purchase, the present-worth expression is similar to that for a cash purchase, except that it has additional items: loan repayments and a tax shield on interest payments. The only way that the lessee can evaluate the cost of a lease is to compare it against the best available estimate of what the cost would be if the lessee owned the equipment.

---

[8] This is known as the Section 179 deduction which allows businesses to deduct the full purchase price of qualifying equipment purchased or financed during the tax year. However, there are the dollar limits of the deduction. The new deduction limits in 2013 are $500,000 on the deduction, and the total amount of equipment purchased for the year cannot exceed $2,000,000. If the cost of your qualifying Section 179 property placed in service in a year is more than $2,000,000, you generally must reduce the dollar limit (but not below zero) by the amount of cost over $2,000,000. If the cost of your Section 179 property placed in service during 2013 is $2,500,000 or more, you cannot take a Section 179 deduction. These limits can change year to year, so you need to visit the IRS website to find the most updated information on the Section 179 deduction.

To lay the groundwork for a more general analysis, we will first consider how to analyze the lease-or-buy decision for a project with a single fixed asset for which the company expects a service life of N periods. Since the net after-tax revenue is the same for both alternatives, we need only consider the incremental cost of owning the asset and the incremental cost of leasing.

- **Buy Option with 100% Debt Financing.** Using the generalized cash flow approach, we may express the incremental cost of owning the asset by borrowing as

$$PW(i)_{Buy} = \ +PW \text{ of loan principal repayment}$$
$$+PW \text{ of after-tax OM costs including debt interest}$$
$$-PW \text{ of tax credit on depreciation}$$
$$-PW \text{ of net proceeds from sale} \qquad (10.4)$$

Note that the acquisition (investment) cost of the asset is offset by the same amount of borrowing at time 0, so that we need to consider only the loan repayment series.

- **Lease Financing.** Suppose that the firm can lease the asset at a constant amount per period. Then the project's incremental cost of leasing becomes

$$PW(i)_{Lease} = PW \text{ of after-tax lease expenses}$$

If the lease does not provide for the maintenance of the equipment leased, then the lessee must assume this responsibility. In that situation, the maintenance term in Eq. (10.4) can be dropped in the calculation of the incremental cost of owning the asset.

- **Decision Rule.** The criterion for the decision to lease as opposed to purchasing the asset thus reduces to a comparison between

$$PW(i)_{Buy} \text{ and } PW(i)_{Lease}$$

Purchasing is preferred if the combined present value of the loan repayment series and the after-tax O&M expense, reduced by the present values of the depreciation tax shield and the net proceeds from disposal of the asset, is less than the present value of the net lease costs.

EXAMPLE 10.8 Lease-or-Buy Decision

The Dallas Electronics Company is considering replacing an old, 1,000-pound-capacity industrial forklift truck. The truck has been used primarily to move goods from production machines into storage. The company is working nearly at capacity and is operating on a two-shift basis six days per week. Dallas management is considering owning or leasing the new truck. The plant engineer has compiled the following data for management.

- The initial cost of a gas-powered truck is $20,000. The new truck would use about 8 gallons of gasoline (in a single shift of 8 hours per day) at a cost of $3.35 per gallon. If the truck is operated 16 hours per day, its expected life will be four years, and an engine overhaul at a cost of $1,500 will be required at the end of two years.

- The Austin Industrial Truck Company has been servicing the old forklift truck, and Dallas would buy the new truck through Austin, which offers a service agreement to users of its trucks. The agreement costs $120 per month and provides for a monthly visit by an experienced service representative to lubricate and tune the truck. Insurance and property taxes for the truck are $650 per year.

- The truck is classified as five-year property under MACRS and could be depreciated accordingly. Dallas can take advantage of the Section 179 tax deduction. In that case, the full purchase amount of $20,000 can be deducted from its gross income. Dallas is in the 40% marginal tax bracket; the estimated resale value of the truck at the end of four years will be 15% of the original cost.

- Austin also has offered to lease a truck to Dallas. Austin will maintain the equipment and will guarantee to keep the truck in serviceable condition at all times. In the event of a major breakdown, Austin will provide a replacement truck at its expense. The cost of the operating lease plan is $850 per month. The contract term is three years at a minimum, with the option to cancel on 30 days' notice thereafter.

- On the basis of recent experience, the company expects that funds committed to new investments should earn at least a 12% rate of return after taxes.

  (a) Determine the cost of owning the truck.
  (b) Determine the cost of leasing the truck.

DISCUSSION: We may calculate the fuel costs for two-shift operation as

$$(8 \text{ gallons/shift})(2 \text{ shifts/day})(\$3.35/\text{gallon}) = \$53.60 \text{ per day}$$

The truck will operate 300 days per year, so the annual fuel cost will be $16,080. However, both alternatives require the company to supply its own fuel, so the fuel cost is not relevant to our decision making. Revenue streams would be the same as fuel costs. Therefore, we may drop these common items from our calculations.

SOLUTION

Given: Preceding cost information, MARR = 12%, and marginal tax rate = 40%.
Find: (a) and (b) Incremental cost of owning versus leasing the truck, (c) Section 179 deduction versus capitalization of the asset

(a) **Cost of owning the truck.** To compare the incremental cost of ownership with the incremental cost of leasing, we make the following additional estimates and assumptions.

Step 1. The preventive-maintenance contract, which costs $120 per month (or $1,440 per year), will be adopted. With annual insurance and taxes of $650, the equivalent present worth of the after-tax O&M is

$$P_1 = (\$1,440 + \$650)(1 - 0.40)(P/A, 12\%, 4) = \$3,809$$

Step 2. The engine overhaul is not expected to increase either the salvage value or the service life of the truck. Therefore, the overhaul cost ($1,500) will be expensed all at once rather than capitalized. The equivalent present worth of this after-tax overhaul expense is

$$P_2 = \$1,500(1 - 0.40)(P/F, 12\%, 2) = \$717$$

Step 3. If Dallas decided to purchase the truck through debt financing, the first step in determining financing costs would be to compute the annual installments of the debt repayment schedule. Assuming that the entire investment of $20,000 is financed at a 10% interest rate, the annual payment would be

$$A = \$20,000(A/P, 10\%, 4) = \$6,309$$

and the interest and principal payment each year is calculated as

Year	Beginning Balance	Interest Charged	Principal Payment	Ending Balance
1	$20,000	$2,000	$4,309	$15,691
2	15,691	1,569	4,740	10,951
3	10,951	1,095	5,214	5,737
4	5,737	573	5,736	0

The equivalent present worth of this loan principal repayment series is

$$P_3 = \$4,309(P/F, 12\%, 1) + \$4,740(P/F, 12\%, 2)$$
$$+ \$5,214(P/F, 12\%, 3) + \$5,736(P/F, 12\%, 4)$$
$$= \$14,983$$

The equivalent present worth of the after-tax interest payment series is

$$P_4 = (1 - 0.40)\left[\begin{array}{l} \$2,000(P/F, 12\%, 1) + \$1,569(P/F, 12\%, 2) \\ +\$1,095(P/F, 12\%, 3) + \$573(P/F, 12\%, 4) \end{array}\right]$$
$$= \$2,508$$

Step 4. With the five-year MACRS depreciation schedule, the tax shield (savings) due to depreciation expenses can be calculated as

n	D_n	Tax Savings $(t_m D_n)$
1	$4,000	$4,000(0.40) = $1,600
2	6,400	6,400(0.40) = 2,560
3	3,840	3,840(0.40) = 1,536
4	1,152*	1,152(0.40) = 461

*Half-year convention.

Therefore, the equivalent present worth of the depreciation tax savings (credit) is

$$P_5 = \$1,600(P/F, 12\%, 1) + \$2,560(P/F, 12\%, 2)$$
$$+ \$1,536(P/F, 12\%, 3) + \$461(P/F, 12\%, 4)$$
$$= \$4,856$$

Step 5. With the estimated salvage value of 15% of the initial investment ($3,000) and with the five-year MACRS depreciation schedule given (total amount of depreciation = $15,392), we compute the net proceeds from the sale of the truck at the end of four years as

Book value	$= \$20,000 - \$15,392 = \$4,608$
Gains (losses)	$= \$3,000 - \$4,608 = (\$1,608)$
Tax savings	$= 0.40(\$1,608) = \$643$
Net proceeds from sale	$= \$3,000 + \$643 = \$3,643$

The present equivalent amount of the net salvage value is

$$P_6 = \$3,643(P/F, 12\%, 4) = \$2,315$$

Step 6. Therefore, the net incremental cost of owning the truck through 100% debt financing is

$$PW(12\%)_{Buy} = P_1 + P_2 + P_3 + P_4 - P_5 - P_6$$
$$= \$3,809 + \$717 + \$14,983 + \$2,508 - \$4,856 - \$2,315$$
$$= \$14,846$$

(b) **Cost of leasing the truck.** How does the cost of acquiring a forklift truck under the lease compare with the cost of owning the truck?

Step 1. The lease payment is also a tax-deductible expense; however, the net cost of leasing has to be computed explicitly on an after-tax basis. The calculation of the annual incremental leasing costs is

Annual lease payments (12 months)	$= \$10,200$
Less 40% taxes	$= 4,080$
Annual net costs after taxes	$= \$6,120$

Step 2. The total net present equivalent incremental cost of leasing is

$$PW(12\%)_{Lease} = \$6,120(P/A, 12\%, 4)$$
$$= \$18,589$$

Step 3. Here, we have assumed that the lease payments occur at the *end* of the period; however, many leasing contracts require the payments to be made at the *beginning* of each period. In the latter situation, we can easily modify the present-worth expression of the lease expense to reflect this cash-flow timing:

$$PW(12\%)_{Lease} = \$6,120 + \$6,120(P/A, 12\%, 3) = \$20,819$$

Purchasing the truck through debt financing would save Dallas either $3,743 or $5,973 in NPW, depending on the lease payment term. Table 10.11 illustrates an alternate way of conducting a buy-versus-lease decision in tabular form using Excel.

TABLE 10.11 Buy versus Lease Decision Based on Generalized Cash Flow Approach (Example 10.8)

	A	B	C	D	E	F
1						
2	**Buy Option**					
3	End of year	**0**	**1**	**2**	**3**	**4**
4						
5	**Operating Activities:**					
6	Maintenance:					
7	- (1 - 0.40)(Maintenance)		(864)	(864)	(864)	(864)
8	- (1 - 0.40)(Property tax)		(390)	(390)	(390)	(390)
9	- (1 - 0.40)(Engine overhaul)			(900)		
10	- (1 - 0.40)(Debt interest)		(1,200)	(941)	(657)	(344)
11	Tax shield:					
12	+ (0.40)(Depreciation)		1,600	2,560	1,536	461
13	**Investing Activities:**					
14	- Capital expenditure	(20,000)				
15	+ Net proceeds from sale					3,643
16	**Financing Activities:**					
17	+ Borrowed funds	20,000				
18	- Repayment of principal		(4,309)	(4,740)	(5,214)	(5,736)
19						
20	Net Cash Flow	$ -	$ (5,163)	$ (5,275)	$ (5,589)	$ (3,230)
21						
22	**Lease Option**					
23	- (1 - 0.40)(Lease payment)		$ (6,120)	$ (6,120)	$ (6,120)	$ (6,120)
24						
25	Net cash flow		$ (6,120)	$ (6,120)	$ (6,120)	$ (6,120)
26						
27	NPW for Buy Option	$ (14,846)				
28	NPW for Lease Option	($18,589)				
29						

COMMENTS: We may wonder here if Dallas should take the Section 179 deduction instead of capitalizing the asset. We can easily answer this question. We already know the cost of owning the truck through capitalizing the purchase cost, which is $14,846. If we go with the Section 179 deduction, there will be no tax shield related to depreciation, and the entire salvage value is subject to taxable gains. With these changes, the cost of owning the truck with the Section 179 deduction is

$$P_5 = \$20,000(1 - 0.40)(P/F, 12\%, 1)$$

$$= \$7,143$$

$$P_6 = \$3,000(1 - 0.40)(P/F, 12\%, 4) = \$1,144$$

$$PW(12\%)_{\text{Buy with Section 179}} = \$3,809 + \$717 + \$14,983 + \$2,508 - \$7,143$$

$$- \$1,144$$

$$= \$13,730$$

Clearly, taking the 179 deduction instead of capitalizing the asset will make buying even more economical. You should not conclude, however, that leasing is *always* more expensive than owning. In many cases, analysis favors a lease option. The interest rate, salvage value, lease payment schedule, and debt financing all have an important effect on decision-making.

SUMMARY

- Identifying and estimating relevant project cash flows is perhaps the most challenging aspect of engineering economic analysis. All cash flows can be organized into one of the following three categories:

  1. Operating activities

  2. Investing activities

  3. Financing activities

- The following types of cash flows are the most common flows a project may generate:

  1. New investment and disposal of existing assets

  2. Salvage value (or net selling price)

  3. Working capital

  4. Working-capital release

  5. Cash revenues or savings

  6. Manufacturing, operating, and maintenance costs

  7. Interest and loan payments

  8. Taxes and tax credits

In addition, although not cash flows, the following elements may exist in a project analysis, and they must be accounted for:

1. Depreciation expenses

2. Amortization expenses

Table 10.1 summarizes these elements and organizes them as operating, investment, or financing.

- The **income statement approach** is typically used in organizing project cash flows. This approach groups cash flows according to whether they are operating, investing, or financing functions.

- The **generalized cash flow approach** (shown in Table 10.10), used to organize cash flows, can be implemented when a project does not change a company's marginal tax rate. The cash flows can be generated more quickly, and the formatting of the results is less elaborate than with the income statement approach. There are also analytical advantages, which we will discover in later chapters. However, the generalized approach is less intuitive and not commonly understood by businesspeople.

PROBLEMS

Generating Net Cash Flows

10.1 A firm is considering purchasing a machine that costs $65,000. It will be used for six years, and the salvage value at that time is expected to be zero. The machine will save $35,000 per year in labor, but it will incur $12,000 in operating and maintenance costs each year. The machine will be depreciated according to five-year MACRS. The firm's tax rate is 40%, and its after-tax MARR is 15%. Should the machine be purchased?

10.2 An automobile-manufacturing company is considering purchasing an industrial robot to do spot welding, which is currently done by skilled labor. The initial cost of the robot is $210,000, and the annual labor savings are projected to be $150,000. If purchased, the robot will be depreciated under MACRS as a five-year recovery property. The robot will be used for seven years, at the end of which time, the firm expects to sell it for $60,000. The company's marginal tax rate is 35% over the project period. Determine the net after-tax cash flows for each period over the project life. Assume MARR = 15%.

10.3 You are considering constructing a luxury apartment building project that requires an investment of $15,500,000, which comprises $12,000,000 for the building and $3,500,000 for land. The building has 50 units. You expect the maintenance cost for the apartment building to be $350,000 the first year and $400,000 the second year, after which it will continue to increase by $50,000 in subsequent years. The cost to hire a manager for the building is estimated to be $85,000 per year. After five years of operation, the apartment building can be sold for $17,000,000. What is the annual rent per apartment unit that will provide a return on investment of 15% after tax? Assume that the building will remain fully occupied during the five years. Assume also that your tax rate is 35%. The building will be depreciated according to 39-year MACRS and will be placed in service in January during the first year of ownership and sold in December during the fifth year of ownership.

10.4 A local delivery company has purchased a delivery truck for $15,000. The truck will be depreciated under MACRS as five-year property. The truck's market value (salvage value) is expected to decrease by $2,500 per year. It is expected that the purchase of the truck will increase its revenue by $10,000 annually. The O&M costs are expected to be $3,000 per year. The firm is in the 40% tax bracket, and its MARR is 15%. If the company plans to keep the truck for only two years, what would be the equivalent present worth?

10.5 A Los Angeles company is planning to market an answering device for people working alone who want the prestige that comes with having a secretary, but who cannot afford one. The device, called Tele-Receptionist, is similar to a voice-mail system. It uses digital-recording technology to create the illusion that a person is operating the switchboard at a busy office. The company purchased a 40,000-ft^2 building and converted it to an assembly plant for $600,000 ($100,000 worth of land and $500,000 worth of building). Installation of the assembly equipment, worth $500,000, was completed on December 31. The plant will begin operation on January 1. The company expects to have a gross annual income of $2,500,000 over the next five years. Annual manufacturing costs and all other operating expenses (excluding depreciation) are projected to be $1,280,000. For depreciation purposes, the assembly-plant building will be classified as a 39-year real property and the assembly equipment as a seven-year MACRS property. The property value of the land and the building at the end of year 5 would appreciate as much as 15% over the initial purchase cost. The residual value of the assembly equipment is estimated to be about $50,000 at the end of year 5. The firm's marginal tax rate is expected to be about 40% over the project period. Determine the project's after-tax cash flows over the period of five years.

10.6 A highway contractor is considering buying a new trench excavator that costs $250,000 and can dig a 3-foot-wide trench at the rate of 16 feet per hour. With the machine adequately maintained, its production rate will remain constant for the first 1,200 hours of operation and then decrease by 2 feet per hour for each additional 400 hours thereafter. The expected average annual use is 400 hours, and maintenance and operating costs will be $50 per hour. The contractor will depreciate the equipment in accordance with a five-year MACRS. At the end of five years, the excavator will be sold for $60,000. Assuming that the contractor's marginal tax rate is 35% per year, determine the annual after-tax cash flow.

10.7 A small children's clothing manufacturer is considering an investment to computerize its management information system for material requirement planning, piece-goods coupon printing, and invoice and payroll services. An outside consultant has been retained to estimate the initial hardware requirement and installation costs. He suggests the following:

Workstation servers	$20,000
(15 PCs, 4 printers)	$65,000
Local area networking system	15,000
System installation and testing	4,000

The expected life of the computer system is five years with no expected salvage value. The proposed system is classified as a five-year property under the MACRS depreciation system. A group of computer consultants needs to be hired to develop various customized software packages to run on the system. Software development costs will be $20,000 and can be expensed during the first tax year. The new system will eliminate two clerks, whose combined annual payroll expenses are $72,000. Additional annual expenses to run this computerized system are expected to be $15,000. Borrowing is not considered an option for this investment, nor is a tax credit available for the system. The firm's expected marginal tax rate over the next six years will be 35%. The firm's interest rate is 13%. Compute the after-tax cash flows over the life of the investment.

10.8 Appliance makers are gearing up to push new energy-efficient systems in the wake of an energy bill that offers tax credits to homeowners who upgrade to electricity-saving appliances. In air conditioning, industry giant Carrier Corporation says it has invested $250 million in developing new heat exchangers—a major component in air-conditioners—that use less energy and are about 20% smaller and 30% lighter than current energy-saving versions. Carrier's current energy-efficient models are almost double the size of its regular central air-conditioning units. The company speculates that this model's bulk may have been a deterrent for homeowners. The new air conditioners were expected to hit the market in the first quarter of 2015.

Cooling efficiency is measured by a standard called SEER: Seasonal Energy Efficiency Ratio. It's similar to the gas mileage system used on cars—the higher the number, the more money you save. Older systems had SEER numbers as low as 8; the new Carrier systems are rated at 18! (Federal standards required a minimum SEER of 13 from January 2006.) Translated into operating costs, this means that for every $100 you used to spend on electricity for cooling, you now can spend just $39.

There are several issues involved in Carrier's pushing more efficient air-conditioning units. First, the market demand is difficult to estimate. Second, it is even more difficult to predict the useful life of the product, as market competition is ever increasing. Carrier's gross margin is about 25%, its operating margin is 8.3%, and its net margin is 7.45%. The expected retail price of the Delux Puron unit is about $4,236. Determine the required sales volume of the new air-conditioning unit to justify the capital investment of $250 million, assuming that Carrier's required return is 15%.

10.9 An asset in the five-year MACRS property class costs $150,000 and has a zero estimated salvage value after six years of use. The asset will generate annual revenues of $320,000 and will require $80,000 in annual labor and $50,000 in annual material expenses. There are no other revenues and expenses. Assume a tax rate of 40%.

(a) Compute the after-tax cash flows over the project life.

(b) Compute the NPW at MARR = 12%. Is the investment acceptable?

10.10 An automaker is considering installing a three-dimensional (3-D) computerized car-styling system at a cost of $230,000 (including hardware and software). With the 3-D computer modeling system, designers will have the ability to view their design from many angles and to fully account for the space required for the engine and passengers. The digital information used to create the computer model can be revised in consultation with engineers, and the data can be used to run milling machines that make physical models quickly and precisely.

The automaker expects to decrease the turnaround time for designing a new automobile model (from configuration to final design) by 22%. The expected savings in dollars is $250,000 per year. The training and operating maintenance cost for the new system is expected to be $50,000 per year. The system has a five-year useful life and can be depreciated according to the five-year MACRS class. The system will have an estimated salvage value of $5,000. The automaker's marginal tax rate is 40%. Determine the annual cash flows for this investment. What is the return on investment for the project?

10.11 A facilities engineer is considering a $55,000 investment in an energy management system (EMS). The system is expected to save $14,000 annually in utility bills for N years. After N years, the EMS will have a zero salvage value. In an after-tax analysis, what would N need to be in order for the investment to earn a 12% return? Assume MACRS depreciation with a three-year class life and a 35% tax rate.

10.12 A corporation is considering purchasing a machine that will save $150,000 per year before taxes. The cost of operating the machine (including maintenance) is $30,000 per year. The machine will be needed for five years, after which it will have a zero salvage value. MACRS depreciation will be used, assuming a three-year class life. The marginal income tax rate is 40%. If the firm wants 15% return on investment after taxes, how much can it afford to pay for this machine?

Investment in Working Capital

10.13 The Newport Company is planning to expand its current spindle product line. The required machinery would cost $520,000. The building that will house the new production facility would cost $1.5 million. The land would cost $350,000, and $250,000 working capital would be required. The product is expected to result in additional sales of $775,000 per year for 10 years, at which time the land can be sold for $500,000, the building for $800,000, and the equipment for $50,000. All of the working capital will be recovered. The annual disbursements for labor, materials, and all other expenses are estimated to be $465,000. The firm's income tax rate is 40%, and any capital gains will be taxed at 35%. The building will be depreciated according to a 39-year property class. The manufacturing facility will be classified as

a seven-year MACRS. The firm's MARR is known to be 15% after taxes.

(a) Determine the projected net after-tax cash flows from this investment. Is the expansion justified?

(b) Compare the IRR of this project with that of a situation with no working capital.

10.14 An industrial engineer proposed the purchase of RFID Fixed-Asset Tracking System for the company's warehouse and weave rooms. The engineer felt that the purchase would provide a better system of locating cartons in the warehouse by recording the locations of the cartons and storing the data in the computer. The estimated investment, annual operating and maintenance costs, and expected annual savings are as follows.

- Cost of equipment and installation: $85,500
- Project life: 6 years
- Expected salvage value: $5,000
- Investment in working capital (fully recoverable at the end of the project life): $15,000
- Expected annual savings on labor and materials: $65,800
- Expected annual expenses: $9,150
- Depreciation method: five-year MACRS

The firm's marginal tax rate is 35%.

(a) Determine the net after-tax cash flows over the project life.

(b) Compute the IRR for this investment.

(c) At MARR = 18%, is the project acceptable?

10.15 The Delaware Chemical Corporation is considering investing in a new composite material. R&D engineers are investigating exotic metal–ceramic and ceramic–ceramic composites to develop materials that will withstand high temperatures, such as those to be encountered in the next generation of jet fighter engines. The company expects a three-year R&D period before these new materials can be applied to commercial products.

The following financial information is presented for management review.

- **R&D cost:** $5 million over a three-year period: $0.5 million at the beginning of year 1; $2.5 million at the beginning of year 2; and $2 million at the beginning of year 3. For tax purposes, these R&D expenditures will be expensed rather than amortized.

- **Capital investment:** $5 million at the beginning of year 4. This investment consists of $2 million in a building and $3 million in plant equipment. The company already owns a piece of land as the building site.
- **Depreciation method:** The building (39-year real property class with the asset placed in service in January) and plant equipment (seven-year MACRS recovery class).
- **Project life:** 10 years after a three-year R&D period.
- **Salvage value:** 10% of the initial capital investment for the equipment and 50% for the building (at the end of the project life).
- **Total sales:** $50 million (at the end of year 4), with an annual sales growth rate of 10% per year (compound growth) during the next five years (year 5 through year 9) and -10% (negative compound growth) per year for the remaining project life.
- **Out-of-pocket expenditures:** 80% of annual sales.
- **Working capital:** 10% of annual sales (considered as an investment at the beginning of each production year and investments fully recovered at the end of the project life).
- **Marginal tax rate:** 40%.
  - (a) Determine the net after-tax cash flows over the project life.
  - (b) Determine the IRR for this investment.
  - (c) Determine the equivalent annual worth for the investment at MARR $= 20\%$.

Effects of Borrowing

10.16 Refer to the data in Problem 10.3. If the firm expects to borrow the initial investment ($15,500,000) at 10% over five years (paying back the loan in equal annual payments of $4,088,861), determine the project's net cash flows.

10.17 In Problem 10.2, to finance the industrial robot, the company will borrow the entire amount from a local bank, and the loan will be paid off at the rate of $30,000 per year, plus 10% on the unpaid balance. Determine the net after-tax cash flows over the project life.

10.18 Refer to the financial data in Problem 10.10. Suppose that 50% of the initial investment of $230,000 will be borrowed from a local bank at an interest rate of 11% over five years (to be paid off in

five equal annual payments). Recompute the after-tax cash flow.

10.19 A special-purpose machine tool set would cost $20,000. The tool set will be financed by a $10,000 bank loan repayable in two equal annual installments at 10% compounded annually. The tool is expected to provide annual (material) savings of $30,000 for two years and is to be depreciated by the MACRS three-year recovery period. The tool will require annual O&M costs in the amount of $5,000. The salvage value at the end of the two years is expected to be $8,000. Assuming a marginal tax rate of 40% and MARR of 15%, what is the net present worth of this project? You may use Table P10.19 as a worksheet for your calculation.

TABLE P10.19

Cash Flow Statement	0	1	2
Operating activities			
Net income		$10,400	$12,019
Depreciation		6,666	4,445
Investment activities			
Investment	$-$20,000		
Salvage			8,000
Gains tax (40%)			
Financial activities			
Borrowed funds	10,000		
Principal repayment	0		
Net cash flow	$-$10,000		

10.20 The A.M.I. Company is considering installing a new process machine for the firm's manufacturing facility. The machine costs $220,000 installed, will generate additional revenues of $85,000 per year, and will save $65,000 per year in labor and material costs. The machine will be financed by a $120,000 bank loan repayable in three equal annual principal installments, plus 9% interest on the outstanding balance. The machine will be depreciated using seven-year MACRS. The useful life of the machine is 10 years, after which it will be sold for $20,000. The combined marginal tax rate is 40%.

(a) Find the year-by-year after-tax cash flow for the project.

(b) Compute the IRR for this investment.

(c) At MARR = 18%, is the project economically justifiable?

10.21 Consider the following financial information about a retooling project at a computer manufacturing company:

- The project costs $2.5 million and has a five-year service life.

- The retooling project can be classified as seven-year property under the MACRS rule.

- At the end of the fifth year, any assets held for the project will be sold. The expected salvage value will be about 10% of the initial project cost.

- The firm will finance 40% of the project money from an outside financial institution at an interest rate of 10%. The firm is required to repay the loan with five equal annual payments.

- The firm's incremental (marginal) tax rate on the investment is 35%.

- The firm's MARR is 18%.

With the preceding financial information,

(a) Determine the after-tax cash flows.

(b) Compute the annual equivalent worth for this project.

10.22 A fully automatic chucker and bar machine is to be purchased for $45,000. The money will be borrowed with the stipulation that it be repaid with six equal end-of-year payments at 12% compounded annually. The machine is expected to provide annual revenue of $13,000 for six years and is to be depreciated by the MACRS seven-year recovery period. The salvage value at the end of six years is expected to be $4,000. Assume a marginal tax rate of 35% and a MARR of 15%.

(a) Determine the after-tax cash flow for this asset over six years.

(b) Determine whether the project is acceptable on the basis of the IRR criterion.

10.23 A manufacturing company is considering acquiring a new injection-molding machine at a cost of $150,000. Because of a rapid change in product mix, the need for this particular machine is expected to last only eight years, after which time the machine is expected to have a salvage value of $10,000. The annual operating cost is estimated to be $11,000. The addition of the machine to the current production facility is expected to generate an annual revenue of $48,000. The firm has only $100,000 available from its equity funds, so it must borrow the additional $50,000 required at an interest rate of 10% per year with repayment of principal and interest in eight equal annual amounts. The applicable marginal income tax rate for the firm is 40%. Assume that the asset qualifies for a seven-year MACRS property class.

(a) Determine the after-tax cash flows.

(b) Determine the NPW of this project at MARR = 14%.

Generalized Cash Flow Method

10.24 Suppose an asset has a first cost of $8,000, a life of five years, a salvage value of $2,000 at the end of five years, and a net annual before-tax revenue of $2,500. The firm's marginal tax rate is 35%. The asset will be depreciated by three-year MACRS.

(a) Using the generalized cash flow approach, determine the cash flow after taxes.

(b) Rework part (a), assuming that the entire investment would be financed by a bank loan at an interest rate of 9%.

(c) Given a choice between the financing methods of parts (a) and (b), show calculations to justify your choice of which is the better one at an interest rate of 9%.

10.25 A construction company is considering acquiring a new earthmover. The purchase price is $110,000, and an additional $25,000 is required to modify the equipment for special use by the company. The equipment falls into the MACRS seven-year classification (the tax life), and it will be sold after five years (the project life) for $50,000. The purchase of the earthmover will have no effect on revenues, but the machine is expected to save the firm $68,000 per year in before-tax operating costs, mainly labor. The firm's marginal tax rate is 40%. Assume that the initial investment is to be financed by a bank loan at an interest rate of 10% payable annually. Determine the after-tax cash flows by using the generalized cash flow approach and the worth of the investment for this project if the firm's MARR is known to be 12%.

10.26 Federal Express (FedEx) is considering adding 18 used Boeing 757 jets by buying the twin engine planes to replace some of its oldest, least-efficient freighters, Boeing 727s. FedEx pays about $10 million each, then FedEx spends about $5 million each to refit the planes to carry cargo. FedEx is required to make a 10% down payment at the time of delivery, and the balance is to be paid over a 10-year period at an interest rate of 12% compounded annually. The actual payment schedule calls for only interest payments over the 10-year period with the original principal amount to be paid off at the end of the 10th year. FedEx expects to generate $45 million per year in fuel savings by adding these aircrafts to its current fleet. The aircraft is expected to have a 15-year service life with a salvage value of 15% of the original purchase price. If the aircrafts are bought, they will be depreciated by the seven-year MACRS property classifications. The firm's combined federal and state marginal tax rate is 38%, and its required minimum attractive rate of return is 18%.

(a) Use the generalized cash-flow approach to determine the cash flow associated with the debt financing.

(b) Is this project acceptable?

Comparing Mutually Exclusive Alternatives

10.27 The Pittsburgh Division of Vermont Machinery, Inc. manufactures drill bits. One of the production processes of a drill bit requires tipping, whereby carbide tips are inserted into the bit to make it stronger and more durable. The tipping process usually requires four or five operators, depending on the weekly workload. The same operators are assigned to the stamping operation, in which the size of the drill bit and the company's logo are imprinted into the bit. Vermont is considering acquiring three automatic tipping machines to replace the manual tipping and stamping operations. If the tipping process is automated, Vermont engineers will have to redesign the shapes of the carbide tips to be used in the machines. The new design requires less carbide, resulting in a savings of material. The following financial data have been compiled.

- **Project life:** Six years.
- **Expected annual savings:** Reduced labor, $56,000; reduced material, $75,000; other benefits (reduction in carpal tunnel syndrome and related problems), $28,000; and reduced overhead, $15,000.
- **Expected annual O&M costs:** $22,000.
- **Tipping machines and site preparation:** Equipment costs (three machines), including delivery, $180,000; site preparation, $20,000.
- **Salvage value:** $30,000 (three machines) at the end of six years.
- **Depreciation method:** Seven-year MACRS.
- **Investment in working capital:** $25,000 at the beginning of the project year. That same amount will be fully recovered at the end of the project year.
- **Other accounting data:** Marginal tax rate of 39% and MARR of 18%.

To raise $200,000, Vermont is considering the following financing options:

- **Option 1:** Use the retained earnings of the tipping machines to finance them.
- **Option 2:** Secure a 12% term loan over six years (to be paid off in six equal annual installments).
- **Option 3:** Lease the tipping machines. Vermont can obtain a six-year financial lease on the equipment (with, however, no maintenance service) for payments of $55,000 at the *beginning* of each year.

(a) Determine the net after-tax cash flows for each financing option.

(b) What is Vermont's present-value cost of owning the equipment by borrowing?

(c) What is Vermont's present-value cost of leasing the equipment?

(d) Recommend the best course of action for Vermont.

10.28 The headquarters building owned by a rapidly growing company is not large enough for the company's current needs. A search for larger quarters revealed two new alternatives that would provide sufficient room, enough parking, and the desired appearance and location. The company now has three options:

- **Option 1:** Lease the new quarters for $144,000 per year.

- **Option 2:** Purchase the new quarters for $800,000, including a $150,000 cost for land.

- **Option 3:** Remodel the current headquarters building.

It is believed that land values will not decrease over the ownership period, but the value of all structures will decline to 10% of the purchase price in 30 years. Annual property tax payments are expected to be 5% of the purchase price. The present headquarters building is already paid for and is now valued at $300,000. The land it is on is appraised at $60,000. The structure can be remodeled at a cost of $300,000 to make it comparable to other alternatives. However, the remodeling will occupy part of the existing parking lot. An adjacent, privately owned parking lot can be leased for 30 years under an agreement that the first year's rental of $9,000 will increase by $500 each year. The annual property taxes on the remodeled property will again be 5% of the present valuation, plus the cost of remodeling. The new quarters are expected to have a service life of 30 years, and the desired rate of return on investments is 12%. Assume that the firm's marginal tax rate is 40% and that the new building and remodeled structure will be depreciated under MACRS using a real-property recovery period of 39 years. If the annual upkeep costs are the same for all three alternatives, which one is preferable?

10.29 An international manufacturer of prepared food items needs 50,000,000 kWh of electrical energy a year, with a maximum demand of 10,000 kW. The local utility company currently charges $0.085 per kWh—a rate considered high throughout the industry. Because the firm's power consumption is so large, its engineers are considering installing a 10,000-kW steam-turbine plant. Three types of plant have been proposed (units in thousands of dollars) and are given in Table P10.29.

The service life of each plant is expected to be 20 years. The plant investment will be subject to

TABLE P10.29

	Plant A	Plant B	Plant C
Average station heat rate (BTU/kWh)	16,500	14,500	13,000
Total investment (boiler/turbine/electrical/structures)	$8,530	$9,498	$10,546
Annual operating cost:			
Fuel	1,128	930	828
Labor	616	616	616
O&M	150	126	114
Supplies	60	60	60
Insurance and property taxes	10	12	14

a 20-year MACRS property classification. The expected salvage value of the plant at the end of its useful life is about 10% of its original investment. The firm's MARR is known to be 12%. The firm's marginal income tax rate is 39%.

(a) Determine the unit power cost ($/kWh) for each plant.

(b) Which plant would provide the most economical power?

Lease-versus-Buy Decisions

10.30 The Prescott Welding Company needs to acquire a new lift truck for transporting its final product to the warehouse. One alternative is to purchase the truck for $45,000, which will be financed by the bank at an interest rate of 12%. The loan must be repaid in four equal installments, payable at the end of each year. Under the borrow-to-purchase arrangement, Prescott Welding would have to maintain the truck at an annual cost of $1,200, also payable at year-end. Alternatively, Prescott Welding could lease the truck under a four-year contract for a lease payment of $12,000 per year. Each annual lease payment must be made at the beginning of each year. The truck would be maintained by the lessor. The truck falls into the five-year MACRS classification, and it has a salvage value of $10,000, which is the expected market value after four years, at which time Prescott Welding plans to replace the truck, irrespective of whether it leases or buys. Prescott Welding has a marginal tax rate of 40% and a MARR of 15%.

(a) What is Prescott Welding's cost of leasing in present worth?

(b) What is Prescott Welding's cost of owning in present worth?

(c) Should the truck be leased or purchased?

Note: This is an operating lease, so the truck would be maintained by the lessor.

10.31 Janet Wigandt, an electrical engineer for Instrument Control, Inc. (ICI), has been asked to perform a lease–buy analysis of a new pin-inserting machine for ICI's PC-board manufacturing that has a project life of four years with annual revenues of $200,000.

- **Buy Option:** The equipment costs $120,000. To purchase it, ICI could obtain a term loan for the full amount at 10% interest, which is payable in four equal end-of-year annual installments. The machine falls into a five-year MACRS property classification. Annual operating costs of $40,000 are anticipated. The machine requires annual maintenance at a cost of $10,000. Because technology is changing rapidly in pin-inserting machinery, the salvage value of the machine is expected to be only $20,000.

- **Lease Option:** Business Leasing, Inc. (BLI) is willing to write a four-year operating lease on the equipment for payments of $44,000 at the beginning of each year. Under this arrangement, BLI will maintain the asset so that the annual maintenance cost of $10,000 will be saved. ICI's marginal tax rate is 40%, and its MARR is 15% during the analysis period.

(a) What is ICI's present-value (incremental) cost of owning the equipment?

(b) What is ICI's present-value (incremental) cost of leasing the equipment?

(c) Should ICI buy or lease the equipment?

10.32 Consider the following lease-versus-borrow-and-purchase problem.

- **Borrow-and-purchase option:**

  1. Jensen Manufacturing Company plans to acquire sets of special industrial tools with a four-year life and a cost of $200,000— delivered and installed. The tools will be depreciated by the MACRS three-year classification.

  2. Jensen can borrow the required $200,000 at a rate of 10% over four years. Four equal end-of-year annual payments would be made in the amount of $63,094 = $200,000(A/P, 10%, 4). The annual interest and principal payment schedule, along with the equivalent present worth of these payments, is

End of Year	Interest	Principal
1	$20,000	$43,094
2	15,961	47,404
3	10,950	52,144
4	5,736	57,358

3. The estimated salvage value for the tool sets at the end of four years is $20,000.

4. If Jensen borrows and buys, it will have to bear the cost of maintenance, which will be performed by the tool manufacturer at a fixed contract rate of $10,000 per year.

- **Lease option:**

    1. Jensen can lease the tools for four years at an annual rental charge of $70,000—payable at the end of each year.

    2. The lease contract specifies that the lessor will maintain the tools at no additional charge to Jensen.

        Jensen's tax rate is 40%. Any gains will also be taxed at 40%.

        (a) What is Jensen's PW of after-tax cash flow of leasing at $i = 15\%$?

        (b) What is Jensen's PW of after-tax cash flow of owning at $i = 15\%$?

10.33 Tom Hagstrom needs a new car for his business. One alternative is to purchase the car outright for $28,000 and to finance the car with a bank loan for the net purchase price. The bank loan calls for 36 equal monthly payments of $881.30 at an interest rate of 8.3% compounded monthly. Payments must be made at the end of each month. The terms of each alternative are

Buy	Lease
$28,000	$696 per month 36-month open-end lease.
	Annual mileage allowed = 15,000 miles

If Tom takes the lease option, he is required to pay $500 for a security deposit, which is refundable at the end of the lease, and $696 a month at the beginning of each month for 36 months. If the car is purchased, it will be depreciated according to a five-year MACRS property classification. The car has a salvage value of $15,400, which is the expected market value after three years, at which time Tom plans to replace the car, irrespective of whether he leases or buys. Tom's marginal tax rate is 28%. His MARR is known to be 13% per year.

(a) Determine the annual cash flows for each option.

(b) Which option is better?

10.34 The Boggs Machine Tool Company has decided to acquire a pressing machine. One alternative is to lease the machine under a three-year contract for a lease payment of $15,000 per year with payments to be made at the beginning of each year. The lease would include maintenance. The second alternative is to purchase the machine outright for $100,000, which involves financing the machine with a bank loan for the net purchase price and amortizing the loan over a three-year period at an interest rate of 12% per year (annual payment = $41,635).

Under the borrow-to-purchase arrangement, the company would have to maintain the machine at an annual cost of $5,000, which is payable at year-end. The machine falls into a five-year MACRS classification and has a salvage value of $50,000, which is the expected market value at the end of year 3, at which time, the company plans to replace the machine, irrespective of whether it leases or buys. Boggs has a tax rate of 40% and a MARR of 15%.

(a) What is Boggs' PW cost of leasing?

(b) What is Boggs' PW cost of owning?

(c) From the financing analysis in parts (a) and (b), what are the advantages and disadvantages of leasing and owning?

10.35 An asset is to be purchased for $25,000. The asset is expected to provide revenue of $10,000 a year and have operating costs of $2,500 a year. The asset is considered to be a seven-year MACRS property. The company is planning to sell the asset at the end of year 5 for $5,000. Given that the company's marginal tax rate is 30% and that it has a MARR of 10% for any project undertaken, answer the following questions.

(a) What is the net cash flow for each year, given that the asset is purchased with borrowed funds at an interest rate of 12% with repayment in five equal end-of-year payments?

(b) What is the net cash flow for each year, given that the asset is leased at a rate of $5,500 payable at beginning of each year (a financial lease)?

(c) Which method (if either) should be used to obtain the new asset?

10.36 Enterprise Capital Leasing Company is in the business of leasing tractors to construction companies. The firm wants to set a three-year lease payment schedule for a tractor purchased at $53,000 from the equipment manufacturer. The asset is classified as a five-year MACRS property. The tractor is expected to have a salvage value of $22,000 at the end of three years' rental. Enterprise will require a lessee to make a security deposit in the amount of $1,500 that is refundable at the end of the lease term. Enterprise's marginal tax rate is 35%. If Enterprise wants an after-tax return of 10%, what lease payment schedule should be set?

10.37 You are considering purchasing industrial equipment to meet the peak demand which will last only two years. The equipment costs $100,000 and has an estimated market value of $40,000 at the end of two years. The expected marginal income tax rate during the project period is known to be 40%. Your firm's interest rate (discount rate) is also 12%. Two types of financing are considered, where revenues and O&M expenses are not affected by the type of financing.

- **Option 1 (Debt financing):** Assuming that the equipment will be financed entirely from a bank loan at 10% over 2 years. You can make an arrangement to pay only the interest each year over the project period by deferring the principal payment until the end of 2 years. For tax purpose, the machine will be depreciated by a straight-line method with zero salvage value over 5 years (no half-year convention).

- **Option 2 (Lease financing):** Now you are considering the possibility of leasing the equipment under a two-year contract for a lease payment of $42,000 per year, payable at the beginning of each year.

Determine the net incremental after-tax cash flows under Option 1.

Determine the net incremental after-tax cash flows under Option 2.

What annual lease payment (before tax) would make the two options economically indifferent?

Short Case Studies

ST10.1 Let's reconsider Problem 10.8. Carrier Corporation, a market leader of air conditioning units, is gearing up to push new energy-efficient systems in the wake of the energy bill passed by Congress. Carrier Corporation says it has invested $8 million in developing new heat exchangers—a major component in air-conditioners—that use less energy and are about 20% smaller and 30% lighter than current energy saving versions. Carrier's current energy-efficient models are almost double the size of its regular central air-conditioning units. The company speculates this model's bulk may have been a deterrent for homeowners. The new air conditioners are expected to hit the market in the second quarter of 2016. Cooling efficiency is measured by a standard called SEER: Seasonal Energy Efficiency Ratio. It's similar to the gas mileage system used on cars—the higher the number, the more money you save. Older systems had SEER numbers as low as 8; the new Carrier systems are rated at 18! (Federal standards will require a minimum SEER of 13 from January 2006). Translated into operating costs, this means that for every $100 you used to spend on electricity for cooling, you now can spend just $39.

There are several issues involved in pushing more-efficient air-conditioning unit by Carrier. First, the size of market demand is difficult to estimate. Second, it is even more difficult to predict the useful product life as market competition is ever increasing. Carrier's marketing department has plans to target sales of the new air-conditioning units to the larger office complexes, and if they are successful there, then the units could be marketed to a wide variety of businesses, including schools, hospitals, and eventually even to households.

The marketing vice president believes that annual sales would be 22,000 units if the units were priced at $8,200 each. The engineering department has estimated that the firm would need a new manufacturing plant; this plant could be built and made ready for production in 2 years, once the "go" decision is made. The plant would require a 25-acre site, and Carrier currently has an option to purchase a suitable tract of land for $2.5 million. Building construction would begin in early 2014 and continue through 2015. The building, which would fall into MACRS 39-year class, would cost an estimated $12 million, and a $2 million payment would be due to the contractor on March 31, 2014, another $6 million on March 31, 2015 and remaining balance of $4 million payable on March 31, 2016.

The necessary manufacturing equipment would be installed late in 2015 and would be paid for on March 31, 2016. The equipment, which would fall into the MACRS 7-year class, would have a cost of $8.5 million, including transportation, plus another $500,000 for installation. To date, the company has spent $8 million on research and development associated with the new technology. The company already expensed $2 million of the R&D costs, and the remaining $6 million has been capitalized and will be amortized over the life of the project. However, if Carrier decides not to go forward with the project, the capitalized R&D expenditures could be written off as expenses on March 31, 2016.

The project would also require an initial investment in net working capital equal to 13% of the estimated sales in the first year. The initial working capital investment would be made on March 31, 2016, and on March 31 of each following year, the net working capital would be increased by an amount equal to 13% of any sales increase expected during the coming year. The project's estimated economic life is 6 years (not counting the construction period). At that time, the land is expected to have a market value of $1.5 million, the building a value of $1.8 million, and the equipment a value of $2.5 million.

The production department has estimated that variable manufacturing costs would total 65% of dollar sales, and that fixed overhead costs, excluding depreciation, would be $7.5 million for the first year of operations. Sales prices and fixed overhead costs, other than depreciation and amortization, are projected to increase with inflation, which is expected to average 5% per year over the 6-year life of the project. (Note that the first year's sales would be $8,200 × 22,000 units = $180.4 million. The second year's sales would be 5% higher than $180.4 million, and so forth.)

Carrier's marginal combined tax rate is 38%; its weighted average cost of capital is 14.5% (meaning that their inflation-adjusted minimum attractive rate of return is 14.5% after tax); and the company's policy, for capital budgeting purposes, is to assume that cash flows occur at the end of each fiscal year (for Carrier, it is March 31). Since the plant would begin operations on April 1, 2016, the first operating cash flows would thus occur on March 31, 2017.

Questions:

1. Determine how you would treat the R&D expenditures.
2. Determine the depreciation schedules for all assets.
3. Determine the taxable gains for each asset at the time of disposal (March 31, 2022).
4. Determine the amount of working capital requirement in each operating period.
5. Develop the project cash flows over the life of investment using an Excel spreadsheet.

Justify the investment based on (1) net present worth criteria and (2) internal rate of return.

ST10.2 As Boeing and other aircraft manufactures are planning to use more aluminum lithium alloys for their future aircrafts, the American Aluminum Company is considering making a major investment of $150 million ($5 million for land, $45 million for buildings, and $100 million for manufacturing equipment and facilities) to develop a stronger, lighter material called aluminum lithium that will make aircraft sturdier and more fuel-efficient. Aluminum lithium, which has been sold commercially for only a few years as an alternative to composite materials, will likely be the material of choice for the next generation of commercial and military aircraft because it is so much lighter than conventional aluminum alloys, which use a combination of copper, nickel, and magnesium to harden aluminum. Another advantage of aluminum lithium is that it is cheaper than composites. The firm predicts that aluminum lithium will account for about 5% of the structural weight of the average commercial aircraft within five years and 10% within 10 years. The proposed plant, which has an estimated service life of 12 years, would have a capacity of about 10 million pounds of aluminum lithium, although domestic consumption of the material is expected to be only 3 million pounds during the first four years, 5 million for the next three years, and 8 million for the remaining life of the plant. Aluminum lithium costs $12 a pound to produce, and the firm would expect to sell it at $17 a pound. The buildings will be depreciated according to the 39-year MACRS real property class, with the buildings placed in service on July 1 of the first year. All manufacturing equipment and facilities will be classified as seven-year MACRS properties.

At the end of the project life, the land will be worth $8 million, the buildings $30 million, and the equipment $10 million. Assuming that the firm's marginal tax rate is 40% and its capital gains tax rate is 35%, determine the following:

(a) The net after-tax cash flows.

(b) The IRR for this investment.

(c) Whether the project is acceptable if the firm's MARR is 15%.

ST10.3 Morgantown Mining Company is considering a new mining method at its Blacksville mine. The method, called longwall mining, is carried out by a robot. Coal is removed by the robot—not by tunneling like a worm through an apple, which leaves more of the target coal than is removed—but rather by methodically shuttling back and forth across the width of the deposit and devouring nearly everything. The method can extract about 75% of the available coal, compared with 50% for conventional mining, which is done largely with machines that dig tunnels. Moreover, the coal can be recovered far more inexpensively. Currently, at Blacksville alone, the company mines 5 million tons a year with 2,200 workers. By installing two longwall robot machines, the company can mine 5 million tons with only 860 workers. (A robot miner can dig more than 6 tons a minute.) Despite the loss of employment, the United Mine Workers union generally favors longwall mines for two reasons: The union officials are quoted as saying, (1) "It would be far better to have highly productive operations that were able to pay our folks good wages and benefits than to have 2,200 shovelers living in poverty," and (2) "Longwall mines are inherently safer in their design." The company projects the financial data given in Table ST10.3 upon installation of the longwall mining.

(a) Estimate the firm's net after-tax cash flows over the project life if the firm uses the unit-production method to depreciate assets. The firm's marginal tax rate is 40%.

(b) Estimate the firm's net after-tax cash flows if the firm chooses to depreciate the robots on the basis of MACRS (seven-year property classification).

ST10.4 National Parts, Inc., an auto-parts manufacturer, is considering purchasing a rapid prototyping system to reduce prototyping time for form, fit, and function applications in automobile parts

TABLE ST10.3

Robot installation (2 units)	$19.3 million
Total amount of coal deposit	50 million tons
Annual mining capacity	5 million tons
Project life	10 years
Estimated salvage value	$0.5 million
Working capital requirement fully recovered at end of project life	$2.5 million
Expected additional revenues:	
Labor savings	$6.5 million
Accident prevention	$0.5 million
Productivity gain	$2.5 million
Expected additional expenses:	
O&M costs	$2.4 million

manufacturing. An outside consultant has been called in to estimate the initial hardware requirement and installation costs. He suggests the following:

- Prototyping equipment: $187,000
- Posturing apparatus: $10,000
- Software: $15,000
- Maintenance: $36,000 per year by the equipment manufacturer
- Resin: Annual liquid polymer consumption of 400 gallons at $350 per gallon
- Site preparation: Some facility changes are required for the installation of the rapid prototyping system (e.g., certain liquid resins contain a toxic substance, so the work area must be well ventilated).

The expected life of the system is six years with an estimated salvage value of $30,000. The proposed system is classified as a five-year MACRS property. A group of computer consultants must be hired to develop customized software to run on the system. Software development costs will be $20,000 and can be expensed during the first tax year. The new system will reduce prototype development time by 75% and material waste (resin) by 25%. This reduction in development time and material waste will save the firm $314,000 and $35,000 annually, respectively.

The firm's expected marginal tax rate over the next six years will be 40%. The firm's interest rate is 20%.

(a) Assuming that the entire initial investment will be financed from the firm's retained earnings (equity financing), determine the after-tax cash flows over the life of the investment. Compute the NPW of this investment.

(b) Assuming that the entire initial investment will be financed through a local bank at an interest rate of 13% compounded annually, determine the net after-tax cash flows for the project. Compute the NPW of the investment.

(c) Suppose that a financial lease is available for the prototype system at $62,560 per year, payable at the beginning of each year. Compute the NPW of the investment with lease financing.

(d) Select the best financing option based on the rate of return on incremental investment.

ST10.5 National Office Automation, Inc. (NOAI) is a leading developer of imaging systems, controllers, and related accessories. The company's product line consists of systems for desktop publishing, automatic identification, advanced imaging, and office information markets. The firm's manufacturing plant in Ann Arbor, Michigan, consists of eight different functions: cable assembly, board assembly, mechanical assembly, controller integration, printer integration, production repair, customer repair, and shipping. The process to be considered is the transportation of pallets loaded with eight packaged desktop printers from printer integration to the shipping department. Several alternatives for minimizing operating and maintenance costs have been examined. The two most feasible alternatives are the following.

- **Option 1:** Use gas-powered lift trucks to transport pallets of packaged printers from printer integration to shipping. The truck also can be used to return printers that must be reworked. The trucks can be leased at a cost of $5,465 per year. With a maintenance contract costing $6,317 per year, the dealer will maintain the trucks. A fuel cost of $1,660 per year is expected. The truck requires a driver for each of the three shifts at a total cost of $58,653 per year for labor. It is estimated that

transportation by truck would cause damages to material and equipment totaling $10,000 per year.

- **Option 2:** Install an automatic guided vehicle system (AGVS) to transport pallets of packaged printers from printer integration to shipping and to return products that require rework. The AGVS, using an electrically powered cart and embedded wire-guidance system, would do the same job that the truck currently does, but without drivers. The total investment costs, including installation, are itemized as in Table ST10.5.

TABLE ST10.5

Vehicle and system installation	$97,255
Staging conveyor	24,000
Power supply lines	5,000
Transformers	2,500
Floor surface repair	6,000
Batteries and charger	10,775
Shipping	6,500
Sales tax	6,970
Total AGVS system cost	**$159,000**

NOAI could obtain a term loan for the full investment amount ($159,000) at a 10% interest rate. The loan would be amortized over five years with payments made at the end of each year. The AGVS falls into the seven-year MACRS classification, and it has an estimated service life of 10 years and no salvage value. If the AGVS is installed, a maintenance contract would be obtained at a cost of $20,000, payable at the beginning of each year. The firm's marginal tax rate is 35% and its MARR is 15%.

(a) Determine the net cash flows for each alternative over 10 years.

(b) Compute the incremental cash flows (option 2–option 1) and determine the rate of return on this incremental investment.

(c) Determine the best course of action based on the rate-of-return criterion.

Note: Assume a zero salvage value for the AGVS.

PART FOUR

HANDLING RISK AND UNCERTAINTY

11

Inflation and Its Impact on Project Cash Flows

Purchasing Power—The Big Mac Index[1] *The Economist* has published "The Big Mac" index annually since 1986, which is considered as an indicator of a country's individual purchasing power. As many countries have different currencies, the standardized Big Mac prices are calculated by converting the average national Big Mac prices with the latest exchange rate to U.S. dollars. For example, the average price of a Big Mac in America in January 2014 was $4.62; in China it was only $2.74 at market exchange rates. So the "raw" Big Mac index says that the yuan was undervalued by 41% at that time. On the other hand, buying a Big Mac in Norway costs $7.80 at market exchange rates; the Big Mac index suggests that the Norwegian krone is almost 70% overvalued.

The index is based on the concept of purchasing power of party, i.e., that the rate of exchange between currencies should reflect the price people pay for the same goods and services from one country to the next. Another important factor

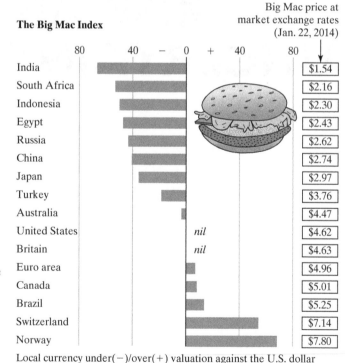

The Big Mac Index

Big Mac price at market exchange rates (Jan. 22, 2014)

Country		Price
India		$1.54
South Africa		$2.16
Indonesia		$2.30
Egypt		$2.43
Russia		$2.62
China		$2.74
Japan		$2.97
Turkey		$3.76
Australia		$4.47
United States	nil	$4.62
Britain	nil	$4.63
Euro area		$4.96
Canada		$5.01
Brazil		$5.25
Switzerland		$7.14
Norway		$7.80

Local currency under(−)/over(+) valuation against the U.S. dollar

[1] "The Big Mac Index—Grease-Proof Taper," *The Economist*, January 25, 2014.

that affects the Big Mac index is inflation, which is never the same between nations. This can erode prices where high inflation exists. The cost of goods and commodity prices may be quite different depending on the nation. For example, the inflation rates and cost of labor per hour among three countries were as follows in January 2014 when the index was published.

Country	Big Mac Price	Exchange Rate	Inflation Rate (January, 2014)	Labor Cost per Hour
Norway	$7.80	6.17 Krone	0	$66
United States	$4.62	$1.00	1.6%	$30
China	$2.74	6.15 Yuan	2.5%	

Clearly, we can see that the higher labor cost in Norway would skew the Big Mac Index even though inflation was virtually nonexistent in the economy.

On a brighter note, we open this chapter with some statistics about engineers' pay rates. The average starting salary of a first-year engineer was $7,500 in 1967. According to a recent survey by the National Association of Colleges and Employers,[2] the average starting salary of a first-year engineer in 2014 was $62,719, which is equivalent to an annual increase of 4.62% over 47 years. During the same period, the general inflation rate remained at 4.27%, indicating that engineers' pay-rates have at least kept up with the rate of inflation.

Up to this point, we have demonstrated how to develop cash flows for an investment project under constant conditions in the general economy. In other words, we have assumed that prices remain relatively unchanged over long periods. As you know from personal experience, that is not a realistic assumption. In this chapter, we define and quantify **inflation** and then go on to apply it in several economic analyses. We will demonstrate inflation's effect on depreciation, borrowed funds, the rate of return of a project, and working capital within the bigger picture of developing projects.

CHAPTER LEARNING OBJECTIVES

After completing this chapter, you should understand the following concepts:

■ How to measure inflation.

■ Conversion from actual dollars to constant dollars or from constant to actual dollars.

[2] *Source*: NACE Salary Survey, National Association of Colleges and Employers, September 2014.

- How to compare the amount of dollars received at different points in time.
- Which inflation rate to use in economic analysis.
- Which interest rate to use in economic analysis (market interest rate versus inflation-free interest rate).
- How to handle multiple inflation rates in project analysis.
- The cost of borrowing and changes in working-capital requirements under inflation.
- How to conduct rate-of-return analysis under inflation.

11.1 Meaning and Measure of Inflation

Historically, the general economy has usually fluctuated in such a way as to exhibit **inflation**, a loss in the purchasing power of money over time. Inflation means that the cost of an item tends to increase over time, or to put it another way, the same dollar amount buys less of an item over time. **Deflation** is the opposite of inflation in that prices usually decrease over time; hence, a specified dollar amount gains in purchasing power. Inflation is far more common than deflation in the real world, so our consideration in this chapter will be restricted to accounting for inflation in economic analyses.

11.1.1 Measuring Inflation

Before we can introduce inflation into an engineering economic problem, we need a means of isolating and measuring its effect. Consumers usually have a relative, if not a precise, sense of how their purchasing power is declining. This sense is based on their experience of shopping for food, clothing, transportation, and housing over the years. Economists have developed a measure called the **Consumer Price Index (CPI)**, which is based on a typical **market basket** of goods and services required by the average consumer. This market basket normally consists of items from eight major groups: (1) food and alcoholic beverages; (2) housing; (3) apparel; (4) transportation; (5) medical care; (6) entertainment; (7) personal care; and (8) other goods and services.

Calculating the Index

The CPI compares the cost of the typical market basket of goods and services in a current month with its cost one month ago, one year ago, or ten years ago. The point in the past with which current prices are compared is called the **base period**. The index value for the base period is set at $100. The original base period used by the Bureau of Labor Statistics (BLS), of the U.S. Department of Labor, for the CPI index was 1967. For example, let us say that, in 1967, the prescribed market basket could have been purchased for $100. Now suppose the same combination of goods and services costs $713[3] in 2014. We can then compute the CPI for 2014 by multiplying the ratio of the current price to the base-period price by 100. In our example, the price index is ($713/$100)100 = 713 which means that the 2014 price of the contents of the market basket is 639% of its base-period price.

---

[3] As of June, 2014.

Updating the Shopping Basket

The revised CPI introduced by the BLS in 1987 includes indexes for two populations: urban wage earners and clerical workers (CPI-W); and all urban consumers (CPI-U). As a result of the revision, both the CPI-W and the CPI-U utilize updated expenditure weights based upon data tabulated from the years 1982, 1983, and 1984 of the Consumer Expenditure Survey and incorporate a number of technical improvements. This method of assessing inflation does not imply, however, that consumers actually purchase the same goods and services year after year. Consumers tend to adjust their shopping practices to changes in relative prices, and they tend to substitute other items for those whose prices have increased greatly in relative terms. We must understand that the CPI does not take into account this sort of consumer behavior, because it is predicated on the purchase of a fixed market basket of the same goods and services, in the same proportions, month after month. For this reason, the CPI is called a **price index** rather than a **cost-of-living index**, although the general public often refers to it as a cost-of-living index.

Price Index for Industrial Goods

The Consumer Price Index is a good measure of the general increase in prices of consumer products. However, it is not a good measure of industrial price increases. In performing engineering economic analysis, the appropriate price indexes must be selected to estimate the price increases of raw materials, finished products, and operating costs. The *Survey of Current Business*, a monthly publication prepared by the BLS, provides the **Producer Price Index (PPI)** for various industrial goods.[4]

Calculating Yearly CPI

Table 11.1 lists the overall CPI, together with several price indexes for certain commodities over a number of years. From the table, we can easily calculate the inflation rate of gasoline from 2013 to 2014 as

$$\frac{304.2 - 301.6}{301.6} = 0.0086 = 0.86\%$$

TABLE 11.1 Selected Price Indexes					
Year	New CPI (Base: 1982–84)	Old CPI (Base: 1967)	Unleaded Gasoline	Sugar	New Cars	Airline Fare
2005	195.3	585.0	194.4	165.2	135.2	236.6
2006	201.6	603.9	220.0	171.5	136.4	247.3
2007	207.3	621.1	238.2	176.8	135.9	251.7
2008	215.3	644.9	277.6	186.6	135.4	282.0
2009	214.5	642.7	175.3	196.4	135.8	248.9
2010	218.1	653.2	237.9	201.2	138.0	278.2
2011	224.9	673.8	301.6	207.8	141.9	304.0
2012	229.6	687.7	311.3	214.7	144.2	304.9
2013	232.9	697.8	301.6	210.9	145.8	312.7
2014	236.4	713.9	304.2	208.6	146.4	342.7

Note: Data for 2014 is for the first half year, a semiannual average.

[4] Both CPI and PPI data are available on the Internet at the BLS home page at http://stats.bls.gov.

Since the inflation rate calculated is positive, the price of gasoline increased at a mere rate of 0.86% over the year 2013, which is one of the least increases in the price of gasoline in recent years.

Average Inflation Rate (*f*)

To account for the effect of varying yearly inflation rates over a period of several years, we can compute a single rate that represents an **average inflation rate**. Since each individual year's inflation rate is based on the previous year's rate, all of these rates have a compounding effect. As an example, suppose we want to calculate the average inflation rate for a two-year period: The first year's inflation rate is 4%, and the second year's rate is 8%—each on a base price of $100.

Step 1. To find the price at the end of the second year, we use the process of compounding:

$$\overbrace{\underbrace{\$100(1 + 0.04)}_{\text{First year}}(1 + 0.08)}^{\text{Second year}} = \$112.32$$

Step 2. To find the average (compound) inflation rate *f*, we establish the following equivalence equation:

$$\$100(1 + f)^2 = \$112.32 \text{ or } \$100(F/P, f, 2) = \$112.32$$

Solving for *f* yields

$$f = \left[\frac{112.32}{100}\right]^{1/2} - 1 = 5.98\%$$

We can say that the price increases in the last two years are equivalent to an average annual percentage rate of 5.98% per year. Note that the average is a geometric (not an arithmetic) average over a several-year period. Our computations are simplified by using a single average rate such as this, rather than a different rate for each year's cash flows.

EXAMPLE 11.1 Average Inflation Rate

Consider again the price increases for the items listed in Table 11.1. Determine the average inflation rate for each item over the nine-year period.

SOLUTION

Let's take the first item, unleaded regular gasoline, for a sample calculation. Since we know the price indices for 2005 and 2014, we can use the appropriate equivalence formula (single-payment compound amount factor or growth formula). We state the problem as follows.

Given: $P = 194.4$, $F = 304.2$, and $N = 2014 - 2005 = 9$.

Find: *f*.

The equation we desire is $F = P(1 + f)^N$, so we have

$$304.2 = 194.4(1 + f)^9$$

$$f = \left(\frac{304.2}{194.4}\right)^{1/9} - 1$$

$$= 5.10\%$$

In a similar fashion, we can obtain the average inflation rates for the remaining items. The answers are

Item	2014 Price Index	2005 Price Index	Average Inflation Rate
New CPI	236.4	195.3	2.14%
Old CPI	713.8	585.0	2.23%
Gasoline	304.2	194.4	5.10%
Sugar	208.6	165.2	2.63%
New Cars	146.4	135.2	0.89%
Airline Fare	342.7	236.6	4.20%

The cost of unleaded gasoline increased most among the items listed in the table, whereas the price of a new car exhibited "most stable" during the same period.

General Inflation Rate $\bar{f}$ versus Specific Inflation Rate f_j

When we use the CPI as a base to determine the average inflation rate, we obtain the **general inflation rate**. We need to distinguish carefully between the general inflation rate and the average inflation rate for specific goods.

- **General inflation rate $\bar{f}$:**   This average inflation rate is calculated on the basis of the CPI for all items in the market basket. The market interest rate is expected to respond to this general inflation rate. We will reserve the symbol $\bar{f}$ strictly for this purpose.
- **Specific inflation rate f_j:**   This rate is based on an index (or the CPI) specific to segment j of the economy. For example, we must often estimate the future cost of a particular item (e.g., labor, material, housing, or gasoline). When we refer to the average inflation rate for just one item, we will drop the subscript j for simplicity. In terms of the CPI, we define the *average general inflation* rate as

$$\text{CPI}_n = \text{CPI}_0 (1 + \bar{f})^n \tag{11.1}$$

or

$$\bar{f} = \left[\frac{\text{CPI}_n}{\text{CPI}_0} \right]^{1/n} - 1 \tag{11.2}$$

where

$\bar{f}$ = The general inflation rate

CPI_n = The consumer price index at the end period n

CPI_0 = The consumer price index for the base period.

In fact, the average inflation rate of 2.14% calculated for the new CPI in Example 11.1 represents the general inflation rate over a nine-year period.

Knowing the CPI values for two consecutive years, we can calculate the *annual general inflation rate* as

$$\bar{f}_n = \frac{\text{CPI}_n - \text{CPI}_{n-1}}{\text{CPI}_{n-1}} \tag{11.3}$$

where $\bar{f}_n$ = the general inflation rate for period n.

As an example, let us calculate the general inflation rate for the year 2009, where $CPI_{2014} = 236.4$ and $CPI_{2013} = 232.9$:

$$\frac{236.4 - 232.9}{215.3} = 1.63\%$$

This was an unusually good rate for the U.S. economy when compared with the average general inflation rate of 4.27%[5] over the last 47 years.

EXAMPLE 11.2 Yearly and Average Inflation Rates

The following table shows a utility company's cost to supply a fixed amount of power to a new housing development.

Year	Cost
0	$504,000
1	538,400
2	577,000
3	629,500

The indices are specific to the utilities industry. Assume that year 0 is the base period, and determine the inflation rate for each period. Then calculate the average inflation rate over the three years listed in the table.

SOLUTION

Given: Annual cost to supply a fixed amount of power.

Find: Annual inflation rate (f_n) and average inflation rate ($\bar{f}$).

The inflation rate during year 1 (f_1) is

$$\frac{(\$538,400 - \$504,000)}{\$504,000} = 6.83\%$$

The inflation rate during year 2 (f_2) is

$$\frac{(\$577,000 - \$538,400)}{\$538,400} = 7.17\%$$

The inflation rate during year 3 (f_3) is

$$\frac{(\$629,500 - \$577,000)}{\$577,000} = 9.10\%$$

The average inflation rate over the three years is

$$f = \left(\frac{629,500}{504,000}\right)^{1/3} - 1 = 0.0769 = 7.69\%$$

Note that, although the average inflation rate[6] is 7.69% for the period taken as a whole, none of the years within the period had that particular rate.

---

[5] To calculate the average general inflation rate from the base period (1967) to 2014, we evaluate

$$\bar{f} = \left[\frac{713.8}{100}\right]^{1/47} - 1 = 4.27\%.$$

[6] Since we obtained this average rate on the basis of costs that are specific to the utility industry, that rate is not the general inflation rate. Rather, it is a specific inflation rate for the utility in question.

11.1.2 Actual versus Constant Dollars

To introduce the effect of inflation into our economic analysis, we need to define several inflation-related terms:[7]

- **Actual (current, nominal) dollars (A_n):** Out-of-pocket dollars paid at the time of purchasing goods and services. Actual dollars are estimates of future cash flows for year n that take into account any anticipated changes in amounts caused by inflationary or deflationary effects. Usually, these amounts are determined by applying an inflation rate to base-year dollar estimates.
- **Constant (real) dollars (A'_n):** Dollars in some base year used to adjust for the effects of inflation. Constant dollars represent constant purchasing power that is independent of the passage of time. In situations where inflationary effects were assumed when cash flows were estimated, the estimates obtained can be converted to constant dollars (base-year dollars) by adjustment with some readily accepted **general inflation rate**. We will assume that the base year is always time 0, unless we specify otherwise.

Conversion from Constant to Actual Dollars

Since constant dollars represent dollar amounts expressed in terms of the purchasing power of the base year, we may find the equivalent dollars in year n by using the general inflation rate $\overline{f}$ in the formula

$$A_n = A'_n(1 + \overline{f})^n = A'_n(F/P, \overline{f}, n) \tag{11.4}$$

where

A'_n = Constant-dollar expression for the cash flow at the end of year n

A_n = Actual-dollar expression for the cash flow at the end of year n

If the future price of a specific cash flow element (j) is not expected to follow the general inflation rate, we will need to use the appropriate average inflation rate f_j applicable to that cash flow element instead of $\overline{f}$. The conversion process is illustrated in Figure 11.1,

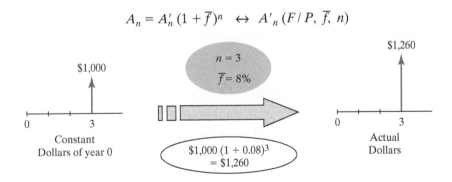

$$A_n = A'_n (1 + \overline{f})^n \leftrightarrow A'_n (F/P, \overline{f}, n)$$

Figure 11.1 Converting a $1,000 cash flow at year 3 estimated in constant dollars of year 0 to its equivalent actual dollars at the same period, assuming an 8% inflation rate.

---

[7] The definitions presented are based on the ANSI Z94 Standards Committee on Industrial Engineering Terminology, *The Engineering Economist*, 1988; 33(2): 145–171.

where $1,260 received in year 3 has a purchasing power of $1,000 in terms of base year 0. Stated differently, if it cost $1,000 to purchase 50 lbs of sugar in year 0, it would cost $1,260 to purchase the same amount of sugar in year 3.

EXAMPLE 11.3 Conversion from Constant to Actual Dollars

Jack Nicklaus won his first Masters Golf Championship in 1963. His prize money was $20,000. Adam Scott won his first Masters in 2013. The prize money was $1,440,000. It seems that Adam's prize money is much bigger. Compare these two prizes in terms of common purchasing power: Whose buying power is greater?

SOLUTION

Given: Two prize monies received in 1963 and 2013, respectively.

Find: Compare these two prizes in terms of purchasing power of 2013.

Both prize monies are in actual dollars at the time of winning. So, if we want to compare the prizes on the basis of the purchasing power of the year 2013, then we need to know the consumer price indexes for 1963 as well as 2013.

Using the old CPI, we find that the CPI for 1963 is 91.7 and the CPI for 2013 is 697.8. The average annual inflation rate between these two periods is 4.14%. To find the purchasing power of Nicklaus's $20,000 in 2013, we simply calculate the expression

$$20,000(1 + 0.0414)^{50} = \$152,023$$

Given inflation and the changing value of money, $20,000 in 1963 is equivalent to $152,023 in terms of 2013 purchasing power. This is clearly much lower than Adam's actual 2013 winning prize, but Jack had 50 years to invest the amount to match Adam's winning prize. Then, at what interest rate would Jack's earning in 1963 match the prize money in 2013?

$$\$20,000(1 + i)^{50} = \$1,440,000$$
$$i = 8.93\%$$

Conversion from Actual to Constant Dollars

This conversion, shown in Figure 11.2, is the reverse process of converting from constant to actual dollars. Instead of using the compounding formula, we use a discounting formula (single-payment present-worth factor):

$$A'_n = \frac{A_n}{(1 + \overline{f})^n} = A_n(P/F, \overline{f}, n) \tag{11.5}$$

Once again, we may substitute f_j for $\overline{f}$ if future prices are not expected to follow the general inflation rate.

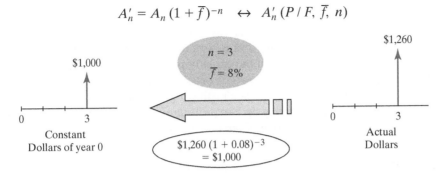

$$A'_n = A_n(1 + \overline{f})^{-n} \quad \leftrightarrow \quad A'_n(P/F, \overline{f}, n)$$

$1,000

Constant
Dollars of year 0

$n = 3$

$\overline{f} = 8\%$

$1,260 (1 + 0.08)^{-3}$
$= $1,000$

$1,260

0   3

Actual
Dollars

Figure 11.2 Conversion from actual to constant dollars. What it means is that $1,260 three years from now will have a purchasing power of $1,000 in terms of base dollars (year 0).

EXAMPLE 11.4 Conversion from Actual to Constant Dollars

The Jagura Creek Fish Company, an aqua-cultural production firm, has negotiated a five-year lease on 20 acres of land, which will be used for fishponds. The annual cost stated in the lease is $20,000 to be paid at the beginning of each of the five years. The general inflation rate is $\overline{f} = 5\%$. Find the equivalent cost in constant dollars during each period.

DISCUSSION: Although the $20,000 annual payments are *uniform*, they are not expressed in constant dollars. Unless an inflation clause is built into a contract, any stated amounts in a contract refer to *actual dollars*.

SOLUTION

Given: Actual dollars and $\overline{f} = 5\%$.
Find: Constant dollars during each period.

Using Eq. (11.5), we can determine the equivalent lease payments in constant dollars as given in Table 11.2.

TABLE 11.2 Conversion from Actual $ to Constant $

End of Period	Cash Flow in Actual $	Conversion at $\overline{f}$	Cash Flow in Constant $	Loss of Purchasing Power
0	$20,000	$(1 + 0.05)^0$	$20,000	0%
1	20,000	$(1 + 0.05)^{-1}$	19,048	4.76
2	20,000	$(1 + 0.05)^{-2}$	18,141	9.30
3	20,000	$(1 + 0.05)^{-3}$	17,277	13.62
4	20,000	$(1 + 0.05)^{-4}$	16,454	17.73

Note that, under the inflationary environment, the lender's receipt of the lease payment in year 5 is worth only 82.27% of the first lease payment.

11.2 Equivalence Calculations under Inflation

In previous chapters, our equivalence analyses took into consideration changes in the **purchasing power** of money implicitly by assuming that all cash flows were given in actual dollars. To factor in changes in **purchasing power** explicitly—that is, inflation—we may use either (1) constant-dollar analysis or (2) actual-dollar analysis. Either method produces the same solution; however, each method requires the use of a different interest rate and procedure. Before presenting the two procedures for integrating interest and inflation, we will give a precise definition of the two interest rates.

11.2.1 Market and Inflation-Free Interest Rates

Two types of interest rates are used in equivalence calculations: the market interest rate and the inflation-free interest rate. The rate to apply depends on the assumptions used in estimating the cash flow.

- **Market interest rate (i):** Interest rate quoted by financial institutions that accounts for both earning and purchasing power. In other words, this rate takes into account the combined effects of the earning value of capital (earning power) and any anticipated inflation or deflation (purchasing power). Virtually all interest rates stated by financial institutions for loans and savings accounts are market interest rates. Most firms use a market interest rate (also known as an **inflation-adjusted MARR**) in evaluating their investment projects.
- **Inflation-free interest rate (i'):** This rate is an estimate of the true earning power of money when the inflation effects have been removed. Commonly known as the **real interest rate**, it can be computed if the market interest rate and the inflation rate are known. As you will see later in this chapter, in the absence of inflation, the market interest rate is the same as the inflation-free interest rate.

In calculating any cash flow equivalence, we need to identify the nature of project cash flows. The three common cases are as follows:

Case 1. All cash flow elements are estimated in constant dollars.

Case 2. All cash flow elements are estimated in actual dollars.

Case 3. Some of the cash flow elements are estimated in constant dollars, and others are estimated in actual dollars.

For case 3, we simply convert all cash flow elements into one type—either constant or actual dollars. Then we proceed with either constant-dollar analysis as for case 1 or actual-dollar analysis as for case 2.

11.2.2 Constant-Dollar Analysis

Suppose that all cash flow elements are already given in constant dollars and that you want to compute the equivalent present worth of the constant dollars (A'_n) occurring in year n. In the absence of an inflationary effect, we should use i' to account for only the earning power of the money. To find the present-worth equivalent of this constant-dollar amount at i', we use

$$P_n = \frac{A'_n}{(1 + i')^n} \tag{11.6}$$

Constant-dollar analysis is common in the evaluation of many long-term public projects, because governments do not pay income taxes. Typically, income taxes are levied on the basis of taxable incomes in actual dollars.

EXAMPLE 11.5 Equivalence Calculation When Flows are Stated in Constant Dollars

The Transco Company is considering making and supplying computer-controlled traffic-signal switching boxes to be used throughout Arizona. Transco has estimated the market for its boxes by examining data on new road construction and on the deterioration and replacement of existing units. The current price per unit is $550; the before-tax manufacturing cost is $450. The start-up investment cost is $250,000. The projected sales and net before-tax cash flows in constant dollars are as follows:

Period	Unit Sales	Net Cash Flow in Constant $
0		−$250,000
1	1,000	100,000
2	1,100	110,000
3	1,200	120,000
4	1,300	130,000
5	1,200	120,000

If Transco managers want the company to earn a 12% inflation-free rate of return (i') before tax on any investment, what would be the present worth of this project?

SOLUTION
Given: Cash flow series stated in constant dollars and $i' = 12\%$.
Find: Present worth of the cash flow series.

Since all values are in constant dollars, we can use the inflation-free interest rate. We simply discount the dollar inflows at 12% to obtain

$$PW(12\%) = -\$250,000 + \$100,000(P/A, 12\%, 5)$$

$$+ \$10,000(P/G, 12\%, 4) + \$20,000(P/F, 12\%, 5)$$

$$= \$163,099 \text{ (in year-0 dollars)}$$

Since the equivalent net receipts exceed the investment, the project can be justified even before considering any tax effects.

11.2.3 Actual-Dollar Analysis

Now let us assume that all cash flow elements are estimated in actual dollars. To find the equivalent present worth of the actual dollar amount (A_n) in year n, we may use either the **deflation method** or the **adjusted-discount method**.

Deflation Method

The deflation method requires two steps to convert actual dollars into equivalent present-worth dollars. First we convert actual dollars into equivalent constant dollars by discounting by the general inflation rate, a step that removes the inflationary effect. Now we can use i' to find the equivalent present worth.

EXAMPLE 11.6 Equivalence Calculation When Cash Flows are in Actual Dollars: Deflation Method

Applied Instrumentation, a small manufacturer of custom electronics, is contemplating an investment to produce sensors and control systems that have been requested by a fruit-drying company. The work would be done under a proprietary contract that would terminate in five years. The project is expected to generate the following cash flows in actual dollars.

n	Net Cash Flow in Actual Dollars
0	−$75,000
1	32,000
2	35,700
3	32,800
4	29,000
5	58,000

(a) What are the equivalent year-0 dollars (constant dollars) if the general inflation rate $(\bar{f})$ is 5% per year?

(b) Compute the present worth of these cash flows in constant dollars at $i' = 10\%$.

SOLUTION

Given: Cash flow series stated in actual dollars, $\bar{f} = 5\%$, and $i' = 10\%$.

Find: (a) Cash flow series converted into constant dollars, and (b) the present worth of the cash flow series.

The net cash flows in actual dollars can be converted to constant dollars by deflating them, again assuming a 5% yearly deflation factor. The deflated (or constant-dollar) cash flows then can be used to determine the NPW at i'.

(a) We convert the actual dollars into constant dollars in Table 11.3.

(b) In Table 11.4, we compute the equivalent present worth of the constant dollars by using $i' = 10\%$.

TABLE 11.3 **Converting Actual Dollars into Constant Dollars**

n	Cash Flows in Actual Dollars	×	Deflation Factor	=	Cash Flows in Constant Dollars
0	−$75,000		1		−$75,000
1	32,000		$(1 + 0.05)^{-1}$		30,476
2	35,700		$(1 + 0.05)^{-2}$		32,381
3	32,800		$(1 + 0.05)^{-3}$		28,334
4	29,000		$(1 + 0.05)^{-4}$		23,858
5	58,000		$(1 + 0.05)^{-5}$		45,445

TABLE 11.4 **Calculating Equivalent Present Worth**

n	Cash Flows in Constant Dollars	×	Discounting	=	Equivalent Present Worth
0	−$75,000		1		−$75,000
1	30,476		$(1 + 0.10)^{-1}$		27,706
2	32,381		$(1 + 0.10)^{-2}$		26,761
3	28,334		$(1 + 0.10)^{-3}$		21,288
4	23,858		$(1 + 0.10)^{-4}$		16,296
5	45,445		$(1 + 0.10)^{-5}$		<u>28,217</u>
					$45,268

Adjusted-Discount Method

The two-step process shown in Example 11.6 can be greatly streamlined by the efficiency of the **adjusted-discount method**, which performs deflation and discounting in one step. Mathematically, we can combine this two-step procedure into one with the formula

$$P_n = \left(\overbrace{\dfrac{\dfrac{A_n}{(1 + \bar{f})^n}}{(1 + i')^n}}^{\text{Conversion to constant dollars}} \right)_{\underbrace{\qquad\qquad}_{\text{discounting}}}$$

$$= \frac{A_n}{(1 + \bar{f})^n(1 + i')^n}$$

$$= \frac{A_n}{[(1 + \bar{f})(1 + i')]^n} \tag{11.7}$$

Since the market interest rate (i) reflects both the earning power and the purchasing power, we have

$$P_n = \frac{A_n}{(1 + i)^n} \tag{11.8}$$

The equivalent present-worth values in Eqs. (11.7) and (11.8) must be equal at year 0. Therefore,

$$\frac{A_n}{(1 + i)^n} = \frac{A_n}{[(1 + \overline{f})(1 + i')]^n}$$

This leads to the following relationship among $\overline{f}$, i', and i:

$$(1 + i) = (1 + \overline{f})(1 + i')$$

Simplifying the terms yields

$$i = i' + \overline{f} + i'\overline{f} \tag{11.9}$$

Equation (11.9) implies that the market interest rate is a function of two terms: i' and $\overline{f}$. Note that without an inflationary effect, the two interest rates are the same: $\overline{f} = 0 \rightarrow i = i'$. As either i' or $\overline{f}$ increases, i also increases. For example, we easily can observe that when prices increase due to inflation, bond rates climb because lenders (i.e., anyone who invests in a money-market fund, a bond, or a certificate of deposit) demand higher rates to protect themselves against erosion in the value of their dollars. If inflation were to remain at 3%, you might be satisfied with an interest rate of 7% on a bond, because your return would more than beat inflation. If inflation were running at 10%, however, you would not buy a 7% bond; you might insist instead on a return of at least 14%. By contrast, when prices are coming down, or at least are stable, lenders do not fear any loss of purchasing power with the loans they make, so they are satisfied to lend at lower interest rates.

In practice, we often approximate the market interest rate (i) simply by adding the inflation rate ($\overline{f}$) to the real interest rate (i') and ignoring the product ($i'\overline{f}$). This practice is fine as long as either i' or $\overline{f}$ is relatively small. With continuous compounding, however, the relationship among i, i', and $\overline{f}$ becomes precisely

$$i' = i - \overline{f} \tag{11.10}$$

So if we assume a nominal APR (market interest rate) of 6% per year compounded continuously and an inflation rate of 4% per year compounded continuously, the inflation-free interest rate is exactly 2% per year compounded continuously.

EXAMPLE 11.7 Equivalence Calculation When Flows Are in Actual Dollars: Adjusted-Discounted Method

Consider the cash flows in actual dollars in Example 11.6. Use the adjusted-discount method to compute the equivalent present worth of these cash flows.

SOLUTION

Given: Cash flow series in actual dollars, $\overline{f} = 5\%$, and $i' = 10\%$.
Find: Equivalent present worth.

First, we need to determine the market interest rate i. With $\overline{f} = 5\%$ and $i' = 10\%$, we obtain

$$i = i' + \overline{f} + i'\overline{f}$$
$$= 0.10 + 0.05 + (0.10)(0.05)$$
$$= 15.5\%$$

Note that the equivalent present worth that we obtain with the adjusted-discount method ($i = 15.5\%$), shown in Table 11.5, gives exactly the same result we obtained in Example 11.6.

TABLE 11.5 Calculating Equivalent Present Worth Using i

n	Cash Flows in Actual Dollars	Multiplied by	=	Equivalent Present Worth
0	−$75,000	1		−$75,000
1	32,000	$(1 + 0.155)^{-1}$		27,706
2	35,700	$(1 + 0.155)^{-2}$		26,761
3	32,800	$(1 + 0.155)^{-3}$		21,288
4	29,000	$(1 + 0.155)^{-4}$		16,296
5	58,000	$(1 + 0.155)^{-5}$		<u>28,217</u>
				$45,268

11.2.4 Mixed-Dollar Analysis

Let us consider another situation in which some cash flow elements are expressed in constant (or today's) dollars and the other elements in actual dollars. In this situation, we can convert all cash flow elements into the same dollar units (either constant or actual). If the cash flow is converted into actual dollars, the market interest rate (i) should be used in calculating the equivalence value. If the cash flow is converted into constant dollars, the inflation-free interest rate (i') should be used.

11.3 Effects of Inflation on Project Cash Flows

We now introduce inflation into some investment projects. We are especially interested in two elements of project cash flows: depreciation expenses and interest expenses. These two elements are essentially immune to the effects of inflation, as they are always given in actual dollars. We will also consider the complication of how to proceed when multiple price indexes have been used to generate various project cash flows.

Because depreciation expenses are calculated on some base-year purchase amount, they do not increase over time to keep pace with inflation. Thus, they lose some of their value to defer taxes, because inflation drives up the general price level and hence, taxable income. Similarly, the selling prices of depreciable assets can increase with the general inflation rate, and because any gains on salvage values are taxable, they can result in increased taxes. Example 11.8 illustrates how a project's profitability changes under an inflationary economy.

EXAMPLE 11.8 Effects of Inflation on Projects with Depreciable Assets

Consider again the automated machining center investment project described in Example 11.1. The summary of the financial facts in the absence of inflation is given in Table 11.6.

TABLE 11.6 **Financial Data for the Automated Machining Center Project**

Item	Description or Data
Project	Automated machining center
Required investment	$125,000
Project life	5 years
Salvage value	$50,000
Depreciation method	7-year MACRS
Annual revenues	$100,000 per year
Annual expenses:	
Labor	$20,000 per year
Material	$12,000 per year
Overhead	$8,000 per year
Marginal tax rate	40%
Inflation-free interest rate (i')	15%

The after-tax cash flow for the automated machining center project was given in Table 11.2, and the net present worth of the project in the absence of inflation was calculated to be $43,152.

What will happen to this investment project if the general inflation rate during the next five years is expected to be 5% annually? Sales and operating costs are assumed to increase accordingly. Depreciation will remain unchanged, but taxes, profits, and thus, cash flow will be higher. The firm's inflation-free interest rate (i') is known to be 15%.

(a) Use the adjusted-discount method to determine the NPW of the project.

(b) Compare this NPW with that of the inflation-free situation.

DISCUSSION: All cash flow elements, except depreciation expenses, are assumed to be in constant dollars. Since income taxes are levied on actual taxable income, we will use actual-dollar analysis, which requires that all cash flow elements be expressed in actual dollars.

- For the purposes of this illustration, all inflationary calculations are made as of year-end.
- Cash flow elements (such as sales, labor, material, overhead, and the selling price of the asset) will be inflated at the same rate as the general inflation rate.[8] For example, whereas annual sales had been estimated at $100,000, under conditions of inflation they become 5% greater in year 1 (or $105,000); 10.25% greater in year 2, and so forth.

Table 11.7 gives the sales conversion from constant to actual dollars for each of the five years of the project life.

TABLE 11.7 Converting Constant Dollars into Actual Dollars

Period	Sales in Constant $	Conversion $\overline{f}$	Sales in Actual $
1	$100,000	$(1 + 0.05)^1$	$105,000
2	100,000	$(1 + 0.05)^2$	110,250
3	100,000	$(1 + 0.05)^3$	115,763
4	100,000	$(1 + 0.05)^4$	121,551
5	100,000	$(1 + 0.05)^5$	127,628

Future cash flows in actual dollars for other elements can be obtained in a similar way. Note that:

- No change occurs in the investment in year 0 or in depreciation expenses, since these items are unaffected by expected future inflation.
- The selling price of the asset is expected to increase at the general inflation rate. Therefore, the salvage value of the machine in actual dollars will be

$$\$50,000(1 + 0.05)^5 = \$63,814$$

This increase in salvage value will also increase the taxable gains, because the book value remains unchanged.

SOLUTION

Given: Financial data in constant dollars, $\overline{f} = 5\%$, and $i' = 15\%$.
Find: The NPW.

Table 11.8 shows after-tax cash flows in actual dollars. Using the adjusted-discount method, we determine the market interest rate at $i = 0.15 + 0.05 + (0.15)(0.05) = 20.75\%$. Then we discount these actual-dollar cash flows at 20.75% to

---

[8] This is a simplistic assumption. In practice, these elements may have price indexes other than the CPI. Differential price indexes will be treated in Example 11.9.

TABLE 11.8 Cash Flow Statement for the Automated Machining Center Project under Inflation (Example 11.8)

	A	B	C	D	E	F	G	H
1	**Income Statement**							
2								
3	End of Year	Inflation Rate	0	1	2	3	4	5
4								
5	Revenue	5%		$ 105,000	$ 110,250	$ 115,763	$ 121,551	$ 127,628
6	Expenses:							
7	Labor	5%		$ 21,000	$ 22,050	$ 23,153	$ 24,310	$ 25,526
8	Materials	5%		$ 12,600	$ 13,230	$ 13,892	$ 14,586	$ 15,315
9	Overhead	5%		$ 8,400	$ 8,820	$ 9,261	$ 9,724	$ 10,210
10	Depreciation	5%		$ 17,863	$ 30,613	$ 21,863	$ 15,613	$ 5,581
11								
12	Taxable income			$ 45,137	$ 35,537	$ 47,595	$ 57,317	$ 70,996
13	Income taxes (40%)			$ 18,055	$ 14,215	$ 19,038	$ 22,927	$ 28,398
14								
15	Net Income			$ 27,082	$ 21,322	$ 28,557	$ 34,390	$ 42,598
16								
17	**Cash Flow Statement**							
18	Operating Activities:							
19	Net income			$ 27,082	$ 21,322	$ 28,557	$ 34,390	$ 42,598
20	Depreciation			$ 17,863	$ 30,613	$ 21,863	$ 15,613	$ 5,581
21	Investing Activities:							
22	Investment		$ (125,000)					
23	Salvage value	5%						$ 63,814
24	Gains tax							$ (12,139)
25								
26	Net Cash Flow (in Actual $)		$ (125,000)	$ 44,945	$ 51,935	$ 50,420	$ 50,003	$ 99,854
27								
28								
29		PW(20.75%) =	$ 38,898					
30		IRR =	32.83%					
31								
32								
33								

=(H23-(C22-SUM(D10:H10)))*0.4

=NPV(21.90%,D26:H26)+C26

=IRR(C26:H26,15%)

determine the NPW. Since NPW = $38,898 > 0, the investment is still economically attractive.

COMMENTS: Note that the NPW in the absence of inflation was $43,152 in Example 11.1. The $4,254 decline just illustrated (known as inflation loss) in the NPW under inflation is due entirely to income tax considerations. The depreciation expense is a charge against taxable income, which reduces the amount of taxes paid and, as a result, increases the cash flow attributable to an investment by the amount of taxes saved. But the depreciation expense under existing tax laws is based on historic cost. As time goes by, the depreciation expense is charged to taxable income in dollars of declining purchasing power; as a result, the "real" cost of the asset is not totally reflected in the depreciation expense. Depreciation costs are thereby understated, and the taxable income is overstated, resulting in higher taxes. In "real" terms, the amount of this additional income tax is $4,254, which is also known as the **inflation tax**. In general, any investment that (for tax purposes) is expensed over time (rather than immediately) is subject to the inflation tax.

11.3.1 Multiple Inflation Rates

As we noted previously, the inflation rate f_j represents a rate applicable to a specific segment j of the economy. For example, if we were estimating the future cost of a piece of machinery, we should use the inflation rate appropriate for that item. Furthermore, we may need to use several rates to accommodate the different costs and revenues in our analysis. The next example introduces the complexity of multiple inflation rates.

EXAMPLE 11.9 Applying Specific Inflation Rates

Let us rework Example 11.8 using different annual indexes (differential inflation rates) in the prices of cash flow components. Suppose that we expect the general rate of inflation, $\overline{f}$, to average 6% per year during the next five years. We also expect that the selling price of the equipment will increase 3% per year, that wages (labor) and overhead will increase 5% per year, and that the cost of materials will increase 4% per year. We expect sales revenue to climb at the general inflation rate. For simplicity, all cash flows and inflation effects are assumed to occur at year's end. Determine the net present worth of this investment using the adjusted-discount method.

SOLUTION

Given: Financial data given in Example 11.8 and multiple inflation rates.
Find: The NPW using the adjusted-discount method.

Table 11.9 summarizes the after-tax cash flows in actual dollars. To evaluate the present worth using actual dollars, we must adjust the original discount rate of 15%,

TABLE 11.9 Cash Flow Statement for the Automated Machining Center Project under Inflation with Multiple Price Indices (Example 11.9)

	A	B	C	D	E	F	G	H
1								
2	**Income Statement**							
3	End of Year	Inflation Rate	0	1	2	3	4	5
4								
5	Revenue	6%		$ 106,000	$ 112,360	$ 119,102	$ 126,248	$ 133,823
6	Expenses:							
7	Labor	5%		$ 21,000	$ 22,050	$ 23,153	$ 24,310	$ 25,526
8	Materials	4%		$ 12,480	$ 12,979	$ 13,498	$ 14,038	$ 14,600
9	Overhead	5%		$ 8,400	$ 8,820	$ 9,261	$ 9,724	$ 10,210
10	Depreciation			$ 17,863	$ 30,613	$ 21,863	$ 15,613	$ 5,581
11								
12	Taxable income			$ 46,257	$ 37,898	$ 51,327	$ 62,562	$ 77,906
13	Income taxes (40%)			$ 18,503	$ 15,159	$ 20,531	$ 25,025	$ 31,162
14								
15	Net income			$ 27,754	$ 22,739	$ 30,796	$ 37,537	$ 46,744
16								
17	**Cash Flow Statement**							
18	Operating Activities:							
19	Net income			$ 27,754	$ 22,739	$ 30,796	$ 37,537	$ 46,744
20	Depreciation			$ 17,863	$ 30,613	$ 21,863	$ 15,613	$ 5,581
21	Investing Activities:							
22	Investment		$ (125,000)					
23	Salvage value	3%						$ 57,964
24	Gains tax							$ (9,799)
25								=-(H23-(-C23-SUM(D10:H10))*0.4
26	Net Cash Flow (in Actual $)		$ (125,000)	$ 45,617	$ 53,352	$ 52,659	$ 53,150	$ 100,490
27								
28								
29			PW(21.90%)=	$ 38,801			=NPV(21.90%,D26:H26)+C26	
30			IRR =	34.15%				
31					=IRR(C26:H26,15%)			
32								
33								

which is an inflation-free interest rate i'. The appropriate interest rate to use is the market interest rate:[9]

$$i = i' + \overline{f} + i'\overline{f}$$
$$= 0.15 + 0.06 + (0.15)(0.06)$$
$$= 21.90\%$$

The equivalent present worth is obtained as

$$\text{PW}(21.90\%) = -\$125,000 + \$45,617(P/F, 21.90\%, 1)$$
$$+\$53,352(P/F, 21.90\%, 2) + \cdots$$
$$+\$100,490(P/F, 21.90\%, 5)$$
$$= \$38,801$$

11.3.2 Effects of Borrowed Funds under Inflation

The repayment of a loan is based on the historical contract amount; the payment size does not change with inflation under the fixed rate loan. Yet inflation greatly affects the value of these future payments, which are computed in year-0 dollars. First, we shall look at how the values of loan payments change under inflation. Interest expenses usually are already stated in the loan contract in actual dollars and need not be adjusted. Under the effect of inflation, the constant-dollar costs of both interest and principal repayments on a debt are reduced. Example 11.10 illustrates the effects of inflation on payments with project financing.

EXAMPLE 11.10 Effects of Inflation on Payments with Financing (Borrowing)

Let us rework Example 11.8 with a debt-to-equity ratio of 0.50, where the debt portion of the initial investment is borrowed at 15.5% annual interest. Assume, for simplicity, that the general inflation rate $\overline{f}$ of 5% during the project period will affect all revenues and expenses (except depreciation and loan payments) and the salvage value of the asset. Determine the NPW of this investment. (Note that the borrowing rate of 15.5% reflects the higher cost of debt financing in an inflationary environment.)

---

[9] In practice, the market interest rate is usually given and the inflation-free interest rate can be calculated when the general inflation rate is known for years in the past or is estimated for some time in the future. In our example, we are considering the opposite situation.

SOLUTION

Given: Cash flow data in Example 11.8, debt-to-equity ratio $= 0.50$, borrowing interest rate $= 15.5\%$, $\overline{f} = 5\%$, and $i' = 15\%$.

Find: The NPW.

For equal future payments, the actual-dollar cash flows for the financing activity are represented by the circles in Figure 11.3. If inflation were to occur, the cash flow (measured in year-0 dollars) would be represented by the shaded circles in the figure. Table 11.10 summarizes the after-tax cash flows in this situation. For simplicity, assume that all cash flows and inflation effects occur at year-end. To evaluate the present worth with the use of actual dollars, we must adjust the original discount rate (MARR) of 15%, which is an inflation-free interest rate i'. The appropriate interest rate to use is thus the market interest rate:

$$i = i' + \overline{f} + i'\overline{f}$$
$$= 0.15 + 0.05 + (0.15)(0.05)$$
$$= 20.75\%$$

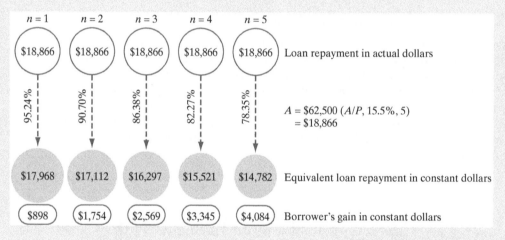

Figure 11.3 Equivalent loan-repayment cash flows measured in year-0 dollars and borrower's gain over the life of the loan (Example 11.10).

Then from Table 11.10, we compute the equivalent present worth of the after-tax cash flow as

$$PW(20.75\%) = -\$62,500 + \$29,953(P/F, 20.75\%, 1) + \cdots$$
$$+ \$82,001(P/F, 20.75\%, 5)$$
$$= \$54,159$$

TABLE 11.10 Cash Flow Statement for the Automated Machining-Center Project under Inflation with Borrowed Funds (Example 11.10)

	A	B	C	D	E	F	G	H
1								
2	**Income Statement**							
3	End of Year	Inflation Rate	0	1	2	3	4	5
4								
5	Revenue	5%		$ 105,000	$ 110,250	$ 115,763	$ 121,551	$ 127,628
6	Expenses:							
7	Labor	5%		$ 21,000	$ 22,050	$ 23,153	$ 24,310	$ 25,526
8	Materials	5%		$ 12,600	$ 13,230	$ 13,892	$ 14,586	$ 15,315
9	Overhead	5%		$ 8,400	$ 8,820	$ 9,261	$ 9,724	$ 10,210
10	Debt interest			$ 9,688	$ 8,265	$ 6,622	$ 4,724	$ 2,532
11	Depreciation			$ 17,863	$ 30,613	$ 21,863	$ 15,613	$ 5,581
12								
13	Taxable income			$ 35,449	$ 27,272	$ 40,973	$ 52,593	$ 68,464
14	Income taxes (40%)			$ 14,180	$ 10,909	$ 16,389	$ 21,037	$ 27,386
15								
16	Net income			$ 21,269	$ 16,363	$ 24,584	$ 31,556	$ 41,078
17								
18	**Cash Flow Statement**							
19	Operating Activities:							
20	Net income			$ 21,269	$ 16,363	$ 24,584	$ 31,556	$ 41,078
21	Depreciation			$ 17,863	$ 30,613	$ 21,863	$ 15,613	$ 5,581
22	Investing Activities:							
23	Investment		$ (125,000)					
24	Salvage value	5%						$ 63,814
25	Gains tax							$ (12,139)
26	Financing Activities:							
27	Borrowed funds		$ 62,500					
28	Principal repayment			$ (9,179)	$ (10,601)	$ (12,244)	$ (14,142)	$ (16,334)
29								
30	Net Cash Flow (in Actual $)		$ (62,500)	$ 29,953	$ 36,375	$ 34,203	$ 33,027	$ 82,001
31								
32								
33			PW(20.75%)=	$ 54,159				
34			IRR =	51.02%				
35								
36								
37								

=-(H24-(C23-SUM(D11:H11)))*0.4

=NPV(20.75%,D30:H30)+C30

=IRR(C30:H30,15%)

COMMENTS: In the absence of debt financing, the project would have a net present worth of $38,898, as shown in Example 11.8. Compared with the preceding result of $54,159, the gain in present worth due primarily to debt financing is $15,261. An inflationary trend decreases the purchasing power of future dollars, which helps long-term borrowers because they can repay a loan with dollars with reduced buying power. That is, the debt-financing cost is reduced in an inflationary environment. In this case, the benefits of financing under inflation have more than offset the *inflation tax* effect on depreciation and the salvage value. The amount of the gain may vary with the interest rate on the loan: The interest rate for borrowing is also generally higher during periods of inflation because it is a market-driven rate.

11.4 Rate-of-Return Analysis under Inflation

In addition to affecting individual aspects of a project's income statement, inflation can have a profound effect on the overall return—that is, the very acceptability—of an investment project. In this section, we will explore the effects of inflation on the rate of return on an investment and present several examples.

11.4.1 Effects of Inflation on Return on Investment

The effect of inflation on the rate of return on an investment depends on how future revenues respond to the inflation. Under inflation, a company is usually able to compensate for increasing material and labor prices by raising its selling prices. However, even if future revenues increase to match the inflation rate, the allowable depreciation schedule (as we have seen) does not increase. The result is increased taxable income and higher income-tax payments. This increase reduces the available constant-dollar after-tax benefits and, therefore, the inflation-free after-tax rate of return (IRR'). The next example will help us to understand this situation.

EXAMPLE 11.11 IRR Analysis with Inflation

The Hartsfield Company, a manufacturer of auto parts, is considering purchasing a set of machine tools at a cost of $30,000. The purchase is expected to generate increased sales of $24,500 per year and increased operating costs of $10,000 per year in each of the next four years. Additional profits will be taxed at a rate of 40%. The asset falls into the three-year MACRS property class for tax purposes. The project has a four-year life with zero salvage value. (All dollar figures represent constant dollars.)

(a) What is the expected internal rate of return?
(b) What is the expected IRR' if the general inflation is 10% during each of the next four years? (Assume that $f_j = \bar{f} = 10\%$.)
(c) If the company requires an inflation-free MARR (or MARR') of 20%, should the company invest in the equipment?

SOLUTION

Given: Financial data in constant dollars, $t_m = 40\%$, and $N = 4$ years.

Find: (a) IRR, (b) IRR' when $\bar{f} = 10\%$, and (c) whether the project can be justified at MARR $= 20\%$.

(a) **Rate-of-return analysis without inflation:** We find the rate of return by first computing the after-tax cash flow using the income statement approach, as illustrated in Table 11.11. The first part of the table shows the calculation of additional sales, operating costs, depreciation, and taxes. The asset will be depreciated fully over four years with no expected salvage value. As we emphasized in Chapter 10,

TABLE 11.11   **Rate-of-Return Calculation in the Absence of Inflation**

	A	B	C	D	E	F
1						
2	**Income Statement**					
3	End of Year	0	1	2	3	4
4						
5	Revenue		$ 24,500	$ 24,500	$ 24,500	$ 24,500
6	Expenses:					
7	O&M		10,000	10,000	10,000	10,000
8	Depreciation		10,000	13,333	4,445	2,222
9	Taxable income		4,500	1,167	10,055	12,278
10	Income taxes (40%)		1,800	467	4,022	4,911
11						
12	Net income		$ 2,700	$ 700	$ 6,033	$ 7,367
13						
14	**Cash Flow Statement**					
15	Operating Activities:					
16	Net income		2,700	700	6,033	7,367
17	Depreciation		10,000	13,333	4,445	2,222
18	Investing Activities:					
19	Machine tools	$ (30,000)				
20	Working capital					
21	Salvage value					
22	Gains tax					
23						
24	Net Cash Flow	$ (30,000)	$ 12,700	$ 14,033	$ 10,478	$ 9,589
25						
26		PW(20%) =	$ 1,016			
27		IRR' =	21.88%			
28						

depreciation is not a cash expense, although it affects taxable income, and thus, cash flow, indirectly. Therefore, we must add depreciation to net income to determine the net cash flow. Thus, if the investment is made, we expect to receive additional annual cash flows of $12,700, $14,033, $10,478, and $9,589. This is a simple investment, so we can calculate the IRR' for the project as

$$PW(i') = -\$30,000 + \$12,700(P/F, i', 1) + \$14,033(P/F, i', 2)$$
$$+\$10,478(P/F, i', 3) + \$9,589(P/F, i', 4)$$
$$= 0$$

Solving for i' yields

$$IRR' = i'^* = 21.88\%$$

The project has an inflation-free rate of return of 21.88%; that is, the company will recover its original investment ($30,000) plus interest at 21.88% each year for each dollar still invested in the project. Since IRR' > MARR' (20%), the company should buy the equipment.

(b) **Rate-of-return analysis under inflation:** With inflation, we assume that sales, operating costs, and the future selling price of the asset will increase. Depreciation will be unchanged, but taxes, profits, and cash flow will be higher. We might think that higher cash flows will mean an increased rate of return. Unfortunately, this is not the case. We must recognize that cash flows for each year are stated in dollars of declining purchasing power. When the net after-tax cash flows are converted to dollars with the same purchasing power as those used to make the original investment, the resulting rate of return decreases. These calculations (assuming an inflation rate of 10% in sales and operating expenses meaning a 10% annual decline in the purchasing power of the dollar) are shown in Table 11.12. For example, whereas additional sales had been $24,500 yearly, under conditions of inflation, they would be 10% greater in year 1 (or $26,950), 21% greater in year 2, and so forth. No change in investment or depreciation expenses will occur, since these items are unaffected by expected future inflation. We have restated the after-tax cash flows (actual dollars) in dollars of a common purchasing power (constant dollars) by deflating them—again assuming an annual deflation factor of 10%. The constant-dollar cash flows then are used to determine the real rate of return. First,

$$PW(i') = -\$30,000 + \$12,336(P/F, i', 1) + \$13,108(P/F, i', 2)$$
$$+ \$10,036(P/F, i', 3) + \$9,307(P/F, i', 4)$$
$$= 0$$

Then, solving for i' yields

$$i'^* = 19.40\%$$

The rate of return for the project's cash flows in constant dollars (year-0 dollars) is 19.40%, which is less than the 21.88% return in the inflation-free case. Since IRR' < MARR', the investment is no longer acceptable.

TABLE 11.12 Rate-of-Return Calculation under Inflation

Income Statement

End of Year	0	1	2	3	4
Revenue		$ 26,950	$ 29,645	$ 32,610	$ 35,870
Expenses:					
O&M		11,000	12,100	13,310	14,641
Depreciation		10,000	13,333	4,445	2,222
Taxable income		5,950	4,212	14,855	19,007
Income taxes (40%)		2,380	1,685	5,942	7,603
Net income		$ 3,570	$ 2,527	$ 8,913	$ 11,404

Cash Flow Statement

	0	1	2	3	4
Operating Activities:					
Net income		3,570	2,527	8,913	11,404
Depreciation		10,000	13,333	4,445	2,222
Investing Activities:					
Machine tools	$ (30,000)		=NPV(20%,C25:F25)+B25		
Working capital					
Salvage value	=NPV(32%,C24:F24)+B24				
Gains tax					
Net Cash Flow (in Actual $)	$ (30,000)	$ 13,570	$ 15,860	$ 13,358	$ 13,626
Net Cash Flow (in Constant $)	$ (30,000)	$ 12,336	$ 13,108	$ 10,036	$ 9,307
	PW(32%)=	$(321.06)		PW(20%) =	$(321.06)
	IRR =	31.34%		IRR' =	19.40%

COMMENTS: We could also calculate the rate of return by setting the PW of the actual dollar cash flows to 0. This would give a value of IRR = 31.34%, but that is an inflation-adjusted IRR. We could then convert to an IRR′ by deducting the amount caused by inflation:

$$i' = \frac{(1 + i)}{(1 + \bar{f})} - 1$$

$$= \frac{(1 + 0.3134)}{(1 + 0.10)} - 1$$

$$= 19.40\%$$

This approach gives the same final result of IRR′ = 19.40%.

11.4.2 Effects of Inflation on Working Capital

The loss of tax savings in depreciation is not the only way that inflation may distort an investment's rate of return. Another source is working-capital drain. Projects requiring increased levels of working capital suffer from inflation because additional cash must be invested to maintain new price levels. For example, if the cost of inventory increases, additional outflows of cash are required to maintain appropriate inventory levels over time. A similar phenomenon occurs with funds committed to accounts receivable. These additional working-capital requirements can significantly reduce a project's rate of return, as the next example illustrates.

EXAMPLE 11.12 Effect of Inflation on Profits with Working Capital

Suppose that, in Example 11.11, a $1,000 investment in working capital is expected and that all of the working capital will be recovered at the end of the project's four-year life. Determine the rate of return on this investment.

SOLUTION
Given: Financial data in Example 11.11 and working-capital requirement = $1,000.
Find: IRR.

Using the data in the upper part of Table 11.13, we can calculate IRR$'$ = IRR of 20.88% in the absence of inflation. Also, PW(20%) = $499. The lower part of the table includes the effect of inflation on the proposed investment. As illustrated in Figure 11.4, working-capital levels can be maintained only by additional infusions of cash; the working-capital drain also appears in the lower part of Table 11.13. For example, the $1,000 investment in working capital made in year 0 will be recovered at the end of the first year, assuming a one-year recovery cycle. However, because of 10% inflation, the required working capital for the second year increases to $1,100. In addition to reinvesting the $1,000 revenues, an additional investment of $100 must be made. This $1,100 will be recovered at the end of the second year. However, the project will need a 10% increase (or $1,210) for the third year, and so forth.

　　As Table 11.13 illustrates, the combined effect of depreciation and working capital is significant. Given an inflationary economy and investment in working capital, the project's IRR$'$ drops to 18.08% or PW(20%) = −$1,074. By using either IRR analysis or NPW analysis, we end up with the same result (as we must): Alternatives that are attractive when inflation does not exist may not be acceptable when inflation does exist.

TABLE 11.13   **Effects of Inflation on Working Capital and After-Tax Rate of Return**

	A	B	C	D	E	F
1	**Case 1: Without Inflation**					
2	**Income Statement**	0	1	2	3	4
3						
4	Revenue		$ 24,500	$ 24,500	$ 24,500	$ 24,500
5	Expenses:					
6	O&M		10,000	10,000	10,000	10,000
7	Depreciation		10,000	13,333	4,445	2,222
8	Taxable income		4,500	1,167	10,055	12,278
9	Income taxes (40%)		1,800	467	4,022	4,911
10						
11	Net Income		$ 2,700	$ 700	$ 6,033	$ 7,367
12	**Cash Flow Statement**					
13	Operating Activities:					
14	Net income		2,700	700	6,033	7,367
15	Depreciation		10,000	13,333	4,445	2,222
16	Investing Activities:					
17	Machine tools	$ (30,000)				
18	Working capital	(1,000)				1,000
19						
20	Net Cash Flow	$ (31,000)	$ 12,700	$ 14,033	$ 10,478	$ 10,589
21						
22		PW(20%) = $	499	IRR' =	20.88%	
23	**Case 2: With Inflation**					
24	**Income Statement**	0	1	2	3	4
25						
26	Revenue		$ 26,950	$ 29,645	$ 32,610	$ 35,870
27	Expenses:					
28	O&M		11,000	12,100	13,310	14,641
29	Depreciation		10,000	13,333	4,445	2,222
30	Taxable income		5,950	4,212	14,855	19,007
31	Income taxes (40%)		2,380	1,685	5,942	7,603
32						
33	Net income		$ 3,570	$ 2,527	$ 8,913	$ 11,404
34	**Cash Flow Statement**					
35	Operating Activities:					
36	Net income		3,570	2,527	8,913	11,404
37	Depreciation		10,000	13,333	4,445	2,222
38	Investing Activities:					
39	Machine tools	$ (30,000)				
40	Working capital	(1,000)	(100)	(110)	(121)	1,331
41						
42	Net Cash Flow (in Actual $)	$ (31,000)	$ 13,470	$ 15,750	$ 13,237	$ 14,957
43	Net Cash Flow (in Constant $)	$ (31,000)	$ 12,245	$ 13,017	$ 9,945	$ 10,216
44						
45		PW(20%) = $ (1,074)		IRR' =	18.08%	
46						

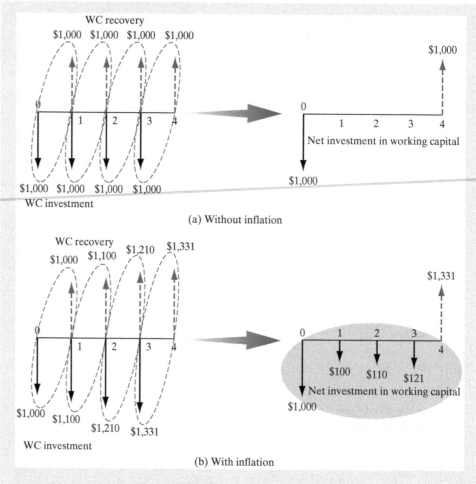

Figure 11.4 Working-capital requirements under inflation: (a) Requirements without inflation and (b) requirements with inflation, assuming a one-year recovery cycle.

SUMMARY

■ The **Consumer Price Index (CPI)** is a statistical measure of change (over time) of the prices of goods and services in major expenditure groups—such as food, housing, apparel, transportation, and medical care—typically purchased by urban consumers. Essentially, the CPI compares the cost of a sample "market basket" of goods and services in a specific period with the cost of the same "market basket" in an earlier reference period, which is designated the **base period**.

■ **Inflation** is the term used to describe a **decline in purchasing power** evidenced in an economic environment of rising prices.

■ **Deflation** is the opposite: an increase in purchasing power evidenced by falling prices.

- The general inflation rate $\bar{f}$ is an average inflation rate based on the CPI. An annual general inflation rate $\bar{f}_j$ can be calculated with the following equation:

$$\bar{f}_n = \frac{CPI_n - CPI_{n-1}}{CPI_{n-1}}$$

- Specific, individual commodities do not always reflect the general inflation rate in their price changes. We can calculate an **average inflation rate** f_j for a specific commodity (j) if we have an index (i.e., a record of historical costs) for that commodity.

- Project cash flows may be stated in one of two forms:

Actual dollars (A_n): Dollars that reflect the inflation or deflation rate.

Constant dollars (A'_n): Year-0 dollars.

- Interest rates used in evaluating a project may be stated in one of two forms:

Market interest rate (i): A rate that combines the effects of interest and inflation when used with an **actual-dollar** analysis.

Inflation-free interest rate (i'): A rate from which the effects of inflation have been removed; this rate is used with a constant-dollar analysis.

- To calculate the present worth of actual dollars, we can use either a two-step or a one-step process:

Deflation Method—two steps:

1. Convert actual dollars by deflating with the general inflation rate $\bar{f}$.

2. Calculate the PW of constant dollars by discounting at i'.

Adjusted-Discount Method—one step (use the market interest rate):

$$P_n = \frac{A_n}{[(1 + \bar{f})(1 + i')]^n}$$
$$= \frac{A_n}{(1 + i)^n}$$

- A number of individual elements of project evaluations can be distorted by inflation. These elements are summarized in Table 11.14.

TABLE 11.14 Effects of Inflation on Project Cash Flows and Return

Item	Effects of Inflation
Depreciation expense	Depreciation expense is charged to taxable income in dollars of declining values; taxable income is overstated, resulting in higher taxes.
Salvage values	Inflated salvage values combined with book values based on historical costs result in higher taxable gains.
Loan repayments	Borrowers repay historical loan amounts with dollars of decreased purchasing power, reducing the cost of financing debt.
Working capital requirement	Known as a *working capital drain*, the cost of working capital increases in an inflationary economy.
Rate of return and NPW	Unless revenues are sufficiently increased to keep pace with inflation, tax effects and a working-capital drain result in a lower rate of return or a lower NPW.

PROBLEMS

Note: *In the problems that follow, the term "market interest rate" represents the "inflation-adjusted MARR" for project evaluation or the "interest rate" quoted by a financial institution for commercial loans.*

Measure of Inflation

11.1 The current gasoline price is $3.50 per gallon, and it is projected to increase next year by 4%, 6% the following year, and 8% the third year. What is the average inflation rate for the projected gasoline price for the next three years?

11.2 The following data indicate the median price of unleaded gasoline during the last 15 years for California residents:

Period	Price
1999	$1.20
2014	$3.80

Assuming that the base period (price index = 100) is 1999, compute the average price index for the unleaded gasoline price for the year 2014.

11.3 The following data indicate the price indexes of medical care services (base period 1982 = 100 between 2010 and 2014:

Period	Price Index
2010	411.21
2011	423.81
2012	440.34
2013	454.01
2014	455.63
2020	?

(a) Assuming that the base period (price index = 100) is reset to the year 2010, compute the average (geometric) price index for medical care services between 2010 and 2014.

(b) If the past trend is expected to continue, how would you estimate the medical care services in 2020?

11.4 For prices that are increasing at an annual rate of 5% the first year and 8% the second year, determine the average inflation rate $(\bar{f})$ over the two years.

11.5 Because of general price inflation in our economy, the purchasing power of the dollar shrinks with the passage of time. If the average general inflation rate is expected to be 8% per year for the foreseeable future, how many years will it take for the dollar's purchasing power to be one-half of what it is now?

Actual versus Constant Dollars

11.6 An annuity provides for 10 consecutive end-of-year payments of $72,000. The average general inflation rate is estimated to be 4% annually, and the market interest rate is 8% annually. What is the annuity worth in terms of a single equivalent amount of today's dollars?

11.7 A company is considering buying a CNC machine. In today's dollars, it is estimated that the maintenance costs for the machine (paid at the end of each year) will be $25,000, $26,000, $28,000, $30,000, and $32,000 for years 1 to 5, respectively. The general inflation rate $(\bar{f})$ is estimated to be 5% per year, and the company will receive 13% return (interest) per year on its invested funds during the inflationary period. The company wants to pay for maintenance expenses in equivalent equal payments (in actual dollars) at the end of each of the five years. Find the amount of the company's payments.

11.8 The following cash flows are in actual dollars:

n	Cash Flow (in Actual $)
0	$25,000
4	35,000
5	45,000
7	55,000

Convert to an equivalent cash flow in constant dollars if the base year is time 0. Keep cash flows at the same point in time—that is, years 0, 4, 5, and 7. Assume that the market interest rate is 14% and that the general inflation rate ($\bar{f}$) is 5% per year.

11.9 The purchase of a car requires a $25,000 loan to be repaid in monthly installments for four years at 9% interest compounded monthly. If the general inflation rate is 4% compounded monthly, find the actual- and constant-dollar value of the 20th payment.

11.10 A series of four annual constant-dollar payments beginning with $10,000 at the end of the first year is growing at the rate of 8% per year. Assume that the base year is the current year ($n = 0$). If the market interest rate is 15% per year and the general inflation rate ($\bar{f}$) is 7% per year, find the present worth of this series of payments, based on

(a) Constant-dollar analysis
(b) Actual-dollar analysis

11.11 Consider the accompanying cash flow diagrams, where the equal-payment cash flow in constant dollars is converted (a) from the equal-payment cash flow in actual dollars, and (b) at an annual general inflation rate of $\bar{f} = 3.8\%$ and $i = 9\%$. What is the amount A in actual dollars equivalent to $A' = \$1,000$ in constant dollars? Assume base year is $n = 0$.

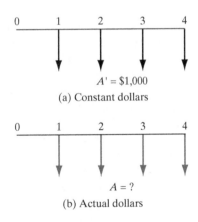

(a) Constant dollars

(b) Actual dollars

Figure P11.11

11.12 A 10-year $1,000 bond pays a nominal rate of 9% compounded semi-annually. If the market interest rate is 12% compounded annually and the general inflation rate is 6% per year, find the actual- and constant-dollar amounts (in time-0 dollars) of the 15th interest payment on the bond.

Equivalence Calculation under Inflation

11.13 If the inflation rate is 4.8% per year and the market interest rate is known to be 10.2% per year, what is the implied real interest (inflation-free) rate in this inflationary economy?

11.14 Suppose that you borrow $42,500 at 11.4% compounded monthly over seven years. Knowing that the 11.4% represents the market interest rate, you realize that the monthly payment in actual dollars will be $736.67. If the average monthly general inflation rate is expected to be 0.6%, determine the equivalent equal monthly payment series in constant dollars.

11.15 The annual fuel costs required to operate a small solid-waste treatment plant are projected to be $2.2 million without considering any future inflation. The best estimates indicate that the annual inflation-free interest rate (i') will be 6% and the general inflation rate ($\bar{f}$) will be 5%. If the plant has a remaining useful life of five years, what is the present equivalent of its fuel costs? Use actual-dollar analysis.

11.16 A series of five constant-dollar (or real-dollar) payments (beginning with $5,000 at the end of the first year) are increasing at the rate of 7% per year. Assume that the average general inflation rate is 5% and the market interest rate is 12% during this inflationary period. What is the equivalent present worth of the series?

11.17 Suppose that you just purchased a used car worth $8,000 in today's dollars. Suppose also that you borrowed $8,000 from a local bank at 9% compounded monthly over two years. The bank calculated your monthly payment at $365.48. Assuming that average general inflation will run at 0.5% per month over the next two years,

(a) Determine the monthly inflation-free interest rate (i') for the bank.
(b) What equal monthly payments (in terms of constant dollars over the next two years) are equivalent to the series of actual payments to be made over the life of the loan?

11.18 A man is planning to retire in 20 years. Money can be deposited at 6% interest compounded monthly, and it is also estimated that the future general inflation ($\bar{f}$) rate will be 4% compounded annually. What amount of end-of month deposit must be made each month until the man retires so that he can make annual withdrawals of $60,000 in terms of today's dollars over the 15 years following his retirement? (Assume that his first withdrawal occurs at the end of the first six months after his retirement.)

11.19 On her 24th birthday, a young woman engineer decides to start saving toward building up a retirement fund that pays 7.2% interest compounded monthly (the market interest rate). She feels that $1,250,000 worth of purchasing power in today's dollars will be adequate to see her through her sunset years after her 65th birthday. Assume a general inflationrate of 5% per year.

(a) If she plans to save by making 432 equal monthly deposits, what should be the amount of her monthly deposit in actual dollars? Assume the first deposit is made at the end of first month.

(b) If she plans to save by making end-of-the-year deposits, increasing by $2,500 over each subsequent year, how much would her first deposit be in actual dollars?

11.20 A father wants to save for his 8-year-old son's college expenses. The son will enter college 10 years from now. An annual amount of $40,000 in today's constant dollars will be required to support the son's college expenses for four years. Assume that these college payments will be made at the beginning of the school year. The future general inflation rate is estimated to be 6% per year, and the market interest rate on the savings account will average 8% compounded annually. Given this information,

(a) What is the amount of the son's freshman-year expense in terms of actual dollars?

(b) What is the equivalent single-sum amount at the present time for these college expenses?

(c) What is the equal amount, in actual dollars, the father must save each year until his son goes to college?

Effects of Inflation on Project Cash Flows

11.21 Consider the following project's after-tax cash flow and the expected annual general inflation rate during the project period.

	Expected	
End of Year	**Cash Flow (in Actual $)**	**General Inflation Rate**
0	−$45,000	
1	32,000	3.5%
2	32,000	4.2
3	32,000	5.5

(a) Determine the average annual general inflation rate over the project period.

(b) Convert the cash flows in actual dollars into equivalent constant dollars with the base year 0.

(c) If the annual inflation-free interest rate is 5%, what is the present worth of the cash flow? Is this project acceptable?

11.22 Gentry Machines, Inc. has just received a special job order from one of its clients. The following financial data have been collected:

• This two-year project requires the purchase of special-purpose equipment for $55,000. The equipment falls into the MACRS five-year class.

• The machine will be sold for $27,000 (today's dollars) at the end of two years.

• The project will bring in additional annual revenue of $114,000 (actual dollars), but it is expected to incur an additional annual operating cost of $53,800 (today's dollars).

• The project requires an investment in working capital in the amount of $12,000 at $n = 0$. In each subsequent year, additional working capital needs to be provided at the general inflation rate. Any investment in working capital will be recovered after the project is terminated.

• To purchase the equipment, the firm expects to borrow $50,000 at 10% over a two-year period. The remaining $5,000 will be taken from the firm's retained earnings. The firm will make equal annual payments of $28,810 (actual dollars) to pay off the loan.

• The firm expects a general inflation of 5% per year during the project period. The firm's marginal tax rate is 40%, and its market interest rate is 18%.

(a) Compute the after-tax cash flows in actual dollars.

(b) What is the equivalent present value of this amount at time 0?

11.23 Hugh Health Product Corporation is considering purchasing a computer system to control plant packaging for a spectrum of health products. The following data have been collected:

- First cost = $120,000 to be borrowed at 9% interest with only interest paid each year and the principal due in a lump sum at end of year 2.
- Economic service life (project life) = 6 years.
- Estimated selling price in year-0 dollars = $15,000.
- Depreciation = five-year MACRS property.
- Marginal income-tax rate = 40% (remains constant).
- Annual revenue = $145,000 (today's dollars).
- Annual expense (not including depreciation and interest) = $82,000 (today's dollars).
- Market interest rate = 18%.

(a) With an average general inflation rate of 5% expected during the project period (which will affect all revenues, expenses, and the salvage value of the computer), determine the cash flows in actual dollars.

(b) Compute the net present value of the project under inflation.

(c) Compute the net present-value loss (gain) due to inflation.

(d) In part (c), how much is the present-value loss (or gain) due to borrowing?

11.24 The Norcross Fiber Company is considering automating its piece-goods screen-printing system at a cost of $20,000. The firm expects to phase out the new printing system at the end of five years due to changes in style. At that time, the firm could scrap the system for $2,000 in today's dollars. The expected net savings due to automation also are in today's dollars (constant dollars):

End of Year	Cash Flow (in Constant $)
1	$22,000
2	19,000
3–5	15,000

The system qualifies as a five-year MACRS property and will be depreciated accordingly. The expected average general inflation rate over the next five years is approximately 5% per year. The firm will finance the entire project by borrowing at 10%. The scheduled repayment of the loan will be:

End of Year	Principal Payment	Interest Payment
1	$6,042	$2,000
2	6,647	1,396
3	7,311	731

The firm's market interest rate for project evaluation during this inflation-ridden time is 20%. Assume that the net savings and the selling price will be responsive to this average inflation rate. The firm's marginal tax rate is known to be 40%.

(a) Determine the after-tax cash flows of this project in actual dollars.

(b) Determine the net present-value reduction (or gains) in profitability due to inflation.

11.25 The J. F. Manning Metal Co. is considering the purchase of a new milling machine during year 0. The machine's base price is $135,000, and it will cost another $15,000 to modify it for special use. This results in a $150,000 cost base for depreciation. The machine falls into the MACRS seven-year property class. The machine will be sold after three years for $80,000 (actual dollars). The use of the machine will require an increase in net working capital (inventory) of $10,000 at the beginning of the project year. The machine will have no effect on revenues, but it is expected to save the firm $80,000 (today's dollars) per year in before-tax operating costs, mainly for labor. The firm's marginal tax rate is 40%, and this rate is expected to remain unchanged over the duration of the project. However, the company expects that the labor cost will increase at an annual rate of 5% but that the working-capital requirement will grow at an annual rate of 8% (caused by inflation). The selling price of the milling machine is not affected by inflation. The general inflation rate is estimated to be 6% per year over the project period. The firm's market interest rate is 20%.

(a) Determine the project cash flows in actual dollars.

(b) Determine the project cash flows in constant (time-0) dollars.

(c) Is this project acceptable?

Rate of Return Analysis under Inflation

11.26 Miami Machine Shops, Ltd. is considering purchasing a vertical drill machine. The machine will cost $62,000 and will have an eight-year service life. The selling price of the machine at the end of eight years is expected to be $5,000 in today's dollars. The machine will generate annual revenues of $22,000 (today's dollars), but the company expects to have an annual expense (excluding depreciation) of $9,500 (today's dollars). The asset is classified as a seven-year MACRS property. The project requires a working-capital investment of $10,000 at year 0. The marginal income tax rate for the firm is averaging 35%. The firm's market interest rate is 18%.

(a) Determine the internal rate of return of this investment.

(b) Assume that the firm expects a general inflation rate of 5%, but that it also expects an 8% annual increase in revenue and working capital and a 6% annual increase in expense caused by inflation. Compute the real (inflation-free) internal rate of return. Is this project acceptable?

11.27 Sonja Jensen is considering the purchase of a fast-food franchise. Sonja will be operating on a lot that is to be converted into a parking lot in six years, but that may be rented in the interim for $800 per month. The franchise and necessary equipment will have a total initial cost of $55,000 and a salvage value of $10,000 (in today's dollars) after six years. Sonja is told that the future annual general inflation rate will be 5%. The projected operating revenues and expenses (in actual dollars) other than rent and depreciation for the business are:

End of Year	Revenues	Expenses
1	$35,000	$16,000
2	38,000	21,000
3	55,000	23,000
4	60,000	32,000
5	70,000	33,000
6	60,000	33,000

Assume that the initial investment will be depreciated under the five-year MACRS and that Sonja's tax rate will be 30%. Sonja can invest her money at a rate of at least 10% in other investment activities during this inflation-ridden period.

(a) Determine the cash flows associated with the investment over its life.

(b) Compute the projected after-tax rate of return (real or inflation-free) for this investment opportunity and justify whether or not it is worth undertaking.

11.28 Suppose you have $10,000 cash that you want to invest. Normally, you would deposit the money in a savings account that pays an annual interest rate of 6%. However, you are now considering the possibility of investing in a bond. Your alternatives are either a nontaxable municipal bond paying 9% or a taxable corporate bond paying 12%. Your marginal tax rate is 30% for both ordinary income and capital gains. You expect the general inflation to be 3% during the investment period. You can buy a high-grade municipal bond costing $10,000 that pays interest of 9% ($900) per year. This interest is not taxable. A comparable high-grade corporate bond is also available that is just as safe as the municipal bond, but it pays an interest rate of 12% ($1,200) per year. This interest is taxable as ordinary income. Both bonds mature at the end of year 5.

(a) Determine the real (inflation-free) rate of return for each bond.

(b) Without knowing your MARR, can you make a choice between these two bonds?

11.29 An airline is considering two types of engines for use in its planes. Each engine has the same life, the same maintenance, and the same repair record.

• Engine A costs $100,000 and uses 50,000 gallons of fuel per 1,000 hours of operation at the average service load encountered in passenger service.

• Engine B costs $200,000 and uses 32,000 gallons of fuel per 1,000 hours of operation at the same service load.

Both engines are estimated to have 10,000 service hours before any major overhaul of the engines is required. If fuel currently costs $5.90 per gallon and its price is expected to increase at the rate of 8% because of inflation, which engine should the firm install for an expected 2,000 hours of operation per year? The firm's marginal income-tax rate is 40%, and the engine will be depreciated on the basis of the unit-of-production method. Assume that the firm's market interest rate is 20%. It is estimated that both engines will retain a market value of 40% of their initial cost (actual dollars) if they are sold on the market after 10,000 hours of operation.

(a) Using the present-worth criterion, which project would you select?

(b) Using the annual-equivalent criterion, which project would you select?

(c) Using the future-worth criterion, which project would you select?

11.30 The Johnson Chemical Company has just received a special subcontracting job from one of its clients. The two-year project requires the purchase of a special-purpose painting sprayer of $60,000. This equipment falls into the MACRS five-year class. After the subcontracting work is completed, the painting sprayer will be sold at the end of two years for $40,000 (actual dollars). The painting system will require an increase of $5,000 in net working capital (for spare-parts inventory, such as spray nozzles). This investment in working capital will be fully recovered after the project is terminated. The project will bring in an additional annual revenue of $120,000 (today's dollars), but it is expected to incur an additional annual operating cost of $60,000 (today's dollars). It is projected that, due to inflation, sales prices will increase at an annual rate of 5%. (This implies that annual revenues will increase at an annual rate of 5%.) An annual increase of 4% for expenses and working-capital requirement is expected. The company has a marginal tax rate of 30%, and it uses a market interest rate of 15% for project evaluation during the inflationary period. The firm expects a general inflation of 8% during the project period.

(a) Compute the after-tax cash flows in actual dollars.

(b) What is the rate of return on this investment (real earnings)?

(c) Is the special subcontracting project profitable?

11.31 The Land Development Corporation is considering purchasing a bulldozer. The bulldozer will cost $100,000 and will have an estimated salvage value of $30,000 at the end of six years. The asset will generate annual before-tax revenues of $80,000 over the next six years. The asset is classified as a five-year MACRS property. The marginal tax rate is 40%, and the firm's market interest rate is known to be 18%. All dollar figures represent constant dollars at time 0 and are responsive to the general inflation rate $\overline{f}$.

(a) With $\overline{f} = 6\%$, compute the after-tax cash flows in actual dollars.

(b) Determine the real rate of return of this project on an after-tax basis.

(c) Suppose that the initial cost of the project will be financed through a local bank at an interest rate of 12% with an annual payment of $24,323 over six years. With this additional condition, answer part (a) again.

(d) In part (a), determine the present-value loss due to inflation.

(e) In part (c), determine how much the project has to generate in additional before-tax annual revenue in actual dollars (equal amount) to make up the loss due to inflation.

Short Case Studies

ST11.1 Suppose a business is considering purchasing an asset with the following financial facts during an inflationary environment: Note that all $ are estimated in terms of purchasing power of the base year "0".

• A machine costing $90,000 will generate annual revenue of $80,000 over three years. The expected annual expenses are $20,000. The machine will be used only for three years, with an estimated

salvage value of $30,000; additional profits will be taxed at a rate of 40%.

• Depreciation (for tax purpose) is assumed to be taken on a straight-line basis over three years with no expected salvage value. For simplicity, assume no half-year convention for the year in which the asset is placed in service or disposed of.

• A $10,000 investment in working capital is required at the beginning of the project year, and any investment in working capital during the project period will be fully recovered at the end of project life.

- The entire capital expenditure ($90,000) will be borrowed at an annual rate of 10%, with the stipulation of interest being paid each year and the principal will be paid in lump sum at the end of 3 years.

(a) Determine the cash flows of this project in *actual dollars* knowing that the general inflation rate is 5% per year during the next three years. Assume that revenues, expenses, salvage value, and working capital requirement will follow the general inflation rate.

(b) Suppose that your inflation-free interest rate is 10%. Determine the net present worth of this project.

ST11.2 Wilson Machine Tools, Inc., a manufacturer of fabricated metal products, is considering purchasing a high-tech computer-controlled milling machine at a cost of $95,000. The cost of installing the machine, preparing the site, wiring, and rearranging other equipment is expected to be $15,000. This installation cost will be added to the cost of the machine to determine the total cost basis for depreciation. Special jigs and tool dies for the particular product also will be required at a cost of $10,000. The milling machine is expected to last 10 years, the jigs and dies only five years. Therefore, another set of jigs and dies has to be purchased at the end of five years. The milling machine will have a $10,000 salvage value at the end of its life, and the special jigs and dies are worth only $300 as scrap metal at any time in their lives. The machine is classified as a seven-year MACRS property, and the special jigs and dies are classified as a three-year MACRS property. With the new milling machine, Wilson expects an additional annual revenue of $80,000 due to increased production. The additional annual production costs are estimated as follows: materials, $9,000; labor, $15,000; energy, $4,500; and miscellaneous O&M costs, $3,000. Wilson's marginal income-tax rate is expected to remain at 35% over the project life of 10 years. All dollar figures represent today's dollars. The firm's market interest rate is 18%, and the expected general inflation rate during the project period is estimated at 6%.

(a) Determine the project cash flows in the absence of inflation.

(b) Determine the internal rate of return for the project in part (a).

(c) Suppose that Wilson expects price increases during the project period: material at 4% per year, labor at 5% per year, and energy and other O&M costs at 3% per year. To compensate for these increases in prices, Wilson is planning to increase annual revenue at the rate of 7% per year by charging its customers a higher price. No changes in salvage value are expected for the machine or the jigs and dies. Determine the project cash flows in actual dollars.

(d) In part (c), determine the real (inflation-free) rate of return of the project.

(e) Determine the economic loss (or gain) in present worth caused by inflation.

ST11.3 Recent biotechnological research has made possible the development of a sensing device that implants living cells on a silicon chip. The chip is capable of detecting physical and chemical changes in cell processes. Proposed uses of the device include researching the mechanisms of disease on a cellular level, developing new therapeutic drugs, and substituting for animals in cosmetic and drug testing. Biotech Device Corporation (BDC) has just perfected a process for mass-producing the chip. The following information has been compiled for the board of directors:

- BDC's marketing department plans to target sales of the device to the larger chemical and drug manufacturers. BDC estimates that annual sales would be 2,000 units if the device were priced at $95,000 per unit (in dollars of the first operating year).

- To support this level of sales volume, BDC would need a new manufacturing plant. Once the "go" decision is made, this plant could be built and made ready for production within one year. BDC would need a 30-acre tract of land that would cost $1.5 million. If the decision were to be made, the land could be purchased on December 31, 2014. The building would cost $5 million and would be depreciated according to the MACRS 39-year class. The first payment of $1 million would be due to the contractor on December 31, 2015, and the remaining $4 million on December 31, 2016.

- The required manufacturing equipment would be installed late in 2016 and would be paid for on December 31, 2016. BDC would have to purchase the equipment at an estimated cost of $8 million, including transportation, plus a further $500,000 for installation. The equipment would fall into the MACRS seven-year class.

- The project would require an initial investment of $1 million in working capital. This investment would be made on December 31, 2016. Then on December 31 of each subsequent year, net working capital would be increased by an amount equal to 15% of any sales increase expected during the coming year. The investments in working capital would be fully recovered at the end of the project year.

- The project's estimated economic life is six years (excluding the two-year construction period). At that time, the land is expected to have a market value of $2 million, the building a value of $3 million, and the equipment a value of $1.5 million. The estimated variable manufacturing costs would total 60% of the dollar sales. Fixed costs, excluding depreciation, would be $5 million for the first year of operations. Since the plant would begin operations on January 1, 2017, the first operating cash flows would occur on December 31, 2017.

- Sales prices and fixed overhead costs (other than depreciation) are projected to increase with general inflation, which is expected to average 5% per year over the six-year life of the project.

- To date, BDC has spent $5.5 million on research and development (R&D) associated with cell implantation. The company has already expensed $4 million in R&D costs. The remaining $1.5 million will be amortized over six years (i.e., the annual amortization expense will be $250,000). If BDC decides not to proceed with the project, the $1.5 million R&D cost could be written off on December 31, 2014.

- BDC's marginal tax rate is 40% and its market interest rate is 20%. Any capital gains will also be taxed at 40%.

(a) Determine the after-tax cash flows of the project in actual dollars.

(b) Determine the inflation-free (real) IRR of the investment.

(c) Would you recommend that the firm accept the project?

12

Project Risk and Uncertainty

Capitalizing on Historic Auto Industry Shift to Lightweight Vehicles[1] Automakers are increasingly choosing aluminum as a cost-effective way to improve the performance, safety, durability, and fuel efficiency of their vehicles. According to automakers, demand for aluminum to produce vehicles—already the second-most-used material utilized to make cars today—is expected to nearly double by 2025. The amount of aluminum body sheet content in North American vehicles is expected to quadruple by 2015 and increase tenfold by 2025, from 2012 levels. In anticipation of this growing demand, Pittsburgh-based Alcoa, one of the U.S.'s biggest sheet aluminum producers, has invested around $670 million at plants in Iowa and Tennessee to meet this projected demand by automakers.

However, *The Wall Street Journal*[2] reports that Western aluminum producers have been battered after they built capacity in anticipation of demand from China that never came. Instead, China developed its own aluminum industry, leaving Alcoa and others to scramble to take smelters out of production. Once all curtailments are finished, Alcoa would have reduced operating smelting capacity by 1.2 million tons, or 28%, since 2007.

What can you say about Alcoa's investment in this capacity expansion project? Certainly, the success of the expansion project is largely dependent on the demand for aluminum products in the auto industry. Meanwhile, the steel industry is readying for battle, and automakers aren't putting all their eggs in the aluminum basket. The steel industry is still the single biggest supplier to the automotive industry, and its organization is working with automakers on specific projects for redesigning parts of their cars with lightweight steel, giving real choices between aluminum and steel. Even if automakers switch to aluminum, it would require extensive overhauls to its

[1] Alcoa is investing approximately $670 million in its automotive "triple play," Alcoa Corporate Website, https://www.alcoa.com/car_truck/en/info_page/in_demand.asp.

[2] "Detroit's Appetite for Aluminum Keeps Growing," *The Wall Street Journal*, June 11, 2014.

Source: iQoncept/Shutterstock

factories and to dealer auto body shops to handle the change to aluminum bodies. That means the projected aluminum demand may not materialize to the degree Alcoa has hoped for. From Alcoa's perspective, how would you factor the fluctuation and uncertainty into the analysis?

In previous chapters, we assumed that cash flows from projects be known with complete certainty; our analysis was concerned with measuring the economic worth of projects and selecting the best ones to invest in. Although that type of analysis can provide a reasonable basis for decision making in many investment situations, we should also consider investment decisions under uncertainty. In practice, management rarely has precise expectations about the future cash flows to be derived from a particular project. In fact, the best that a firm can reasonably expect to do is to estimate the range of possible future costs and benefits and the relative chances of achieving a reasonable return on the investment. We use the term **risk** to describe an investment project whose cash flow is not known in advance with absolute certainty, but for which an array of alternative outcomes and their probabilities (odds) are known. We will also use the term **project risk** to refer to variability in a project's NPW. A greater project risk usually means a greater variability in a project's net-present-worth (NPW), or simply that the *risk is the potential for loss.* This chapter begins by exploring the origins of project risk.

609

CHAPTER LEARNING OBJECTIVES

After completing this chapter, you should understand the following concepts:

- How to describe the nature of project risk.
- How to conduct a sensitivity analysis of key input variables.
- How to conduct a break-even analysis.
- How to develop a net-present-worth probability distribution.
- How to compare mutually exclusive risky alternatives.
- How to develop a risk simulation model.
- How to make a sequential investment decision with a decision tree.

12.1 Origins of Project Risk

The decision to make a major capital investment requires information about cash flow over the life of a project. The profitability estimate of an investment depends on these cash flow estimations, which are generally subject to a great deal of uncertainty. For a new product introduction, the factors to be estimated include the total market for the product; the market share that the firm can attain; the growth in the market; the cost of producing the product, including labor and materials; the selling price; the life of the product; the cost and life of the equipment needed; and the effective tax rates. Many of these factors are subject to substantial uncertainty. In practice, a common approach is to make single-number "best estimates" for each of the uncertain factors and then to calculate measures of profitability, such as the NPW or rate of return for the project. This approach, however, has two drawbacks:

1. No guarantee can ever ensure that the "best estimates" will match actual values.
2. No provision is made to measure the risk associated with an investment or the project. In particular, most managers have no way of determining either the probability that a project will lose money or the probability that it will generate large profits.

Because cash flows can be rather difficult to estimate accurately, project managers frequently consider a range of possible values for cash flow elements. If a range of values for individual cash flows is possible, it follows that a range of values for the NPW of a given project is also possible. Clearly, the analyst will want to gauge the probability and reliability of individual cash flows and, consequently, the level of certainty about the overall project worth.

12.2 Methods of Describing Project Risk

We may begin analyzing project risk by first determining the uncertainty inherent in a project's cash flows. We can do this analysis in a number of ways, which range from making informal judgments to calculating complex economic and statistical quantities. In this section, we will introduce three methods of describing project risk: (1) sensitivity analysis; (2) break-even analysis; and (3) scenario analysis. Each method will be explained with reference to a single example, involving the Boston Metal Company.

12.2.1 Sensitivity (What-if) Analysis

One way to glean a sense of the possible outcomes of an investment is to perform a sensitivity analysis. This kind of analysis determines the effect on the NPW of variations in the input variables (such as revenues, operating cost, and salvage value) used to estimate after-tax cash flows. A **sensitivity analysis** reveals how much the NPW will change in response to a given change in an input variable. In calculating cash flows, some items have a greater influence on the final result than others. In some problems, the most significant item may be easily identified. For example, the estimate of sales volume is often a major factor in a problem in which the quantity sold varies with the alternatives. In other problems, we may want to locate the items that have an important influence on the final results so that they can be subjected to special scrutiny.

Sensitivity analysis is sometimes called "**what-if**" analysis, because it answers questions such as "What if incremental sales are only 1,000 units, rather than 2,000 units? Then what will the NPW be?" Sensitivity analysis begins with a base-case situation, which is developed by using the most likely values for each input. We then change the specific variable of interest by several specified percentage points above and below the most likely value, while holding other variables constant. Next, we calculate a new NPW for each of the values we obtained. A convenient and useful way to present the results of a sensitivity analysis is to plot **sensitivity graphs**. The slopes of the lines show how sensitive the NPW is to changes in each of the inputs: The steeper the slope, the more sensitive the NPW is to a change in a particular variable. Sensitivity graphs identify the crucial variables that affect the final outcome most. Example 12.1 illustrates the concept of sensitivity analysis.

EXAMPLE 12.1 Sensitivity Analysis

Boston Metal Company (BMC), a manufacturer of various fabricated metal parts, must decide whether to enter the competition to become the supplier of transmission-housings for Gulf Electric, a company that produces the housings in its own in-house manufacturing facility but has almost reached its maximum production capacity. Therefore, Gulf is looking for an outside supplier. To compete, BMC must design a new fixture for the production process and purchase a new forge. The available details for this purchase are as follows.

- The new forge would cost $125,000. This total includes retooling costs for the transmission-housings.

- If BMC gets the order, it may be able to sell as many as 2,000 units per year to Gulf Electric for $50 each, in which case variable production costs,[3] such as direct labor and direct material costs, will be $15 per unit. The increase in fixed costs,[4] other than depreciation, will amount to $10,000 per year.

- The firm expects that the proposed transmission-housings project will have about a five-year product life. The firm also estimates that the amount ordered by Gulf

---

[3] These are expenses that change in direct proportion to the change in volume of sales or production, as defined in Section 8.3.

[4] These are expenses that do not vary with the volume of sales or production. For example, property taxes, insurance, depreciation, and rent are usually fixed expenses.

Electric in the first year will be ordered in each of the subsequent four years. (Due to the nature of contracted production, the annual demand and unit price would remain the same over the project after the contract is signed.)

- The initial investment can be depreciated on a MACRS basis over a seven-year period, and the marginal income-tax rate is expected to remain at 40%. At the end of five years, the forge is expected to retain a market value of about 32% (or $40,000) of the original investment.

- On the basis of this information, the engineering and marketing staffs of BMC have prepared the cash flow forecasts shown in Table 12.1. Since the NPW is positive ($40,169) at the 15% opportunity cost of capital (MARR), the project appears to be worth undertaking.

What Makes BMC Managers Uneasy: BMC's managers are uneasy about this project, because too many uncertain elements have not been considered in the analysis.

- If BMC decided to take on the project, it would have to invest in the forging machine to provide Gulf Electric with some samples as a part of the bidding process. If Gulf Electric were not to like BMC's sample, BMC would stand to lose its entire investment in the forging machine.

- If Gulf were to like BMC's sample, and if it were overpriced, BMC would be under pressure to bring the price in line with competing firms. Even the possibility that BMC would get a smaller order must be considered, as Gulf may utilize its overtime capacity to produce some extra units. Finally, BMC is not certain about its projections of variable and fixed costs.

Recognizing these uncertainties, the managers want to assess the various possible future outcomes before making a final decision. Put yourself in BMC's management position, and describe how you may address the uncertainty associated with the project. In doing so, perform a sensitivity analysis for each variable and develop a sensitivity graph.

DISCUSSION: Table 12.1 shows BMC's expected cash flows—but that they will indeed materialize cannot be assumed. In particular, BMC is not very confident in its revenue forecasts. The managers think that if competing firms enter the bidding, BMC will lose a substantial portion of the projected revenues by not being able to increase its bidding price. Before undertaking the project, the company needs to identify the key variables that will determine whether it will succeed or fail. The marketing department has estimated revenue as

$$\text{Annual revenue} = (\text{Product demand})(\text{unit price})$$

$$= (2{,}000)(\$50) = \$100{,}000$$

The engineering department has estimated variable costs, such as those of labor and materials, at $15 per unit. Since the projected sales volume is 2,000 units per year, the total variable cost is $30,000.

After first defining the unit sales, unit price, unit variable cost, fixed cost, and salvage value, we conduct a sensitivity analysis with respect to these key input variables. This is done by varying each of the estimates by a given percentage and determining what effect the variation in that item will have on the final results. If the effect is large, the result is sensitive to that item.

TABLE 12.1   **Cash Flow Statement for BMC's Transmission-housings Project (Base Case)**

	A	B	C	D	E	F	G
1							
2	**Input Data (Base):**				**Sensitivity Analysis:**		
3							
4	Unit price ($)	50		Category		% Change	
5	Demand	2,000		Unit price		0%	
6	Var. cost ($/unit)	15		Demand		0%	
7	Fixed cost ($)	10,000		Var. cost (unit)		0%	
8	Salvage ($)	40,000		Fixed cost		0%	
9	Tax rate (%)	40%		Salvage		0%	
10	MARR (%)	15%					
11				**Output (NPW)**		$40,169	
12							
13		0	1	2	3	4	5
14	**Income Statement**						
15	Revenues:						
16	Unit price		$ 50	$ 50	$ 50	$ 50	$ 50
17	Demand (units)		2,000	2,000	2,000	2,000	2,000
18	Sales revenue		$ 100,000	$ 100,000	$ 100,000	$ 100,000	$ 100,000
19	Expenses:						
20	Unit variable cost		$ 15	$ 15	$ 15	$ 15	$ 15
21	Variable cost		30,000	30,000	30,000	30,000	30,000
22	Fixed cost		10,000	10,000	10,000	10,000	10,000
23	Depreciation		17,863	30,613	21,863	15,613	5,581
24							
25	Taxable income		$ 42,137	$ 29,387	$ 38,137	$ 44,387	$ 54,419
26	Income taxes (40%)		16,855	11,755	15,255	17,755	21,768
27							
28	Net income		$ 25,282	$ 17,632	$ 22,882	$ 26,632	$ 32,651
29							
30	**Cash Flow Statement**						
31	Operating Activities:						
32	Net income		25,282	17,632	22,882	26,632	32,651
33	Depreciation		17,863	30,613	21,863	15,613	5,581
34	Investing Activities:						
35	Investment	(125,000)					
36	Salvage						40,000
37	Gains tax						(2,613)
38							
39	Net cash flow	$ (125,000)	$ 43,145	$ 48,245	$ 44,745	$ 42,245	$ 75,619
40							

SOLUTION

Given: After-tax cash flow statement based on most likely estimates.

Find: Input variables to which the NPW is most sensitive.

Sensitivity analysis: We begin the sensitivity analysis with a consideration of the base-case situation, which reflects the best estimate (expected value) for each input variable. In developing Table 12.2, we changed a given variable by 20% in 5%

TABLE 12.2 Sensitivity Analysis for Five Key Input Variables (Example 12.1)

	A	B	C	D	E	F	G	H	I	J
1										
2						Percent (%) Deviation from the Base				
3	Variables					Base				
4		-20%	-15%	-10%	-5%	0%	5%	10%	15%	20%
5										
6	Unit price	40.00	42.50	45.00	47.50	50.00	52.50	55.00	57.50	60.00
7	NPW	(57)	9,999	20,055	30,111	40,169	50,225	60,281	70,337	80,393
8										
9	Demand	1,600	1,700	1,800	1,900	2,000	2,100	2,200	2,300	2,400
10	NPW	12,010	19,049	26,088	33,130	40,169	47,208	54,247	61,286	68,325
11										
12	Variable cost	12.00	12.75	13.50	14.25	15.00	15.75	16.50	17.25	18.00
13	NPW	52,236	49,219	46,202	43,186	40,169	37,152	34,135	31,118	28,101
14										
15	Fixed cost	8,000	8,500	9,000	9,500	10,000	10,500	11,000	11,500	12,000
16	NPW	44,191	43,185	42,179	41,175	40,169	39,163	38,157	37,151	36,145
17										
18	Salvage value	32,000	34,000	36,000	38,000	40,000	42,000	44,000	46,000	48,000
19	NPW	37,782	38,378	38,974	39,573	40,169	40,765	41,361	41,957	42,553
20										

increments above and below the base-case value and calculated new NPWs, while other variables were held constant. The values for both sales and operating costs were the expected (or base-case) values, and the resulting $40,169 is the base-case NPW. Now we ask a series of "what-if" questions: What if sales are 20% below the expected level? What if operating costs rise? What if the unit price drops from $50 to $45? Table 12.2 summarizes the results of varying the values of the key input variables.

Sensitivity graph: Next, we construct a sensitivity graph for five of the transmission-housing project's key input variables. (See Figure 12.1.) We plot the base-case NPW on the ordinate of the graph at the value 1.0 on the abscissa (or 0% deviation). Then we reduce the value of product demand to 0.95 of its base-case value and recompute the NPW with all other variables held at their base-case value. We repeat the process by either decreasing or increasing the relative deviation from the base case. The lines for the variable unit price, variable unit cost, fixed cost, and salvage value are obtained in the same manner. In Figure 12.1, we see that the project's NPW is (1) very sensitive to changes in product demand and unit price, (2) fairly sensitive to changes in variable costs, and (3) relatively insensitive to changes in the fixed cost and the salvage value.

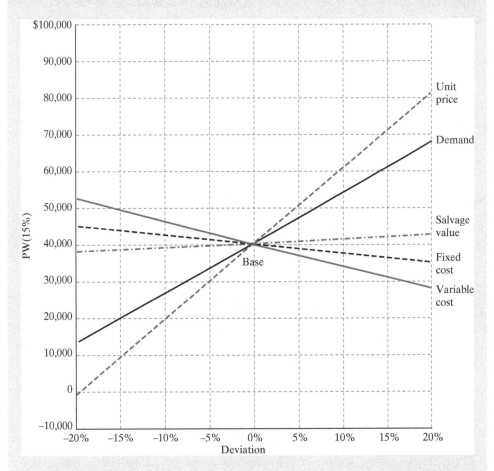

Figure 12.1 Sensitivity graph for BMC's transmission-housings project (Example 12.1).

Graphic displays (such as the one in Figure 12.1) provide a useful means to communicate the relative sensitivities of the different variables to the corresponding NPW value. However, sensitivity graphs do not explain any interactions among the variables or the likelihood of realizing any specific deviation from the base case. Certainly, it is conceivable that an answer might not be very sensitive to changes in either of two items, but it might be very sensitive to combined changes in them.

12.2.2 Break-Even Analysis

When we perform a sensitivity analysis of a project, we are asking how serious the effect of lower revenues or higher costs will be on the project's profitability. Managers sometimes prefer to ask instead how much sales can decrease below forecasts before the project begins to lose money. This type of analysis is known as **break-even analysis**. In other words, break-even analysis is a technique for studying the effect of variations in output on a firm's NPW (or other measures). We will present an approach to break-even analysis based on the project's cash flows.

To illustrate the procedure of break-even analysis based on NPW, we use the generalized cash-flow approach we discussed in Section 10.4. We compute the PW of cash inflows as a function of an unknown variable (say, x), perhaps annual sales. For example,

$$\text{PW of cash inflows} = f(x)_1$$

Next, we compute the PW of cash outflows as a function of x:

$$\text{PW of cash outflows} = f(x)_2$$

NPW is, of course, the difference between these two numbers. Accordingly, we look for the break-even value of x that makes

$$f(x)_1 = f(x)_2$$

Note that this break-even value is similar to that used to calculate the internal rate of return when we want to find the interest rate that makes the NPW equal zero. The break-even value is also used to calculate many other similar "cutoff values" at which a choice changes.

EXAMPLE 12.2 Break-Even Analysis

Through the sensitivity analysis in Example 12.1, BMC's managers become convinced that the NPW is quite sensitive to changes in annual sales volume. Determine the break-even NPW value as a function of that variable.

SOLUTION
Given: Cash flow statement based on most-likely estimates.
Find: Determine the break-even sales volume.

The analysis is shown in Table 12.3, in which the revenues and costs of the BMC transmission-housings project are set out in terms of an unknown amount of annual sales X.

TABLE 12.3 Break-Even Analysis with Unknown Annual Sales (Example 12.2)

	0	1	2	3	4	5
Cash inflow:						
Net salvage						37,389
Revenue:						
$X(1 - 0.4)(\$50)$		30X	30X	30X	30X	30X
Depreciation credit 0.4		7,145	12,245	8,745	6,245	2,230
Cash outflow:						
Investment	−125,000					
Variable cost:						
$-X(1 - 0.4)(\$15)$		−9X	−9X	−9X	−9X	−9X
Fixed cost:						
$-(1 - 0.4)(\$10,000)$		−6,000	−6,000	−6,000	−6,000	−6,000
Net cash flow	−125,000	21X + 1,145	21X + 6,245	21X + 2,745	21X + 245	21X + 33,617

617

We calculate the PWs of cash inflows and outflows as follows.

- PW of cash inflows:

$$\begin{aligned}
\text{PW}(15\%)_{\text{Inflow}} &= (\text{PW of after-tax net revenue}) \\
&\quad + (\text{PW of net salvage value}) \\
&\quad + (\text{PW of tax savings from depreciation}) \\
&= 30X(P/A, 15\%, 5) + \$37,387(P/F, 15\%, 5) \\
&\quad + \$7,145(P/F, 15\%, 1) + \$12,245(P/F, 15\%, 2) \\
&\quad + \$8,745(P/F, 15\%, 3) + \$6,245(P/F, 15\%, 4) \\
&\quad + \$2,230(P/F, 15\%, 5) \\
&= 30X(P/A, 15\%, 5) + \$44,490 \\
&= 100.5650X + \$44,490
\end{aligned}$$

- PW of cash outflows:

$$\begin{aligned}
\text{PW}(15\%)_{\text{Outflow}} &= (\text{PW of capital expenditure}) \\
&\quad + (\text{PW of after-tax expenses}) \\
&= \$125,000 + (9X + \$6,000)(P/A, 15\%, 5) \\
&= 30.1694X + \$145,113
\end{aligned}$$

The NPW of cash flows for the BMC is thus

$$\begin{aligned}
\text{PW}(15\%) &= 100.5650X + \$44,490 \\
&\quad - (30.1694X + \$145,113) \\
&= 70.3956X - \$100,623
\end{aligned}$$

In Table 12.4, we compute the PW of the inflows and the PW of the outflows as a function of demand (X).

The NPW will be just slightly positive if the company sells 1,430 units. Precisely calculated, the zero-NPW point (break-even volume) is 1,429.43 units:

$$\begin{aligned}
\text{PW}(15\%) &= 70.3956X - \$100,623 \\
&= 0 \\
X_b &= 1,430 \text{ units}
\end{aligned}$$

In Figure 12.2, we have plotted the PWs of the inflows and outflows under various assumptions about annual sales. The two lines cross when sales are 1,430 units, the point at which the project has a zero NPW. Again we see that, as long as sales are greater or equal to 1,430, the project has a positive NPW.

TABLE 12.4 Determination of Break-Even Volume Based on Project's NPW
(Example 12.2)

Demand (X)	PW of Inflow (100.5650X + 44,490)	PW of Outflow (30.1694X + 145,113)	NPW (70.3956X − 100,623)
0	$ 44,490	$ 145,113	$ (100,623)
500	94,773	160,198	(65,425)
1,000	145,055	175,282	(30,227)
1,429	188,197	188,225	(28)
1,430	188,298	188,255	43
1,500	195,338	190,367	4,970
2,000	245,620	205,452	40,168
2,500	295,903	220,537	75,366

Break-even volume = 1,430 units

Figure 12.2 Break-even analysis based on net cash flow (Example 12.2).

COMMENTS: Instead of using the analytical approach described, we may use the Goal Seek function for the Excel worksheet developed in Table 12.1. All we have to do is set the Output (NPW) value (cell $F$11) to zero by changing the demand deviation (cell $F$6). As shown in Figure 12.3, to break-even, the demand could be as low as 28.55% below the base (or 1,429 units).

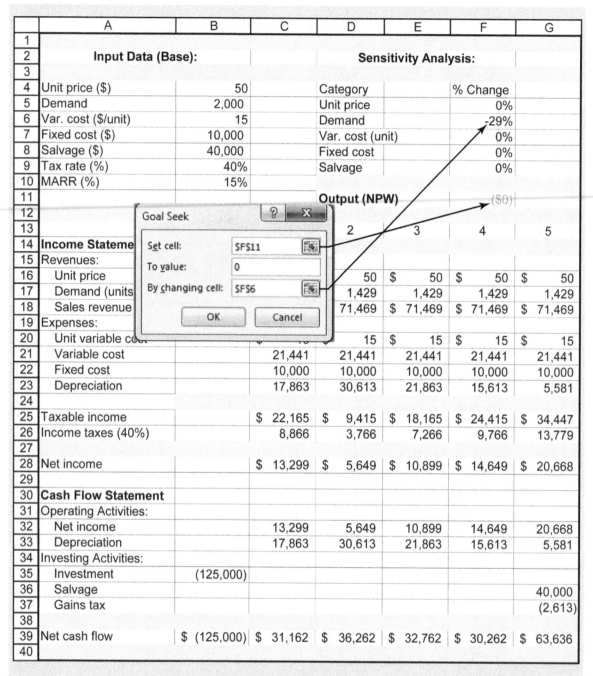

	A	B	C	D	E	F	G
1							
2	**Input Data (Base):**			**Sensitivity Analysis:**			
3							
4	Unit price ($)	50		Category		% Change	
5	Demand	2,000		Unit price		0%	
6	Var. cost ($/unit)	15		Demand		-29%	
7	Fixed cost ($)	10,000		Var. cost (unit)		0%	
8	Salvage ($)	40,000		Fixed cost		0%	
9	Tax rate (%)	40%		Salvage		0%	
10	MARR (%)	15%					
11				Output (NPW)		($0)	
12							
13				2		3	4	5
14	**Income Statement**						
15	Revenues:						
16	Unit price			50	$ 50	$ 50	$ 50
17	Demand (units)			1,429	1,429	1,429	1,429
18	Sales revenue			71,469	$ 71,469	$ 71,469	$ 71,469
19	Expenses:						
20	Unit variable cost			15	$ 15	$ 15	$ 15
21	Variable cost		21,441	21,441	21,441	21,441	21,441
22	Fixed cost		10,000	10,000	10,000	10,000	10,000
23	Depreciation		17,863	30,613	21,863	15,613	5,581
24							
25	Taxable income		$ 22,165	$ 9,415	$ 18,165	$ 24,415	$ 34,447
26	Income taxes (40%)		8,866	3,766	7,266	9,766	13,779
27							
28	Net income		$ 13,299	$ 5,649	$ 10,899	$ 14,649	$ 20,668
29							
30	**Cash Flow Statement**						
31	Operating Activities:						
32	Net income		13,299	5,649	10,899	14,649	20,668
33	Depreciation		17,863	30,613	21,863	15,613	5,581
34	Investing Activities:						
35	Investment	(125,000)					
36	Salvage						40,000
37	Gains tax						(2,613)
38							
39	Net cash flow	$ (125,000)	$ 31,162	$ 36,262	$ 32,762	$ 30,262	$ 63,636
40							

Goal Seek dialog box:
Set cell: $F$11
To value: 0
By changing cell: $F$6
OK Cancel

Figure 12.3 Break-even analysis using Excel's Goal Seek function.

12.2.3 Scenario Analysis

Although both sensitivity and break-even analyses are useful, they have limitations. Often, it is difficult to specify precisely the relationship between a particular variable and the NPW. The relationship is further complicated by interdependencies among the variables.

Holding operating costs constant while varying unit sales may ease the analysis, but in reality, operating costs do not behave in this manner. Yet, it may complicate the analysis too much to permit movement in more than one variable at a time.

Scenario analysis is a technique that considers the sensitivity of NPW both to changes in key variables and to the range of likely values of those variables. For example, the decision maker may examine two extreme cases: a "worst-case" scenario (low unit sales, low unit price, high variable cost per unit, high fixed cost, and so on) and a "best-case" scenario. The NPWs under the worst and the best conditions are then calculated and compared with the expected (or base-case) NPW. Example 12.3 illustrates a plausible scenario analysis for BMC's transmission-housings project.

EXAMPLE 12.3 Scenario Analysis

Consider again BMC's transmission-housings project first presented in Example 12.1. Assume that the company's managers were asked to come up with the range of estimates for each key input variable: a worst case, a most-likely case, and a best case. For example, they regard a drop in unit sales below 1,600 or a rise above 2,400 as extremely unlikely. Thus, a decremental annual sale of 400 units defines the lower bound (the worst-case scenario), whereas an incremental annual sale of 400 units defines the upper bound (the best-case scenario). (Remember that the most likely value was 2,000 in annual unit sales.) Discuss the worst- and best-case scenarios, assuming that the unit sales for all five years are equal.

DISCUSSION: To carry out the scenario analysis, we ask the marketing and engineering staffs to give optimistic (best-case) and pessimistic (worst-case) estimates for the key variables. Then we use the worst-case variable values to obtain the worst-case NPW and the best-case variable values to obtain the best-case NPW. Then, we use the Excel worksheet developed in Table 12.1 to recalculate the NPW value associated with the set of variables for each scenario.

SOLUTION

Given: Three plausible scenarios assuming a set of values for key input variables.
Find: NPW for each scenario.

The results of our analysis are summarized as follows.

Variable Considered	Worst-Case Scenario	Most-Likely Case Scenario	Best-Case Scenario
Unit demand	1,600	2,000	2,400
Unit price ($)	48	50	53
Variable cost ($)	17	15	12
Fixed cost ($)	11,000	10,000	8,000
Salvage value ($)	30,000	40,000	50,000
PW(15%)	−$5,856	$40,169	$104,295

We see that the base case produces a positive NPW, the worst case produces a negative NPW, and the best case produces a large positive NPW. Still, by just looking at the results in the table, it is not easy to interpret the scenario analysis or to make a decision based on it. For example, we could say that there is a chance of losing money on the project, but we do not yet have a specific likelihood for that possibility. Clearly, we need estimates of the probabilities of occurrence of the worst case, the best case, the base (most likely) case, and all the other possibilities.

The need to estimate probabilities leads us directly to our next step: developing a probability distribution (or, put another way, the probability that the variable in question takes on a certain value). If we can predict the effects on the NPW of variations in the parameters, why should we not assign a probability distribution to the possible outcomes of each parameter and combine these distributions in some way to produce a probability distribution for the possible outcomes of the NPW? We shall consider this issue in the next two sections.

12.3 Probability Concepts for Investment Decisions

In this section, we will assume that the analyst has available the probabilities (likelihoods) of future events from either previous experience with a similar project or a market survey. The use of probability information can provide management with a range of possible outcomes and the likelihood of achieving different goals under each investment alternative.

12.3.1 Assessment of Probabilities

We begin by defining terms related to probability, such as *random variable*, *probability distribution*, and *cumulative probability distribution*. Quantitative statements about risk are given as numerical probabilities or as likelihoods (odds) of occurrence. Probabilities are given as decimal fractions in the interval from 0.0 to 1.0. An event or outcome that is certain to occur has a probability of 1.0. As the probability of an event approaches 0, the event becomes increasingly less likely to occur.

Risk Analysis

Risk analysis refers to conducting analysis to identify the degree of the uncertainty of forecasted future cash flows streams, and perform statistical analysis to determine the probability of a project's success or failure. One of the important steps in risk analysis is to assign probabilities of the various outcomes related to a given course of action and possible future economic states.

For example, airlines need to properly hedge fuel price in ever-chaning market dynamics, while PC makers must factor in the possibility of reduced revenues due to a global recession. Risk analysis allows professionals to identify and mitigate risks, but not avoid them completely. Proper risk analysis often includes mathematical and statistical software programs such as @RISK or Crystal Ball.

Random Variables

A **random variable** is a parameter or variable that can have more than one possible value (though not simultaneously). The value of a random variable at any one time is unknown until the event occurs, but the probability that the random variable will have a specific value is known in advance. In other words, associated with each possible value of the random variable is a likelihood (or probability) of occurrence. For example, when your college team plays a football game, only two events regarding the outcome of the game are possible: win or lose. The outcome is a random variable dictated largely by the strength of your opponent.

To indicate random variables, we will adopt the convention of a capital italic letter (for example, X). To denote the situation in which the random variable takes a specific value, we will use a lowercase italic letter (for example, x). Random variables are classified as either discrete or continuous:

- Any random variables that take on only isolated (countable) values are **discrete random variables**.
- **Continuous random variables** may have any value within a certain interval.

For example, the outcome of a game should be a discrete random variable. By contrast, suppose you are interested in the amount of beverages sold on a given day that the game is played. The quantity (or volume) of beverages sold will depend on the weather conditions, the number of people attending the game, and other factors. In this case, the quantity is a continuous random variable—a random variable that takes a value from a continuous set of values.

Probability Distributions

For a discrete random variable, the probability of each random event needs to be assessed. For a continuous random variable, the probability function needs to be defined, as the event takes place over a continuous domain. In either case, a range of probabilities for each feasible outcome exists. Together, these probabilities make up a **probability distribution**.

Probability assessments may be based on past observations or historical data if the trends that were characteristic of the past are expected to prevail in the future. Forecasting weather or predicting the outcome of a game in many professional sports is done on the basis of compiled statistical data. Any probability assessments based on objective data are called **objective probabilities**. However, in many real investment situations, no objective data are available to consider. In these situations, we assign **subjective probabilities** that we think are appropriate to the possible states of nature. As long as we act consistently with our beliefs about the possible events, we may reasonably account for the economic consequences of those events in our profitability analysis.

For a continuous random variable, we usually try to establish a range of values; that is, we try to determine a **minimum value** (L) and a **maximum value** (H). Next, we determine whether any value within these limits might be *more likely* to occur than the other values; in other words, does the distribution have a **mode** (M_O) or a most frequently occurring value?

- If the distribution does have a mode, we can represent the variable by a **triangular distribution (or Beta distribution)**, such as that shown in Figure 12.4.
- If we have no reason to assume that one value is any more likely to occur than any other, perhaps the best we can do is to describe the variable as a **uniform distribution**, as shown in Figure 12.5.

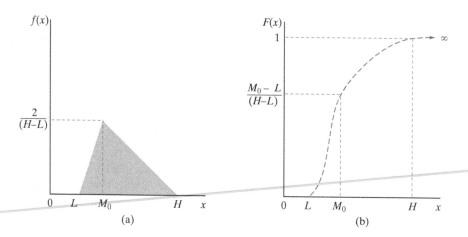

Figure 12.4 A triangular probability distribution: (a) Probability function and (b) cumulative probability distribution.

These two distributions are frequently used to represent the variability of a random variable when the only information we have is its minimum, its maximum, and whether the distribution has a mode. For example, suppose the best judgment of the analyst was that the sales revenue could vary anywhere from \$2,000 to \$5,000 per day, and any value within the range is equally likely. This judgment about the variability of sales revenue could be represented by a uniform distribution.

For BMC's transmission-housings project, we can treat two input variables, the product demand (X) and the unit price (Y) as discrete random variables whose values cannot be predicted with certainty at the time of decision-making. Let us assume the probability distributions in Table 12.5. We see that the product demand with the highest probability is 2,000 units, whereas the unit sale price with the highest probability is \$50. These, therefore, are the most likely values. We also see a substantial probability that a unit demand other than 2,000 units will be realized. When we use only the most likely values in an economic analysis, we are in fact ignoring these other outcomes.

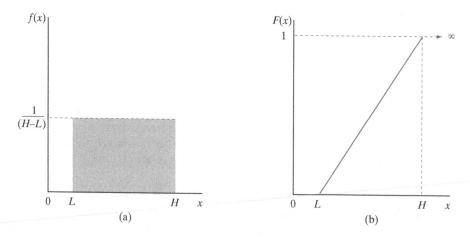

Figure 12.5 A uniform probability distribution: (a) Probability function and (b) cumulative probability distribution.

TABLE 12.5 **Probability Distributions for Unit Demand (X) and Unit Price (Y) for BMC's Project***

Product Demand (X)		Unit Sale Price (Y)	
Units (x)	$P(X = x)$	Unit Price (y)	$P(Y = y)$
1,600	0.20	$48	0.30
2,000	0.60	50	0.50
2,400	0.20	53	0.20

*X and Y are independent random variables.

Cumulative Distribution

As we have observed in the previous section, the probability distribution provides information regarding the probability that a random variable will assume some value x. We can use this information, in turn, to define the cumulative distribution function. The **cumulative distribution** function gives the probability that the random variable will attain a value smaller than or equal to some specific value x. A common notation for the cumulative distribution is

$$F(x) = P(X \le x) = \begin{cases} \sum_{j=1}^{j} p_j & \text{(for a discrete random variable)} \\ \int_{L}^{x} f(x)dx & \text{(for a continuous random variable)} \end{cases}$$

where p_j is the probability of occurrence of the x_jth value of the discrete random variable and $f(x)$ is a probability function for a continuous variable. With respect to a continuous random variable, the cumulative distribution rises continuously in a smooth (rather than stairwise) fashion.

Example 12.4 illustrates the method by which probabilistic information can be incorporated into our analysis. Again, BMC's transmission-housings project will be used. In the next section, we will show you how to compute some composite statistics with the use of all the data.

EXAMPLE 12.4 Cumulative Probability Distributions

Suppose that the only parameters subject to risk are the number of unit sales (X) to Gulf Electric each year and the unit sales price (Y). From experience in the market, BMC assesses the probabilities of outcomes for each variable as shown in Table 12.5. Determine the cumulative probability distributions for these random variables.

SOLUTION
Given: Discrete probability distributions for X and Y.
Find: Cumulative distributions for X and Y.

Consider the demand probability distribution (X) given in Table 12.5 for BMC's transmission-housings project:

Unit Demand (X)	Probability $P(X = x)$
1,600	0.2
2,000	0.6
2,400	0.2

If we want to know the probability that demand will be less than or equal to any particular value, we can use the following cumulative probability function:

$$F(x) = P(X \le x) = \begin{cases} 0 & x < 1,600 \\ 0.2 & 1,600 \le x < 2,000 \\ 0.8 & 2,000 \le x < 2,400 \\ 1.0 & 2,400 \le x \end{cases}$$

For example, if we want to know the probability that the demand will be less than or equal to 2,000, we can examine a single value, namely $x = 2,100$, and we will find that the probability is 80%.

We can find the cumulative distribution for Y in a similar fashion. Figures 12.6 and 12.7 illustrate probability and cumulative probability distributions for X and Y, respectively.

Figure 12.6 Probability and cumulative probability distributions for random variable X (annual sales).

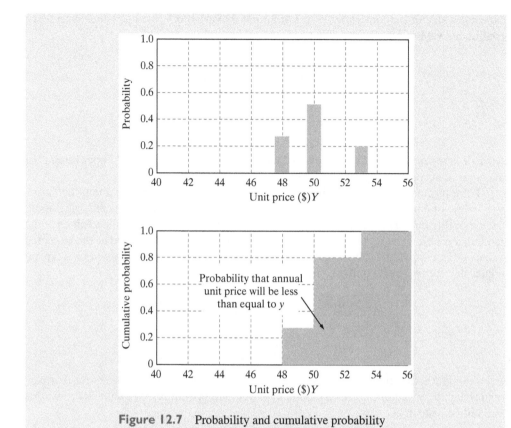

Figure 12.7 Probability and cumulative probability distributions for random variable Y from Table 12.5.

12.3.2 Summary of Probabilistic Information

Although knowledge of the probability distribution of a random variable allows us to make a specific probability statement, a single value that may characterize the random variable and its probability distribution is often desirable. Such a quantity is the **expected value** of the random variable. We also want to know something about how the values of the random variable are dispersed about the expected value (i.e., the **variance**). In investment analysis, this dispersion information is interpreted as the degree of project risk. The expected value indicates the weighted average of the random variable, and the variance captures the variability of the random variable.

Measure of Expectation

The **expected value** (also called the **mean**) is a weighted average value of the random variable, where the weighting factors are the probabilities of occurrence. All distributions (discrete and continuous) have an expected value. We will use $E[X]$ (or μ_X) to denote the expected value of random variable X. For a random variable X that has either discrete or continuous values, we compute the expected value with the formula

$$E[X] = \mu_X = \begin{cases} \displaystyle\sum_{j=1}^{J} (p_j)x_j & \text{(discrete case)} \\ \displaystyle\int_{L}^{H} xf(x)dx & \text{(continuous case)} \end{cases} \tag{12.1}$$

where J is the number of discrete events and L and H are the lower and upper bounds of the continuous probability distribution.

The expected value of a distribution gives us important information about the "average" value of a random variable, such as the NPW, but it does not tell us anything about the variability on either side of the expected value. Will the range of possible values of the random variable be very small, and will all the values be located at or near the expected value? For example, the following table represents the temperatures recorded on a typical winter day for two cities in the United States.

Location	Low	High	Average
Atlanta	15°F	67°F	41°F
Seattle	36°F	48°F	42°F

Even though both cities had almost identical mean (average) temperatures on that particular day, they had different variations in extreme temperatures. We shall examine this variability issue next.

Measure of Variation—Risk

Another measure needed when we are analyzing probabilistic situations is a measure of the risk due to the variability of the outcomes. Among the various measures of the variation of a set of numbers that are used in statistical analysis are the **range**, the **variance**, and the **standard deviation**. The variance and the standard deviation are used most commonly in the analysis of financial risk. We will use Var[X] or σ_X^2 to denote the variance, and σ_X to denote the standard deviation of random variable X. (If there is only one random variable in an analysis, we normally omit the subscript.)

The variance tells us the degree of spread (or dispersion) of the distribution on either side of the mean value. As the variance increases, the spread of the distribution increases: the smaller the variance, the narrower is the spread about the expected value.

To determine the variance, we first calculate the deviation of each possible outcome x_j from the expected value $(x_j - \mu)$. Then we raise each result to the second power and multiply it by the probability of x_j occurring (i.e., p_j). The summation of all these products serves as a measure of the distribution's variability. For a random variable that has only discrete values, the equation for computing the variance is[5]

$$\text{Var}[X] = \sigma_X^2 = \sum_{j=1}^{J} (x_j - \mu)^2(p_j) \tag{12.2}$$

---

[5] For a continuous random variable, we compute the variance as

$$\text{Var}[X] = \int_{L}^{H} (x - \mu)^2 f(x)dx$$

where p_j is the probability of occurrence of the jth value of the random variable (x_j) and μ is as defined by Eq. (12.1).

To be most useful, any measure of risk should have a definite value (unit). One such measure is the **standard deviation**, which we may calculate by taking the positive square root of Var[X] measured in the same units as X:

$$\sigma_X = \sqrt{\text{Var}[X]} \qquad (12.3)$$

The standard deviation is a probability-weighted deviation (more precisely, the square root of the sum of squared deviations) from the expected value. Thus, it gives us an idea of how far above or below the expected value the actual value is likely to be. For most probability distributions, the actual value will be observed within the $\pm 3\sigma$ range.

In practice, the calculation of the variance is somewhat easier if we use the formula

$$\begin{aligned}
\text{Var}[X] &= \Sigma p_j x_j^2 - (\Sigma p_j x_j)^2 \\
&= E[X^2] - (E[X])^2 \qquad (12.4)
\end{aligned}$$

The term $E[X^2]$ in Eq. (12.4) is interpreted as the mean value of the squares of the random variable (i.e., the actual values squared). The second term is simply the mean value squared. Example 12.5 illustrates how we compute measures of variation.

EXAMPLE 12.5 Calculation of Mean and Variance

Consider once more BMC's transmission-housings project. Unit sales (X) and unit price (Y) are estimated as in Table 12.5. Compute the means, variances, and standard deviations for the random variables X and Y.

SOLUTION
Given: Probability distributions for X and Y.
Find: Compute the mean and variance for each distribution.

For the product demand variable (X), we have

x_j	p_j	$x_j(p_j)$	$(x_j - E[X])^2$	$(x_j - E[X])^2(p_j)$
1,600	0.20	320	$(-400)^2$	32,000
2,000	0.60	1,200	0	0
2,400	0.20	480	$(400)^2$	32,000
		$EX = 2,000$		$\text{Var}[X] = 64,000$
				$\sigma_X = 252.98$

For the variable unit price (Y), we have

y_j	p_j	$y_j(p_j)$	$(y_j - E[Y])^2$	$(y_j - E[Y])^2(p_j)$
$48	0.3	$14.40	$(-2)^2$	1.20
50	0.50	25.00	$(0)^2$	0
53	0.20	10.60	$(3)^2$	1.80
		$E[Y] = 50.00$		$\text{Var}[Y] = 3.00$
				$\sigma_Y = 1.73$

12.3.3 Joint and Conditional Probabilities

Thus far, we have not looked at how the values of some variables can influence the values of others. It is, however, entirely possible—indeed, it is likely—that the values of some parameters will be dependent on the values of others. We commonly express these dependencies in terms of conditional probabilities. An example is product demand, which probably will be influenced by unit price.

We define a **joint probability** as

$$P(x, y) = P(X = x | Y = y)P(Y = y) \tag{12.5}$$

where $P(X = x | Y = y)$ is the **conditional probability** of observing x given $Y = y$, and $P(Y = y)$ is the **marginal probability** of observing $Y = y$. Certainly, important cases exist in which a knowledge of the occurrence of event X does *not* change the probability of an event Y. That is, if X and Y are **independent**, then the joint probability is simply

$$P(x, y) = P(x)P(y) \tag{12.6}$$

The concepts of joint, marginal, and conditional distributions are best illustrated by numerical examples.

Suppose that BMC's marketing staff estimates that, for a given unit price of $48, the conditional probability that the company can sell 1,600 units is 0.10. The probability of this joint event (unit sales = 1,600 and unit sales price = 48) is

$$P(x, y) = P(1{,}600, \$48)$$
$$= P(x = 1{,}600 | y = \$48)P(y = \$48)$$
$$= (0.10)(0.30)$$
$$= 0.03$$

TABLE 12.6 **Assessments of Conditional and Joint Probabilities**

Unit Price Y	Probability	Unit Sales X	Conditional Probability	Joint Probability
		1,600	0.10	0.03
$48	0.30	2,000	0.40	0.12
		2,400	0.50	0.15
		1,600	0.10	0.05
50	0.50	2,000	0.64	0.32
		2,400	0.26	0.13
		1,600	0.50	0.10
53	0.20	2,000	0.40	0.08
		2,400	0.10	0.02

We can obtain the probabilities of other joint events in a similar fashion; these are presented in Table 12.6.

From Table 12.6, we can see that the unit demand (X) ranges from 1,600 to 2,400 units, the unit price (Y) ranges from $48 to $53, and nine joint events are possible. These joint probabilities must sum to unity, as in Table 12.7.

The marginal distribution for x can be developed from the joint event by fixing x and summing over y, as given in Table 12.8.

TABLE 12.7 **Joint Probabilities**

Joint Event (xy)	P(x, y)
(1,600, $48)	0.03
(2,000, $48)	0.12
(2,400, $48)	0.15
(1,600, $50)	0.05
(2,000, $50)	0.32
(2,400, $50)	0.13
(1,600, $53)	0.10
(2,000, $53)	0.08
(2,400, $53)	0.02
	Sum = 1.00

TABLE 12.8 Marginal Probabilities

x_j	$P(x_j) = \Sigma_y P(x, y)$
1,600	$P(1,600, \$48) + P(1,600, \$50) + P(1,600, \$53) = 0.18$
2,000	$P(2,000, \$48) + P(2,000, \$50) + P(2,000, \$53) = 0.52$
2,400	$P(2,400, \$48) + P(2,400, \$50) + P(2,400, \$53) = 0.30$

This marginal distribution tells us that 52% of the time we can expect to have a demand of 2,000 units and 18% and 30% of the time we can expect to have a demand of 1,600 and 2,400, respectively.

12.3.4 Covariance and Coefficient of Correlation

When two random variables are not independent, we need some measure of their dependence on each other. The parameter that tells the degree to which two variables (X, Y) are related is the covariance $\text{Cov}(X, Y)$, denoted by σ_{XY}. Mathematically, we define

$$
\begin{aligned}
\text{Cov}(X, Y) &= \sigma_{XY} \\
&= E\{(X - E[X])(Y - E[Y])\} \\
&= E(XY) - E(X)E(Y) \\
&= \rho_{XY}\sigma_X\sigma_Y
\end{aligned}
\tag{12.7}
$$

where ρ_{XY} is the coefficient of correlation between X and Y. It is clear that, if X tends to exceed its mean whenever Y exceeds its mean, $\text{Cov}(X, Y)$ will be positive. If X tends to fall below its mean whenever Y exceeds its mean, then $\text{Cov}(X, Y)$ will be negative. The sign of $\text{Cov}(X, Y)$, therefore, reveals whether X and Y vary directly or inversely with one another. We can rewrite Eq. (12.7) in terms of ρ_{XY}:

$$
\rho_{XY} = \frac{\text{Cov}(X, Y)}{\sigma_X\sigma_Y}
\tag{12.8}
$$

The value of ρ_{XY} can vary within the range of -1 and $+1$, with $\rho_{XY} = 0$ indicating no correlation between the two random variables. As shown in Table 12.9, the coefficient of correlation between the product demand (X) and the unit price (Y) is negatively correlated as $\rho_{XY} = -0.439$, indicating that as the firm lowers the unit price it tends to generate a higher demand.

TABLE 12.9 Calculating the Correlation Coefficient between Two Random Variables X and Y

(x, y)	$(x - E[X])$	$(y - E[Y])$	$p(x, y)$	$(x - E[X])(y - E[Y])$	$p(x, y) \times (x - E[X])(y - E[Y])$
(1,600, 48)	(1,600 − 2,000)	(48 − 50)	0.03	800	24
(2,000, 48)	(2,000 − 2,000)	(48 − 50)	0.12	0	0
(2,400, 48)	(2,400 − 2,000)	(48 − 50)	0.15	−800	−120
(1,600, 50)	(1,600 − 2,000)	(50 − 50)	0.05	0	0
(2,000, 50)	(2,000 − 2,000)	(50 − 50)	0.32	0	0
(2,400, 50)	(2,400 − 2,000)	(50 − 50)	0.13	0	0
(1,600, 53)	(1,600 − 2,000)	(53 − 50)	0.10	−1,200	−120
(2,000, 53)	(2,000 − 2,000)	(53 − 50)	0.08	0	0
(2,400, 53)	(2,400 − 2,000)	(53 − 50)	0.02	1,200	24

$$\text{Cov}(X, Y) = -192$$

$$\rho_{XY} = \frac{\text{Cov}(X, Y)}{\sigma_X \sigma_Y}$$

$$= \frac{-192}{(\sqrt{64,000} \times \sqrt{3.00})}$$

$$= -0.439$$

12.4 Probability Distribution of NPW

After we have identified the random variables in a project and assessed the probabilities of the possible events, the next step is to develop the project's NPW distribution.

12.4.1 Procedure for Developing an NPW Distribution

We will consider the situation in which all the random variables used in calculating the NPW are *independent*. To develop the NPW distribution, we may follow these steps:

Step 1. Express the NPW as functions of unknown random variables.
Step 2. Determine the probability distribution for each random variable.
Step 3. Determine the joint events and their probabilities.
Step 4. Evaluate the NPW equation at these joint events.
Step 5. Rank the NPW values in increasing order of NPW.

These steps are illustrated in Example 12.6.

EXAMPLE 12.6 Procedure for Developing an NPW Distribution

Consider again BMC's transmission-housings project first set forth in Example 12.1. Use the unit demand (X) and price (Y) given in Table 12.5, and develop the NPW distribution for the BMC project. Then calculate the mean and variance of the NPW distribution.

SOLUTION

Given: Probability distributions for X and Y.
Find: NPW probability distribution.

Using the generalized cash flow approach discussed in Chapter 10, we can summarize the after-tax cash flow for the BMC's transmission-housings project as functions of random variables X and Y. As shown in Table 12.10, we can compute the PW of cash inflows as

$$PW(15\%) = 0.6XY(P/A, 15\%, 5) + \$44,490$$
$$= 2.0113XY + \$44,490$$

The PW of cash outflows is

$$PW(15\%) = \$125,000 + (9X + \$6,000)(P/A, 15\%, 5)$$
$$= 30.1694X + \$145,113$$

Thus, the NPW is

$$PW(15\%) = 2.0113X(Y - \$15) - \$100,623$$

TABLE 12.10 After-Tax Cash Flow as a Function of Unknown Unit Demand (X) and Unit Price (Y) (Example 12.6)

Item	0	1	2	3	4	5
Cash inflow:						
Net salvage						37,387
Revenue $X(1 - 0.4)Y$		0.6XY	0.6XY	0.6XY	0.6XY	0.6XY
Depreciation credit:						
0.4 (depreciation)		7,145	12,245	8,745	6,245	2,230
Cash outflow:						
Investment	−125,000					
Variable cost $-X(1 - 0.4)(15)$		−9X	−9X	−9X	−9X	−9X
Fixed cost $-(1 - 0.4)(10,000)$		−6,000	−6,000	−6,000	−6,000	−6,000
Net cash flow	−125,000	0.6X(Y − 15) +1,145	0.6X(Y − 15) +6,245	0.6X(Y − 15) +2,745	0.6X(Y − 15) +245	0.6X(Y − 15) +33,617

If the product demand X and the unit price Y are independent random variables, then PW(15%) also will be a random variable. To determine the NPW distribution, we need to consider all the combinations of possible outcomes.[6] The first possibility is the event in which $X = 1,600$ and $Y = \$48$. Since X and Y are considered to be independent random variables, the probability of this joint event is

$$P(X = 1,600, Y = \$48) = P(X = 1,600)P(Y = \$48)$$
$$= (0.20)(0.30)$$
$$= 0.06$$

With these values as input, we compute the possible NPW outcome as

$$PW(15\%) = 2.0113X(Y - \$15) - \$100,623$$
$$= 2.0113(1,600)(\$48 - \$15) - \$100,623$$
$$= \$5,574$$

Eight other outcomes are possible; they are summarized with their joint probabilities in Table 12.11 and depicted in Figure 12.8.

The NPW probability distribution in Table 12.11 indicates that the project's NPW varies between \$5,574 and \$82,808, but that no loss occurs under any of the circumstances examined. On the one hand, from the cumulative distribution, we further observe that there is a 0.38 probability that the project would realize an NPW less than that forecast for the base case (\$40,168). On the other hand, there is a 0.32 probability that the NPW will be greater than this value. Certainly, the probability distribution

TABLE 12.11 The NPW Probability Distribution with Independent Random Variables

Event No.	x	y	$P(X = x, Y = y)$	Cumulative Joint Probability	NPW
1	1,600	$48.00	0.06	0.06	$5,574
2	1,600	50.00	0.10	0.16	12,010
3	1,600	53.00	0.04	0.20	21,664
4	2,000	48.00	0.18	0.38	32,123
5	2,000	50.00	0.30	0.68	40,168
6	2,000	53.00	0.12	0.80	52,236
7	2,400	48.00	0.06	0.86	58,672
8	2,400	50.00	0.10	0.96	68,326
9	2,400	53.00	0.04	1.00	82,808

[6] If X and Y are dependent random variables, the joint probabilities developed in Table 12.6 should be used.

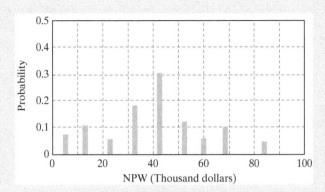

Figure 12.8 The NPW probability distribution when X and Y are independent (Example 12.6).

provides much more information on the likelihood of each possible event than does the scenario analysis presented in Section 12.2.3.

We have developed a probability distribution for the NPW by considering random cash flows. As we observed, a probability distribution helps us to see what the data imply in terms of project risk. Now we can learn how to summarize the probabilistic information—the mean and the variance.

- For BMC's transmission-housings project, we compute the expected value of the NPW distribution as shown in Table 12.12. Note that this expected value happens to be the same as the most likely value of the NPW distribution.

TABLE 12.12 Calculation of the Mean of the NPW Distribution

Event No.	x	y	$P(x, y)$	Cumulative Joint Probability	NPW	Weighted NPW
1	1,600	$48.00	0.06	0.06	$5,574	$334
2	1,600	50.00	0.10	0.16	12,010	1,201
3	1,600	53.00	0.04	0.20	21,664	867
4	2,000	48.00	0.18	0.38	32,123	5,782
5	2,000	50.00	0.30	0.68	40,168	12,050
6	2,000	53.00	0.12	0.80	52,236	6,268
7	2,400	48.00	0.06	0.86	58,672	3,520
8	2,400	50.00	0.10	0.96	68,326	6,833
9	2,400	53.00	0.04	1.00	82,808	3,312

$$E[\text{NPW}] = \$40,168$$

- We obtain the variance of the NPW distribution, assuming independence between X and Y and using Eq. (12.2), as shown in Table 12.13. We could obtain the same result more easily by using Eq. (12.4).

COMMENTS: We can obtain the mean and variance of the NPW analytically by using the properties of the product of random variables. Let $W = XY$, where X and Y are random variables with known means and variances (μ_X, σ_X^2) and (μ_Y, σ_Y^2), respectively. If X and Y are independent of each other, then

$$E(W) = E(XY) = \mu_X\mu_Y \tag{12.9}$$

and

$$\text{Var}(W) = \text{Var}(XY) = \mu_X^2\sigma_Y^2 + \mu_Y^2\sigma_X^2 + \sigma_X^2\sigma_Y^2 \tag{12.10}$$

The expected net present worth is then

$$E[\text{PW}(15\%)] = 2.0113E(XY) - (2.0113)(15)E(X) - 100{,}623$$

$$= 2.0113(2{,}000)(50) - (2.0113)(15)(2{,}000) - 100{,}623$$

$$= \$40{,}168$$

Now let $Z = Y - 15$. Then $E(Z) = E(Y) - 15 = \mu_Y - 15 = 50 - 15 = 35$ and $\text{Var}[Z] = \text{Var}[Y] = \sigma_Y^2 = 3$. So

TABLE 12.13 Calculation of the Variance of the NPW Distribution

Event No. (i)	x	y	$P(X = x, Y = y)$	NPW	$(\text{NPW} - E[\text{NPW}])^2$	$P_i(\text{NPW}_i - E[\text{NPW}])^2$
1	1,600	$48.00	0.06	$5,574	1,196,769,744	71,806,185
2	1,600	50.00	0.10	12,010	792,884,227	79,288,423
3	1,600	53.00	0.04	21,664	342,396,536	13,695,861
4	2,000	48.00	0.18	32,123	64,725,243	11,650,544
5	2,000	50.00	0.30	40,168	0	0
6	2,000	53.00	0.12	52,236	145,631,797	17,475,816
7	2,400	48.00	0.06	58,672	342,396,536	20,543,792
8	2,400	50.00	0.10	68,326	792,884,227	79,288,423
9	2,400	53.00	0.04	82,808	1,818,132,077	72,725,283
					$\text{Var}[\text{NPW}] =$	366,474,326
					$\sigma =$	$19,144

$$
\begin{aligned}
\text{Var}[PW(15\%)] &= \text{Var}[2.0113X(Y - 15) - 100{,}623] \\
&= \text{Var}[2.0113XZ] \\
&= (2.0113)^2 \, \text{Var}[XZ] \\
&= (2.0113)^2 (\mu_X^2 \sigma_Z^2 + \mu_Z^2 \sigma_X^2 + \sigma_X^2 \sigma_Z^2) \\
&= (2.0113)^2 [(2{,}000^2)(3) + (35^2)(64{,}000) + (64{,}000)(3)] \\
&= 366{,}474{,}326
\end{aligned}
$$

Note that the mean and variance thus calculated are exactly the same as in Tables 12.12 and 12.13, respectively. Note also that if X and Y are correlated with each other, then Eq. (12.10) cannot be used.[7]

12.4.2 Aggregating Risk over Time

In the previous section, we developed an NPW equation by aggregating all cash flow components over time. One approach to estimating the amount of risk present in a particular investment opportunity is to determine the mean and variance of cash flows in each period; then we may be able to aggregate the risk over the project life in terms of net present worth. We have two cases.

- **Independent random variables:** In this case,

$$
E[PW(i)] = \sum_{n=0}^{N} \frac{E(A_n)}{(1 + i)^n} \tag{12.11}
$$

and

$$
\text{Var}[PW(i)] = \sum_{n=0}^{N} \frac{\text{Var}(A_n)}{(1 + i)^{2n}} \tag{12.12}
$$

where

A_n = cash flow in period n

$E(A_n)$ = expected cash flow in period n

$\text{Var}(A_n)$ = variance of the cash flow in period n

In defining Eq. (12.12), we are also assuming the independence of cash flows, meaning that knowledge of what will happen to one particular period's cash flow will not allow us to predict what will happen to cash flows in other periods.

- **Dependent random variables:** In case we cannot assume a statistical independence among cash flows, we need to consider the degree of dependence among them. The

---

[7] Analytical treatment for the products of random variables including the correlated case is given by Chan S. Park and Gunter Sharp-Bette, *Advanced Engineering Economics*, New York, John Wiley, 1990 (Chapter 10).

expected-value calculation will not be affected by this dependence, but the project variance will be calculated as

$$\text{Var}[\text{PW}(i)] = \sum_{n=0}^{N} \frac{\text{Var}(A_n)}{(1 + i)^{2n}} + 2\sum_{n=0}^{N} \sum_{s=n+1}^{N} \frac{\rho_{ns}\sigma_n\sigma_s}{(1 + i)^{n+s}} \tag{12.13}$$

where ρ_{ns} = correlation coefficient (degree of dependence) between A_n and A_s. The value of ρ_{ns} can vary within the range from -1 to $+1$, with $\rho_{ns} = -1$ indicating perfect negative correlation and $\rho_{ns} = +1$ indicating perfect positive correlation. The result $\rho_{ns} = 0$ implies that no correlation exists between A_n and A_s. If $\rho_{ns} > 0$, we can say that A_n and A_s are positively correlated, meaning that if the actual realization of cash flow for A_n is higher than its expected value, it is likely that you also will observe a higher cash flow than its expected value for A_s. If $\rho_{ns} < 0$, the opposite relation exists.

Present Worth Distribution

One overriding question is what NPW distribution looks like if the periodic cash flow (A_n) takes some form of random variable. The normal distribution shown in Figure 12.9 has the very important property that the distribution of a sum of a large number of *independent* variables is approximately normal.[8] This is the central limit theorem. Since PW(i) is the sum of discounted random cash flows (A_n), it follows that (under certain conditions) PW(i)

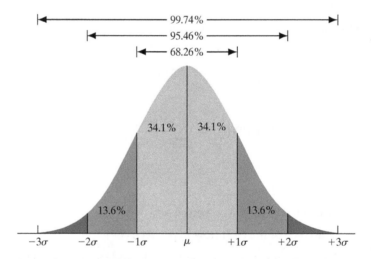

- The area under the normal curve equals 1.0 or 100%.
- Of the area under the curve, 99.74% is within $\pm 3\sigma$ of the mean (μ), indicating that the probability is 99.74% that the actual outcome will be within the range $\mu - 3\sigma$ to $\mu + 3\sigma$.
- The larger the value of σ, the greater the probability that the actual outcome will vary widely from the expected or most-likely estimate.
- Tables exist for finding the probability of other ranges as shown in Appendix C.

Figure 12.9 Probability ranges for a normal distribution.

[8] A number of CLTs for dependent random variables have been developed, and for some special situations, a normal approximation may be also valid.

is also closely approximated by a normal distribution with the mean and variance as calculated in Eqs. (12.11) and (12.12). A normal distribution is a bell-shaped probability distribution with many practical applications found in the real world. If a probability distribution is normal, the actual cash flow will be within ± 1 standard deviation (σ) of the expected cash flow 68.26% of the time. Figure 12.9 illustrates this point, and it also shows the situation for $\pm 2\sigma$ and $\pm 3\sigma$. Example 12.7 illustrates the mechanics involved in calculating the mean and variance of a project's net present worth.

EXAMPLE 12.7 Aggregation of Risk over Time

Consider the following financial data for an investment project, where the only random component is the operating expense in each period:

- Investment required = $10,000
- Project life = 3 years
- Expected salvage value = $0
- Annual operating revenue = $20,000
- Annual operating expenses are random variables with the following means and variances:

n	X_1	X_2	X_3
Mean	$9,000	$13,000	$15,000
Variance	250,000	490,000	1,000,000

- Depreciation method is alternative MACRS (3 years) which is basically a straight-line method with a half-year convention.

n	1	2	3
D_n	$1,667	$3,333	$1,667

- Tax rate = 40%
- Discount rate (or MARR) = 12%

Determine the expected net present worth and the variance of the NPW assuming the following two situations: (a) X_i are independent random variables and (b) X_i are dependent random variables.

SOLUTION

Given: Cash flow data and input probability distributions.
Find: NPW distribution.

Step 1. Calculate the net proceeds from disposal of the equipment at the end of the project life:

Salvage value = 0

$$\text{Book value} = 10,000 - \sum_{n=1}^{3} D_n = \$3,333$$

$$\text{Taxable gain or loss} = 0 - \$3,333 = (\$3,333)$$

$$\text{Loss credit} = 0.4(\$3,333) = \$1,333$$

$$\text{Net proceeds from sale} = \text{Salvage value} + \text{Loss credit} = \$1,333$$

Step 2. Construct a generalized cash flow table (Table 12.14) as a function of the random variable X_n.

The means and variances of random variables (X_1, X_2, & X_3) are given as

$$E[X_1] = 9,000 \qquad E[X_2] = 13,000 \qquad E[X_3] = 15,000$$

$$\text{Var}[X_1] = 250,000 \quad \text{Var}[X_2] = 490,000 \quad \text{Var}[X_3] = 1,000,000$$

TABLE 12.14 Project's After-Tax Cash Flows

	Cash Flow			
Description	**0**	**1**	**2**	**3**
Investment	−10,000			
Revenue × (0.6)		12,000	12,000	12,000
−O&M × (0.6)		$-0.6X_1$	$-0.6X_2$	$-0.6X_3$
D_n × (0.4)		667	1,334	667
Net proceeds from sale				1,333
Net cash flow	**−10,000**	**12,667**	**13,334**	**14,000**
		$-\mathbf{0.6X_1}$	$-\mathbf{0.6X_2}$	$-\mathbf{0.6X_3}$

The net present worth is then

$$PW(12\%) = \left[-10,000 + \frac{12,667}{1.12} + \frac{13,334}{1.12^2} + \frac{14,000}{1.12^3} \right]$$

$$-0.6\left[\frac{X_1}{1.12} + \frac{X_2}{1.12^2} + \frac{X_3}{1.12^3} \right]$$

$$= 21,905 - 0.6\left[\frac{X_1}{1.12} + \frac{X_2}{1.12^2} + \frac{X_3}{1.12^3} \right]$$

Case 1. Independent Cash Flows: Using Eqs. (12.11) and (12.12), we obtain

$$E[PW(12\%)] = 21{,}905 - 0.6\left[\frac{E[X_1]}{1.12} + \frac{E[X_2]}{1.12^2} + \frac{E[X_3]}{1.12^3}\right]$$

$$= 21{,}905 - 17{,}446$$

$$= \$4{,}459$$

$$Var[PW(12\%)] = \left(\frac{-0.6}{1.12}\right)^2 Var[X_1] + \left(\frac{-0.6}{1.12^2}\right)^2 Var[X_2]$$

$$+ \left(\frac{-0.6}{1.12^3}\right)^2 Var[X_3]$$

$$= 71{,}747.45 + 112{,}103.39 + 182{,}387.20$$

$$= 366{,}240$$

$$\sigma[PW(12\%)] = \$605.18$$

Case 2. Dependent Cash Flows: If the random variables X_1, X_2 and X_3 are partially correlated with the correlation coefficients $\rho_{12} = 0.3$, $\rho_{13} = 0.5$, and $\rho_{23} = 0.4$, respectively (see Figure 12.10), then the expected value and the variance of the NPW can be calculated with Eq. (12.13):

$$E[PW(12\%)] = 21{,}905 - 0.6\left[\frac{E[X_1]}{1.12} + \frac{E[X_2]}{1.12^2} + \frac{E[X_3]}{1.12^3}\right]$$

$$= 21{,}905 - 17{,}446$$

$$= \$4{,}459$$

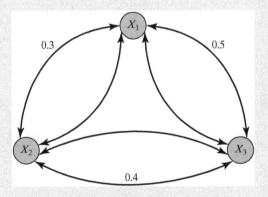

Figure 12.10 A correlation diagram.

$$\text{Var}[\text{PW}(12\%)] = (\text{Original variance}) + (\text{Covariance terms})$$

$$= 366{,}240$$

$$+2\left(\frac{-0.6}{1.12}\right)\left(\frac{-0.6}{1.12^2}\right)\rho_{12}\sigma_1\sigma_2$$

$$+2\left(\frac{-0.6}{1.12}\right)\left(\frac{-0.6}{1.12^3}\right)\rho_{13}\sigma_1\sigma_3$$

$$+2\left(\frac{-0.6}{1.12^2}\right)\left(\frac{-0.6}{1.12^3}\right)\rho_{23}\sigma_2\sigma_3$$

$$= 366{,}240$$

$$+2\left(\frac{0.6^2}{1.12^3}\right)(0.3)(500)(700)$$

$$+2\left(\frac{0.6^2}{1.12^4}\right)(0.5)(500)(1{,}000)$$

$$+2\left(\frac{0.6^2}{1.12^5}\right)(0.4)(700)(1{,}000)$$

$$= 366{,}240 + 53{,}810.59 + 114{,}393.25 + 114{,}393.25$$

$$= 648{,}837$$

$$\sigma[\text{PW}(12\%)] = \$805.50$$

Note that the variance increases significantly when at least some of the random variables are positively correlated.

COMMENTS: How is information such as the preceding used in decision making? Most probability distributions are completely described by six standard deviations—three above, and three below, the mean. As shown in Figure 12.11, the NPW of this project would almost certainly fall between $2,643 and $6,275 for the independent case and between $2,043 and $6,876 for the dependent case. In either situation, the NPW below 3σ from the mean is still positive, so we may say that the project is quite safe. If that figure were negative, it would then be up to the decision maker to determine whether it is worth investing in the project, given its mean and standard deviation.

If we assume a normal probability distribution, we could find the probability that the project's NPW will be greater than $5,000. With case 1,

$$\Pr(X \geq x) = \Pr(\text{PW}(12\%) \geq \$5{,}000)$$

$$= 1 - \Pr(\text{PW}(12\%) \leq \$5{,}000)$$

$$= 1 - \Phi\left(\frac{X - \mu}{\sigma}\right) = 1 - \Phi\left(\frac{5{,}000 - 4{,}459}{605.18}\right)$$

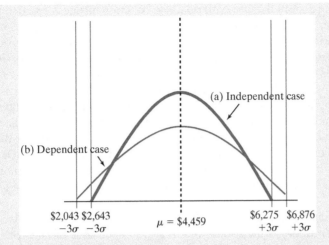

Figure 12.11   NPW distributions with $\pm 3\sigma$: (a) Independent case and (b) Dependent case.

$$= 1 - \Phi(0.89)$$

$$= 1 - 0.81317$$

$$= 18.68\%$$

Similarly, we can find the probability that the project will lose money (or the project's NPW will be negative), which turns out to be 0.

$$\Pr(X \leq x) = \Pr(PW(12\%) \leq 0)$$

$$= \Phi\left(\frac{X - \mu}{\sigma}\right) = \Phi\left(\frac{0 - 4,459}{605.18}\right) = \Phi(-7.3681)$$

$$= 1 - \Phi(7.3681)$$

$$= 0$$

See Appendix C for procedures on how to use the normal probability distribution.

12.4.3 Decision Rules for Comparing Mutually Exclusive Risky Alternatives

Once the expected value has been located from the NPW distribution, it can be used to make an accept–reject decision in much the same way that a single NPW is used when a single possible outcome for an investment project is considered.

Expected-Value Criterion Using Risk-Adjusted Discount Rate

The decision rule is called the **expected-value criterion**, and using it, we may accept a single project if its expected NPW value is positive. In the case of mutually exclusive alternatives, we select the one with the highest expected NPW. The use of the expected NPW has an advantage over the use of a point estimate, such as the likely value, because it includes all possible cash flow events and their probabilities.

In using the expected value critierion, one common business practice is to adjust the discount rate to incorporate the amount of risk in project. This approach is commonly known as the **risk-adjusted discount rate method.** What makes the risk-adjusted discount rate approach appealing in practical application is its simplicity. Most financial officers (CFOs) have at least a rough idea of how an investment's required rate of return should vary with perceived risk. For example, they know from the historical data that, over many years, common stocks hae yielded an average annual return about 7% higher than the return on government bonds (risk-free return). If the present return on government bonds is 8%, it is plausible to expect an investment that is about as risky as common stocks to yield a return of about 15%. Similary, CFOs know that an investment promsing a return of 40% is attractive unless its risk is extraordinarily high. Granted, such reasoning is impresice; nonetheless, it does lend some objectivity to risk assessment.

The justification for the use of the expected-value criterion is also based on the **law of large numbers**, which states that if many repetitions of an experiment are performed, the average outcome will tend toward the expected value. This justification may seem to negate the usefulness of the criterion, since, in project evaluation, we are most often concerned with a single, nonrepeatable "experiment" (i.e., an investment alternative). However, if a firm adopts the expected-value criterion as a standard decision rule for *all* of its investment alternatives, then (over the long term) the law of large numbers predicts that accepted projects tend to meet their expected values. Individual projects may succeed or fail, but the average project tends to meet the firm's standard for economic success.

Mean-and-Variance Criterion

The expected-value criterion is simple and straightforward to use, but it fails to reflect the variability of the outcome of an investment in any explicit manner. Certainly, we can enrich our decision by incorporating information on variability along with the expected value. Since the variance represents the dispersion of the distribution, it is desirable to minimize it. Basically, the decision rule is as follows:

- The smaller the variance, the less the variability (the potential for loss) is associated with the NPW. Therefore, when we compare mutually exclusive projects, we may select the alternative with the smaller variance if its expected value is the same as or larger than those of other alternatives.
- In cases where preferences are not clear cut, the ultimate choice depends on the decision maker's trade-offs (**utility function**)—how far he or she is willing to take the variability to achieve a higher expected value. In other words, the challenge is to decide what level of risk you are willing to accept and then, having decided on your tolerance for risk, to understand the implications of that choice.
- Since the risk is explicitly considered in terms of variance, the interest rate to use in discounting the cash flows should be the cost of capital (not the risk-adjusted discounted rate used in the expected value criterion). (We will discuss this discount rate issue in Chapter 15.)

Example 12.8 illustrates some of the critical issues that need to be considered in evaluating mutually exclusive risky projects.

EXAMPLE 12.8 Comparing Risky Mutually Exclusive Projects

With ever-growing concerns about air pollution, the greenhouse effect, and increased dependence on oil imports in the United States, Green Engineering has developed a prototype conversion unit that allows a motorist to switch from gasoline to compressed natural gas (CNG) or vice versa. Driving a car equipped with Green's conversion kit is not much different from driving a conventional model. A small dial switch on the dashboard controls which fuel is to be used. Four different configurations are available, according to the type of vehicle: compact size, mid-size, large size, and trucks. In the past, Green has built a few prototype vehicles powered by alternative fuels other than gasoline but has been reluctant to go into higher volume production without more evidence of public demand.

As a result, Green Engineering initially would like to target one market segment (one configuration model) in offering the conversion kit. Green Engineering's marketing group has compiled the potential NPW distribution at MARR = 10% (Table 12.15) for each different configuration when marketed independently.

Evaluate the expected return and risk for each model configuration, and recommend which one, if any, should be selected.

SOLUTION

Given: Discrete probability distributions as shown in Table 12.5.
Find: Mean and variance for each model; select the best model.

For model 1, we calculate the mean and variance of the NPW distribution as

$$E[\text{NPW}]_1 = \$1,000(0.35) + \$2,000(0.40)$$
$$+ \$3,000(0.20) + \$4,000(0.05)$$
$$= \$1,950$$

TABLE 12.15 Green Engineering's Potential NPW Distribution

Event (NPW) (unit: thousands)	Probabilities			
	Model 1	Model 2	Model 3	Model 4
$1,000	0.35	0.10	0.40	0.20
1,500	0	0.45	0	0.40
2,000	0.40	0	0.25	0
2,500	0	0.35	0	0.30
3,000	0.20	0	0.20	0
3,500	0	0	0	0
4,000	0.05	0	0.15	0
4,500	0	0.10	0	0.10

$$Var[NPW]_1 = 1,000^2(0.35) + 2,000^2(0.40)$$
$$+ 3,000^2(0.20) + 4,000^2(0.05) - (1,950)^2$$
$$= 747,500$$

In a similar manner, we compute the other values as

Configuration	E[NPW]	Var[NPW]
Model 1	$1,950	747,500
Model 2	2,100	915,000
Model 3	2,100	1,190,000
Model 4	2,000	1,000,000

The results are plotted in Figure 12.12. If we make a choice solely on the basis of the expected value, we may select either model 2 or model 3, because they have the highest expected NPW.

If we consider the variability along with the expected NPW, however, the correct choice is not obvious. We will first eliminate those alternatives that are clearly inferior to others:

- **Model 2 versus Model 3.** We see that models 2 and 3 have the same mean of $2,100, but model 2 has a much lower variance. In other words, model 2 dominates model 3, so we eliminate model 3 from further consideration.

- **Model 2 versus Model 4.** Similarly, we see that model 2 is preferred over model 4, because model 2 has a higher expected value and a smaller variance; model 2 again dominates model 4, so we eliminate model 4. In other words, model 2 dominates both models 3 and 4 as we consider the variability of NPW.

- **Model 1 versus Model 2.** Even though the mean-and-variance rule has enabled us to narrow down our choices to only two models (models 1 and 2), it does not indicate what the ultimate choice should be. In other words, comparing

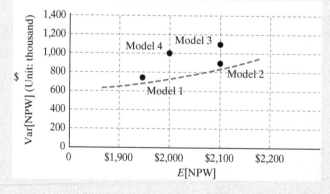

Figure 12.12 Mean–variance chart showing project dominance. Both model 3 and model 4 are dominated by model 2.

models 1 and 2, we see that $E[\text{NPW}]$ increases from \$1,950 to \$2,100 at the price of a higher $\text{Var}[\text{NPW}]$, which increases from 747,500 to 915,000. The choice, then, will depend on the decision maker's trade-offs between the incremental expected return (\$150) and the incremental risk (167,500). We cannot choose between the two simply on the basis of mean and variance, so we must resort to other probabilistic information.[9]

12.5 Risk Simulation

In the previous sections, we examined analytical methods of determining the NPW distributions and computing their means and variances. As we saw in Section 12.4.1, the NPW distribution offers numerous options for graphically presenting to the decision maker probabilistic information, such as the range and likelihoods of occurrence of possible levels of NPW. Whenever we can adequately evaluate the risky investment problem by analytical methods, it is generally preferable to do so. However, many investment situations are such that we cannot solve them easily by analytical methods, especially when many random variables are involved. In these situations, we may develop the NPW distribution through computer simulation.

12.5.1 Computer Simulation

Before we examine the details of risk simulation, let us consider a situation in which we wish to train a new astronaut for a future space mission. Several approaches exist for training this astronaut. One (somewhat unlikely) possibility is to place the trainee in an actual space shuttle and to launch him or her into space. This approach certainly would be expensive; it also would be extremely risky, because any human error made by the trainee could have tragic consequences. As an alternative, we can place the trainee in a flight simulator designed to mimic the behavior of the actual space shuttle in flight. The advantage of this approach is that the astronaut trainee learns all the essential functions of space operation in a simulated space environment. The flight simulator generates test conditions approximating operational conditions, and any human errors made during training cause no harm to the astronaut or to the equipment being used.

The use of computer simulation is not restricted to simulating a physical phenomenon such as the flight simulator. In recent years, techniques for testing the results of some investment decisions before they are actually executed have been developed. As a result, many phases of business investment decisions have been simulated with considerable success. In fact, we can analyze BMC's transmission-housings project by building a simulation model. The general approach is to assign a subjective (or objective) probability distribution to each unknown factor and to combine all these distributions into a probability distribution for the project's profitability as a whole. The essential idea is that if we can simulate the actual state of nature for unknown investment variables on a computer, we may be able to obtain the resulting NPW distribution.

---

[9] As we seek further refinement in our decision under risk, we may consider expected-utility theory or stochastic dominance rules, which are beyond the scope of our text. (See Park, C. S., and Sharp-Bette, G. P., *Advanced Engineering Economics* [New York: John Wiley, 1990], Chapters 10 and 11.)

The unit demand (X) in our BMC's transmission-housing project was one of the random variables in the problem. We can know the exact value for this random variable only after the project is implemented. Is there any way to predict the actual value before we make any decision about the project?

The following logical steps are often suggested for a computer program that simulates investment scenarios:

Step 1. Identify all the variables that affect the measure of investment worth (e.g., NPW after taxes).

Step 2. Identify the relationships among all the variables. The relationships of interest here are expressed by the equations or the series of numerical computations by which we compute the NPW of an investment project. These equations make up the model we are trying to analyze.

Step 3. Classify the variables into two groups: the parameters whose values are known with certainty and the random variables for which exact values cannot be specified at the time of decision making.

Step 4. Define distributions for all the random variables.

Step 5. Perform Monte Carlo sampling (see Section 12.5.3) and describe the resulting NPW distribution.

Step 6. Compute the distribution parameters and prepare graphic displays of the results of the simulation.

Figure 12.13 illustrates the logical steps involved in simulating a risky investment project. The risk simulation process has two important advantages compared with the analytical approach discussed in Section 12.4.

1. The number of variables that can be considered is practically unlimited, and the distributions used to define the possible values of each random variable can be of any type and any shape. The distributions can be based on statistical data if they are available or, more commonly, on subjective judgment.

2. The method lends itself to sensitivity analyses. By defining some factors that have the most significant effect on the resulting NPW values and using different distributions (in terms of either shape or range) for each variable, we can observe the extent to which the NPW distribution is changed.

12.5.2 Model Building

In this section, we shall present some of the procedural details related to the first three steps (model building) outlined in Section 12.5.1. To illustrate the typical procedure involved, we will work with the investment setting for BMC's transmission-housings project described in Example 12.6.

The initial step is to define the measure of investment worth and the factors that affect that measure. For our presentation, we choose the measure of investment worth as an after-tax NPW computed at a given interest rate i. In fact, we are free to choose any measure of worth, such as annual worth or future worth. In the second step, we must divide into two groups all of the variables that we listed in step 1 as affecting the NPW. One group consists of all of the parameters for which values are known. The second group includes

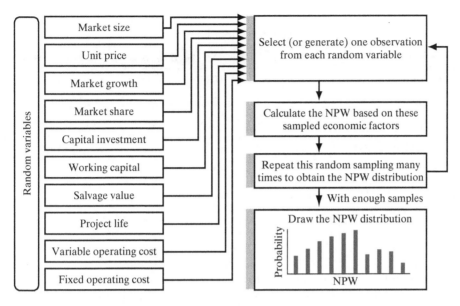

Figure 12.13 Logical steps involved in simulating a risky investment.

all remaining parameters for which we do not know exact values at the time of analysis. The third step is to define the relationships that tie together all of the variables. These relationships may take the form of a single equation or several equations.

EXAMPLE 12.9 Developing a Simulation Model

Consider again BMC's transmission-housings project, as set forth in Example 12.6. Identify the input factors related to the project and develop the simulation model for the NPW distribution.

DISCUSSION: For the BMC project, the variables that affect the NPW are the investment required, unit price, demand, variable production cost, fixed production cost, tax rate, and depreciation expenses, as well as the firm's interest rate. Some of the parameters that might be included in the known group are the investment cost and interest rate (MARR). If we have already purchased the equipment or have received a price quote, then we also know the depreciation amount. In addition, assuming that we are operating in a stable economy, we would probably know the tax rates for computing income taxes due.

The group of parameters with unknown values would usually include all of the variables relating to costs and future operating expense and to future demand and sales prices. These are the random variables for which we must assess the probability distributions.

For simplicity, we classify the input parameters or variables for BMC's transmission-housings project as given in Table 12.16.

TABLE 12.16 **Classification of Input Parameters**

Assumed to Be Known Parameters	Assumed to Be Unknown Parameters
MARR	Unit price
Tax rate	Demand
Depreciation amount	Salvage value
Investment amount	
Project life	
Fixed production cost	
Variable production cost	

Note that, unlike the situation in Example 12.6, here we treat the salvage value as a random variable. With these assumptions, we are now ready to build the NPW equation for the BMC project.

SOLUTION

Given: Input probability distributions.
Find: Mathematical expression for an NPW distribution.

Recall that the basic investment parameters assumed for BMC's five-year project in Example 12.6 were as follows:

- Investment = $125,000
- Marginal tax rate = 0.40
- Annual fixed cost = $10,000
- Variable unit production cost = $15/unit
- MARR(i) = 15%
- Annual depreciation amounts:

n	D_n
1	$17,863
2	30,613
3	21,863
4	15,613
5	5,575

The after-tax annual revenue is expressed in terms of functions of product demand (X) and unit price (Y):

$$R_n = XY(1 - t_m) = 0.6XY$$

The after-tax annual expenses excluding depreciation are also expressed as a function of product demand (X):

$$E_n = \text{(Fixed cost + variable cost)}(1 - t_m)$$
$$= (\$10,000 + 15X)(0.60)$$
$$= \$6,000 + 9X$$

Then the net after-tax cash revenue is

$$V_n = R_n - E_n$$
$$= 0.6XY - 9X - \$6,000$$

The present worth of the net after-tax cash inflow from revenue is

$$\sum_{n=1}^{5} V_n(P/F, 15\%, n) = [0.6X(Y - 15) - \$6,000](P/A, 15\%, 5)$$
$$= 0.6X(Y - 15)(3.3522) - \$20,113$$

The present worth of the total depreciation credits is

$$\sum_{n=1}^{5} D_n t_m(P/F, i, n) = 0.40[\$17,863(P/F, 15\%, 1) + \$30,613(P/F, 15\%, 2)$$
$$+ \$21,863(P/F, 15\%, 3) + 15,613(P/F, 15\%, 4)$$
$$+ \$5,575(P/F, 15\%, 5)$$
$$= \$25,901$$

Since the total depreciation amount is \$91,527, the book value at the end of year 5 is \$33,473 (\$125,000 − \$91,527). Any salvage value greater than this book value is treated as a gain taxed at the rate t_m. In our example, the salvage value is considered to be a random variable. Thus, the amount of taxable gains (losses) also becomes a random variable. Therefore, the net salvage value after tax adjustment is

$$S - (S - \$33,473)t_m = S(1 - t_m) + 33,473t_m$$
$$= 0.6S + \$13,389$$

Then the equivalent present worth of this amount is

$$(0.6S + \$13,389)(P/F, 15\%, 5) = (0.6S + \$13,389)(0.4972)$$

Now the NPW equation can be summarized as

$$PW(15\%) = -\$125,000 + 0.6X(Y - 15)(3.3522) - \$20,113 + \$25,901$$
$$+ (0.6S + \$13,389)(0.4972)$$
$$= -\$112,555 + 2.0113X(Y - 15) + 0.2983S$$

Note that the NPW function is now expressed in terms of the three random variables X, Y, and S.

12.5.3 Monte Carlo Sampling

With some variables, we may base the probability distribution on objective evidence gleaned from the past if the decision maker feels that the same trend will continue to operate in the future. If not, we may use subjective probabilities as discussed in Section 12.3.1. Once we specify a distribution for a random variable, we need to determine how to generate samples from that distribution. **Monte Carlo sampling** is a simulation method in which a random sample of outcomes is generated for a specified probability distribution. In this section, we shall discuss the Monte Carlo sampling procedure for an *independent* random variable.

Random Numbers

The sampling process is the key part of the analysis. It must be done such that the sequence of values sampled will be distributed in the same way as the original distribution. To accomplish this objective, we need a source of independent, identically distributed, uniform random numbers between 0 and 1. Toward that end, we can use a table of random numbers, but most digital computers have programs available to generate "equally likely (uniform)" random decimals between 0 and 1. We will use $U(0,1)$ to denote such a statistically reliable uniform random-number generator, and we will use U_1, U_2, U_3, ... to represent uniform random numbers generated by this routine. (In Microsoft Excel, the RAND function can be used to generate such a sequence of random numbers.)

Sampling Procedure

For any given random numbers, the question is, "How are they used to sample a distribution in a simulation analysis?" The first task is to convert the distribution into its corresponding cumulative frequency distribution. Then the random number generated is set equal to its numerically equivalent percentile and is used as the entry point on the $F(x)$-axis of the cumulative-frequency graph. The sampled value of the random variable is the x value corresponding to this cumulative percentile entry point.

This method of generating random values works, because choosing a random decimal between 0 and 1 is equivalent to choosing a random percentile of the distribution. Then the random value is used to convert the random percentile to a particular value. The method is general and can be used for any cumulative probability distribution—either continuous or discrete.

EXAMPLE 12.10 Monte Carlo Sampling

In Example 12.9, we developed an NPW equation for BMC's transmission-housings project as a function of three random variables—demand (X), unit price (Y), and salvage value (S):

$$PW(15\%) = -\$112,555 + 2.0113X(Y - 15) + 0.2983S$$

- For random variable X, we will assume the same discrete distribution as defined in Table 12.5.
- For random variable Y, we will assume a triangular distribution with $L = \$48$, $H = \$53$, and $M_O = \$50$.
- For random variable S, we will assume a uniform distribution with $L = \$30,000$ and $H = \$50,000$.

With the random variables (X, Y, and S) distributed as just set forth, and assuming that these random variables are *mutually independent* of each other, we need three uniform random numbers to sample one realization from each random variable. We determine the NPW distribution on the basis of 200 iterations.

DISCUSSION: As outlined in Figure 12.14, a simulation analysis consists of a series of repetitive computations of the NPW. To perform the sequence of repeated simulation trials, we generate a sample observation for each random variable in the model and substitute all of the values into the NPW equation. Each trial requires that we use a different random number in the sequence to sample each distribution. Thus, if three random variables affect the NPW, we need three random numbers for each trial. After each trial, the computed NPW is stored in the computer. Each value of the NPW computed in this manner represents one state of nature. The trials are continued until a sufficient number of NPW values is available to define the NPW distribution.

SOLUTION

Given: Mathematical expression for an NPW distribution, demand (X) = discrete random variable, price (Y) = triangular distribution, and salvage value (S) = uniform distribution.

Find: Develop an NPW probability distribution based on 200 Monte Carlo samplings. Suppose the following three uniform random numbers are generated for the first iteration: $U_1 = 0.12135$ for X, $U_2 = 0.82592$ for Y, and $U_3 = 0.86886$ for S.

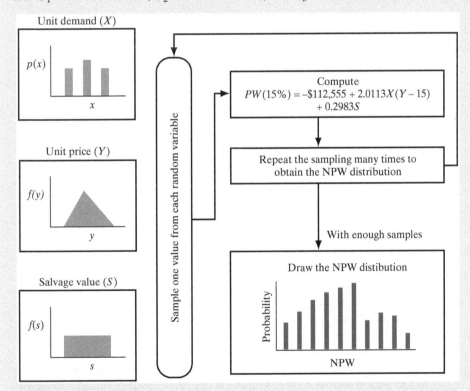

Figure 12.14 A logical sequence of a Monte Carlo simulation to obtain the NPW distribution for BMC's transmission-housings project.

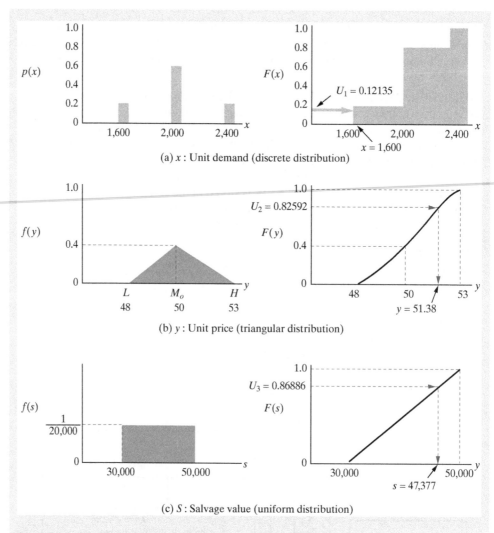

Figure 12.15 Illustration of a sampling scheme for discrete and continuous random variables.

- **Demand (X):** The cumulative distribution for X is already given in Example 12.4. To generate one sample (observation) from this discrete distribution, we first find the cumulative probability function, as depicted in Figure 12.15(a). In a given trial, suppose the computer gives the random number 0.12135. We then enter the vertical axis at the 12.135th percentile (the percentile numerically equivalent to the random number), read across to the cumulative curve, and then read down to the x-axis to find the corresponding value of the random variable X; this value is 1,600, and it is the value of x that we use in the NPW equation. In the next trial, we sample another value of x by obtaining another random number, entering the ordinate at the numerically equivalent percentile and reading the corresponding value of x from the x-axis.

- **Price (Y):** Assuming that the unit-price random variable can be estimated by the three parameters, its probability distribution is shown in Figure 12.15(b). Note that Y takes a continuous value (unlike the assumption of discreteness in Table 12.5). The sampling procedure is again similar to that in the discrete situation. Using the random number $U = 0.82592$, we can approximate $y = \$51.38$ by performing a linear interpolation.
- **Salvage (S):** With the salvage value (S) distributed uniformly between \$30,000 and \$50,000, and a random number $U = 0.86886$, the sample value is $s = \$30,000 + (50,000 - 30,000)0.86886$, or $s = \$47,377$. The sampling scheme is shown in Figure 12.15(c).

Now we can compute the NPW equation with these sample values, yielding

$$PW(15\%) = -\$112,555 + 2.0113(1,600)(\$51.3841 - \$15)$$
$$+ 0.2983(\$47,377)$$
$$= \$18,665$$

This result completes the first iteration of NPW_1 computation.

For the second iteration, we need to generate another set of three uniform random numbers (suppose they are 0.72976, 0.79885, and 0.41879) to generate the respective sample from each distribution and to compute $NPW_2 = \$44,752$. If we repeat this process for 200 iterations, we obtain the NPW values listed in Table 12.17.

By ordering the observed data by increasing NPW value and tabulating the ordered NPW values, we obtain the frequency distribution shown in Table 12.18. Such a tabulation results from dividing the entire range of computed NPWs into a series of subranges (20 in this case) and then counting the number of computed values that fall in each of the 20 intervals. Note that the sum of all the observed frequencies (column 3) is the total number of trials.

TABLE 12.17 Observed NPW Values (\$) for BMC's Simulation Project

18,665	44,752	41,756	75,804	67,508
74,177	47,664	43,106	17,394	35,841
46,400	36,237	12,686	47,947	48,411
50,365	64,287	44,856	47,970	47,988
42,071	43,421	36,991	59,551	9,788
50,147	46,010	11,777	71,202	76,259
20,937	48,250	33,837	37,378	7,188
47,875	14,891	48,362	35,894	75,360
38,726	73,987	45,346	62,508	36,738

(continued)

TABLE 12.17 *(continued)*

| 44,849 | 46,076 | 48,374 | 47,287 | 70,702 |
| 47,168 | 42,348 | 8,213 | 6,689 | 14,415 |

No.	U_1	U_2	U_3	X	Y	S	NPW
1	0.12135	0.82592	0.86886	1,600	$51.38	$47,377	$18,665
2	0.72976	0.79885	0.41879	2,000	$51.26	$38,376	$44,752
⋮	⋮	⋮	⋮	⋮	⋮	⋮	⋮
200	0.57345	0.75553	0.61251	2,000	$51.08	$42,250	$45,204

41,962	45,207	46,487	77,414	67,723
79,089	51,846	76,157	49,960	17,091
15,438	37,214	49,542	45,830	46,679
36,393	15,714	44,899	48,109	76,593
34,927	42,545	45,452	48,089	43,163
78,925	39,179	38,883	80,994	45,202
18,807	74,707	44,787	5,944	6,377
45,811	44,608	42,448	34,584	48,677
10,303	35,597	37,910	62,624	62,060
44,813	36,629	75,111	15,739	43,621
34,288	49,116	76,778	46,067	62,853
9,029	42,485	51,817	44,096	35,198
36,345	75,337	14,034	18,253	41,865
48,089	64,934	81,532	74,407	57,465
32,969	46,529	19,653	10,250	31,269
33,300	47,552	19,139	48,664	46,410
33,457	44,605	35,160	51,496	34,247
46,657	37,249	44,171	34,189	36,673
47,567	62,654	65,062	34,519	45,204

Column 4 simply expresses the frequencies of column 3 as a fraction of the total number of trials. At this point, all we have done is arrange the 200 numerical values of NPW into a table of relative frequencies.

TABLE 12.18 Simulated NPW Frequency Distribution for BMC's Transmission-Housings Project

Cell No.	Cell Interval	Observed Frequency	Relative Frequency	Cumulative Frequency
1	$5{,}944 \leq NPW \leq \$9{,}723$	8	0.04	0.04
2	$9{,}723 < NPW \leq 13{,}502$	6	0.03	0.07
3	$13{,}502 < NPW \leq 17{,}281$	9	0.05	0.12
4	$17{,}281 < NPW \leq 21{,}061$	10	0.05	0.17
5	$21{,}061 < NPW \leq 24{,}840$	0	0.00	0.17
6	$24{,}840 < NPW \leq 28{,}620$	0	0.00	0.17
7	$28{,}620 < NPW \leq 32{,}399$	2	0.01	0.18
8	$32{,}399 < NPW \leq 36{,}179$	20	0.10	0.28
9	$36{,}179 < NPW \leq 39{,}958$	18	0.09	0.37
10	$39{,}958 < NPW \leq 43{,}738$	13	0.07	0.43
11	$39{,}958 < NPW \leq 43{,}738$	39	0.19	0.63
12	$47{,}517 < NPW \leq 51{,}297$	28	0.14	0.77
13	$51{,}297 < NPW \leq 55{,}076$	3	0.02	0.78
14	$55{,}076 < NPW \leq 58{,}855$	1	0.01	0.78
15	$58{,}855 < NPW \leq 62{,}635$	6	0.03	0.81
16	$62{,}635 < NPW \leq 66{,}414$	6	0.03	0.84
17	$66{,}414 < NPW \leq 70{,}194$	4	0.02	0.86
18	$70{,}194 < NPW \leq 73{,}973$	4	0.02	0.88
19	$73{,}973 < NPW \leq 77{,}752$	18	0.09	0.97
20	$77{,}752 < NPW \leq 81{,}532$	5	0.03	1.00

12.5.4 Simulation Output Analysis

After a sufficient number of repetitive simulation trials has been run, the analysis is essentially completed. The only remaining tasks are to tabulate the computed NPW values in order to determine the expected value and to make various graphic displays that will be useful to management.

Interpretation of Simulation Results

Once we obtain an NPW frequency distribution (such as that shown in Table 12.17), we need to make the assumption that the actual relative frequencies of column 4 in the table are representative of the probability of having an NPW in each range. That is, we assume that the relative frequencies we observed in the sampling are representative of the proportions we would have obtained had we examined all the possible combinations.

This sampling is analogous to polling the opinions of voters about a candidate for public office. We could speak to every registered voter if we had the time and resources, but a simpler procedure would be to interview a smaller group of persons selected with an unbiased sampling procedure. If 60% of this scientifically selected sample supports the candidate, it probably would be safe to assume that 60% of all registered voters support the candidate. Conceptually, we do the same thing with simulation. As long as we ensure that a sufficient number of representative trials has been made, we can rely on the simulation results.

Once we have obtained the probability distribution of the NPW, we face the crucial question, "How do we use this distribution in decision making?" Recall that the probability distribution provides information regarding the probability that a random variable will attain some value x. We can use this information, in turn, to define the cumulative distribution, which expresses the probability that the random variable will attain a value smaller than or equal to some x [i.e., $F(x) = P(X \leq x)$]. Thus, if the NPW distribution is known, we can also compute the probability that the NPW of a project will be negative. We use this probabilistic information in judging the profitability of the project.

With the assurance that 200 trials was a sufficient number for the BMC project, we may interpret the relative frequencies in column 4 of Table 12.18 as probabilities. The NPW values range between $5,944 and $81,532, thereby indicating no loss for any situation. The NPW distribution has an expected value of $44,245 and a standard deviation of $18,585.

Creation of Graphic Displays

Using the output data (the relative frequency) in Table 12.18, we can create the distribution in Figure 12.16(a). A picture such as this can give the decision maker a feel for the

Figure 12.16 Simulation result for BMC's transmission-housings project based on 200 iterations.

ranges of possible NPWs, for the relative likelihoods of loss versus gain, for the range of NPWs that are most probable, and so on.

Another useful display is the conversion of the NPW distribution to the equivalent cumulative frequency, as shown in Figure 12.16(b). Usually, a decision maker is concerned with the likelihood of attaining at least a given level of NPW. Therefore, we construct the cumulative distribution by accumulating the areas under the distribution as the NPW decreases. The decision maker can use Figure 12.16(b) to answer many questions—for example, "What is the likelihood of making at least a 15% return on investment (i.e., the likelihood that the NPW will be at least 0)?" In our example, this probability is virtually 100%.

Dependent Random Variables

All of our simulation examples have involved independent random variables. We must recognize that some of the random variables affecting the NPW may be related to one another. If they are, we need to sample from distributions of the random variables in a manner that accounts for any such dependencies. For example, in the BMC project, both the demand and the unit price are not known with certainty. Each of these parameters would be on our list of variables for which we need to describe distributions, but they could be related inversely. When we describe distributions for these two parameters, we have to account for that dependency. This issue can be critical, as the results obtained from a simulation analysis can be misleading if the analysis does not account for dependencies. The sampling techniques for these dependent random variables are found in many textbooks on simulation as well using risk simulation software, such as Crystal Ball or @RISK.

12.5.5 Risk Simulation with Oracle Crystal Ball[10]

One practical way to conduct a risk simulation is to use an Excel-based program such as Crystal Ball or @RISK.[11] All we have to do is to develop a worksheet for a project's cash-flow statement. Then we identify the random variables in the cash flow elements. With Crystal Ball, we have a variety of probability distributions to choose from to describe our beliefs about the random variables of interest. We illustrate how to conduct a risk simulation on Crystal Ball, using the same financial data described in Example 12.9. Crystal Ball uses Monte Carlo simulation to show you all possible outcomes. Running an analysis with Crystal Ball involves three simple steps.

> **Step 1. Create a Cash Flow Statement with Excel.** Crystal Ball is an add-in to Microsoft Excel. As an add-in, Crystal Ball becomes seamlessly integrated with your spreadsheet, adding risk analysis to your existing models. Table 12.19 is an Excel spreadsheet in which the cash flow entries are a function of the input variables. Here, what we are looking for is the NPW of the project when we assigned specific values to the random variables. As in Example 12.9, we treat the unit price (Y), the demand (X), and the salvage value (S) as random variables.

[10] A risk simulation software developed by Oracle, a leading spreadsheet-based application suite for predictive modeling, forecasting, simulation, and optimization.
[11] A risk simulation software developed by Palisade Corporation, Ithaca, NY, 14850.

TABLE 12.19 An Excel Worksheet for Risk Simulation Using Crystal Ball Software

	A	B	C	D	E	F	G
1							
2	Require Crystal Ball software to Properly Display the Worksheet						
3	Crystal Ball Simulation for Example 12.9						
4							
5							
6	Input Data (Base):				Type of Distributions for the Random Variables:		
7							
8	Unit Price ($)	50.33			Category		Distribution
9	Demand	2000					
10	Var. cost ($/unit)	15			Unit price (Y)		Triangular(48,50,53)
11	Fixed cost ($)	10000			Demand (X)		Discrete (1600,2000,2400)
12	Salvage ($)	40000			Salvage (S)		Uniform(30000,50000)
13	Tax rate (%)	40%					
14	MARR (%)	15%					
15					Output (NPW)	$41,508	
16							
17		0	1	2	3	4	5
18	Income Statement						
19	Revenues:						
20	Unit price		$ 50.33	$ 50.33	$ 50.33	$ 50.33	$ 50.33
21	Demand (units)		2000	2000	2000	2000	2000
22	Sales revenue		$ 100,666	$ 100,666	$ 100,666	$ 100,666	$ 100,666
23	Expenses:						
24	Unit variable cost		$ 15	$ 15	$ 15	$ 15	$ 15
25	Variable cost		30,000	30,000	30,000	30,000	30,000
26	Fixed cost		10,000	10,000	10,000	10,000	10,000
27 28	Depreciation		17,863	30,613	21,863	15,613	5,581
29	Taxable income		$ 42,803	$ 30,053	$ 38,803	$ 45,053	$ 55,085
30 31	Income taxes (40%)		17,121	12,021	15,521	18,021	22,034
32	Net income		$ 25,682	$ 18,032	$ 23,282	$ 27,032	$ 33,051
33							
34	Cash Flow Statement						
35	Operating Activities:						
36	Net income		25,682	18,032	23,282	27,032	33,051
37	Depreciation		17,863	30,613	21,863	15,613	5,581
38	Investing Activities:						
39	Investment	(125,000)					
40	Salvage						40,000
41	Gains tax						(2,613)
42							
43	Net Cash Flow	$ (125,000)	$ 43,545	$ 48,645	$ 45,145	$ 42,645	$ 76,019
44							

Step 2. Define Uncertainty. We start by replacing uncertain values in our spreadsheet model with Crystal Ball's probability distribution functions. What we have to do is to simply let the cursor point to the cell and then go to the top of the menu bar and click on Define Assumption. This will bring out the

distribution gallery. Then pick the type of probability distribution and enter the required distribution parameters, as shown here.

Random Variable	Probability Distribution	Distribution Parameters Required by Crystal Ball		Cell Location in Table 12.19
Y	Triangular	Minimum	48.00	B8
		Likeliest	50.00	
		Maximum	53.00	
X	Discrete	Value	Probability	B9
		1,600	0.20	
		2,000	0.60	
		2,400	0.20	
S	Uniform	Minimum	30,000	B12
		Maximum	50,000	

Step 3. Define Forecast. We select our output cell—the "bottom line" cell whose value we are interested in. In our case, this is the net present value for the transmission-housing project located at cell F15. To designate cell F15 as the output cell, we move the cursor to cell F15 and select the Define Forecast command in the tool bar. Then the cell formula in F15 will look like

=NPV(B14,C43:G43)+B43

Step 4. Simulate. Now we are ready to simulate. Click the Start button and watch. Crystal Ball recalculates our spreadsheet model hundreds or thousands of times. We can specify the number of iterations (trials to run) through the Run Preferences command. There is no limit to the number of different scenarios we can look at in our simulations. Each time, Crystal Ball samples random values from the Crystal Ball functions we entered and records the resulting outcome. The result: a look at a whole range of possible outcomes, including the probabilities that they will occur! Almost instantly, you will see what critical situations to seek out—or avoid! With Crystal Ball, we can answer questions like, "What is the probability of NPW exceeding $50,000?" or "What are the chances of losing money on this investment?"

Step 5. Analyzing the Simulation Result Screen. Crystal Ball provides a wide range of graphing options for interpreting and presenting the simulation results. Crystal Ball also gives us a full statistical report on our simulations, as well as access to all the data generated. The Create Reports function includes cumulative graphs, Tornado charts for sensitivity analysis, histograms, and summary statistics. As shown in Figure 12.17, based on 5000 iterations, we find that the NPW ranges between $3,591 and $85,149 with the mean value of $40,995 and the standard deviation of $18,730. These results are comparable to those obtained in Example 12.9.

Figure 12.17 The NPW distribution created by Crystal Ball based on 5000 trial runs.
Source: Oracle Corporation.

12.6 Decision Trees and Sequential Investment Decisions

Most investment problems that we have discussed so far involved only a single decision at the time of investment (i.e., an investment is accepted or rejected) or a single decision that entails a different decision option, such as "make a product in-house" or "farm out," and so on. Once the decision is made, there are no later contingencies or decision options to follow up on. However, certain classes of investment decisions cannot be analyzed in a single step. As an oil driller, for example, you have an opportunity to bid on an offshore lease. Your first decision is whether to bid. But if you decide to bid and eventually win the contract, you will have another decision regarding whether to drill immediately or run more seismic (drilling) tests. If you drill a test well that turns out to be dry, you have another decision whether to drill another well or drop the entire drilling project. In a sequential decision problem such as this, in which the actions taken at one stage depend on actions taken in earlier stages, the evaluation of investment alternatives can be quite complex. Certainly, all of these future options must be considered when one is evaluating the feasibility of bidding at the initial stage. In this section, we describe a general approach for dealing with more complex decisions that is useful both for structuring the decision problems and for finding a solution to them. The approach utilizes a decision tree—a graphic tool for describing the actions available to the decision maker, the events that can occur, and the relationship between the actions and events.

12.6.1 Structuring a Decision-Tree Diagram

To illustrate the basic decision-tree concept, let's consider an investor named Bill Henderson who wants to invest some money in the financial market. He wants to choose between a highly speculative growth stock (d_1) and a very safe U.S. Treasury bond (d_2). Figure 12.18 illustrates this situation with the decision point represented by a square box ($\square$) or decision node. The alternatives are represented as branches emanating from the node. Suppose Bill were to select some particular alternative, say, "invest in the stock." There are three chance events that can happen—each representing a potential return on investment. These events are shown in Figure 12.18 as branches emanating from a circular node ($\bigcirc$). Note that the branches represent uncertain events over which Bill has no control. However, Bill can assign a probability to each chance event and enter it beside each branch in the decision tree. At the end of each branch is the conditional monetary transaction associated with the selected action and given event.

Relevant Net After-Tax Cash Flow

Once the structure of the decision tree is determined, the next step is to find the relevant cash flow (monetary value) associated with each of the alternatives and the possible chance outcomes. As we have emphasized throughout this text, the decision has to be made on an after-tax basis. Therefore, the relevant monetary value should be on that basis. Since the costs and revenues occur at different points in time over the study period, we also need to convert the various amounts on the tree's branches to their equivalent lump-sum amounts, say, net present value. The conditional net present value thus represents the profit associated with the decisions and events along the path from the first part of the tree to the end.

Rollback Procedure

To analyze a decision tree, we begin at the end of the tree and work backwards—the **rollback** procedure. In other words, starting at the tips of the decision tree's branches and working toward the initial node, we use the following two rules.

1. For each chance node, we calculate the expected monetary value (EMV). This is done by multiplying probabilities by conditional profits associated with branches emanating from that node and summing these conditional profits. We then place the EMV in the node to indicate that it is the expected value calculated over all branches emanating from that node.

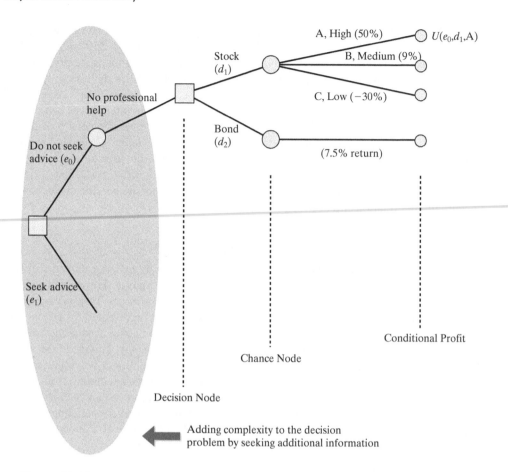

Figure 12.18 Structure of a typical decision tree (Bill's investment problem).

2. For each decision node, we select the one with the highest EMV (or minimum cost). Then those decision alternatives which are not selected are eliminated from further consideration. On the decision-tree diagram, we draw a mark across the non-optimal decision branches, indicating that they are not to be followed.

Example 12.11 illustrates how Bill could transform his investment problem into a tree format using numerical values.

EXAMPLE 12.11 Bill's Investment Problem in a Decision-Tree Format

Suppose Bill has $50,000 to invest in the financial market for one year. His choices have been narrowed to two options.

- **Option 1:** Buy 1,000 shares of a technology stock at $50 per share that will be held for one year. Since this is a new initial public offering (IPO), there is not much research information available on the stock; hence, there will be a brokerage fee of $100 for this size of transaction (for either buying or selling stocks). For simplicity, assume that the stock is expected to provide a return at any one of three different levels: a high level (A) with a 50% return ($25,000), a medium level (B) with a 9% return ($4,500), or a low level (C) with a 30% loss (−$15,000). Assume also that the probabilities of these occurrences are assessed at 0.25, 0.40, and 0.35, respectively. No stock dividend is anticipated for such a growth-oriented company.

- **Option 2:** Purchase a $50,000 U.S. Treasury bond, which pays interest at an effective annual rate of 7.5% ($3,750). The interest earned from the Treasury bond is nontaxable income. However, there is a $150 transaction fee for either buying or selling the bond.

Bill's dilemma is which alternative to choose to maximize his financial gain. At this point, Bill is not concerned about seeking some professional advice on the stock before making a decision. We will assume that any long-term capital gains will be taxed at 20%. Bill's minimum acceptable rate of return is known to be 5% after taxes. Determine the payoff amount at the tip of each branch.

SOLUTION

Given: $50,000 to invest in either stock or bond.

Find: Which investment option is better?

Figure 12.19(a) shows the costs and revenues on the branches transformed to their present equivalents.

Option 1:

(a) With a 50% return (a real winner) over a one-year holding period,

- The net cash flow associated with this event at period 0 will include the amount of investment and the brokerage fee. Thus, we have

$$\text{Period 0: } (-\$50,000 - \$100) = -\$50,100$$

- When you sell stock, you need to pay another brokerage fee. However, any investment expenses (such as brokerage commissions) can be included in determining the cost basis for investment. The taxable capital gains will be calculated as ($75,000 − $50,000 − $100 − $100) = $24,800. Therefore, the net cash flow at period 1 would be

$$\text{Period 1: } (+\$75,000 - \$100) - 0.20(\$24,800) = \$69,940$$

- Then the conditional net present value of this stock transaction is

$$PW(5\%) = -\$50,100 + \$69,940(P/F, 5\%, 1) = \$16,510$$

This amount of $16,510 is entered at the tip of the corresponding branch. This procedure is repeated for each possible branch, and the resulting amounts are shown in Figure 12.19(b).

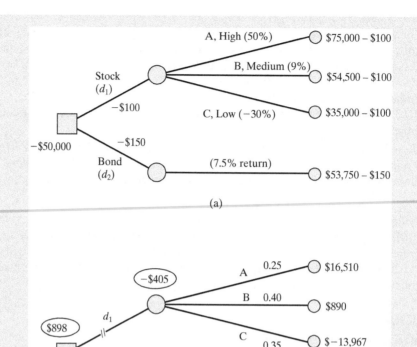

Figure 12.19 Decision tree for Bill's investment problem:
(a) Relevant cash flows (before tax) and (b) net present worth for
each decision path.

(b) With a 9% return, we have
- Period 0: −$50,100
- Period 1: (+$54,500 − $100) − 0.20($54,500 − $50,000 − $200)
  = $53,540
- PW(5%) = −50,100 + 53,540(P/F, 5%, 1) = $890

(c) With a 30% loss,[12] we have
- Period 0: −$50,100
- Period 1: (+$35,000 − $100) − 0.20($35,000 − $50,000 − $200)
  = $37,940
- PW(5%) = −$50,100 + $37,940(P/F, 5%, 1) = −$13,967

---

[12] If Bill has taxable gains from other investment, he can offset the loss from these gains, so the loss will reduce the gain taxes–tax savings.

Option 2: Since the interest income on the U.S. government bond will not be taxed, there will be no capital-gains tax. Considering only the brokerage commission, we find that the relevant cash flows would be

- Period 0: $(-\$50,000 - \$150) = -\$50,150$
- Period 1: $(+\$53,750 - \$150) = \$53,600$
- $PW(5\%) = -\$50,150 + \$53,600(P/F, 5\%, 1) = \$898$

Figure 12.19(b) shows the complete decision tree for Bill's investment problem. Now Bill can calculate the expected monetary value (EMV) at each chance node. The EMV of option 1 represents the sum of the products of the probabilities of high, medium, and low returns and the respective conditional profits (or losses):

$$EMV = \$16,510(0.25) + \$890(0.40) - \$13,967(0.35) = -\$405$$

For option 2, the EMV is simply

$$EMV = \$898$$

In Figure 12.19, the expected monetary values are shown in the event nodes. Bill must choose which action to take, and this would be the one with the highest EMV, namely, option 2, with EMV = $898. This expected value is indicated in the tree by putting $898 in the decision node at the beginning of the tree. Note that the decision tree uses the idea of maximizing expected monetary value developed in the previous section. In addition, the mark ‖ is drawn across the nonoptimal decision branch (option 1), indicating that it is not to be followed. In this simple example, the benefit of using a decision tree may not be evident. However, as the decision problem becomes more complex, the decision tree becomes more useful in organizing the information flow needed to make the decision. This is true in particular if Bill must make a sequence of decisions, rather than a single decision, as we next illustrate.

12.6.2 Worth of Obtaining Additional Information

In this section, we introduce a general method for evaluating whether it is worthwhile to seek more information. Most of the information that we can obtain is imperfect in the sense that it will not tell us exactly which event will occur. Such imperfect information may have some value if it improves the chances of making a correct decision—that is, if it improves the expected monetary value. The problem is whether the reduced uncertainty is valuable enough to offset its cost. The gain is in the improved efficiency of decisions that may become possible with better information.

We use the term *experiment* in a broad sense in what follows. An experiment may be a market survey to predict sales volume for a typical consumer product, a statistical sampling of the quality of a product, or a seismic test to give a well-drilling firm some indications of the presence of oil.

The Value of Perfect Information

Let us take Bill's prior decision to purchase U.S. Treasury bonds as a starting point. How do we determine whether further strategic steps would be profitable? We could do more to obtain additional information about the potential return on an investment in stock, but such steps cost money. Thus, we have to balance the monetary value of reducing uncertainty with the cost of obtaining additional information. In general, we can evaluate the worth of a given experiment only if we can estimate the reliability of the resulting information. In our stock investment problem, an expert's opinion may be helpful in deciding whether to abandon the idea of investing in a stock. This can be of value, however, only if Bill can say beforehand how closely the expert can estimate the performance of the stock.

The best place to start a decision improvement process is with a determination of how much we might improve our incremental profit by removing uncertainty completely. Although we probably couldn't ever really obtain such certainty, the value of perfect information is worth computing as an upper bound to the value of additional information.

We can easily calculate the value of perfect information. First, we merely note the mathematical difference of the incremental profit from an optimal decision based on perfect information and the original decision to purchase Treasury bonds made without foreknowledge of the actual return on the stock. This difference is called **opportunity loss** and must be computed for each potential level of return on the stock. Then we compute the **expected opportunity loss (EOL)** by weighting each potential loss by the assigned probability associated with that event. What turns out in this expected opportunity loss is exactly the **expected value of perfect information (EVPI)**. In other words, if we had perfect information about the performance of the stock, we should have made a correct decision about each situation, resulting in no regrets (i.e., no opportunity losses). Example 12.12 illustrates the concept of opportunity loss and how it is related to the value of perfect information.

EXAMPLE 12.12 Expected Value of Perfect Information

Reluctant to give up a chance to make a larger profit with the stock option, Bill may wonder whether to obtain further information before acting. As with the decision to purchase U.S. Treasury bonds, Bill needs a single figure to represent the expected value of perfect information (EVPI) about investing in the stock. Calculate the EVPI on the basis of the financial data in Example 12.11 and Figure 12.19(b).

SOLUTION

Given: Decision tree as described in Figure 12.19.
Find: Expected value of perfect information.

As regards stock, the decision of whether or not to purchase may hinge on the return on the stock during the next year—the only unknown, subject to a probability distribution. The opportunity loss associated with the decision to purchase bonds is shown in Table 12.20.

- For example, the conditional net present value of $16,510 is the net profit if the return is high. Also, recall that, without Bill's having received any new information, the indicated action (the prior optimal decision) was to select option 2 "Purchase Treasury bonds" with an expected net present value of $898.

- However, if we know that the stock will be a definite winner (yield a high return), then the prior decision to "Purchase Treasury bonds" is inferior to option 2 "Invest

TABLE 12.20 Opportunity Loss Cost Associated with Investing in Bonds

Potential Return	Probability	Decision Option Option 1: Invest in Stock	Option 2: Invest in Bonds	Optimal Choice with Perfect Information	Opportunity Loss Associated with Investing in Bonds
High (A)	0.25	$16,510	$898	Stock	$15,612
Medium (B)	0.40	890	898	Bond	0
Low (C)	0.35	−13,967	898	Bond	0
Expected Value		−$405	$898		**$ 3,903**

in stock" to the extent of $16,510 - $898 = $15,612$. With either a medium or low return, however, there is no opportunity loss, as a decision strategy with and without perfect information is the same. Again, an average weighted by the assigned chances is used, the result is

$$\text{EVPI} = (0.25)(\$15,612) + (0.40)(\$0) + (0.35)(\$0) = \$3,903$$

Or we could calculate the EVPI by the difference between the best profit we can expect from each potential return from stock investment (known as the expected profit with perfect information, EPPI) and the profit we can expect without perfect information. That is

$$\text{EVPI} = \text{EPPI} - \text{EMV}$$
$$= \$16,510(0.25) + \$898(0.40) + \$898(0.35) - \$898$$
$$= \$3,903$$

This EVPI, representing the expected amount that could be gained in incremental profit from having perfect information, places an upper limit on the sum Bill would be willing to pay for additional information.

Updating Conditional Profit (or Loss) after Paying a Fee to the Expert

With a relatively large EVPI in Example 12.12, there is a huge potential that Bill can benefit by obtaining some additional information on the stock. Bill can seek an expert's advice by receiving a detailed research report on the stock under consideration. This report, available at a nominal fee (say, $200) will suggest either of two possible outcomes:

1. The report may come up with a "buy" rating on the stock after concluding that business conditions were favorable.
2. The report may not recommend any action or may take a neutral stance after concluding that business conditions were unfavorable or that the firm's business model on which the decision is based was not well received in the financial marketplace.

Bill's alternatives are either to pay for this service before making any further decision or simply to make his own decision without seeing the report.

If Bill seeks advice before acting, he can base his decision upon the stock report. This problem can be expressed in terms of a decision tree, as shown in Figure 12.20. The upper part of the tree shows the decision process if no advice is sought. This is the same as Figure 12.19 with probabilities of 0.25, 0.40, and 0.35 for high, medium, and low returns, respectively, an expected loss of $405 for investing in the stock, and an expected profit of $898 from purchasing Treasury bonds.

The lower part of the tree, following the branch "Seek advice," displays the results and the subsequent decision possibilities. Using the analyst's research report, Bill can expect two possible events: the expert's recommendation and the actual performance of the stock. After reading either of the two possible report outcomes, Bill must make a decision about whether to invest in the stock. Since Bill has to pay a nominal fee ($200) to get the report, the profit or loss figures at the tips of the branches must also be updated. For the stock investment option, if the actual return proves to be very high (50%), the resulting cash flows would be

$$\text{Period 0: } (-\$50,000 - \$100 - \$200) = -\$50,300$$

$$\text{Period 1: } (+\$75,000 - \$100) - (0.20)(\$25,000 - \$400) = \$69,980$$

$$\text{PW}(5\%) = -\$50,300 + \$69,980(P/F, 5\%, 1) = \$16,348$$

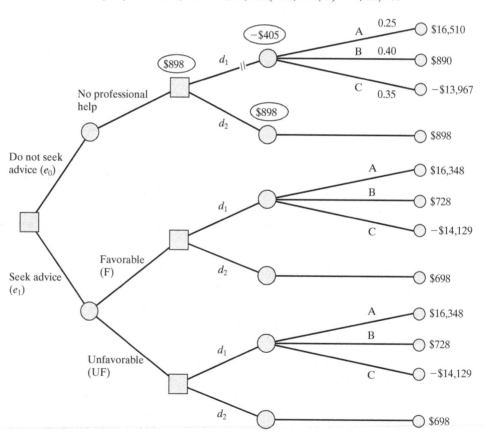

Figure 12.20 Decision tree for Bill's investment problem with an option of getting professional advice.

Since the cost of obtaining the report ($200) will be a part of the total investment expenses (along with the brokerage commissions), the taxable capital gains would be adjusted accordingly. With a medium return, the equivalent net present value would be $729. With a low return, there would be a net loss in the amount of $14,129.

Similarly, Bill can compute the net profit for option 2 as

Period 0: $(-\$50,000 - \$150 - \$200) = -\$50,350$

Period 1: $(+\$53,750 - \$150) = \$53,600$

$$PW(5\%) = -\$50,350 + \$53,600(P/F, 5\%, 1) = \$698$$

Once again, note that gains earned from the U.S. government bond will not be taxed.

12.6.3 Decision Making after Having Imperfect Information

In Figure 12.18, the analyst's recommendation precedes the actual investment decision. This allows Bill to change his prior probabilistic assessments regarding the performance of the stock. In other words, if the analyst's report has any bearing on Bill's reassessments of the stock's performance, then Bill must change his prior probabilistic assessments accordingly. But what types of information should be available and how should the prior probabilistic assessments be updated?

Conditional Probabilities of the Expert's Predictions Given a Potential Return on the Stock

Suppose that Bill knows an expert (stock analyst) who will charge a fee in the amount of $200 to provide Bill with a report on the stock that is either favorable or unfavorable. This research report is not infallible, but it is known to be pretty reliable. From past experience, Bill estimates that

- when the stock under consideration is relatively highly profitable (A), there is an 80% chance that the report will be favorable;
- when the stock return is medium (B), there is a 65% chance that the report will be favorable; and
- when the stock return is relatively unprofitable (C), there is a 20% chance that the report will be favorable. Bill's estimates are summarized in Table 12.21.

TABLE 12.21 Conditional Probabilities of the Expert's Predictions

What the Report Will Say	Given Level of Stock Performance		
	High (A)	Medium (B)	Low (C)
Favorable (F)	0.80	0.65	0.20
Unfavorable (UF)	0.20	0.35	0.80

These conditional probabilities can be expressed in a tree format, as shown in Figure 12.21. For example, if the stock indeed turned out to be a real winner, the probability that the report would have predicted the same (with a favorable recommendation) is 0.80, and the probability that the report would have issued an "unfavorable

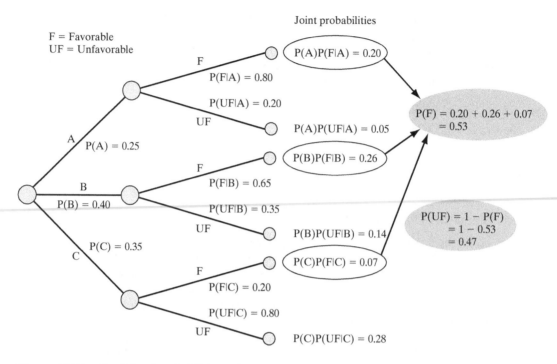

Figure 12.21 Conditional probabilities and joint probabilities expressed in a tree format.

recommendation" is 0.20. Such probabilities would reflect past experience with the analyst's service of this type, modified perhaps by Bill's own judgment. In fact, these probabilities represent the reliability or accuracy of the expert opinion. It is a question of counting the number of times the expert's forecast was "on target" in the past and the number of times the expert "missed the mark" for each of the three actual levels of return on the stock. With this past history of the expert's forecasts, Bill can evaluate the economic worth of the service. Without such a history, no specific value can be attached to seeking the advice.

Joint and Marginal Probabilities

Suppose now, before seeing the research report, Bill has some critical information on (1) all the probabilities for the three levels of stock performance (prior probabilities), and (2) with what probability the report will say that the return on the stock is favorable or unfavorable—given the three levels of the stock performance in the future. A glance at the sequence of events in Figure 12.20, however, shows that these probabilities are not the ones required to find "expected" values of various strategies in the decision path. What is really needed are the probabilities pertaining to what the report will say about the stock—favorable (a recommendation to buy) or unfavorable (no recommendation or a neutral stance)—and the conditional probabilities of each of the three potential returns after the report is seen. In other words, the probabilities of Table 12.21 are not directly useful in Figure 12.20. Rather, we need the *marginal* probabilities of the "favorable" and "unfavorable" recommendations.

To remedy this situation, the probabilities must be put in a different form: We must construct a joint probabilities table. To start with, we have the original (prior) probabilities assessed by Bill before seeing the report: a 0.25 chance that the stock will result in a high return (A), a 0.40 chance of a medium return (B), and a 0.35 chance of a loss (C). From these and the conditional probabilities of Table 12.21, the joint probabilities of Figure 12.21 can be calculated. Thus, the joint probability of both a prediction of a favorable stock market (F) and a high return on the stock (A) is calculated by multiplying the conditional probability of a favorable prediction—given a high return on the stock (which is 0.80 from Table 12.21)—by the probability of a high return (A):

$$P(A, F) = P(F \mid A)P(A) = (0.80)(0.25) = 0.20$$

Similarly,

$$P(A, UF) = P(UF \mid A)P(A) = (0.20)(0.25) = 0.05$$

$$P(B, F) = P(F \mid B)P(B) = (0.65)(0.40) = 0.26$$

$$P(B, UF) = P(UF \mid B)P(B) = (0.35)(0.40) = 0.14$$

and so on.

The marginal probabilities of return in Table 12.22 are obtained by summing the values across the columns. Note that these are precisely the original probabilities of a high, medium, and low return, and they are designated prior probabilities because they were assessed before any information from the report was obtained.

In understanding Table 12.22, it is useful to think of it as representing the results of 100 past situations identical to the one under consideration. The probabilities then represent the frequency with which the various outcomes occurred. For example, in 25 of the 100 cases, the actual return on the stock turned out to be high; and in these 25 high-return cases, the report predicted a favorable condition in 20 instances [that is $P(A, F) = 0.20$] and predicted an unfavorable one in five instances.

TABLE 12.22   Joint as Well as Marginal Probabilities

Potential Level of Return	What Report Will Say Joint Probabilities Favorable (F)	Unfavorable (UF)	Prior Marginal Probabilities of Return
High (A)	0.20	0.05	**0.25**
Medium (B)	0.26	0.14	**0.40**
Low (C)	0.07	0.28	**0.35**
Marginal probabilities	**0.53**	**0.47**	**1.00**

Similarly, the marginal probabilities of prediction in the bottom row of Table 12.22 can be interpreted as the relative frequency with which the report predicted favorable and unfavorable conditions. For example, the survey predicted favorable conditions 53 out of 100 times: 20 of these times when stock returns actually were high, 26 times when returns were medium, and seven times when returns were low. These marginal probabilities of what the report will say are important to our analysis, for they give us the probabilities associated with the information received by Bill before the decision to invest in the stock is made. The probabilities are entered beside the appropriate branches in Figure 12.20.

Determining Revised Probabilities

We still need to calculate the probabilities for the branches that represent "high return," "medium return," and "low return" in the lower part of Figure 12.20. We cannot use the values 0.25, 0.40, and 0.35 for these events, as we did in the upper part of the tree, because those probabilities were calculated prior to seeking the advice. At this point on the decision tree, Bill will have received information from the analyst, and the probabilities should reflect that information. (If Bill's judgment has not changed even after seeing the report, then there will be no changes in the probabilities.) The required probabilities are the conditional probabilities for the various levels of return on the stock—given the expert's opinion. Thus, using the data from Table 12.22, we can compute the probability $P(A|F)$ of seeing a high return (A), given the expert's buy recommendation (F), directly from the definition of conditional probability:

$$P(A|F) = P(A, F)/P(F) = 0.20/0.53 = 0.38$$

The probabilities of a medium and a low return—given a buy recommendation—are, respectively,

$$P(B|F) = P(B, F)/P(F) = 0.26/0.53 = 0.49$$

and

$$P(C|F) = P(C, F)/P(F) = 0.07/0.53 = 0.13$$

These probabilities are called *revised* (or *posterior*) probabilities, since they come after the inclusion of the information received from the report. To understand the meaning of the preceding calculations, think again of Table 12.22 as representing 100 past identical situations. Then, in 53 cases [since $P(F) = 0.53$], 20 actually resulted in a high return. Hence, the posterior probability of a high return is 20/53 = 0.38, as calculated.

The posterior probabilities of other outcomes after seeing an unfavorable recommendation can be calculated similarly:

$$P(A|UF) = P(A, UF)/P(UF) = 0.05/0.47 = 0.11$$

$$P(B|UF) = P(B, UF)/P(UF) = 0.14/0.47 = 0.30$$

$$P(C|UF) = P(C, UF)/P(UF) = 0.28/0.47 = 0.59$$

EXAMPLE 12.13 Decision Making after Getting Some Professional Advice

On the basis of Figure 12.22, (a) determine how much the research report is worth to Bill. Should Bill seek an expert's advice (research report)? (b) What is the expected value of sample information (EVSI)?

SOLUTION

Given: Decision tree as described in Figure 12.22.
Find: Should Bill seek an expert advice?

(a) All of the necessary information is now available, and Figure 12.22 can be analyzed, starting from the right and working backward. The expected values are shown in the circles. For example, follow the branches "Seek Advice (e_1)," "Favorable (Buy recommendation)," and "Invest in the Stock (d_1)." The expected value of $4,733 shown in the circle at the end of these branches is calculated as

$$0.38(\$16,348) + 0.49(\$729) + 0.13(-\$14,129) = \$4,733.$$

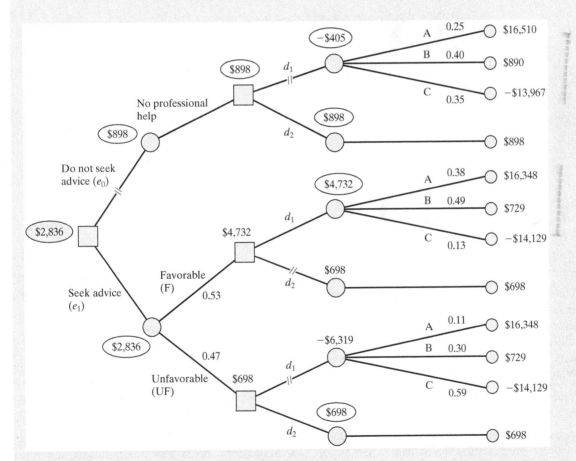

Figure 12.22 Decision tree for Bill's investment problem with an option of having professional advice.

Since Bill can expect a profit of $4,733 from investing in the stock after receiving a buy recommendation, the investment in Treasury bonds ($698) becomes a less profitable alternative. Thus, it can be marked out ($\neq$) of the "Purchase Treasury bonds" branch to indicate that it is not optimal.

There is an expected loss of $6,319 if the report gives an unfavorable recommendation, but Bill goes ahead with buying the stock. In this situation, the "Treasury bonds" option becomes a better choice, and the "Buy stock" branch is marked out. The part of the decision tree relating to seeking the advice is now reduced to two chance events (earning $4,733 and earning $698), and the expected value in the circular node is calculated as

$$0.53(\$4,733) + 0.47(\$698) = \$2,836$$

Thus, if the expert's opinion is taken and Bill acts on the basis of the information received, the expected profit is $2,836. Since this amount is greater than the $898 profit that would be obtained in the absence of the expert's advice, we conclude that it is worth spending $200 to receive some additional information from the expert.

(b) EVSI – What is the worth of the advice?

The expected value of sample information is simply how much expected profit can improve with the advice.

$$\text{EVSI} = (\$2,836 + \$200) - \$898 = \$2,138$$

Note that we need to add the report fee of $200 to determine the expected profit with the sample information. In this case, the professional advice is worth $2,138, and you are only paying $200 to get this advice—a real bargain.

In summary, the optimal decision is as follows: (1) Bill should seek professional advice at the outset; (2) If the expert's report indicates a buy recommendation, go ahead and invest in the stock; (3) If the report says otherwise, invest in the Treasury bonds.

SUMMARY

■ Often, cash flow amounts and other aspects of an investment project analysis are uncertain. Whenever such uncertainty exists, we are faced with the difficulty of **project risk**: The possibility that an investment project will not meet our minimum requirements for acceptability and success.

■ Three of the most basic tools for assessing project risk are as follows:

1. **Sensitivity analysis:** A means of identifying those project variables which, when varied, have the greatest effect on the acceptability of the project.

2. **Break-even analysis:** A means of identifying the value of a particular project variable that causes the project to exactly break even.

3. **Scenario analysis:** A means of comparing a "base-case" or expected project measurement (such as the NPW) with one or more additional scenarios (such as the best case and the worst case) to identify the extreme and most likely project outcomes.

■ Sensitivity, break-even, and scenario analyses are reasonably simple to apply, but they also are somewhat simplistic and imprecise in cases where we must deal with multifaceted project uncertainty. **Probability concepts** allow us to further refine the analysis of project risk by assigning numerical values to the likelihood that project variables will have certain values.

■ The end goal of a probabilistic analysis of project variables is to produce an NPW distribution from which we can extract such useful information as (1) the **expected NPW value**, (2) the extent to which other NPW values vary from or are clustered around the expected value (the **variance**), and (3) the best- and worst-case NPWs.

■ Our real task is not try to find "risk-free" projects; they don't exist in real life. The challenge is to decide what level of risk we are willing to assume and then, having decided on our tolerance for risk, to understand the implications of our choice.

■ **Risk simulation**, in general, is the process of modeling reality to observe and weigh the likelihood of the possible outcomes of a risky undertaking.

■ **Monte Carlo sampling** is a specific type of randomized sampling method in which a random sample of outcomes is generated for a specified probability distribution. Because Monte Carlo sampling and other simulation techniques often rely on generating a significant number of outcomes, they can be more conveniently performed on the computer than manually.

■ The **decision tree** is another technique that can facilitate investment decision making when uncertainty prevails, especially when the problem involves a sequence of decisions. Decision-tree analysis involves the choice of a decision criterion—say, to maximize expected profit. If possible and feasible, an experiment is conducted, and the prior probabilities of the states of nature are revised on the basis of the experimental results. The expected profit associated with each possible decision is computed, and the act with the highest expected profit is chosen as the optimum action.

PROBLEMS

Sensitivity Analysis

12.1 Green Management Company is considering the acquisition of a new eighteen-wheeler.

• The truck's base price is $80,000, and it will cost another $20,000 to modify it for special use by the company.

• This truck falls into the MACRS five-year class. It will be sold after three years for $30,000.

• The truck purchase will have no effect on revenues, but it is expected to save the firm $45,000 per year in before-tax operating costs, mainly in leasing expenses.

• The firm's marginal tax rate (federal plus state) is 40%, and its MARR is 15%.

(a) Is this project acceptable, based on the most likely estimates given in the problem?

(b) If the truck's base price is $95,000, what would be the required savings in leasing so that the project would remain profitable?

12.2 The Wellington Construction Company is considering acquiring a new earthmover. The mover's basic price is $90,000, and it will cost another $18,000 to modify it for special use by the company. This earthmover falls into the MACRS five-year class. It will be sold after four years for $30,000. The purchase of the earthmover will have no effect on revenues, but it is expected to save the firm $35,000 per year in before-tax operating costs, mainly labor. The firm's marginal tax rate (federal plus state) is 40%, and its MARR is 10%.

(a) Is this project acceptable, based on the most likely estimates given?

(b) Suppose that the project will require an increase in net working capital (spare-parts inventory) of $5,000, which will be recovered at the end of year 4. Taking this new requirement into account, would the project still be acceptable?

(c) If the firm's MARR is increased to 18% and with the working capital requirement from (b) not in effect, what would be the required savings in labor so that the project remains profitable?

12.3 The Minnesota Metal Forming Company has just invested $500,000 of fixed capital in a

TABLE P12.4

	Height			
	2 Floors	**3 Floors**	**4 Floors**	**5 Floors**
First cost	$500,000	$750,000	$1,250,000	$2,000,000
Lease revenue	199,100	169,200	149,200	378,150
Net resale value	600,000	900,000	2,000,000	3,000,000

manufacturing process that is estimated to generate an after-tax annual cash flow of $200,000 in each of the next five years. At the end of year 5, no further market for the product and no salvage value for the manufacturing process is expected. If a manufacturing problem delays the start-up of the plant for one year (leaving only four years of process life), what additional after-tax cash flow will be needed to maintain the same internal rate of return as would be experienced if no delay occurred?

12.4 A real-estate developer seeks to determine the most economical height for a new office building, which will be sold after five years. The relevant net annual revenues and salvage values on after-tax basis are as given in Table P12.4.

(a) The developer is uncertain about the interest rate (i) to use, but is certain that it is in the range from 5% to 30%. For each building height, find the range of values of i for which that building height is the most economical.

(b) Suppose that the developer's interest rate is known to be 15%. What would be the cost (in terms of net present value) of a 10% overestimation of the resale value? (In other words, the true value was 10% lower than that of the original estimate.)

12.5 A special-purpose milling machine was purchased four years ago for $20,000. It was estimated at that time that this machine would have a life of 10 years, a salvage value of $1,000, and a cost of removal of $1,500. These estimates are still good. This machine has annual operating costs of $2,000, and its current book value is $13,000 (based on "alternative MACRS," a straight-line depreciation with a half-year convention with zero salvage value). If the machine is retained for its entire

10-year life, the remaining annual depreciation schedule would be $2,000 for years 5 through 10. A new machine that is more efficient will reduce operating costs to $1,000, but it will require an investment of $12,000. The life of the new machine is estimated to be six years with a salvage value of $2,000. The new machine would fall into the five-year MACRS property class. An offer of $6,000 for the old machine has been made, and the purchaser would pay for removal of the machine. The firm's marginal tax rate is 40%, and its required minimum rate of return is 10%.

(a) What incremental cash flows will occur at the end of years 0 through 6 as a result of replacing the old machine? Should the old machine be replaced now?

(b) Suppose that the annual operating costs for the old milling machine would increase at an annual rate of 5% over the remaining service life of the machine. With this change in future operating costs for the old machine, would the answer in part (a) change?

(c) What is the minimum salvage value for the old machine at the year 0 so that both alternatives are economically equivalent?

12.6 A company is currently paying a sales representative $0.60 per mile to drive her car for company business. The company is considering supplying the representative with a car, which would involve the following: A car costs $25,000, has a service life of 5 years, and a market value of $5,000 at the end of that time. Monthly storage costs for the car are $200, and the cost of fuel, tires, and maintenance is 20 cents per mile. The car will be depreciated by MACRS, using a recovery period of 5 years (20%, 32%, 19.20%, 11.52%, and 11.52%). The firm's marginal tax rate is 40%.

(a) Suppose that the required travel by the sales person would be 30,000 miles per year. Which option would be better at $i = 15\%$?

(b) What annual mileage must a salesman travel by car for the cost of the two methods of providing transportation to be equal if the interest rate is 15%?

12.7 A small manufacturing firm is considering purchasing a new boring machine to modernize one of its production lines. Two types of boring machines are available on the market. The lives of machine A and machine B are eight years and 10 years, respectively. The machines have the following receipts and disbursements:

Item	Machine A	Machine B
First cost	$6,000	$8,500
Service life	8 years	10 years
Salvage value	$500	$1,000
Annual O&M costs	$700	$520
Depreciation (MACRS)	7 years	7 years

Use a MARR (after tax) of 10% and a marginal tax rate of 30%, and answer the following questions.

(a) Which machine would be most economical to purchase under an infinite planning horizon? Explain any assumption that you need to make about future alternatives.

(b) Determine the break-even annual O&M costs for machine A so that the annual equivalent cost of machine A is the same as that of machine B.

(c) Suppose that the required service life of the machine is only five years. The salvage values at the end of the required service period are estimated to be $3,000 for machine A and $3,500 for machine B. Which machine is more economical?

12.8 The management of Langdale Mill is considering replacing a number of old looms in the mill's weave room. The looms to be replaced are two 86-inch President looms, sixteen 54-inch President looms, and twenty-two 72-inch Draper X-P2 looms. The company may either replace the old looms with new ones of the same kind or buy 21 new shutterless Pignone looms (Table P12.8). The first alternative requires purchasing 40 new President and Draper looms and scrapping the old looms. The second alternative involves scrapping the 40 old looms, relocating 12 Picanol looms, and constructing a concrete floor, plus purchasing the 21 Pignone looms and various related equipment.

The firm's MARR is 18%, set by corporate executives who feel that various investment opportunities available for the mills will guarantee a rate of return on investment of at least 18%. The mill's marginal tax rate is 40%.

(a) Perform a sensitivity analysis on the project's data, varying the operating revenue, labor cost, annual O&M cost, and MARR. Assume that each of these variables can deviate from its base-case expected value by $\pm 10\%$, by $\pm 20\%$, and by $\pm 30\%$.

(b) From the results of part (a), prepare sensitivity diagrams and interpret the results.

12.9 Mike Lazenby, an industrial engineer at Energy Conservation Service, has found that the anticipated profitability of a newly developed water-heater temperature-control device can be measured by present worth with the formula

$$\text{NPW} = 4.028V(2X - \$11) - 77,860$$

where V is the number of units produced and sold and X is the sales price per unit. Mike also has found that the value of the parameter V could occur anywhere over the range from 1,000 to 6,000 units and that of the parameter X anywhere between $20 and $45 per unit. Develop a sensitivity graph as a function of the number of units produced and the sales price per unit.

12.10 A local U.S. Postal Service office is considering purchasing a 4,000-pound forklift truck, which will be used primarily for processing incoming as well as outgoing postal packages. Forklift trucks traditionally have been fueled by either gasoline, liquid propane gas (LPG), or diesel fuel. Battery-powered electric forklifts, however, are increasingly popular in many industrial sectors due to the economic and environmental benefits that accrue from their use. Therefore, the postal service is interested in comparing forklifts that use the four different types of fuel

TABLE P12.8

Description	Alternative 1	Alternative 2
Machinery/related equipment	$2,119,170	$1,071,240
Removal cost of old looms/site preparation	26,866	49,002
Salvage value of old looms	62,000	62,000
Annual sales increase with new looms	7,915,748	7,455,084
Annual labor	261,040	422,080
Annual O&M	1,092,000	1,560,000
Depreciation (MACRS)	7 years	7 years
Project life	8 years	8 years
Salvage value	169,000	54,000

(Table P12.10). The purchase costs as well as annual operating and maintenance costs are provided by a local utility company and the Lead Industries Association. Annual fuel and maintenance costs are measured in terms of number of shifts per year, where one shift is equivalent to eight hours of operation.

The postal service is unsure of the number of shifts per year, but it expects it should be somewhere between 200 and 260 shifts. Since the U.S. Postal Service does not pay income taxes, no depreciation or tax information is required. The U.S. government uses 10% as an interest rate for any project evaluation of this nature. Develop a sensitivity graph that shows how the choice of alternatives changes as a function of number of shifts per year.

Break-Even Analysis

12.11 Susan Campbell is thinking about going into the motel business near Disney World in Orlando, Florida. The cost to build a motel is $2,200,000. The lot costs $600,000. Furniture and furnishings cost $400,000 and should be recovered in eight years (seven-year MACRS property), while the motel building should be recovered in 39 years (39-year MACRS real property placed in service on January 1st). The land will appreciate at an annual rate of 5%

TABLE P12.10

	Electrical Power	LPG	Gasoline	Diesel Fuel
Life expectancy	7 years	7 years	7 years	7 years
Initial cost	$29,739	$21,200	$20,107	$22,263
Salvage value	$3,000	$2,000	$2,000	$2,200
Maximum shifts per year	260	260	260	260
Fuel consumption/shift	31.25 kWh	11 gal	11.1 gal	7.2 gal
Fuel cost/unit	$0.10/kWh	$3.50/gal	$2.24/gal	$2.45/gal
Fuel cost per shift	$3.125	$38.50	$24.86	$17.64
Annual maintenance (fixed cost)	$500	$1,000	$1,000	$1,000
Variable cost/shift	$4.5	$7	$7	$7

over the project period, but the building will have a zero salvage value after 25 years. When the motel is full (100% capacity), it takes in (receipts) $4,000 per day, 365 days per year. Exclusive of depreciation, the motel has fixed operating expenses of $230,000 per year. The variable operating expenses are $170,000 at 100% capacity, and these vary directly with percent capacity down to zero at 0% capacity. If the interest is 10% compounded annually, at what percent capacity over 25 years must the motel operate in order for Susan to break even? (Assume that Susan's tax rate is 31%.)

12.12 A plant engineer wishes to know which of two types of lightbulbs should be used to light a warehouse. The bulbs that are currently used cost $45.90 per bulb and last 14,600 hours before burning out. The new bulb (at $60 per bulb) provides the same amount of light and consumes the same amount of energy, but it lasts twice as long. The labor cost to change a bulb is $16.00. The lights are on 19 hours a day, 365 days a year. If the firm's MARR is 15%, what is the maximum price (per bulb) the engineer should be willing to pay to switch to the new bulb? (Assume that the firm's marginal tax rate is 40%.)

12.13 Robert Cooper is considering purchasing a piece of business rental property containing stores and offices at a cost of $250,000. Cooper estimates that annual disbursements (other than income taxes) will be about $12,000. The property is expected to appreciate at the annual rate of 5%. Cooper expects to retain the property for 20 years once it is acquired. Then it will be depreciated as a 39-year real-property class (MACRS), assuming that the property will be placed in service on January 1st. Cooper's marginal tax rate is 30% and his MARR is 15%. What would be the minimum annual total of rental receipts that would make the investment break even?

12.14 Two different methods of solving a production problem are under consideration. Both methods are expected to be obsolete in six years. Method A would cost $80,000 initially and have annual operating costs of $22,000 per year. Method B would cost $52,000 and costs $17,000 per year to operate. The salvage value realized would be $20,000 with Method A and $15,000 with Method B. Investments in both methods are subject to a five-year MACRS property class. The firm's marginal income tax rate

is 40%. The firm's MARR is 20%. What would be the required additional annual revenue for Method A such that an engineer would be indifferent to choosing one method over the other?

12.15 The Rocky Mountain Publishing Company is considering introducing a new morning newspaper in Denver. Its direct competitor charges $0.25 at retail with $0.05 going to the retailer. For the level of news coverage the company desires, it determines the fixed cost of editors, reporters, rent, pressroom expenses, and wire-service charges to be $300,000 per month. The variable cost of ink and paper is $0.10 per copy, but advertising revenues of $0.05 per paper will be generated. To print the morning paper, the publisher has to purchase a new printing press, which will cost $600,000. The press machine will be depreciated according to a seven-year MACRS class. The press machine will be used for 10 years, at which time its salvage value would be about $100,000. Assume 300 issues per year, a 40% tax rate, and a 13% MARR. How many copies per day must be sold to break even at a retail selling price of $0.25 per paper?

Probabilistic Analysis

12.16 A corporation is trying to decide whether to buy the patent for a product designed by another company. The decision to buy will mean an investment of $8 million, and the demand for the product is not known. If demand is light, the company expects a return of $1.3 million each year for three years. If demand is moderate, the return will be $2.5 million each year for four years, and high demand means a return of $4 million each year for four years. It is estimated the probability of a high demand is 0.4, and the probability of a light demand is 0.2. The firm's (risk-free) interest rate is 12%. Calculate the expected present worth of the patent. On this basis, should the company make the investment? (All figures represent after-tax values.)

12.17 Consider the following investment cash flows over a two-year life:

Period	Cash Flow	Mean (μ)	Variance (σ^2)
0	A_0	−$800	0
1	A_1	400	60^2
2	A_2	700	65^2

(a) Compute the mean and variance of PW of this project at $i = 12\%$ if

  1. A_1 and A_2 are mutually independent.

  2. A_1 and A_2 are partially correlated with $\rho_{12} = 0.25$.

(b) In part (a)(1), if random variables (A_1 and A_2) are normally distributed with the mean and variance as specified in the table, compute the probability that the PW will be negative. (See Appendix C.)

12.18 Consider the following investment cash flows over a 2-year life:

Period	Cash Flow	Mean (μ)	Variance (σ^2)
0	A_0	$-\$800$	90^2
1	A_1	500	90^2
2	A_2	600	90^2

(a) Compute the mean and variance of PW of this project at $i = 12\%$ if

  1. A_0, A_1 and A_2 are mutually independent.

  2. A_1 and A_2 are partially correlated with $\rho_{12} = 0.4$, but $\rho_{01} = \rho_{02} = 0.25$

  3. A_0, A_1, and A_2 are perfectly positively correlated.

(b) In part (a)(1), if random variables (A_0, A_1 and A_2) are normally distributed with the mean and variance specified in the table, compute the probability that the PW will be greater than $120.

12.19 A financial investor has an investment portfolio worth $325,000. A bond in the portfolio will mature next month and provide him with $22,500 to reinvest. The choices have been narrowed down to the following two options.

• **Option 1:** Reinvest in a foreign bond that will mature in one year. This will entail a brokerage fee of $162. For simplicity, assume that the bond

will provide interest of $2,625, $2,225, or $1,825 over the one-year period and that the probabilities of these occurrences are assessed to be 0.20, 0.45, and 0.35, respectively.

• **Option 2:** Reinvest in a $22,500 certificate with a savings-and-loan association. Assume that this certificate has an effective annual rate of 8.4%.

(a) Which form of reinvestment should the investor choose in order to maximize his expected financial gain?

(b) If the investor can obtain professional investment advice from Salomon Brothers,Inc.,what would be the maximum amount the investor should pay for this service?

12.20 The Kellogg Company is considering the following investment project and has estimated all cost and revenues in constant dollars. The project requires the purchase of a $9,000 asset, which will be used for only two years (the project life).

• The salvage value of this asset at the end of two years is expected to be $4,000.

• The project requires an investment of $2,000 in working capital, and this amount will be fully recovered at the end of the project year.

• The annual revenue, as well as general inflation, are discrete random variables but can be described by the probability distributions in Table P12.20 (both random variables are statistically independent).

• The investment will be classified as a three-year MACRS property (tax life).

• It is assumed that the revenues, salvage value, and working capital are responsive to the general inflation rate.

• The revenue and inflation rate dictated during the first year will prevail over the remainder of the project period.

TABLE P12.20

Annual Revenue (X)	Probability	General Inflation Rate (Y)	Probability
$10,000	0.30	3%	0.25
20,000	0.40	5%	0.50
30,000	0.30	7%	0.25

TABLE P12.21 Comparison of Mutually Exclusive Projects

First Cost	Project 1 ($1,000)		Project 2 ($800)	
	Probability	Revenue	Probability	Revenue
	0.2	$2,000	0.3	$1,000
Net revenue, given in PW	0.6	3,000	0.4	2,500
	0.2	3,500	0.3	4,500

- The marginal income tax rate for the firm is 40%. The firm's inflation-free interest rate (i') is 10%.

(a) Determine the NPW as a function of X and Y.

(b) In part (a), compute the expected NPW of this investment.

(c) In part (a), compute the variance of the NPW of the investment.

Comparing Risky Projects

12.21 A manufacturing firm is considering two mutually exclusive projects, both of which have an economic service life of one year with no salvage value. The initial cost and the net year-end revenue for each project are given in Table P12.21.

Assume that both projects are statistically independent of each other.

(a) If you are an expected-value maximizer, which project would you select?

(b) If you also consider the variance of the project, which project would you select?

12.22 A business executive is trying to decide whether to undertake one of two contracts or neither one. He has simplified the situation somewhat and feels that it is sufficient to imagine that the contracts provide alternatives as given in Table P12.22.

TABLE P12.22

Contract A		Contract B	
NPW	Probability	NPW	Probability
$100,000	0.2	$40,000	0.3
50,000	0.4	10,000	0.4
0	0.4	−10,000	0.3

(a) Should the executive undertake either one of the contracts? If so, which one? What would he do if he made decisions with an eye toward maximizing his expected NPW?

(b) What would be the probability that contract A would result in a larger profit than that of contract B?

12.23 Two alternative machines are being considered for a cost-reduction project.

- Machine A has a first cost of $60,000 and a salvage value (after tax) of $22,000 at the end of six years of service life. The probabilities of annual after-tax operating costs of this machine are estimated as follows:

Annual O&M Costs	Probability
$5,000	0.20
8,000	0.30
10,000	0.30
12,000	0.20

- Machine B has an initial cost of $35,000, and its estimated salvage value (after tax) at the end of four years of service is negligible. The annual after-tax operating costs are estimated to be as follows:

Annual O&M Costs	Probability
$8,000	0.10
10,000	0.30
12,000	0.40
14,000	0.20

The MARR on this project is 10%. The required service period of these machines is estimated to be 12 years, and no technological advance in either machine is expected.

(a) Assuming independence, calculate the mean and variance for the equivalent annual cost of operating each machine.

(b) From the results of part (a), calculate the probability that the annual cost of operating machine A will exceed the cost of operating machine B.

12.24 Two mutually exclusive investment projects are under consideration. It is assumed that the cash flows are statistically independent random variables with means and variances estimated as given in Table P12.24.

(a) For each project, determine the mean and standard deviation of the NPW, using an interest rate of 15%.

(b) On the basis of the results of part (a), which project would you recommend?

Decision-Tree Analysis

12.25 Your company must decide whether or not to introduce a new product. The sales of the product will be either at a high level (a success) or quite low (a flop). The conditional value for this decision is as follows:

	States of Nature	
Decision	**High Sales (H)**	**Low Sales (L)**
Introduce product	$4M	−$2M
Do not introduce product	0	0
Probability	0.3	0.7

Your company has the option to conduct a market survey to better ascertain the true market demand.

The survey will cost $0.2M. The survey does not provide perfect information about the sales, but there are three possible outcomes from the survey: (1) the survey may predict success (high sales) for the new product; (2) the survey may predict that the new product will fail (low sales); or (3) the survey result may be inconclusive. You are given the following notation and information:

Conditional Probabilities of Survey Predictions, Given Potential Sales

	Potential Level of Sales	
Market Survey Results (Survey prediction)	**High Sales (H)**	**Low Sales (L)**
Survey predicts high sales (S)	0.4	0.1
Survey inconclusive (I)	0.4	0.5
Survey predicts low sales (F)	0.2	0.4

To evaluate the economic worth of the survey, answer the following questions.

(a) Based on expected monetary value, what is your decision before taking the survey?

(b) What is the expected value of perfect information before taking the survey?

(c) Draw the complete decision tree diagram with the survey option.

(d) What is the expected value of perfect information after you have taken the survey?

(e) What is the worth of the survey information?

(f) What is the best course of action?

12.26 Delta College's campus police are quite concerned with ever-growing weekend parties taking place at the various dormitories on the campus, where alcohol is commonly served to underage college students. According to reliable information, on

TABLE P12.24

	Project A		Project B	
End of Year	**Mean**	**Variance**	**Mean**	**Variance**
0	−$5,000	$1,000^2$	−$10,000	$2,000^2$
1	4,000	$1,000^2$	6,000	$1,500^2$
2	4,000	$1,500^2$	8,000	$2,000^2$

a given Saturday night, one may observe a party to take place 60% of the time. Police Lieutenant Shark usually receives a tip regarding student drinking that is to take place in one of the residence halls the next weekend. According to Officer Shark, this tipster has been correct 40% of the time when the party is planned. The tipster has been also correct 80% of the time when the party does not take place. (That is, the tipster says that no party is planned.) If Officer Shark does not raid the residence hall in question at the time of the supposed party, he loses 10 career progress points. (The police chief gets complete information on whether there was a party only after the weekend.) If he leads a raid and the tip is false, he loses 50 career progress points, whereas if the tip is correct, he earns 100 points. Certainly he could raid always, or never, or do the opposite of what the tip says.

(a) What is the probability that no party is actually planned even though the tipster says that there will be a party?

(b) If the lieutenant wishes to maximize his expected career progress points, what should he do before getting any tip?

(c) What is the EVPI (in terms of career points)?

12.27 As a plant manager of a corporation, you are trying to decide whether to open a new factory outlet store, which would cost about $500,000. Success of the outlet store depends on demand in the new region. If demand is high, you expect to gain $1 million per year; if demand is average, the gain is $500,000; and if demand is low, you lose $80,000. From your knowledge of the region and your product, you feel that the chances are 0.4 that sales will be average and equally likely that they will be high or low (0.3, respectively). Assume that the firm's MARR is known to be 15%, and the marginal tax rate will be 40%. Also, assume that the salvage value of the store at the end of 15 years will be about $100,000. The store will be depreciated under a 39-year property class.

(a) If the new outlet store will be in business for 15 years, should you open it? How much would you be willing to pay to know the true state of nature?

(b) Suppose a market survey is available at $1,000, with the following reliability (the values shown were obtained from past experience, where actual demand was compared with predictions made by the survey):

Actual	Survey Prediction		
Demand	Low	Medium	High
Low	0.75	0.20	0.05
Medium	0.20	0.60	0.20
High	0.05	0.25	0.70

Determine the strategy that maximizes the expected payoff after taking the market survey. In doing so, compute the EVPI after taking the survey. What is the true worth of the sample information?

12.28 Jay Olsen, a writer of novels, just has completed a new thriller novel. A movie company and a TV network both want exclusive rights to market his new title. If he signs with the network, he will receive a single lump sum of $900, but if he signs with the movie company, the amount he will receive depends on how successful the movie is at the box office. (All $ units are in thousands.)

• TV Network: $900

• Movie

	Event		
	Small box office (S)	Medium box office (M)	Large box office (L)
Payouts (in $)	$250	$1,000	$3,200
Probabilities	0.30	0.60	0.1

(a) Which option would you recommend based on the expected monetary value (EMV) criterion? (Assume that he is a risk-neutral person interested in maximizing the expected monetary value.)

(b) How much would he be willing to pay to know the true state of nature?

(c) Jay can send his novel to a prominent movie critic to assess the potential box office success. From his past experience, the movie critic's reliability of predicting the box office success is as follows. Favorable prediction means that it is highly likely the movie would be a box office success. It will cost $20 to get his novel evaluated by the movie critic.

Critic will say	Given Actual Demand		
	Small box office (S)	Medium box office (M)	Large box office (L)
Favorable (F)	0.20	0.40	0.70
Unfavorable (UF)	0.80	0.60	0.30

For example, when the true state of nature is "S," the movie critic will say "F" with a 20% probability.

Determine Jay's strategy that maximizes the expected payoff after receiving the movie critic's report. In doing so, compute the EVPI after taking the survey. What is the true worth of the market survey?

12.29 A machine shop specializing in industrial pumps is experiencing a substantial backlog, and the shop's management is considering three courses of action:

- Arrange for subcontracting (A)
- Construct new facilities (B)
- Do nothing (C)

The correct choice depends largely upon demand, which may be low, medium, or high. By consensus, management estimates the respective demand probabilities as 0.2, 0.5, and 0.3. The management estimates the profits when choosing from the three alternatives (A, B, and C) under the differing probable levels of demand. These profits are already expressed in terms of NPW.

Event	NPW (in $100,000)		
	Low	Medium	High
A	10	50	90
B	−120	25	200
C	20	40	60

(a) Which option would you recommend based on the expected monetary value (EMV) criterion? (Assume that you are a risk-neutral person so that you are interested in maximizing the expected monetary value.)

(b) How much would you be willing to pay to know the true state of nature?

(c) Suppose a market survey can be undertaken with the following reliability (values obtained from past experience where actual demand was compared with predictions made by the market survey). Favorable prediction means that there would be sufficient demand for the product in the market place.

Given: Actual Demand	Survey Prediction of Demand	
	Favorable (F)	Unfavorable (UF)
Low	0.30	0.70
Medium	0.40	0.60
High	0.75	0.25

Determine the strategy that maximizes the expected payoff after taking the market survey. In doing so, compute the EVPI after taking the survey. What is the true worth of the market survey?

12.30 An acquisition editor is trying to decide whether or not to publish a highly technical manuscript he has received. In making a decision he feels that it is sufficient to imagine that there are just two states of nature which are

- S1, the book will sell just an average of 500 copies per year (primarily from library adoptions);
- S2, the average annual sales will be 3,000 copies.

The prior probabilities that he assigns to these states of nature are 0.50 and 0.50, respectively. The discounted profits if he publishes the book are projected to be ($30,000, loss), and +$40,000, respectively. Of course, the discounted profit is $0 if he does not publish it.

To gain additional information, the publisher can send the manuscript to a reviewer. From previous experience with this reviewer, the publisher feels that the conditional probabilities of the state of nature given this response from the reviewer are respectively as follows:

State of Actual Sales	What the reviewer would say	
	Like (Z1)	Dislike (Z2)
S1	0.2	0.8
S2	0.6	0.4

Should the publisher publish the book? To answer this question, do the following:

(a) What is the prior optimal act (before sending out the manuscript to the reviewer)?

(b) Calculate the EVPI before sending the manuscript out for review.

(c) What is the posterior optimal act? Draw the decision tree.

(d) What is the EVPI after finding out that the reviewer liked the manuscript?

(e) What is the EVPI after finding out that the reviewer disliked the manuscript?

(f) Calculate the EVSI.

(g) Suppose that it costs $4,500 to get the manuscript reviewed. Is it worth doing it? What is the ENGS?

Short Case Studies

ST12.1 In Virginia's six-number lottery (or lotto) players pick six numbers from 1 to 44. The winning combination is determined by a machine that looks like a popcorn machine, except that it is filled with numbered table-tennis balls. The Virginia lottery drawing offered the prizes shown in Table ST12.1, assuming that the first prize is not shared with a ticket price of $1. Common among regular lottery players is this dream: waiting until the jackpot reaches an astronomical sum and then buying every possible number, thereby guaranteeing a winner. Sure, it would cost millions of dollars, but the payoff would be much greater. Is it worth trying? How do the odds of winning the first prize change as you increase the number of tickets purchased?

ST12.2 The city of Opelika was having a problem locating land for a new sanitary landfill when the Alabama Energy Extension Service offered the solution of burning the solid waste to generate steam. At the same time, Uniroyal Tire Company seemed to be having a similar problem disposing of solid waste in the form of rubber tires. It was determined that there would be about 200 tons per day of waste to be burned, including municipal and industrial waste. The city is considering building a waste-fired steam plant, which would cost $6,688,800. To finance the construction cost, the city will issue resource-recovery revenue bonds in the amount of $7,000,000 at an interest rate of 11.5%. Bond interest is payable annually. The differential amount between the actual construction costs and the amount of bond financing

$$(\$7,000,000 - \$6,688,800 = \$311,200)$$

will be used to settle the bond discount and expenses associated with the bond financing. The expected life of the steam plant is 20 years. The expected salvage value is estimated to be about $300,000. The expected labor costs are $335,000 per year. The annual operating and maintenance costs (including fuel, electricity, maintenance, and water) are expected to be $175,000. The plant would generate 9,360 pounds of waste, along with 7,200 pounds of waste after incineration, which will have to be disposed of as landfill. At the present rate of $19.45 per pound, this will cost the city a total of $322,000 per year.

The revenues for the steam plant will come from two sources: (1) sales of steam and (2) tipping fees for disposal. The city expects 20% downtime per year for the waste-fired steam plant. With an input

TABLE ST12.1 Virginia Lottery Prizes

Number of Prizes	Prize Category	Total Amount
1	First prize	$27,007,364
228	Second prizes ($899 each)	204,972
10,552	Third prizes ($51 each)	538,152
168,073	Fourth prizes ($1 each)	168,073
	Total winnings	**$27,918,561**

of 200 tons per day and 3.01 pounds of steam per pound of refuse, a maximum of 1,327,453 pounds of steam can be produced per day. However, with 20% downtime, the actual output would be 1,061,962 pounds of steam per day. The initial steam charge will be approximately $4.00 per thousand pounds, which will bring in $1,550,520 in steam revenue the first year and every year thereafter. The tipping fee is used in conjunction with the sale of steam to offset the total plant cost. It is the goal of the Opelika steam plant to phase out the tipping fee as soon as possible. The tipping fee will be $20.85 per ton in the first year of plant operation and will be phased out at the end of the eighth year. The scheduled tipping fee assessment is as follows:

Year	Tipping Fee
1	$976,114
2	895,723
3	800,275
4	687,153
5	553,301
6	395,161
7	208,585

(a) At an interest rate of 10%, would the steam plant generate sufficient revenue to recover the initial investment?

(b) At an interest rate of 10%, what would be the minimum charge (per thousand pounds) for steam sales to make the project break even?

(c) Perform a sensitivity analysis to determine the input variable of the plant's downtime.

ST12.3 A local telephone company is considering installing a new phone line for a new row of apartment complexes. Two types of cables are being examined: conventional copper wire and fiber optics. Transmission by copper-wire cables, although cumbersome, involves much less complicated and less expensive support hardware than do fiber optics. The local company may use five different types of copper wire cables: 100 pairs, 200 pairs, 300 pairs, 600 pairs, and 900 pairs per cable. In calculating the cost of cable, the following equation is used:

$$\text{Cost} = [\text{cost per foot} + \text{cost per pair}$$
$$(\text{number of pairs})](\text{length})$$

where

$$\text{22-gauge copper wire} = \$1.692 \text{ per foot}$$

and

$$\text{Cost per pair} = \$0.013 \text{ per pair}$$

The annual cost of the cable as a percentage of the initial cost is 18.4%. The life of the system is 30 years. In fiber optics, a cable is referred to as a ribbon. One ribbon contains 12 fibers, grouped in fours; therefore, one ribbon contains three groups of 4 fibers. Each group can produce 672 lines (equivalent to 672 pairs of wires), and since each ribbon contains three groups, the total capacity of the ribbon is 2,016 lines. To transmit signals via fiber optics, many modulators, waveguides, and terminators are needed to convert the signals from electric currents to modulated light waves. Fiberoptic ribbon costs $15,000 per mile. At each end of the ribbon, three terminators are needed, one for each group of 4 fibers, at a cost of $30,000 per terminator. Twenty-one modulating systems are needed at each end of the ribbon, at a cost of $12,092 for a unit in the central office and $21,217 for a unit in the field. Every 22,000 feet, a repeater is required to keep the modulated light waves in the ribbon at an intensity that is intelligible for detection. The unit cost of this repeater is $15,000. The annual cost, including income taxes for the 21 modulating systems, is 12.5% of the initial cost of the units. The annual cost of the ribbon itself is 17.8% of the initial cost. The life of the whole system is 30 years. (All figures represent after-tax costs.)

(a) Suppose that the apartments are located 5 miles from the phone company's central switching system and that about 2,000 telephones will be required. This would require either 2,000 pairs of copper wire or one fiberoptic ribbon and related hardware. If the telephone company's interest rate is 15%, which option is more economical?

(b) In part (a), suppose that the apartments are located 10 miles or 25 miles from the phone company's central switching system. Which option is more economically attractive under each scenario?

ST12.4 Capstone Turbine Corporation is the world's leading provider of microturbine based MicroCHP (combined heat and power) systems for clean, continuous, distributed-generation electricity. The MicroCHP unit is a compact turbine generator that delivers electricity onsite or close to the point where it is needed. Designed to operate on a variety of gaseous and liquid fuels, this form of distributed-generation technology first debuted in 1998. The microturbine is designed to operate on demand or continuously for up to one year between recommended maintenance (filter cleaning/replacement). The generator is cooled by airflow into the gas turbine, thus eliminating the need for liquid cooling. It can make electricity from a variety of fuels—natural gas, kerosene, diesel oil, and even waste gases from landfills, sewage plants, and oilfields.

Capstone's focus applications include combined heat and power, resource recovery of waste fuel from wellhead and biogas sites, power quality and reliability, and hybrid electric vehicles. And, unlike traditional backup power, this solution can support everyday energy needs and generate favorable payback. With the current design, which has a 60-KW rating, one of Capstone's generators would cost about $84,000. The expected annual expenses, including capital costs as well as operating costs, would run close to $19,000. These expenses yield an annual savings of close to $25,000 compared with the corresponding expenses for a conventional generator of the same size.

Capstone is considering a full-scale production of the system unit for residential use. The project requires an initial investment of $55 million. The company has prepared the financial data in Table ST12.4 related to the project.

The initial investment can be depreciated on a seven-year MACRS, and the project is expected to have an economic service life of eight years. The firm's marginal tax rate is 35%, and its MARR is known to be 15%. Capstone will conduct the economic analysis based on the following implicit model of cash flow.

- $n = 0$: Cash flow = −$55 million
- $n = 1$ to 8: Cash flow = (revenue − costs − depreciation)(1 − tax rate) + depreciation

- Revenues = market size × unit price × $(1+\text{growth rate})^{(n-1)}$
- Costs = (market size) × variable unit cost × $(1 + \text{growth rate})^{(n-1)}$ + Fixed cost (excluding depreciation)
- $n = 8$: Adjust the cash flow in year 8 to reflect the gains tax (loss credit) and the salvage value

Assume that each random variable in Table ST12.4 is triangularly distributed with the parameters specified. Using your choice of software (such as Excel, @Risk, or Crystal Ball), develop the NPW distribution of the project.

(a) Document the procedure to obtain your NPW distribution.

(b) Compute the mean and variable of the NPW distribution.

(c) Determine the probability the actual NPW would be below the expected NPW.

ST12.5 A firm is considering whether to market a new type of computer game during the upcoming Christmas season. The profit contribution of this computer game will depend on the extent of the demand θ, where θ denotes the number of thousands of copies of the computer game to be sold. It is estimated that the net profit of marketing the computer game will be approximately $2\theta^2 - 500,000$ dollars. Assume that the firm's initial prediction of demand distribution on θ is a uniform density on the interval $0 \le \theta \le M$. In terms of M determine the following.

a. For what value of the demand θ should the firm market the new type of toy?

b. What is the expected value of θ?

c. How much should the firm be willing to pay to find out exactly what the demand θ will be?

TABLE ST12.4

Random Variables	Low	Most Likely	High
Annual Market size (units)	5,000	8,000	10,000
Growth rate (annual)	3%	5%	8%
Unit price	$80,000	$84,000	$86,000
Unit variable cost	$56,000	$60,000	$65,000
Fixed cost (annual) excluding depreciation	$5,000,000	$8,000,000	$9,000,000
Salvage value	$6,000,000	$7,000,000	$8,000,000

13

Real-Options Analysis

Travel is Adventure. Prepare For the Unexpected. Option thinking is not new, it is everywhere. Our whole life consists of options. Have you ever thought about buying trip insurance for protection of your trip? Whether you are flying halfway across the country or halfway around the world, trip insurance is offered to protect you against unforeseen circumstances that can disrupt your plans. The reason why you would be interested in this type of protection is that you want to minimize the downside risk–in case of cancellation or change in your travel plan, you would like to recover all your prepaid expenses such as flight and hotel. If there is no change in your plan, you simply lose your insurance premium. The bottom line is that you would like to have flexibility in your travel plan through the trip protection insurance. In the option world, buying trip insurance is equivalent to buying a put option such that you would exercise your put option only in case of travel disruption. If not, you simply let the option expire. Here the insurance premium is viewed as an option premium to have the right to cancel the trip.

In a business investment setting, managers can and do obtain valuable information after a project is launched, and their informed actions can make a big difference. The more options you have, the more flexible you can be to changing market dynamics. The greater the uncertainty and flexibility,

Source: Mark Carrel/Fotolia

the greater the value of management's options. Essentially, option thinking is a structured and explicit way of thinking about investment possibilities, actions, and decisions in order to be able to react to future conditions.

An article in *BusinessWeek* magazine[1] took a closer look at a new trend in corporate planning known as "the real-options revolution in decision making." In a nutshell, real-options analysis simply says that companies benefit by keeping their options open. For example, suppose a company is deciding whether to fund a large R&D project that could either make or lose a lot of money. A traditional calculation of net present worth (NPW), which discounts projected costs and revenues into today's dollars, examines the project as a whole and concludes that it is a no go. But a real-options analysis breaks the project into stages and concludes that it makes sense to fund the first stage at least. Real-options analysis rewards flexibility—and that is what makes it better than NPW, one of today's advanced financial decision-making tools. In this chapter, we will examine the fundamentals of the real-options decision framework and discuss how we may apply its logic to approach solving capital-investment decision problems under uncertainty.

CHAPTER LEARNING OBJECTIVES

After completing this chapter, you should understand the following concepts:

- How the financial option can be used to hedge the market risk.
- How the financial options are priced.
- What features of financial options would be useful in managing risk for real assets.
- How the various real-options models are structured.
- How the flexibility of various real options are valued.

13.1 Risk Management: Financial Options

Let us start with an ordinary "option," which is just the *opportunity* to buy or sell stock at a specific price within a certain period. The two basic types of financial options are a *call option* and a *put option*. The call option gives the investor the right to buy a stock at a specified price within a specified period. The put option gives the investor the right to sell a stock at a specified price within a particular period.

[1] "Exploiting Uncertainty: The 'Real-Options' Revolution in Decision-Making." *BusinessWeek*, June 7, 1999, p. 119.

13.1.1 Features of Financial Options

Financial options have several important features. First we review important terminologies used in financial options.

Terminology

- ***The contracting parties.*** The party granting the right is referred to as the *option seller* (or option writer). The party purchasing the right is referred to as the *option buyer*. In the option world, the option buyer is said to have a *long* position in the option, while the option seller is said to have a *short* position in the option. The process is shown in Figure 13.1

- ***The right or obligation.*** An option to buy a financial asset at a specified price is a call-option on the asset. The call buyer has the *right* to purchase the asset; the call-option seller has the *obligation* to sell. An option to sell an asset at a specified price is a put option on the asset. Consequently, the put-option seller has the *obligation* to buy and the put-option buyer has the *right* to sell.

- ***The underlying asset*** (S) dictates the value of the financial option. Typically, the underlying asset is a financial asset, such as equities (stocks), equity indexes, and commodities.

- ***The strike*** or ***exercise price*** (K) gives the value at which the investor can purchase or sell the underlying security (financial asset).

- The ***maturity*** (T) of the option defines the period within which the investor can buy or sell the stock at the strike price. An option that can be exercised earlier than its maturity date is called an **American option**. An option that can be exercised only at the maturity date is a **European option**.

- ***The option premium*** (C) is the price that an investor has to pay to own the option. The **intrinsic value** of an option is the value the option would have if it were exercised immediately. For a call option, it is max $\{S - K, 0\}$. For a put option it is max $\{K - S, 0\}$. The difference between C and the intrinsic value is known as **time value** of the option.

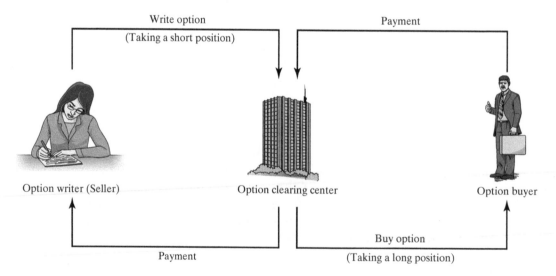

Figure 13.1 Process of buying and selling a financial option.

Payoffs

The payoff of an option depends on the price of the underlying asset at the exercise date (S_T). Three terms—*at the money*, *in the money*, and *out of the money*—identify where the current stock price (S) is relative to the strike price (K).

- An option that is in the money is an option that, if exercised, would result in a positive value to the investor.
- An option that is out of the money is an option that, if exercised, would result in a negative value to the investor.
- You will exercise your option when the option is in the money. If not, you let the option expire. Your loss is limited to the option premium itself.

Option Positions

In every option contract, there are two sides: the one who holds the option (the long position) and the one who issues the option (the short position). Thus, there are four possible positions: a long position in a call option, a long position in a put option, a short position in a call option, and a short position in a put option. In general, the call and put options have the following characteristics:

- The greater the difference between the exercise price and the actual current price of the item, the cheaper the premium, because there is less of a chance that the option will be exercised.
- The closer the expiration date of an option that is *out of the money* (the market price is higher than the strike price), the cheaper the price.
- The more time there is until expiration, the larger the premium, because the chance of reaching the strike price is greater and the carrying costs are more.
- Call and put options move in opposition. Call options rise in value as the underlying market prices go up. Put options rise in value as market prices go down. However, the call (c) and put (p) option premiums are related to

$$c + Ke^{-rf} = p + S_0$$

This equation is known as the **put-call parity** in financial option.

We will consider two types of investors: one who thinks that the share price of the stock will continue to go up (a call-option buyer) and the other who thinks that the share price of the stock will go down (a put-option buyer). Let's examine how much profit each investor would make if his or her prediction proves to be true.

13.1.2 Buy Call Options When You Expect the Price to Go Up

An investor purchases a call option when he or she is bullish on a company and thinks that the stock price will rise within a specific amount of time—usually three, six, or nine months. For this right, you pay the call-option seller a fee (called a "premium"), which is forfeited if you do not exercise the option before the agreed-upon expiration date.

EXAMPLE 13.1 **Profit from Call Option**

You buy a call option to purchase 100 shares (or one contract) of Google stock.

- Strike price = $625
- Expiration date = January 22, 2016
- Current share price = $579.11 (September 11, 2014)

- Option premium (price of an option to buy one share) = $45.60
- The initial investment (one contract) = 100 × $45.60 = $4,560

Suppose that at the expiration of the option, the share price is $700. What would be the net gain from exercising the call option?

SOLUTION

Given: Option parameters given in the problem statement.
Find: Profit from exercising call option.

At the expiration, the option exercised for a net gain of

$$(\$700 - \$625) \times 100 - \$4,560 = \$2,940$$

As shown in Figure 13.2, you will exercise your option when the stock price is higher than $625. Otherwise, your option is worthless, and you let the option expire and take a $45.60 loss per share. Clearly, the investor is betting that the stock price will be much higher than $625 by January 2016. (Otherwise he or she would not own the call option.)

COMMENTS: As we mentioned earlier, there are two sides to every option contract. So, we would be interested in when the investor who has taken a short position (has sold or written the call option) makes profit. As the call-option writer receives cash up front (premium paid by the option buyer) but has the potential liabilities later, the profit or loss is the reverse of that for the purchase of the call option. Figure 13.3 shows the writer's profit or loss as a function of the final stock price at expiration.

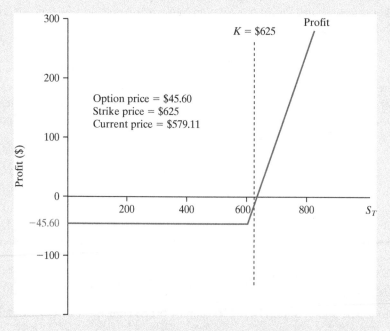

Figure 13.2 Profit from buying a call option on one share of Google stock.

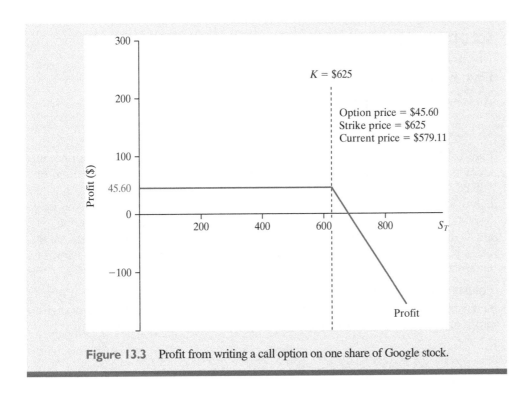

Option price = $45.60
Strike price = $625
Current price = $579.11

Figure 13.3   Profit from writing a call option on one share of Google stock.

13.1.3 Buy Put Options When You Expect the Price to Go Down

One way to protect the value of financial assets from a significant downside risk is to buy put-option contracts. When an investor purchases a put, he or she is betting that the underlying investment is going to decrease in value within a certain amount of time. This means that if the value does go down, you will *exercise the put*. Buying a put-option contract is equivalent to purchasing a home insurance policy. You pay an annual premium for protection in case, say, your house burns to the ground. If that happens, you will exercise the put and get the money from the insurance company. If it never happens during the contract period, your loss is limited to the cost of the premium.

EXAMPLE 13.2 **Profit from Put Option**

You buy a put option to sell 100 shares of Google stock.

- Strike price = $580
- Expiration date = January 22, 2016
- Current share price = $579.11 (September 11, 2014)
- Option premium (price of an option to buy one share) = $58.30
- The initial investment (one contract) = 100 × $58.30 = $5,830

Suppose that at the expiration of the option, the share price is $500. What would be the net gain from exercising the put option?

SOLUTION

Given: Option parameters give in the problem statement.
Find: Profit from exercising put option.

At the expiration, the option exercised for a net gain of

$$(\$580 - \$500) \times 100 - \$5,830 = \$2,170$$

Figure 13.4 illustrates how an investor might profit from buying a put option on Google stock. Once again, with a strike price of $580 and an option premium of $58.30, you would exercise your put only when the share price of the stock dipped below $580. At $521.70, you will break even; between $580 and $521.70, you will have a partial loss by recovering some of your premium. Once again, the reason you are buying a put option with a strike price of $580 is that the current price of the stock is too high to sustain that value in the future, and so you want to protect the value of the stock at the current level.

COMMENTS: As in the case of a call option, the option writer's profit or loss is the reverse of that for the purchase of the option. Figure 13.5 illustrates the put writer's profit or loss as a function of the final stock price at option expiration.

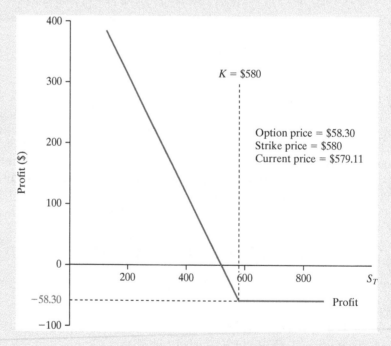

Figure 13.4 Profit from buying a put option on one share of Google stock.

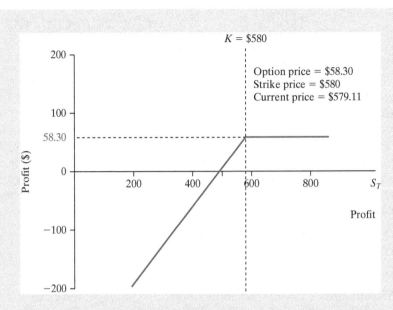

Figure 13.5 Profit from writing a put option on one share of Google stock.

13.2 Option Strategies

Now that we have described how the basic financial options work, we will review two basic strategies on how to hedge the financial risk when we use these options. The first strategy is to reduce the capital that is at risk by buying call options, and the second strategy is to buy a put option to hedge the market risk.

13.2.1 Buying Calls to Reduce Capital That Is at Risk

Once again, buying a call gives the owner a right, but not an obligation. The risk for the call buyer is limited to the premium paid for the call (the price of the call), plus commissions. The value of the call tends to increase as the price of the underlying stock rises. This gain will increasingly reflect a rise in the value of the underlying stock when the market price moves above the option's strike price. As an investor, you could buy the underlying security (stock), or you could buy call options on the underlying security. As an American call buyer, you have three options: (1) Hold the option to maturity and exercise at the strike price; (2) trade for profit before the option expires (known as exercising your option); and (3) let the option expire if doing so is advantageous to you. Example 13.3 illustrates how you might buy a call to participate in the upward movement of a stock while limiting your downside risk.

EXAMPLE 13.3 How to Buy a Call to Participate in the Upward Movement of a Stock while Limiting Your Downside Risk

Gilead Sciences, Inc. (GILD) is trading at $106. Instead of spending $53,000 for 500 shares of GILD stock, an investor could purchase a five-month call with a $110-strike price for $8.23. By doing this, the investor is saying that he or she anticipates that GILD will rise above the strike of $110 (which is where GILD can be purchased no matter how high GILD has risen), plus another $8 (the option premium), or $118 by expiration. Without considering commissions and taxes, discuss the amount of risk presented in each of the following scenarios, as opposed to purchasing 500 shares:

- GILD is trading at or above $118 by expiration.
- GILD is trading between $100 and $118 at expiration.
- GILD is trading at or below $100 at expiration.

DISCUSSION: Each call contract represents 100 shares of stock, so five calls could be bought in place of 500 shares of stock. The cost of five calls at $8 is $4,000 (5 calls × $8 × 100). Thus, instead of spending $53,000 on stock, the investor needs to spend only $4,000 for the purchase of the five calls. The balance of $49,000 could then be invested in short-term instruments. This investor has unlimited profit potential when GILD rises above $118. The risk for the option buyer is limited to the premium paid, which in this example is $4,000.

SOLUTION

Given: Buying five GILD five-month $110 calls at $8 versus buying 500 shares of GILD at $106

Find: Profit or loss for three possible scenarios at expiration.

(a) GILD is at or above $118 by expiration.

- If GILD is at $120 at expiration, the option will be worth the difference between the strike and the current price of the stock:

$120 (current price) − $110 (strike price) = $10 (current option value)

The option could be exercised, and a 25% return would be earned on the initial investment. That is,

$10 (current option value) − $8 (premium paid for option)

= $2 (profit if option is sold)

- Had the stock been purchased at $106 (at a cost of $53,000) and had it risen to $120, it would now be worth $60,000. This would be a 13.21% increase in value over the original cost of $53,000. But the call buyer spent only $4,000 and earned 25% on his options. Table 13.1 compares both scenarios.

TABLE 13.1 Comparison Between Purchasing Stock versus Purchasing Options

Had Stock Been Purchased				
Stock Purchase Price	Initial Cost of Stock 500 Shares	Stock Price at Expiration	Value of Stock at Expiration[*]	Change in Stock Value
$106	$53,000	$120	$60,000	$7,000
$106	$53,000	$115	$57,500	$4,500
$106	**$53,000**	**$90**	**$45,000**	**($8,000)**

Had Call Options Been Purchased				
Option Price per Contract	Initial Total Cost, 5 Option Contracts	Options Value Per Contract at Expiration	Total Value of Options	Change in Options Value
$8	$4,000	$10	$5,000	$1,000[†]
$8	$4,000	$5	$2,500	($1,500)[†]
$8	**$4,000**	**0**	**0**	**($4,000)[†]**

*Plus dividends if any.

†Plus interest earned on cash not used, cost of stock less call premium.

(b) GILD is between $110 and $118 at expiration.

- The investor's option will still hold some value if GILD is between $110 and $118 but not enough to break even on the position. The option still can be sold to recoup some of the cost. For example, suppose GILD is at $115 on the last day of trading, usually the third Friday of the expiration month. Then the option can be sold to close out the position through the last trading day of the call. What happened is that GILD did rise in value, but not as much as anticipated. The option that cost $8 is now worth just five points. Instead of letting the option expire, you can sell the call and recoup some of your losses:

$$\$8 \text{ (Cost)} - \$5 \text{ (Sale)} = \$3 \text{ (Net loss excluding commissions)}$$

- If just the stock had been bought and it rose to $115, $4,500 would have been earned, while the holder of calls would have lost $1,500. However, the holder of calls would have been earning interest on $49,000, which would offset some of the loss in the options.

(c) GILD is at or below $110 at expiration.

- Suppose GILD is now at $90 and the option has expired worthless. The premium that was paid for the calls has been lost. However, had the stock been bought, the investor would now be in a losing stock position, hoping to break even. By purchasing a call he or she had limited capital at risk. Now the investor still has most of the money that would have gone into buying the stock, plus interest. Thus, the investor can make another investment decision.

- If the investor had purchased GILD at $106 and the stock did not move as he or she anticipated, the investor would have had two choices: sell the stock and, after commission costs, incur some losses; or hold onto it and hope that it rises over the long term. Had the investor bought the options and been wrong, the options would expire worthless and the loss would be limited to the premium paid ($4,000). Figure 13.6 shows the variation of the profit or loss with the final stock price for both investment strategies.

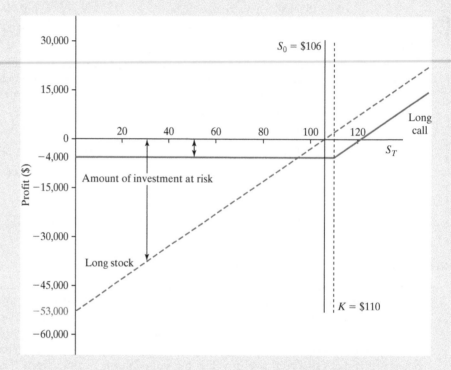

Figure 13.6 Profit or loss as a function of the final stock price for each investment strategy.

13.2.2 Protective Puts as a Hedge[2]

People insure their valuable assets, but most investors have not realized that many of their stock positions also can be insured. That is exactly what a *protective put* does. Typically, by paying a relatively small premium (compared with the market value of the stock), an investor knows that no matter how far the stock drops, it can be sold at the strike price of the put anytime up until the put expires. Buying puts against an existing stock position or simultaneously purchasing stock and puts can supply the insurance needed to overcome the uncertainty of the marketplace. Although a protective put may not be suitable for all investors, it can provide the protection needed to invest in individual stocks in volatile markets, because it ensures limited downside risk and unlimited profit potential for the life of the option.

[2]This section is based on the materials prepared by the Chicago Board Options Exchange (CBOE). These materials are available on the company's website at http://www.cboe.com.

Buying a protective put involves the purchase of one put contract for every 100 shares of stock already owned or purchased. Purchasing a put against stock is similar to purchasing insurance. The investor pays a premium (the cost of the put) to insure against a loss in the stock position. No matter what happens to the price of the stock, the put owner can sell it at the strike price at any time prior to expiration.

EXAMPLE 13.4 How to Use a Protective Put as Insurance

Suppose you are investing in Qualcomm (QCOM) stock. QCOM is trading at $76 per share, and the QCOM six-month, $75-put contract can be purchased at $4.40. To take a look at what happens to a protective put position as the underlying stock (QCOM) moves up or down, compare the amount of protection anticipated under the following two possible scenarios at expiration (do not consider commissions and taxes for now):

- **Option 1:** Buy QCOM at $76, without owning a put for protection.
- **Option 2:** Buy QCOM at $76, and buy a QCOM $75-put contract.

SOLUTION

Given: (a) option 1 and (b) option 2.

Find: Compare buying QCOM stock with buying QCOM with a protective put.

(a) Option 1: Buy QCOM at $76.
If stock is bought at $76 per share, the investor begins to lose money as soon as the stock drops below the purchase price. The entire $76 purchase price is at risk.

- **Upside potential:** If the price increases, the investor benefits from the entire increase without incurring the cost of the put premium or insurance.
- **Downside risk:** When only the stock is bought, there is no protection or insurance. The investor is at risk of losing the entire investment.

(b) Option 2: Buy QCOM at $76 and buy a QCOM $75-put contract.
A six-month put with a strike price of $75 can be bought for $4.40 or $440 per contract ($4.40 × 100). This put can be considered insurance "with a $1 discount," because the stock is purchased at $76 and an "at-the-money" put with the strike price ($75) is purchased. If the stock drops below $75, the put or insurance will begin to offset any loss in QCOM (less the cost of the put). If QCOM remains at $75 or above, the put will expire worthless and the premium will be lost. If just the stock had been bought, the investor would begin to profit as soon as the stock rose above $76. However, the investor would have no protection from the risk of the stock declining in value. Owning a put along with stock ensures limited risk while increasing the break-even on the stock by the cost of the put—still allowing for unlimited profit potential above the break-even.

	Buy QCOM	Buy QCOM and Six-Month $75 Put
Stock Cost	$76.00	$76.00
Put Cost	0.00	$4.40
Total Cost	$76.00	$80.40
Risk	$76.00	$5.40

- **Upside potential:** This strategy gives an investor the advantage of having downside protection without limiting upside potential above the total cost of the position (or $80.40). The only disadvantage is that the investor will not begin to profit until the stock rises above $80.40.
- **Downside risk:** No matter how low QCOM falls, buying the six-month put with a 75-strike price gives the investor the right to sell QCOM at $75 up until expiration. The downside risk is only $5.40, which is the total cost for this position—$80.40 less $75 (the strike price). Figure 13.7 shows the profit pattern for a long position in a stock combined with a long position in a put.

COMMENTS: If you buy QCOM $70 put (at $2.45 premium) instead of $75 put, your behavior is viewed as if you want some downside protection on a stock position but are willing to have a deductible in exchange for a lower insurance cost. The potential volatility of the equity markets can be of great concern to investors. The purchase of a protective put can give the investor the comfort level needed to purchase individual securities. This strategy is actually more conservative than purchasing stock. As long as a put is held against a stock position, there is limited risk; you know where the stock can be sold. The only disadvantages are that money cannot be made until the stock moves above the combined cost of the stock and the put, and that the put has a finite life. Once the stock rises above the total cost of the position, however, an investor has the potential for unlimited profit.

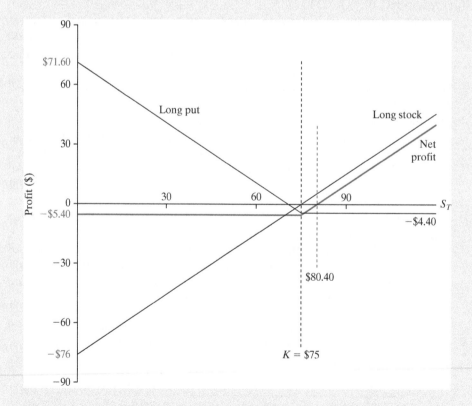

Figure 13.7 Potential profit or loss with long position in a stock combined with long position in a put as a function of stock price at the expiration of option.

13.3 Option Pricing

In this section, we illustrate the conceptual foundation of how to determine the price of an option. We know exactly the value of a financial option at maturity. However, our interest is not in the value at maturity, but rather in the value *today*. Here, we are basically looking at the pricing of a European option on a share of stock that pays no dividend. In this simplified approach to option pricing, we are going to envision our continuous-process world as a series of snapshots. In what follows, we will review three different approaches to valuing an option. The first approach is to use the replicating-portfolio concept, the second approach is to use the risk-free financing concept, and the third approach is to use the risk-neutral concept. All approaches yield the same result, so the choice depends on the preference of the analyst.

13.3.1 Replicating-Portfolio Approach with a Call Option

Let us suppose that today, day 0, the price of a particular share of stock is $300. Let us suppose further that tomorrow it could rise or fall by 5% (to either $315 or $285, respectively). Now consider the value of a one-day call option on this share of stock. Suppose that the exercise price of this one-day call option is $300. The value of the call option at maturity—day 1—is as shown in Figure 13.8. However, if you were thinking about buying this call option, you would be interested in knowing its value today. The issue is how we calculate that value. First we will define the symbols to use in pricing a financial option:

Δ = the number of shares of the underlying asset to purchase

b = the amount of cash borrowed or bond purchased at the risk-free rate

R = 1 plus the risk free rate, $(1 + r)$

S_0 = the value of the underlying asset today

uS = the upward movement in the value of the underlying asset in the future at some time t

dS = the downward movement in the value of the underlying asset in the future at some time t

K = the strike (or exercise) price of the option

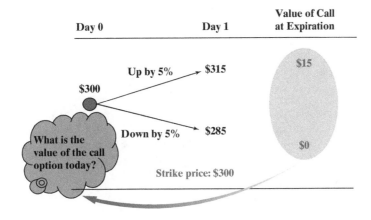

Figure 13.8 A simplified approach to a one-day call option.

C = the value of the call option

C_u = the upward movement in the value of the call option

C_d = the downward movement in the value of the call option

We will value the option by creating an arbitrage portfolio that contains the option. An **arbitrage (risk-free) portfolio** is a portfolio that earns a sure return. Let's form such a portfolio out of two risky assets: the share of stock and the call option on the share of the stock. We will create a portfolio such that the gains made on one of the assets will be exactly offset by losses on the other. As Figure 13.9 indicates, creating a portfolio in which we are selling (writing) one call-option against Δ shares of stock could form such a portfolio. In option language, we are *long* Δ shares of stock and *short* (selling) one call option. (Note that you [the writer] have already been paid for selling the option at day 0. So on day 1, you are concerned only about what the option buyer will do with his or her position.)

As illustrated in Figure 13.8, if at expiration the share price is higher than the exercise price, then the owner of the call option benefits at the expense of the seller: The payoff to the option owner is $15 ($315 − $300). By contrast, the payoff (expense) to the seller of the option is the reverse. If the share price at expiration is less than the exercise price, the call option is worthless: The payoff to both parties is zero. Therefore, this approach is equivalent to buying an insurance policy to protect any potential loss caused by changes in the price of a share.

- On day 1, if the stock price moves up from $300 to $315, the value of the shares is 315Δ and the value of the option is 15, so that the total value of this arbitrage portfolio is $315\Delta - 15$. If the stock price moves down from $300 to $285, the value of shares will be 285Δ, and the value of the option is zero. Therefore, the total value of the portfolio is 285Δ. Here our objective is to select the value of Δ such that the total value of the portfolio is the same regardless of the value of the share of stock at option expiration. Hence, we have formed a portfolio that has no risk. Here, Δ refers to the number of shares necessary to form the arbitrage portfolio. This number is easily found by equating

$$315\Delta - 15 = 285\Delta$$

Thus,

$$\Delta = 0.50$$

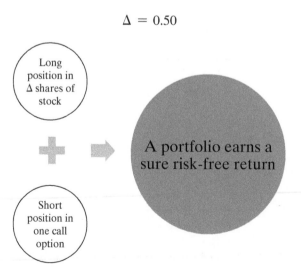

Long position in Δ shares of stock

Short position in one call option

A portfolio earns a sure risk-free return

Figure 13.9 Forming an arbitrage portfolio.

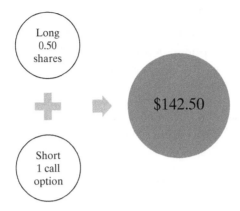

Figure 13.10 Creating an arbitrage portfolio.

A riskless portfolio (Figure 13.10) is

Long: 0.50 shares

Short: 1 call option

Then, if the stock price moves up to $315, the value of the portfolio is

$$\$315 \times 0.5 - 15 = \$142.50$$

If the stock price moves down to $285, the value of the portfolio is

$$\$285 \times 0.5 = \$142.50$$

Clearly, regardless of whether the stock moves up or down, the value of the portfolio will be $142.50 at the end of day 1.

- On day 0, the value of the share is $300, but we do not know the option value. However, we know the value of the arbitrage portfolio: We also know that on day 1 the value of the portfolio is $142.50. Therefore, we conclude that

$$(300 \times 0.50 - C)_{\text{day 0}} < (142.50)_{\text{day 1}}$$

To equate these two option values, it is necessary to discount day-1 values to day-0 values. So, we can express the present value of the $142.50 to be received one day earlier as

$$300 \times 0.50 - C = \frac{142.50}{(1 + r)}$$

where r is a one-day interest rate. Since the arbitrage portfolio is riskless, the interest rate used is the risk-free interest rate—say, 6%. The interest rate for one day is $(1/365) \times 0.06 = 0.00016$, and the preceding equation becomes

$$300 \times 0.50 - C = \frac{142.50}{(1 + 0.00016)}$$

$$C = \$7.52$$

Solving for C, we obtain the value of the one-day call option on day 0 as $7.52.

Now we can summarize what we have demonstrated in this simple example.

- Assuming no **arbitrage opportunities,**[3] the portfolio consisting of the stock and the option can be set up such that we know with certainty what the value of the portfolio will be at the expiration date.
- Because there is no uncertainty in the portfolio's value, the portfolio has no risk premium; hence, discounting can be done at the risk-free rate.
- As long as a portfolio consisting of shares of stock plus a short position in call options is set up, the value of this portfolio at expiration will be the same in both the up state and the down state.
- In essence, this portfolio mitigates all risk associated with the underlying asset's price movement.
- Because the investor has hedged against all risk, the appropriate discount rate to account for the time value of money is the risk-free rate!

13.3.2 Risk-Free Financing Approach

Another way of pricing financial options is to use the risk-free financing approach, which also is based on the arbitrage-portfolio concept illustrated in Section 13.3.1. As before, let's assume that the value of an asset today is $300 and it is known that in one day the price of the asset will go up by $u = 1.05$ with probability p or down by $d = 0.95$ with probability $(1 - p)$ (i.e., $uS = 315 and $dS = 285). Here, we are interested in knowing the value today of a European call option with an exercise price of $300 that expires on day 1.

We want to create a replicating portfolio that consists of the underlying asset (a stock in this example) together with a risk-free bond instead of a call option. As shown in Figure 13.11, we are creating a portfolio that consists of Δ shares of stock and b dollars'

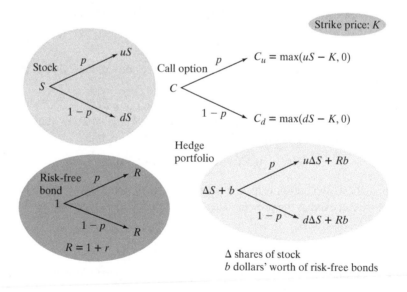

Figure 13.11 Creating a replicating portfolio by owning Δ shares of stock and b dollars worth of a risk-free asset such as a U.S. Treasury bond.

[3]Arbitrage opportunity: attempting to profit by exploiting price differences of identical or similar financial instruments, on different markets or in different forms.

worth of risk-free bonds. On day 1, we know that the value of the up-branch will be $u\Delta S + Rb$, and the down-branch will be $d\Delta S + Rb$. This replicating portfolio should have the same value with the call option; otherwise investors will profit through an arbitrage opportunity. Therefore, we can set:

$$u\Delta S + Rb = C_u$$

$$d\Delta S + Rb = C_d$$

Solving for ΔS and b respectively, we obtain:

$$\Delta S = \frac{C_u - C_d}{u - d} \tag{13.1}$$

$$b = \frac{C_u - u\Delta S}{R} = \frac{uC_d - dC_u}{R(u - d)} \tag{13.2}$$

Using Eqs. (13.1) and (13.2) yields:

$$\Delta S = \frac{C_u - C_d}{u - d} = \frac{15 - 0}{1.05 - 0.95} = \$150$$

$$b = \frac{C_u - u\Delta S}{R} = \frac{uC_d - dC_u}{R(u - d)} = \frac{15 - 1.05(150)}{1.00016} = -\$142.48$$

$$C = \Delta S + b = \$150.00 - \$142.48 = \$7.52$$

These equations tell us that a replicating portfolio needs to be formed with $150 worth of stock financed in part by $142.48 at the risk-free rate of 6% interest and that the option value at day 0 should be $7.52. Note that this is exactly the same value we obtained in Section 13.3.1.

13.3.3 Risk-Neutral Probability Approach

Let us now assume a risk-neutral world in which to value the option. Recall from Section 13.3.2 that the replicating portfolio hedged against all risks associated with the underlying asset's movement. Because all risk has been mitigated, this approach is equivalent to valuing the option in a risk-neutral world.

As before, let us define p and $(1 - p)$ as the *objective* probabilities of obtaining the up state (uS) and down state (dS) volatiles, respectively, as shown in Figure 13.12. Then the value of a replicating portfolio will change according to these probabilities. Since no opportunity for arbitrage exists, the following equality for the asset, based on Eqs. (13.1) and (13.2), must hold:

$$\Delta S + b = \frac{C_u - C_d}{u - d} + \frac{uC_d - dC_u}{R(u - d)} = \frac{1}{R}\left(\frac{R - d}{u - d}C_u + \frac{u - R}{u - d}C_d\right) \tag{13.3}$$

Now let $q = (R - d)/(u - d)$ in Eq. (13.3). Then we can rewrite Eq. (13.3) as

$$C = \Delta S + b = \frac{1}{R}[qC_u + (1 - q)C_d] \tag{13.4}$$

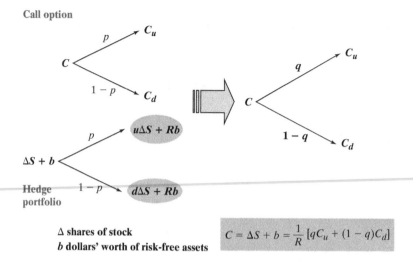

Figure 13.12 Risk-neutral probability approach to valuing an option.

Notice that the option value is found by taking the expected value of the option with probability q and then discounting this value according to the risk-free rate. Here, the probability (q) is a **risk-neutral probability**, and the objective probability (p) never enters to the calculation. In other words, we do not need a risk-adjusted discount rate in valuing an option, either.

If we assume continuous *discounting*, then the risk-neutral probability of an up movement q and the value of a call option can be calculated as:

$$q = \begin{cases} \dfrac{R - d}{u - d}, & \text{discrete compounding} \\[3mm] \dfrac{e^{rt} - d}{u - d}, & \text{continuous compounding} \end{cases} \tag{13.5}$$

and

$$C = \Delta S + b = \frac{1}{e^{rt}}[qC_u + (1 - q)C_d]$$

$$= e^{-rt}[qC_u + (1 - q)C_d] \tag{13.6}$$

Note that t represents the unit of time expressed in years. If the duration is exactly one month, then $t = \frac{1}{12}$.

In a risk-neutral world and discrete compounding, the following equality for the asset must hold:

$$q = \frac{R - d}{u - d} = \frac{1.00016 - 0.95}{1.05 - 0.95}$$

$$= 0.5016$$

$$C = \Delta S + b = \frac{1}{R}[qC_u + (1 - q)C_d]$$

$$= \frac{1}{1.00016}[0.5016(\$15) + (1 - 0.5016)(\$0)]$$

$$= \$7.52$$

Once again, we obtain the same result as the other approaches gave. The risk-neutral valuation approach is the most popular method due to its ease of implementation and computational simplicity. Therefore, we will use that approach in most of our option valuations.

13.3.4 Put-Option Valuation

A put option is valued the same way as a call option, except with a different payoff function. Under a risk-neutral approach with continuous compounding, the valuation formulas are

$$q = \frac{e^{rt} - d}{u - d} \tag{13.7}$$

$$C = e^{-rt}[qP_u + (1 - q)P_d] \tag{13.8}$$

where P_u and P_d are the put-option values associated with up and down movements, respectively.

EXAMPLE 13.5 A Put-Option Valuation

Suppose the value of an asset today is $50, and it is known that in one year the asset price will go up by $u = 1.2$ or down by $d = 0.8$ (i.e., $uS = \$60$ and $dS = \$40$, respectively). What is the value of a European put option with an exercise price of $55 that expires in $T = 1$ year? Assume a risk-free rate of 6% compounded continuously.

SOLUTION
Given: $u = 1.2, d = 0.8, T = 1, r = 6\%, S_0 = \$50$, and $K = \$55$.
Find: The risk-neutral probabilities and the put-option value.

From Eqs. 13.7 and 13.8, we first determine q and C_{put}:

$$q = \frac{e^{0.06} - 0.8}{1.2 - 0.8} = 0.6546$$

$$C_{put} = e^{-0.06}[(0)(0.6546) + (15)(0.3454)]$$

$$= \$4.88$$

The process is illustrated in Figure 13.13.

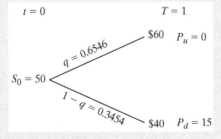

Figure 13.13
Binomial lattice model with risk-neutral probabilities.

13.3.5 Two-Period Binomial Lattice Option Valuation

The examples in the previous section assume a one-period movement of the underlying asset. Let's see what happens when we let the option run for two days. Continuing to assume that the share price can move up or down by 5% every day, we find that the distribution of share prices over the three days is as shown in Figure 13.14:

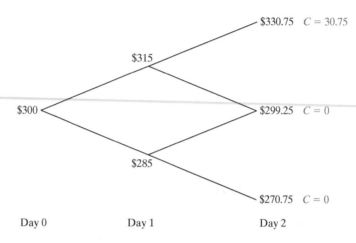

Figure 13.14 Stock prices in a two-step tree.

Our objective is to determine the option price at day 0, which is the initial node of the tree. In doing so, we first must determine the values of the option for the share prices that could exist on day 1 (i.e., $315 and $285). Then we use these values to determine what the option will be worth on day 0. Using the risk-neutral approach, we will use the value of q calculated earlier (page 682), that is:

$$q = \frac{R - d}{u - d} = \frac{1.00016 - 0.95}{1.05 - 0.95} = 0.5016$$

- On day 1, if the value of the share is $315, the call option value would be:

$$C_{\text{day 1,\$315}} = \frac{1}{1.00016}[0.5016(\$30.75) + (1 - 0.5016)(\$0)]$$

$$= \$15.4217$$

- On day 1, if the value of the share is $285, the value of the call would be zero, since the value of the call will be zero regardless of whether the value of the share rises to $299.25 or falls to $270.75. (Note that our strike price is still $300 per share.) This is shown in Figure 13.15.
- On day 0, the relevant lattice only involves the option prices at day 1. Therefore, continuing to use 6% as the relevant annualized rate for a one-day interest, we obtain

$$C_{\text{day 0}} = \frac{1}{1.00016}[0.5016(\$15.4217) + (1 - 0.5016)(\$0)]$$

$$= \$7.73$$

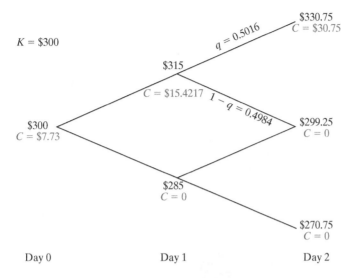

K = $300

$315
C = $15.4217

q = 0.5016

$330.75
C = $30.75

$300
C = $7.73

1 − q = 0.4984

$299.25
C = 0

$285
C = 0

$270.75
C = 0

Day 0   Day 1   Day 2

Figure 13.15   Two-period binomial lattice and option prices with the upper number at each node being the stock price and the lower number being the option price at that node.

so the value of the call option would be $7.73. Note that the value of the option increases from $7.52 to $7.73 merely by extending the maturity of the option for one more period.

13.3.6 Multiperiod Binomial Lattice Model

If we can value a two-day option, we can calculate a three-day, four-day, or in-general (*n*-day) option. The logic is exactly the same: We solve iteratively from expiration to period 0; the only thing that changes is the magnitude of the problem. The general form of an *n*-period binomial lattice is shown in Figure 13.16. The stock price can be visualized as moving from node to node in a rightward direction. As before, the probability of an upward movement from any node is p, and the probability of downward movement is $1 - p$.

The binomial lattice models described in Figures 13.16(a) or (b) are still unrealistically simple if we consider only a few time steps in a year. When binomial trees are used in practice, the life of the option is typically divided into 30 or more steps. In each time step, there is a binomial stock price movement. This means that 2^{30} or about 1.073 billion possible stock-price paths are possible in a year, resulting in a price distribution very close to a lognormal [Figure 13.16(c)]. Because the binomial model is multiplicative in nature, the stock price will never become negative. Percentage changes in the stock price in a short period of time are also normally distributed. If we define

$$\mu = \text{Expected yearly growth rate}$$

$$\sigma = \text{Volatility of yearly growth rate}$$

then the mean of the percentage change in time Δt is $\mu \Delta t$, and the standard deviation of the percentage change is $\sigma \sqrt{\Delta t}$. To capture the movement of the stock price, we need to select values for u and d and the probability p. If a period length of Δt is chosen as a time

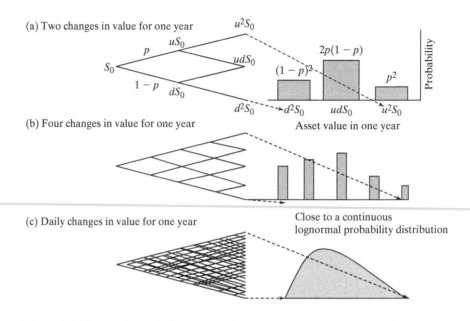

(a) Two changes in value for one year

(b) Four changes in value for one year

(c) Daily changes in value for one year

Figure 13.16 The binominal representation of asset price over one year: (a) with only two changes in value, (b) with four changes in value, and (c) with daily changes in value.

step, which is small compared to 1 (one year), the parameters of the binomial lattice can be selected as a function of the stock volatility σ:

$$p = \frac{1}{2} + \frac{1}{2}\left(\frac{\mu}{\sigma}\right)\sqrt{\Delta t}$$

$$u = e^{\sigma\sqrt{\Delta t}}$$

$$d = \frac{1}{u} = e^{-\sigma\sqrt{\Delta t}} \tag{13.9}$$

With this choice, the binomial model will closely match the values of the expected growth rate of the stock μ and its variance σ^2. The closeness of the match improves if Δt is made smaller, becoming exact as Δt goes to zero.

EXAMPLE 13.6 Construction of a Binomial Lattice Model

Consider a stock with starting price of $430 and with the parameters $\mu = 25\%$ and $\sigma = 50\%$ for a historical yearly volatility. We wish to make a binomial model based on weekly periods. Construct a binomial lattice to show how the stock price would move for four weeks.

SOLUTION
Given: $\mu = 25\%$, $\sigma = 50\%$, and $\Delta t = 1/52$.
Find: p, u, d, and the binomial lattice for four weeks.

Using Eq. (13.9), we get

$$u = e^{0.50/\sqrt{52}} = 1.07180$$

$$d = \frac{1}{u} = 0.9330$$

and

$$p = \frac{1}{2} + \frac{1}{2}\left(\frac{0.25}{0.50}\right)\sqrt{\frac{1}{52}} = 0.53467$$

The binomial lattice for the next four weeks is shown in Figure 13.17. The joint probabilities to reach the final nodes can be calculated as

- Price node ($567.44): $(0.53467)^4 = 0.0817$
- Price node ($493.96): $(0.53467)^3 (0.46533) \times 4 = 0.2845$
- Price node ($430.00): $(0.53467)^2 (0.46533)^2 \times 6 = 0.3714$
- Price node ($374.32): $(0.53467)(0.465433)^3 \times 4 = 0.2155$
- Price node ($325.85): $(0.46533)^4 = 0.0469$

The mean and variance of the final stock price are also determined as

$$\text{Mean } (\mu_T) = \$567.44 \times 0.0817 + \$493.96 \times 0.2845 + \ldots + \\ \$325.85 \times 0.0469 \\ = \$442.54$$

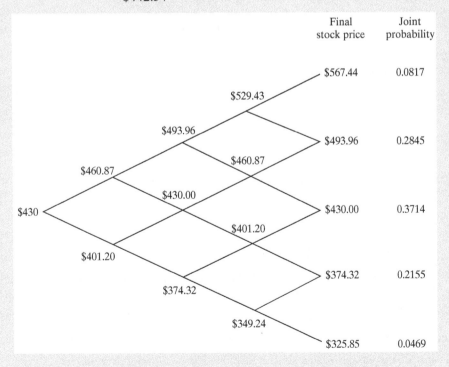

Figure 13.17 Binomial lattice for Google stock for a period of four weeks.

$$\text{Variance } (\sigma_T^2) = (567.44 - 442.54)^2 (0.0817)$$
$$+ \ldots (325.85 - 442.54)^2 (0.0469)$$
$$= (61.05)^2$$

13.3.7 Black–Scholes Option Model

As shown in Figure 13.16(c), if we take a sufficiently small time interval and the expiration time becomes long, we can approximate the resulting share distribution by a lognormal function, and the option value at the current point can be calculated with the Black–Scholes option model. In 1973, Black and Scholes developed an option-pricing model based on *risk-free arbitrage*, which means that, over a short time interval, an investor is able to replicate the future payoff of the stock option by constructing a portfolio involving the stock and a risk-free asset. The model provides a trading strategy in which the investor is able to profit with a portfolio return equal to the risk-free rate. The Black–Scholes model is a continuous-time model and assumes that the resulting share distribution at expiration would be distributed lognormally.

European Call Option

A standard call option gives its holder the right, but not the obligation, to buy a fixed number of shares at the exercise price (K) on the maturity date. If the current price of the stock is S_0, the Black–Scholes formula for the price of the call is

$$C_{\text{call}} = S_0 N(d_1) - Ke^{-rT}N(d_2) \tag{13.10}$$

where

$$d_1 = \frac{\ln(S_0/K) + (r + \sigma^2/2)T}{\sigma\sqrt{T}}$$

$$d_2 = \frac{\ln(S_0/K) + (r - \sigma^2/2)T}{\sigma\sqrt{T}} = d_1 - \sigma\sqrt{T}$$

$N(\bullet) = $ the standard cumulative normal distribution function

$T = $ the time to maturity

$r = $ the risk-free rate of return

$\sigma = $ the volatility of the stock return.

The model is independent of the expected rate of return and the risk preference of investors. The advantage of the model is that all the input variables are observable except for the variance of the return, which easily can be estimated from historical stock price data.

European Put Option

We can value a put option in a similar fashion. The new option formula is:

$$C_{\text{put}} = Ke^{-rT}N(-d_2) - S_0 N(-d_1) \tag{13.11}$$

where

$$d_1 = \frac{\ln(S_0/K) + (r + \sigma^2/2)T}{\sigma\sqrt{T}}$$

$$d_2 = \frac{\ln(S_0/K) + (r - \sigma^2/2)T}{\sigma\sqrt{T}} = d_1 - \sigma\sqrt{T}$$

Note that, in developing the preceding continuous-time model, Black and Scholes address only the valuation of a European option, which pays no dividend. For an American option, we still need to use the *discrete-time valuation* given in the binomial lattice approach. Of course, we can consider the effect of paying a cash dividend in the model, but we will not address this embellishment or others, as our focus is not the evaluation of financial options.

EXAMPLE 13.7 Option Valuation under a Continuous-Time Process

Consider a stock currently trading at $40. For a strike price of $44, you want to price both a call and a put option that mature two years from now. The volatility of the stock σ is 40%, and the risk-free interest rate is 6%.

SOLUTION
Given: $S_0 = 40$, $K = \$44$, $r = 6\%$, $T = 2$ years, and $\sigma = 40\%$.
Find: C_{call} and C_{put}

The call-option value is calculated as

$$d_1 = \frac{\ln(40/44) + (0.06 + {}^{(0.4)^2}/_2)2}{0.4\sqrt{2}} = 0.3265$$

$$d_2 = 0.3265 - 0.4\sqrt{2} = -0.2392$$

$$N(d_1) = N(0.3265) = 0.628$$

$$N(d_2) = N(-0.2392) = 0.405$$

$$C_{\text{call}} = 40(0.628) - 44e^{-0.06(2)}(0.405)$$

$$= \$9.30$$

The put-option value is calculated as

$$N(-d_1) = N(-0.3265) = 0.372$$

$$N(-d_2) = N(0.2392) = 0.595$$

$$C_{\text{put}} = 44e^{-0.06(2)}(0.595) - 40(0.372)$$

$$= \$8.32$$

COMMENTS: Note that the put-option premium is smaller than the call-option premium, indicating that the upside potential is higher than the downside risk. The evaluation of the Black–Scholes formula can be tedious but can be automated with Excel. Figure 13.18 demonstrates a typical worksheet design to evaluate the Black–Scholes formulas.

	A	B	C	D	E	F	G
1							
2	**Black-Scholes Model**						
3							
4	Stock price (S_0) or Project Value (V_0)	40		=B4*EXP(-B9*B7)*B15-B5*EXP(-B8*B7)*B16			
5	Exercise Price or Investment Cost (K)	44					
6	Volatility (σ)	0.4		=B5*EXP(-B8*B7)*B18-B4*EXP(-B9*B7)*B17			
7	Time to Investment Decision (T)	2					
8	Risk-free Rate (r)	0.06					
9	Dividend Yield (ω)	0					
10	Call Value	9.30		=(LN(B4/B5)+(B8-B9+(B6^2)/2)*B7)/(B6*SQRT(B7))			
11	Put Value	8.32					
12				=B13-B6*SQRT(B7)			
13	d_1	0.326488561		=NORMSDIST(B13)			
14	d_2	-0.239196864					
15	$N(d_1)$	0.627972629		=NORMSDIST(B14)			
16	$N(d_2)$	0.405476467		=NORMSDIST(-B13)			
17	$N(-d_1)$	0.372027371					
18	$N(-d_2)$	0.594523533		=NORMSDIST(-B14)			
19							

Figure 13.18 Excel worksheet to evaluate the Black–Scholes formulas.

13.4 Real-Options Analysis

So far, we have discussed the conceptual foundation for pricing financial options. The idea is, "Can we apply the same logic to value the real assets?" To examine this possibility, we will explore a new way of thinking about corporate investment decisions. The idea is based on the premise that any corporate decision to invest or divest real assets is simply an option, giving the option holder a right to make an investment without any obligation to act. The decision maker therefore has more flexibility, and the value of this operating flexibility should be taken into consideration. Let's consider the following scenario as a starting point for our discussion.

13.4.1 How is Real Options Analysis Different?

Let's say a company is deciding whether to fund a large Internet project that could make or lose lots of money—most likely, lose it. A traditional calculation of net present value, which discounts projected costs and revenues into today's dollars, examines the project as a whole and concludes that it's a no go. But a real-options analysis breaks the project into stages and concludes that it makes sense to fund at least the value of the first stage. Here is how it works.

Step 1. **Evaluate each stage of the project separately:** Say the first stage, setting up a website, has a net present value of $-\$5$ million. The second stage, an e-commerce venture to be launched in one year, is tough to value, but let's say that the best guess of its net present value is $-\$30$ million.

Step 2. **Understand your options:** Setting up the website gives you the opportunity—but not the obligation—to launch the e-commerce venture later. In a year, you will know better whether that opportunity is worth pursuing. If it's not, all you have lost is the investment in the website. However, the second stage could be immensely valuable.

Step 3. **Reevaluate the project using an options mindset:** In the stock market, formulas such as the Black and Scholes calculate how much you should pay for an option to buy. Think of the first stage of your Internet project as buying such an option—risky and out of the money, but cheap.

Step 4. **Go figure:** Taking into account the limited downside of building a website and the huge (albeit iffy) opportunities it creates, we see that a real-options analysis could give the overall project a present value of, say, $70 million. So the no go changes to a go. Next, we will see how the logical procedure just outlined can be applied to address the investment risk inherent in strategic business decisions.

13.4.2 A Conceptual Framework for Real Options in Engineering Economics

In this section, we will conceptualize how the financial options approach can be used to value the flexibility associated with a real investment opportunity. A decision maker with an opportunity to invest in real assets can be viewed as having a *right, but not an obligation*, to invest. He or she therefore owns a real option similar to a simple call option on a stock:

- An investment opportunity can be compared to a call option on the present value of the cash flows (not counting the investment cost) arising from that investment (V).
- The investment outlay that is required to acquire the assets is the exercise price (I).
- The time to maturity is the time it takes to make the investment decision or the time until the opportunity disappears.

The analogy between a call option on stocks and a real option in capital budgeting is shown in Figure 13.19. The value of the real option at expiration depends on the value of the asset and would influence the decision as to whether to exercise the option. The decision maker would exercise the real option only if doing so were favorable.

Concept of Real Call Option

The NPW of an investment obtained with the options framework in order to capture strategic concerns is called the *expanded* (or *strategic*) NPW (ENPW or SNPW) and is equivalent to the option value calculated by the Black–Scholes model in Eq. (13.10). The value of the right to undertake the investment now is (R_{min}), which is the payoff if the option is exercised immediately. If $V > I$, the payoff is $V - I$, and if $V < I$, the payoff is 0. The true value of the option (R) that one needs to find is the ENPW of the real option

Figure 13.19 The analogy between a call option on a stock and a real option.

(Figure 13.20). Since one would undertake the investment later only if the outcome is favorable, the ENPW is greater than the conventional NPW. The value of the flexibility associated with the option to postpone the investment is the difference between the ENPW and the conventional NPW. This is the value of real-option or the value of flexibility defined as

$$\text{Value of "real-option"} = \text{NPW with option} - \text{NPW without option}$$
$$= \text{Expanded NPW} - \text{NPW without option} \qquad (13.12)$$

| • Value of "Real Option"
• Value of flexibility
• Real-option premium | = | • NPW with option
• Expanded NPW (ENPW)
• Strategic NPW (SNPW) | − | • NPW without option
• Conventional NPW |

Figure 13.20 Definition of value of real-option, commonly known as "value of flexibility" or "real-option premium."

Concept of a Real Put Option

The concept of a real put option is important from a strategic perspective. A put option gives its owner the right to dispose of an asset when it is favorable to do so. The put works like a guarantee or insurance when things go bad. An early-abandonment decision can be viewed as a simple put option. The option to abandon a project early may have value, as when an asset has a higher resale value in a secondary market than its use value. The put guarantees that the use value (V) of an asset does not fall below its market resale value (I). If it does, the option holder will exercise the put. In most instances, it is not possible to make an exact comparison between a standard put option on a stock and a real put option in capital budgeting.

Now we will explain the main conceptual real-options decision framework with a simple numerical example, assuming that we can resolve the uncertainty about the project cash flows by just waiting, say one year.

EXAMPLE 13.8 A Conceptual Real-Options Decision Framework When You Can Resolve Uncertainty by Waiting

Consider a project that requires a $200 million investment and expects to last three years with the following cash flows:

Demand	Probability	Annual Cash Flow
Good	0.25	$250 million
Moderate	0.30	$100 million
Poor	0.45	$ 35 million

The cost of capital that accounts for the market risk, known as the risk-adjusted discount rate, is 10%, and a risk-free interest rate is known to be 6%. Can we defer the project by one year and then implement it only if demand is either moderate or good?

SOLUTION

Given: Three demand scenarios, required investment = $200 million, risk-free rate = 6%, risk-adjusted discount rate = 10%, and $T = 1$ year.

Find: Option value for deferring the investment for one year.

Conventional Approach: We will determine the acceptability of the project on the basis of the expected NPW as

$$E[A_n] = 0.25(\$250) + (0.30)(\$100) + 0.45(\$35)$$

$$= \$108.25$$

$$E[PW(10\%)]_{\text{Benefits}} = \$108.25(P/A, 10\%, 3)$$

$$= \$269.20$$

$$E[PW(10\%)] = \$269.20 - \$200$$

$$= \$69.20 > 0$$

Since the expected NPW is positive, the project would be considered for immediate action. Note that we need to use the risk-adjusted discount rate, which reflects the risk inherent in estimating the project cash flows. If we proceed immediately with the project, its expected NPW is $69.20 million. However, the project is very risky. If demand is good, NPW = $421.72 million; if demand is moderate, NPW = $48.69 million; and if demand is poor, NPW = −$112.96 million. (See Figure 13.21.) Certainly, we are uneasy about the prospect of realizing a poor demand, even though its expected NPW is positive.

Real-Options Approach: Suppose we have an option to defer the investment decision by *one year*, and then we will gain additional information regarding demand.

Figure 13.21 Cash flow diagrams associated with demand scenarios in Example 13.8.

If demand is low, we will not implement the project. If we wait, the up-front cost and cash flows will stay the same, except they will be shifted ahead by a year. The following table depicts the demand scenarios:

Demand Scenarios	Future Cash Flows					NPW at $n = 0$	Probability
	0	1	2	3	4		
Good	▶	−$200	$250	$250	$250	$376.51	25%
Moderate	▶	−$200	$100	$100	$100	$37.40	30%
Poor	▶	0	0	0	0	$0	45%

Note that our assumption here is that there is no cost of delaying the investment in terms of the initial investment of $200 million. If there is, we simply increase the initial investment by the amount of cost of delaying. In finding the NPW, *we will use two different discount rates*. We will use 6% to discount the cost of the project at the risk-free rate, since the cost is known. Then we will discount the operating cash flows at the cost of capital. Under the various scenarios, we have

$$PW_{Good} = -\$200(P/F, 6\%, 1) + \$250(P/A, 10\%, 3)(P/F, 10\%, 1)$$

$$= \$376.51$$

$$PW_{Moderate} = -\$200(P/F, 6\%, 1) + \$100(P/A, 10\%, 3)(P/F, 10\%, 1)$$
$$= \$37.40$$
$$PW_{Poor} = \$0$$

Next, we calculate the project's expected NPW if we wait by using the preceding three scenarios with their given probabilities:

$$E[NPW]_{wait} = (0.25)(\$376.51) + 0.30(\$37.40) + 0.45(\$0)$$
$$= \$105.35$$

This is exactly what we call the expanded NPW (ENPW). Now we can compare the option to wait against the conventional NPW analysis. Note that the option to wait is higher ($105.35 million versus $69.20 million):

$$\text{Value of real-option} = \$105.35 - \$69.20 = \$36.15$$

In other words, the waiting option is worth $36.15 million. If we implement the project today, we gain $69.20 million but could lose the future project, worth $105.35 million for a net loss of $36.15 million on the differential project. Therefore, we should wait and decide next year whether to implement the project based on the observed demand. Note also that the cash flows are less risky under the option to wait, since we can avoid the low cash flows. Also, the cost of implementing the project may not be risk free. Given the change in risk, perhaps we should use different rates to discount the cash flows. Another important observation is that the ENPW is equivalent to the EVPI concept we discussed in Chapter 12, as this is the value that we can attain if we have the perfect information about the future.

COMMENTS: The option to wait resembles a financial call option: We get to "buy" the project for $200 million in one year if the projected value is greater than $200 million after one year. This is like a call option with an exercise price of $200 million and an expiration date of one year. If we use the Black–Scholes model to value the option to wait, the required inputs are

$K = I$, which is the exercise price = cost to implement the project = $200 million
r = risk-free rate = 6%
T = time to maturity = 1 year
$S_0 = V$ = current stock price = estimated current value of the project
σ^2 = variance of stock return = estimated variance on project return

In general, we need to estimate two unknown parameters: V and σ^2. Estimating V is equivalent to matching the current stock price in a financial option. In a real option, V is the value of the project today, which is found by computing the present value of all of the project's future expected cash flows at its cost of capital (or risk-adjusted discount rate). Estimating σ^2 will be more difficult, but the procedures will be discussed in Section 13.6.

In the following sections, we will demonstrate how we can come up with these values and use them in calculating the real-option premium.

13.5 Simple Real-Option Models

Most common types of real options can be classified into three categories, as summarized in Figure 13.22. Some of the unique features of each option will be examined with numerical examples.

13.5.1 Option to Defer Investment

The option to defer an investment is similar to a call option on stock. Suppose that you have a new product that is currently selling well in the United States and your firm is considering expanding the market to China. Because of many uncertainties and risks in the Chinese market (pricing, competitive pressures, market size, and logistics), the firm is thinking about hiring a marketing firm in China that can conduct a test market for the product. Any new or credible information obtained through the test market will determine whether or not the firm will launch the product. If the market is ready for the product, the firm will execute the expansion. If the market is not ready for the product, then the firm may wait another year or walk away and abandon the expansion plan altogether. In this case, the firm

Common Real Options

Real Option Category	Real Option Type	Description	Examples
Invest/ grow	Scale up	Well-positioned businesses can scale up later through cost-effective sequential investments as the market grows.	• High technology • R&D intensive • Multinational • Strategic acquisition
	Switch up	A flexibility option to switch products or processes on plants given a shift in the underlying price or demand of inputs or outputs.	• Small-batch goods producers • Utilities • Farming
	Scope up	Investments in proprietary assets in one industry enables a company to enter another industry cost effectively. Link and leverage.	• Companies with lock-in • De facto standard bearers
Defer/ learn	Study/ start	Delay the investment until more information or skill is acquired.	• Natural-resource companies • Real-estate development
Disinvest/ shrink	Scale down	Shrink or shut down a project partway through if new information changes the expected payoffs.	• Capital-intensive industries • Financial services • New-product introduction • Airframe order cancellations
	Switch down	Switch to more cost-effective and flexible assets as new information is obtained.	• Small-batch goods producers • Utilities
	Scope down	Limit the scope of (or abandon) operations in a related industry when there is no further potential in a business opportunity.	• Conglomerates

Figure 13.22 Types of real options.
Source: "Get Real—Using Real Options in Security Analysis," *Frontiers of Finance*, Volume 10, by Michael J. Mauboussin. Credit Suisse First Boston, June 23, 1999.

will lose only the cost associated with testing the market. An American expansion option will provide the firm with an estimate of the value to spend in the market research phase.

EXAMPLE 13.9 Delaying Investment: Value of Waiting— Real Call Option

A firm is preparing to manufacture and sell a new brand of golf balls. This ball will be marketed for recreational purposes (not USGA-approved for competition). Consider the following financial information regarding the golf ball project:

- The investment costs are estimated at $50M today, and a "most likely" estimate for net cash inflows is $12M per year for the next five years.
- Due to the high uncertainty in the demand for this new type of golf balls, the volatility of cash inflows is estimated at 50%.
- It is assumed that a two-year "window of opportunity" exists for the investment decision.
- If the firm delays the investment decision, the investment costs are expected to increase 10% per year.
- The firm's risk-adjusted discount rate (MARR) is 12%, and the risk-free rate is 6%.

Should the firm invest in this project today? If not, is there value associated with delaying the investment decision for two years?

SOLUTION

Given: Estimated project cash flows, required investment = $50 million, risk-free rate = 6%, risk-adjusted discount rate = 12%, $\sigma = 50\%$, and $T = 2$ years.

Find: Option value for deferring the investment for two years.

Conventional Approach: Should the Firm Invest Today? If the traditional NPW criterion is used, the decision is not to undertake the investment today, as it has a negative NPW of $6.74 million:

$$PW(12\%) = -\$50M + \$12M(P/A, 12\%, 5)$$
$$= -\$6.74M$$

Real-Options Approach: Is there value to waiting two years? From an options approach, the investment will be treated as an opportunity to wait and then to undertake the investment two years later if events are favorable. As shown in Figure 13.23, this delay option can be viewed as a call option. If the value V of the project is greater than the investment cost I two years from now, then the option to undertake the project should be exercised. If not, it expires worthless. Therefore,

$$\text{Delay option} = \max[V_2 - I_2, 0]$$

where V_2 is a random variable that is dependent upon market demand.

Now, assume that one can use the Black–Scholes formula[4] to value the opportunity to undertake this investment in year 2. The opportunity to wait can be considered

---

[4]One can arrive at approximately the same value with the binomial model by dividing the one-year time to maturity into sufficiently small time intervals.

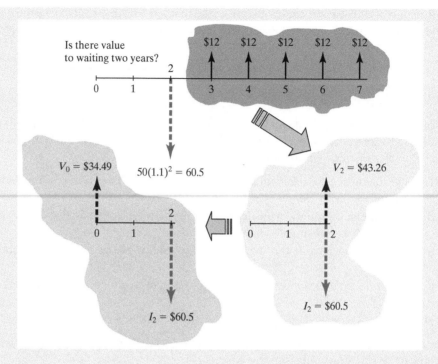

Figure 13.23 Transforming cash-flow data to obtain input parameters V_0 and I_2 for option valuation.

as a European call option on the present value of the future cash flows, $V_0 = \$34.49$ million, with an exercise price of $I = \$60.5$ million expiring two years from now ($T = 2$). Since $r = 6\%$ and $\sigma = 50\%$, using the Black–Scholes formula, we obtain $d_1 = -0.2715$, $d_2 = -0.9786$, $N(d_1) = 0.3930$, and $N(d_2) = 0.1639$. The value of the real option is therefore:

$$C = \text{ENPW} = 34.49\,(0.3930)$$
$$- 60.5e^{-(0.06)2}\,(0.1639)$$
$$= 13.5549 - 8.7941$$
$$= \$4.76$$
$$\text{Value of real-option} = \$4.76 - (-\$6.74)$$
$$= \$11.50\text{ M}$$

Thus, the value of retaining the flexibility of having the delay option is worth $11.50M. The ability to wait provides a decision maker a strategic NPW of $4.76M instead of a negative NPW of $6.74M. On the one hand, if the market for the product turns out to be favorable, the company will exercise the real option in the money and undertake the investment in year 2. On the other hand, if the market two years later turns out to be unfavorable, the decision will be not to undertake the investment. In summary, you could pay up to $11.50M to have this delay option.

COMMENTS: What exactly does the $4.76M delay value imply? Suppose the firm's project portfolio consists of ten projects, including this golf ball project. Then if the other nine projects have a net present value of $100M, the value of the firm using standard NPW still would be $100M, because the golf ball project would not be accepted, since it has a negative NPW. If the option to invest in the golf balls is included in the valuation, then the value of the firm's portfolio is $104.76M. Therefore, the delay option that the firm possesses is worth an additional $4.76M.

13.5.2 Patent and License Valuation

A patent or license provides a firm the right, but not the obligation, to develop a product (or land) over some prescribed time interval. The right to use the patent has value if and only if the expected benefits (V) exceed the projected development costs (I), which is max $\{V - I, 0\}$. This right is considered a real option that can be used to value the worth of a license to a firm.

EXAMPLE 13.10 Option Valuation for Patent Licensing

A technology firm is contemplating purchasing a patent on a new type of digital-TV converter. The patent would provide the firm with three years of exclusive rights to the digital-TV converter technology. Estimates of market demand show net revenues of $50M per annum for seven years. The estimated cost of production is $200M. Assume that MARR = 12%, $r = 6\%$, and the volatility due to market uncertainty related to product demand is 35%. Determine the value of the patent.

SOLUTION

- **Conventional Approach: Should the Firm Purchase the License Now?** If the traditional NPW criterion is used, the NPW without option would be:

$$PW(12\%) = -\$200M + \$50M(P/A, 12\%, 7)$$

$$= \$28.19M$$

- **Real Options Approach:** The patent licensing provides the right to use the digital-TV converter technology anytime over the next three years. Therefore,

$$V_3 = \$50M(P/A, 12\%, 7) = \$228.19M$$

$$V_0 = \$228.19M(P/F, 12\%, 3) = \$162.42M$$

$$I_3 = \$200M$$

$$T = 3$$

$$r = 6\%$$

$$\sigma = 35\%$$

Substituting these values into Eq. (13.10), we obtain the Black–Scholes value of this call option: $36.95M.

$$\text{Value of real-option} = \$36.95M - \$28.19M = \$8.76M$$

With the real-options analysis, the firm now has an upper limit on the value of the patent. At the most, the firm should pay $8.76M for it, and the firm can use this value as part of the negotiation process.

13.5.3 Growth Option—Option to Expand

A growth option occurs when an initial investment is required to support follow-on investments, such as (1) an investment in phased expansion, (2) a Web-based technology investment, and (3) an investment in market positioning. The amount of loss from the initial investment represents the call-option premium. Making the initial investment provides the option to invest in any follow-on opportunity. For example, suppose a firm plans on investing in an initial small-scale project. If events are favorable, then the firm will invest in a large-scale project. The growth option values the flexibility to invest in the large-scale project if events are favorable. The loss on the initial project is viewed as the option premium.

EXAMPLE 13.11 Valuation of a Growth Option

A firm plans to market and sell its product in two markets: locally and regionally. The local market will require an initial investment, followed by some estimated cash inflows for three years. If events are favorable, the firm will invest and sell the product regionally, expecting benefits for four years. Figure 13.24 details the investment opportunities related to the growth options of this project.

Assume that the firm's risk-adjusted discount rate is 12% and the risk-free interest rate is 6%. Determine the value of this growth option.

SOLUTION
The NPW of each phase is

$$\text{PW}(12\%)_{\text{Small}, 0} = -\$30 + \$10(P/F, 12\%, 1) + \$12(P/F, 12\%, 2)$$
$$+ \$14(P/F, 12\%, 3)$$
$$= -\$1.54M \text{ [in year 0]}$$
$$\text{PW}(12\%)_{\text{Large}, 3} = -\$60 + \$16(P/F, 12\%, 1) + \$18(P/F, 12\%, 2)$$
$$+ \$20(P/F, 12\%, 3) + \$20(P/F, 12\%, 4)$$
$$= -\$4.42M \text{ [in year 3]}$$

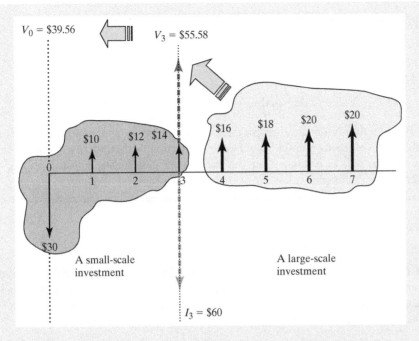

Figure 13.24 Cash flows associated with the two investment opportunities: a small-scale investment followed by a large-scale investment. The large-scale investment is contingent upon the small-scale investment.

Therefore, the total net present worth of the two-phased investment project is

$$\text{PW}(12\%)_{\text{Total, 0}} = -\$1.54\text{M} - \$4.42\text{M}(P/F, 12\%, 3)$$

$$= -\$1.54\text{M} - 3.14\text{M}$$

$$= -\$4.68\text{M} < 0$$

After investing in the small-scale project, the firm is not obligated to invest in the large-scale phase; hence, it is an option. The value of the two investment opportunities can be calculated in two steps.

Step 1. The option to expand can be valued as a European call option with the use of the Black–Scholes equation (Eq. 13.10). The option inputs are

$$V_{\text{Large, 3}} = \$55.58M$$

$$V_{\text{Large, 0}} = \$55.58M(P/F, 12\%, 3) = \$39.56M$$

$$I_3 = 60M$$

$$T = 3$$

$$r = 6\%$$

$$\sigma = 40\%$$

$$C = \$7.54\text{M}$$

Step 2. The total value of the two investment opportunities is

$$ENPW = -\$1.54M + \$7.5M = \$6M$$

Step 3. The value of real-option is

$$\text{Value of real-option} = \$6M - (-\$4.68) = \$10.68M$$

Because the option premium for these combined investment opportunities exceeds $6M, the investments should be pursued. Even though the initial investment in the local market will lose money, the benefits of the large-scale investment later will offset the initial losses.

13.5.4 Scale-Up Option

A firm has the right to scale up an investment *at any time* if the initial project is favorable. In other words, a firm has the option to increase its investment in a project in return for increased revenues. In a way, this is just a growth option, but it will be valued on a binomial basis. Example 13.12 illustrates how we value a scale-up option.

EXAMPLE 13.12 A Scale-Up Option Valuation Using the Binomial Lattice Approach

A firm has undertaken a project in which the firm has the option to invest in additional manufacturing and distribution resources (or scale up). The project's current value is $V_0 = \$10M$. Anytime over the next three years, the firm can invest an additional $I = \$3M$ and receive an expected 30% increase in net cash flows and, therefore, a 30% increase in project value. The risk-free interest rate is 6% compounded annually and the volatility of the project's value is 30%. Use a binomial lattice with a one-year time increment to value the scale-up option as an American option. (You can scale up anytime you see fit to do so.)

SOLUTION

We will solve the problem in two steps, the first of which is to determine how the value of the underlying asset changes over time and the second of which is to create the option valuation lattice.

Step 1. Binomial Approach: Lattice Evolution of the Underlying Asset. First, we determine the following parameters to build a binomial lattice that shows how the value of the project changes over time:

$$u = e^{0.30(1)} = 1.35$$

$$d = \frac{1}{u} = 0.74$$

$$q = \frac{1.06 - 0.74}{1.35 - 0.74} = 0.5241$$

The equity binomial lattice looks like the one in Figure 13.25; the values are created in a forward multiplication of up and down factors, from left to right.

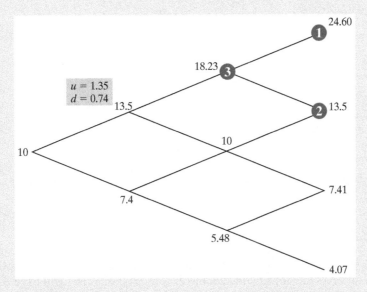

Figure 13.25   Event tree: Lattice evolution of the underlying value of a project.

Step 2. Binomial Approach: Option Valuation Lattice.   Once you obtain the lattice evolution of the underlying project, we can create the option valuation lattice in two steps: the valuation of the terminal nodes and the valuation of the intermediate nodes. The calculation proceeds in a backward manner, starting from the terminal nodes: The nodes at the end of the lattice are valued first, going from right to left.

All the required calculations and steps are shown in Figure 13.26. To illustrate, we will take a sample circled node (❶) in the figure:

• Node ❶ reveals a value of $24.60, which is the value the firm can expect to achieve without any scaling up. Now, with the scale-up option, the project value can increase 30% at the expense of $3 million of investment, for a new value of

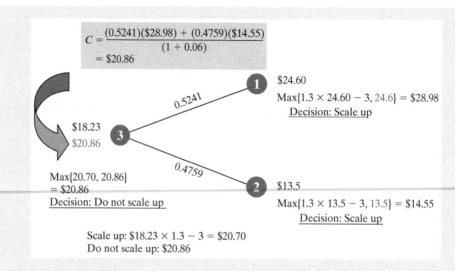

Figure 13.26 Decision tree diagram illustrating the process of reaching a decision at node ❸.

$1.3(24.60) - 3 = \$28.98$ million. Since this figure is larger than $24.60, the firm will opt for scaling up the operation.

- Similarly, the firm will reach the same decision for node ❷, which yields a scale-up result of $14.55.
- Moving on to node ❸, we see that the firm again has one of two options: to scale up the operation at that point rather than in the future, or to keep the option open for the future in the hope that when the market is up, the firm will have the ability to execute the option. Mathematically, we can state the options as

$$\text{Scale-up now } (n = 2): 1.3(\$18.23) - \$3 = \$20.70$$

$$\text{Keep the option open: } C = \frac{(0.5241)(\$28.98) + (0.4759)(\$14.55)}{(1 + 0.06)} = \$20.86$$

Thus, the decision is to keep the option open. Finally at year 0, we compute

$$C = \frac{(0.5241)(\$15.29) + (0.4759)(\$7.66)}{(1 + 0.06)}$$

$$= \$11.00$$

The end-result option value on the binomial tree shown in Figure 13.27 represents the value of the project with the scale-up option. The value of the scale-up option itself is

$$\text{Option value} = \$11.00 - \$10 = \$1M$$

Therefore, the option to expand the project in the future is worth $1,000,000.

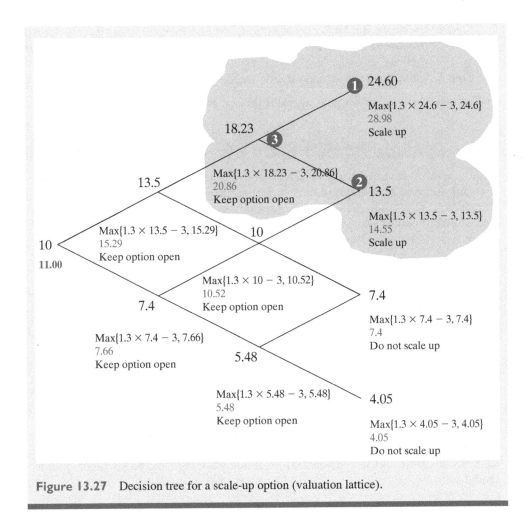

Figure 13.27 Decision tree for a scale-up option (valuation lattice).

13.5.5 Compound Options

Many large-scale project initiatives are taken place over several phases where management can decide to expand, scale back, or abandon the project after gaining new information to resolve uncertainty. For example, a capital investment project divided into multiple phases, including permitting, design, engineering, and construction, can either be terminated or continued into the next stage, depending on market conditions at the end of each phase. These are compound options, where exercising one option generates another, thereby making the value of one option contingent upon the value of another option. A sequential compound option exists when a project has multiple phases and later phases depend on the success of earlier phases. The logical steps to use in handling this sequential compound-option problem are as follows:

Step 1. Calculate the initial underlying asset lattice by first calculating the up and down factors and evolving the present value of the future cash flow over the planning horizon.

Step 2. Calculate the longer term option first and then the shorter term option, because the value of a sequential compound option is based on earlier options.

Step 3. Calculate the combined-option value.

Example 13.13 illustrates how we value the compound option.

EXAMPLE 13.13 **A Real-World Way to Manage Real Options[5]**

Suppose a firm is considering building a new chemical plant. Permits and preparation for the plant cost $60 million. At the end of year 1, the firm has the right to invest $400 million on the design phase. Upon completion of the design, the firm has another right to invest $800 million in building the plant over the next two years. The firm's risk-adjusted discount rate is 10.83%. If the plant existed today, its market value would be $1 billion. The volatility of the plant's value (σ) is 18.23%. The risk-free interest rate is known to be 8%. Determine the NPW of the project, the option values at each stage of the decision point, and the combined-option value of the project.

STRATEGY: Note that there are three investment phases: the permit-and-preparation phase (phase 0), which provides an option to invest in the design phase; (2) the design phase (phase 1), which provides the option to invest in plant construction; and (3) the construction phase (phase 2). Each phase depends upon a decision made during the previous phase. Therefore, investing in phase 2 is contingent upon investing in phase 1, which in turn is contingent upon the results of phase 0. The first task at hand is to develop an event tree or a binomial lattice tree to see how the project's value changes over time. We need two parameters: upward movement u and downward movement d. Since the time unit is one year, $\Delta t = 1$; and since $\sigma = 0.1823$, we compute u and d as

$$u = e^{\sigma\sqrt{\Delta t}} = e^{0.1823\sqrt{1}} = 1.2$$

$$d = \frac{1}{u} = 0.833$$

The lattice evolution of the underlying project value will look like the event tree shown in Figure 13.28.

SOLUTION

(a) **Traditional NPW calculation:** Using a traditional, discounted cash-flow model, we may calculate the present value of the expected future cash flows, discounted at 10.83%, as

$$PW(10.83\%)_{\text{Investment}} = \$60 + \frac{\$400}{\$1.1083} + \frac{\$800}{1.1083^3}$$

$$= \$1,008.56 \text{ million}$$

---

[5] This example is based on the article "A Real World Way to Manage Real Options," by Tom Copeland, *Harvard Business Review*, March 2004.

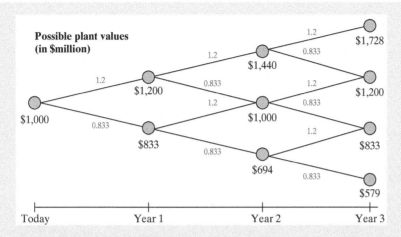

Figure 13.28 The event tree that illustrates how the project's value changes over time.

$$PW(10.83\%)_{\text{Value of investment}} = \$1,000 \text{ million}$$

$$NPW = \$1,000 - \$1,008.56$$

$$= -\$8.56 < 0 \text{ (reject)}$$

Since the NPW is negative, the project would be a no-go one. Note that we used a cost of capital k of 10.83% in discounting the expected cash flows. This cost of capital represents a risk-adjusted discount rate.

(b) **Real-options analysis:** Using a process known as backward induction, we may proceed to create the option valuation lattice in two steps: the valuation of the terminal nodes and the valuation of the intermediate nodes. In our example, we start with the nodes at year 3.

Step 1. Valuing Phase 2 Options: We pretend that we are at the end of year 3, at which time the project's value would be one of the four possible values ($1,728, $1,200, $833, and $579), as shown in Figure 13.29. Clearly, the project values at the first three nodes exceed the investment cost of $800 million, so our decision should be to invest in phase 2. Only if we reach the last node ($579) will we walk away from the project.

Step 2. What to Do in Year 2: In order to determine the option value at each node at year 2, we first need to calculate the risk-neutral probabilities. For node B, the option value comes out to be $699 million:

$$u = e^{\sigma\sqrt{\Delta t}} = e^{0.1823\sqrt{1}} = 1.2$$

$$d = \frac{1}{u} = 0.833$$

$$R = 1 + r = 1 + 0.08 = 1.08$$

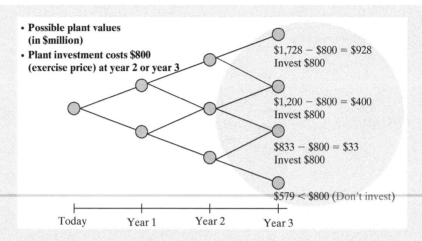

- **Possible plant values (in $million)**
- **Plant investment costs $800 (exercise price) at year 2 or year 3**

$1,728 − $800 = $928
Invest $800

$1,200 − $800 = $400
Invest $800

$833 − $800 = $33
Invest $800

$579 < $800 (Don't invest)

Today   Year 1   Year 2   Year 3

Figure 13.29   Valuing phase 2 options.

$$q = \frac{R - d}{u - d} = \frac{1.08 - 0.833}{1.2 - 0.833} = 0.673$$

$$1 - q = 1 - 0.673 = 0.327$$

$$C = \frac{1}{1 + 0.08}[0.673(\$928) + 0.327(\$400)]$$

$$= \$699$$

Thus, keeping the option open is more desirable than exercising it (committing $800 million), as the net investment value is smaller than the option value ($1,440 − $800 = $640).

The option values for the rest of the nodes at the end of year 2 are calculated in a similar fashion and are shown in Figure 13.30. Note that in year 2, regardless of which node we arrive at, we should not exercise the option.

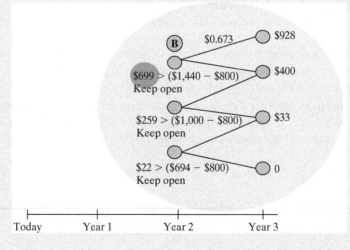

B $0.673 $928

$699 > ($1,440 − $800)
Keep open $400

$259 > ($1,000 − $800)
Keep open $33

$22 > ($694 − $800)
Keep open   0

Today   Year 1   Year 2   Year 3

Figure 13.30   Sequential compound options (valuation lattice).

Step 3. **What to Do in Year 1:**   Moving on to the nodes at year 1, we see that, at node C in Figure 13.31, the value of executing the option is $514 − $400 = $114 million, and

$$\text{Max}\{\$514 - \$400, 0\} = \$114$$

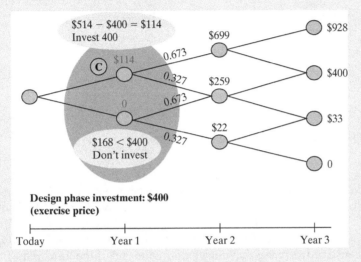

Design phase investment: $400
(exercise price)

Figure 13.31   Valuing phase 1 options.

Keep in mind that the value $514 million comes from the option valuation lattice from year 2:

$$\frac{1}{1 + 0.08}[(0.673)(\$699) + (0.327)(\$259)] = \$514$$

To realize the option value of $514 million at node C, you need to invest $400 million. Therefore, the value of continuing the project is $514 − $400 = $114 million. Note that at the end of first period, the second option expires. Therefore, it must be exercised at a cost of $400 million or be left unexercised (at no cost). If it is exercised, the payouts are directly dependent not on the value of the underlying project ($1,200 million) but on the value provided by the option to invest at the next stage.

Step 4. **Standing at Year 0:**   We can estimate the present value of the compound option by recognizing that we can either keep the first option open or exercise it at a cost of $60 million. As shown in Figure 13.32, the value of that option is determined by

$$C = \frac{1}{1.08}[(0.673)(\$114) + (0.327)(0)]$$

$$= \$71.039$$

$$\text{Max}\{\$71 - \$60, 0\} = \$11$$

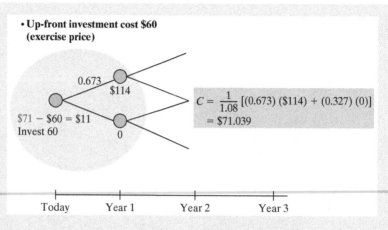

Figure 13.32 The value of phase 0 options.

We can interpret $71.039 million as the compound option value of a project that has a present value of $1,000 million today, has a standard deviation of 18.23% per year, and requires completing three investments: $60 million for the first stage, $400 million for a design stage, and $800 million for a construction phase that must start by the end of the third year. If the start-up cost is greater than $71.039 million, the project would be rejected; otherwise, it would be accepted. Figure 13.33 illustrates how we go about making various investment decisions along the paths of the event tree.

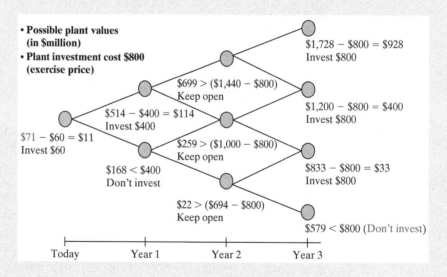

Figure 13.33 Sequential compound options (combined lattice).

Now we can summarize what we have done:

- The ENPW is $71.039 million. Since the required initial investment is only $60 million, it is worth keeping the option by initiating the permit and preparation work. Overall, the real option premium for this compound option is

$$\text{Value of real-option} = \$71.039M - (-\$8.56M)$$

$$= \$79.60M > \$60M$$

13.6 Estimating Volatility at the Project Level

Up to this point, we assumed that the volatility of project cash flows is known to us or can be estimated somehow. Unfortunately, the volatility of cash flows is probably one of the most difficult input parameters to estimate in any real-options analysis. The volatility that we need for the binomial tree is the volatility of the value of project's yearly growth rate.

13.6.1 Mathematical Relationship between σ and σ_T

One approach we will adopt is to use the mathematical relationship[6] between σ (project volatility) and the parameters for the project value distribution at T. The mean and variance of the project value distribution at T can be expressed in terms of the project volatility (σ), V_0, r, and T, as in Eq. (13.13):

$$\mu_T = V_0 \cdot e^{rT}$$
$$\sigma_T^2 = V_0^2 \cdot e^{2rT} \cdot (e^{\sigma^2 T} - 1) \tag{13.13}$$

where

V_0 = project value if it is available today

μ_T = mean of the project value distribution at period T

σ_T^2 = variance of the project value distribution at period T

Or, we can rewrite Eq. (13.13) as

$$\sigma = \sqrt{\frac{\ln\left(\left(\frac{\sigma_T}{\mu_T}\right)^2 + 1\right)}{T}} \tag{13.14}$$

Figure 13.34 illustrates these mathematical relations. What this relationship suggests is if we can come up with the V_T distribution, then we may be able to estimate the project volatility through Eq. (13.14) that could be used in our binomial lattice.

---

[6] *Investment under Uncertainty*. Dixit, A. K. and R. S. Pindyck, Princeton University Press, Princeton, NJ, 1994.

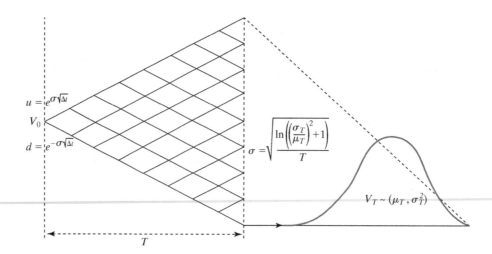

Figure 13.34 Mathematical relationships between σ and σ_T. If Δt is small compared to 1, V_T is log-normally distributed.

13.6.2 Estimating V_T Distribution

To estimate the project value distribution at the option expiration, we may use either an analytical approach or the Monte Carlo simulation to combine various random elements in project cash flows. For most applications, simulation approach would be more practical than the analytical approach. In doing so, we may use the following procedure.

Step 1. Estimate the project cash flows over the project life $(T + n)$ if the project would be undertaken at the end of option life T. Typically, each cash flow component may be estimated in terms of the three-point estimate (low, most-likely, and high).

Step 2. Obtain the V_T distribution by aggregating the project cash flow at each period and then discount them at a risk-free rate, as we are dealing with cash-flow distributions rather than an estimated cash-flow series.

Step 3. Compute the mean (μ_T) and variance (σ_T^2) of the V_T distribution using Eqs. (12.11) through (12.13) if the mean and variance of cash flow for each period is known. If cash flow estimates are given in terms of individual probability distribution, use software, such as Crystal Ball or @RISK to obtain the statics of the V_T distribution.

Step 4. Compute σ by using Eq. (13.14).

Figure 13.35 summarizes the process that we have described, and Example 13.14 gives a numerical example.

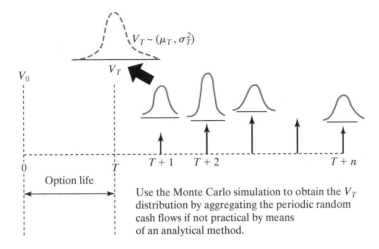

Figure 13.35 The process of obtaining the V_T distribution by aggregating the periodic random cash-flow series. A risk simulation software such as Crystal Ball or @RISK may be used.

EXAMPLE 13.14 Estimating the Project Volatility for a Simple Deferral Option

If a project under consideration is deferred for two years, the required investment would be $35 million, and it would be expected to last three years with the following cash flows in thousand dollars.

	A_n		
Period (n)	Pessimistic Estimate	Most-Likely Estimate	Optimistic Estimate
0			
1	(Delayed for two years)		
2			
3	$2,000	$16,000	$30,000
4	$4,500	$20,000	$40,000
5	$3,000	$18,000	$38,000

The cost of capital that accounts for the market risk (known as the risk-adjusted discount rate) is 10%, and a risk-free interest rate is known to be 6%. Estimate the project volatility to be used in option calculation.

STRATEGY: The estimating procedure that is common in practice is to make three-point estimates as described in the problem statement. The meaning of these three estimates is

- Optimistic estimate H: Everything will go as well as reasonably possible.
- Pessimistic estimate L: Everything will go as poorly as reasonably possible.
- Most-likely estimate (M_o): The projected amount of net cash flow is most likely to occur.

Then, the three estimated are used as the upper bound, the lower bound, and the mode of the corresponding cash-flow probability distribution, respectively. The probability distribution itself could be assumed to be a beta distribution with a standard deviation of one-sixth of the spread between the upper and lower bounds (the optimistic and pessimistic estimates). With these assumptions, the mean and variance become explicit functions of the bounds and mode. Given the estimates of the bounds and mode of cash-flow distribution of A_n at the end of period n, we find:

$$E[A_n] = \frac{L + 4M_o + H}{6} \tag{13.15}$$

and

$$\text{Var}[A_n] = \left(\frac{H - L}{6}\right)^2 \tag{13.16}$$

The model of an underlying beta distribution has been assumed primarily for convenience, but we could use any other distribution, such as a triangular distribution.

SOLUTION

Given: Project cash flow estimates for each period over the project life, risk-free rate $= 6\%$, and $T = 2$.

Find: μ_T, σ_T^2, project volatility (σ), and option premium.

Analytical Approach:

$n = 3$: Using Eqs. (13.15) and (13.16), we find that

$$E[A_3] = \frac{\$2,000 + 4(\$16,000) + \$30,000}{6} = \$16,000$$

$$\text{Var}[A_3] = \left(\frac{30,000 - 2,000}{6}\right)^2 = 21,777,778$$

Similarly, we can calculate the means and variances for other periods as

	A_n				
Period (n)	**Pessimistic Estimate**	**Most-Likely Estimate**	**Optimistic Estimate**	**E[A_n]**	**Var[A_n]**
0					
1		(Delayed for two years)			
2					
3	$2,000	$16,000	$30,000	16,000	21,777,778
4	$4,500	$20,000	$40,000	20,750	35,006,944
5	$3,000	$18,000	$38,000	18,833	34,027,778

Using Eqs. (12.11) and (12.12) with the assumption of a statistical independence[7] among the cash flows, we compute the project value distribution at period 1 (V_2) at a risk-free discount rate[8] of 6% as

$$E[V_2] = \mu_1 = \frac{\$16,000}{(1 + 0.06)} + \frac{\$20,750}{(1 + 0.06)^2} + \frac{\$18,833}{(1 + 0.06)^3} = \$49,374$$

$$Var[V_2] = \sigma_1^2 = \frac{21,777,778}{(1 + 0.06)^2} + \frac{35,006,944}{(1 + 0.06)^4} + \frac{34,027,778}{(1 + 0.06)^6} = (8,432)^2$$

Using Eq. (13.14), we estimate the project's volatility (σ) as

$$\sigma = \sqrt{\frac{\ln\left(\left(\frac{\sigma_T}{\mu_T}\right)^2 + 1\right)}{T}} = \sqrt{\frac{\ln\left(\left(\frac{8,432}{49,374}\right)^2 + 1\right)}{2}}$$

$$= 0.1199$$

Then, we may use 11.99% as the project volatility in real-options analysis.

Simulation Approach: If the cash flow estimates (A_n) are given in terms of individual probability distribution, we may use Crystal Ball to develop the V_2 distribution. Once we obtain $E[V_2]$ and $Var[V_2]$, the remaining procedure will be the same as before.

Option Value Calculation: Finally, we may calculate the option premium by either binomial lattice (discrete process assumption) or Black and Scholes (continuous-process assumption).

First, we need to determine the value of V_0. Here it is not logical to use the value $E[V_2]$, as it was based on a risk-free rate. Recall that we are not dealing with a V_0 distribution directly—instead we are using a point estimate, ignoring the variance figure. In that case, we need to use a risk-adjusted discount rate to determine the V_0 value.

$$V_0 = \left(\frac{\$16,000}{(1 + 0.10)} + \frac{\$20,750}{(1 + 0.10)^2} + \frac{\$18,833}{(1 + 0.10)^3}\right)(P/F, 10\%, 2)$$

$$= \$37,888$$

Or, the conventional NPW is

$$\text{Conventional NPW} = V_0 - I = \$37,888 - \$35,000$$

$$= \$2,888$$

- **Option Value Calculation by Black–Scholes Model:** Substituting the estimated parameters into Eq. (13.10), we find that the option value is $8,954:

$$V_0 = \$37,888, K = \$35,000, r = 0.06, T = 2, \sigma^2 = (0.1199)^2 = 0.0144$$

[7] If the cash flows are correlated dependent random variables, we use Eq. (12.13) to calculate the variance of the project value distribution.

[8] Once again, here we need to use a risk-free rate, as any randomness in project cash flows is reflected in the project value distribution.

$$d_1 = \frac{\ln(37,888/35,000) + (0.06 + {}^{0.0144}/_2)(2)}{0.1199\sqrt{2}}$$

$$= 1.26008$$

$$d_2 = 1.26008 - 0.1199\sqrt{2}$$

$$= 1.090504$$

$$N(d_1) = 0.896178$$

$$N(d_2) = 0.862254$$

$$C_{\text{call}} = 37,888(0.896178) - 35,000e^{-0.06(2)}(0.862254)$$

$$= \$7,188$$

- **Option Value Calculation by Binomial Model:** If we use the discrete version of the binomial lattice model, we need to create an event tree that models the potential values of the underlying risky asset. This tree contains no decision nodes and simply models the evolution of the underlying asset. Defining T as the number of years per upward movement and σ^2 as the annualized volatility of the underlying project value, we can determine the up and down movements with Eq. (13.10):

$$\text{Up movement} = u = e^{\sigma\sqrt{\Delta t}}$$

$$\text{Down movement} = d = \frac{1}{u}$$

Substituting numerical values into these formulas with $\Delta t = 1$ yields

$$u = e^{\sigma\sqrt{\Delta t}} = e^{0.1199\sqrt{1}} = 1.1274$$

$$d = \frac{1}{u} = \frac{1}{1.1274} = 0.8870$$

When we add decision points to an event tree, it becomes a decision tree (Figure 13.36).

To determine the value of the project with the flexibility to defer it, we work backward through the decision tree (Figure 13.37), using the risk-neutral probability approach at each node:

$$R = 1 + r = 1 + 0.06 = 1.06$$

$$q = \frac{R - d}{u - d} = \frac{1.06 - 0.8870}{1.1274 - 0.8870} = 0.7196$$

$$1 - q = 1 - 0.7196 = 0.2804$$

$$C = \$7,100$$

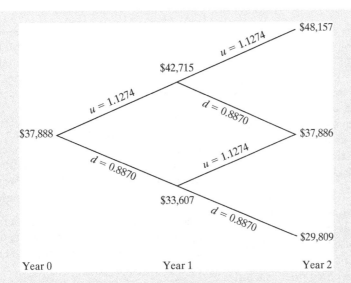

Figure 13.36 Event tree: Binomial lattice.

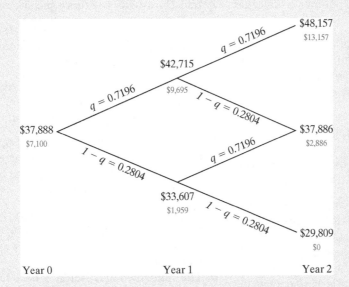

Figure 13.37 Option value calculation by binomial model.

Note that this amount ($7,100) is calculated under the binomial model, which is different from the value ($7,188) obtained under the Black–Scholes model earlier. This is not surprising, as the discrete model with $T = 2$ is a rough approximation of the continuous version. Since the option itself is worth $7,188 million, the value of real-option is

$$\text{Value of real-option} = \$7,188 - \$2,888 = \$4,300$$

Clearly, it makes sense to defer the investment project for two years in this case.

SUMMARY

In this chapter, we have introduced a new tool to evaluate a risk associated with strategic investment decision problems. The tool is known as "real options" and is one of the most recent developments in corporate finance. Since the use of real options is based on financial options on stock, we reviewed some fundamentals of option valuation techniques and demonstrated how the financial option theory can be extended to evaluate the risk associated with real assets.

■ An option to buy a financial asset at a specified price is a **call option** on the asset. The call-option buyer has the *right* to purchase the asset, whereas the call-option seller has the *obligation* to sell. An option to sell an asset at a specified price is a **put option** on the asset. Consequently, the put-option seller has the *right* to sell; the put-option buyer has the *obligation* to buy.

■ The value of the call tends to increase as the price of the underlying stock rises. This gain will increasingly reflect a rise in the value of the underlying stock when the market price moves above the option's strike price.

■ Buying puts against an existing stock position or simultaneously purchasing stock and puts can supply the insurance needed to overcome the uncertainty of the marketplace.

■ There are three conceptual approaches to valuing option premiums: the arbitrage-portfolio approach with a call option, the risk-free financing approach, and the risk-neutral probabilistic approach. All three approaches lead to the same option valuation, so the choice is dependent on the preference of the analyst.

■ As analytical approaches to option valuation, we use a discrete version of the binominal lattice model and the continuous version of the

Black–Scholes model. As the number of time steps gets larger, the value calculated with the binomial lattice model approaches the closed-form solution to the Black–Scholes model.

■ Real-options analysis is the process of valuing managerial strategic and operating flexibilities. Real options exist when managers can influence the size and risk of a project's cash flows by taking different actions during the project's life in response to changing market dynamics.

■ The single most important characteristic of an option is that it does not obligate its owner to take any action. It merely gives the owner the right to buy or sell an asset.

■ Financial options have an underlying asset that is traded—usually a security such as a stock.

■ Real options have an underlying asset that is not a security—for example, a project or a growth opportunity—and it is not traded.

■ The payoffs for financial options are specified in the contract, whereas real options are found or created inside of projects. Their payoffs vary.

■ Among the different types of real-option models are the option to defer investment, an abandonment option, follow-on (compound) options, and the option to adjust production.

■ One of the most critical parameters in valuing real options is the volatility of the return on the project and we have shown an analytical process of estimating the parameter.

■ The fundamental difference between the traditional NPW approach and real-options analysis is in how they treat managing the project risk: The traditional NPW approach is to avoid risk whenever possible, whereas the real-options approach is to manage the risk.

PROBLEMS

Financial Options

13.1 What would be the price of a six-month European call option on a non-dividend-paying stock when the stock price is $120, the strike price is $150, and the risk-free interest rate is 6% per year? Assume a volatility of 45% for this stock.

13.2 What would be the price of a two-month European put option on a non-dividend-paying stock when the stock price is $68.50, the strike price is $70, and the risk-free interest rate is 5% per annum? Assume a volatility of 60% for this stock.

13.3 Assume $S_0 = \$55$, the exercise price $K = \$45$, call-option premium $= \$10$, and put-option premium $= \$2$. Draw the corresponding expiration payoff diagrams for long-call, short-call, long-put, and short-put option positions.

13.4 You are given the following information. A call option's premium is currently $4, the exercise price equals $52, the risk free rate is 5%, volatility is 40%, and the time to expiration is nine months. What must the current stock price be equal to?

13.5 Use a binomial lattice with the following attributes to value a European call option.

- Current underlying asset value of 60
- Exercise price of 60
- Volatility of 30%
- Risk-free rate of 5% compounded continuously
- Time to expiration equal to 18 months
- A two-period lattice

13.6 Use a binomial lattice with the following attributes to value an American put option.

- Current underlying asset value of 40
- Exercise price of 45
- Volatility of 30%
- Risk-free rate of 5% compounded continuously
- Time to expiration equal to three years
- A three-period lattice

13.7 Consider the following purchasing options on Facebook (FB) stock.

- The current price to buy one share of Facebook stock is $76.33.
- The stock does not pay dividends.
- The risk-free interest rate, compounded continuously, is 6%.
- A European call option on one share of FB stock with a strike price of K that expires in one year costs $12.
- A European put option on one share of FB stock with a strike price of K that expires in one year costs $9.45.

Determine the strike price, K, if the volatility is estimated to be 35.85%.

13.8 An investor has a portfolio consisting of the following assets and instruments.

- A long-call option with $K = \$40$ and a call premium of $3 at the time of purchase.
- A short-put option with $K = \$45$ and a put premium of $4 at the time of purchase.
- Two short-call options with $K = \$35$ and a call premium of $5 at the time of purchase.
- Two short-stock positions that cost $40 per share at the time of purchase.

Assume that each of these contracts has the same expiration date, and ignore the time value of money. If the stock price at expiration is $S_T = \$60$, what is the net profit of this portfolio?

13.9 A put-option premium is currently $4, with $S_0 = \$30$, $K = \$32$, and $T = 6$ months. Calculate the intrinsic value and time premium for this put option. In addition, explain *why* the time to contract maturity and the underlying asset volatility affect a put option's time premium.

13.10 An investor has $10,000 that she wants to invest in Apple Corporation. She can either go long in Apple's stock at a spot price of $100 per share on September 17, 2015 or purchase option contracts for $10 per share with an exercise price $K = \$110$ per share, which expires on January 2016. Determine the payoff for both investment strategies assuming $S_T = \$95$, $S_T = \$115$, and $S_T = \$140$. Which strategy would you recommend?

13.11 Let $S_0 = \$100$, the risk-free rate equal 5%, and the volatility equal 35%. Answer the following, assuming continuous compounding.

(a) Use the three discrete-time approaches to value a European call option with a two time-period tree with an exercise price equal to $105 and $T = 18$ months. To maintain a riskless hedge, how many shares of stock should be held long today and in nine months?

(b) Value a European put option assuming the underlying asset with $K = \$100$, $T = 1$ year, and use a four time-period tree.

(c) Value an American put option with the same attributes as in part b.

13.12 Put together a Black–Scholes option calculator in Excel to answer the following.

(a) What is the call-option value with $S_0 = \$45$, $K = \$48$, $r = 6\%$, $T = 15$ months, and volatility $= 40\%$?

(b) What is the put-option value with $S_0 = \$60$, $K = \$65$, $r = 6\%$, $T = 18$ months, and volatility $= 20\%$?

(c) What is the put-option value with $S_0 = \$38$, $K = \$40$, $r = 6\%$, $T = 3$ months, and volatility $= 60\%$?

(d) What is the call-option value with $S_0 = \$100$, $K = \$95$, $r = 8\%$, $T = 3$ years, and volatility $= 40\%$?

13.13 The Rio-Coffee Farm, a Brazilian coffee producer, has an exclusive contract to supply coffee beans to Seattle's Best Company (SBC). The contract states that SBC will take delivery of 1,000,000 lbs of coffee beans in one year at the market price. It will cost Rio-Coffee $1,000,000 to provide 1,000,000 lbs of beans, and today's market price is $1.25 per lb. Rio-Coffee has decided to hedge as follows (both options are one-year, European):

- Buy 1,000,000 $1.25-strike put options for $50,000
- Sell 1,000,000 $1.40-strike call options for $38,000

Seattle's Best believes the market price in one year will be somewhere between $1.00 and $1.50 per lb. Which interval represents the range of possible profit

one year from now for Rio-Coffee? The continuously compounded risk-free interest rate is 6%.

Real Options Analysis

13.14 A company is planning to undertake an investment of $2 million to upgrade one of its products for an emerging market. The market is highly volatile, but the company owns a product patent that will protect it from competitive entry until the next year. Because of the uncertainty of the demand for the upgraded product, there is a chance that the market will be in favor of the company. The present value of expected future cash flows is estimated to be $1.9 million. Assume a risk-free interest rate of 8% and a standard deviation of 40% per annum for the PV of future cash flows. What is the value of delaying the investment rather than canceling the product?

13.15 A mining firm has the opportunity to purchase a license on a plot of land to mine for gold. Consider the following financial information:

- The investment cost to mine is $140M.
- It costs $1,115 per ounce to mine gold.
- The spot price of gold is $1,222 per ounce.
- The licensing agreement provides exclusive rights for three years.
- The historical volatility of gold prices is 20%.
- The estimated gold reserve on the plot of land is 1.5M ounces.
- The firm's MARR is 12% and $r = 6\%$.

Determine the maximum amount the firm should pay for a license to mine for gold on the property in question, or equivalently, what is the value of the gold mine today?

Growth Options

13.16 You are considering an investment in a Christmas tree farm over three years. The farm already has trees planted a year ago, and these trees grow each year by the following factors.

Year	1	2	3
Growth	1.6	1.5	1.4

The price of the trees follows a binomial lattice with $u = 1.25$ and $d = 0.80$. As the trees grow older, the value of the investment in general will increase.

However, if the trees get too big, they are less attractive, so a three-year rotation appears to be a typical lifecycle for this type of Christmas tree. The interest rate (risk-free) is constant at continuous 6%. It costs $400,000 each year—payable at the beginning of the year—to lease the forestland. The initial value of the one-year old trees is $1 million (assuming they were harvested immediately). You can cut the trees at the beginning of the Christmas season (say, December 1) of any year and then not pay rent after that.

(a) It costs $400,000 each year, payable at the beginning of the year, to lease the forest land. You can cut the trees at the beginning of Christmas season (Say, December 1) of any year and then not pay rent after that. Find the best cutting policy with this rent.

(b) Suppose there is no rent. Find the best cutting policy without rent.

Deferral Options

13.17 You currently own a piece of land and are contemplating the construction of either a small shopping mall with 10 stores or a large shopping mall with 30 stores. The current market value of one store is $60,000, and the volatility surrounding this value is 20%. The cost today of constructing one store within a 10-store mall is $35,000, and the cost today of constructing one store within a 30-store mall is $55,000. You have the option of delaying the construction decision for two years. However, if you do so, today's construction costs are expected to increase 5% per year. Assume the risk-free rate is 6% compound continuously. What is the value of postponing the construction decision for two years?

Switching Options

13.18 A firm has invested in and is currently receiving benefits from project A. The current value of project A is $4M. In next five years, the firm will have the option to use most of the same equipment from project A and switch to project B. Switching over would entail a $2M investment cost. The expected net cash inflow of project B is $1M per annum for 10 years. What is the total value of this investment scenario? Assume that MARR = 12%, $r = 6\%$, and $\sigma = 50\%$.

R&D Options

13.19 A pharmaceutical company needs to estimate the maximum amount to spend on R&D for a new type of diet drug. It is estimated that three years of R&D spending will be required to develop and test market the drug. After the initial three years, an investment in manufacturing and production will be required in year 4. It is estimated today that net cash inflows for six years will be received from sales of the drug. Figure P13.19 is a cash flow diagram summarizing the firm's estimates.

Assuming a MARR of 12%, $\sigma = 50\%$, and a risk-free interest rate of 6%, determine whether or not the required spending on R&D for this diet drug can be justified.

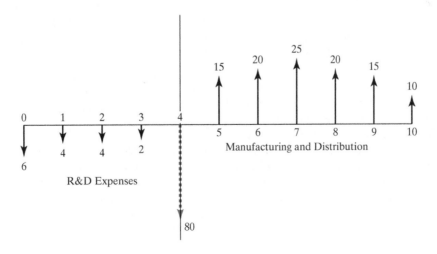

Figure P13.19

Abandonment Option

13.20 A firm is considering purchasing equipment to manufacture a new product. The equipment will cost $3M, and expected net cash inflows are $0.35M indefinitely. If market demand for the product is low, then over the next five years the firm will have the option of discarding the equipment on a secondary market for $2.2M. Assume that MARR = 12%, $\sigma = 50\%$, and $r = 6\%$. What is the value of this investment opportunity for the firm?

Scale-Down Options

13.21 A firm that has undertaken a project has the option to sell some of its equipment and facilities and to sublet out some of the project workload. The project's current value is $V_0 = \$10$ M. Anytime over the next three years, the firm can sell off $4M in resources but receive an expected 20% decrease in net cash flows (and therefore, a 20% decrease in project value). Let $r = 6\%$ and $\sigma = 30\%$. A binomial lattice will be used with a one-year time increment. Determine the value of this scale-down investment opportunity.

Expansion–Contraction Options

13.22 Suppose a large manufacturing firm decides to hedge against risk through the use of strategic options. Specifically, it has the option to choose between two strategies.

- Expanding its current manufacturing operations.
- Contracting its current manufacturing operations at any time within the next two years.
- Neither expanding not contracting.

Suppose that the firm has a current operating structure whose static valuation of future profitability based on a discounted cash-flow model is found to be $100 million. Suppose also that the firm estimates the implied volatility of the returns on the projected future cash flows to be 15%. The risk-free interest rate is 5%. Finally, suppose that the firm has the option to contract 10% of its current operations at any time over the next two years, creating an additional $25 million in savings after the contraction. The expansion option will increase the firm's operations by 30% with a $20 million implementation cost.[9]

(a) Show the binomial lattice of the underlying asset over two years.

(b) What is the value of retaining the option to choose between both alternatives (i.e., consider both the expanding option and the contracting option together)?

Compound Options

13.23 A firm relies on R&D to maintain profitability. The firm needs to determine the maximum amount to invest today (or invest in phase I) for its three-phased project.

- **Phase I:** Research. (Invest R_0 today.)
- **Phase II:** Development. (Invest I_1 one year from now.)
- **Phase III:** Implementation. (Invest I_2 in facilities, manpower, etc., three years from today.)

The three-phased investment cash flows are as given in Figure P13.23.

Using a MARR of 12%, $\sigma = 50\%$, and $r = 6\%$, determine the best investment strategy for the firm.

13.24 The J&J Construction Company is evaluating an investment project to build a golf resort complex on a particular land site. J&J can obtain a one-year option to buy the required parcel of land, and if the land is purchased, the price would be $3,000,000. The land option (if purchased) would expire one year from now. J&J has two years from today to make a decision on whether to build the golf resort complex and start operations. The cost of the building and initial operating capital is $12,000,000. They estimate that the land could be sold for 90% of its purchase price if they decided not to build the resort complex. Given the following criteria:

- The value of the complex if it existed today would be worth $V = \$19,500,000$.
- Volatility of the project return is 25%.
- Risk-free interest rate is 5% continuous per year.

What would the company be willing to pay for the combined value of the land option and related prefeasibility studies (i.e., environmental impact assessment, etc.) before actual construction of the golf resort complex?

---

[9] Adapted from Johnathan Mun, *Real Options Analysis*, New York: John Wiley, 2002, p. 182.

Figure P13.23

13.25 A firm has decided to hedge itself through the use of strategic options. It has the option to choose among three strategies:

- It has the option to contract 10% of its current operations at any time over the next three years, thereby creating an additional $25 million in savings after this contraction.

- It has the option to expand its current operation by 30% (resulting in 30% increase in project value) with a $20 million implementation cost.

- It has the option to abandon its current operations and then sell its intellectual property for $100 million.

The value of the current operation is $100 million, the implied volatility of project return is 15%, and the risk-free rate is 5%.

(a) What is the value of just having the contract option?

(b) What is the value of just having the abandonment option?

(c) What is the value of just having the expansion option?

(d) What is the combined value of having all three options together?

Short Case Studies

ST13.1 A Silicon Valley start-up company faces an opportunity to invest in an R&D project that will revolutionize the way consumers use telephones, internet and TV. Suppose a project has two phases, where the first phase (R&D) has a one-year expiration that costs $50 million. The second phase (testing and product development) last three years and requires $70 million. Due to fast-changing technology in TV market, the volatility of the project return can vary in each phase of investment - 20% during the first phase and 30% during the second phase. The risk-free rate for the next five years is 6%. If the product is available today, the value of the project would be $100 million.

(a) Determine the underlying asset lattice (how the project value changes over time).

(b) Determine the value of this compound option.

ST13.2 A drug company is considering developing a new drug. Due to the uncertain nature of the drug's progress in development, market demand, success in human and animal testing, and FDA approval, management has decided that it will create a strategic abandonment option. That is, anytime within the next two years, management can review the progress of the R&D effort and decide whether to terminate the development of the drug. If the program is terminated, the firm can sell off its intellectual property rights to the drug to another pharmaceutical firm.

- The present value of the expected future cash flows (discounted at an appropriate market-adjusted rate) is $150 million.
- A Monte Carlo simulation indicates that the volatility of the returns on future cash flows is 30%.
- The risk-free rate on a riskless asset for the same time frame is 5%.
- The drug's patent is worth $100 million if sold within the next two years.

Given that $S = \$150$, $\sigma = 30\%$, $T = 2$, and the risk-free interest rate $(r) = 5\%$, determine the value of the abandonment option if the following statements are true.

(a) It is an American option using a binomial lattice approach.

(b) Suppose that the R&D program was indeed successful at the end of two years, and the present value of the expected future cash flows discounted at an appropriate market risk-adjusted discount rate is found to be $400 million. The implied volatility of the returns on the projected future cash flows to be 35%. The risk-free rate is 7% for the next two years. Suppose that the firm has the option to expand and double its operations by acquiring its competitor for a sum of $250 million at any time over the next two years. What is the total value of the project to this firm, assuming you account for this expansion option?

ST13.3 The pharmaceutical industry is composed of both large and small firms competing for new research, the introduction of new products, and the sales of existing products. Performing research and development for new drugs can be financially risky, as few researched drugs actually make it to the market. Sometimes, large firms will enter into agreements with smaller firms to conduct research. There are benefits to both parties in these situations. The biggest advantage is that the large firm does not need to invest in-house for required expertise and facilities to develop a new type of drug. In essence, large firms contract out their research work to smaller firms for an investment amount smaller than what the large

firms would have to spend if they conducted the new research internally. On their part, small firms receive much-needed funds to carry out their research.

Merck Co. and Genetics, Inc. are publicly traded large and small pharmaceutical firms, respectively. Merck Co. is considering entering into an agreement with Genetics. The agreement would require Merck to give $4M to Genetics to develop a new drug over the next two years. By doing so, Merck would obtain the right to acquire Genetics for $60 per share three years from today. If the new drug is successful, Genetics' share price is expected to double or even triple. However, if the new drug is unsuccessful, Genetics' share price is expected to decrease in value significantly. Genetics has a current stock price of $30 per share, has an expected share price volatility of 50% (including the riskiness surrounding the new-drug research), and has 1.2 million shares outstanding.

(a) Is the initial $4M that Merck is agreeing to pay Genetics justified? (Quantify your answer, and assume the risk-free rate is 6%.)

(b) Assume that Merck could also buy Genetics today for $40 per share. Compare buying Genetics today versus entering into the aforesaid agreement. Briefly discuss the advantages and disadvantages of both alternatives. If you utilize the binomial lattice approach, then only use a *one-period lattice* for your computations.

ST13.4 GlaxoSmithKline is a world leading research-based pharmaceutical company with a powerful combination of skills and resources that provides a platform for delivering strong growth in today's rapidly changing healthcare environment. GlaxoSmithKline's mission is to improve the quality of human life by enabling people to do more, feel better and live longer. Headquartered in the UK and with operations based in the US, the new company is one of the industry leaders, with an estimated seven per cent of the world's pharmaceutical market.

GlaxoSmithKline intends to continue developing innovative and cost-effective solutions in prevention, diagnosis and drug therapy to the ultimate benefit of patients. Keeping with its history of product development, GlaxoSmithKline announced recently that it has received FDA clearance for its Elecsys proBNP (pro-B-type natriuretic peptide) assay, the first fully automated test for use in the diagnosis of congestive heart failure (CHF). Now, laboratories can have results ready to report to physicians in as few as 18 minutes, using an automated platform that reduces technologist time and minimizes human errors.

GlaxoSmithKline now believes that it has the possibility to start a new project based on its Elecsys proBNP research, which could result in a test for the diagnosis of hardening of the arteries. No such test currently exists and GlaxoSmithKline believes the successful development of such a product would result in great financial reward.

- **Research Phase:** The Company will hire 10 scientists and will pay each $80,000/year. Also there is an overhead cost of $160,000/year. The project requires a $3M initial equipment investment for the research phase, which will take about 4 years. This equipment will be depreciated with the straight-line method over 4 years. In the research phase there will be a cost of $500,000, $400,000, $400,000, $400,000 respectively for material cost each year.

- **Development Phase:** In year 5 GlaxoSmithKline can pursue approval from appropriate agencies. This development phase will require 4 scientists at the previous pay level. The development phase will require the same equipment and also new equipment is required for development which costs $2M. This equipment is depreciated by the straight-line method over 4 years. Assume both sets of equipment will have no salvage value after they're fully depreciated. During the 4-year development phase initial overhead cost is $1M/year for legal and lobbying fees. The material cost for development phase will be $350,000 for each year.

- **Manufacturing Phase:** The research and development phases will give GlaxoSmithKline the opportunity to implement the new product in year 9. Manufacturing phase requires a $100M equipment and machine investment which is to be depreciated according to the straight-line method over 10 years. (Once again, assume a zero salvage value of the assets.) Also, for marketing and distribution GlaxoSmithKline will spend $4M except for the first year, which is $10M. Labor cost in this phase will be $1M. The overhead, which will also include management, legal costs, will be $8M for each year. GlaxoSmithKline expects revenue of $50M/year for the foreseeable future.

Note that all cash flow figures above represent expected values. The risk free rate is known to be 6%, and the risk-adjusted discount rate is 12%. The firm's income tax rate is 40%. At this point, the firm has no clear idea on the volatility of the project return, even though its stock volatility is known to be around 15%. Formulate GlaxoSmithKline's decision process using the compound option framework, and then determine the desired level of project return volatility at minimum that allows the firm to initiate the research phase.

Note: If you need to make some specific additional assumptions in approaching this problem, do so but spell them out clearly at the beginning of your analysis.

FIVE

SPECIAL TOPICS IN ENGINEERING ECONOMICS

14

Replacement Decisions

Replacing an Old Absorption Chiller   University of California at San Francisco (UCSF) Medical Center, one of the nation's leading centers of health sciences, research, and patient care, spent $7.7 million to replace the 1980s steam-driven chiller, which chills water that is piped throughout the hospital for needed air conditioning. Absorption chillers use heat rather than electricity as their energy source, mainly natural gas. Absorption cooling is most frequently used to air condition large commercial buildings, such as hospitals.

The old chiller was unreliable, inefficient, required a lot of maintenance and had an inadequate cooling capacity. The new 600-ton electric centrifugal chiller uses more electricity, but UCSF expected to save more than 65M kBtu of energy per year, which translates to a projected annual savings of $1.3 million in utility costs, and more than $100,000 in maintenance costs. UCSF predicted that the $7.7 million investment would be recovered within less than six years.

Another benefit from this replacement project is that UCSF saves almost 411,000 kilowatt hours and more than 658,000 therms of energy each year, savings equivalent to avoiding nearly 8 million pounds of carbon dioxide, powering 921 homes for one year, or removing 551 cars from the road.[1]

n Chapters 7 through 9, we presented methods that helped us choose the best among a number of investment alternatives. The problems we examined in those chapters concerned primarily *profit-adding projects*. However, economic analysis is also frequently performed on projects with existing facilities or *profit-maintaining projects*—those projects whose primary purpose is not to increase sales, but rather, simply to maintain ongoing operations. In practice, profit-maintaining projects less frequently involve the comparison of new machines. Instead, the problem often facing management is whether to buy new and more efficient equipment or to continue to use existing equipment. This class of decision problems is known as the **replacement problem**. In this chapter, we examine the basic concepts and techniques related to replacement analysis.

CHAPTER LEARNING OBJECTIVES

After completing this chapter, you should understand the following concepts:

- What makes the replacement decision problems differ from the other capital investment decisions.

- What types of financial information should be collected to conduct a typical replacement decision problem.

- How to compare a defender with a challenger on the basis of opportunity cost concept.

- How to determine the economic service life for any given asset.

- How to determine the optimal time to replace a defender.

- How to consider the tax effects in replacement analysis.

14.1 Replacement Analysis Fundamentals

We will examine three aspects of the replacement problem: (1) approaches to comparing defender and challenger; (2) the determination of economic service life; and (3) replacement analysis when the required service period is long. Initially, we will present the

---

[1] San Francisco, June 26, 2013, ARUP, http://www.arup.com/News/2013_06_June/26_June_2013_Chiller_replacement_project_recognised.aspx.

replacement concepts and logics on a before-tax basis, and then revisit these replacement problems later, taking income taxes into account.

14.1.1 Basic Concepts and Terminology

The replacement projects are decision problems involving the replacement of existing obsolete or worn-out assets. The continuation of operations is dependent on these assets. The failure to make an appropriate decision results in a slowdown or shutdown of the operations. The question is, "When existing equipment should be replaced with more efficient equipment?"

Defender and Challenger

Replacement problem has given rise to the use of the terms **defender** and **challenger**, terms commonly used in the boxing world. In every boxing class, the current defending champion is constantly faced with a new challenger. In replacement analysis, the defender is the existing machine (or system), and the challenger is the best available replacement equipment.

An existing piece of equipment will be removed at some future time—either when the task it performs is no longer necessary or when the task can be performed more efficiently by newer and better equipment. The question is not *whether* the existing piece of equipment will be removed but *when* it will be removed. A variation of this question is why we should replace existing equipment at the current time rather than postponing replacement of the equipment by repairing or overhauling it. Another aspect of the defender–challenger comparison concerns deciding exactly which equipment is the best challenger. If the defender is to be replaced by the challenger, we would generally want to install the very best of the possible alternatives.

Current Market Value

The most common problem encountered in considering the replacement of existing equipment is the determination of what financial information is actually relevant to the analysis. Often, a tendency to include irrelevant information in the analysis is apparent. To illustrate this type of decision problem, let us consider Example 14.1.

EXAMPLE 14.1 Information Relevant to Replacement Analysis

Macintosh Printing, Inc. purchased a $20,000 printing machine two years ago. The company expected this machine to have a five-year life and a salvage value of $5,000. The company spent $5,000 last year on repairs, and current operating costs are running at the rate of $8,000 per year. Furthermore, the anticipated salvage value of the machine has been reduced to $2,500 at the end of its remaining useful life. In addition, the company has found that the current machine has a market value of $10,000 today. If the company desires to replace the current printing machine with a new one, the equipment vendor will allow the company this full amount as a trade-in on a new machine. What values for the defender are relevant to our analysis?

SOLUTION

Given: Original cost, market value, and trade-in allowance.

Find: Cost information relevant to replacement analysis.

In this example, three different dollar amounts relating to the defender are presented:

1. **Original cost.** The printing machine was purchased for $20,000.
2. **Market value.** The company estimates the old machine's market value at $10,000.
3. **Trade-in allowance.** This is the same as the market value. (In other problems, however, it could be different from the market value.)

COMMENTS: In this example and in all defender analyses, the relevant information is the **current market value** of the equipment. The original cost, repair cost, and trade-in value are irrelevant. A common misconception is that the trade-in value is the same as the current market value of the equipment and thus could be used to assign a suitable current value to the equipment. This is not always true, however. For example, a car dealer typically offers a trade-in value on a customer's old car to reduce the price of a new car. Would the dealer offer the same value on the old car if he or she were not also selling the new one? The answer is, "Not generally." In many instances, the trade-in allowance is inflated to make the deal look good, and the price of the new car is also inflated to compensate for the dealer's trade-in cost. In this type of situation, the trade-in value does not represent the true value of the item, so we should not use it in economic analysis.[2]

Sunk Costs

As mentioned in Section 3.4.3, a **sunk cost** is any past cost that is unaffected by any future investment decision. In Example 14.1, the company spent $20,000 to buy the machine two years ago. Last year, $5,000 more was spent on the machine. The total accumulated expenditure is $25,000. If the machine were sold today, the company could get only $10,000 back (Figure 14.1). It is tempting to think that the company would lose $15,000 in addition to the cost of the new machine if the old machine were to be sold and replaced with a new one. This is an incorrect way of doing economic analysis, however. In a proper engineering economic analysis, only future costs should be considered; past or sunk costs should be ignored. Thus, *the value of the defender that should be used in a replacement analysis is its current market value, not what it cost when it was originally purchased and not the cost of repairs that already have been made to the machine.*

Sunk costs are money that is gone, and no present action can recover them. They represent past actions—the results of decisions made in the past. In making economic decisions at the present time, one should consider only the possible outcomes of the various decisions and pick the one with the best possible future results. Using sunk costs in arguing one option over the other would only lead to more bad decisions.

---

[2] If we do make the trade, however, the actual net cash flow at the time of the trade, properly used, is certainly relevant.

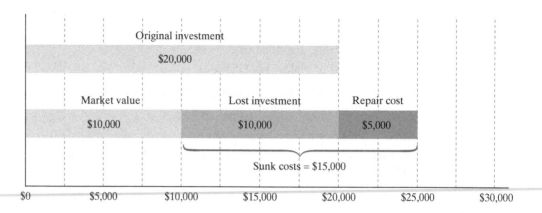

Figure 14.1 Sunk cost associated with an asset's disposal, as described in Example 14.1.

Operating Costs

The driving force for replacing existing equipment is that it becomes more expensive to operate with time. The total cost of operating a piece of equipment may include repair and maintenance costs, wages for the operators, energy consumption costs, and costs of materials. Increases in any one or a combination of these cost items over a period of time may impel us to find a replacement for the existing asset. The challenger is usually newer than the defender and often incorporates improvements in design and newer technology. As a result, some or all of the cost items for the challenger are likely to be less expensive than those for the defender.

We will call the sum of the various cost items related to the operation of an asset the **operating costs**. As is illustrated in the sections that follow, keeping the defender involves a lower initial cost than purchasing the challenger but requires higher annual operating costs. Usually, operating costs increase over time for both the defender and the challenger. In many instances, the labor costs, material costs, and energy costs are the same for the defender and the challenger and do not change with time. It is the repair and maintenance costs that increase and cause the operating costs to increase each year as an asset ages.

When repair and maintenance costs are the only cost items that differ between the defender and the challenger on a year-by-year basis, we need to include only those costs in the operating costs used in the analysis. Regardless of which cost items we choose to include in the operating costs, it is essential that the same items be included for both the defender and the challenger. For example, if energy costs are included in the operating costs of the defender, they also should be included in the operating costs of the challenger. A more comprehensive discussion of the various types of costs incurred in a complex manufacturing facility was provided in Chapter 8.

14.1.2 Opportunity Cost Approach to Comparing Defender and Challenger

Although replacement projects are a subcategory of the mutually exclusive categories of project decisions we studied in Chapter 5, they do possess unique characteristics that allow us to use specialized concepts and analysis techniques in their evaluation. We consider a

EXAMPLE 14.2 Replacement Analysis Using the Opportunity Cost Approach

Consider again Example 14.1. The company has been offered a chance to purchase another printing machine for $15,000. Over its three-year useful life, the machine will reduce the usage of labor and raw materials sufficiently to cut operating costs from $8,000 to $6,000. It is estimated that the new machine can be sold for $5,500 at the end of year 3. If the new machine were purchased, the old machine would be sold to another company rather than to be traded in for the new machine.

Suppose that the firm will need either machine (old or new) for only three years and that it does not expect a new, superior machine to become available on the market during this required service period. Assuming that the firm's interest rate is 12%, decide whether replacement is justified now.

SOLUTION
Given: Financial data for both the defender and challenger.
Find: Whether or not to replace the defender.

- **Option 1: Keep the defender.** If the decision is to keep the defender, the opportunity cost approach treats the $10,000 current salvage value of the defender as an investment cost. The annual operating cost for the next three years will be $8,000 per year, and the defender's salvage value three years from today will be $2,500. The cash flow diagram for the defender is shown in Figure 14.2(a).
- **Option 2: Replace the defender with the challenger.** The new machine costs $15,000. The annual operating cost of the challenger is $6,000. The salvage value of the challenger three years later will be $5,500. The actual cash-flow diagram for this option is shown in Figure 14.2(b). We calculate the net present worth and annual equivalent cost for each of the two options as follows.

$$PW(12\%)_D = \$10,000 + \$8,000(P/A, 12\%, 3)$$
$$-\$2,500(P/F, 12\%, 3)$$
$$= \$27,435$$
$$AEC(12\%)_D = PW(12\%)_D(A/P, 12\%, 3)$$
$$= \$11,423$$
$$PW(12\%)_C = \$15,000 + \$6,000(P/A, 12\%, 3) - \$5,500(P/F, 12\%, 3)$$
$$= \$25,496$$
$$AEC(12\%)_C = PW(12\%)_C(A/P, 12\%, 3)$$
$$= \$10,615$$

Because of the annual difference of $808 in favor of the challenger, the replacement should be made now.

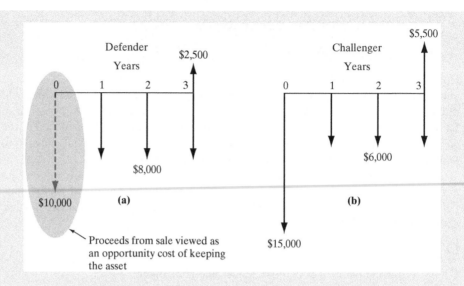

Figure 14.2 Comparison of defender and challenger based on the opportunity cost approach (Example 14.2).

COMMENTS: Recall that we assumed the same service life for both the defender and the challenger. In general, however, old equipment has a relatively short remaining life compared with new equipment, so that assumption is too simplistic. When the defender and challenger have unequal lifetimes, we must make an assumption in order to obtain a common analysis period. A typical assumption is that, after the initial decision, we make **perpetual replacements** with assets similar to the challenger. Certainly, we can still use PW analysis with actual cash flows, but that would require evaluating *infinite* cash-flow streams.

basic approach to analyzing replacement problems commonly known as the **opportunity cost approach**.

The key issue is how to treat the proceeds from the sale of the old equipment. In fact, if you decide to keep the old machine, this potential sales receipt is forgone. *The opportunity cost approach views the net proceeds from the sale as the opportunity cost of keeping the defender.* In other words, we consider the salvage value as a cash outflow for the defender (or an investment required in order to keep the defender).

If our analysis in Example 14.2 showed instead that the defender should not be replaced now, we still need to address the question of whether the defender should be kept for one or two years and then should be replaced with the challenger. This is a valid question that requires more data on market values over time. We address the situation later, in Section 14.3. Also, when comparing the defender with challenger, we must do so based on the most favorable life of each asset. This most favorable life is defined as the economic service life of the asset. In the next section, we discuss how to determine such an economic service life of equipment.

14.2 Economic Service Life

Perhaps you have seen a 20-year-old automobile that is still in service. Provided that it receives the proper repair and maintenance, almost anything can be kept operating for an extended period of time. If it's possible to keep a car operating for an almost indefinite period, why aren't more old cars spotted on the streets? Two reasons are that some people may get tired of driving the same old car, and other people may want to keep a car as long as it will last but realize that repair and maintenance costs will become excessive.

In general, we need to consider economically how long an asset should be held once it is placed in service. For instance, a truck-rental firm that frequently purchases fleets of identical trucks may wish to arrive at a policy decision on how long to keep each vehicle before replacing it. If an appropriate life span is computed, a firm could stagger a schedule of truck purchases and replacements to smooth out annual capital expenditures for its overall truck purchases.

As we discussed in Chapter 6, the costs of owning and operating an asset can be divided into two categories: **capital costs** and **operating costs**. Capital costs have two components: the initial investment and the salvage value at the time of disposal of the asset. The initial investment for the challenger is simply its purchase price. For the defender, we should treat the opportunity cost as its initial investment. We will use N to represent the length of time in years the asset will be kept, I to denote the initial investment, and S_N to designate the salvage value at the end of the ownership period of N years.

Capital (Ownership) Costs

Recall that the annual equivalent of capital costs, which is called the capital recovery cost (see Section 6.2), over the period of N years can be calculated with

$$CR(i) = I(A/P, i, N) - S_N(A/F, i, N)$$
$$= (I - S_N)(A/P, i, N) + iS_N \tag{14.1}$$

Generally speaking, as an asset becomes older, its salvage value becomes smaller. As long as the salvage value is less than the initial cost, the capital recovery cost is a decreasing function of N. In other words, the longer we keep an asset, the lower the capital recovery cost becomes. If the salvage value is equal to the initial cost no matter how long the asset is kept, the capital recovery cost is constant.

Operating Cost

As described earlier, the operating costs of an asset include operating and maintenance (O&M) costs, labor costs, material costs, and energy consumption costs. Labor costs, material costs, and energy costs are often constant for the same equipment from year to year if the usage of the equipment remains constant. However, O&M costs tend to increase as a function of the age of the asset. Because of the increasing trend of the O&M costs, the total operating costs of an asset usually increase as well as the asset ages. We use OC_n to represent the total operating costs in year n of the ownership period and $OC(i)$ to represent the annual equivalent of the operating costs over a life span of N years. Then $OC(i)$ can be expressed as

$$OC(i) = \underbrace{\left(\sum_{n=1}^{N} OC_n(P/F, i, n) \right)}_{\text{Equivalent total operating cost at } n = 0} (A/P, i, N) \tag{14.2}$$

As long as the annual operating costs increase with the age of the equipment, $OC(i)$ is an increasing function of the life of the asset. If the annual operating costs are the same from year to year, $OC(i)$ is constant and equal to the annual operating costs no matter how long the asset is kept.

Annual Equivalent Cost (AEC)

The total annual equivalent costs of owning and operating an asset $[AEC(i)]$ are a summation of the capital recovery costs and the annual equivalent of operating costs of the asset:

$$AEC(i) = CR(i) + OC(i) \tag{14.3}$$

The economic service life of an asset is defined as the period of useful life that minimizes the annual equivalent costs of owning and operating the asset. On the basis of the foregoing discussions, we need to find the value of N that minimizes AEC as expressed in Eq. (14.3). If $CR(i)$ is a decreasing function of N and $OC(i)$ is an increasing function of N, as is often the case, AEC will be a convex function of N with a unique minimum point (see Figure 14.3). In this book, we assume that AEC has a unique minimum point. Three special situations are worth considering:

1. If the salvage value of the asset is constant and equal to the initial cost and the annual operating cost increases with time, then AEC is an increasing function of N and attains its minimum at $N = 1$. In this case, we should try to replace the asset as soon as possible.

2. If, however, the annual operating cost is constant and the salvage value is less than the initial cost and decreases with time, then AEC is a decreasing function of N. In this case, we would try to delay the replacement of the asset as long as possible.

3. Finally, if the salvage value is constant and equal to the initial cost and the annual operating costs are constant, then AE also will be constant. In this case, when to replace the asset does not make any economic difference.

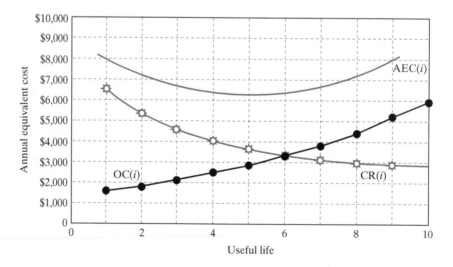

Figure 14.3 A schematic illustrating the trends of capital recovery cost (ownership cost), annual operating cost, and total annual equivalent cost.

Replacement Strategy

If a new asset is purchased and operated for the length of its economic life, the annual equivalent cost is minimized. If we further assume that a new asset of identical price and features can be purchased repeatedly over an indefinite period, we would always replace this kind of asset at the end of its economic life. By replacing assets perpetually according to its economic life, we obtain the minimum AEC stream over an indefinite period. However, if the identical-replacement assumption cannot be made, we will have to use the methods to be covered in Section 14.3 to carry out a replacement analysis. The next example explains the computational procedure for determining an asset's economic service life.

EXAMPLE 14.3 Economic Service Life of a Lift Truck

Suppose a company has a forklift but is considering purchasing a new electric-lift truck that would cost $18,000 and have operating costs of $3,000 in the first year. For the remaining years, operating costs increase each year by 15% over the previous year's operating costs. Similarly, the salvage value declines each year by 20% from the previous year's salvage value. The lift truck has a maximum life of eight years. The firm's required rate of return is 12% before tax. Find the economic service life of this new machine.

DISCUSSION: For an asset whose revenues are either unknown or irrelevant, we compute its economic life on the basis of the costs for the asset and its year-by-year salvage values. To determine an asset's economic service life, we need to compare the options of keeping the asset for one year, two years, three years, and so forth. The option that results in the lowest annual equivalent cost (AEC) gives the economic service life of the asset.

SOLUTION

Given: $I = \$18,000$, O&M cost, salvage value of the asset over eight years, and $i = 12\%$,
Find: Economic service life.

- $n = 1$: **One-year replacement cycle.** In this case, the machine is bought, used for one year, and sold at the end of year 1. The annual equivalent cost for this option is

$$\text{AEC}(12\%)_{n=1} = \overbrace{(\$18,000 - \$14,400)(A/P, 12\%, 1) + 0.12(\$14,400)}^{\text{capital cost}}$$
$$+ \underbrace{\$3,000}_{\text{operating cost}}$$
$$= \$8,760$$

- Note that $(A/P, 12\%, 1) = (F/P, 12\%, 1)$ and the annual equivalent cost is the equivalent cost at the end of year 1, since $N = 1$. Because we are dealing with the annual equivalent cost, we have treated cost items with a positive sign, while the salvage value has a negative sign.

- $n = 2$: **Two-year replacement cycle.** In this case, the truck will be used for two years and disposed of at the end of year 2. The operating cost in year 2 is 15% higher than that in year 1, and the salvage value at the end of year 2 is 20% lower than that at the end of year 1. The cash flow diagram for this option is shown in Figure 14.4. The annual equivalent cost over the two-year period is

$$
\text{AEC}(12\%)_{n=2} = \overbrace{[(\$18,000 - \$11,520)(A/P, 12\%, 2) + 0.12(\$11,520)]}^{\text{capital cost}}
$$
$$
+ \underbrace{\$3,000(P/A_1\ 15\%, 12\%, 2)(A/P, 12\%, 2)}_{\text{operating cost}}
$$

$$
= \$8,429
$$

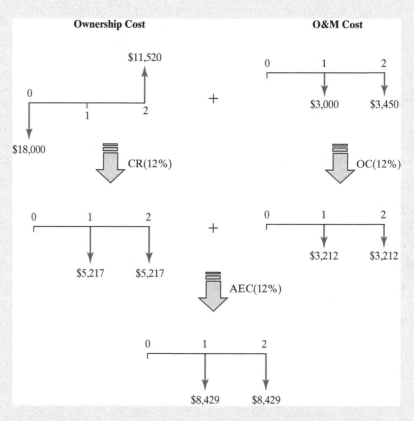

Figure 14.4 Annual equivalent cost (AEC) for owning and operating the asset for two years.

TABLE 14.1 Economic Service Life Calculations for a Lift Truck

	A	B	C	D	E	F
1						
2		Annual changes in MV		-20%		
3		Annual increases in O&M		15%		
4		Interest rate		12%		
5						
6	n	Market Value	O&M Costs	CR(12%)	OC(12%)	AEC(12%)
7						
8	0	$18,000				
9	1	$14,400	$3,000	$5,760	$3,000	$8,760
10	2	$11,520	$3,450	$5,217	$3,212	$8,429
11	3	$9,216	$3,968	$4,763	$3,436	$8,199
12	4	$7,373	$4,563	$4,384	$3,672	$8,055
13	5	$5,898	$5,247	$4,065	$3,920	$7,985
14	6	$4,719	$6,034	$3,797	$4,180	$7,977
15	7	$3,775	$6,939	$3,570	$4,454	$8,024
16	8	$3,020	$7,980	$3,378	$4,740	$8,118
17						
18						=D16+E16
19	=B15*(1+$D$2)					
20			=C15*(1+$D$3)		=-PMT($D$4,A16,NPV($D$4,$C$9:C16),0)	
21						
22			=-PMT($D$4,A16,$B$8,-B16)			
23						

- Similarly, we can find the annual equivalent costs for the options of keeping the asset for three, four, five, six, seven, and eight years. The computed annual equivalent costs for each of these options are shown in Table 14.1.

From the preceding calculated AEC values for $n = 1, \ldots, 8$, we find that AEC(12%) is smallest when $n = 6$. If the truck were to be sold after six years, it would have an annual equivalent cost of $7,977 per year. If it were to be used for a period other than six years, the annual equivalent costs would be higher than $7,977. We conclude that the economic service life of the truck is six years. By replacing the assets perpetually according to an economic life of six years, we obtain the minimum annual equivalent cost stream. Figure 14.5 illustrates this concept. Of course, we should envision a long period of required service for this kind of asset.

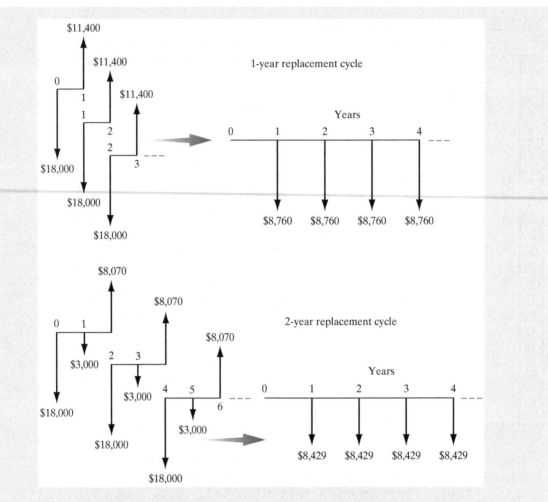

Figure 14.5 Conversion of an infinite number of replacement cycles to infinite AE cost streams (Example 14.3).

14.3 Replacement Analysis when the Required Service is Long

Once we understand how the economic service life of an asset is determined, the next question is when to replace the defender. If *now* is not the right time, when *is* the optimal time to replace the defender? Before presenting an analytical approach to answer this question, we consider several important assumptions.

14.3.1 Required Assumptions and Decision Frameworks

In deciding when to replace the defender, we need to consider the following three factors (Figure 14.6):

- Planning horizon (study period)
- Technology
- Relevant cash flow information

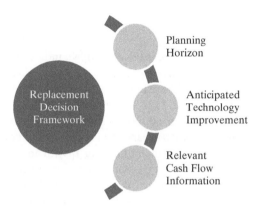

Figure 14.6 Elements of replacement decision framework.

Planning Horizon (Study Period)

By the planning horizon, we simply mean the service period required by the defender and a sequence of future challengers. The infinite planning horizon is used when we are simply unable to predict when the activity under consideration will be terminated. In other situations, it may be clear that the project will have a definite and predictable duration. In these cases, the replacement policy should be formulated more realistically on the basis of a finite planning horizon.

Technology

Predictions of technological patterns over the planning horizon refer to the development of types of challengers that may replace those under study. A number of possibilities exist in predicting purchase cost, salvage value, and operating cost that are dictated by the efficiency of a new machine over the life of an existing asset.

- If we assume that all future machines will be the same as those now in service, we are implicitly saying that no technological progress in the area will occur.
- In other cases, we may explicitly recognize the possibility of machines becoming available in the future that will be significantly more efficient, reliable, or productive than those currently on the market. (Personal computers are a good example.) This situation leads to the recognition of technological change and obsolescence.

Clearly, if the best available machine gets better and better over time, we should certainly investigate the possibility of delaying an asset's replacement for a couple of years—a viewpoint that contrasts with the situation in which technological change is unlikely.

Relevant Cash Flows over the Life of an Asset

Many varieties of predictions can be used to estimate the patterns of revenue, cost, and salvage value over the life of an asset. Sometimes revenue is constant, but costs increase, while salvage value decreases over the life of a machine. In other situations, a decline in revenue over the life of a piece of equipment can be expected. The specific situation will determine whether replacement analysis is directed toward cost minimization (with constant revenue) or profit maximization (with varying revenue). We formulate a replacement policy for an asset whose salvage value does not increase with age.

Decision Frameworks

To illustrate how a decision framework is developed, we indicate a replacement sequence of assets by the notation $(j_0, n_0), (j_1, n_1), (j_2, n_2), \ldots, (j_K, n_k)$. Each pair of numbers (j, n) indicates a type of asset and the lifetime over which that asset will be retained. The defender, asset 0, is listed first; if the defender is replaced now, $n_0 = 0$. A sequence of pairs may cover a finite period or an infinite period. For example, the sequence $(j_0, 2), (j_1, 5), (j_2, 3)$ indicates retaining the defender for two years, then replacing the defender with an asset of type j_1 and using it for five years, and then replacing j_1 with an asset of type j_2 and using it for three years. In this situation, the total planning horizon covers 10 years $(2 + 5 + 3)$. The special case of keeping the defender for n_0 periods followed by infinitely repeated purchases and the use of an asset of type j for n^* years is represented by $(j_0, n_0), (j, n^*)_\infty$. This sequence covers an infinite period, and the relationship is illustrated in Figure 14.7.

Decision Criterion

Although the economic life of the defender is defined as the additional number of years of service which minimizes the annual equivalent cost (or maximizes the annual equivalent revenue), that is *not* necessarily the *optimal* time to replace the defender. The correct replacement time depends on data on the challenger as well as on data on the defender.

As a decision criterion, the AEC method provides a more direct solution when the planning horizon is infinite. When the planning horizon is finite, the PW method is more

Figure 14.7 Types of typical replacement decision frameworks.

convenient to use. We will develop the replacement decision procedure for both situations. We begin by analyzing an infinite planning horizon without technological change. Even though a simplified situation such as this is not likely to occur in real life, the analysis of this replacement situation introduces methods that will be useful in analyzing infinite-horizon replacement problems with technological change.

14.3.2 Replacement Strategies under the Infinite Planning Horizon

Consider a situation in which a firm has a machine that is in use in a process which is expected to continue for an indefinite period. Presently, a new machine will be on the market that is (in some ways) more effective for the application than the defender is. The problem is when (if at all) the defender should be replaced with the challenger.

Under the infinite planning horizon, the service is required for a very long time. Either we continue to use the defender to provide the service, or we replace the defender with the best available challenger for the same service requirement. In this case, we may apply the following procedure in replacement analysis.

1. Compute the economic lives of both the defender and the challenger. Let's use N_D^* and N_C^* to indicate the economic lives of the defender and the challenger, respectively. The annual equivalent costs for the defender and the challenger at their respective economic lives are indicated by AEC_D^* and AEC_C^*.

2. Compare AEC_D^* and AEC_C^*.
   - If $AEC_D^* > AEC_C^*$, it is more costly to keep the defender than to replace it with the challenger. Thus, the challenger should replace the defender now.
   - If $AEC_D^* < AEC_C^*$, it costs less to keep the defender than to replace it with the challenger. Thus, the defender should *not* be replaced now. The defender should continue to be used at least for the duration of its economic life if there are no technological changes over that life.

3. If the defender should not be replaced now, when should it be replaced? First we need to continue to use it until its economic life is over. Then we should calculate the cost of running the defender for one more year after its economic life. This approach is called **marginal analysis**; that is, we calculate the incremental cost of operating the defender for just one more year. In other words, we want to see whether the cost of extending the use of the defender for an additional year exceeds the savings resulting from delaying the purchase of the challenger.
   - If this marginal cost is greater than AEC_C^*, the defender should be replaced at the end of its economic life. Otherwise, we should calculate the cost of running the defender for the second year after its economic life.
   - If the marginal cost for the second year is bigger than AEC_C^*, the defender should be replaced one year after its economic life. The process should be continued until we find the optimal replacement time. Here, we have assumed that the best available challenger does not change. Figure 14.8 summarizes the general procedure to conduct a replacement analysis under an infinite planning horizon.

Note that this procedure might be applied dynamically. For example, it may be performed annually for replacement analysis. Whenever there are updated data on the costs of the defender or new challengers available on the market, the new data should be used in the procedure. Example 14.4 illustrates the procedure.

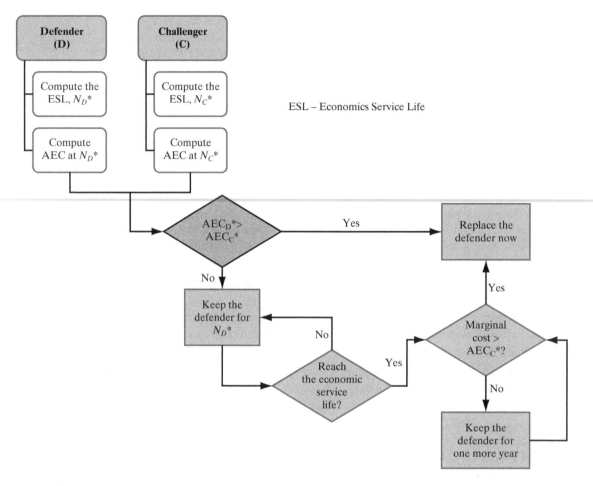

Figure 14.8 Overall replacement strategies under an infinite planning horizon.

EXAMPLE 14.4 Replacement Analysis under an Infinite Planning Horizon

The Advanced Electrical Insulator Company is considering replacing a broken inspection machine which has been used to test the mechanical strength of electrical insulators with a newer and more efficient one.

- **Option 1:** If repaired, the old machine can be used for another six years. However, the firm can sell it now to another firm in the industry for $5,000. If the machine is kept, it will require an immediate $1,200 overhaul to restore it to operable condition. The overhaul will neither extend the service life originally estimated nor increase the value of the inspection machine. The operating costs are estimated at $2,000 during the first year, and these are expected to increase by $1,500 per year thereafter. Future market values are expected to decline by 25% each year over the previous year's value.

- **Option 2:** The new machine costs $10,000 and will have operating costs of $2,200 in the first year, increasing by 20% per year thereafter. The expected salvage value is $6,000 after one year and will decline 15% each year. The physical life of the new machine would last eight years.

The company requires a rate of return of 15% before tax. Find the economic life for each option, *and* determine when the defender should be replaced.

SOLUTION

Given: Financial data for both defender and challenger and $i = 15\%$.

Find: (a) Economic service lives for both defender and challenger, (b) when to replace the defender.

(a) **Economic service life.**

- **Defender:** If the company retains the inspection machine, it is in effect deciding to overhaul the machine and invest the machine's current market value in that alternative. The opportunity cost of the machine is $5,000. Because an overhaul costing $1,200 is also needed to make the machine operational, the total initial cash outlay in the machine is $5,000 + $1,200 = $6,200. Other market (salvage) value and O&M cost data for the defender are summarized in Table 14.2. We can calculate the annual equivalent costs if the defender is to be kept for one year, two years, three years, and so forth, with the equation

$$\text{AEC}(15\%)_N = (\$5,000 + \$1,200)(A/P, 15\%, N) + \$2,000$$
$$+ \$1,500(A/G, 15\%, N)$$
$$- \$5,000(1 - 0.25)^N(A/F, 15\%, N)$$

TABLE 14.2 Economic Service Life Calculation for the Defender (Example 14.4)

	A	B	C	D	E	F
1						
2		Annual changes in MV		-25%		
3		Annual increases in O&M		$1,500		
4		Interest rate		15%		
5						
6	n	Market Value	O&M Costs	CR(15%)	OC(15%)	AEC(15%)
7						
8	0	$5,000	$1,200			
9	1	$3,750	$2,000	$2,000	$3,380	$5,380
10	2	$2,813	$3,500	$1,767	$3,436	$5,203
11	3	$2,109	$5,000	$1,582	$3,886	$5,469
12	4	$1,582	$6,500	$1,435	$4,410	$5,844
13	5	$1,187	$8,000	$1,316	$4,942	$6,258
14	6	$890	$9,500	$1,220	$5,463	$6,682
15						
16						
17						
18						
19		=B13*(1+$D$2)	=C13+$D$3	=-PMT($D$4,A14,$B$8,-B14)		=D14+E14
20						
21				=-PMT($D$4,A14,NPV($D$4,$C$9:C14)+$C$8,0)		
22						

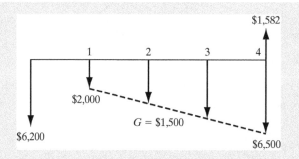

Figure 14.9 Cash flow diagram for defender when $N = 4$ years (Example 14.4).

$$AEC(15\%)_N = \overbrace{[\$6{,}200 - \$5{,}000(1 - 0.25)^N](A/P, 15\%, N)}^{\text{CR}(15\%)}$$
$$+ 0.15\{\$5{,}000(1 - 0.25)^N\}$$
$$\underbrace{- \$2{,}000 + \$1{,}500(A/G, 15\%, N)}_{\text{OC}(15\%)}$$

For example, with $N = 4$ years, the annual cash flows looks like the one in Figure 14.9, and its annual equivalent costs are $AEC(15\%)_4 = \$5{,}844$. As shown in Table 14.2, we get the lowest AEC value when $N = 2$ years. Thus, the defender's economic life is two years. Using the notation we defined in the procedure, we have

$$N_D^* = 2 \text{ years}$$

$$AEC_D^* = \$5{,}203$$

- **Challenger:** The economic life of the challenger can be determined with the same procedure we used in Example 14.3. The equation is

$$AEC(15\%)_N = \$10{,}000(A/P, 15\%, N)$$
$$+\$2{,}200(P/A_1, 20\%, 15\%, N)(A/P, 15\%, N)$$
$$-\$6{,}000(1 - 0.15)^{N-1}(A/F, 15\%, N).$$

We can also develop an Excel worksheet similar to the defender, as shown in Table 14.3, to determine the economic service life. Clearly, the economic life of the challenger is four years; that is,

$$N_C^* = 4 \text{ years}$$

Thus,

$$AEC_C^* = \$5{,}625$$

(b) **Should the defender be replaced now?** Since $AEC_D^* = \$5{,}203 < AEC_C^* = \$5{,}625$, the defender should not be replaced for now. If there are no technological advances in the next few years, the defender should be used for at least $N_D^* = 2$ more years. However, it is not necessarily best to replace the defender right at the high point of its economic life, either. It depends on what type of challenger would be available in the future.

TABLE 14.3   Economic Service Life Calculation for the Challenger (Example 14.4)

	A	B	C	D	E	F
1						
2		Annual changes in MV		-15%		
3		Annual increases in O&M		20%		
4		Interest rate		15%		
5						
6	*n*	Market Value	O&M Costs	CR(12%)	OC(12%)	AEC(12%)
7						
8	0	$10,000				
9	1	$6,000	$2,200	$5,500	$2,200	$7,700
10	2	$5,100	$2,640	$3,779	$2,405	$6,184
11	3	$4,335	$3,168	$3,131	$2,624	$5,756
12	4	$3,685	$3,802	$2,765	$2,860	$5,625
13	5	$3,132	$4,562	$2,519	$3,113	$5,631
14	6	$2,662	$5,474	$2,338	$3,382	$5,721
15	7	$2,263	$6,569	$2,199	$3,670	$5,869
16	8	$1,923	$7,883	$2,088	$3,977	$6,066
17						
18						=D16+E16
19		=B15*(1+$D$2)	=C15*(1+$D$3)			
20				=-PMT($D$4,A16,NPV($D$4,$C$9:C16),0)		
21		=-PMT($D$4,A16,$B$8,-B16)				
22						

(c) **When should the defender be replaced?**   If we need to find the answer to this question today, we have to calculate the cost of keeping and using the defender for the third year from today. That is, what is the cost of not selling the defender at the end of year 2, using it for the third year, and replacing it at the end of year 3? The following cash flows are related to this question:

- Opportunity cost at the end of year 2: equal to the market value then, or $2,813.
- Operating cost for the third year: $5,000.
- Salvage value of the defender at the end of year 3: $2,109.

Figure 14.10 represents these cash flows. The cost of using the defender for one more year from the end of its economic life is

$$\$2,813 \times 1.15 + \$5,000 - \$2,109 = \$6,126$$

Figure 14.10   Marginal cost of keeping the defender for one more year beyond its economic service life.

> Since $AEC_C^* = \$5,625$, it is more expensive to keep the defender for the third year than to replace it with the challenger. Accordingly, we conclude that we should replace the defender at the end of year 2. If this one-year cost is still smaller than AEC_C^*, we need to calculate the cost of using the defender for the fourth year and then compare that cost with the AEC_C^* of the challenger.

Handling Unequal Service Life Problem in Replacement Analysis

In replacement analysis, it is common for a defender and its challenger to have different economic service lives. The annual-equivalent approach is frequently used, but it is important to know that we use the AEC method in replacement analysis, not because we have to deal with the problem of unequal service lives, but rather because the AEC approach provides some computational advantage for a special class of replacement problems.

In Chapter 5, we discussed the general principle for comparing alternatives with unequal service lives. In particular, we pointed out that use of the AEC method relies on the concept of repeatability of projects and one of two assumptions: an infinite planning horizon or a common service period. In defender–challenger situations, however, repeatability of the defender cannot be assumed. In fact, by virtue of our definition of the problem, we are not repeating the defender, but replacing it with its challenger, which is an asset that in some way constitutes an improvement over the current equipment. Thus, the assumptions we made for using an annual cash flow analysis with unequal service-life alternatives are not valid in the usual defender–challenger situation.

The complication—the unequal-life problem—can be resolved if we recall that the replacement problem at hand is not *whether* to replace the defender but *when* to do so. When the defender is replaced, it will always be by the challenger—the best available equipment. An identical challenger can then replace the challenger repeatedly. In fact, we really are comparing the following two options in replacement analysis:

1. **Replace the defender now.** The cash flows of the challenger will be used from today and will be repeated because an identical challenger will be used if replacement becomes necessary again in the future. This stream of cash flows is equivalent to a cash flow of AEC_C^* each year for an infinite number of years.

2. **Replace the defender, say, x years later.** The cash flows of the defender will be used in the first x years. Starting in year $x + 1$, the cash flows of the challenger will be used indefinitely.

The annual-equivalent cash flows for the years beyond year x are the same for these two options. We need only to compare the annual-equivalent cash flows for the first x years to determine which option is better. This is why we can compare AEC_D^* with AEC_C^* to determine whether now is the time to replace the defender.

14.3.3 Replacement Strategies under the Finite Planning Horizon

If the planning period is finite (for example, eight years), a comparison based on the AEC method over a defender's economic service life does not generally apply. The procedure for solving such a problem with a finite planning horizon is to establish all "reasonable" replacement patterns and then use the PW value for the planning period to select the most economical pattern. To illustrate this procedure, consider Example 14.5.

EXAMPLE 14.5 Replacement Analysis under the Finite Planning Horizon (PW Approach)

Consider again the defender and the challenger in Example 14.4. Suppose that the firm has a contract to perform a given service, using the current defender or the challenger for the next eight years. After the contract work, neither the defender nor the challenger will be retained. What is the best replacement strategy?

SOLUTION

Given: AEC values for both the defender and the challenger over eight years, planning horizon = 8 years, and $i = 15\%$.

Find: Most economical replacement scenario.

Recall again the annual equivalent costs for the defender and challenger under the assumed holding periods (a boxed number denotes the minimum AEC value at $N_D^* = 2$ and $N_C^* = 4$, respectively):

	Annual Equivalent Cost ($)	
n	**Defender**	**Challenger**
1	5,380	7,700
2	$\boxed{5{,}203}$	6,184
3	5,469	5,756
4	5,844	$\boxed{5{,}625}$
5	6,258	5,631
6	6,682	5,721

Many replacement scenario options would fulfill an eight-year planning horizon, as shown in Figure 14.11. Of these options, six appear to be the most likely by inspection. These options are listed, and the present equivalent cost for each option is calculated, as follows.

- **Option 1:** $(j_0, 0), (j, 4), (j, 4)$

$$PW(15\%)_1 = \$5,625(P/A, 15\%, 8)$$

$$= \$25,241$$

- **Option 2:** $(j_0, 1), (j, 4), (j, 3)$

$$PW(15\%)_2 = \$5,380(P/F, 15\%, 1)$$

$$+ \$5,625(P/A, 15\%, 4)(P/F, 15\%, 1)$$

$$+ \$5,756(P/A, 15\%, 3)(P/F, 15\%, 5)$$

$$= \$25,177$$

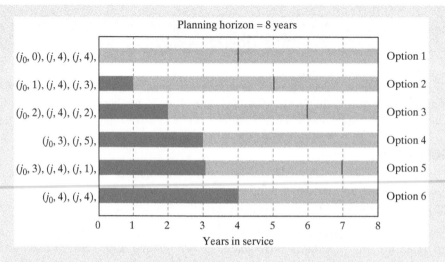

Figure 14.11 Some likely replacement patterns under a finite planning horizon of eight years (Example 14.5).

- **Option 3:** $(j_0, 2), (j, 4), (j, 2)$

$$PW(15\%)_3 = \$5,203(P/A, 15\%, 2)$$
$$+ \$5,625(P/A, 15\%, 4)(P/F, 15\%, 2)$$
$$+ \$6,184(P/A, 15\%, 2)(P/F, 15\%, 6)$$
$$= \$24,948$$

- **Option 4:** $(j_0, 3), (j, 5)$

$$PW(15\%)_4 = \$5,469(P/A, 15\%, 3)$$
$$+ \$5,631(P/A, 15\%, 5)(P/F, 15\%, 3)$$
$$= 24,898 \leftarrow \text{minimum cost}$$

- **Option 5:** $(j_0, 3), (j, 4), (j, 1)$

$$PW(15\%)_5 = \$5,469(P/A, 15\%, 3)$$
$$+ \$5,625(P/A, 15\%, 4)(P/F, 15\%, 3)$$
$$+ \$7,700(P/F, 15\%, 8)$$
$$= \$25,563$$

- **Option 6:** $(j_0, 4), (j, 4)$

$$PW(15\%)_6 = \$5,844(P/A, 15\%, 4)$$
$$+ \$5,625(P/A, 15\%, 4)(P/F, 15\%, 4)$$
$$= \$25,866$$

An examination of the present equivalent cost of a planning horizon of eight years indicates that the least-cost solution among six decision options considered appears to be option 4: Retain the defender for three years, purchase the challenger, and keep it for five years.

COMMENTS: In this example, we examined only six decision options that were likely to lead to the best solution, but it is important to note that several other possibilities have not been looked at. To explain, consider Figure 14.12, which shows a graphical representation of various replacement strategies under a finite planning horizon. For example, the replacement strategy $[(j_0, 2), (j, 3), (j, 3)]$ (shown as a solid line in the figure) is certainly feasible, but we did not include it in the previous computation. Naturally, as we extend the planning horizon, the number of possible decision options can easily multiply. To make sure that we indeed find the optimal solution for such a problem, an optimization technique such as dynamic programming can be used.[3]

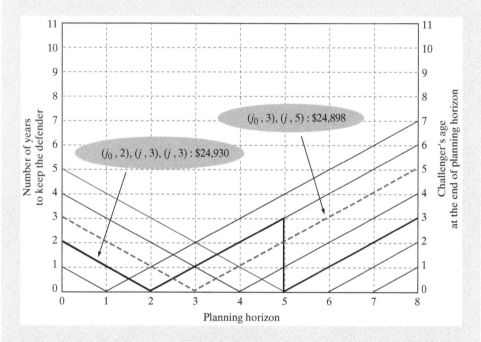

Figure 14.12 Graphical representations of replacement strategies under a finite planning horizon (Example 14.5).

[3] F. S. Hillier and G. S. Lieberman, *Introduction to Operations Research*, 10th ed. New York: McGraw-Hill, 2014.

14.3.4 Consideration of Technological Change

Thus far, we have defined the challenger simply as the best available replacement for the defender. It is more realistic to recognize that the replacement decision often involves an asset now in use versus a candidate for replacement—that is, in some way, an improvement on the current asset. This, of course, reflects technological progress that is ongoing continually. Future models of a machine are likely to be more efficient than a current model. In most areas, technological change appears as a combination of gradual advances in effectiveness; the occasional technological breakthrough, however, can revolutionize the character of a machine.

The prospect of improved future challengers makes a current challenger a less desirable alternative. By retaining the defender, we may have an opportunity to acquire an improved challenger later. If this is the case, the prospect of improved future challengers may affect a current decision between a defender and its challenger. It is difficult to forecast future technological trends in any precise fashion. However, in developing a long-term replacement policy, we need to take technological change into consideration.

14.4 Replacement Analysis with Tax Considerations

Up to this point, we have covered various concepts and techniques that are useful in replacement analysis. In this section, we illustrate how to extend those concepts and techniques to consider the effects of taxes on replacement analysis, using the same examples (Example 14.1 through 14.4) with the following additional information:

- Depreciation schedules for both the defender and the challenger
- Taxable gains and gain taxes at the time of asset disposal
- All operating cash flows on after-tax basis
- MARR based on after-tax, usually lower than the one on before-tax basis.

EXAMPLE 14.6 Net Proceeds from the Disposal of an Old Machine

Suppose that, in Example 14.1, the $20,000 capital expenditure was set up to be depreciated on a seven-year MACRS (allowed annual depreciation: $2,858, $4,898, $3,498, $2,498, $1,786, $1,784, and $892). If the firm's marginal income-tax rate is 40%, determine the taxable gains (or losses) and the net proceeds from disposal of the old printing machine.

SOLUTION

Given: Depreciation schedule, $S = \$10,000$, and $t_m = 40\%$.
Find: Market value to be used in replacement analysis.

First we need to find the current book value of the old printing machine. The original cost minus the accumulated depreciation, calculated with the half-year convention (if the machine is sold now), is

$$\$20,000 - (\$2,858 + \$4,898/2) = \$14,693$$

So we compute the following:

$$\text{Allowed book value} = \$14,693$$
$$\text{Current market value} = \$10,000$$
$$\text{Losses} = \$4,693$$
$$\text{Tax savings} = \$4,693(0.40) = \$1,877$$
$$\text{Net proceeds from the sale} = \$10,000 + \$1,877$$
$$= \$11,877$$

This calculation is illustrated in Figure 14.13.

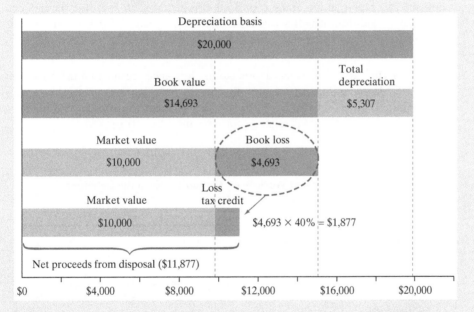

Figure 14.13 Net proceeds from the sale of the old printing machine—defender (Example 14.6).

EXAMPLE 14.7 After-Tax Replacement Analysis Using the Opportunity Cost Approach

Suppose, in Example 14.2, that the new machine would fall into the same seven-year MACRS class as the old machine. The firm's *after-tax* interest rate (MARR) is 10%, and the marginal income tax rate is 40%. Use the opportunity cost approach to decide whether replacement of the old machine is justified now.

SOLUTION
Given: Financial information for both defender and challenger, $t_m = 40\%$, and $i = 10\%$.
Find: Whether or not to replace the defender now.

- **Option 1: Keep the defender.** Table 14.4 shows the worksheet format the company uses to analyze a typical replacement project with the generalized cash-flow approach. Each line is numbered, and a line-by-line description is given.

  Lines 1 to 4: If the old machine is kept, the depreciation schedule would be ($3,498, $2,498, and $1,786). Following the half-year convention, it is assumed that the asset will be retired at the end of three years; thus, the depreciation for year three is (0.5)($1,786) = $893. This results in total depreciation in the amount of $14,645 and a remaining book value of $20,000 − $14,645 = $5,355.

  Lines 5 to 6: Repair costs in the amount of $5,000 were already incurred before the replacement decision. This is a sunk cost and should not be considered in the analysis. If a repair in the amount of $5,000 is required to keep the defender in serviceable condition, it will show as an expense in year 0. If the old machine is retained for the next three years, the before-tax annual O&M costs are as shown in line 6.

  Lines 7 to 10: Recall that the depreciation allowances result in a tax reduction equal to the depreciation amount multiplied by the tax rate. The operating expenses are multiplied by the factor of (1 − the tax rate) to obtain the after-tax O&M. For a situation in which the asset is retained for now, the AEC(10%) is $7,524.

TABLE 14.4 Replacement Worksheet: Option 1—Keep the Defender (Example 14.7)

	A	B	C	D	E	F	G
1							
2	*n*	-2	-1	0	1	2	3
3	**Financial Data**						
4							
5	(1) Depreciation		$ 2,858	$ 4,898	$ 3,498	$ 2,498	$ 893
6	(2) Book value	$ 20,000	17,142	12,244	8,746	6,248	5,355
7	(3) Salvage value						2,500
8	(4) Loss from sale						(2,855)
9	(5) Repair cost		5,000				
10	(6) O&M costs				8,000	8,000	8,000
11							
12	**Cash Flow Statement**						
13	(7) Opportunity cost			(11,877)			
14	(8) Net proceeds from sale						3,642
15	(9) O&M costs × (0.60)				(4,800)	(4,800)	(4,800)
16	(10) Depreciation × (0.40)				1,399	999	357
17							
18	Net cash flow			$ (11,877)	$ (3,401)	$ (3,801)	$ (801)
19							
20			PW(10%) =	$ (18,712)		AEC(10%) =	$7,791
21							
22					=PMT(12%,3,C20)		

TABLE 14.5 Replacement Worksheet: Option 2—Replace the Defender (Example 14.7)

	A	B	C	D	E	F	G
1							
2		n	-2	-1	0	1	2	3
3	Financial Data						
4							
5	(1) Depreciation				$ 2,144	$ 3,674	$ 1,312
6	(2) Book value			$ 15,000	12,856	9,182	7,870
7	(3) Salvage value						5,500
8	(4) Loss from sale						(2,370)
9	(5) Repair cost						
10	(6) O&M costs				6,000	6,000	6,000
11							
12	Cash Flow Statement						
13	(7) Opportunity cost			(15,000)			
14	(8) Net proceeds from sale						6,448
15	(9) O&M costs × (0.60)				(3,600)	(3,600)	(3,600)
16	(10) Depreciation × (0.40)				858	1,470	525
17							
18	Net cash flow			$ (15,000)	$ (2,742)	$ (2,130)	$ 3,373
19							
20			PW(10%) =	$ (16,720)		AEC(10%) =	$6,723	
21							

- **Option 2: Replace the defender.** Table 14.5 shows an Excel worksheet based on the generalized cash-flow approach. As before, each line is numbered, and a line-by-line description of the table is given.

  Line 1: The depreciation schedule, along with the book values for the new machine (seven-year MACRS), is shown. Again, the depreciation amount of $1,312 in year 3 reflects the half-year convention.

  Line 2: The purchase price of the new machine, including installation and freight charges (if any), will be the depreciation base for the challenger. The book values are also shown here.

  Line 4: With the salvage value estimated at $5,500, we expect a loss ($2,370 = $7,870 − $5,500) on the sale of the new machine at the end of year 3.

  Line 6: The O&M costs for the new machine are listed.

  Lines 7 to 10: The after-tax calculation for both O&M costs and depreciation tax credit are the same as those in Table 14.4. For a situation in which the asset is replaced with the challenger, the AEC(10%) is $6,723.

Because of the annual difference of $801 in favor of the challenger, the replacement should be made now. Figure 14.14 illustrates the cash flows related to these decision options.

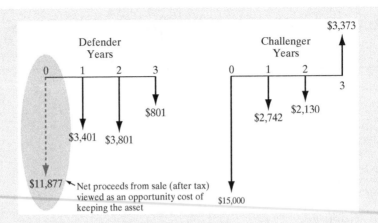

Figure 14.14 Comparison of defender and challenger on the basis of the opportunity cost approach (Example 14.7).

EXAMPLE 14.8 Economic Service Life of a Lift Truck on After-Tax Basis

Consider again Example 14.3, using the following additional data. The asset belongs to a five-year MACRS property class with the following annual depreciation allowances: 20%, 32%, 19.20%, 11.52%, 11.52%, and 5.76%. The firm's marginal tax rate is 40%, and its after-tax MARR is 9%. Find the economic service life of this new machine.

DISCUSSION: To determine an asset's economic service life, we first list the gains or losses that will be realized if the truck were to be disposed of at the end of each operating year. In doing so, we need to compute the book values at the end of each operating year, assuming that the asset would be disposed of at that time. Recall that, with the half-year convention, the book value (for the MACRS property) at the end of the year is based on its disposal during the year. As summarized in Table 14.6, these values provide a basis for identifying the net proceeds from sale at the end of an assumed holding period.

SOLUTION

Given: Financial data as in Example 14.3, depreciation schedule, $t_m = 40\%$, and $i = 9\%$.

Find: Economic service life.

Two approaches may be used to find the economic life of an asset: (a) the generalized cash-flow approach and (b) the tabular approach.

(a) **Generalized cash-flow approach.** Since we have only a few cash-flow elements (O&M, depreciation, and salvage value), an efficient way to obtain the after-tax cash flow is to use the generalized cash-flow approach discussed in Section 10.4.

TABLE 14.6 Calculation of Net Proceeds from Sale as a Function of Holding Period (Example 14.8)

	0	1	2	3	4	5	6	7	8
Allowed Annual									
Depreciation Amount					Holding Period				
D1		$3,600	$3,600	$3,600	$3,600	$3,600	$3,600	$3,600	$3,600
D2			$2,880	$5,760	$5,760	$5,760	$5,760	$5,760	$5,760
D3				$1,728	$3,456	$3,456	$3,456	$3,456	$3,456
D4					$1,037	$2,074	$2,074	$2,074	$2,074
D5						$1,037	$2,074	$2,074	$2,074
D6							$1,036	$1,036	$1,036
D7								$0	$0
D8									$0
Total Depreciation		$3,600	$6,480	$11,088	$13,853	$15,927	$18,000	$18,000	$18,000
Book Value	$18,000	$14,400	$11,520	$6,912	$4,147	$2,073	$0	$0	$0
Market Value	$18,000	$14,400	$11,520	$9,216	$7,373	$5,898	$4,719	$3,775	$3,020
Taxable Gains		$0	$0	$2,304	$3,226	$3,825	$4,719	$3,775	$3,020
Net Proceeds from Sale		$14,400	$11,520	$8,294	$6,082	$4,368	$2,831	$2,265	$1,812

TABLE 14.7 After-Tax Cash Flow Calculation for Owning and Operating the Asset for Two Years

	A	B	C	D	E	F
1						
2	*n*	-2	-1	0	1	2
3	**Financial Data**					
4	(1) Depreciation				$ 3,600	$ 2,880
5	(2) Book value			$ 18,000	14,400	11,520
6	(3) Salvage value					11,520
7	(4) Loss from sale					-
8	(5) O&M costs				3,000	3,450
9						
10	**Cash Flow Statement**					
11	(6) Investment cost			(18,000)		
12	(7) Net proceeds from sale					11,520
13	(8) O&M costs × (0.60)				(1,800)	(2,070)
14	(9) Depreciation × (0.40)				1,440	1,152
15						
16	Net cash flow			$ (18,000)	$ (360)	$ 10,602
17						
18		PW(9%) =	$ (9,407)		AEC(9%) =	$5,347
19						

For example, we can develop an Excel worksheet, as shown in Table 14.7, to summarize the cash flows for two-year ownership.

We can continue to generate yearly after-tax entries for the asset's remaining physical life, as we know the forecasted O&M costs and the net proceeds from the sale of the asset as a function of holding period. Similarly, the annual equivalent costs for the subsequent years can be computed as

Holding Period	AEC(9%)
1	$5,580
2	$5,347
3	$5,149
4	$5,013
5	$4,935
6	$4,901
7	$4,910
8	$4,959

If the truck were to be sold after six years, it would have a minimum annual cost of $4,901 per year. Graphically it is shown in Figure 14.15, and this is the life

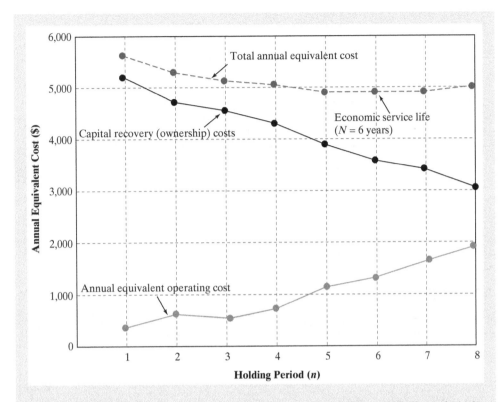

Figure 14.15 Economic service life obtained by finding the minimum AEC (Example 14.8). Note that we treat the cost items with a positive sign.

that is most favorable for comparison purposes. By replacing the asset perpetually according to an economic life of six years, we obtain the minimum infinite AE cost stream.

(b) **Tabular approach.** The tabular approach separates the annual cost elements into two parts, one associated with the capital recovery of the asset and the other associated with operating the asset. In computing the capital recovery cost, we need to determine the after-tax salvage values at the end of each holding period, as calculated previously in Figure 14.19. Then, we compute the total annual equivalent cost of the asset for any given year's operation using Eq. (14.3).

If we examine the equivalent annual costs itemized in Table 14.8 (columns I and J), we see that, as the asset ages, the equivalent annual O&M costs increase. At the same time, capital recovery costs decrease with prolonged use of the asset. The combination of decreasing capital recovery costs and increasing annual O&M costs results in the total annual equivalent cost taking on a form similar to that depicted in Figure 14.15.

TABLE 14.8 Economic Service Life Calculation Based After-Tax (Example 14.8)

	A	B	C	D	E	F	G	H	I	J	K
1											
2	Tax Rate	40%									
3	MARR	9%									
4	Holding	Permitted Annual Depreciation Amounts over the								Total	Book
5	Period	Holding Period								Depreciation	Value
6	(N)	1	2	3	4	5	6	7	8		
7	0										$18,000
8	1	$3,600								$3,600	$14,400
9	2	$3,600	$2,880							$6,480	$11,520
10	3	$3,600	$5,760	$1,728						$11,088	$6,912
11	4	$3,600	$5,760	$3,456	$1,037					$13,853	$4,147
12	5	$3,600	$5,760	$3,456	$2,074	$1,037				$15,927	$2,073
13	6	$3,600	$5,760	$3,456	$2,074	$2,074	$1,036			$18,000	$0
14	7	$3,600	$5,760	$3,456	$2,074	$2,074	$1,036	$0		$18,000	$0
15	8	$3,600	$5,760	$3,456	$2,074	$2,074	$1,036	$0	$0	$18,000	$0
16											
17	Holding	Annual O&M Costs over the Holding Period								Total PW of	Total PW of A/T
18	Period									Total PW of	of A/T
19	(N)	1	2	3	4	5	6	7	8	O&M Costs	O&M Costs
20	0										
21	1	$3,000								$2,752	$1,651
22	2	$3,000	$3,450							$5,656	$3,394
23	3	$3,000	$3,450	$3,968						$8,720	$5,232
24	4	$3,000	$3,450	$3,968	$4,563					$11,953	$7,172
25	5	$3,000	$3,450	$3,968	$4,563	$5,247				$15,363	$9,218
26	6	$3,000	$3,450	$3,968	$4,563	$5,247	$6,034			$18,961	$11,376
27	7	$3,000	$3,450	$3,968	$4,563	$5,247	$6,034	$6,939		$22,757	$13,654
28	8	$3,000	$3,450	$3,968	$4,563	$5,247	$6,034	$6,939	$7,980	$26,761	$16,057
29											
30	Holding	Expected			Net A/T	A/T Operating Costs (in PW)					Total
31	Period	Market	Taxable	Gains	Market	over the Holding Period			OC(9%)	CR(9%)	AEC(9%)
32	(N)	Value	Gains	Tax	Value	O&M Costs	Tax Shield	Total OC			
33	0	$18,000									
34	1	$14,400	$0	$0	$14,400	$1,651	$1,321	$330	$360	$5,220	$5,580
35	2	$11,520	$0	$0	$11,520	$3,394	$2,291	$1,103	$627	$4,720	$5,347
36	3	$9,216	$2,304	$922	$8,294	$5,232	$3,794	$1,438	$568	$4,581	$5,149
37	4	$7,373	$3,226	$1,290	$6,083	$7,172	$4,622	$2,550	$787	$4,226	$5,013
38	5	$5,898	$3,825	$1,530	$4,368	$9,218	$5,185	$4,033	$1,037	$3,898	$4,935
39	6	$4,719	$4,719	$1,888	$2,831	$11,376	$5,702	$5,675	$1,265	$3,636	$4,901
40	7	$3,775	$3,775	$1,510	$2,265	$13,654	$5,702	$7,952	$1,580	$3,330	$4,910
41	8	$3,020	$3,020	$1,208	$1,812	$16,057	$5,702	$10,355	$1,871	$3,088	$4,959
42											
43	Sample cell formulas:										
44	cell K13:=$K$7-J13				cell D39:=C39*$B2			cell G39:=NPV($B$3,B13:I13)*$B2			
45	cell K26:==J26*(1-$B2)				cell E39:=B39-D39			cell I39:=-PMT($B$3,A26,H39,0,0)			
46	cell C39:=B39-K113				cell F39:=K26			cell J39: =-PMT($B$3,A26,$B$33,-E39)			
47											

EXAMPLE 14.9 After-Tax Replacement Analysis under the Infinite Planning Horizon

Recall Example 14.4, in which the Advanced Electrical Insulator Company is considering replacing a broken inspection machine. Let's assume the following additional data:

- The old machine has been fully depreciated, so it has zero book value. The machine could be used for another six years.
- The new machine falls into the five-year MACRS property class and will be depreciated accordingly.

The marginal income tax rate is 40%, and the MARR is 12% after taxes. Find the useful life for each option presented in Example 14.4, *and* decide whether the defender should be replaced now or later.

SOLUTION

Given: Financial data as provided in Example 14.4, depreciation schedules for both the defender and the challenger, $i = 12\%$, and $t_m = 40\%$.

Find: When to replace the defender.

Economic service life:

- **Defender.** The defender is fully depreciated, so that all salvage values can be treated as ordinary gains and taxed at 40%. The after-tax market values are thus as shown in cells B21 through B27 in Table 14.9.

  If the company retains the inspection machine, it is in effect deciding to overhaul the machine and invest the machine's current market value (after taxes) in that alternative. Although the company will make no physical cash flow transaction, it is withholding the market value of the inspection machine (the opportunity cost) from the investment. The after-tax O&M costs are as shown in cells C21 through C27 in Table 14.9.

  Using the current year's market value as the investment required to retain the defender, we obtain the AEC as a function of holding period as shown in Table 14.9, indicating that the remaining useful life of the defender is two years *in the absence of future challengers*. The overhaul (repair) cost of $1,200 in year 0 can be treated as a deductible operating expense for tax purposes as long as it does not add value to the property. (Any repair or improvement expenses that increase the value of the property must be capitalized by depreciating them over the estimated service life.)

- **Challenger.** Because the challenger will be depreciated over its tax life, we must determine the book value of the asset at the end of each period to compute the after-tax salvage value. This is done in rows 4 through 15 in Table 14.10. With the after-tax O&M values computed in rows 17 through 28 in that table, we are now ready to find the economic service life of the challenger by generating AEC value entries. These calculations are summarized in rows 30 through 41 in Table 14.10. The economic life of the challenger is five years, with an AEC(12%) value of $3,541.

TABLE 14.9 Economics of Retaining the Defender for N More Years (Example 14.9)

	A	B	C	D	E	F
1						
2		Annual changes in MV		-25%		
3		Annual changes in O&M		$1,500		
4		Tax rate		40%		
5		Interest rate (After tax)		12%		
6						
7	n	Forecasted		Book	Taxable	
8		Market Value	O&M Costs	Value	Gains	
9						
10	0	$5,000	$1,200	$0	$5,000	
11	1	$3,750	$2,000	$0	$3,750	
12	2	$2,813	$3,500	$0	$2,813	=B12-D12
13	3	$2,109	$5,000	$0	$2,109	
14	4	$1,582	$6,500	$0	$1,582	
15	5	$1,187	$8,000	$0	$1,187	
16	6	$890	$9,500	$0	$890	
17					=-PMT($D$5,A22,$B$21,-B22))	
18		After-Tax	After-Tax			
19	N	Market Value	O&M Costs	CR(12%)	OC(12%)	AEC(12%)
20					=D10+E10	
21	0	$3,000	$720			
22	1	$2,250	$1,200	$1,110	$2,006	$3,116
23	2	$1,688	$2,100	$979	$2,051	$3,030
24	3	$1,266	$3,000	$874	$2,332	$3,206
25	4	$949	$3,900	$789	$2,660	$3,449
26	5	$712	$4,800	$720	$2,997	$3,717
27	6	$534	$5,700	$664	$3,330	$3,994
28						
29	=E11*(1-$D$4))	=C13*(1-$D$4)				
30						
31			=-PMT($D$5,A23,NPV($D$5,$C$22:C23)+$C$21,0)			
32						
33						

Optimal time to replace the defender:

Since the AEC value for the defender's remaining useful life (two years) is $3,030, which is less than $3,541, the decision will be to keep the defender for now. Of course, the defender's remaining useful life of two years does not imply that the defender should actually be kept for two years before the company switches to the challenger. The reason for this is that the defender's remaining useful life of two years was calculated without considering what type of challenger would be available in the future.

TABLE 14.10 Economics of Owning and Operating the Challenger for N More years (Example 14.9)

	A	B	C	D	E	F	G	H	I	J	K
1											
2	Tax Rate	40%									
3	MARR	12%									
4	Holding	Permitted Annual Depreciation Amounts over the Holding Period								Total	Book
5	Period										
6	(N)	1	2	3	4	5	6	7	8	Depreciation	Value
7	0										$10,000
8	1	$2,000								$2,000	$8,000
9	2	$2,000	$1,600							$3,600	$6,400
10	3	$2,000	$3,200	$960						$6,160	$3,840
11	4	$2,000	$3,200	$1,920	$576					$7,696	$2,304
12	5	$2,000	$3,200	$1,920	$1,152	$576				$8,848	$1,152
13	6	$2,000	$3,200	$1,920	$1,152	$1,152	$576			$10,000	$0
14	7	$2,000	$3,200	$1,920	$1,152	$1,152	$576	$0		$10,000	$0
15	8	$2,000	$3,200	$1,920	$1,152	$1,152	$576	$0	$0	$10,000	$0
16											
17	Holding	Annual O&M Costs over the Holding Period								Total PW of	Total PW of A/T
18	Period										
19	(N)	1	2	3	4	5	6	7	8	O&M Costs	O&M Costs
20	0										
21	1	$2,200								$1,964	$1,179
22	2	$2,200	$2,640							$4,069	$2,441
23	3	$2,200	$2,640	$3,168						$6,324	$3,794
24	4	$2,200	$2,640	$3,168	$3,802					$8,740	$5,244
25	5	$2,200	$2,640	$3,168	$3,802	$4,562				$11,329	$6,797
26	6	$2,200	$2,640	$3,168	$3,802	$4,562	$5,474			$14,102	$8,461
27	7	$2,200	$2,640	$3,168	$3,802	$4,562	$5,474	$6,569		$17,073	$10,244
28	8	$2,200	$2,640	$3,168	$3,802	$4,562	$5,474	$6,569	$7,883	$20,257	$12,154
29											

	A	B	C	D	E	F	G	H	J	K	(Total)
30	Holding	Expected			Net A/T	A/T Operating Costs (in PW) over the Holding Period			OC(12%)	CR(12%)	Total AEC(12%)
31	Period	Market	Taxable	Gains	Market						
32	(N)	Value	Gains	Tax	Value	O&M Costs	Tax Shield	Total OC			
33	0										
34	1	$6,000	($2,000)	($800)	$6,800	$1,179	$714	$464	$520	$4,400	$4,920
35	2	$5,100	($1,300)	($520)	$5,620	$2,441	$1,224	$1,217	$720	$3,266	$3,986
36	3	$4,335	$495	$198	$4,137	$3,794	$2,008	$1,786	$744	$2,937	$3,681
37	4	$3,685	$1,381	$552	$3,133	$5,244	$2,428	$2,816	$927	$2,637	$3,564
38	5	$3,132	$1,980	$792	$2,340	$6,797	$2,705	$4,092	$1,135	$2,406	$3,541
39	6	$2,662	$2,662	$1,065	$1,597	$8,461	$2,952	$5,509	$1,340	$2,235	$3,575
40	7	$2,263	$2,263	$905	$1,358	$10,244	$2,952	$7,292	$1,598	$2,057	$3,654
41	8	$1,923	$1,923	$769	$1,154	$12,154	$2,952	$9,202	$1,852	$1,919	$3,772
42											

43	Sample cell formulas:						
44	cell K13:=$K$7-J13			cell D39:=C39*$B$2		cell G39:=NPV($B$3,B13:I13)*$B$2	
45	cell K26:==J26*(1-$B$2)			cell E39:=B39-D39		cell I39:=-PMT($B$3,A26,H39,0,0)	
46	cell C39:=B39-K113			cell F39:=K26		cell J39: =-PMT($B$3,A26,10000,-E39)	
47							

When a challenger's financial data is available, we need to enumerate all replacement timing possibilities.

The possible replacement cash patterns associated with each of these alternatives are shown in Figure 14.16, assuming that the cost and efficiency of the current challenger remain unchanged in future years. From the figure, we observe that on an annual basis the cash-flows after the remaining physical life of the defender are the same.

Before we evaluate the economics of various replacement-decision options, recall the AEC values for the defender and the challenger under the assumed service lives (a boxed figure denotes the economic service life as calculated in Figures 14.23 and 14.24).

	Annual Equivalent Cost	
n	Defender	Challenger
1	$3,116	$4,920
2	3,030	3,986
3	3,206	3,681
4	3,449	3,564
5	3,717	3,541
6	3,994	3,575
7	—	3,654

Instead of using the marginal analysis in Example 14.4, we will use PW analysis, which requires an evaluation of infinite cash-flow streams. (The result is the same under both analyses.) Immediate replacement of the defender by the challenger is equivalent to computing the PW for an infinite cash-flow stream of $-\$3,541$. If we use the capitalized equivalent-worth approach of Chapter 5 ($CE(i) = A/i$), we obtain the following results.

- $n = 0$:

$$PW(12\%)_{n_0=0} = (1/0.12)(\$3,541)$$

$$= \$29,508$$

Suppose we retain the old machine n more years and then replace it with the new one. Now we will compute $PW(i)_{n_0=n}$.

- $n = 1$:

$$PW(12\%)_{n_0=1} = \$3,116(P/F, 12\%, 1) + \$29,508(P/F, 12\%, 1)$$

$$= \$29,129$$

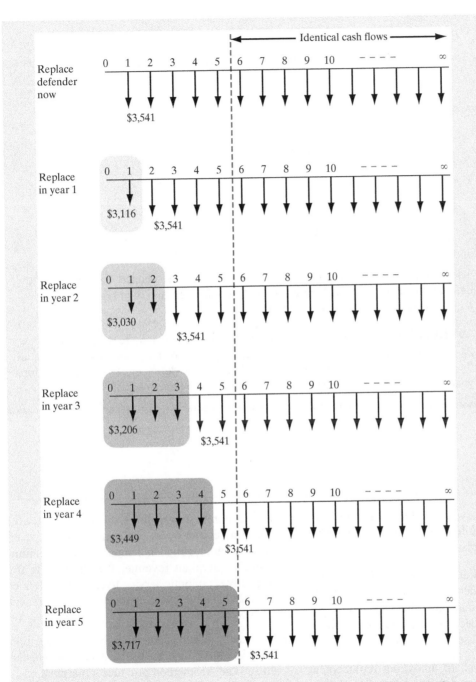

Figure 14.16 Equivalent annual cash-flow streams when the defender is kept for *n* years followed by infinitely repeated purchases of the challenger every four years (Example 14.9).

- $n = 2$:

$$PW(12\%)_{n_0=2} = \$3,030(P/A, 12\%, 2) + \$29,508(P/F, 12\%, 2)$$
$$= \boxed{\$28,644}$$

- $n = 3$:

$$PW(12\%)_{n_0=3} = \$3,206(P/A, 12\%, 3) + \$29,508(P/F, 12\%, 3)$$
$$= \$28,703$$

- $n = 4$:

$$PW(12\%)_{n_0=4} = \$3,449(P/A, 12\%, 4) + \$29,508(P/F, 12\%, 4)$$
$$= \$29,228$$

and so forth.

The present worth of $28,644 represents the net cost associated with retaining the defender for two years, replacing it with the challenger, and then replacing the challenger every five years for an indefinite period.

COMMENTS: In our example, it happens to be that we are keeping the defender over its economic service life, but we should not expect to reach the same conclusion all the time. There are many situations where keeping the defender for a couple of more years beyond its economic service life would be more economical.

SUMMARY

- In replacement analysis, the **defender** is an existing asset; the **challenger** is the best available replacement candidate.

- The **current market value** is the value to use in preparing a defender's economic analysis. **Sunk costs**—past costs that cannot be changed by any future investment decision—should not be considered in a defender's economic analysis.

- The basic approach to analyzing replacement problems is the **opportunity cost approach**. The opportunity cost approach views the net proceeds from the sale of the defender as an opportunity cost of keeping the defender. That is, instead of deducting the salvage value from the purchase cost of the challenger, we consider the net proceeds as an investment required to keep the asset.

- The **economic service life** is the remaining useful life of a defender *or* a challenger that results in the minimum equivalent annual cost or maximum annual equivalent revenue. We should use the respective economic service lives of the defender and the challenger when conducting a replacement analysis.

- Ultimately, in replacement analysis, the question is not *whether* to replace the defender, but *when* to do so. The AE method provides a marginal basis on which to make a year-by-year decision about the best time to replace the defender. As a general decision criterion, the PW method provides a more direct solution to a variety of replacement problems with either an infinite or a finite planning horizon or a technological change in a future challenger.

- The role of **technological change** in improving assets should be evaluated in making long-term replacement plans. If a particular item is undergoing rapid, substantial technological improvements, it may be prudent to delay replacement (to the extent where the loss in production does not exceed any savings from improvements in future challengers) until a desired future model is available.

- Conceptually, all replacement analyses should be done on after-tax basis. However, the replacement decisions are often carried out on the basis of before-tax. It is simpler and the ultimate replacement decisions are not likely changed by the consideration of taxes in many cases. This is because we may be able to reflect the tax effects by adjusting the discount rate.

PROBLEMS

Sunk Costs, Opportunity Costs, and Cash Flows

14.1 Consider the following two options related to one of the old but special machine tools in your machine shop.

- **Option 1:** You continue to use the old machine tool that was bought four years ago for $12,000. It has been fully depreciated but can be sold for $2,000. If kept, it could be used for three more years with proper maintenance and with some extra care. No salvage value is expected at the end of three years. The maintenance costs would run $10,000 per year for the old machine tool.

- **Option 2:** You purchase a brand-new machine tool at a price of $15,000 to replace the present equipment. Because of the nature of the product manufactured, it also has an expected economic life of three years and will have a salvage value of $5,000 at the end of that time. With the new machine tool, the expected operating and maintenance costs (with the scrap savings) amount to $3,000 each year for three years.

(a) For the old machine tools, what would be the amount of sunk cost that should be recognized in replacement analysis?

(b) What is the opportunity cost of retaining the old machine tool now?

(c) What is the net incremental benefit (or loss) in present value associated with replacing the old machine tool at an interest rate of 15%?

14.2 The Columbus Electronics Company is considering replacing a 1,000-pound-capacity forklift truck that was purchased three years ago at a cost of $15,000. The diesel-operated forklift was originally expected to have a useful life of eight years and a zero estimated salvage value at the end of that period. The truck has not been dependable and is frequently out of service while awaiting repairs. The maintenance expenses of the truck have been rising steadily and currently amount to about $3,000 per year. The truck could be sold for $6,000. If retained, the truck will require an immediate $1,500 overhaul to keep it in operating condition. This overhaul will neither extend the originally estimated service life nor increase the value of the truck. The updated annual operating costs, engine overhaul cost, and market values over the next five years are estimated as given in Table P14.2.

A drastic increase in operating costs during the fifth year is expected due to another overhaul, which will again be required to keep the truck in operating condition. The firm's MARR is 15%.

(a) If the truck is to be sold now, what will be its sunk cost?

(b) What is the opportunity cost of not replacing the truck now?

(c) What is the equivalent annual cost of owning and operating the truck for two more years?

(d) What is the equivalent annual cost of owning and operating the truck for five years?

14.3 Komatsu Cutting Technologies is considering replacing one of its CNC machines with one that is newer and more efficient. The firm purchased the CNC machine 10 years ago at a cost of $135,000. The machine had an expected economic life of 12 years at the time of purchase and an expected salvage value of $12,000 at the end of the 12 years. The original salvage estimate is still good, and the machine has a remaining useful life of two years. The firm can sell this old machine now to another

TABLE P14.2

n	O&M	Depreciation	Engine Overhaul	Market Value
−3	—	—	—	—
−2	—	$3,000	—	—
−1	—	4,800	—	—
0	—	2,880	$1,500	$6,000
1	$3,000	1,728	—	4,000
2	3,500	1,728	—	3,000
3	3,800	864	—	1,500
4	4,500	0	—	1,000
5	4,800	0	5,000	0

firm in the industry for $30,000. The new machine can be purchased for $165,000, including installation costs. It has an estimated useful (economic) life of eight years. The new machine is expected to reduce cash operating expenses by $30,000 per year over its eight-year life, at the end of which the machine is estimated to be worth only $5,000. The company has a MARR of 12%.

(a) If you decided to retain the old machine, what is the opportunity (investment) cost of retaining the old asset?

(b) Compute the cash flows associated with retaining the old machine in years one and two.

(c) Compute the cash-flows associated with purchasing the new machine in years one through eight.

(d) If the firm needs the service of these machines for an indefinite period and no technology improvement is expected in future machines, what will be your decision?

14.4 Air Links, a commuter airline company, is considering replacing one of its baggage-handling machines with a newer and more efficient one. The current book value of the old machine is $50,000, and it has a remaining useful life of five years. The salvage value expected from scrapping the old machine at the end of five years is zero, but the company can sell the machine now to another firm in the industry for $10,000. The new baggage-handling machine has a purchase price of $120,000 and an estimated useful life of seven years. It has an estimated salvage value of $30,000 and is expected to realize economic savings on electric power usage, labor, and repair costs and also to reduce the amount of damaged luggage. In total, an annual savings of $50,000 will be realized if the new machine is installed. The firm uses a 15% MARR. Using the opportunity cost approach,

(a) What is the initial cash outlay required for the new machine?

(b) What are the cash flows for the defender in years zero through five?

(c) Should the airline purchase the new machine?

14.5 Duluth Medico purchased a digital image-processing machine three years ago at a cost of $50,000. The machine had an expected life of eight years at the time of purchase and an expected salvage value of $5,000 at the end of the eight years. The old machine has been slow at handling the increased business volume, so management is considering replacing the machine. A new machine can be purchased for $75,000, including installation costs. Over its five-year life, the machine will reduce cash operating expenses by $30,000 per year. Sales are not expected to change. At the end of its useful life, the machine is estimated to be worthless. The old machine can be sold today for $10,000. The firm's interest rate for project justification is known to be 15%. The firm does not expect a better machine (other than the current challenger) to be available for the next five years. Assuming that the economic service life of the new machine, as well as the remaining useful life of the old machine, is five years,

(a) Determine the cash flows associated with each option (keeping the defender versus purchasing the challenger).

(b) Should the company replace the defender now?

14.6 The Northwest Manufacturing Company is currently manufacturing one of its products on a hydraulic stamping-press machine. The unit cost of the product is $12, and 3,000 units were produced and sold for $19 each during the past year. It is expected that both the future demand of the product and the unit price will remain steady at 3,000 units per year and $19 per unit.

- The old machine has a remaining useful life of three years. The old machine could be sold on the open market now for $5,500. Three years from now, the old machine is expected to have a salvage value of $1,200.

- The new machine would cost $36,500, and the unit manufacturing cost on the new machine is projected to be $11. The new machine has an expected economic service life of five years and an expected salvage value of $6,300.

- The appropriate MARR is 12%. The firm does not expect a significant improvement in technology, and it needs the service of either machine for an indefinite period.

(a) Compute the cash flows over the remaining useful life of the old machine if the firm decides to retain it.

(b) Compute the cash flows over the economic service life if the firm decides to purchase the new machine.

(c) Should the new machine be acquired now?

Economic Service Life

14.7 A new asset is available for $200,000. O&M costs are $20,000 each year for the first five years, $22,000 in year six, $25,000 in year seven, and $28,000 in year eight. Salvage values are estimated to be $130,000 after one year and will decrease at the rate of 20% per year thereafter. At a MARR of 12%, determine the economic service life of the asset.

14.8 A firm is considering replacing a machine that has been used to make a certain kind of packaging material. The new, improved machine will cost $31,000 installed and will have an estimated economic life of 10 years with a salvage value of $2,500. Operating costs are expected to be $1,000 per year throughout the service life of the machine. The old machine (still in use) had an original cost of $25,000 four years ago, and at the time it was purchased, its service life (physical life) was estimated to be seven years with a salvage value of $5,000. The old machine has a current market value of $7,700. If the firm retains the old machine, its updated market values and operating costs for the next four years will be as given in Table P14.8.

The firm's MARR is 12%.

TABLE P14.8

Year-End	Market Value	Book Value	Operating Costs
0	$7,700	$7,810	
1	4,300	5,578	$3,200
2	3,300	3,347	3,700
3	1,100	1,116	4,800
4	0	0	5,850

(a) Working with the updated estimates of market values and operating costs over the next four years, determine the remaining useful life of the old machine.

(b) Determine whether it is economical to make the replacement now.

(c) If the firm's decision is to replace the old machine, when should it do so?

14.9 General Foods Company is considering replacing an old food dispenser which was purchased 10 years ago at a cost of $25,000. It has a current market value of $4,000. If the firm spends $1000 to recalibrate the dispenser, it would extend the life of the dispenser for another three years with following estimated market values and operating costs:

End of Year	Market Value	Overhaul	O&M Cost
0	$4,000	$1,000	
1	$3,000		$3,500
2	$2,250		$4,500
3	$1,688		$5,500

A computer-controlled new dispenser can be installed for $12,000. This dispenser would have an estimated economic life of 5 years, with an annual equivalent cost of $5,900. (Note that I calculated the economic service life of the challenger here.) The firm's MARR is 12%. Ignore any income tax effects for this problem.

(a) What is the remaining economic service life of the old machine with overhaul?

(b) What is your decision – replace the defender now?

(c) In (b), if now is not the right time to replace the defender, when to replace it?

(d) In order to reach the conclusion in (c), what assumption(s) do you need in your replacement analysis?

14.10 A special-purpose machine is to be purchased at a cost of $15,000. Table P14.10 shows the expected annual operating and maintenance cost and the salvage values for each year of the machine's service.

(a) If the interest rate is 10%, what is the economic service life for this machine?

(b) Repeat part (a) using $i = 15\%$.

| TABLE P14.10

Year of Service	O&M Costs	Market Value
1	$2,500	$12,800
2	3,200	8,100
3	5,300	5,200
4	6,500	3,500
5	7,800	0

Replacement Decisions with an Infinite Planning Horizon and No Technological Change

14.11 A special-purpose turnkey stamping machine was purchased four years ago for $20,000. It was estimated at that time that this machine would have a life of 10 years, a salvage value of $3,000, and a removal cost of $1,500. These estimates are still good. The

machine has annual operating costs of $2,000. A new machine that is more efficient will reduce the operating costs to $1,000, but it will require an investment of $20,000 plus $1,000 for installation. The life of the new machine is estimated to be 12 years, a salvage of $2,000, and a removal cost of $1,500. An offer of $6,000 has been made for the old machine, and the purchaser is willing to pay for removal of the machine. Find the economic advantage of replacing or of continuing with the present machine. State any assumptions that you make. The firm's interest rate is 12% for any project justification on a before-tax basis.

14.12 A five-year-old defender has a current market value of $4,000 and expected O&M costs of $3,000 this year, increasing by $1,500 per year. Future market values are expected to decline by $1,000 per year. The machine can be used for another three years. The challenger costs $6,000 and has O&M costs of $2,000 per year, increasing by $1,000 per year. The machine will be needed for only three years, and the salvage value at the end of that time is expected to be $2,000. The MARR is 15%.

(a) Determine the annual cash flows for retaining the old machine for three years.

(b) Determine whether now is the time to replace the old machine. First show the annual cash flows for the challenger.

14.13 The Greenleaf Company is considering purchasing a new set of air-electric quill units to replace an obsolete one. The machine currently being used for the operation has a market value of zero. However, it is in good working order, and it will last for at least an additional five years. The new quill units will perform the operation with so much more efficiency that the firm's engineers estimate that labor, material, and other direct costs will be reduced $3,000 a year if the units are installed. The new set of quill units costs $10,000 delivered and installed, and its economic life is estimated to be five years with zero salvage value. The firm's MARR is 13%.

(a) What investment is required to keep the old machine?

(b) Compute the cash flow to use in the analysis for each option.

(c) If the firm uses the internal-rate-of-return criterion, should the firm buy the new machine on that basis?

14.14 The Wu Lighting Company is considering replacing an old, relatively inefficient vertical drill machine that was purchased seven years ago at a cost of $10,000. The machine had an original expected life of 12 years and a zero estimated salvage value at the end of that period. The divisional manager reports that a new machine can be bought and installed for $12,000. Furthermore, over its five-year life, the machine will expand sales from $10,000 to $11,500 a year and will reduce the usage of labor and raw materials sufficiently to cut annual operating costs from $7,000 to $5,000. The new machine has an estimated salvage value of $2,000 at the end of its five-year life. The old machine's current market value is $1,000; the firm's MARR is 15%.

(a) Should the new machine be purchased now?

(b) What price of the new machine would make the two options equal?

14.15 The Advanced Robotics Company is faced with the prospect of replacing its old call-switching system, which has been used in the company's headquarters for 10 years. This particular system was installed at a cost of $100,000, and it was assumed that it would have a 15-year life with no appreciable salvage value. The current annual operating costs for the old system are $20,000, and these costs would be the same for the rest of its life. A sales representative from North Central Bell is trying to sell the company a computerized switching system that would require an investment of $200,000 for installation. The economic life of this computerized system is estimated to be 10 years, with a salvage value of $18,000, and the system will reduce annual operating costs to $5,000. No detailed agreement has been made with the sales representative about the disposal of the old system. Determine the ranges of resale value associated with the old system that would justify installation of the new system at a MARR of 14%.

14.16 A company is currently producing chemical compounds by a process that was installed 10 years ago at a cost of $100,000. It was assumed that the process would have a 20-year life with a zero salvage

value. The current market value of the process, however, is $60,000, and the initial estimate of its economic life is still good. The annual operating costs associated with the process are $18,000. A sales representative from the U.S. Instrument Company is trying to sell a new chemical-compound-making process to the company. This new process will cost $200,000, have a service life of 10 years, a salvage value of $20,000, and reduce annual operating costs to $4,000. Assuming that the company desires a return of 12% on all investments, should it invest in the new process?

14.17 Eight years ago, a lathe was purchased for $45,000. Its operating expenses were $8,700 per year. An equipment vendor offers a new machine for $53,500. An allowance of $8,500 would be made for the old machine when the new one is purchased. The old machine is expected to be scrapped at the end of five years. The new machine's economic service life is five years with a salvage value of $12,000. The new machine's O&M cost is estimated to be $4,200 for the first year, increasing at an annual rate of $500 thereafter. The firm's MARR is 12%. What option would you recommend?

14.18 The New York Taxi Cab Company has just purchased a new fleet of models for the year 2000. Each brand-new cab cost $20,000. From past experience, the company estimates after-tax cash returns for each cab as

$$A_n = \$65,800 - 30,250(1 + 0.15)^{n-1}$$
$$S_n = \$20,000(1 - 0.15)^n$$

where A_n stands for net after-tax cash flows from the cab's operation during period n, and S_n stands for the after-tax salvage value of the cab at the end of period n. The management views the replacement process as a constant and infinite chain.

(a) If the firm's MARR is 10% and it expects no major technological and functional change in future models, what is the optimal period (constant replacement cycle) to replace its cabs? (Ignore inflation.)

(b) What is the internal rate of return for a cab if it is retired at the end of its economic service life? What is the internal rate of return for a sequence

TABLE P14.32

Year End	Book Value	Salvage Value	Operating Costs
1	$8,000	$5,300	$1,500
2	4,800	3,900	2,100
3	2,880	2,800	2,700
4	1,728	1,800	3,400
5	576	1,400	4,200
6	0	600	4,900

14.33 Given the following data:

$$I = \$30,000$$

$$S_n = 22,000 - 2,000n$$

$$B_n = 30,000 - 3,000n$$

$$O\&M_n = 3,000 \times (1 + 0.15)^{n-1}$$

and

$$t_m = 0.40$$

where

I = Asset purchase price
S_n = Market value at the end of year n
B_n = Book value at the end of year n
$O\&M_n$ = O&M cost during year n
t_m = Marginal tax rate

(a) Determine the economic service life of the asset if $i = 12\%$.

(b) Determine the economic service life of the asset if $i = 20\%$.

(c) Assume that $i = 0$ and determine the economic service life of the asset mathematically (i.e., use the calculus technique for finding the minimum point, as described in Chapter 8).

14.34 Redo Problem 14.10 with the following additional information.

- For tax purposes, the entire cost of $15,000 can be depreciated according to a five-year MACRS property class.
- The firm's marginal tax rate is 40%.
- The firm's after-tax MARR is 12%.

14.35 The Quintana Electronic Company is considering purchasing new robot-welding equipment to perform operations currently being performed by less efficient equipment. The new machine's purchase price is $150,000 delivered and installed. A Quintana industrial engineer estimates that the new equipment will produce savings of $30,000 in labor and other direct costs annually when compared with the current equipment. He estimates the proposed equipment's economic life at 10 years with a zero salvage value. The current equipment is in good working order and will last, physically, for at least 10 more years. Quintana Company expects to pay income taxes of 40%, and any gains also will be taxed at 40%. Quintana uses a 10% discount rate for analysis performed on an after-tax basis. Depreciation of the new equipment for tax purposes is computed on the basis of a seven-year MACRS property class.

(a) Assuming that the current equipment has zero book value and zero salvage value, should the company buy the proposed equipment?

(b) Assuming that the current equipment is being depreciated at a straight-line rate of 10%, has a book value of $72,000 (cost, $120,000; accumulated depreciation, $48,000), and zero net salvage value today, should the company buy the proposed equipment?

(c) Assuming that the current equipment has a book value of $72,000, a salvage value today of $45,000, and if the current equipment is retained for 10 more years, its salvage value will be zero, should the company buy the proposed equipment?

(d) Assume that the new equipment will save only $15,000 a year, but that its economic life is expected to be 12 years. If other conditions are as described in part (a), should the company buy the proposed equipment?

14.36 The Quintana Company decided to purchase the equipment described in Problem 14.35 (hereafter called "model A" equipment). Two years later, even better equipment (called "model B") came onto the market, making model A obsolete, with no resale value. The model B equipment costs $300,000 delivered and installed, but it is expected to result in annual savings of $75,000 over the cost of operating the Model A equipment. The economic life of model B is estimated to be 10 years with a zero salvage value. (Model B also is classified as a seven-year MACRS property.)

(a) What action should the company take?

(b) If the company decides to purchase the model B equipment, a mistake must have been made, because good equipment (bought only two years previously) is being scrapped. How did this mistake come about?

14.37 Redo Problem 14.11 with the following additional information.

- The current book value of the old machine is $6,248, and the old asset has been depreciated as a seven-year MACRS property.

- The new asset is also classified as a seven-year MACRS property.

- The company's marginal tax rate is 30%, and the firm uses 10% after-tax MARR.

14.38 Redo Problem 14.12 with the following additional information.

- The old machine has been fully depreciated.

- The new machine will be depreciated under a three-year MACRS class.

- The marginal tax rate is 40%, and the firm's after-tax MARR is 12%.

14.39 Redo Problem 14.13 with the following additional information.

- The current book value of the old machine is $4,000, and the annual depreciation charge is $800 if the firm decides to keep the old machine for the additional five years.

- The new asset is classified as a seven-year MACRS property.

- The company's marginal tax rate is 40%, and the firm uses a 10% after-tax MARR.

14.40 Redo Problem 14.14 with the following additional information.

- The old machine has been fully depreciated.

- The new machine will be depreciated under a seven-year MACRS class.

- The marginal tax rate is 40%, and the firm's after-tax MARR is 12%.

14.41 Redo Problem 14.15 the following additional information.

- The old switching system has been fully depreciated.

- The new system falls into a five-year MACRS property class.

- The company's marginal tax rate is 40%, and the firm uses a 10% after-tax MARR.

14.42 Five years ago, a conveyor system was installed in a manufacturing plant at a cost of $35,000. It was estimated that the system, which is still in operating condition, would have a useful life of eight years with a salvage value of $3,000. It was also estimated that if the firm continues to operate the system, its market values and operating costs for the next three years would be as follows.

Year End	Market Value	Book Value	Operating Cost
0	$11,500	$15,000	—
1	5,200	11,000	$4,500
2	3,500	7,000	5,300
3	1,200	3,000	6,100

A new system can be installed for $43,500; it would have an estimated economic life of 10 years with a salvage value of $3,500. Operating costs are expected to be $1,500 per year throughout the service life of the system. The firm's MARR is 18%. The system belongs to the seven-year MACRS property class. The firm's marginal tax rate is 35%.

(a) Decide whether to replace the existing system now.

(b) If the decision is to replace the existing system, when should replacement occur?

14.43 Redo Problem 14.16 with the following additional information.

- The old machine has been depreciated on a straight-line basis.
- The new machine will be depreciated under a seven-year MACRS class.
- The marginal tax rate is 40%, and the firm's after-tax MARR is 10%.

14.44 Redo Problem 14.17 with the following additional information.

- The old machine has been fully depreciated.
- The new machine will be depreciated under a seven-year MACRS class.
- The marginal tax rate is 35%, and the firm's after-tax MARR is 10%.

14.45 Redo Problem 14.19 with the following additional information.

- The old machine has been depreciated according to a seven-year MACRS.
- The new machine also will be depreciated under a seven-year MACRS class.

- The marginal tax rate is 40%, and the firm's after-tax MARR is 10%.

14.46 Redo Problem 14.20 with the following additional information.

- **Option 1:** The old sprayer has been fully depreciated. The new sprayer is classified as having a seven-year MACRS recovery period.
- **Option 2:** The larger capacity sprayer is classified as a seven-year MACRS property.

The company's marginal tax rate is 40%, and the firm uses a 12% after-tax MARR.

14.47 A six-year-old computer numerical control (CNC) machine that originally cost $8,000 has been fully depreciated, and its current market value is $1,500. If the machine is kept in service for the next five years, its O&M costs and salvage value are estimated as given in Table P14.47

It is suggested that the machine be replaced by a new CNC machine of improved design at a cost of $6,000. It is believed that this purchase will completely eliminate breakdowns and the resulting cost of delays and that operation and repair costs will be reduced $200 a year from what they would be with the old machine. Assume a five-year life for the challenger and a $1,000 terminal salvage value. The new machine falls into a five-year MACRS property class. The firm's MARR is 12%, and its marginal tax rate is 30%. Should the old machine be replaced now?

TABLE P14.47

End of Year	O&M Costs		Salvage Value
	Operation and Repairs	Delays Due to Breakdowns	
1	$1,300	$600	$1,200
2	1,500	800	1,000
3	1,700	1,000	500
4	1,900	1,200	0
5	2,000	1,400	0

14.48 Redo Problem 14.24 with the following additional information.

- The old asset has been depreciated according to a five-year MACRS.
- The new asset also will be depreciated under a five-year MACRS class.
- The marginal tax rate is 30%, and the firm's after-tax MARR is 11%.

14.49 The University Resume Service has just invested $8,000 in a new desktop publishing system. From past experience, the owner of the company estimates its after-tax cash returns as

$$A_n = \$8,000 - \$4,000(1 + 0.15)^{n-1}$$
$$S_n = \$6,000(1 - 0.3)^n$$

where A_n stands for the net after-tax cash flows from operation of the system during period n and S_n stands for the after-tax salvage value at the end of period n.

(a) If the company's MARR is 12%, compute the economic service life of the system.

(b) Explain how the economic service life varies with the interest rate.

TABLE P14.49

Year-End	Market Value	Book Value	Operating Costs
0	$7,700	$,7810	
1	4,300	5,578	$3,200
2	3,300	3,347	3,700
3	1,100	1,116	4,800
4	0	0	5,850

Short Case Studies

ST14.1 Five years ago, a conveyor system was installed in a manufacturing plant at a cost of $35,000. It was estimated that the system, which is still in operating condition, would have a useful life of eight years, with a salvage value of $3,000. If the firm continues to operate the system, the system's estimated market values and operating costs for the next three years are as follows:

End of Year	Market Value	Overhaul	O&M Cost
0	$11,500	$15,000	
1	$5,200	$11,000	$4,500
2	$3,500	$7,000	$5,300
3	$1,200	$3,000	$8,000

A new system can be installed for $43,500. This system would have an estimated economic life of 10 years, with a salvage value of $3,500. The operating costs for the new system are expected to be $1,500 per year throughout its service life. The firm's MARR is 18%. Ignore any income tax effects for this problem.

(a) What is the remaining economic service life of the old machine?

(b) What is your decision—replace the defender now?

(c) In (b), if now is not the right time to replace the defender, when should it be replaced?

(d) In order to reach the conclusion in (c), what assumption(s) do you need in your replacement analysis?

ST14.2 The National Woodwork Company, a manufacturer of window frames, is considering replacing a conventional manufacturing system with a flexible manufacturing system (FMS). The company cannot produce rapidly enough to meet demands. One manufacturing problem that has been identified is that the present system is expected to be useful for another five years but will require an estimated $105,000 per year in maintenance, which will increase $10,000 each year as parts become scarcer. The current market value of the existing system is $140,000, and the machine has been fully depreciated.

The proposed system will reduce or entirely eliminate set-up times. Each window can be made as it is ordered by the customer, who phones the order into the head office where details then are fed into the company's main computer. These manufacturing details then are dispatched to computers on the manufacturing floor, which are, in turn, connected to a computer that controls the proposed FMS. This system eliminates the warehouse space and material-handling time that are needed when the conventional system is used.

Before the FMS is installed, the old equipment will be removed from the job shop floor at an estimated cost of $100,000. This cost includes needed electrical work on the new system. The proposed FMS will cost $1,200,000. The economic life of the machine is expected to be 10 years, and the salvage value is expected to be $120,000. The change in window styles has been minimal in the past few decades and is expected to continue to remain stable in the future. The proposed equipment falls into the seven-year MACRS category. The total annual savings will be $664,243: $12,000 attributed to a reduction in the number of defective windows, $511,043 from the elimination of 13 workers, $100,200 from the increase in productivity, and $41,000 from the near elimination of warehouse space and material handling. The O&M costs will be only $45,000, increasing by $2,000 per year. The National Woodwork's MARR is about 15%, and the expected marginal tax rate over the project years is 40%.

(a) What assumptions are required to compare the conventional system with the FMS?

(b) With the assumptions defined in part (a), should the FMS be installed now?

ST14.3 In 2 × 4 and 2 × 6 lumber production, significant amounts of wood are present in sideboards produced after the initial cutting of logs. Instead of processing the sideboards into wood chips for the paper mill, the Union Camp Company uses an "edger" to reclaim additional lumber, resulting in savings for the company. An edger is capable of reclaiming lumber by any of the following three methods: (1) removing rough edges; (2) splitting large sideboards; and (3) salvaging 2 × 4 lumber from low-quality 4 × 4 boards. Union Camp Company's engineers have discovered that a significant reduction in production costs could be achieved simply by replacing the original edger with a newer, laser-controlled model.

The old edger was placed in service 12 years ago and is fully depreciated. Any machine scrap value would offset the removal cost of the equipment. No market exists for this obsolete equipment. The old edger needs two operators. During the cutting operation, the operator makes edger settings, using his or her judgment. The operator has no means of determining exactly what dimension of lumber could be recovered from a given sideboard and must guess at the proper setting to recover the highest grade of lumber. Furthermore, the old edger is not capable of salvaging good-quality 2 × 4s from poor-quality 4 × 4s. The defender can continue in service for another five years with proper maintenance. The following are the financial data for the old edger:

Current market value:	$0
Current book value:	$0
Annual maintenance cost:	$2,500 in year one, increasing at a rate of 15% each year over the previous year's cost
Annual operating costs (labor and power):	$65,000

The new edger has numerous advantages over its defender, including laser beams that indicate where cuts should be made to obtain the maximum yield. The new edger requires a single operator. Labor savings will be reflected in lower operating and maintenance costs of $35,000 a year. Table ST14.3 gives the estimated costs and the depreciation methods associated with the new edger.

Twenty-five percent of the total mill volume passes through the edger. A 12% improvement in yield is expected to be realized with the new edger, resulting

TABLE ST14.3

Estimated Cost	
Equipment	$ 35,700
Equipment installation	21,500
Building	47,200
Conveyor modification	14,500
Electrical (wiring)	16,500
Subtotal	**$135,400**
Engineering	7,000
Construction management	20,000
Contingency	16,200
Total	**$178,600**
Useful life of new edger	10 years
Salvage value:	
Building (tear down)	$0
Equipment	10% of the original cost
Annual O&M costs	$35,000
Depreciation Methods	
Building	39-year MACRS
Equipment and installation	7-year MACRS

in an improvement of $(0.25)(0.12) = 3\%$ in the total mill volume. The annual savings due to the improvement in productivity is thus expected to be $57,895.

(a) Should the defender be replaced now if the mill's MARR and marginal tax rate are 16% and 40%, respectively?

(b) If the defender will eventually be replaced by the current challenger, when is the optimal time to perform the replacement?

ST14.4 The Tiger Construction Company purchased its current bulldozer (a Caterpillar D8H) and placed it in service six years ago. Since the purchase of the Caterpillar, new technology has produced changes in machines resulting in an increase in productivity of approximately 20%. The Caterpillar worked in a system with a fixed (required) production level to maintain overall system productivity. As the Caterpillar aged and logged more downtime, more hours had to be scheduled to maintain the required production. Tiger is considering purchasing a new bulldozer (a Komatsu K80A) to replace the Caterpillar. Table ST14.4 contains data which has been collected by Tiger's civil engineer.

(a) A civil engineer notices that both machines have different working hours and hourly production capacities. To compare the different units of capacity, the engineer needs to devise a combined index that reflects each machine's productivity as well as actual operating hours. Develop such a combined productivity index for each period.

(b) Adjust the operating and maintenance costs by the index you have come up with.

(c) Compare the two alternatives. Should the defender be replaced now?

(d) If the following price indices were estimated for the next five project years, should the defender be replaced now?

TABLE ST14.4

Defender (Caterpillar D8H)		Challenger (Komatsu K80A)
Useful life	Not known	Not known
Purchase price	—	$400,000
Salvage value if kept for		
0 year	$75,000	$400,000
1 year	60,000	300,000
2 years	50,000	240,000
3 years	30,000	190,000
4 years	30,000	150,000
5 years	10,000	115,000
Fuel use (gallon/hour)	11.30	16
Maintenance costs:		
1	$46,800	$35,000
2	46,800	38,400
3	46,800	43,700
4	46,800	48,300
5	46,800	58,000
Operating hours (hours/year):		
1	1,800	2,500
2	1,800	2,400
3	1,700	2,300
4	1,700	2,100
5	1,600	2,000
Productivity index	1.00	1.20
Other relevant information:		
Fuel cost ($/gallon)		$3.20
Operator's wages ($/hour)		$23.40
Market interest rate (MARR)		15%
Marginal tax rate		40%
Depreciation methods	Fully	MACRS
	Depreciated	(5-year)

	Forecasted Price Index			
Year	**General Inflation**	**Fuel**	**Wage**	**Maintenance**
0	100	100	100	100
1	108	110	115	108
2	116	120	125	116
3	126	130	130	124
4	136	140	135	126
5	147	150	140	128

Source: Data from AHS UAI annual report.

15

Capital-Budgeting Decisions

Laredo Petroleum Holdings, Inc.[1] Laredo, an independent energy company with headquarters in Tulsa, Oklahoma, announced that its Board of Directors has approved a $1 billion capital budget for 2014, excluding acquisitions. Approximately 84% of the capital budget is focused on drilling and completion activities, approximately 13% for facilities and 3% for land, tests, and other capital. This development-focused capital program is expected to drive an approximate 30% annual growth in production volumes from the Company's Permian Basin properties.

The 2014 capital budget of $1 billion, which is expected to be funded from internally-generated cash flow and borrowings under the Company's senior secured credit facility, is allocated as follows.

Source: Mark Payne/Shutterstock

[1]Laredo Petroleum Announces 2014 Capital Budget of $1 Billion, Laredo Petroleum, Inc., December 9, 2013.

	($ in millions)
Horizontal Drilling	$ 545
Vertical Drilling	240
Nonoperated	55
Drilling and Completion	840
Facilities	130
Land and Seismic	20
Other	10
Total	$1,000

To maximize efficiency, rigs drilling development wells will operate in concentrated geographic areas where the Company has started building out production corridors to maximize the cost savings from common facilities. The Company's facilities capital expenditures in 2014 are expected to focus on building out the core infrastructure that will enable the Company's transition to full-scale hydrocarbon manufacturing. Laredo is building multiple production corridors that will support development drilling for hundreds of wells are expected to benefit the future cost structure of the Company.

We may be interested in knowing how Laredo identified these projects among other competing projects. Laredo also stated that they will use the internal funds to finance these large-scale projects. Because of the size of the financing involved, the firm financing method will affect the firm's capital structure, and this in turn affects the firm's financial risk. The firm's cost of capital will tend to increase during the project period, as well.

In this chapter, we present the basic framework of **capital budgeting**, which involves investment decisions related to fixed assets. Here, the term **capital budget** includes planned expenditures on fixed assets; **capital budgeting** encompasses the entire process of analyzing projects and deciding whether they should be included in the capital budget. In previous chapters, we focused on how to evaluate and compare investment projects—the analysis aspect of capital budgeting. In this chapter, we focus on the budgeting aspect. Proper capital-budgeting decisions require a choice of the method of project financing, a schedule of investment opportunities, and an estimate of the minimum attractive rate of return (MARR).

CHAPTER LEARNING OBJECTIVES

After completing this chapter, you should understand the following concepts:

- How a corporation raises its capital to finance a project.
- How to determine the cost of debt.
- How to determine the cost of equity.
- How to determine the marginal cost of capital.
- How to determine the MARR in project evaluation.
- How to create an optimal project portfolio under capital rationing.

15.1 Methods of Financing

In previous chapters, we focused on problems relating to investment decisions. In reality, investment decisions are not always independent of the source of finance. For convenience however, in economic analysis, investment decisions are usually separated from financing decisions: First the investment project is selected, and then the source of financing is considered. After the source is chosen, appropriate modifications to the investment decision are made.

We have also assumed that the assets employed in an investment project are obtained with the firm's own capital (retained earnings) or from short-term borrowings. In practice, this arrangement is not always attractive or even possible. If the investment calls for a significant infusion of capital, the firm may raise the needed capital by issuing stock. Alternatively, the firm may borrow the funds by issuing bonds to finance such purchases. In this section, we will first discuss how a typical firm raises new capital from external sources. Then we will discuss how external financing affects after-tax cash flows and how the decision to borrow affects the investment decision.

The two broad choices a firm has for financing an investment project are **equity financing** and **debt financing**.[2] We will look briefly at these two options for obtaining external investment funds and also examine their effects on after-tax cash flows.

15.1.1 Equity Financing

Equity financing can take one of two forms: (1) the use of retained earnings otherwise paid to stockholders or (2) the issuance of stock. Both forms of equity financing use funds invested by the current or new owners of the company.

Until now, most of our economic analyses presumed that companies had cash on hand to make capital investments; implicitly, we were dealing with cases of financing by retained earnings. If a company had not reinvested these earnings, it might have paid them to the company's owners—the stockholders—in the form of a dividend, or it might have kept these earnings on hand for future needs.

If a company does not have sufficient cash on hand to make an investment and does not wish to borrow in order to fund the investment, financing can be arranged by selling

---

[2]A hybrid financing method, known as *lease financing*, was discussed in Section 10.4.3.

common stock to raise the required funds. To do this, the company has to decide how much money to raise, the type of securities to issue (common stock or preferred stock), and the basis for pricing the issue.

Once the company has decided to issue common stock, it must estimate **flotation costs**—the expenses it will incur in connection with the issue, such as investment bankers' fees, lawyers' fees, accountants' costs, and the cost of printing and engraving. Usually, an investment banker will buy the issue from the company at a discount—below the price at which the stock is to be offered to the public. (The discount usually represents the *flotation costs*.) If the company is already publicly owned, the offering price commonly will be based on the existing market price of the stock. If the company is going public for the first time, no established price will exist, so investment bankers have to estimate the expected market price at which the stock will sell after the stock issue. Example 15.1 illustrates how the flotation cost affects the cost of issuing common stock.

EXAMPLE 15.1 Equity Financing by Issuing Common Stock

Scientific Sports, Inc. (SSI), a golf club manufacturer, has developed a new metal club (driver). The club is made out of titanium, which delivers optimal aerodynamics for faster swing speeds and promotes a faster ball speed and low spin for extraordinary distance. The company expects to acquire considerable market penetration with this new product. To produce it, the company needs a new manufacturing facility, which will cost $10 million. The company decided to raise this $10 million by selling common stock. The firm's current stock price is $30 per share. Investment bankers have informed management that the new public issue must be priced at $28 per share because of decreasing demand, which will occur as more shares become available on the market. The flotation costs will be 6% of the issue price, so SSI will net $26.32 per share. How many shares must SSI sell to net $10 million after flotation expenses?

SOLUTION

Given: $10 million to raise, price of share = $28, and flotation cost = 6%.
Find: Number of shares to be sold.

Let X be the number of shares to be sold. Then total flotation cost will be

$$(0.06)(\$28)(X) = 1.68X$$

To net $10 million, we must have

$$\text{Sales proceeds} - \text{flotation cost} = \text{Net proceeds}$$
$$28X - 1.68X = \$10,000,000$$
$$26.32X = \$10,000,000$$
$$X = 379,940 \text{ shares}$$

Now we can figure out the flotation cost for issuing the common stock. The cost is

$$1.68(379,940) = \$638,300$$

15.1.2 Debt Financing

The second major type of financing a company can select is **debt financing**, which includes both short-term borrowing from financial institutions and the sale of long-term bonds, wherein money is borrowed from investors for a fixed period. As we have seen in Chapter 10 with debt financing, the interest paid on the loans or bonds is treated as an expense for income-tax purposes. Since interest is a tax-deductible expense, companies in high tax brackets may incur lower after-tax financing costs with a debt. In addition to influencing the borrowing interest rate and tax bracket, a loan-repayment method can affect financing costs.

When the debt-financing option is used, we need to separate the interest payments from the repayment of the loan for our analysis. The interest-payment schedule depends on the repayment schedule established at the time the money is borrowed. One of the primary methods of debt-financing is through issuing bonds. Bond financing does not involve the partial payment of principal: Only interest is paid each year (or semiannually). The principal is paid in a lump sum when the bond matures. (See Section 4.6.3 for bond terminologies and valuation.) Bond financing is similar to equity financing in that flotation costs are involved when bonds are issued.

Example 15.2 illustrates how these different methods can affect the cost of issuing bonds or term loans.

EXAMPLE 15.2 Debt Financing by Issuing Bonds

Consider again Example 15.1. Suppose SSI has instead decided to raise the $10 million by debt financing. SSI could issue a mortgage bond. The flotation cost is 1.8% of the $10 million issue. The company's investment bankers have indicated that a five-year bond issue with a face value of $1,000 can be sold at $985 per share. The bond would require annual interest payments of 12%.

SOLUTION

Given: $10 million to raise, par value $= \$1,000$, market price $= \$985$, coupon rate $= 12\%$ interest payable annually, and flotation cost $= 1.8\%$.
Find: Number of bonds to be sold.

(a) To net $10 million, SSI would have to sell

$$\frac{\$10,000,000}{1 - 0.018} = \$10,183,300$$

worth of bonds and pay $183,300 in flotation costs. Since the $1,000 bond will be sold at a 1.5% discount, the total number of bonds to be sold would be

$$\frac{\$10,183,300}{\$985} = 10,338.38 \text{ (or 10,339)}$$

(b) For the bond financing, the annual interest is equal to

$$\$10,338,380(0.12) = \$1,240,606$$

Only the interest is paid each period; thus, the principal amount owed remains unchanged. The principal and interest components of each annual payment are summarized in Table 15.1.

TABLE 15.1 Debt Financing Through Issuing Bonds (Example 15.2)

	0	1	2	3	4	5
Bond financing: No principal repayments until end of life						
Beginning balance		$10,338,380	$10,338,380	$10,338,380	$10,338,380	$10,338,380
Interest owed		1,240,606	1,240,606	1,240,606	1,240,606	1,240,606
Repayment						
Interest payment		(1,240,606)	(1,240,606)	(1,240,606)	(1,240,606)	(1,240,606)
Principal payment						(10,338,380)
Ending balance	$10,338,380	$10,338,380	$10,338,380	$10,338,380	$10,338,380	0

15.1.3 Capital Structure

The ratio of total debt to total capital, generally called the **debt ratio** or **capital structure**, represents the percentage of the total capital provided by borrowed funds. For example, a debt ratio of 0.4 indicates that 40% of the capital is borrowed and the remaining funds are provided from the company's equity (retained earnings or stock offerings). This type of financing is called **mixed financing**.

Borrowing affects a firm's capital structure, and firms must determine the effects of a change in the debt ratio on their market value before making an ultimate financing decision. Even if debt financing is attractive, you should understand that companies do not simply borrow funds to finance projects. A firm usually establishes a **target capital structure** or **target debt ratio** after considering the effects of various financing methods. This target may change over time as business conditions vary, but a firm's management always strives to achieve the target whenever individual financing decisions are considered. On the one hand, if the actual debt ratio is below the target level, any new capital will easily be raised by issuing debt. On the other hand, if the debt ratio is currently above the target, expansion capital will be raised by issuing stock. Figure 15.1 illustrates the basic concepts of capital structure and degree of financial risk associated with source of capital.

How does a typical firm set the target capital structure? This is a rather difficult question to answer, but we can list several factors that affect the capital-structure policy.

- First, capital-structure policy involves a trade-off between risk and return. As you take on more debt for business expansion, the inherent business risk[3] also

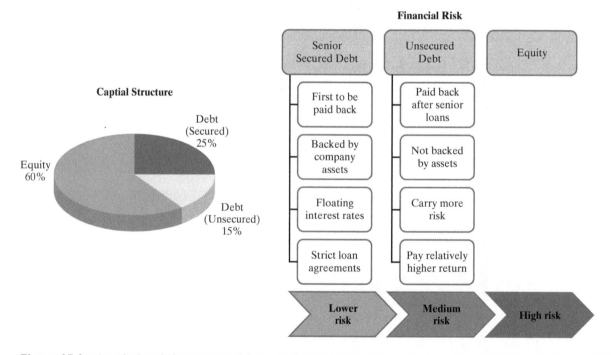

Figure 15.1   A typical capital structure and financial risk associated with each source of capital.

[3]Unlike equity financing, in which dividends are optional, debt interest and principal (face value) must be repaid on time. Also, uncertainty is involved in making projections of future operating income as well as expenses. In bad times, debt can be devastating, but in good times, the tax deductibility of interest payments increases profits to owners.

increases, but investors view business expansion as a healthy indicator for a corporation with higher expected earnings. When investors perceive higher business risk, the firm's stock price tends to be depressed. By contrast, when investors perceive higher expected earnings, the firm's stock price tends to increase. The optimal capital structure is thus the one that strikes a balance between business risks and expected future earnings. The greater the firm's business risk, the lower is its optimal debt ratio.

- Second, a major reason for using debt is that interest is a deductible expense for business operations, which lowers the effective cost of borrowing. Dividends paid to common stockholders, however, are not deductible. If a company uses debt, it must pay interest on this debt, whereas if it uses equity, it pays dividends to its equity investors (shareholders). A company needs $1 in before-tax income to pay $1 of interest, but if the company is in the 35% tax bracket, it needs $1/(1 - 0.35) = $1.54 of before-tax income to pay a $1 dividend.

- Third, financial flexibility—the ability to raise capital on reasonable terms from the financial market—is an important consideration. Firms need a steady supply of capital for stable operations. When money is tight in the economy, investors prefer to advance funds to companies with a healthy capital structure (lower debt ratio).

These three elements (business risk, taxes, and financial flexibility) are major factors that determine the firm's optimal capital structure. Example 15.3 illustrates how a typical firm finances a large-scale engineering project by maintaining the predetermined capital structure.

EXAMPLE 15.3 Project Financing Based on an Optimal Capital Structure

Consider again SSI's $10 million venture project in Example 15.1. Suppose that SSI's optimal capital structure calls for a debt ratio of 0.5. After reviewing SSI's capital structure, the investment banker convinced management that it would be better off (in view of current market conditions) to limit the stock issue to $5 million and to raise the other $5 million as debt by issuing bonds. Because the amount of capital to be raised in each category is reduced by half, the flotation cost would also change. The flotation cost for common stock would be 8.1%, whereas the flotation cost for bonds would be 3.2%. As in Example 15.2, the five-year, 12% bond will have a par value of $1,000 and will be sold for $985.

Assuming that the $10 million capital would be raised from the financial market, the engineering department has detailed the following financial information.

- The new venture will have a five-year project life.

- The $10 million capital will be used to purchase land for $1 million, a building for $3 million, and equipment for $6 million. The plant site and building are already available, and production can begin during the first year. The building falls into a 39-year MACRS property class and the equipment into a seven-year MACRS class. At the end of year five, the salvage value of each asset is as follows: the land $1.5 million, the building $2 million, and the equipment $2.5 million.

- For common stockholders, an annual cash dividend in the amount of $2 per share is planned over the project life. This steady cash-dividend payment is deemed necessary to maintain the market value of the stock.
- The unit production cost is $50.31 (material, $22.70; labor and overhead (excluding depreciation), $10.57; and tooling, $17.04).
- The unit price is $250, and SSI expects an annual demand of 20,000 units.
- The operating and maintenance cost, including advertising expenses, would be $600,000 per year.
- An investment of $500,000 in working capital is required at the beginning of the project; the amount will be fully recovered when the project terminates.
- The firm's marginal tax rate is 40%, and this rate will remain constant throughout the project period.

(a) Determine the after-tax cash flows for this investment with external financing.

(b) Is this project justified at an interest rate of 20%?

DISCUSSION: As the amount of financing and flotation costs change, we need to recalculate the number of shares (or bonds) to be sold in each category. For a $5 million common-stock issue, the flotation cost increases to 8.1%.[4] The number of shares to be sold to net $5 million is 5,000,000/(0.919)(28) = 194,311 shares (or $5,440,708). For a $5 million bond issue, the flotation cost is 3.2%. Therefore, to net $5 million, SSI has to sell 5,000,000/(0.968)(985) = 5,243.95 units of $1,000 par value. This implies that SSI is effectively borrowing $5,243,948, upon which figure the annual bond interest will be calculated. The annual bond interest payment is $5,243,948(0.12) = $629,274.

SOLUTION

Given: Financial data related to the SSI investment project.

Find: Net cash flow with financing activities.

(a) **After-tax cash flows.** Table 15.2 summarizes the after-tax cash flows for the new venture. The following calculations and assumptions were used in developing the table.

- Revenue: $250 × 20,000 = $5,000,000 per year.
- Costs of goods: $50.31 × 20,000 = $1,006,200 per year.
- Bond interest: $5,243,948 × 0.12 = $629,274 per year.
- Depreciation: Assuming that the building is placed in service in January, the first year's depreciation percentage is 2.4573%. Therefore, the allowed depreciation amount is $3,000,000 × 0.024573 = $73,718. The percentages for the remaining years would be 2.5641% per year (or $76,923) and year five is back to $73,718 due to the mid-month convention. Equipment is depreciated according to a seven-year MACRS.

---

[4]Flotation costs are higher for small issues than for large ones due to the existence of fixed costs. Certain costs must be incurred regardless of the size of the issue, so the percentage of flotation costs increases as the size of the issue gets smaller.

TABLE 15.2 Effects of Project Financing and Maintaining Target Capital Structure on Project Cash Flows (Example 15.3)

	A	B	C	D	E	F	G
1							
2	**Income Statement**						
3	End of Year	0	1	2	3	4	5
4							
5	Revenue		$5,000,000	$5,000,000	$5,000,000	$5,000,000	$5,000,000
6	Expenses:						
7	Cost of goods		$1,006,200	$1,006,200	$1,006,200	$1,006,200	$1,006,200
8	O&M		$600,000	$600,000	$600,000	$600,000	$600,000
9	Bond interest		$629,274	$629,274	$629,274	$629,274	$629,274
10	Depreciation:						
11	Building		73,718	76,923	76,923	76,923	73,718
12	Equipment		$857,400	$1,469,400	$1,049,400	$749,400	$267,900
13							
14	Taxable income		$1,833,408	$1,218,203	$1,638,203	$1,938,203	$2,422,908
15	Income taxes (40%)		$733,363	$487,281	$655,281	$775,281	$969,163
16							
17	Net income		$1,100,045	$730,922	$982,922	$1,162,922	$1,453,745
18							
19	**Cash Flow Statement**						
20	Operating Activities:						
21	Net income		$1,100,045	$730,922	$982,922	$1,162,922	$1,453,745
22	Depreciation		$931,118	$1,546,323	$1,126,323	$826,323	$341,618
23	Investment Activities:						
24	Land	$(1,000,000)					$1,500,000
25	Building	$(3,000,000)					$2,000,000
26	Equipment	$(6,000,000)					$2,500,000
27	Working capital	$(500,000)					$500,000
28	Gains tax						$(308,682)
29	Financing Activities:						
30	Common stock	$5,000,000					$(5,440,708)
31	Bond	$5,000,000					$(5,243,948)
32	Cash dividend		$(388,622)	$(388,622)	$(388,622)	$(388,622)	$(388,622)
33							
34	Net Cash Flow	$(500,000)	$1,642,541	$1,888,623	$1,720,623	$1,600,623	$(3,086,597)
35							
36	PW(20%)=	$2,707,529					
37			=NPV(20%,C34:G34)+B34					
38							

- Gains tax:

Property	Salvage Value	Book Value	Gains (Losses)	Gains Tax
Land	$1,500,000	$1,000,000	$500,000	$200,000
Building	2,000,000	2,621,795	(621,795)	(248,718)
Equipment	2,500,000	1,606,500	893,500	357,400
				$308,682

- Cash dividend: 194,311 shares $\times$ \$2 = \$388,622.
- Common stock: When the project terminates and the bonds are retired, the debt ratio is no longer 0.5. If SSI wants to maintain the constant capital structure (0.5), SSI would have to repurchase the common stock in the amount of \$5,440,708 at the prevailing market price. In developing Table 15.2, we assumed that this repurchase of common stock had taken place at the end of project year. In practice, a firm may or may not repurchase the common stock. As an alternative means of maintaining the desired capital structure, the firm may use this extra debt capacity, released to borrow for other projects.
- Bond: When the bonds mature at the end of year five, the total face value in the amount of \$5,243,948 must be paid to the bondholders.

(b) Measure of project worth. The NPW for this project is then

$$PW(20\%) = -\$500,000 + \$1,642,541(P/F, 20\%, 1) + \dots$$

$$= -\$3,086,597(P/F, 20\%, 5)$$

$$= \$2,707,529$$

The investment is not simple, and it is also a mixed investment. The RIC at MARR of 20% is 327%. Even though the project requires a significant amount of cash expenditure at the end of its life, it still appears to be a very profitable one.

In Example 15.3, we neither discussed the cost of capital required to finance this project nor explained the relationship between the cost of capital and the MARR. In the remaining sections, these issues will be discussed. As we will see in Section 15.3, we can completely ignore the detailed cash flows related to project financing if we adjust our discount rate according to the capital structure, namely, by using the weighted cost of capital.

15.2 Cost of Capital

In most of the capital-budgeting examples in earlier chapters, we assumed that the firms under consideration were financed entirely with equity funds. In those cases, the cost of capital may have represented the firm's required return on equity. However, most firms finance a substantial portion of their capital budget with long-term debt (bonds), and many also use preferred stock as a source of capital. In these cases, a firm's cost of capital must reflect the average cost of the various sources of long-term funds that the firm uses, not only the cost of equity. In this section, we will discuss the ways in which the cost of each individual type of financing (retained earnings, common stock, preferred stock, and debt) can be estimated[5] given a firm's target capital structure.

---

[5]Estimating or calculating the cost of capital in any precise fashion is a very difficult task.

15.2.1 Cost of Equity

Whereas debt and preferred stocks are contractual obligations that have easily determined costs, it is not easy to measure the cost of equity. In principle, the cost of equity capital involves an **opportunity cost**. In fact, the firm's after-tax cash flows belong to the stockholders. Management may either pay out these earnings in the form of dividends or retain the earnings and reinvest them in the business. If management decides to retain the earnings, an opportunity cost is involved: Stockholders could have received the earnings as dividends and invested the money in other financial assets. Therefore, the firm should earn on its retained earnings at least as much as the stockholders themselves could earn in alternative, but comparable, investments.

When investors are contemplating buying a firm's stock, they have two things in mind: (1) cash dividends and (2) gains (appreciation of shares) at the time of sale. From a conceptual standpoint, investors determine market values of stocks by discounting expected future dividends at a rate that takes into account any future growth. Since investors seek growth companies, a desired growth factor for future dividends is usually included in the calculation.

To illustrate, let's take a simple numerical example. Suppose investors in the common stock of ABC Corporation expect to receive a dividend of $5 by the end of the first year. The future annual dividends will grow at an annual rate of 10%. Investors will hold the stock for two more years and will expect the market price of the stock to rise to $120 by the end of the third year. Given these hypothetical expectations, ABC expects that investors would be willing to pay $100 for this stock in today's market. What is the required rate of return k_r on ABC's common stock? We may answer this question by solving the following equation for k_r:

$$\$100 = \frac{\$5}{(1 + k_r)} + \frac{\$5(1 + 0.1)}{(1 + k_r)^2} + \frac{\$5(1 + 0.1)^2 + \$120}{(1 + k_r)^3}$$

In this case, $k_r = 11.44\%$. This implies that if ABC finances a project by retaining its earnings or by issuing additional common stock at the going market price of $100 per share, it must realize at least 11.44% on new investment just to provide the minimum rate of return perceived by the investors. Therefore, 11.44% is the specific cost of equity that should be used in calculating the weighted-average cost of capital. Because flotation costs are involved in issuing new stock, the cost of equity will increase. If investors view ABC's stock as risky and therefore are willing to buy the stock at a price lower than $100 (but with the same expectations), the cost of equity will also increase. Now we can generalize the preceding result in two different viewpoints.

Cost of Equity Calculation Based on Sources of Capital

One approach is to identify the specific cost incurred in raising the capital by the source of equity.

- **Cost of Retained Earnings:**   Let's assume the same hypothetical situation for ABC. Recall that ABC's retained earnings belong to holders of its common stock. If ABC's current stock is traded for a market price of P_0 with a first-year dividend[6] of D_1 but is growing at the annual rate of g thereafter, the specific cost of retained earnings for an infinite period of holding (stocks will change hands over the years, but it does not matter who holds the stock) can be calculated as

$$P_0 = \frac{D_1}{(1 + k_r)} + \frac{D_1(1 + g)}{(1 + k_r)^2} + \frac{D_1(1 + g)^2}{(1 + k_r)^2} + \cdots$$

$$= \frac{D_1}{1 + k_r \sum_{n=0}^{\infty}\left[\dfrac{(1 + g)}{(1 + k_r)}\right]^n}$$

$$= \frac{D_1}{1 + k_r}\left[\frac{1}{1 - \dfrac{1 + g}{1 + k_r}}\right]$$

where $g < k_r$.

Solving for k_r, we obtain

$$k_r = \frac{D_1}{P_0} + g \tag{15.1}$$

- If we use k_r as the discount rate for evaluating the new project, it will have a positive NPW only if the project's IRR exceeds k_r. Therefore, any project with a positive NPW calculated at k_r induces a rise in the market price of the stock. Hence, by definition, k_r is the rate of return required by shareholders and should be used as the cost of the equity component in calculating the weighted average cost of capital.
- **Issuing New Common Stock:** Again, because flotation costs are involved in issuing new stock, we can modify the cost of retained earnings k_r by

$$k_e = \frac{D_1}{P_0(1 - f_c)} + g \tag{15.2}$$

where k_e is the cost of common equity and f_c is the flotation cost as a percentage of the stock price.

Either calculation is deceptively simple, because in fact, several ways are available to determine the cost of equity. In reality, the market price fluctuates constantly—as do a firm's future earnings. Thus, future dividends may not grow at a constant rate, as the model indicates. For a stable corporation with moderate growth, however, the cost of equity as calculated by evaluating either Eq. (15.1) or Eq. (15.2) serves as a good approximation.

- **Cost of Preferred Stock:** A preferred stock is a hybrid security in the sense that it has some of the properties of bonds and other properties that are similar to common stock. Like bondholders, holders of preferred stock receive a fixed annual dividend. In fact, many firms view the payment of the preferred dividend as an obligation just

---

[6]When we check the stock listings in the newspaper, we do not find the expected first-year dividend D_1. Instead, we find the dividend paid out most recently, D_0. So if we expect growth at a rate g (the dividend at the end of one year from now), D_1 may be estimated as $D_1 = D_0(1 + g)$.

like interest payments to bondholders. It is therefore relatively easy to determine the cost of preferred stock. For the purposes of calculating the weighted average cost of capital, the specific cost of a preferred stock will be defined as

$$k_p = \frac{D^*}{P^*(1 - f_c)} \tag{15.3}$$

- where D^* is the fixed annual dividend, P^* is the issuing price, and f_c is as defined in Eq. (15.2).
- **Cost of Equity:** Once we have determined the specific cost of each equity component, we can determine the weighted-average cost of equity (i_e) for a new project. We have

$$i_e = \left(\frac{c_r}{c_e}\right)k_r + \left(\frac{c_c}{c_e}\right)k_e + \left(\frac{c_p}{c_e}\right)k_p \tag{15.4}$$

where c_r is the amount of equity financed from retained earnings, c_c is the amount of equity financed from issuing new stock, c_p is the amount of equity financed from issuing preferred stock, and $c_r + c_c + c_p = c_e$. Example 15.4 illustrates how we may determine the cost of equity.

EXAMPLE 15.4 Determining the Cost of Equity

The Alpha Corporation needs to raise $10 million for plant modernization. Alpha's target capital structure calls for a debt ratio of 0.4, indicating that $6 million has to be financed from equity.

Source	Amount	Fraction of Total Equity
Retained earnings	$1 million	0.167
New common stock	4 million	0.666
Preferred stock	1 million	0.167

- Alpha is planning to raise $6 million from the following equity sources:
- Alpha's current common stock price is $40, which is the market price that reflects the firm's future plant modernization. Alpha is planning to pay an annual cash dividend of $5 at the end of the first year, and the annual cash dividend will grow at an annual rate of 8% thereafter.
- Additional common stock can be sold at the same price of $40, but there will be 12.4% flotation costs.
- Alpha can issue $100 par preferred stock with a 9% dividend. (This means that Alpha will calculate the dividend on the basis of the par value, which is $9 per share.) The stock can be sold on the market for $95, and Alpha must pay flotation costs of 6% of the market price.

Determine the cost of equity to finance the plant modernization.

SOLUTION

Given: $6 million to raise by equity, $D_1 = \$5$, $g = 8\%$, $D^* = 9$, $P^* = \$95$, and $P_0 = \$40$.

Find: Cost of equity (i_e).

We will itemize the cost of each component of equity.

- Cost of retained earnings: With $D_1 = \$5$, $g = 8\%$, and $P_0 = \$40$,

$$k_r = \frac{5}{40} + 0.08 = 20.5\%$$

- Cost of new common stock: With $D_1 = \$5$, $g = 8\%$, and $f_c = 12.4\%$,

$$k_e = \frac{5}{40(1 - 0.124)} + 0.08 = 22.27\%$$

- Cost of preferred stock: With $D^* = \$9$, $P^* = \$95$, and $f_c = 0.06$,

$$k_p = \frac{9}{95(1 - 0.06)} = 10.08\%$$

- Cost of equity: With $\dfrac{c_r}{c_e} = 0.167$, $\dfrac{c_c}{c_e} = 0.666$, and $\dfrac{c_p}{c_e} = 0.167$,

$$i_e = (0.167)(0.205) + (0.666)(0.2227) + (0.167)(0.1008)$$
$$= 19.96\%$$

Cost of Equity Calculation Based on Financial Market

An alternative (but more practical approach) is to determine the cost of equity from the investors' point of view.

- **Rate of Return Expectation on Risky Assets:** What rate of return can stockholders expect to earn on retained earnings? This question is difficult to answer, but the value sought is often regarded as the rate of return stockholders require on a firm's common stock. If a firm cannot invest retained earnings in order to earn at least the rate of return on equity, it should pay these funds to its stockholders and let them invest directly in other assets that do provide that rate of return. In general, the expected return on any risky asset is composed of three factors:[7]

$$\begin{pmatrix} \text{Expected return} \\ \text{on risky asset} \end{pmatrix} = \begin{pmatrix} \text{Risk-free} \\ \text{interest rate} \end{pmatrix} + \begin{pmatrix} \text{Inflation} \\ \text{premium} \end{pmatrix} + \begin{pmatrix} \text{Risk} \\ \text{premium} \end{pmatrix}$$

This equation says that the owner of a risky asset should expect to earn a return from three sources:

---

[7] An excellent discussion of this subject matter is found in Robert C. Higgins, *Analysis for Financial Management*, 5th ed. Boston: Irwin/McGraw-Hill, 1998.

- Compensation from the opportunity cost incurred in holding the asset, known as the risk-free interest rate.
- Compensation for the declining purchasing power of the investment over time, known as the inflation premium.
- Compensation for bearing risk, known as the risk premium.

Fortunately, we do not need to treat the first two terms as separate factors, because together they equal the expected return on a default-free bond, such as a government bond. In other words, owners of government bonds expect a return from the first two sources but not the third—a state of affairs we may express as

$$\begin{pmatrix} \text{Expected return} \\ \text{on risky asset} \end{pmatrix} = \begin{pmatrix} \text{Interest rate on} \\ \text{government bond} \end{pmatrix} + \begin{pmatrix} \text{Risk} \\ \text{premium} \end{pmatrix}$$

- **Cost of Equity Calculation:** The cost of equity is the risk-free cost of debt (e.g., 20-year U.S. Treasury bills around 6%) plus a premium for taking a risk as to whether a return will be received. The risk premium is the average return on the market, typically the mean of Standard & Poor's 500 large U.S. stocks—or S&P 500 (say, 12.5%)—less the risk-free cost of debt. This premium is multiplied by *beta* (β), which is an approximate measure of stock price volatility that quantifies risk. β measures one firm's stock price relative to the market stock price as a whole. A number greater than unity means that the stock is *more* volatile than the market (on average); a number less than unity means that the stock is *less* volatile than the market (on average). Values for β for most publicly traded stocks are commonly found in various sources, such as Value-Line.[8] The following formula quantifies the cost of equity (i_e):

$$i_e = r_f + \beta[r_M - r_f] \tag{15.5}$$

Here, r_f is the risk-free interest rate (commonly referenced to the U.S. Treasury bond yield, adjusted for inflation) and r_M is the market rate of return (commonly referenced to the average return on S&P 500 stock index funds, adjusted for inflation). Example 15.5 illustrates how we may determine the cost of equity.

EXAMPLE 15.5 Determining the Cost of Equity by the Financial Market

Alpha Corporation needs to raise $10 million for plant modernization. Alpha's target capital structure calls for a debt ratio of 0.4, indicating that $6 million has to be financed from equity.

- Alpha is planning to raise $6 million from the financial market.
- Alpha's β is known to be 1.99, which is greater than unity, indicating that the firm is perceived as more risky than the market average.

[8]*Value Line Reports* are presently available for over 5,000 public companies, and the number is growing. The *Value Line Reports* contain the following information: (1) total assets; (2) total liabilities; (3) total equity; (4) long-term debt as a percentage of capital; (5) equity as a percentage of capital; (6) financial strength (which is used to determine the interest rate); (7) β, and (8) return on invested capital.

• The risk-free interest rate is 6%, and the average market return is 13%. (All of these interest rates are adjusted to reflect inflation in the economy.)

Determine the cost of equity to finance the plant modernization.

SOLUTION

Given: $r_M = 13\%$, $r_f = 6\%$, and $\beta = 1.99$.
Find: i_e.

$$i_e = 0.06 + 1.99(0.13 - 0.06)$$

$$= 19.93\%$$

COMMENTS: In this example, we purposely selected the value of β to approximate the cost of equity derived from Example 15.4. What does this 19.93% represent? If Alpha finances the project entirely from its equity funds, the project must earn at least a 19.93% return on investment.

Note that the cost of equity is almost always higher than the cost of debt. This is because the U.S. Tax Code allows the deduction of interest expense but does not allow the deduction of the cost of equity, which could be considered more subjective and complicated.

15.2.2 Cost of Debt

Now let us consider the calculation of the specific cost that is to be assigned to the debt component of the weighted-average cost of capital. The calculation is relatively straight-forward and simple. As we said in Section 15.1.2, the two types of debt financing are term loans and bonds. Because the interest payments on both are tax deductible, the effective cost of debt will be reduced.

To determine the after-tax cost of debt (i_d), we evaluate the expression

$$i_d = \left(\frac{c_s}{c_d}\right)k_s(1 - t_m) + \left(\frac{c_b}{c_d}\right)k_b(1 - t_m) \tag{15.6}$$

where c_s is the amount of the short-term loan, k_s is the before-tax interest rate on the term loan, t_m is the firm's marginal tax rate, k_b is the before-tax interest rate on the bond, c_b is the amount of bond financing, and $c_s + c_b = c_d$.

As for bonds, a new issue of long-term bonds incurs flotation costs. These costs reduce the proceeds to the firm, thereby raising the specific cost of the capital financed. For example, when a firm issues a $1,000 par bond, but nets only $940, the flotation cost will be 6%. Therefore, the effective after-tax cost of the bond component will be higher than the nominal interest rate specified on the bond. We will examine this problem with a numerical example.

EXAMPLE 15.6 Determining the Cost of Debt

Consider again Example 15.4, and suppose that Alpha has decided to finance the remaining $4 million by securing a term loan and issuing 20-year $1,000 par bonds under the following conditions:

Source	Amount	Fraction	Interest Rate	Flotation Cost
Term loan	$1 million	0.25	12% per year	
Bonds	3 million	0.75	10% per year	6%

If the bond can be sold to net $940 (after deducting the 6% flotation cost), determine the cost of debt to raise $4 million for the plant modernization. Alpha's marginal tax rate is 38%, and it is expected to remain constant in the future.

SOLUTION

Given: $4 million to raise by debt, $k_b = 10\%$, and $k_s = 12\%$.
Find: Cost of debt (i_d).

First, we need to find the effective after-tax cost of issuing the bond with a flotation cost of 6%. The before-tax specific cost is found by solving the equivalence formula

$$\$940 = \frac{\$100}{(1 + k_b)} + \frac{\$100}{(1 + k_b)^2} + \cdots + \frac{\$100 + \$1,000}{(1 + k_b)^{20}}$$

$$= \$100(P/A, k_b, 20) + \$1,000(P/F, k_b, 20)$$

Solving for k_b, we obtain $k_b = 10.74\%$. Note that the cost of the bond component increases from 10% to 10.74% after the 6% flotation cost is taken into account.

The after-tax cost of debt is the interest rate on debt multiplied by $(1 - t_m)$. In effect, the government pays part of the cost of debt, because interest is tax deductible. Now we are ready to compute the after-tax cost of debt as

$$i_d = (0.25)(0.12)(1 - 0.38) + (0.75)(0.1074)(1 - 0.38)$$

$$= 6.85\%$$

15.2.3 Calculating the Cost of Capital

With the specific cost of each financing component determined, we are ready to calculate the tax-adjusted weighted-average cost of capital based on total capital. Then we will define the marginal cost of capital that should be used in project evaluation.

Weighted-Average Cost of Capital

Assuming that a firm raises capital on the basis of the target capital structure and that the target capital structure remains unchanged in the future, we can determine a **tax-adjusted weighted-average cost of capital** (or simply stated, the **cost of capital**). This cost of capital represents a composite index reflecting the cost of raising funds from different sources. The cost of capital is defined as

$$k = \frac{i_d c_d}{V} + \frac{i_e c_e}{V} \tag{15.7}$$

where

c_d = total debt capital (such as bonds) in dollars

c_e = total equity capital in dollars

$V = c_d + c_e$

i_e = average equity interest rate per period, taking into account all equity sources

i_d = after-tax average borrowing interest rate per period, taking into account all debt sources

k = tax-adjusted weighted-average cost of capital

Note that the cost of equity is already expressed in terms of after-tax cost, because any return to holders of either common stock or preferred stock is made after payment of income taxes. Figure 15.2 summarizes the process of determining the cost of capital.

Marginal Cost of Capital

Now that we know how to calculate the cost of capital, we can ask, "Could a typical firm raise unlimited new capital at the same cost?" The answer is no. As a firm tries to attract more new dollars, the cost of raising each additional dollar will rise at some point. As this occurs, the weighted-average cost of raising each additional new dollar also rises. Thus, the **marginal cost of capital** is defined as the cost of obtaining another dollar of

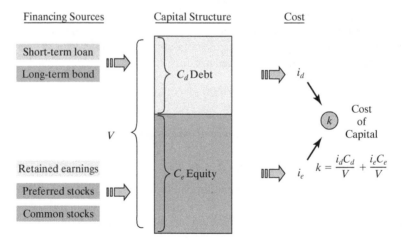

Figure 15.2 Process of calculating the cost of capital. Recall that there are two different ways of determining the cost of equity—one by the traditional approach and the other by the financial market, commonly known as "capital asset pricing model."

new capital, and the marginal cost rises as more and more capital is raised during a given period. In evaluating an investment project, we are using the concept of the marginal cost of capital. The formula to find the marginal cost of capital is exactly the same as Eq. (15.6); however, the costs of debt and equity in that equation are the interest rates on new debt and equity not outstanding (or combined) debt or equity. In other words, we are interested in the marginal cost of capital—specifically, to use it in evaluating a new investment project. The rate at which the firm has borrowed in the past is less important for this purpose. Example 15.7 works through the computations for finding the cost of capital (k).

EXAMPLE 15.7 Calculating the Marginal Cost of Capital

Consider again Examples 15.5 and 15.6. The marginal income tax rate (t_m) for Alpha is expected to remain at 38% in the future. Assuming that Alpha's capital structure (debt ratio) also remains unchanged, determine the cost of capital (k) for raising $10 million in addition to existing capital.

SOLUTION
Given: $c_d =$ $4 million, $c_e =$ $6 million, $V =$ $10 million, $i_d = 6.85\%$, and $i_e = 19.93\%$.
Find: k. Using Eq. (15.7), we calculate

$$k = \frac{(0.0685)(4)}{10} + \frac{(0.1993)(6)}{10}$$

$$= 14.70\%$$

This 14.70% would be the marginal cost of capital that a company with the given financial structure would expect to pay to raise $10 million.

15.3 Choice of Minimum Attractive Rate of Return

Thus far, we have said little about what interest rate or minimum attractive rate of return (MARR) is suitable for use in a particular investment situation. Choosing the MARR is a difficult problem; no single rate is always appropriate. In this section, we will discuss briefly how to select a MARR for project evaluation. Then we will examine the relationship between capital budgeting and the cost of capital.

15.3.1 Choice of MARR when Project Financing is Known

In Chapter 10, we focused on calculating after-tax cash flows, including situations involving debt financing. When cash flow computations reflect interest, taxes, and debt repayment, what is left is called **net equity flow**. If the goal of a firm is to maximize the wealth of its stockholders, why not focus only on the after-tax cash flow to equity, instead of on the flow to all suppliers of capital? Focusing on only the equity flows will permit us to use the cost of equity as the appropriate discount rate. In fact, we have implicitly assumed that all after-tax cash flow problems, in which financing flows are explicitly stated in earlier chapters, represent net equity flows, so the MARR used represents the **cost of equity** (i_e). Example 15.8 illustrates project evaluation by the net equity flow method.

EXAMPLE 15.8 Project Evaluations by Net Equity Flow

Suppose the Alpha Corporation, which has the capital structure described in Example 15.7, wishes to install a new set of machine tools, which is expected to increase revenues over the next five years. The tools require an investment of $150,000 to be financed with 60% equity and 40% debt. The equity interest rate (i_e), which combines the two sources of common and preferred stocks, is 19.96%. Alpha will use a 12% short-term loan to finance the debt portion of the capital ($60,000), with the loan to be repaid in equal annual installments over five years. Depreciation is according to a MACRS over a three-year property class life, and zero salvage value is expected. Additional revenues and operating costs are expected to be as follows:

n	Revenues ($)	Operating Cost
1	$68,000	$20,500
2	73,000	20,000
3	79,000	20,500
4	84,000	20,000
5	90,000	20,500

The marginal tax rate (combined federal, state, and city rate) is 38%. Evaluate this venture by using net equity flows at $i_e = 19.96\%$.

SOLUTION

Given: Project cash flow information and $i_e = 19.96\%$.
Find: Net equity flow and justify the project, based on $i_e = 19.96\%$.

The calculations are shown in Table 15.3. The NPW and IRR calculations are

$$PW(19.96\%) = -\$90,000 + \$34,541(P/F, 19.96\%, 1)$$
$$+\$43,854(P/F, 19.96\%, 2) + \$29,893(P/F, 19.96\%, 3)$$
$$+\$28,540(P/F, 19.96\%, 4) + \$27,124(P/F, 19.96\%, 5)$$
$$= \$11,285$$
$$IRR = 25.91\% > 19.96\%$$

The internal rate of return for this cash flow is 25.91%, which exceeds $i_e = 19.96\%$. Thus, the project would be profitable.

TABLE 15.3 New Equity Flow Method When Source of Project Financing is Known (Example 15.8)

	A	B	C	D	E	F	G
1							
2	**Income Statement**						
3	End of Year	0	1	2	3	4	5
4							
5	Revenue		$ 68,000	$ 73,000	$ 79,000	$ 84,000	$ 90,000
6	Expenses:						
7	Operating cost		$ 20,500	$ 20,000	$ 20,500	$ 20,000	$ 20,500
8	Debt interest		$ 7,200	$ 6,067	$ 4,797	$ 3,376	$ 1,783
9	Depreciation		$ 50,000	$ 66,667	$ 22,222	$ 11,111	$ -
10							
11	Taxable income		$ (9,700)	$ (19,734)	$ 31,481	$ 49,513	$ 67,717
12	Income taxes (38%)		$ (3,686)	$ (7,499)	$ 11,963	$ 18,815	$ 25,732
13							
14	Net income		$ (6,014)	$ (12,235)	$ 19,518	$ 30,698	$ 41,985
15							
16	**Cash Flow Statement**						
17	Operating Activities:						
18	Net income		$ (6,014)	$ (12,235)	$ 19,518	$ 30,698	$ 41,985
19	Depreciation		$ 50,000	$ 66,667	$ 22,222	$ 11,111	$ -
20	Investment Activities:						
21	Milling machine	$ (150,000)					
22	Salvage value						$ -
23	Gains Tax						$ -
24	Financing Activities:						
25	Borrowed funds	$ 60,000					
26	Principal repayment		$ (9,445)	$ (10,578)	$ (11,847)	$ (13,269)	$ (14,861)
27							
28	Net Cash Flow	$ (90,000)	$ 34,541	$ 43,854	$ 29,893	$ 28,540	$ 27,124
29							
30							
31	PW(19.96%)=	$ 11,285					
32	IRR =	25.91%					
33							

COMMENT: In this problem, we assumed that the Alpha Corporation would be able to raise the additional equity funds at the same rate of 19.96%, so this 19.96% can be viewed as the marginal cost of capital.

15.3.2 Choice of MARR when Project Financing is Unknown

You might well ask, "Why, if we use the cost of equity (i_e) exclusively, of what use is the k?" The answer to this question is that, by using the value of k, we may evaluate investments without explicitly treating the debt flows (both interest and principal). In this case, we make a tax adjustment to the discount rate by employing the effective after-tax cost of debt. This approach recognizes that the net interest cost is effectively transferred from the tax collector to the creditor in the sense that there is a dollar-for-dollar reduction in taxes up to the debt interest payments. Therefore, debt financing is treated implicitly. The method would be appropriate when debt financing is not identified with individual investments, but rather it enables the company to engage in a set of investments. (Except where financing flows are explicitly stated, all previous examples in this book have implicitly assumed the more realistic and appropriate situation wherein debt financing is not identified with individual investment. Therefore, the MARRs represent the weighted cost of capital [k].) Example 15.9 illustrates this concept.

EXAMPLE 15.9 Project Evaluation by Marginal Cost of Capital

In Example 15.8, suppose that Alpha Corporation has not decided yet how the $150,000 will be financed. However, Alpha believes that the project should be financed according to its target capital structure with a debt ratio of 40%. Using k, find the NPW and IRR.

SOLUTION

Given: Cash flow information, debt ratio = 0.40, amount of financing required = $150,000 and $k = 14.7\%$.

Find: Justify the project based on marginal cost of capital.

By not accounting for the cash flows related to debt financing, we calculate the after-tax cash flows as shown in Table 15.4. Notice that when we use this procedure,

TABLE 15.4 Cost of Capital Method When Project Financing is Unknown (Example 15.9)

	A	B	C	D	E	F	G
1							
2	**Income Statement**						
3	End of Year	0	1	2	3	4	5
4							
5	Revenue		$ 68,000	$ 73,000	$ 79,000	$ 84,000	$ 90,000
6	Expenses:						
7	Operating cost		$ 20,500	$ 20,000	$ 20,500	$ 20,000	$ 20,500
8	Depreciation		$ 50,000	$ 66,667	$ 22,222	$ 11,111	$ -
9							
10	Taxable income		$ (2,500)	$ (13,667)	$ 36,278	$ 52,889	$ 69,500
11	Income taxes (38%)		$ (950)	$ (5,193)	$ 13,786	$ 20,098	$ 26,410
12							
13	Net income		$ (1,550)	$ (8,474)	$ 22,492	$ 32,791	$ 43,090
14							
15	**Cash Flow Statement**						
16	Operating Activities:						
17	Net income		$ (1,550)	$ (8,474)	$ 22,492	$ 32,791	$ 43,090
18	Depreciation		$ 50,000	$ 66,667	$ 22,222	$ 11,111	$ -
19	Investment Activities:						
20	Milling machine	$(150,000)					
21	Salvage value						$ -
22	Gains tax						$ -
23							
24	Net Cash Flow	$(150,000)	$ 48,450	$ 58,193	$ 44,714	$ 43,902	$ 43,090
25							
26				=NPV(14.7%,C24:G24)+B24			
27	PW(14.7%)=	$ 13,175					
28	IRR =	18.47%		=IRR(B24:G24,15%)			
29							

interest and the resulting tax shield are ignored in deriving the net incremental after-tax cash flow. In other words, no cash flow is related to any financing activity. Thus, taxable income is overstated, as are income taxes. To compensate for these overstatements, the discount rate is reduced accordingly. The implicit assumption is that the tax overpayment is exactly equal to the reduction in interest implied by i_d.

The flow at time 0 is simply the total investment ($150,000 in this example). Recall that Alpha's k was calculated to be 14.70%. The internal rate of return for the after-tax flow in the last line of Table 15.4 is calculated as 18.47% > 14.70%. Since the IRR exceeds the value of k, the investment would be profitable. Here, we evaluated the after-tax flow by using the value of k, and we reached the same conclusion about the desirability of the investment as we did in Example 15.8.

COMMENTS: The net equity flow and the cost-of-capital methods usually lead to the same accept/reject decision for independent projects (assuming the same amortization schedule for debt repayment, such as term loans) and usually rank projects identically for mutually exclusive alternatives. Some differences may be observed as special financing arrangements may increase (or even decrease) the attractiveness of a project by manipulating tax shields and the timing of financing inflows and payments.

To summarize, in cases where the exact debt-financing and repayment schedules are known, we recommend the use of the net equity flow method. The appropriate MARR would be the cost of equity, i_e. If no specific assumption is made about the exact instruments that will be used to finance a particular project (but we do assume that the given capital-structure proportions will be maintained), we may determine the after-tax cash flows without incorporating any debt cash flows. Then we use the marginal cost of capital (k) as the appropriate MARR.

15.3.3 Choice of MARR under Capital Rationing

It is important to distinguish between the cost of capital (k), as calculated in Section 15.2.3, and the MARR (i) used in project evaluation under **capital rationing**—situations in which the funds available for capital investment are not sufficient to cover all potentially acceptable projects. When investment opportunities exceed the available money supply, we must decide which opportunities are preferable. Obviously, we want to ensure that all of the selected projects are more profitable than the best rejected project (or at least as profitable as the worst accepted project), which is the best opportunity forgone and whose value is called the **opportunity cost**. When a limit is placed on capital, the MARR is assumed to be equal to this opportunity cost, which is usually greater than the marginal cost of capital. In other words, the value of i represents the corporation's time-value trade-offs and partially reflects the available investment opportunities. Thus, there is nothing illogical about borrowing money at k and evaluating cash flows according to the different rate i. Presumably, the money will be invested to earn a rate i or greater. In the next example, we will provide guidelines for selecting a MARR for project evaluation under capital rationing.

A company may borrow funds to invest in profitable projects, or it may return to (invest in) its **investment pool** any unused funds until they are needed for other investment activities. Here, we may view the borrowing rate as a marginal cost of capital (k), as defined in Eq. (15.7). Suppose that all available funds can be placed in investments yielding a return equal to l, which is the **lending rate**. We view these funds as an investment pool. The firm may withdraw funds from this pool for other investment purposes, but if left in the pool, the funds will earn at the rate l, which is the opportunity cost. The MARR thus is related to either the borrowing interest rate or the lending interest rate. To illustrate the relationship among the borrowing interest rate, the lending interest rate, and the MARR, let us define the following variables:

$$k = \text{Borrowing rate (or the cost of capital)}$$

$$l = \text{Lending rate (or the opportunity cost)}$$

$$i = \text{MARR}$$

Generally (but not always), we might expect k to be greater than or equal to l. We must pay more for the use of someone else's funds than we can receive for "renting out" our own funds (unless we are running a lending institution). Then we will find that the appropriate MARR would be between l and k. The concept for developing a discount rate (MARR) under capital rationing is best understood by a numerical example.

EXAMPLE 15.10 Determining an Appropriate MARR as a Function of the Budget

The Sand Hill Corporation has identified six investment opportunities that will last one year. The firm drew up a list of all potentially acceptable projects. Table 15.5 shows each project's net cash flows over one-year life, and IRR. Then it ranks the projects according to their IRR, listing the highest IRR first.

TABLE 15.5 Sand Hill's Investment Opportunities with Estimated IRRs

Project	Cash Flow A0	Cash Flow A1	IRR
1	−$10,000	$12,000	20%
2	−10,000	11,500	15
3	−10,000	11,000	10
4	−10,000	10,800	8
5	−10,000	10,700	7
6	−10,000	10,400	4

Suppose $k = 10\%$, which remains constant for the budget amount up to $60,000, and $l = 6\%$. Assuming that the firm has available (a) $40,000, (b) $60,000, and (c) $0 for investments, what is the reasonable choice for the MARR in each case?

SOLUTION

Given: Capital budget and project profiles.

Find: Correct MARR to use as a function of budget available.

We will consider the following steps to determine the appropriate discount rate (MARR) under a capital-rationing environment.

Step 1. Develop the firm's cost-of-capital schedule as a function of the capital budget. For example, the cost of capital can increase as the amount of financing increases. Also, determine the firm's lending rate if any unspent money is lent out or remains invested in the company's investment pool.

Step 2. Plot this investment opportunity schedule by showing how much money the company could invest at different rates of return as shown in Figure 15.3.

(a) If the firm has $40,000 available for investing, it should invest in projects 1, 2, 3, and 4. Clearly, it should not borrow at 10% to invest in either project 5 or project 6. In these cases, the best rejected project is project 5. The worst accepted project is project 4. If you view the opportunity cost as the cost associated with accepting the worst project, the MARR could be 8%.

(b) If the firm has $60,000 available, it should invest in projects 1, 2, 3, 4, and 5. It could lend the remaining $10,000 rather than invest these funds in project 6. For this new situation, we have MARR $= l = 6\%$.

(c) If the firm has no funds available, it probably would borrow to invest in projects 1 and 2. The firm might also borrow to invest in project 3, but it would be indifferent to this alternative unless some other consideration was involved. In this case, MARR $= k = 10\%$; therefore, we can say that $l \leq$ MARR $\leq k$ when we have complete certainty about future investment opportunities. Figure 15.4 illustrates the concept of selecting a MARR under capital rationing.

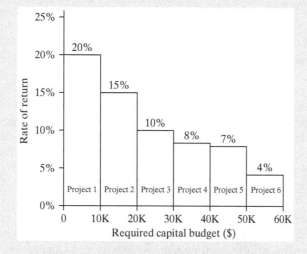

Figure 15.3   An investment opportunity schedule ranking alternatives by the rate of return (Example 15.10).

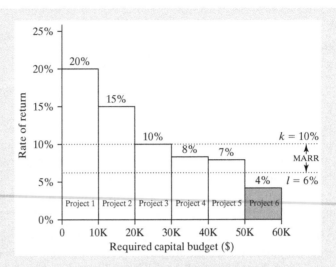

Figure 15.4 A range of MARRs as a function of a budget under capital rationing (Example 15.10): If you have an unlimited budget, MARR = lending rate; if you have no budget but are allowed to borrow, MARR = 10%.

COMMENTS: In this example, for simplicity, we assumed that the timing of each investment is the same for all competing proposals—say, period 0. If each alternative requires investments over several periods, the analysis will be significantly complicated, as we have to consider both the investment amount and its timing in selecting the appropriate MARR.[9]

Now we can generalize what we have learned. If a firm finances investments through borrowed funds, it should use MARR = k; if the firm is a lender, it should use MARR = l. A firm may be a lender in one period and a borrower in another; consequently, the appropriate rate to use may vary from period to period. In fact, whether the firm is a borrower or a lender may well depend on its investment decisions.

In practice, most firms establish a single MARR for all investment projects. Note the assumption that we made in Example 15.10: **Complete certainty** about investment opportunities was assumed. Under highly uncertain economic environments, the MARR generally would be much greater than k, which is the firm's cost of capital.

It is important to recognize that one of the components in calculating k is the cost of equity, which we assume an average risk of the firm. However, a project is conceived as much higher risky one, then we need to adjust the MARR to reflect this additional risk. For example, if $k = 10\%$, a MARR of 15% would not be considered excessive. Few firms are willing to invest in projects earning only slightly more than their cost of capital because of elements of *risk* in the project. This way of incorporating the risk element in MARR is known as a **risk-adjusted discount rate** as we discussed in Chapter 13.

---

[9]This is certainly beyond the scope of any introductory engineering economics text, but can be found in C. S. Park and G. P. Sharp-Bette, *Advanced Engineering Economics*. New York: John Wiley, 1990.

If the firm has a large number of current and future opportunities that will yield the desired return, we can view the MARR as the minimum rate at which the firm is willing to invest, and we can also assume that proceeds from current investments can be reinvested to earn at the MARR. Furthermore, *if we choose the "do-nothing" alternative, all available funds are invested at the MARR.* In engineering economics, we also normally separate the risk issue from the concept of MARR. As seen in Chapter 12, we treat the effects of risk explicitly when we must.

15.4 Capital Budgeting

In this section, we will examine the process of deciding whether projects should be included in the capital budget. In particular, we will consider decision procedures that should be applied when we have to evaluate a set of multiple investment alternatives for which we have a limited capital budget.

15.4.1 Evaluation of Multiple Investment Alternatives

In Chapters 5, 6, and 7, we learned how to compare two or more mutually exclusive projects. Now we shall extend the comparison techniques to a set of multiple decision alternatives that are not necessarily mutually exclusive. Here, we distinguish a **project** from an **investment alternative**, which is a decision option. For a single project, we have two investment alternatives: to accept or reject the project. For two independent projects, we can have four investment alternatives: (1) to accept both projects; (2) to reject both projects; (3) to accept only the first project; and (4) to accept only the second project. As we add interrelated projects, the number of investment alternatives to consider grows exponentially.

To perform a proper capital-budgeting analysis, a firm must group all projects under consideration into decision alternatives. This grouping requires the firm to distinguish between projects that are independent of one another and projects that are dependent on one another in order to formulate alternatives correctly.

Independent Projects

An **independent project** is a project that may be accepted or rejected without influencing the accept–reject decision of another independent project. For example, the purchase of a milling machine, office furniture, and a forklift truck constitutes three independent projects. Only projects that are economically independent of one another can be evaluated separately. (Budget constraints may prevent us from selecting one or more of several independent projects; this external constraint does not alter the fact that the projects are independent.)

Dependent Projects

In many decision problems, several investment projects are related to one another such that the acceptance or rejection of one project influences the acceptance or rejection of others. Two such types of dependencies are mutually exclusive projects and contingent projects. We say that two or more projects are **contingent** if the acceptance of one requires the acceptance of the other. For example, the purchase of a computer printer is dependent upon the purchase of a computer, but the computer may be purchased without purchasing the printer.

15.4.2 Formulation of Mutually Exclusive Alternatives

We can view the selection of investment projects as a problem of selecting a single decision alternative from a set of mutually exclusive alternatives. Note that each investment project is an investment alternative, but that a single investment alternative may entail a whole group of investment projects. The common method of handling various project relationships is to arrange the investment projects so that the selection decision involves only mutually exclusive alternatives. To obtain this set of mutually exclusive alternatives, we need to enumerate all of the feasible combinations of the projects under consideration.

Independent Projects

With a given number of independent investment projects, we can easily enumerate mutually exclusive alternatives. For example, in Table 15.6, we consider two projects (A and B) where we have four decision alternatives, including a do-nothing alternative.

In our notation, X_j is a decision variable associated with investment project j. If $X_j = 1$, project j is accepted; if $X_j = 0$, project j is rejected. Since the acceptance of one of these alternatives will exclude any other, the alternatives are mutually exclusive.

TABLE 15.6 **Formulating Mutually Exclusive Investment Alternatives with Independent Projects**

Alternative	Description	X_A	X_B
1	Reject A, Reject B	0	0
2	Accept A, Reject B	1	0
3	Reject A, Accept B	0	1
4	Accept A, Accept B	1	1

Mutually Exclusive Projects

Suppose we are considering two independent sets of projects (A and B). Within each independent set are two mutually exclusive projects (A1, A2) and (B1, B2). The selection of either A1 or A2, however, is also independent of the selection of any project from the set (B1, B2). In other words, you can select A1 and B1 together, but you cannot select A1 and A2 together. For this set of investment projects, the mutually exclusive alternatives are as given in Table 15.7.

Note that with two independent sets of mutually exclusive projects, we have nine different decision alternatives.

Contingent Projects

Suppose the acceptance of C is contingent on the acceptance of both A and B, and the acceptance of B is contingent on the acceptance of A. Then the number of decision alternatives can be formulated as given in Table 15.8. Thus, we can easily formulate a set of mutually exclusive investment alternatives with a limited number of projects that are independent, mutually exclusive, or contingent merely by arranging the projects in a logical sequence.

One difficulty with the enumeration approach is that, as the number of projects increases, the number of mutually exclusive alternatives increases exponentially. For example, for 10 independent projects, the number of mutually exclusive alternatives is

TABLE 15.7 **Formulating Mutually Exclusive Investment Alternatives with Dependent Projects**

Alternative	(X_{A1}, X_{A2})	(X_{B1}, X_{B2})
1	(0, 0)	(0, 0)
2	(1, 0)	(0, 0)
3	(0, 1)	(0, 0)
4	(0, 0)	(1, 0)
5	(0, 0)	(0, 1)
6	(1, 0)	(1, 0)
7	(0, 1)	(1, 0)
8	(1, 0)	(0, 1)
9	(0, 1)	(0, 1)

2^{10} (or 1,024). For 20 independent projects, 2^{20} (or 1,048,576) alternatives exist. As the number of decision alternatives increases, we may have to resort to mathematical programming to find the solution to an investment problem. Fortunately, in real-world business, the number of engineering projects to consider at any one time is usually manageable, so the enumeration approach is a practical one.

TABLE 15.8 **Formulating Decision Options with Contingent Projects**

Alternative	X_A	X_B	X_C
1	0	0	0
2	1	0	0
3	1	1	0
4	1	1	1

15.4.3 Capital-Budgeting Decisions with Limited Budgets

Recall that capital rationing refers to situations in which the funds available for capital investment are not sufficient to cover all the projects. In such situations, we enumerate all investment alternatives as before, but we eliminate from consideration any mutually exclusive alternatives that exceed the budget. The most efficient way to proceed in a capital-rationing situation is to select the group of projects that maximizes the total NPW of future cash flows over required investment outlays.

Optimal Capital Budget If Projects Can Be Accepted in Part

We will first consider the situation where projects can be funded partially. Example 15.11 will illustrate the process of determining the optimal capital budget and an appropriate MARR to be used in project selection.

EXAMPLE 15.11 Four Energy-Saving Projects under Budget Constraints

The facilities department at an electronic instrument firm has four energy-efficiency projects under consideration:

> **Project 1: Electrical.** This project requires replacing the existing standard-efficiency motors in the air-conditioners and exhaust blowers of a particular building with high-efficiency motors.
>
> **Project 2: Building envelope.** This project involves coating the inside surface of existing fenestration in a building with low-emissivity solar film.
>
> **Project 3: Air-conditioning.** This project requires the installation of heat exchangers between a building's existing ventilation and relief air ducts.
>
> **Project 4: Lighting.** This project requires the installation of specular reflectors and removing a building's existing ceiling grid-lighting troffers.

These projects require capital outlays in the $50,000 to $140,000 range and have useful lives of about eight years. The facilities department's first task was to estimate the annual savings that could be realized by these energy-efficiency projects. Currently, the company pays 7.80 cents per kilowatt-hour (kWh) for electricity and $4.85 per thousand cubic feet (MCF). Assuming that the current energy prices would continue for the next eight years, the company has estimated the cash flow and the IRR for each project as follows.

Project	Investment	Annual O&M Cost	Annual Savings (Energy)	Annual Savings (Dollars)	IRR
1	$46,800	$1,200	151,000 kWh	$11,778	15.43%
2	104,850	1,050	513,077 kWh	40,020	33.48%
3	135,480	1,350	6,700,000 CF	32,495	15.95%
4	94,230	942	470,738 Wh	36,718	34.40%

Because each project could be adopted in isolation, at least as many alternatives as projects are possible. For simplicity, assume that all projects are independent (as opposed to being mutually exclusive), that they are equally risky, and that their risks are all equal to those of the firm's average existing assets.

(a) Determine the optimal capital budget for the energy-saving projects.

(b) With $250,000 approved for energy improvement funds during the current fiscal year, the department does not have sufficient capital on hand to undertake all four projects without any additional allocation from headquarters. Enumerate the total number of decision alternatives and select the best alternative.

DISCUSSION: The NPW calculation cannot be shown yet, as we do not know the marginal cost of capital. Therefore, our first task is to develop the **marginal-cost-of-capital (MCC)** schedule, which is a graph that shows how the cost of capital changes as more and more new capital is raised during a given year. The graph in Figure 15.5 is

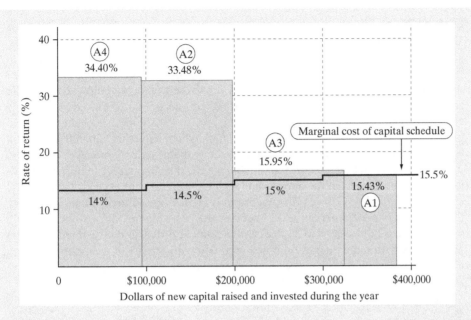

Figure 15.5 Combining the marginal-cost-of-capital schedule and investment opportunity schedule curves to determine a firm's optimal capital budget (Example 15.11).

the company's marginal-cost-of-capital schedule. The first $100,000 would be raised at 14%, the next $100,000 at 14.5%, the next $100,000 at 15%, and any amount over $300,000 at 15.5%. We then plot the IRR data for each project as the **investment opportunity schedule (IOS)** shown in the graph. The IOS shows how much money the firm could invest at different rates of return.

SOLUTION

Given: IRRs for four different energy projects.
Find: Optimal capital budget.

Consider project A4. Its IRR is 34.40%, and it can be financed with capital that costs only 14%. Consequently, it should be accepted. Projects A2 and A3 can be analyzed similarly; all are acceptable, because the IRR exceeds the marginal cost of capital. Project A1, by contrast, should be rejected, because its IRR is less than the marginal cost of capital. Therefore, the firm should accept the three projects A4, A2, and A3, which have rates of return in excess of the cost of capital that would be used to finance them if we end up with a capital budget of $334,560. This should be the amount of the *optimal capital budget*.

In Figure 15.5, even though two rate changes occur in the marginal cost of capital in funding project A3 (the first change is from 14.5% to 15%, the second from 15% to 15.5%), the accept–reject decision for A3 remains unchanged, as its rate of return exceeds the marginal cost of capital.

What would happen if the MCC cut through project A3? For example, suppose that the marginal cost of capital for any project raised above $300,000 would cost 16% instead of 15.5%, thereby causing the MCC schedule to cut through project A3. Should we then accept project A3? If we can take project A3 in part, we would take on only part of it up to 74.49%.

COMMENTS: Now, if projects cannot be funded partially, how do we compare these alternatives? Consider again Figure 15.5. If we take project A1 first, it would be acceptable, because its 15.43% return would exceed the 14% cost of capital used to finance it. Why couldn't we do this? The answer is that we are seeking to maximize the excess of returns over costs (or the area above the MCC) but below the IOS. We accomplish that by accepting the most profitable projects first. This logic leads us to conclude that, as long as the budget permits, project A4 should be selected first and project A2 second. This will consume $199,080, which leaves us $50,920 in unspent funds. The question is, "What are we going to do with this leftover money?" Certainly, it is not enough to take project A3 in full, but we can take project A1 in full. Full funding for project A1 will fit the budget, and the rate of return still exceeds the marginal cost of capital (15.43% > 15%). Accordingly, unless the leftover funds earn more than 15.43% interest, alternative 12 becomes the best (Figure 15.6). A more logical way to address this issue is to enumerate the number of feasible investment decision alternatives within the budget limit and come up with the list of mutually exclusive alternatives. This will be discussed in the next section.

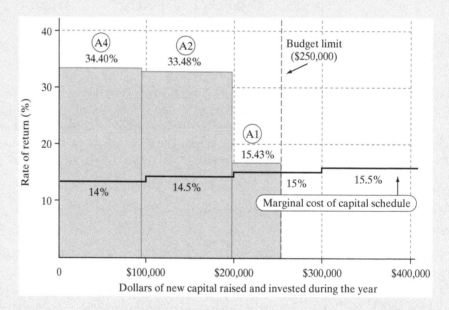

Figure 15.6   The appropriate cost of capital to be used in the capital-budgeting process for decision alternative 12, with a $250,000 budget limit (Example 15.11) is 15%.

Optimal Capital Budget If Projects Cannot Be Accepted in Part

If projects cannot be partially funded, we first need to enumerate the number of feasible investment decision alternatives within the budget limit. Now, how do we compare these alternatives as the marginal cost of capital changes for each one? One way we can compare these alternatives is to apply the incremental rate of return approach discussed in Chapter 7. Example 15.12 will illustrate the process of identifying the best alternative when the marginal cost of capital changes as a function of budget.

EXAMPLE 15.12 Interdependent Projects with a Limited Budget

Consider six investment projects as shown in Table 15.9 and their interdependencies as follows:

- Projects A and B are mutually exclusive.
- Projects C and F are independent projects.
- Project D is contingent on project C.
- Project E is contingent on project B.

TABLE 15.9 Cash Flows for Six Interdependent Projects

Project	Cash Flows			
	0	1	2	3
A	−$100	50	50	50
B	−$200	50	100	200
C	−$100	30	80	120
D	−$300	200	120	50
E	−$200	100	100	150
F	−$150	30	70	80

The following also indicates the marginal cost of capital (MCC) as a function of budget:

Amount of Financing	$100	$200	$300	$400	$500	$600	$700	$800	$900	$1000
k	12%	12.5%	13%	13.5%	14%	14.5%	15%	16%	17%	20%

(a) Formulate the entire list of mutually exclusive decision alternatives.
(b) What is the optimal capital budget if projects cannot be accepted in part? Which projects would be funded under the optimal capital budget? What is the appropriate MARR?
(c) If the firm is allowed to borrow up to $800, which projects would be funded? What is the appropriate MARR?

SOLUTION

Given: IRRs for six different projects with interdependencies as specified and the marginal cost of capital as a function of budget.

Find: (a) Number of mutually exclusive alternatives, (b) Optimal capital budget, and (c) an appropriate MARR with a budget limit of $600.

DISCUSSION: If we calculate the IRR for each project as in Table 15.10, we know that project F is not a good one as its IRR is less than 12.5% (cost of capital to borrow $150). Therefore, we can eliminate the project from the mutually exclusive list.

TABLE 15.10 IRRs for Six Interdependent Projects

	Project	Cash Flows				IRR
		0	**1**	**2**	**3**	
1	A	−100	50	50	50	23%
2	B	−200	50	100	200	27%
3	C	−100	30	80	120	44%
4	D	−300	200	120	50	14%
5	E	−200	100	100	150	31%
6	F	−150	30	70	80	8%

(a) **Generation of mutually exclusive alternatives:**

As shown in Table 15.11, by dropping project F, the mutually exclusive list reduces to 12 alternatives. If we borrow $800, we would be able to fund the most expensive alternative. The marginal cost of capital (k) for this amount is 16%. Now if we calculate the IRR for each alternative on the list, we see that all the alternatives would

TABLE 15.11 List of Mutually Exclusive Alternatives

ALT	Projects	0	1	2	3	IRR	k	PW(k)
1	0	0	0	0	0	0%		$ -
2	A	−100	50	50	50	23%	12.0%	$20.09
3	C	−100	30	80	120	44%	12.0%	$75.97
4	B	−200	50	100	200	27%	12.5%	$63.92
5	AC	−200	80	130	170	35%	12.5%	$93.22
6	BC	−300	80	180	320	33%	13.0%	$133.54
7	BE	−400	150	200	350	29%	13.5%	$126.79
8	CD	−400	230	200	170	25%	13.5%	$74.16
9	CBE	−500	180	280	470	32%	14.0%	$190.58
10	ACD	−500	280	250	220	25%	14.0%	$86.47
11	BCD	−600	280	300	370	26%	14.5%	$119.85
12	BECD	−800	380	400	520	27%	16.0%	$157.99

provide enough return to cover the borrowing cost. (If there is any alternative that does not meet this, we can further drop the alternative from the list.)

(b) **Optimal capital budget and the projects to be funded:**

- **Total investment approach:** If we want to adopt the net present worth maximization, we need to calculate the NPW for each alternative at its appropriate cost of capital. For example, to find the net present worth for ALT 6 which requires $300, we look up the MCC table to see how much it would cost to finance the amount. It is 13%. Then, we use 13% to calculate the NPW, which is $133. We do the same for the rest of alternatives, and find that ALT 9 is the best alternative. The optimal capital budget is $500.

- **Incremental Rate of Return Approach:** To determine the optimal capital budget, we need to identify the best alternative out of 12 alternatives. Since the MARR is a function of projects selected, we may conduct the incremental analysis to determine the best alternative. In doing so, we need to do a series of pairwise comparisons as shown in Table 15.12.

TABLE 15.12 Incremental Analysis to Determine the Best Alternative

Incremental	Cash Flow				ΔIRR	MARR (k)	Selection
	0	**1**	**2**	**3**			
ALT 2 - ALT 1	−100	50	50	50	23.4%	12.0%	ALT 2
ALT 3 - ALT 2	0	−20	30	70	176.6%	12.0%	ALT 3
ALT 4 - ATL 3	−100	20	20	80	7.6%	12.5%	ALT 3
ALT 5 - ALT 3	−100	50	50	50	23.4%	12.5%	ALT 5
ALT 6 - ALT 5	−100	0	50	150	29.0%	13.0%	ALT 6
ALT 7 - ALT 6	−100	70	20	30	11.9%	13.5%	ALT 6
ALT 8 - ALT 6	−100	150	20	−150	−49.0%	13.5%	ALT 6
ATL 9 - ALT 6	−200	100	100	150	31.4%	14.0%	ALT 9
ALT 9 - ALT 10	0	−100	30	250	73.8%	14.0%	ALT 9
ALT 11 - ALT 9	−100	100	20	−100	−58.8%	14.5%	ALT 9
ALT 12 - ALT 9	−300	200	120	50	14.4%	16.0%	ALT 9

Note: When we are comparing ALT 9 with ALT 10, we need to create an incremental investment by subtracting ATL 10 from ALT 9. See Example 7.11.

- For example, ALT4 versus ALT 3 – the higher cost investment ALT 4 will cost additional $100 over ALT 3. According to the cost of capital schedule, this additional $100 will be financed at the marginal cost of capital of 12.5% if we decided to go with ALT 4.

- Incremental analysis (ALT 4 – ALT 3):

$$(-\$100, \$20, \$20, \$80) \rightarrow \Delta IRR = 7.6\% < 12.5\%$$

- So, it is not worth spending additional $100 to fund ALT 4. Then we drop ALT 4 from the list. We go to the next alternative on the list, which ALT 5.
- Incremental analysis (ALT 5 – ALT 3):

$$(-\$100, \$50, \$50, \$50) \rightarrow \Delta IRR = 23.4\% > 12.5\%$$

So, we prefer ALT 5 over ALT 3. This process continues until no alternative left to compare. It turns out that ALT 9 will be the best alternative, which requires $500 investment with the marginal cost of capital of 14%. The optimal capital budget is depicted in Figure 15.7.

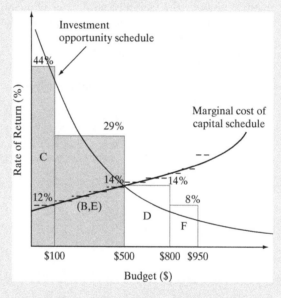

Figure 15.7 The optimal capital budget (Example 15.12).

(c) With a budget limit of $800, which projects to fund? What is the appropriate MARR?

If you are allowed to borrow up to $800, you would only do so up to $500, as the best portfolio still remains at (CBE), and your MARR should be 14%.

SUMMARY

- Methods of financing fall into two broad categories:

  1. **Equity financing** uses retained earnings or funds raised from an issuance of stock to finance a capital investment.

  2. **Debt financing** uses money raised through loans or by an issuance of bonds to finance a capital investment.

- Companies do not simply borrow funds to finance projects. Well-managed firms usually establish a **target capital structure** and strive to maintain the **debt ratio** when individual projects are financed.

- The cost-of-capital formula is a composite index reflecting the cost of funds raised from different sources. The formula is

$$k = \frac{i_d c_d}{V} + \frac{i_e c_e}{V}, V = c_d + c_e$$

- The selection of an appropriate MARR depends generally upon the **cost of capital**—the rate the firm must pay to various sources for the use of capital.

  - The **cost of equity** (i_e) is used when debt-financing methods and repayment schedules are known explicitly.
  - The **cost of capital** (k) is used when exact financing methods are unknown, but a firm keeps its capital structure on target. In this situation, a project's after-tax cash flows contain no debt cash flows, such as principal and interest payment.

- The marginal cost of capital is defined as the cost of obtaining another dollar of new capital. The marginal cost rises as more and more capital is raised during a given period.

- Without a capital limit, the choice of MARR is dictated by the availability of financing information:

  1. In cases where the exact debt-financing and repayment schedules are known, we recommend the use of the net equity flow method. The proper MARR would be the cost of equity, i_e.

  2. If no specific assumption is made about the exact instruments that will be used to finance a particular project (but we do assume that the given capital-structure proportions will be maintained), we may determine the after-tax cash flows without incorporating any debt cash flows. Then we use the marginal cost of capital (k) as the proper MARR. If borrowing is allowed, a proper choice of MARR is summarized in Table 15.13.

- Under a highly uncertain economic environment, the MARR generally would be much greater than k, which is the firm's cost of capital with normal risk, as the risk premium increases.

- Under conditions of capital rationing, the selection of the MARR is more difficult, but generally, the following possibilities exist:

  - The cost of capital used in the capital-budgeting process is determined at the intersection of the **IOS** and **MCC** schedules. If the cost of capital at

TABLE 15.13 How to Choose MARR when Borrowing is permitted

Conditions	MARR
A firm borrows some capital from lending institutions at the borrowing rate k and some from its investment pool at the lending rate l.	$l < \text{MARR} < k$
A firm borrows all capital from lending institutions at the borrowing rate k.	$\text{MARR} = k$
A firm borrows all capital from its investment pool at the lending rate l.	$\text{MARR} = l$

the intersection is used, then the firm will make correct accept–reject decisions, and its level of financing and investment will be optimal. This view assumes that the firm can invest and borrow at the rate where the two curves intersect.

- If a strict budget is placed in a capital-budgeting problem and no projects can be taken in part, all feasible investment decision scenarios need to be enumerated. Depending upon each such scenario, the cost of capital will also likely change. The task is then to find the best-investment scenario in light of a changing-cost-of-capital environment. As the number of projects to consider increases, we may eventually resort to a more advanced technique, such as a mathematical programming procedure.

PROBLEMS

Methods of Financing

15.1 The Optical World Corporation, a manufacturer of peripheral vision storage systems, needs $10 million to market its new robotics-based vision systems. The firm is considering two financing options: common stock and bonds. If the firm decides to raise the capital through issuing common stock, the flotation costs will be 6%, and the share price will be $25. If the firm decides to use debt financing, it can sell a 10-year, 12% bond with a par value of $1,000. The bond flotation costs will be 1.9%.

(a) For equity financing, determine the flotation costs and the number of shares to be sold to net $10 million.

(b) For debt financing, determine the flotation costs and the number of $1,000 par value bonds to be sold to net $10 million. What is the required annual interest payment?

15.2 Consider a project whose initial investment is $500,000 financed at an interest rate of 9% per year. Assuming that the required repayment period is six years, determine the repayment schedule by identifying the principal as well as the interest payments for each of the following methods:

(a) Equal repayment of the principal

(b) Equal repayment of the interest

(c) Equal annual installments

15.3 A chemical plant is considering purchasing a computerized control system. The initial cost is $200,000, and the system will produce net savings of $100,000 per year. If purchased, the system will be depreciated under MACRS as a five-year recovery property. The system will be used for four years, at the end of which time the firm expects to sell it for $30,000. The firm's marginal tax rate on this investment is 35%. Any capital gains will be taxed at the same income tax rate. The firm is considering purchasing the computer control system either through its retained earnings or through borrowing from a local bank. Two commercial banks are willing to lend the $200,000 at an interest rate of 10%, but each requires different repayment plans (Table P15.3). Bank A requires four equal annual principal payments with interest calculated on the basis of the unpaid balance. Bank B offers a payment plan, with five equal annual payments.

TABLE P15.3

Repayment Plan of Bank A			
End of Year		**Principal**	**Interest**
1		$50,000	$20,000
2		50,000	15,000
3		50,000	10,000
4		50,000	5,000
Repayment Plan of Bank B			
End of Year	**Principal**	**Interest**	**Total**
1	$32,759	$20,000	$52,759
2	36,035	16,724	52,759
3	39,638	13,121	52,759
4	43,602	9,157	52,759
5	47,963	4,796	52,759

(a) Determine the cash flows if the computer control system is to be bought through its retained earnings (equity financing).

(b) Determine the cash flows if the asset is financed through either bank A or bank B.

(c) Recommend the best course of financing the project. (Assume that the firm's MARR is known to be 10%.)

15.4 The Edison Power Company currently owns and operates a coal-fired combustion turbine plant that was installed 20 years ago. Because of degradation of the system, 65 forced outages occurred during the last year alone and two boiler explosions during the last seven years. Edison is planning to scrap the current plant and install a new, improved gas-turbine plant that produces more energy per unit of fuel than typical coal-fired boilers produce.

The new 50-MW gas-turbine plant, which runs on gasified coal, wood, or agricultural wastes, will cost Edison $65 million. Edison wants to raise the capital from three financing sources: 45% common stock, 10% preferred stock (which carries a 6% cash dividend when declared), and 45% borrowed funds. Edison's investment banks quote the following flotation costs:

Financing Source	Flotation Costs	Selling Price	Par Value
Common stock	4.6%	$32/share	$10
Preferred stock	8.1	55/share	15
Bond	1.4	980	1,000

(a) What are the total flotation costs to raise $65 million?

(b) How many shares (both common and preferred) or bonds must be sold to raise $65 million?

(c) If Edison makes annual cash dividends of $2 per common share and annual bond interest payments are at the rate of 12%, how much cash should Edison have available to meet both the equity and debt obligation? (Note that whenever a firm declares cash dividends to its common stockholders, the preferred stockholders are entitled to receive dividends of 6% of par value.)

Cost of Capital

15.5 Calculate the after-tax cost of debt under each of the following conditions:

(a) Interest rate, 12%; tax rate, 25%

(b) Interest rate, 14%; tax rate, 34%

(c) Interest rate, 15%; tax rate, 40%

15.6 The Sweeney Paper Company is planning to sell $10 million worth of long-term bonds with an 11% interest rate. The company believes that it can sell the $1,000 par value bonds at a price that will provide a yield to maturity of 13% to its investors. The flotation costs will be 1.9%. If Sweeney's marginal tax rate is 35%, what is its annual after-tax cost of debt?

15.7 The Mobil Appliance Company's earnings, dividends, and stock price are expected to grow at an annual rate of 12%. Mobil's common stock is currently traded at $18 per share. Mobil's last cash dividend was $1.00, and its expected cash dividend for the end of this year is $1.12. Determine the cost of retained earnings (k_r).

15.8 Refer to Problem 15.7, and suppose that Mobil wants to raise capital to finance a new project by issuing new common stock. With the new project, the cash dividend is expected to be $1.10 at the end of the current year, and its growth rate is 10%. The stock now sells for $18, but new common stock can be sold to net Mobil $15 per share.

(a) What is Mobil's flotation cost, expressed as a percentage?

(b) What is Mobil's cost of new common stock (k_e)?

15.9 The Callaway Company's cost of equity is 22%. Its before-tax cost of debt is 13%, and its marginal tax rate is 40%. The firm's capital structure calls for a debt-to-equity ratio of 45%. Calculate Callaway's cost of capital.

15.10 The Delta Chemical Corporation is expected to have the capital structure for the foreseeable future as given in Table P15.10.

The flotation costs are already included in each cost component. The marginal income tax rate (t_m) for Delta is expected to remain at 40% in the future.

(a) Determine the cost of capital (k).

(b) If the risk-free rate is known to be 6% and the average return on the S&P 500 is about 12%, determine the cost of equity with $\beta = 1.2$ based on the capital-asset pricing principle.

(c) Determine the cost of capital on the basis of the cost of equity obtained in part (b).

TABLE P15.10

Source of After-Tax Financing	Percent of Total Funds	Before-Tax Cost	Cost
Debt	30%		
Short term	10	14%	
Long term	20	12	
Equity	70%		
Common stock	70		30%

15.11 Matsushita Machining (MPM) is considering acquiring a new precision cutting machine at a cost of $120,000. The need for this particular machine is expected to last only five years, after which the machine is expected to have a salvage value of $30,000. The annual operating cost is estimated at $20,000. The addition of this machine to the current production facility is expected to generate an additional revenue of $70,000 annually and will be depreciated in the seven-year MACRS property class. The income tax rate applicable to Charleston is 36%. The initial investment will be financed with 60% equity and 40% debt. The before-tax debt interest rate, which combines both short-term and long-term financing, is 12% with the loan to be repaid in equal annual installments. The equity interest rate (i_e), which combines the two sources of common and preferred stocks, is 18%.

(a) Evaluate this investment project by using net equity flows.

(b) Evaluate this investment project by using k.

15.12 The Huron Development Company is considering buying an overhead pulley system. The new system has a purchase price of $100,000, an estimated useful life and MACRS class life of five years, and an estimated salvage value of $30,000. It is expected to allow the company to economize on electric power usage, labor, and repair costs, as well as to reduce the number of defective products. A total annual savings of $45,000 will be realized if the new pulley system is installed. The company is in the 30% marginal tax bracket. The initial investment will be financed with 40% equity and 60% debt. The before-tax debt interest rate, which combines both short-term and long-term financing, is 15% with the loan to be repaid in equal annual installments over the project life. The equity interest rate (i_e), which combines the two sources of common and preferred stocks, is 20%.

(a) Evaluate this investment project by using net equity flows.

(b) Evaluate this investment project by using k.

15.13 Consider the two mutually exclusive machines given in Table P15.13.

TABLE P15.13

	Machine A	Machine B
Initial investment	$40,000	$60,000
Service life	6 years	6 years
Salvage value	$4,000	$8,000
Annual O&M cost	$8,000	$10,000
Annual revenues	$20,000	$28,000
MACRS property	5 year	5 year

The initial investment will be financed with 70% equity and 30% debt. The before-tax debt interest rate, which combines both short-term and long-term financing, is 10% with the loan to be repaid in equal annual installments over the project life. The equity interest rate (i_e), which combines the two sources of common and preferred stock, is 15%. The firm's marginal income tax rate is 35%.

(a) Compare the alternatives using $i_e = 15\%$. Which alternative should be selected?

(b) Compare the alternatives using k. Which alternative should be selected?

(c) Compare the results obtained in parts (a) and (b).

Capital Budgeting

15.14 The DNA Corporation, a biotech engineering firm, has identified seven R&D projects for funding. Each project is expected to be in the R&D stage for three to five years. The IRR figures shown in Table P15.14 represent the royalty income from selling the R&D results to pharmaceutical companies.

DNA Corporation can raise only $100 million. DNA's borrowing rate is 18%, and its lending rate is 12%. Which R&D projects should be included in the budget?

15.15 Consider the following investment projects and their interdependencies:

- Projects A and F are mutually exclusive.
- Projects C and D are independent projects.
- Project B is contingent on Project C.
- Project E is contingent on project F.

The following indicates the cost of capital as a function of budget:

TABLE P15.14

Project	Investment Type	Required IRR
1. Vaccines	$15M	22%
2. Carbohydrate chemistry	25M	40
3. Antisense	35M	80
4. Chemical synthesis	5M	15
5. Antibodies	60M	90
6. Peptide chemistry	23M	30
7. Cell transplant/gene therapy	19M	32

Amount of Financing	$100	$200	$400	$600	$800	$1,000
k	14.5%	16%	17.5%	20%	21%	23%

TABLE P15.15

Project	0	1	2	3	IRR
		Cash Flows			
A	−$100	40	60	40	18.82%
B	−$200	150	100	30	24.73%
C	−$100	30	80	60	28.54%
D	−$300	200	120	150	28.25%
E	−$200	100	100	150	31.44%
F	−$150	50	70	80	14.63%

(a) Formulate the entire list of mutually exclusive decision alternatives.

(b) What is the optimal capital budget? What is the appropriate MARR for capital budgeting purpose?

(c) If the firm has a budget limit placed at $800, which projects would be funded? What is the appropriate MARR?

Short Case Studies

ST15.1 Games, Inc., is a publicly traded company that makes computer software and accessories. Games's stock price over the last five years is plotted in Figure ST15.1a, Games's earnings per share for the last five years are shown in Figure ST15.1b, and Games's dividends per share for the last five years are shown in Figure ST15.1c. The company currently has 1,000,000 shares outstanding and a long-term debt of $12,000,000. The company also paid $1,200,000 in interest expenses last year, has other assets of $5,000,000, and had earnings before taxes of $3,500,000 last year.

Figure ST15.1a   Stock price per share.

Figure ST15.1b   Earnings per share.

Figure ST15.1c Dividends per share.

Games has decided to manufacture a new product. In order to make the new product, Games will need to invest in a new piece of equipment that costs $10,000,000. The equipment is classified as a seven-year MACRS property and is expected to depreciate 30% per year. Equipment installation will require 20 employees working for two weeks and charging $50 per hour each. Once the equipment has been installed, the facility is expected to remain operational for two years.

Games intends to maintain its current debt-to-equity ratio and therefore plans on borrowing the appropriate amount today to cover the purchase of the equipment. The interest rate on the loan will be equal to its current cost of debt. The loan will require equal annual interest payments over its life (i.e., the loan rate times the principal borrowed for each year).

The principal will be repaid in full at the end of year two. Games plans on issuing new stock to cover the equity portion of the investment. The underwriter of the new stock issue charges an 11% flotation cost.

Games estimates that its new product will acquire 20% of all market share within the United States. Even though this product has never been produced before, Games has identified an older product that should have market attributes similar to those of the new product. The older product's unit market sales for the entire United States for the last five years are shown in Figure ST15.1d. It is estimated that the new product will sell for $20 per unit (in today's dollars). Costs are estimated at 80% of the selling price for the first year and 60% of the selling price for the second year. Inflation is estimated at 10% per annum for the new product.

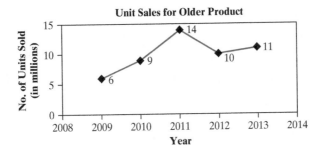

Figure ST15.1d Total unit sales in the United States for the older product.

Due to the large size of the investment required to manufacture the new product, Games's analysts predict that once the decision to accept the project is made, investors will revalue the company's stock price. Because you are the chief engineering economist for Games, you have been requested by upper management to determine how manufacturing this new product will change Games's stock price. Management would like answers to the following questions. (*Note*: The first three questions are asked in a sequence that will assist you in answering the last two questions.)

(a) Games's investors usually use the corporate value model (CVM) to determine the total market value of the company. In its most basic form, the CVM states that a firm's market value is nothing more than the present value of its expected future net cash flows plus the value of its assets. Using this logic, what must investors currently assess the present value of Games's future net cash flows to be (not including the investment in the new product)?

(b) Determine Games's tax rate.

(c) Determine an appropriate MARR to use in the analysis when the financing source is known. Now do the same when the financing source is unknown.

(d) Assuming that the new-product venture is accepted and the new piece of equipment is disposed of at the end of year two, what is a most-likely estimate for Games's stock price?

(e) Now assume that the new piece of equipment is not disposed of in year two and that Games has not decided how the $10,000,000 for the equipment will be financed. Instead, assume that the revenues generated from the new product will continue indefinitely. Estimate a pessimistic, a most-likely, and an optimistic estimate for Games's stock price.[10]

ST15.2 The National Food Processing Company is considering investing in plant modernization and plant expansion. These proposed projects would be completed in two years, with varying requirements of money and plant engineering. Management is willing to use the somewhat uncertain data in Table ST15.2 in selecting the best set of proposals.

The resource limitations are as follows:

- First-year expenditures: $450,000
- Second-year expenditures: $420,000
- Engineering hours: 11,000 hours

The situation requires that a new or modernized production line be provided (project 1 or project 2 which are mutually exclusive). The numerical control (project 3) is applicable only to the new line. The company obviously does not want to both buy (project 6) and build (project 5) raw-material processing facilities; it can, if desirable, rely on the present supplier as an independent firm. Neither the maintenance-shop project (project 4) nor the delivery-truck purchase (project 7) is mandatory.

TABLE ST15.2

No	Project	Investment Year 1	Investment Year 2	Engineering IRR	Engineering Hours
1	Modernize production line	$300,000	0	30%	4,000
2	Build new production line	100,000	$300,000	43%	7,000
3	Provide numerical control for new production line	0	200,000	18%	2,000
4	Modernize maintenance shops	50,000	100,000	25%	6,000
5	Build raw-material processing plant	50,000	300,000	35%	3,000
6	Buy present subcontractor's facilities for raw-material processing	200,000	0	20%	600
7	Buy a new fleet of delivery trucks	70,000	10,000	16%	0

[10]Dr. Luke Miller.

(a) Enumerate all possible mutually exclusive alternatives without considering the budget and engineering-hour constraints.

(b) Identify all feasible mutually exclusive alternatives.

(c) Suppose that the firm's marginal cost of capital will be 14% for raising the required capital up to $1 million. Which projects would be included in the firm's budget?

ST15.3 Consider the investment projects given in Table ST15.3.

Suppose that you have only $3,500 available at period 0. Neither additional budgets nor borrowing is allowed in any future budget period. However, you can lend out any remaining funds (or available funds) at 10% interest per period.

(a) If you want to maximize the future worth at period 3, which projects would you select? What is that future worth (the total amount available for lending at the end of period 3)? No partial projects are allowed.

TABLE ST15.3

	Project Cash Flows			
n	A	B	C	D
0	−$2,000	−$3,000	−$1,000	0
1	1,000	4,000	1,400	−$1,000
2	1,000	0	−100	1,090
3	1,000	0	0	0
$i*$	23.38%	33.33%	30.91%	9%

(b) Suppose in part (a) that at period 0 you are allowed to borrow $500 at an interest rate of 13%. The loan has to be repaid at the end of year one. Which project would you select to maximize your future worth at period 3?

(c) Considering a lending rate of 10% and a borrowing rate of 13%, what would be the most reasonable MARR for project evaluation?

ST15.4 The American Chemical Corporation (ACC) is a multinational manufacturer of industrial chemical products. ACC has made great progress in reducing energy costs and has implemented several cogeneration projects in the United States and Puerto Rico, including the completion of a 35-megawatt (MW) unit in Chicago and a 29-MW unit in Baton Rouge. The division of ACC being considered for one of its more recent cogeneration projects is a chemical plant located in Texas. The plant has a power usage of 80 million kilowatt hours (kWh) annually. However, on the average, it uses 85% of its 10-MW capacity, which would bring the average power usage to 68 million kWh annually. Texas Electric currently charges $0.09 per kWh of electric consumption for the ACC plant, a rate that is considered high throughout the industry.

Because ACC's power consumption is so large, the purchase of a cogeneration unit would be desirable. Installation of the unit would allow ACC to generate its own power and to avoid the annual $6,120,000 expense to Texas Electric. The total initial investment cost would be $10,500,000, including $10,000,000 for the purchase of the power unit itself (which is a gas-fired 10-MW Allison 571), engineering, design, and site preparation, and $500,000 for the purchase of interconnection equipment (such as poles and distribution lines) that will be used to interface the co-generator with the existing utility facilities.

ACC is considering two financing options:

- ACC could finance $2,000,000 through the manufacturer at 10% for 10 years and the remaining $8,500,000 through issuing common stock. The flotation cost for a common-stock offering is 8.1%, and the stock will be priced at $45 per share.

- Investment bankers have indicated that 10-year 9% bonds could be sold at a price of $900 for each $1,000 bond. The flotation costs would be 1.9% to raise $10.5 million.

(a) Determine the debt-repayment schedule for the term loan from the equipment manufacturer.

(b) Determine the flotation costs and the number of common stocks to sell to raise the $8,500,000.

(c) Determine the flotation costs and the number of $1,000 par value bonds to be sold to raise $10.5 million.

ST15.5 After ACC management decided to raise the $10.5 million by selling bonds (Problem ST15.4), the company's engineers estimated the operating costs of the cogeneration project. The annual cash flow is composed of many factors: maintenance, standby power, overhaul costs, and miscellaneous expenses. Maintenance costs are projected to be approximately $500,000 per year. The unit must be overhauled every three years at a cost of $1.5 million. Miscellaneous expenses, such as the cost of additional personnel and insurance, are expected to total $1 million. Another annual expense is that for stand-by power, which is a service provided by the utility in the event of a cogeneration unit trip or scheduled maintenance outage. Unscheduled outages are expected to occur four times annually with each averaging two hours in duration at an annual expense of $6,400. Overhauling the unit takes approximately 100 hours and occurs every three years, requiring another power cost of $100,000. Fuel (spot gas) will be consumed at a rate of $8,000 BTU per kWh, including the heat recovery cycle. At $2.00 per million BTU, the annual fuel cost will reach $1,280,000. Due to obsolescence, the expected life of the cogeneration project will be 12 years, after which Allison will pay ACC $1 million for the salvage of all equipment.

Revenue will be incurred from the sale of excess electricity to the utility company at a negotiated rate. Since the chemical plant will consume (on average) 85% of the unit's 10-MW output, 15% of the output will be sold at $0.04 per kWh, bringing in an annual revenue of $480,000. ACC's marginal tax rate (combined federal and state) is 36%, and the company's minimum required rate of return for any cogeneration project is 27%. The anticipated costs and revenues are summarized in Table ST15.5.

(a) If the cogeneration unit and other connecting equipment could be financed by issuing corporate bonds at an interest rate of 9% compounded annually, with the flotation expenses as indicated in Problem ST15.4, determine the net cash flow from the cogeneration project.

(b) If the cogeneration unit can be leased, what would be the maximum annual lease amount that ACC is willing to pay?

TABLE ST15.5

Initial investment:	
Cogeneration unit, engineering, design, and site preparation (15-year MACRS class)	$10,000,000
Interconnection equipment (5-year MACRS class)	500,000
Salvage after 12 years' use	1,000,000
Annual expenses:	
Maintenance	500,000
Miscellaneous (additional personnel and insurance)	1,000,000
Stand-by power	6,400
Fuel	1,280,000
Other operating expenses:	
Overhaul every 3 years	1,500,000
Stand-by power during overhaul	100,000
Revenues:	
Sale of excess power to Texas Electric	480,000

CHAPTER 16

Economic Analysis in the Service Sector

New NHTSA Study Shows Motor Vehicle Crashes Have $871 Billion Economic and Societal Impact on U.S. Citizens The U.S. Department of Transportation's National Highway Traffic Safety Administration (NHTSA)[1] recently released a new study that underscores the high economic toll and societal impact of motor vehicle crashes in the United States. The price tag for crashes comes at a heavy burden for Americans at $871 billion in economic loss and societal harm. This includes $277 billion in economic costs—nearly $900 for each person living in the United States based on calendar year 2010 data—and $594 billion in harm from the loss of life and the pain and decreased quality of life due to injuries.

Source: Leks052/Shutterstock

[1] *New NHTSA Study Shows Motor Vehicle Crashes Have $871 Billion Economic and Societal Impact on U.S. Citizens*, NHTSA 19–14, National Highway Traffic Safety Administration, May 28, 2014.

The economic cost of motor vehicle crashes in the U.S. is the equivalent of 1.9 percent of the $14.96 trillion Gross Domestic Product (GDP) in 2010. Factors contributing to the price tag include productivity losses, property damage, medical and rehabilitation costs, congestion costs, legal and court costs, emergency services, insurance administration costs, and the costs to employers, among others. Overall, nearly 75 percent of these costs are paid through taxes, insurance premiums, and congestion-related costs such as travel delay, excess fuel consumption, and increased environmental impacts. These costs, borne by society rather than individual crash victims, totaled over $200 billion.

As the annual economic loss of road accidents is so staggering, reducing road accidents is a major policy goal in the majority of the U.S. These states believe that road accidents caused by mechanical defects of the vehicles can be significantly reduced by periodic mechanical inspections. Fewer than 20 states now require safety inspections, down from a peak of about 31.

According to PennLive report,[2] Pennsylvania garages go through this ritual on some 11 million vehicles a year—at a cost to motorists of hundreds of millions of dollars—but the commonwealth is bucking a national trend in which a majority of states no longer require safety inspections. PennLive further reported that neighboring New Jersey and Washington, D.C., dropped their inspection programs, both citing beliefs that there is no conclusive evidence that they are necessary. Still, a report by Cambridge Systematics Inc., commissioned by the Pennsylvania Department of Transportation, says emphatically: "The results of the research clearly demonstrate that the Vehicle Safety Inspection program in Pennsylvania is effective and saves lives." Without inspections, the report said, the number of fatal crashes could be expected to increase by between 127 and 169 each year. How do we conduct such a cost-benefit study for a public program?

Up to this point, we have focused our analysis on the economic issues related to the manufacturing sector of the U.S. economy. The main reason for doing this is that many engineers will be working in that sector. However, an increasing number of engineers are now seeking their careers in the service sector; such as health care, financial institutions, transportation and logistics, and government. In this chapter, we will present some unique features that must be considered in evaluating investment projects in the service sector.

CHAPTER LEARNING OBJECTIVES

After completing this chapter, you should understand the following concepts:

- How to price service.
- How to conduct cost benefit analysis in the public sector.

---

[2] *"Despite other states' moves, Pennsylvania's vehicle inspection program going strong,"* PennLive, January 10, 2011 (http://www.pennlive.com/midstate/index.ssf/2011/01/despite_other_states_moves_pen.html).

■ How to conduct cost effectiveness analysis.

■ How to evaluate investment projects in the health care industry.

16.1 What Is the Service Sector?[3]

The service sector of the U.S. economy dominates both gross domestic product (GDP) and employment. It is also the fastest-growing part of the economy and the one offering the most fertile opportunities to engineers to improve their productivity. For example, service activities now approach 80% of U.S. employment, far outstripping sectors such as manufacturing (14%) and agriculture (2%), as shown in Figure 16.1.

The mere scale of the service sector makes it a critical element of the U.S. economy, employing, as it does, many millions of workers producing trillions of dollars in economic value. The service sector has expanded far beyond the traditional consumer or institutional service industries, and currently includes distributive services such as transportation, communications, and utilities; the rapidly expanding producer services, such as finance, insurance, real estate, and advertising; wholesale and retail trade and sales; the nonprofit sector, including health, philanthropy, and education; and government. The use of technology in the service sector promotes deskilling (i.e., automation). However, secure and reliable U.S. services provide much of the key infrastructure on which the whole nation, and indeed much of the world's commerce, depends. Tremendous new challenges have arisen as changes in processes and operations are developed to better assure the safety and reliability of critical services. These changes all lead to significant investment in infrastructures requiring detailed economic analysis.

16.1.1 Characteristics of the Service Sector

What makes the evaluation of a service sector project unique compared with one from the manufacturing sector? Some of the unique features are summarized as follows.

- Services are generally intangible. They have sometimes been defined as "anything of economic value that cannot be held or touched."

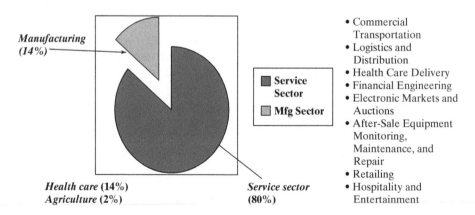

Figure 16.1 Contribution of the service sector to the U.S. gross domestic product (GDP).

[3] National Science Foundation Program, "Exploratory Research on Engineering the Service Sector (ESS)," NSF 02–029.

- It is usually impossible to build inventories of services. Either the demand for the service must be backlogged, or enough resources need to be provided to meet an acceptable fraction of the demand as it arises.
- Services are more dynamic and responsive to demand than are manufactured products. This means that variability and risk are more central issues in service industries. Indeed, the management of financial risk is an important service in itself.
- Many services (examples are medical treatment and equipment repair) require a diagnostic step to design the service as part of its delivery. Coproduction (i.e., active collaboration between the server and the customer) is also required in many settings.
- Service products are usually less standardized and less subject to design specifications than are manufactured goods, because the outputs are tailored to customer needs as they are delivered. This also makes it harder to distinguish service product design from product manufacture and delivery.
- The dimensions of service quality are more subtle and subjective than those of physical products. Not only are the parameters of services more difficult to express, but customers' perceptions play a much greater role in deciding what is satisfactory or valuable.
- Most service operations are more labor intensive than the production of goods.
- Compared with goods industries, the service economy has a much greater fraction of its operations performed by governments and institutions.
- Information technology is central to service industries. Often, it is the only significant means of multiplying human output.

Certainly, our objective is not to address all aspects of the service sector, but only some of the common economic analysis issues confronted by its engineers.

16.1.2 Difficulty of Pricing Service

Improving service can be many different things. For example, for a delivery business such as United Parcel Service (UPS) or FedEx, anything that reduces the total time taken from pickup to delivery is considered an improvement of service. For an airline, on-time departures and arrivals, a reduction in the number of mishandled pieces of checked-in baggage, and a speedy check-in at the gate could all be considered important service parameters, because they make airline passengers happy, which in turn will translate into more business volume. Accordingly, one of the critical questions related to improvement in service is, "What do service providers gain by improving their own service to suppliers and customers?" If we can quantify improvements in service in terms of dollars, the economic analysis is rather straightforward: If the net increase in revenue due to improvements in service exceeds the investment required, the project can be justified. For this type of decision problem, we can simply use any of the measures of investment worth discussed in Chapter 5 through Chapter 7.

Improving service is one thing, and putting a price on services is another thing. This is far more difficult than pricing products, because the benefits of services are less tangible and service companies often lack well-documented standard unit-production costs to go by. When a company designs an after-service contract to go with equipment sales, customers may be segmented according to their service needs rather than their size, industry, or type of equipment. Companies then develop the pricing, contracting, and monitoring capabilities to support the cost-effective delivery of the service.[4] For example, when customers

---

[4] *Source:* Russell G. Bundschuh and Theodore M. Dezvane, "How to Make After-Sales Services Pay Off," *The McKinsey Quarterly*, 2003, No. 4, pp. 3–13.

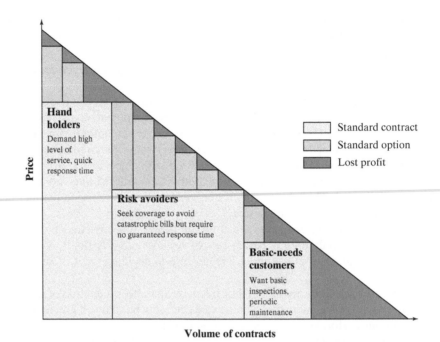

Figure 16.2 How to capture lost profits. By segmenting customers into three categories with appropriate service options, companies can capture much of the lost profit. Options may include guaranteed response time, remote equipment monitoring, or extended warranty.

are segmented according to the service level they need, they tend to fall into one of at least three common categories. The "basic-needs customers" want a standard level of service with basic inspections and periodic maintenance. The "risk avoiders" are looking for coverage to avoid big bills but care less about other elements, such as response times. The "hand holders" need high levels of service, often with quick and reliable response times, and are willing to pay for the privilege. Therefore, to maximize return, companies need to capture tremendous value from their service businesses by taking a more careful, fact-based approach to designing and pricing services (Figure 16.2).

The types of service problems that we are interested in this chapter, however, are (1) those involving economic decisions by the public sector and (2) those encountered by nonprofit organizations (such as hospitals) or public policy makers on health care. In the next section, we will review some of the common decision tools adopted by the public sector (such as government agencies and the military). The economic issues in the health-care service industry will be discussed in Section 16.5.

16.2 Economic Analysis in the Public Sector

In earlier chapters, we have focused attention on investment decisions in the private sector; the primary objective of these investments was to increase the wealth of corporations. In the public sector, federal, state, and local governments spend hundreds of billions of dollars annually on a wide variety of public activities, such as the vehicle inspection project described in the chapter opener. In addition, governments at all levels regulate the behavior of individuals and businesses by influencing the use of enormous quantities of productive resources. How can public decision makers determine whether their decisions, which affect the use of these productive resources, are in fact in the best public interest?

Many civil engineers work on public-works areas such as highway construction, airport construction, and water projects. In the vehicle inspection program presented at the beginning of the chapter, each prevention measure requires a different level of investment and produces a different degree of benefits. One of the most important aspects of airport expansion is to quantify the cost of airport delays in dollar terms. In other words, planners ask, "What is the economic benefit of reducing airport delays?" From the airline's point of view, taxiing and arrival delays mean added fuel costs. For the airport, delays mean lost revenues in landing and departure fees. From the public's point of view, delays mean lost earnings, as people then have to spend more time on transportation. Comparing the investment costs with the potential benefits, an approach known as benefit–cost analysis is an important feature of economic analysis.

16.2.1 What is Benefit–Cost Analysis?

Benefit–cost analysis is a decision-making tool that is used to systematically develop useful information about the desirable and undesirable effects of public projects. In a sense, we may view benefit–cost analysis in the public sector as profitability analysis in the private sector. In other words, benefit–cost analysis attempts to determine whether the social benefits of a proposed public activity outweigh the social costs. Usually, public investment decisions involve a great deal of expenditure, and their benefits are expected to occur over an extended period of time. Examples of benefit–cost analysis include studies of (1) public transportation systems; (2) environmental regulations on noise and pollution; (3) public-safety programs; (4) education and training programs; (5) public-health programs; (6) flood control; (7) water resource development; and (8) national defense programs.

The three typical aims of benefit–cost analyses are (1) to maximize the benefits for any given set of costs (or budgets); (2) to maximize the net benefits when both benefits and costs vary; and (3) to minimize costs to achieve any given level of benefits (often called "cost-effectiveness" analysis). Three types of decision problems, each having to do with one of these aims, will be considered in this chapter.

16.2.2 Framework of Benefit–Cost Analysis

To evaluate public projects designed to accomplish widely differing tasks, we need to measure the benefits or costs in the same units in all projects so that we have a common perspective by which to judge the different projects. In practice, this means expressing both benefits and costs in monetary units, a process that often must be performed without accurate data. In performing benefit–cost analysis, we define **users** as the public and **sponsors** as the government.

The general framework for benefit–cost analysis can be summarized as follows:

Step 1. Identify all users' benefits expected to arise from the project.

Step 2. Quantify these benefits in dollar terms as much as possible, so that different benefits may be compared against one another and against the costs of attaining them.

Step 3. Identify sponsors' costs.

Step 4. Quantify these costs in dollar terms as much as possible to allow for comparisons.

Step 5. Determine the equivalent benefits and costs during the base period; use an interest rate appropriate for the project.

Step 6. Accept the project if the equivalent users' benefits exceed the equivalent sponsor's costs.

We can use benefit–cost analysis to choose among such alternatives as allocating funds for the construction of a mass-transit system, a dam with irrigation, highways, or an air-traffic control system. If the projects are on the same scale with respect to cost, it is merely a question of choosing the project for which the benefits exceed the costs by the greatest amount. The steps just outlined are for a single (or independent) project evaluation. As in the case of the internal-rate-of-return criterion, in comparing mutually exclusive alternatives, an incremental benefit–cost ratio must be used. Section 16.2.3 illustrates this important issue in detail.

16.2.3 Valuation of Benefits and Costs

In the abstract, the framework we just developed for benefit–cost analysis is no different from the one we have used throughout this text to evaluate private investment projects. The complications (as we shall discover in practice) arise in trying to identify and assign values to all the benefits and costs of a public project.

Users' Benefits

To begin a benefit–cost analysis, we identify all project **benefits** (favorable outcomes) and **disbenefits** (unfavorable outcomes) to the user. We should also consider the indirect consequences resulting from the project—the so-called **secondary effects**. For example, the construction of a new highway will create new businesses such as gas stations, restaurants, and motels (benefits), but it will divert some traffic from the old road. As a consequence, some businesses would be lost (disbenefits). Once the benefits and disbenefits are quantified, we define the users' benefits as:

$$\text{Users' Benefits (B)} = \text{benefits} - \text{disbenefits}$$

In identifying users' benefits, we should classify each one as a **primary benefit**—a benefit that is directly attributable to the project—or a **secondary benefit**—a benefit that is indirectly attributable to the project. As an example, at one time the U.S. government was considering building a superconductor research laboratory in Texas. If the project ever were to materialize, it would bring many scientists and engineers, along with other supporting population, to the region. Primary national benefits might include the long-term benefits that could accrue as a result of various applications of the research to United States businesses. Primary regional benefits might include economic benefits created by the research laboratory activities, which would generate many new supporting businesses. Secondary benefits might include the creation of new economic wealth as a consequence of a possible increase in international trade, and an increase in the incomes of various regional producers attributable to a growing population.

The reason for making this distinction is that it may make our analysis more efficient: If primary benefits alone are sufficient to justify project costs, we can save time and effort by *not* quantifying the secondary benefits.

Sponsor's Costs

We determine the cost to the sponsor by identifying and classifying the expenditures required and any savings (or revenues) to be realized. The sponsor's costs should include both capital investment and annual operating costs. Any sales of products or services that take place upon completion of the project will generate some revenues—for example, toll revenues on highways. These revenues reduce the sponsor's costs. Therefore, we calculate the sponsor's costs by combining these cost elements:

Sponsor's Costs = capital cost + operating and maintenance costs − revenues.

Social Discount Rate

As we learned in Chapter 15, the selection of an appropriate MARR for evaluating an investment project is a critical issue in the private sector. In public project analyses, we also need to select an interest rate, called the **social discount rate**, to determine equivalent benefits as well as the equivalent costs. The selection of a social discount rate in public project evaluation is as critical as the selection of a MARR in the private sector.

Ever since present-value calculations were initiated to evaluate public water resources and related land-use projects in the 1930s, a tendency to use relatively low rates of discount when compared with those existing in markets for private assets has persisted. During the 1950s and into the 1960s, the rate for water resource projects was 2.63%, which for much of that period was even below the yield on long-term government securities. The persistent use of a lower interest rate for water resource projects is a political issue. In recent years, with the growing interest in performance budgeting and systems analysis in the 1960s, the tendency on the part of government agencies has been to examine the appropriateness of the discount rate in the public sector in relation to the efficient allocation of resources in the economy as a whole. Two views of the basis for determining the social discount rate prevail.

1. **Projects without private counterparts.** *The social discount rate should reflect only the prevailing government borrowing rate.* Projects such as dams designed purely for flood control, access roads for noncommercial uses, and reservoirs for community water supply may not have corresponding private counterparts. In those areas of government activity where benefit–cost analysis has been employed in evaluation, the rate of discount traditionally used has been the cost of government borrowing. In fact, water resource project evaluations follow this view exclusively.

2. **Projects with private counterparts.** *The social discount rate should represent the rate that could have been earned had the funds not been removed from the private sector.* For public projects that are financed by borrowing at the expense of private investment, we may focus on the opportunity cost of capital in alternative investments in the private sector to determine the social discount rate. In the case of public capital projects, similar to some in the private sector that produce a commodity or a service (such as electric power) to be sold on the market, the rate of discount employed would be the average cost of capital as discussed in Chapter 15. The reasons for using the private rate of return as the opportunity cost of capital in projects similar to those in the private sector are:

   1. To prevent the public sector from transferring capital from higher yielding to lower yielding investments.

   2. To force public-project evaluators to employ market standards in justifying projects.

The Office of Management and Budget (OMB) holds the second view. Since October 1992, OMB[5] has required that constant-dollar benefit–cost analyses of proposed investments and regulations should report net present value and other outcomes determined using a real discount rate of 7% (inflation-free rate). This rate approximates the marginal pretax rate of return on an average investment in the private sector in recent years. This is equivalent to about 10% market interest rate if we consider 3% inflation in the market place. Significant changes in this rate will be reflected in future updates of the OMB Circular.

16.2.4 Quantifying Benefits and Costs

Now that we have defined the general framework for benefit–cost analysis and discussed the appropriate discount rate, we will illustrate the process of quantifying the benefits and costs associated with a public project.[6] Table 16.1 attests that many states in the United States employ inspection systems for motor vehicles.[7] Critics often claim that these programs lack efficacy and have a poor benefit-to-cost ratio in terms of reducing fatalities, injuries, accidents, and pollution.

To illustrate the process of quantifying the costs and benefits for such a vehicle inspection program, we will consider a motor-vehicle inspection program initiated by the state of Pennsylvania.

TABLE 16.1 Passenger Safety Inspection Programs in the U.S.

State/Territory	Random	Mandatory	One Year	Two Year	Resale
Delaware		•		•	
District of Columbia		•	•		
Hawaii		•	•		
Illinois	•				
Iowa	•				
Louisiana		•	•		
Maine	•	•	•		
Maryland					•
Massachusetts		•		•	•
Mississippi		•	•		
Missouri		•		•	•
New Hampshire		•	•		

(Continued)

[5] Circular No. A-94 Revised, Office of Management and Budget, Whitehouse website, http://www.whitehouse.gov/sites/default/files/omb/assets/a94/a094.pdf.

[6] General analysis framework based on Loeb, P. D. and Gilad, B., "The Efficacy and Cost Effectiveness of Vehicle Inspection," *Journal of Transport Economics and Policy*, May, 1984, pp. 145–164.

[7] Pennsylvania's Vehicle Safety Inspection Program Effectiveness Study, Final Report, Pennsylvania Department of Transportation, March, 2009.

TABLE 16.1 (*Continued*)

State/Territory	Random	Mandatory	One Year	Two Year	Resale
New Jersey	•	•		•	•
New York		•	•		
North Carolina		•	•		
Ohio	•				
Oregon	•				
Pennsylvania		•	•		
Vermont		•	•		
Virginia		•	•		
West Virginia		•	•		

Source: Data from American Association of Motor Vehicle Administrators Fast Track to Vehicle Services Facts, 2003.

Elements of Benefits and Costs

Using Figure 16.3 as road map, the state of Pennsylvania identified the primary and secondary benefits of its new inspection program as follows:

- **Users' Benefits**
  Primary benefits.   Deaths and injuries related to motor-vehicle accidents impose financial costs on individuals and society. Preventing such costs through the inspection program has the following primary benefits.
  1. Retention of contributions to society that might be lost due to an individual's death.
  2. Retention of productivity that might be lost while an individual recuperates from an accident.

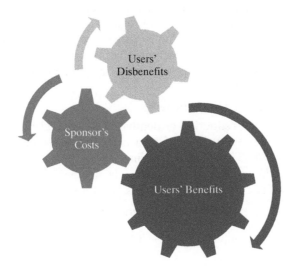

Figure 16.3   Elements of Benefits and Costs.

3. Savings of medical, legal, and insurance services.

4. Savings on property replacement or repair costs.

Secondary benefits.   Some secondary benefits are not measurable (e.g., the avoidance of pain and suffering); others can be quantified. Both types of benefits should be considered.

1. Savings of income of families and friends of accident victims who might otherwise be tending to the victims.

2. Avoidance of air and noise pollution and savings on fuel costs.

3. Savings on enforcement and administrative costs related to the investigation of accidents.

4. Avoidance of pain and suffering.

- **Users' Disbenefits**

  1. Cost of spending time to have a vehicle inspected (including travel time), as opposed to devoting that time to an alternative endeavor (opportunity cost).

  2. Cost of inspection fees.

  3. Cost of repairs that would not have been made if the inspection had not been performed.

  4. Value of time expended in repairing the vehicle (including travel time).

  5. Cost in time and direct payment for reinspection.

- **Sponsor's Costs**

  1. Capital investments in inspection facilities.

  2. Operating and maintenance costs associated with inspection facilities. (These include all direct and indirect labor, personnel, and administrative costs.)

- **Sponsor's Revenues or Savings**

  1. Inspection fee.

Valuation of Benefits and Costs

The aim of benefit–cost analysis is to maximize the equivalent value of all benefits minus that of all costs (expressed either in present values or in equivalent annual values). This objective is in line with promoting the economic welfare of citizens. In general, the benefits of public projects are difficult to measure, whereas the costs are more easily determined. For simplicity, we will attempt only to quantify the primary users' benefits and sponsor's costs on an annual basis.

- **Calculation of Primary Users' Benefits**

  1. **Benefits due to the reduction of deaths.**   The equivalent value of the average income stream lost by victims of fatal accidents[8] was estimated at \$6,000,000 per death victim and \$1,314,665 per severe injuries, in 2010 dollars. The state estimated that the inspection program would reduce the number of annual fatal accidents by 127 in deaths and 114.3 in severe injuries, resulting in a potential savings of

  $$\text{Deaths} - (127)(\$5,800,000) = \$736,600,000$$

  $$\text{Severe injuries} - (114.30)(\$1,314,665) = \$150,266,210$$

---

[8] These estimates were based on the total average income that these victims could have generated had they lived. The average value on human life was calculated by considering several factors, such as age, sex, and income group.

2. **Benefits due to the reduction of damage to property.** The average cost of damage to property per accident was estimated at $2,784. This figure includes the cost of repairs for damages to the vehicle, the cost of insurance, the cost of legal and court administration, the cost of police accident investigation, and the cost of traffic delay due to accidents. The state had 121,312 reported crashes in 2010 and estimated that accidents would be reduced by 23,450 per year, and that about 63% of all accidents would result in damage to property only. Therefore, the estimated annual value of benefits due to a reduction in property damage was estimated at

$$\$2,784(23,450)(0.63) = \$41,129,424$$

The overall annual primary benefits are estimated as

Value of reduction in fatalities	
Deaths	$736,600,000
Severe Injuries	$150,266,210
Value of reduction in property damage	41,129,424
Total	**$927,995,634**

- **Calculation of Primary Users' Disbenefits**
  1. **Opportunity cost associated with time spent bringing vehicles in for inspection.** This cost is estimated as

$$C_1 = \text{(number of cars inspected)}$$
$$\times \text{(average duration involved in travel)}$$
$$\times \text{(average wage rate)}$$

With an estimated average duration of 1.02 travel-time hours per car, an average wage rate of $21.77 per hour, and 10.9 million inspected cars per year, we obtain

$$C_1 = 10,900,000(1.02)(\$21.77)$$
$$= \$242,038,860$$

  2. **Cost of inspection fee.** This cost may be calculated as

$$C_2 = \text{(inspection fee)} \times \text{(number of cars inspected)}$$

Assuming that an inspection fee of $23 is to be paid for each car, the total annual inspection cost is estimated as

$$C_2 = (\$23)(10,900,000)$$
$$= \$250,700,000$$

  3. **Opportunity cost associated with time spent waiting during the inspection process.** This cost may be calculated by the formula

$$C_3 = \text{(average waiting time in hours)}$$
$$\times \text{(average wage rate per hour)}$$
$$\times \text{(number of cars inspected)}$$

With an average waiting time of 33 minutes (or 0.5618 hours),

$$C_3 = (0.5618)(\$21.77)(10,900,000) = \$133,307,000$$

4. Vehicle usage costs for the inspection process. These costs are estimated as

$$C_4 = \text{(number of inspected cars)}$$
$$\times \text{(vehicle operating cost per mile)}$$
$$\times \text{(average round-trip miles to inspection station)}$$

Assuming a $0.55 operating cost per mile and 20 round-trip miles, we obtain

$$C_4 = 10,900,000(\$0.55)(20) = \$119,900,000$$

The overall primary annual disbenefits are estimated as

Item	Amount
C_1	$242,038,860
C_2	250,700,000
C_3	133,307,000
C_4	119,900,000
Total disbenefits	**$745,945,860**
or	**$68 per vehicle**

- **Calculation of Primary Sponsor's Costs**
  PennDOT does not own the inspection facilities, so there is no direct capital expenditure for these needs. However, the agency is responsible for regulating the inspection stations and certifying the inspectors, so annual operating expenditure should be a part of the primary sponsor's cost. Here we assume an annual operating expenditure of $25 million.
- **Calculation of Primary Sponsor's Revenues**
  If PennDOT runs the inspection facilities, the sponsor's costs can be offset to a large degree by annual inspection revenues, which must be subtracted to avoid double counting. In this case, annual inspection revenues are the same as the direct cost of inspection incurred by the users (C_2), which was calculated as $250,700,000. However, the PennDOT certified inspection stations collect the inspection fees, so there is no revenue to offset the sponsor's cost.

Reaching a Final Decision

Finally, a discount rate of 6% was deemed appropriate, because the state of Pennsylvania finances most state projects by issuing a 6% long-term tax-exempt bond. The streams of costs and benefits are already discounted in order to obtain their present and annual equivalent values.

From the state's estimates, the primary benefits of inspection are valued at $927,995,634, which are compared with the primary disbenefits of inspection, which total $745,945,860. Therefore, the users' net benefits are

$$\text{Users' net benefits} = \$927,995,634 - \$745,945,860$$
$$= \$182,049,774$$

The sponsor's net costs are

$$\text{Sponsor's net costs} = \$25,000,000$$

Since all benefits and costs are expressed in annual equivalents, we can use these values directly to compute the degree of benefits that exceeds the sponsor's costs:

$$\$182,049,774 - \$25,000,000 = \$157,049,774 \text{ per year}$$

This positive AE amount indicates that, if all estimates of the benefits and costs were reasonable, the Pennsylvania inspection system would be economically justifiable. We can assume the AE amount would have been even greater had we also factored in secondary benefits such as the potential savings from reduction in minor and moderate injuries from crash accidents. Also, for simplicity, we have not explicitly considered vehicle growth in the state of Pennsylvania. For a complete analysis, this growth factor must be considered to account for all related benefits and costs in equivalence calculations.

16.2.5 Difficulties Inherent in Public Project Analysis

As we observed in the motor-vehicle inspection program in the previous section, public benefits are very difficult to quantify in a convincing manner. For example, consider the valuation of a saved human life in any category. Conceptually, the total benefit associated with saving a human life may include the avoidance of the insurance administration costs as well as legal and court costs. In addition, the average potential income lost because of premature death (taking into account age and sex) must be included. Obviously, the difficulties associated with any attempt to put precise numbers on human life are insurmountable.

Consider the following example. A few years ago, a 50-year-old business executive was killed in a plane accident. The investigation indicated that the plane was not properly maintained according to federal guidelines. The executive's family sued the airline, and the court eventually ordered the airline to pay $5,250,000 to the victim's family. The judge calculated the value of the lost human life assuming that if the executive had lived and worked in the same capacity until his retirement, his remaining lifetime earnings would have been equivalent to $5,250,000 at the time of award. This is an example of how an individual human life was assigned a dollar value, but clearly any attempt to establish an average amount that represents the general population is controversial. We might even take exception to this individual case: Does the executive's salary adequately represent his worth to his family? Should we also assign a dollar value to their emotional attachment to him, and if so, how much?

Now consider a situation in which a local government is planning to widen a typical municipal highway to relieve chronic traffic congestion. Knowing that the project will be financed by local and state taxes and that many out-of-state travelers also are expected to benefit, should the planner justify the project solely on the benefits to local residents? Which point of view should we take in measuring the benefits—the municipal level, the state level, or both? It is important that any benefit measure be performed from the appropriate *point of view*.

In addition to valuation and point-of-view issues, many possibilities for tampering with the results of benefit–cost analyses exist. Unlike private projects, many public projects are undertaken because of political pressure rather than on the basis of their economic benefits alone. In particular, whenever the benefit–cost ratio becomes marginal or less than unity, a potential to inflate the benefit figures to make the project look good exists.

16.3 Benefit-Cost Ratios

An alternative way of expressing the worthiness of a public project is to compare the user's benefits (B) with the sponsor's cost (C) by taking the ratio B/C. In this section, we shall define the benefit–cost (B/C) ratio and explain the relationship between it and the conventional NPW criterion.

16.3.1 Definition of Benefit-Cost Ratio

For a given benefit–cost profile, let B and C be the present values of benefits and costs defined respectively by

$$B = \sum_{n=0}^{N} b_n (1 + i)^{-n} \tag{16.1}$$

and

$$C = \sum_{n=0}^{N} c_n (1 + i)^{-n} \tag{16.2}$$

where

b_n = benefit at the end of period n, $b_n \geq 0$

c_n = expense at the end of period n, $c_n \geq 0$

$A_n = b_n - c_n$

N = project life

i = sponsor's interest rate (discount rate)

The sponsor's costs (C) consist of the equivalent capital expenditure (I) and the equivalent annual operating costs (C') accrued in each successive period. (Note the sign convention we use in calculating a benefit–cost ratio. Since we are using a ratio, all benefits and cost flows are expressed in positive units. Recall that in previous equivalent-worth calculations our sign convention was to explicitly assign "+" for cash inflows and "−" for cash outflows.) Let's assume that a series of initial investments is required during the first K periods, while annual operating and maintenance costs accrue in each subsequent period. Then the equivalent present value for each component is

$$I = \sum_{n=0}^{K} c_n (1 + i)^{-n} \tag{16.3}$$

$$C' = \sum_{n=K+1}^{N} c_n (1 + i)^{-n} \tag{16.4}$$

and

$$C = I + C'$$

The B/C ratio is defined as

$$\text{BC}(i) = \frac{B}{C} = \frac{B}{I + C'}, \quad I + C' > 0 \tag{16.5}$$

If we are to accept a project, $\text{BC}(i)$ must be greater than unity.

Note that we must express the values of B, C', and I in present-worth equivalents. Alternatively, we can compute these values in terms of annual equivalents and use them in calculating the B/C ratio. The resulting B/C ratio is not affected.

EXAMPLE 16.1 Benefit–Cost Ratio for the Indian River Lagoon–South (IRL-S) Project[9]

Martin and St. Lucie counties—Florida's "Treasure Coast"—encompass one of the state's most productive and most threatened estuarine treasures, the Indian River Lagoon and St. Lucie Estuary. A huge effort, called the Indian River Lagoon–South (IRL-S) Restoration Project, is underway to reverse the damaging effects of pollution and unnaturally large fresh-water discharges into these ecologically vital water bodies. The purpose of the project is

- To restore natural water flow to the Florida Everglades.
- To enhance water supply to South Florida region.
- To provide continuous flood protection.

Figure 16.4 Indian River Lagoon–South, Florida.

[9] Indian River Lagoon–South, Facts & Information, http://www.evergladesplan.org/index.aspx.

After a series of feasibility studies and public hearings, the project was authorized in the Water Resources Development Act (WRDA) of 2007 with a price tag of $1.2 billion and a project completion set for 2015. The U.S. Army Corps of Engineers estimated the costs and benefits of the project as given in Figure 16.5. Some of the financial highlights for the restoration project is as follows:

- Estimated initial cost for construction: $1,207,288,000
- Investment cost during construction: $103,987,398
- Annual recurring operating and maintenance, repair costs: $6,144,700
- Estimated annual benefits: $159,000,000
- Discount rate used: 5 5/8%
- Project life: 39 years

The U.S. Army Corps of Engineers further detailed the costs and benefits of the project as given in Figure 16.5. Compute the benefit–cost ratio for the project based on these estimates.

SOLUTION

Given: Project's benefits and cost series as shown in Figure 16.5, $i = 5\ 5/8\%$, $N = 39$, and $K = 5$.

Find: B, C, I, C', and B/C ratio.

$$B = \$159(P/A, 5.625\%, 25)(P/F, 5.625\%, 5)$$
$$= \$1,602.65$$
$$C' = \$6.144(P/A, 5.625\%, 25)(P/F, 5.625\%, 5)$$
$$= \$61.93$$

Figure 16.5 Project cash flows related to IRL-S Project.

$$I = \$12.155 + \$345(P/F, 5.625\%, 1) + \$165(P/F, 5.625\%, 2)$$

$$+ \cdots + \$15.133(P/F, 5.625\%, 5)$$

$$= \$1,046.54$$

$$C = I + C'$$

$$= \$1,046.54 + \$61.93$$

$$= \$1,108.47$$

Using Eq. (16.5), we can compute the B/C ratio as

$$BC(5.625\%) = \frac{\$1,602.65}{\$1,046.54 + \$61.93}$$

$$= 1.45$$

The B/C ratio exceeds unity, so the users' benefits exceed the sponsor's costs.

COMMENTS: Home to more than 4,300 species of plants and animals and supporting an annual economic contribution of more than \$730 million, the lagoon region will benefit from careful restoration and protection of these water bodies. On what basis did the U.S. Army Corps of Engineers estimate annual benefits of \$159 million to justify the IRL-S Project? Some of the specific benefit features are as follows:

- Annual benefits of \$159 million after project is completed.
- 12,000 acres of aboveground storage.
- 9,000 acres of manmade wetlands.
- 90,000 acres of natural areas, including 53,000 acres of restored wetlands, providing additional water storage.
- 90 acres of artificial submerged habitat created for aquatic vegetation.
- 922 acres of submerged aquatic vegetation restored.
- 9 million cubic yards of removed muck.
- 41% long-term reduction in phosphorus.
- 26% long-term reduction in nitrogen.
- 2,650 acres of benthic habitat created in St. Lucie River and Estuary.
- 889 acres of restored oyster habitat.
- \$6.1 million in improved agricultural productivity through improved freshwater supplies.

16.3.2 Profitability Index (Net B/C Ratio)

A variation to the benefit-cost ratio called the profitability index (commonly called the **net B/C ratio**) considers only the initial capital expenditure as a cash outlay, and annual net benefits are used:

$$PI(i) = \frac{B - C'}{I} = \frac{B'}{I} > 1, \quad I > 0 \tag{16.6}$$

The decision rule has not changed—the ratio must still be greater than one. It can be easily shown that a project with $BC(i) > 1$ will always have $PI(i) > 1$ as long as both C and I are > 0, as they must represent the inequalities in the decision rules to maintain the stated senses. The magnitude of $BC(i)$ generally will be different than that for $PI(i)$, but the magnitudes are irrelevant for making decisions. All that matters is whether the ratio exceeds the threshold value of one. However, some analysts prefer to use $PI(i)$, because it indicates the net benefit (B') expected per dollar invested.

EXAMPLE 16.2 **Profitability Index for the IRL-S Project**

Reconsider Example 16.1. Compute the profitability index for the IRL-S project at $i = 5.625\%$.

SOLUTION

Given: Project's benefits and cost series as shown in Figure 16.5, $i = 5\ 5/8\%$, $N = 39$, and $K = 5$.
Find: $PI(i)$

$$B = \$1,602.65$$
$$C' = \$61.93$$
$$I = \$1,046.54$$

and

$$PI(i) = \frac{B - C'}{I}$$

$$= \frac{\$1,602.65 - \$61.93}{\$1,046.54}$$

$$= 1.47 > 1$$

Basically, the IRL-S project is expected to generate $1.47 benefits per $1 invested. Therefore the project can be justified, which is the same conclusion reached under the benefit–cost ratio analysis.

16.3.2 Relationship among B/C Ratio, Profitability Index, and NPW

The *B/C* ratio yields the same decision for a project as does the NPW criterion. Recall that the $BC(i)$ criterion for project acceptance can be stated as

$$\frac{B}{I + C'} > 1 \tag{16.7}$$

If we multiply both sides of the equation by the term $(I + C')$ and transpose $(I + C')$ to the left-hand side, we have

$$B > (I + C')$$

$$B - (I + C') > 0$$

$$\mathrm{PW}(i) = B - C > 0 \qquad (16.8)$$

which is the same decision rule[10] as that which accepts a project by the NPW criterion. The same results can be obtained for the profitability index. This implies that we could use either the benefit–cost ratio or the profitability index in evaluating private projects instead of using the NPW criterion. Or we could use the NPW criterion in evaluating public projects. Either approach will signal consistent project selection. Recall that, in Example 16.1,

$$\mathrm{PW}(5.625\%) = B - C = \$1,602.65 - \$1,108.47 = \$494.18 > 0$$

Clearly, the project would thus be acceptable under the NPW criterion.

16.3.3 Comparing Mutually Exclusive Alternatives: Incremental Analysis

Let us now consider how we choose among mutually exclusive public projects. As we explained in Chapter 7, we must use the incremental investment approach in comparing alternatives based on any relative measure, such as IRR, BC(i), or PI(i).

To apply incremental analysis, we compute the incremental differences for each term (B, I, and C'), and take the B/C ratio on the basis of these differences. To use BC(i) on incremental investment, we may proceed as follows.

Step 1. If one or more alternatives have B/C ratios greater than unity, eliminate any alternatives with a B/C ratio less than that.

Step 2. Arrange the remaining alternatives in increasing order of the denominator $(I + C')$. Thus, the alternative with the smallest denominator should be the first (j), the alternative with the second smallest denominator should be second (k), and so forth.

Step 3. Compute the incremental differences for each term (B, I, and C') for the paired alternatives (j, k) in the list:

$$\Delta B = B_k - B_j$$

$$\Delta I = I_k - I_j$$

$$\Delta C' = C'_k - C'_j$$

Step 4. Compute the BC(i) on incremental investment by evaluating

$$\mathrm{BC}(i)_{k-j} = \frac{\Delta B}{\Delta I + \Delta C'} \qquad (16.9)$$

If $\mathrm{BC}(i)_{k-j} > 1$, select an alternative k. Otherwise, select an alternative j.

---

[10] We can easily verify a similar relationship between the net B/C ratio and the NPW criterion.

Step 5. Compare the alternative selected with the next one on the list by computing the incremental benefit–cost ratio.[11] Continue the process until you reach the bottom of the list. The alternative selected during the last pairing is the best one.

We may modify this decision procedure when we encounter the following situations:

- If $\Delta I + \Delta C' = 0$, we cannot use the benefit–cost ratio, because the equation implies that both alternatives require the same initial investment and operating expenditure. When this happens, we simply select the alternative with the largest B value.
- In situations where public projects with unequal service lives are to be compared but the projects can be repeated, we may compute all component values (B, C', and I) on an annual basis and use them in incremental analysis.

EXAMPLE 16.3 Incremental Benefit–Cost Ratios

Consider four design alternatives in terms of length of the light-rail line: A1, A2, A3, and A4. Each design alternative has the same service life of 30 years with negligible salvage value. The anticipated annual benefits from travel-time savings, accident prevention, emission reduction, and costs from operating and maintaining the rail system (along with the initial investment costs) are as follows (all units in million $).

Design Options	Investment Cost	Annual Benefits	Annual O&M cost
A1	$ 700	$ 80	$ 20
A2	$ 950	$ 120	$ 30
A3	$ 1,300	$ 160	$ 40
A4	$ 2,000	$ 210	$ 50

Knowing that these are mutually exclusive alternatives, which would be the best alternative at a 5% interest rate? Show the sequence of calculations that would be required to produce the correct results using the B/C ratio on incremental investment.

SOLUTION
Given: I, B, C', and $i = 5\%. N = 30$ years.
Find: Which design option to choose?

---

[11] If we use the PI(i) as a basis, we need to order the alternatives in increasing order of I and compute the PI on the incremental investment.

Calculation of B/C ratios: First, we need to calculate I, B, and C' for each alternative at 5% interest over 30 years. They are summarized as

Design Options	I	B	C'	BC(5%)
A1	$700	$1,229.80	$307.45	1.22
A2	$950	$1,844.69	$461.17	1.31
A3	$1,300	$2,459.59	$614.90	1.28
A4	$2,000	$3,228.21	$768.62	1.17

Since the BC(i) values for each project are greater than unity, all of the design options are acceptable on an individual basis.

Incremental analysis: With the projects being mutually exclusive, we must use the principle of incremental analysis. Obviously, if we attempt to rank the projects according to the size of the B/C ratio, we will observe a different project preference. For example, if we use the BC(i) on the total investment, we see that A2 appears to be the most desirable and A4 the least desirable, but this proves to be wrong. By computing the incremental B/C ratios, we will select a project that is consistent with the NPW criterion.

- **A1 versus A2.** With the do-nothing alternative, we first drop from consideration any project that has a B/C ratio smaller than unity, so the first incremental comparison is between A1 and A2:

$$\text{BC(5\%)}_{2-1} = \frac{\$1,844.69 - \$1,229.80}{(\$950 - \$700) + (\$461.17 - \$307.45)}$$

$$= 1.52 > 1$$

Since the ratio is greater than unity, we prefer A2 to A1. Therefore, A2 becomes the "current best" alternative.

- **A3 versus A2.** Next, we must determine whether the incremental benefits to be realized from A3 would justify the additional expenditure. Therefore, we need to compare A2 and A3 as

$$\text{BC(5\%)}_{3-2} = \frac{\$2,459.59 - \$1,844.69}{(\$1,300 - \$950) + (\$614.90 - \$461.17)}$$

$$= 1.22 > 1$$

The incremental B/C ratio again exceeds unity; therefore, we prefer A3 over A2.

- **A4 versus A3.** Next, comparing A3 with A4,

$$\text{BC(5\%)}_{4-3} = \frac{\$3,228.21 - \$2,459.59}{(\$2,000 - \$1,300) + (\$768.62 - \$614.90)}$$

$$= 0.90 < 1$$

The incremental *B/C* ratio is less than unity; therefore, we prefer A3 over A4. With no further projects to consider, A3 becomes our final choice. Figure 16.6 graphically illustrates elimination of inferior alternatives.

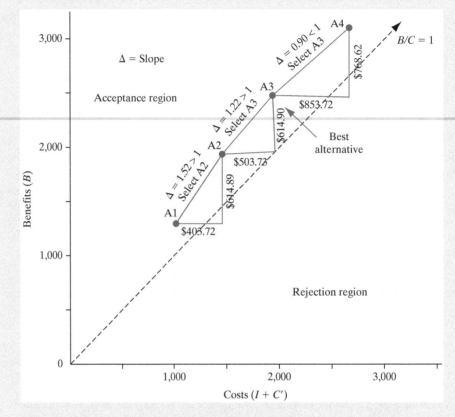

Figure 16.6 Incremental benefit–cost analyses. The slope of 1.52 for line A2–A1 indicates a $\Delta B/\Delta C$ ratio > 1. This is also true for line A3–A2, but not for line A4–A3, indicating A3 will be the ultimate choice.

EXAMPLE 16.4 Incremental Profitability Indices

Reconsider Example 16.3. Select the best alternative based on the profitability index criterion.

SOLUTION

Given: *I*, *B*, *C'*, and $i = 5\%$. $N = 30$ years.

Find: Which design option to choose based on PI?

By comparing the first pair of projects on this list, we obtain

- **A1 versus A2.** With the do-nothing alternative, we first drop from consideration any project that has a PI smaller than unity, so the first incremental comparison is between A1 and A2:

$$\text{PI}(5\%)_{2-1} = \frac{(\$1,844.69 - \$1,229.80) - (\$461.17 - \$307.45)}{(\$950 - \$700)}$$

$$= 1.84 > 1 \rightarrow \text{select A2}.$$

Since the ratio is greater than unity, we prefer A2 to A1. Therefore, A2 becomes the "current best" alternative.

- **A3 versus A2:** Next, a comparison of A2 and A3 yields a preference of A3.

$$\text{PI}(5\%)_{3-2} = \frac{(\$2,459.59 - \$1,844.69) - (\$614.90 - \$461.17)}{(\$1,300 - \$950)}$$

$$= 1.32 > 1 \rightarrow \text{select A3}.$$

- **A4 versus A3:** Finally, you will find a comparison between A3 and A4 results in favoring A3.

$$\text{PI}(5\%)_{4-3} = \frac{(\$3,228.21 - \$2,459.59) - (\$768.62 - \$614.90)}{(\$2,000 - \$1,300)}$$

$$== 0.88 < 1 \rightarrow \text{select A3}.$$

Therefore, A3 becomes the best choice by the PI criterion.

16.4 Analysis of Public Projects Based on Cost-Effectiveness

In evaluating public investment projects, we may encounter situations where competing alternatives have the same goals, but the effectiveness with which those goals can be met may or may not be measurable in dollars. In these situations, we compare decision alternatives directly on the basis of their **cost effectiveness**. Here, we judge the effectiveness of an alternative in dollars or some nonmonetary measure by the extent to which that alternative, if implemented, will attain the desired objective. The preferred alternative is then either the one that produces the maximum effectiveness for a given level of cost or the one that produces the minimum cost for a fixed level of effectiveness.

16.4.1 Cost-Effectiveness Studies in the Public Sector

A typical cost-effectiveness analysis procedure in the public sector involves the following steps.

Step 1. Establish the goals to be achieved by the analysis.

Step 2. Identify the restrictions imposed on achieving the goals, such as those having to do with the budget or with weight.

Step 3. Identify all the feasible alternatives for achieving the goals.

Step 4. Identify the social interest rate to use in the analysis.

Step 5. Determine the equivalent life-cycle cost of each alternative, including R&D costs, testing costs, capital investment, annual operating and maintenance costs, and the salvage value of the item under consideration.

Step 6. Determine the basis for developing the cost-effectiveness index. Two approaches may be used (Figure 16.7): (1) the fixed-cost approach and (2) the fixed-effectiveness approach. If the fixed-cost approach is used, determine the amount of effectiveness obtained at a given cost. If the fixed-effectiveness approach is used, determine the cost required to obtain the predetermined level of effectiveness.

Step 7. Compute the cost-effectiveness ratio for each alternative based on the criterion selected in step 6.

Step 8. Select the alternative with the maximum cost-effective index.

When either the cost or the level of effectiveness is clearly stated in achieving the declared program objective, most cost-effectiveness studies will be relatively straightforward. If that is not the case, however, the decision maker must come up with his or her own program objective by fixing either the cost required in the program or the level of effectiveness to be achieved. The next section shows how such a problem can easily evolve in many public projects or military applications.

16.4.2 A Cost-Effectiveness Case Study[12]

To illustrate the procedures involved in cost-effectiveness analysis, we shall present an example of how the U.S. Army selected the most cost-effective program for providing on-time delivery of time-sensitive, high-priority cargos and key personnel to the battle staging grounds.

Figure 16.7 Types of Cost-Effectiveness Analysis.

[12] This case study is provided by Dr. George C. Prueitt of CAS, Inc.

Statement of the Problem

The U.S. Army has been engaged in recent conflicts that have transformed from the traditional set-piece battles into operations in distributed locations that require rapid sequences of activities. To support success in these new operations, the Army needs timely delivery of cargos and personnel in the distributed locations. Consequently, the dispersion of forces has created a condition of unsecured landlines of communication (LOC), and aerial delivery is considered as one of the practical solutions to the problem. The time-sensitive nature of these priority cargos makes it most preferable to have direct delivery from intermediate staging bases (ISB) to the forward brigade combat teams (BCT).

The transportation network typically involves a long-haul move (a fixed-wing aerial movement) and a short-haul distribution (by helicopter or truck). The long haul usually stops at the closest supporting airfield, but with the appropriate asset, a single, direct movement could be made. The long-haul portion has been the bigger problem—trucks take too long and are vulnerable to hostilities, the U.S. Air Force (USAF) assets are not always available or appropriate, and existing helicopters do not have the necessary range to fly. Furthermore, the use of the larger USAF aircraft results in inefficient loads, as the high-priority cargos use only a small fraction of the capacity of the aircraft. The Army is investigating how to best correct the problem. They are considering three options to handle the long haul: (1) Use the U.S. Air Force (C-130J) assets; (2) procure a commercially available future cargo aircraft (FCA); or (3) use additional helicopters (CH-47).

Defining the Goals

The solution will be a "best value" selection, providing the best capability for the investment required. The selection process must address three operational perspectives.

1. **A "micro" examination:** For a given specific battlefield arrangement (available transportation network, supported organizations and locations), the decision metric is the time required to delivery specific critical cargos and the solutions' comparative operating costs. This metric is a suitability measure to identify the system that best and most quickly meets the timely delivery requirement.

2. **A "macro" examination:** Given a broader, more prolonged theater support requirement, the decision metric becomes, "What percentage of critical cargos can be delivered?" and "What is the cost of procuring all of the transportation assets needed to create the delivery network combinations (fixed-wing airplanes, helicopters, and trucks) to meet those delivery percentages?" This metric is a feasibility measure to determine the number of assets needed to meet the quantities of cargos to be delivered.

3. Given relatively comparable fixed-wing aircraft fleets (differing quantities for equal capability), what are the 15-year life-cycle costs associated with those alternatives? This is an affordability metric, to ensure that all life-cycle costs are properly identified and can be met within the Army budget.

Description of Alternatives

As mentioned earlier, the Army is considering three alternatives. They are:

- **Alternative 1: Use the USAF C-130J assets.** The USAF C-130J aircraft has been proposed as an alternative for the long-haul task. The C-130J is a new acquisition, and will be considered in two employment concepts: first, in a direct movement pattern;

second, as part of a standard USAF Scheduled Tactical Air Re-supply (STAR) route (analogous to a bus route). Current operations almost exclusively use the STAR route method when C-130 aircraft are employed. It is assumed that transportation planning and theater priorities would be sufficiently high to permit the C-130J to be employed in the more direct delivery method and are not limited solely to STAR route operations.

• **Alternative 2: Use future cargo aircraft.** Another long-haul alternative is to procure one of two commercially available products capable of delivering these cargos on a timely and efficient basis.

• **Alternative 3: Use additional CH-47 helicopters.** Looking first at the long-haul task, most of today's deliveries are provided by either the CH-47 helicopter or the C-130 aircraft in conjunction with another vehicle. The CH-47 is being used to perform this mission because it is the "best available" Army asset. Unfortunately, it is expensive to operate, and its range limitations make it an inefficient method of carrying out the mission. Furthermore, the long distances between the ISB to the forward units are causing the helicopters to accumulate flight hours more rapidly than planned in their service-life projections, and this has generated a significant increase in their maintenance requirements.

These long-haul delivery alternatives are described in Table 16.2.

In addition to the long-haul task, the short-haul distribution legs will usually need a complementary ground or helicopter asset to make the final delivery when the fixed-wing assets cannot land close enough to the BCT because of a lack of improved landing strips or runways of sufficient length. This creates a greater burden on the transportation network to ensure that adequate ground or rotary-wing assets are available to complete the mission. The combinations of long-haul alternatives and final delivery systems create the series of network transportation options described in Table 16.3. Figure 16.8 provides a descriptive portrayal of these transportation network options.

TABLE 16.2 Descriptions of Alternatives for Transportation Network Long-Haul Requirement

Alternative	Description	Comment
Alternative 1: C-130J	Procure additional C-130Js and maximize their use to perform the long-haul task of the intratheater delivery of critical, time-sensitive cargo and key personnel. Minimize requirement for additional CH-47s.	C-130Js will use both USAF Scheduled Tactical Air Resupply (STAR) routes and more direct routes. The C-130J has longer runway requirements than smaller FCA aircraft does.
Alternative 2: Army FCA	Procure a variant of the FCA and maximize its use to perform the long-haul task of the intratheater delivery of mission-critical, time-sensitive cargo and key personnel with minimal use of CH-47s.	In some situations, the FCA will be able to land either closer to or at the BCT locations. Two versions of the FCA (alternative 2A and alternative 2B) have been included in this examination.
Alternative 3: More CH-47s	Procure additional CH-47s to perform the long-haul task of the intratheater delivery of critical, time-sensitive cargo and key personnel.	This alternative will include those additional CH-47s needed to carry support systems for the CH-47s that are used to move cargo over extended distances.

TABLE 16.3   Descriptions of Transportation Network Options

Main Alternatives	Network Option	Description
	NO_1	ALT 1 (C-130) direct from ISB to C-130 Supportable Forward Airfield, then truck to BCT
Alternative 1	NO_2	ALT 1 (C-130) direct from ISB to C-130 Supportable Forward Airfield, then CH-47 to BCT
	NO_3	ALT 1 (C-130) STAR route from ISB to C-130 Supportable Forward Airfield (stop nearest to BCT), then truck to BCT
	NO_4	ALT 1 (C-130) STAR route from ISB to C-130 Supportable Forward Airfield (stop nearest to BCT), then CH-47 to BCT
	NO_5	ALT 2A direct from ISB to FCA Supportable Forward Airfield, then truck to BCT
	NO_6	ALT 2A direct from ISB to FCA Supportable Forward Airfield, then CH-47 to BCT
Alternative 2	NO_7	ALT 2A direct from ISB to BCT
	NO_8	ALT 2B direct from ISB to FCA Supportable Forward Airfield, then truck to BCT
	NO_9	ALT 2B direct from ISB to FCA Supportable Forward Airfield, then CH-47 to BCT
	NO_{10}	ALT 2B direct from ISB to BCT
Alternative 3	NO_{11}	ALT 3 (CH-47) direct from ISB to BCT, with multiple refueling stops en route

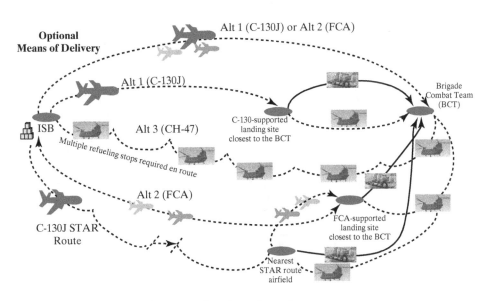

Figure 16.8   Transportation-network option schematics.

Micro Analysis

To perform a cost-effectiveness analysis, the micro comparisons were made by selecting the delivery time as the performance metric (vertical axis in Figure 16.9) and one-way operating cost as the other metric (horizontal axis). Lower cost and shorter delivery times cause the direction of preference to be in the lower left-hand corner of the figure. The respective operating costs (per flight hour) for each system were: C-130J at $3,850, FCA 2A at $2,800, FCA 2B at $1,680, and CH-47 at $4,882.

Transportation network option 10 (NO_{10}) based on alternative 2B FCA and NO_7 based on alternative 2A FCA—each flying directly from ISB to BCT—had the shortest times to accomplish the mission. Their respective operating costs pushed them into the most favored regions on the graph.

Not surprisingly, the next quickest option was NO_2 based on alternative 1 aircraft flying directly to a C-130 supportable forward airfield with a CH-47 delivering for the last leg; however, the higher operating costs of the C-130J (over the FCA alternatives) and the CH-47 made its operating cost more than double NO_{10}.

For the remaining options, NO_3 and NO_4 (using the C-130J on a STAR route) required the longest delivery time, although the operating costs were comparable to the most favored solutions. NO_{11} which is the CH-47–based option, had the highest operating costs. Figure 16.9 summarizes the results for all 11 transportation-network options.

Macro Analysis

The macro analysis was performed with the criterion of capability to deliver high-priority cargos in two simultaneous combat operations over a 30-day period. The minimum acceptable on-time delivery rate is 80% of all required shipments. The requirements were derived from one operation with 10 days of high-intensity combat followed by 20 days of stability operations and a second operation consisting of 30 days of low-intensity stability operations. Current transportation-asset allocations were used to establish a "baseline" capability.

Figure 16.9 Assessment results from micro cost-effectiveness analysis.

The baseline included 28 CH-47 helicopters and 10 C-130J aircraft for this specific mission and additional quantities of Army systems (other aircraft and trucks) to do the final cargo distributions. These assets were capable of satisfying about 12% of the total on-time delivery requirement. The shortfall between 12% and the 80% standard represents the capability "gap" that additional assets are required to fill.

To complete the macro analysis, additional quantities of alternatives 1, 2A, 2B, and 3 were incrementally added, and the resultant increase in on-time deliveries was computed. The cost to procure the additional quantities was determined with the following unit prices: alternative 1 at $75.6M each, alternative 2A at $34.5M, alternative 2B at $29.2M, and alternative 3 at $26.1M. Figure 16.10 shows the changes in percent-on-time delivery versus the procurement cost of the respective alternatives. There is a significant procurement cost differentially between alternatives, with alternatives 2A and 2B being the most cost effective.

Life-Cycle Cost for Each Alternative

The third part of the case study is to perform the life-cycle cost (LCC) analysis for the various alternatives. Table 16.4 lists the cost items that are considered in this LCC analysis. In fact, these cost items represent the common decision elements mandated by the U.S. Department of Defense, along with appropriate rules and assumptions, to ensure that each alternative is evaluated for comparable cost and affordability. First, an LCC estimate is developed for specific quantities of each primary long-haul aircraft in the alternatives. The number of aircraft is chosen to establish equal fleet capabilities (not necessarily fleet size). Second, to determine the number of aircraft required for each alternative, the U.S. Army uses the equal-effectiveness standard of 80% on-time delivery of high-priority cargo and key personnel. Specifically,

- Alternatives 2A and 2B use a fleet size of 56 as derived from the macro analysis and with additional fleet-support considerations.

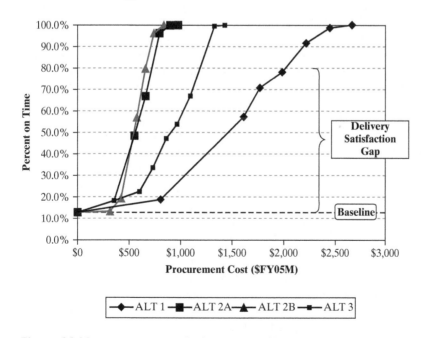

Figure 16.10 Assessment results from macro-effectiveness analysis.

TABLE 16.4 Definitions of Life-Cycle Cost Variables

Life-Cycle Cost Element	Cost Variable	Description
1	RDT&E	Government and contractor costs associated with the research, development, test, and evaluation (RDT&E) of the material system.
2	Procurement	Costs resulting from the production and initial fielding of the material system in the Army's operational inventory.
3	Military construction	Cost of all military construction projects associated with a material system. No military construction money is anticipated to be needed for any of the FCA alternatives.
4	Military personnel	Cost of military personnel associated with the development and operation of the material system.
5	O&M	O&M funded cost associated with development, production, fielding, operation, and support of the material system for 20 years.

- Alternative 1, the USAF C-130J, uses 40 aircraft, because it was assumed that all additional fleet-support requirements would be addressed with the USAF larger program. Similarly,
- Alternative 3 uses 92 CH-47F as its mission requirement with no additional fleet support requirements, as the 92 would be added to the existing fleet of about 400 CH-47 aircraft.

Cost estimates were developed for the respective alternatives. In a simple form, these estimates provide the future estimated costs for procurement and operations. The procurement costs include both RDT&E and procurement requirements. The operating cost includes operations and maintenance (O&M) and military personnel costs. There were some nominal military construction costs, but they were equal for all alternatives.

A 15-year operating cycle (from first procurement through last operation) is assumed in developing these cost estimates. The projected costs for future years are provided in Table 16.5.

For most of the procurements by the U.S. Department of Defense, future inflation factors must be specified in the analysis. For this case study, an annual inflation rate of 2% is used. In terms of interest rate, an inflation-adjusted discount rate of 8% was used to calculate the LCC for each alternative. The sample LCC calculation for ALT1 in Table 16.5 is

$$\text{PW}(8\%)_{\text{ALT 1}} = (\$702.8 + \$25.0) + (\$796.5 + 51.9)(P/F, 8\%, 1)$$

$$+ \cdots + \$10.6(P/F, 8\%, 15)$$

$$= \$3,726.50$$

Figure 16.11 illustrates the cost component of each alternative's LCC. The figure shows that alternatives 2A and 2B, with 56 aircraft, have both lower procurement and operations and maintenance costs than either alternative 1, which is the C-130J alternative with its quantity of 40 aircraft, or alternative 3, which is the CH-47F alternative based on 92 helicopters. As anticipated earlier, the CH-47F alternative turns out to be the most expensive option due to its higher operating costs, even though its procurement of

TABLE 16.5 Life-Cycle Cost Estimates for Four Alternatives

	A	B	C	D	E	F	G	H	I
1									
2		ALT1		ALT 2A		ALT 2B		ALT 3	
3	Year	Proc	O&M	Proc	O&M	Proc	O&M	Proc	O&M
4									
5	0	$ 702.80	$ 25.00	$ 257.50	$ 4.80	$ 221.10	$ 3.30	$ 620.20	$ 10.80
6	1	$ 860.27	$ 56.10	$ 529.90	$ 14.70	$ 450.30	$ 12.50	$ 654.40	$ 82.00
7	2	$ 877.47	$ 88.40	$ 800.80	$ 43.00	$ 687.20	$ 35.00	$ 451.40	$ 163.20
8	3	$ 447.51	$ 122.00	$ 431.30	$ 64.00	$ 376.10	$ 52.60	$ 145.20	$ 224.10
9	4		$ 151.60	$ 206.10	$ 105.10	$ 189.20	$ 80.70	$ 351.00	$ 251.40
10	5		$ 154.70		$ 114.30		$ 80.50	$ 262.60	$ 258.80
11	6		$ 157.80		$ 120.60		$ 81.50		$ 262.30
12	7		$ 160.90		$ 128.40		$ 86.80		$ 274.70
13	8		$ 164.10		$ 132.20		$ 90.50		$ 287.50
14	9		$ 167.40		$ 139.90		$ 95.60		$ 301.20
15	10		$ 170.80		$ 141.40		$ 98.30		$ 315.50
16	11		$ 174.20		$ 145.80		$ 102.50		$ 324.40
17	12		$ 145.80		$ 129.60		$ 106.50		$ 362.10
18	13		$ 110.00		$ 119.80		$ 89.60		$ 248.80
19	14		$ 72.60		$ 106.20		$ 77.60		$ 86.70
20	15		$ 33.60		$ 100.90		$ 74.20		$ 35.40
21									
22	PW(8%)	$3,726.39		$2,773.53		$2,278.84		$4,123.38	
23									
24	Note 1: All costs in million $.					=NPV(8%,F6:F9)+F5+NPV(8%,G6:G20)+G5			
25	Note 2: "Proc" means procurement cost estimate.								
26	Note 3: "O&M" includes operations, maintenance, and manpower costs.								
27									

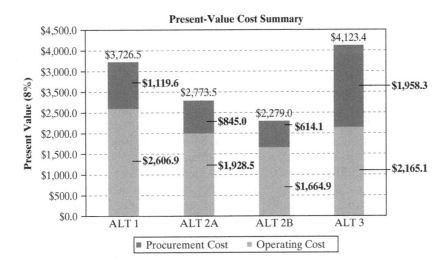

Figure 16.11 Alternatives present value (8%) for procurement and operations.

92 helicopters is less expensive than 40 C-130J aircraft. The total operating costs differ in contribution by type of aircraft. The fixed-wing aircraft estimates are based on 600 hours of flying time per year, while the helicopters are scheduled for less than 200 hours of flying time. However, because the airspeed and flying range for the helicopter are much lower than for the fixed-wing aircraft, more CH-47F helicopters are required to provide the same level of capability. Alternative 1, using the C-130J as the primary aircraft, requires fewer total aircraft than alternatives 2A and 2B, but with nearly twice the unit cost, it is a more expensive option. In conclusion, for the alternatives considered, the commercially available candidates for the FCA (alternatives 2A and 2B) provide the best opportunity for the Army to satisfy its requirement to delivery of mission-critical, time-sensitive cargo and key personnel from an ISB directly to a BCT, with alternative 2B having a slight edge over 2A for the metrics used.

16.5 Economic Analysis in Health-Care Service

The health-care service industry alone constitutes 14% to 15% of GDP when all of its dimensions are included. Clearly, health care is one of the largest, most research-intensive service industries in the United States. Accordingly, its medical knowledge base is expanding at a staggering pace, and many Americans enjoy unparalleled advancements in medical science and technology. At the same time, the nation's health-care system consistently fails to deliver high-quality care: Variation in both access to and delivery of care is considerable, errors are widespread, costs are spiraling, and few resources are devoted to optimizing system operations and improving the delivery of care.

16.5.1 Economic Evaluation Tools

Economic evaluation can be used to inform decision making and can provide information to assist in answering the following questions:

- What services do we provide (or improve), when, and at what level?
- How do we provide such services?
- Where do we provide the services?
- What are the costs associated with providing or improving the services?

The following three methods of economic evaluation are related to health service.

- **Cost-effectiveness analysis.** This technique is used in health economics to compare the financial costs of therapies whose outcomes can be measured purely in terms of their health effects (e.g., years of life saved or ulcers healed). Cost-effectiveness analysis (CEA) is the most commonly applied form of economic analysis in health economics. However, it does not allow comparisons to be made between courses of action that have different health effects.
- **Cost-utility analysis.** This technique is similar to CEA in that there is a defined outcome and the cost to achieve that outcome is measured in money. The outcome is measured in terms of survival and quality of life (for example, quality-adjusted life years—QALYs). CEA can indicate which one of a number of alternative interventions represents the best value for money, but it is not as useful when comparisons need to be made across different areas of health care, since the outcome measures used may

be very different. As long as the outcome measure is life years saved or gained, a comparison can still be made, but even in such situations, CEA remains insensitive to the *quality-of-life* dimension. In order to know which areas of health care are likely to provide the greatest benefit in improving health status, a cost-utility analysis[13] needs to be undertaken using a common currency for measuring the outcomes across health-care areas.

- **Cost–benefit analysis.** If information is needed as to which interventions will result in overall resource savings, a cost–benefit analysis (CBA) has to be performed, although, like a cost–utility analysis, a cost–benefit analysis has its own drawbacks. In CBA, the benefit is measured as the associated economic benefit of an intervention; hence, both costs and benefits are expressed in money, and the CBA may ignore many intangible (but very important) benefits that are difficult to measure in monetary terms (e.g., relief of pain and anxiety). Even though the virtue of this analysis is that it enables comparisons to be made between schemes in very different areas of health care, the approach is not widely accepted for use in health economics.

16.5.2 Cost–Effectiveness Analysis in the Health-Care Sector

In cost–effectiveness analysis (CEA), outcomes are reported in a single unit of measurement. CEA compares the costs and health effects of an intervention to assess whether it is worth doing from an economic perspective. First of all, CEA is a specific type of economic analysis in which all costs are related to a single, common effect. Decision makers can use it to compare different resource allocation options in like terms. A general misconception is that CEA is merely a means of finding the least expensive alternative or getting the "most bang for the buck." In reality, CEA is a comparison tool; it will not always indicate a clear choice, but it will evaluate options quantitatively and objectively on the basis of a defined model. CEA was designed to evaluate health-care interventions, but the methodology can be used for non-health economic applications as well. CEA can compare any resource allocation with measurable outcomes.

What Constitutes a Cost?

In CEA, it is common to distinguish between the direct costs and the indirect costs associated with an intervention. Some interventions may also result in intangibles, which are difficult to quantify but should be included in the cost profile. Examples of these different kinds of costs are as follows.

- **Direct cost.** Drugs, medical staff time, medical equipment, transport, and out-of-pocket expenses by the patients.
- **Indirect costs.** Loss of productive time by the patients during the intervention.
- **Intangibles.** Pain and suffering, and adverse effects from the intervention.

It is essential to specify which costs are included in a CEA and which are not to ensure that the findings are not subject to misinterpretation.

---

[13] The cost-utility approach is commonly used in comparing different health benefit outcomes.

Cost–Effectiveness Ratio

The cost-effectiveness ratio is simply the sum of all costs, divided by the sum of all health effects:

$$\text{Cost-effectiveness ratio} = \frac{\Sigma(\text{all costs})}{\Sigma(\text{all measured health effects})} \qquad (16.10)$$

The benefits are not measured in terms of just dollars, but in a ratio that incorporates both health outcomes and dollars.

Cost-effectiveness ratios should be related to the size of relevant budgets to determine the most cost-effective strategies. CEAs compare several program strategies and rank them by cost-effectiveness ratios. An analysis of two screening interventions might show you that one costs $10,000 per life year gained while the other costs $40,000 per life year gained. The first intervention requires monthly screening, and the second requires biannual screening. Realizing that compliance is a greater problem with monthly screening, the decision maker would implement the most appropriate coverage strategy for the population in question. Sometimes, the analysis compares an option against a baseline option, such as "do nothing" or "give usual care." The last two are valid strategic options.

Discounting

There is often a significant time lag between the investment of health-service resources and the arrival of the associated health gain. In general, we prefer to receive benefits now and pay costs in the future. In order to reflect this preference in economic evaluation, costs are discounted.

Independent versus Mutually Exclusive Intervention Programs

When we use a CEA, we need to distinguish between those interventions which are completely independent and those which are dependent. Two (or more) interventions are said to be **independent** if the costs and effects of one neither affect nor are affected by the costs and effects of the other. Two (or more) interventions are dependent if the implementation of one results in changes to the costs and effects of the other.

- **Independent interventions:** Using CEA with independent-intervention programs requires that cost-effectiveness ratios be calculated for each program and placed in rank order. Then as many programs as are budget permitting are selected with the first ranked project getting the top priority. If the first ranked project is fully funded and there are still some budget funds left, then the second ranked project gets the opportunity for funding, and so forth.
- **Mutually exclusive interventions.** In reality, the likelihood is that choices will have to be made between different treatment regimes for the same condition and between different dosages or treatments versus prophylaxis (i.e., mutually exclusive interventions). In this case, incremental cost-effectiveness ratios (ΔCERs) should be used:

$$\Delta\text{CER}_{2-1} = \frac{\text{cost of P2} - \text{cost of P1}}{\text{effects of P2} - \text{Effects of P1}}$$

The alternative interventions are ranked according to their effectiveness on the basis of securing the maximum effect rather than considering cost.

Example 16.5 illustrates the process of performing a cost-effectiveness analysis for independent programs and Example 16.6 considers mutually exclusive alternatives.

EXAMPLE 16.5 Cost–Effectiveness Analysis for Independent Interventions[14]

Consider the three interventions shown in Table 16.6 for different patient groups. Each intervention has as an alternative "doing nothing." Determine which project(s) would be funded as a function of budget.

TABLE 16.6 Cost-Effectiveness of Three Independent Intervention Programs

Program	Cost ($)	Health Effect (Life Years Gained)	Cost-Effectiveness Ratio
A	100,000	1,200	83.33
B	120,000	1,350	88.89
C	150,000	1,850	81.08

SOLUTION

Given: Cost and health effect for three drug-intervention programs.

Find: Which project to fund as a function of available budget.

According to CEA, program C should be given priority over program A, since it has a lower cost-effectiveness ratio (CER), but in order to decide which program to implement, the extent of the resources available must be considered. Clearly, the choice of independent intervention is a function of the budget that is available to implement. For example, with $200,000, we will go with program C, and the remaining $50,000 will be available for funding (up to 50%) of program A. Table 16.7 summarizes the scope of project funding as a function of the available budget.

TABLE 16.7 Choices of Program as a Function of Budget

Budget Available ($)	Programs to Be Implemented
Less than $150,000	As much of C as budget allows
$150,000	100% of C
$150,000–$250,000	C and as much of A as budget allows
$250,000	C and A
$250,000–$370,000	C and A, and as much of B as budget allows
$370,000	All three programs: A, B, C

[14] This section is based on the article "What Is Cost-effectiveness?" by Ceri Phillips and Guy Thompson, Vol. 1, No. 3, *Hayward Medical Communications*. Copyright © 2001; on the Internet at www.evidence-based-medicine.co.uk.

EXAMPLE 16.6 Cost–Effectiveness Analysis for Mutually Exclusive Intervention Programs

Consider the following five mutually exclusive drug intervention programs with the costs and health effects as calculated in Table 16.8. Select the best intervention program.

TABLE 16.8 Mutually Exclusive Intervention Programs

Program	Cost	Effects (Life Years Gained)	Cost Effectiveness (Dollars Per) (Life Years Gained)	Incremental Cost-Effectiveness Ratio
P1	$125,000	1,300	96.15	96.15
P2	$100,000	1,500	66.67	−125
P3	$160,000	2,000	80.00	120
P4	$140,000	2,200	63.63	−100
P5	$170,000	2,600	65.38	75

SOLUTION

Given: Five drug intervention programs.
Find: Incremental CER and which program to fund out of five programs.

- If money is no object, P5 is clearly the best alternative, as the number of life years gained is most significant.
- The least-effective intervention (P1) has the same average CER as its incremental cost-effectiveness ratio (ICER), because it is compared with the alternative of "doing nothing":

$$\Delta CER_{1-0} = \frac{125,000 - 0}{1,300 - 0}$$

$$= 96.15$$

- A comparison between P1 and P2 yields

$$\Delta CER_{2-1} = \frac{\text{Cost of P2} - \text{Cost of P1}}{\text{Effect of P2} - \text{Effect of P1}}$$

$$= \frac{100,000 - 125,000}{1,500 - 1,300}$$

$$= -125 < 96.15$$

The negative ICER for P2 means that adopting P2 rather than P1 results in an improvement in life years gained and a reduction in costs. It also indicates that P2 dominates P1. Hence, we can eliminate P1 at this stage of the analysis.

- A comparison between P2 and P3 yields

$$\Delta CER_{3-2} = \frac{160,000 - 100,000}{2,000 - 1,500}$$

$$= 120 > 66.67$$

The ICER for P3 works out to be 120, which means that it costs $120 to generate two times each additional life year gained compared with P2. Thus, there is no clear dominance between P2 and P3.

- A comparison between P4 and P3 yields

$$\Delta CER_{4-3} = \frac{140,000 - 160,000}{2,200 - 2,000}$$

$$= -100 < 80$$

P4 is more effective than P3 as the incremental cost-effectiveness ratio becomes negative. As P4 dominates P3, we can eliminate P3 from the analysis. Having excluded P1 and P3, we now recalculate for P2, P4, and P5, as shown in Table 16.9.

TABLE 16.9 Remaining Mutually Exclusive Alternatives after Eliminating More Costly and Less Effective Programs

Program	Cost	Effects (Life Years Gained)	Cost Effectiveness (Dollars per) (Life Years Gained)	Incremental Cost-Effectiveness Ratio
P2	$100,000	1,500	66.67	66.67
P4	$140,000	2,200	63.63	57.14
P5	$170,000	2,600	65.38	75.00

- A comparison between P2 and P4 yields

$$\Delta CER_{4-2} = \frac{140,000 - 100,000}{2,200 - 1,500}$$

$$= \$57.14 < 66.67$$

Thus, P2 is dominated by P4, since the latter is more effective, and it costs less to produce an additional unit ($57.14 compared with $66.67). The dominated alternative is then excluded, and the ICERs are recalculated again (Table 16.10).

- Finally, a comparison between P4 and P5 yields

$$\Delta CER_{5-4} = \frac{170,000 - 140,000}{2,600 - 2,200}$$

$$= 75 > 63.63$$

TABLE 16.10 Remaining Mutually Exclusive Alternatives after Eliminating All Dominated Programs

Program	Cost	Effects (Life Years Gained)	Cost Effectiveness (Dollars per) (Life Years Gained)	Incremental Cost-Effectiveness Ratio
P4	$140,000	2,200	63.63	63.63
P5	$170,000	2,600	65.38	75.00

No clear dominance exists between P4 and P5. As shown in Figure 16.12, these two programs are therefore the ones that deserve funding consideration. In deciding between them, the size of the available budget must be brought to bear on the matter.

- If the available budget is $140,000, all patients should receive intervention P4, whereas if the available budget is $170,000, all patients should receive the more effective P5.

- If the budget is, say, $150,000, then since the cost difference between P4 and P5 is $30,000 and the budget surplus is $10,000, it is possible to switch one-third of the patients to P5 and still remain within budget.

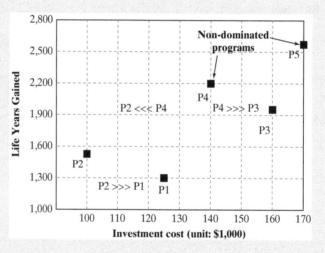

Figure 16.12 Cost-effectiveness diagram—life years gained.

16.5.3 Cost-Utility Analysis[15]

Cost-utility analysis expresses the value for money in terms of a single type health outcome. Here we are measuring the incremental cost to gain an extra quality-adjusted life-year (QALY). This approach incorporates both increases in survival time (extra life-years)

[15] Introduction to Drug Utilization Research, World Health Organization, 2003. http://apps.who.int/medicinedocs/en/d/Js4876e/5.4.html.

Figure 16.13 Quality-Adjusted Life-Years (QALY) per person per year.

and changes in quality of life (with or without increased survival) into one measure. The specific features of this analytic measurement are as follows:

- An increased quality of life is expressed as a utility value on a scale of 0 (dead) to one (perfect quality of life), as shown in Figure 16.13. An increased duration of life of one year (without change in quality of life), or an increase in quality of life from 0.5 to 0.7 utility units for five years, would both result in a gain of one QALY. This allows for easy comparison across different types of health outcome, but still requires value judgments to be made about increases in the quality of life (utility) associated with different health outcomes.

  For example, a particular treatment option produces 0.80 QALYs per person a year, then aggregate QALYs produced for 1,000 patients over 15 years would be calculated as

  $$0.80(1,000)(15) = 12,000 \text{ QALYs}.$$

The use of **incremental cost-utility ratios** enables the cost of achieving a health benefit by treatment with a drug to be assessed against similar ratios calculated for other health interventions. The incremental cost-utility ratios are the same method as the incremental cost-effectiveness ratios (ICER) as we discussed in the previous section—simply use utility instead of natural units to measure outcomes.

EXAMPLE 16.6 Cost-Utility Analysis of Two Different Treatment Options

Table 16.11 is a summary of findings from a simulation model to estimate the cost-utility of annual, repeated projected 15-year costs and quality-adjusted life years saved by lung cancer screening and treatment with and without smoking cessation in New York.[16]

[16] *Source*: "A Cost-Utility Analysis of Lung Cancer Screening and the Additional Benefits of Incorporating Smoking Cessation Interventions," Andrea C. Villanti et al., *PLOS*, August 07, 2013.

TABLE 16.11 Projected 15-year costs and quality-adjusted life years saved by lung cancer screening and treatment with and without smoking cessation using stage shifts from the State of New York

		Treatment Options	
	Base: Screening	Option 1: Screening + Light smoking cessation intervention	Option 2: Screening + Intensive smoking cessation intervention
Additional treatment cost over the base over 15 years	$27,824,282,242	$1,361,556,665	$3,212,191,737
Additional QALYs saved over the base over 15 years	985,284	273,566	930,754
Cost per QALY saved	$28,240	$23,185	$16,198

Source: "A Cost-Utility Analysis of Lung Cancer Screening and the Additional Benefits of Incorporating Smoking Cessation Interventions," Andrea C. Villanti, etc., PLOS, August 07, 2013.

The model assumed that screenings would be performed over 15 years in a high risk hypothetical cohort of 18 million adults between age 50 and 64 with 30+ pack-years of smoking history. In the base case, the lung cancer screening intervention cost $27.8 billion over 15 years and yielded 985,284 quality-adjusted life years (QALYs) gained for a cost-utility ratio of $28,240 per QALY gained. Adding smoking cessation to these annual screenings resulted in increases in both the costs and QALYs saved, reflected in cost-utility ratios ranging from $16,198 per QALY gained to $23,185 per QALY gained. Which treatment option would you recommend?

SOLUTION

Given: Cost and utility data for two additional treatment options.

Find: Incremental cost-utility ratio. What is the incremental cost-utility ratio for the more expansive treatment option, Option 2, for 15 years?

- The difference in cost between the two treatment options:
- Δcost = $3,212,191,737 − $1,361,556,665 = $1,850,635,072
- The difference in utility production of the two treatment options—how many extra QALYs are produced by Option 2 treatment, over 15 years:

  Δutility = 930,754 − 273,566 = 657,188 QALYs over 15 years

- An incremental cost-utility ratio for Option 2 over Option 1:

$$ICER_{\text{Option 2-Option 1}} = \frac{\text{Change in cost}}{\text{Change in outcome (utility)}}$$

$$= \frac{\$1,850,635,072}{657,188}$$

$$= \$2,816 \text{ per extra QALY gained by Option 2}$$

Clearly, Option 2 is more cost-effective treatment option, if there is no budget restrictions. Figure 16.14 summarizes the incremental QALYs gained over the base screenings for each treatment option.

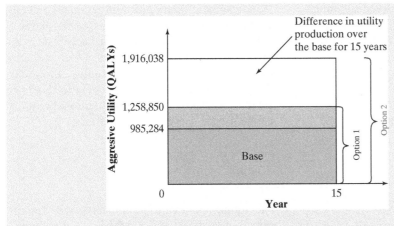

Figure 16.14 Incremental utility produced over 15 years by two treatment options over the base screening.

SUMMARY

■ The service sector accounts for about 80% of U.S. GDP. Health care and government services account for almost 30% of the total GDP.

■ **Benefit–cost analysis** is commonly used to evaluate public projects; several facets unique to public-project analysis are neatly addressed by benefit–cost analysis.

1. Benefits of a nonmonetary nature can be quantified and factored into the analysis.

2. A broad range of project users distinct from the sponsor can and should be considered; benefits and disbenefits to *all* these users can and should be taken into account.

■ Difficulties involved in public-project analysis include the following:

1. Identifying all the users of the project.

2. Identifying all the benefits and disbenefits of the project.

3. Quantifying all the benefits and disbenefits in dollars or some other unit of measure.

4. Selecting an appropriate interest rate at which to discount benefits and costs to a present value.

■ The *B/C* ratio is defined as

$$BC(i) = \frac{B}{C} = \frac{B}{I + C'}, \quad I + C' > 0$$

The decision rule is that if $BC(i) > 1$ then the project is acceptable.

■ The profitability index (or the net *B/C* ratio) is defined as

$$PI(i) = \frac{B - C'}{I} = \frac{B'}{I}, \quad I > 0$$

The PI expresses the net benefit expected per dollar invested. The same decision rule applies as for the B/C ratio, $PI(i) > 1$.

■ The **cost-effectiveness method** allows us to compare projects on the basis of cost and nonmonetary effectiveness measures. We may either maximize effectiveness for a given cost criterion or minimize cost for a given effectiveness criterion.

■ In health economics, cost-effectiveness analysis is the most widely used economic evaluation method. The cost-effectiveness ratio is defined as

Cost-effectiveness ratio

$$= \frac{\Sigma(\text{all costs})}{\Sigma(\text{all measured health effects})}$$

If independent intervention programs are evaluated, we can rank the programs on the basis of their cost-effectiveness ratios and accept as many programs as the budget permits. When we compare mutually exclusive intervention programs, the incremental cost-effectiveness ratio should be used to determine whether an additional cost can be justified to increase health effects.

PROBLEMS

Valuation of Benefits and Costs

16.1 The state of Michigan is considering a bill that would ban the use of road salt on highways and bridges during icy conditions. Road salt is known to be toxic, costly, corrosive, and caustic. Chevron Chemical Company produces a calcium magnesium acetate (CMA) deicer and sells it for $650 a ton as Ice-B-Gon. Road salts, by contrast, sold for an average of $54 a ton. Michigan needs about 500,000 tons of road salt each year. (The state spent $27 million on road salt annually.) Chevron estimates that each ton of salt on the road costs $650 in highway corrosion, $525 in rust on vehicles, $150 in corrosion to utility lines, and $100 in damages to water supplies, for a total of $1,425. Unknown salt damage to vegetation and soil surrounding areas of highways has occurred. Michigan would ban road salt (at least on expensive steel bridges or near sensitive lakes) if state studies support Chevron's cost claims.

(a) What would be the users' benefits and sponsor's costs if a complete ban on road salt were imposed in Michigan?

(b) How would you go about determining the salt damages (in dollars) to vegetation and soil?

16.2 A public school system in Ohio is considering the adoption of a four-day school week as opposed to the current five-day school week in high schools. The community is hesitant about the plan, but the superintendent of the school system envisions many benefits associated with the four-day system, Wednesday being the "day off." The following pros and cons have been cited.

• Experiments with the four-day system indicate that the "day off" in the middle of the week will cut down on both teacher and pupil absences.

• The longer hours on school days will require increased attention spans, which is not an appropriate expectation for younger children.

• The reduction in costs to the federal government should be substantial, as the number of lunches served would be cut by approximately 20%.

• The state bases its expenditures on its local school systems largely on the average number of pupils attending school in the system. Since the number of absences will decrease, state expenditures on local systems should increase.

• Older students might want to work on Wednesdays. Unemployment is a problem in this region, however, and any influx of new job seekers could aggravate an existing problem. Community centers, libraries, and other public areas also may experience increased usage on Wednesdays.

• Parents who provide transportation for their children will see a savings in fuel costs. Primarily, only those parents whose children live less than 2 miles from the school would be involved. Children living more than 2 miles from school are eligible for free transportation provided by the local government.

• Decreases in both public and private transportation should result in fuel conservation, decreased pollution, and less wear on the roads. Traffic congestion should ease on Wednesdays in areas where congestion caused by school traffic is a problem.

• Working parents will be forced to make child-care arrangements (and possibly payments) for one weekday per week.

• Students will benefit from wasting less time driving to and from school; Wednesdays will be available for study, thus taking the heavy demand off most nights. Bused students will spend far less time per week waiting for buses.

- The local school board should see some ease in funding problems. The two areas most greatly affected are the transportation system and nutritional programs.

(a) For this type of public study, what do you identify as the users' benefits and disbenefits?

(b) What items would be considered as the sponsor's costs?

(c) Discuss any other benefits or costs associated with the four-day school week.

16.3 The Electric Department of the City of Tallahassee, Florida, operates generating and transmission facilities serving approximately 140,000 people in the city and surrounding Leon County. The city has proposed the construction of a $300 million, 235-MW Circulating Fluidized-Bed Combustor (CFBC) at the Arvah B. Hopkins Station to power a turbine generator that is currently receiving steam from an existing boiler fueled by gas or oil. Among the advantages associated with the use of CFBC systems are the following:

- A variety of fuels can be burned, including inexpensive low-grade fuels with high ash and a high sulfur content.

- The relatively low combustion temperatures inhibit the formation of nitrogen oxides. Acid-gas emissions associated with CFBC units would be expected to be significantly lower than emissions from conventional coal-fueled units.

- The sulfur-removal method, low combustion temperatures, and high-combustion efficiency characteristic of CFBC units result in solid wastes, which are physically and chemically more amenable to land disposal than the solid wastes resulting from conventional coal-burning boilers equipped with flue-gas desulfurization equipment.

On the basis of the Department of Energy's (DOE's) projections of growth and expected market penetration, the demonstration of a successful 235-MW unit could lead to as much as 41,000 MW of CFBC generation being constructed by the year 2017. The proposed project would reduce the city's dependency on oil and gas fuels by converting its largest generating unit to coal-fuel capability. Consequently, substantial reductions in local acid–gas emissions could be realized in comparison to the permitted emissions associated with oil fuel. The city has requested a $50 million cost share from the DOE. Under the Clean Coal Technology Program, cost sharing is considered attractive because the DOE cost share would largely offset the risk of using such a new technology. To qualify for cost-sharing money, the city has to address the following questions for the DOE.

(a) What is the significance of the project at local and national levels?

(b) What items would constitute the users' benefits and disbenefits associated with the project?

(c) What items would constitute the sponsor's costs?

Put yourself in the city engineer's position, and respond to these questions.

Benefit–Cost Analyses

16.4 In September 2015, a 3.45 megawatt cooling, heating, and power (CHP) system was placed in into operation for Southeast Regional Hospital with 360 beds. The CHP system was part of a new 20,000 sq. ft. central utility plant that provided the 800,000 sq. ft. health-care facility with cooling, heating and power. By centralizing all of the utilities in one location, the hospital accrued significant energy savings, as well as improved maintenance efficiency. The financial facts related to the project are summarized as

- Installed cost: $2.5 million
- Annual energy cost savings: $786,000
- Annual O&M costs: $233,000
- Project service life: 15 years
- Salvage value: $200,000

At an interest rate of 8%, what is the net savings per dollar invested?

16.5 A city government is considering two types of town-dump sanitary systems. Design A requires an initial outlay of $400,000 with annual operating and maintenance costs of $50,000 for the next 15 years; design B calls for an investment of $300,000 with annual operating and maintenance costs of $80,000 per year for the next 15 years. Fee collections from the residents would again be $85,000 per year. The interest rate is 8%, and no salvage value is associated with either system.

(a) Using the benefit–cost ratio BC(i), which system should be selected?

(b) If a new design (design C), which requires an initial outlay of $350,000 and annual operating and maintenance costs of $65,000, is proposed, would your answer in part (a) change?

16.6 The City of Boston is considering adding new buses for its current mass-transit system that links from the Hartsfield International Airport to major city destinations on nonstop basis. The total investment package is worth $8 million and expected to last 10 years with a $750,000 salvage value. The annual operating and maintenance costs for buses would be $2 million during the first year, and will grow by 5% each year over the previous year thereafter. If the system is used for 600,000 trips per year, what would be the fair price to charge per trip to have a profitability index of 1? Assume that the City of Boston uses 5% interest rate for any city-sponsored projects.

16.7 A local county is considering purchasing some dump trucks for the trash pickups. Each truck will cost $55,000 and have an operating and maintenance cost that starts at $18,000 during the first year and increases by $3,000 per year thereafter. Assume the salvage value is $12,000 at the end of 5 years and the interest rate is 10%. To meet the equivalent annual cost of owning and operating each truck over 5-year planning horizon, what is the required annual trash collection fees?

16.8 Consider three investment projects, A1, A2, and A3. Each project has the same service life, and the present worth of each component value (B, I, and C') is computed at 10% as follows:

TABLE P16.8

	Projects		
	A1	**A2**	**A3**
I	$5,000	$20,000	$14,000
B	12,000	35,000	21,000
C'	4,000	8,000	1,000
PW(i)	$3,000	$7,000	$6,000

(a) If all three projects are independent, which projects would be selected based on BC(i)?
(b) If the three projects are mutually exclusive, which project would be the best alternative? Show the sequence of calculations that would

be required to produce the correct results. Use the B/C ratio on incremental investment.

16.9 The U.S. government is considering building apartments for government employees working in a foreign country and living in locally owned housing. A comparison of two possible buildings is given in Table P16.9.

TABLE P16.9

	Building X	Building Y
Original investment by government agencies	$8,000,000	$12,000,000
Estimated annual maintenance costs	240,000	180,000
Savings in annual rent now being paid to house employees	1,960,000	1,320,000

Assume the salvage or sale value of the apartments to be 60% of the first investment. Use 10% and a 20-year study period to compute the B/C ratio on incremental investment, and make a recommendation. (Assume no do-nothing alternative.)

16.10 The U.S. Department of Interior is planning to build a dam and construct a hydroelectric power plant. In addition to producing electric power, this project will provide flood control, irrigation, and recreational benefits. The estimated benefits and costs expected to be derived from the three alternatives under consideration are listed as follows:

TABLE P16.10

	Design Alternatives		
	A	**B**	**C**
Initial investment	$100 M	$160M	$220M
Annual recurring benefits	$20M	$40M	$60M
Annual recurring costs	$8M	$18M	$25M
Salvage value	0	0	0

The interest rate is 5%, and the life of each project is estimated to be 40 years.

(a) Find the benefit–cost ratio for each alternative.

(b) Select the best alternative, according to BC(i).

(c) Select the best alternative, according to PI(i).

16.11 Three public-investment alternatives with the same service life are available: A1, A2, and A3. Their respective total benefits, costs, and first costs are given in present worth as follows:

TABLE P16.11

	Proposals		
Present worth	**A1**	**A2**	**A3**
I	100	300	200
B	400	700	500
C'	100	200	150

Assuming no do-nothing alternative, which project would you select on the basis of the benefit–cost ratio BC(i) on incremental investment?

16.12 A local city that operates automobile parking facilities is evaluating a proposal that it will erect and operate a structure for parking in the city's downtown area. Three designs for a facility to be built on available sites have been identified in Table P16.12 (all dollar figures are in thousands).

At the end of the estimated service life, whichever facility had been constructed would be torn down, and the land would be sold. It is estimated that the proceeds from the resale of the land will be equal to the cost of clearing the site. If the city's interest rate is known to be 10%, which design alternative would be selected on the basis of the benefit–cost criterion?

16.13 The federal government is planning a hydro-electric project for a river basin. In addition to producing electric power, this project will provide flood control, irrigation, and recreational benefits. The estimated benefits and costs expected to be derived from the three alternatives under consideration are listed in Table P16.13.

The interest rate is 10%, and the life of each of the projects is estimated to be 50 years.

(a) Find the benefit–cost ratio for each alternative.

(b) Select the best alternative on the basis of BC(i).

TABLE P16.12

	Design A	Design B	Design C
Cost of site	$240	$180	$200
Cost of building	2,200	700	1,400
Annual fee collection	830	750	600
Annual maintenance cost	410	360	310
Service life (years)	30	30	30

TABLE P16.13

	Decision Alternatives		
	A	**B**	**C**
Initial cost	$8,000,000	$10,000,000	$15,000,000
Annual benefits or costs:			
Power sales	$1,000,000	$1,200,000	$1,800,000
Flood-control savings	250,000	350,000	500,000
Irrigation benefits	350,000	450,000	600,000
Recreation benefits	100,000	200,000	350,000
O&M costs	200,000	250,000	350,000

16.14 Two different routes are under consideration for a new interstate highway:

TABLE P16.14

	Length of Highway	First Cost	Annual Upkeep
The "long" route	22 miles	$21 million	$140,000
Transmountain shortcut	10 miles	$45 million	$165,000

For either route, the volume of traffic will be 400,000 cars per year. These cars are assumed to operate at $0.25 per mile. Assume a 40-year life for each road and an interest rate of 10%. Determine which route should be selected.

16.15 The government is considering undertaking four projects. These projects are mutually exclusive, and the estimated present worth of their costs and the present worth of their benefits are shown in millions of dollars in Table P16.15. All of the projects have the same duration.

TABLE P16.15

Projects	PW of Benefits	PW of Costs
A1	$40	$85
A2	150	110
A3	70	25
A4	120	73

Assuming no do-nothing alternative, which alternative would you select? Justify your choice by using a benefit–cost ($BC(i)$) analysis on incremental investment.

Cost-Effectiveness Analysis

16.16 Table P16.16 summarizes the costs of treatment of a disease based on two different antibiotics and associated health benefits (effectiveness). Which treatment option is the best?

16.17 Table P16.17 summarizes cervical-cancer treatment options and their health effectiveness. Find the best strategy for treating cervical cancer.

TABLE P16.16

Type of Treatment	Cost of Treating 100 Patients	Effectiveness (Percent Successful Treatment of Infections)
Antibiotic A	$12,000	75%
Antibiotic B	$13,500	80%
Antibiotic C	$14,800	82%

TABLE P16.17

Strategy	Cost	Marginal Cost	Effectiveness	Marginal Effectiveness	CE Ratio
Nothing	$0	—	0 years	—	—
Simple	$5,000	$5,000	5 years	5 years	$1,000/yr
Complex	$50,000	$45,000	5.5 years	0.5 years	$90,000/yr

Short Case Studies

ST16.1 Fast growth in the population of the city of Orlando and in surrounding counties—Orange County in particular—has resulted in insurmountable traffic congestion. The county has few places to turn for extra money for road improvements—except new taxes. County officials have said that the money they receive from current taxes is insufficient to widen overcrowded roads, improve roads that don't meet

TABLE ST16.1

District	Project	Type of Improvement	Construction Cost	Annual O&M	Annual Benefits
	I-1	Four-lane	$ 980,000	$ 9,800	$313,600
	I-2	Four-lane	3,500,000	35,000	850,000
I	I-3	Four-lane	2,800,000	28,000	672,000
	I-4	Four-lane	1,400,000	14,000	490,000
	I-5	Realign	2,380,000	47,600	523,600
II	II-1	Four-lane	5,040,000	100,800	1,310,400
	II-2	Four-lane	2,520,000	50,400	831,600
	II-3	Four-lane	4,900,000	98,000	1,021,000
	II-4	Realign	1,365,000	20,475	245,700
III	III-1	Four-lane	2,100,000	31,500	567,000
	III-2	Two-lane	1,170,000	17,550	292,000
	III-3	Four-lane	1,120,000	16,800	358,400
	III-4	Four-lane	2,800,000	56,000	980,000
IV	IV-1	Reconstruct	1,690,000	33,800	507,000
	IV-2	Widen	975,000	15,900	273,000
	IV-3	Widen	1,462,500	29,250	424,200

modern standards, and pave dirt roads. State residents now pay 12 cents in taxes on every gallon of gas. Four cents of that goes to the federal government, 4 cents to the state, 3 cents to the county in which the tax is collected, and 1 cent to the cities.

The county commissioner has suggested that the county get the money by tacking an extra penny-a-gallon tax onto gasoline, bringing the total federal and state gas tax to 13 cents a gallon. This new tax would add about $2.6 million a year to the road-construction budget. The extra money would have a significant impact. With the additional revenue, the county could sell a $24 million bond issue. It would then have the option of spreading that amount among many smaller projects or concentrating on a major project.

Assuming that voters would approve a higher gas tax, county engineers were asked to prepare a priority list outlining which roads would be improved with the extra money. The road engineers also computed the possible public benefits associated with each road-construction project; they accounted for a possible reduction in travel time, a reduction in the accident rate, land appreciation, and savings in the operating costs of vehicles. Table ST16.1 is a list of the projects and their characteristics.

Assume a 20-year planning horizon and an interest rate of 10%. Which projects would be considered for funding in parts (a) and (b)?

(a) Due to political pressure, each district will have the same amount of funding, say, $6 million.

(b) The funding will be based on tourist traffic volumes. Districts I and II combined will get $15 million, and Districts III and IV combined will get $9 million. It is desirable to have at least one four-lane project from each district.

ST16.2 The City of Portland Sanitation Department is responsible for the collection and disposal of all solid waste within the city limits. The city must collect and dispose of an average of 300 tons of garbage each day. The city is considering ways to improve the current solid-waste collection and disposal system.

- The current system uses Dempster Dumpmaster frontend loaders for collection and incineration or landfill for disposal. Each collecting vehicle has a load capacity of 10 tons (or 24 cubic yards) and dumping is automatic. The incinerator in use was manufactured in 1942 and was designed to incinerate 150 tons per 24 hours. A natural-gas afterburner has been added in an effort to reduce air pollution. However, the incinerator still does not meet state air-pollution requirements, and it is operating under a permit from the Oregon State Air and Water Pollution Control Board. Prison-farm labor is used for the operation of the incinerator. Because the capacity of the incinerator is relatively low, some trash is not incinerated, but it is taken to the city landfill. The trash landfill is located approximately 11 miles, and the incinerator approximately 5 miles, from the center of the city. The mileage and costs in person-hours for delivery to the disposal sites are excessive; a high percentage of empty vehicle miles and person-hours are required, because separate methods of disposal are used and the destination sites are remote from the collection areas. The operating cost for the present system is $905,400, including $624,635 to operate the prison-farm incinerator, $222,928 to operate the existing landfill, and $57,837 to maintain the current incinerator.

- The proposed system locates a number of portable incinerators, each with 100-ton-per-day capacity for the collection and disposal of refuse waste collected for three designated areas within the city. Collection vehicles will also be staged at these incineration-disposal sites together with the plant and support facilities that are required for incineration, fueling and washing of the vehicles, a support building for stores, and shower and locker rooms for collection and site crew personnel. The pickup-and-collection procedure remains essentially the same as in the existing system. The disposal-staging sites, however, are located strategically in the city on the basis of the volume and location of wastes collected, thus eliminating long hauls and reducing the number of miles the collection vehicles must retravel from pickup to disposal site.

Four variations of the proposed system are being considered, containing one, two, three, and four incinerator-staging areas, respectively. The type of

incinerator is a modular prepackaged unit that can be installed at several sites in the city. Such units exceed all state and federal standards for exhaust emissions. The city of Portland needs 24 units, each with a rated capacity of 12.5 tons of garbage per 24 hours. The price per unit is $137,600, which means a capital investment of about $3,304,000. The plant facilities, such as housing and foundation, were estimated to cost $200,000 per facility based on a plan incorporating four incinerator plants strategically located around the city. Each plant would house eight units and be capable of handling 100 tons of garbage per day. Additional plant features, such as landscaping, were estimated to cost $60,000 for site 1 and other sites cost some additional expenses.

The annual operating cost of the proposed system would vary according to the type of system configuration. It takes about 1.5 to 1.7 million cubic feet (MCF) of fuel to incinerate 1 ton of garbage. The conservative 1.7-MCF figure was used for total cost. This means a fuel cost of $4.25 per ton of garbage at a cost of $2.50 per MCF. Electric requirements at each plant will be 230 kW per day, which means a $0.48-per-ton cost for electricity if the plant is operating at full capacity. The maintenance cost of each plant was estimated to be $1.19 per ton. Since three plants will require fewer transportation miles, it is necessary to consider the savings accruing from this operating advantage. Three plant locations will save 6.14 miles per truck per day on the average. At an estimated cost of $0.30 per mile, this would mean that an annual savings of $6,750 is realized on minimum trips to the landfill disposer for a total annual savings in transportation of $15,300. Savings in labor are also realized because of the shorter routes, which permit more pickups during the day. The annual savings from this source are $103,500. Table ST16.2 summarizes all costs, in thousands of dollars, associated with the present and proposed systems.

A bond will be issued to provide the necessary capital investment at an interest rate of 8% with a maturity date 20 years in the future. The proposed systems are expected to last 20 years with negligible salvage values. If the current system is to be retained, the annual O&M costs would be expected to increase at an annual rate of 10%. The city will use the bond interest rate as the interest rate for any public-project evaluation.

TABLE ST16.2

Item	Present System	Costs for Proposed Systems Site Number			
		1	**2**	**3**	**4**
Capital costs:					
Incinerators		$3,302	$3,302	$3,302	$3,302
Plant facilities		600	900	1,260	1,920
Annex buildings		91	102	112	132
Additional features		60	80	90	100
Total		**$4,053**	**$4,384**	**$4,764**	**$5,454**
Annual O&M costs	$905.4	$342	$480	$414	$408
Annual savings:					
Pickup transportation		$13.2	$14.7	$15.3	$17.1
Labor		87.6	99.3	103.5	119.40

(a) Determine the operating cost of the current system in terms of dollars per ton of solid waste.

(b) Determine the economics of each solid-waste disposal alternative in terms of dollars per ton of solid waste.

ST16.3 Because of a rapid growth in population, a small town in Pennsylvania is considering several options to establish a wastewater treatment facility that can handle upto a flow of 2 million gallons per day (MGD). The town has five treatment options available.

- **Option 1: No action.**   This option will lead to continued deterioration of the environment. If growth continues and pollution results, fines imposed (as high as $10,000 per day) would soon exceed construction costs.

- **Option 2: Land-treatment facility.**   This option will provide a system for land treatment of the wastewater to be generated over the next 20 years and will require the utilization of the most land for treatment of the wastewater. In addition to the need to find a suitable site, pumping of the wastewater for a considerable distance out of town will be required. The land cost in the area is $3,000 per acre. The system will use spray irrigation to distribute wastewater over the site. No more than 1 inch of wastewater can be applied in one week per acre.

- **Option 3: Activated sludge-treatment facility.** This option will provide an activated sludge-treatment facility at a site near the planning area. No pumping will be required, and only seven acres of land will be needed for construction of the plant, at a cost of $7,000 per acre.

- **Option 4: Trickling filter-treatment facility.** Provide a trickling filter-treatment facility at the same site selected for the activated sludge plant of option 3. The land required will be the same as that for option 3. Both facilities will provide similar levels of treatment but will be using different units.

- **Option 5: Lagoon-treatment system.**   Utilize a three-cell lagoon system for treatment. The lagoon system requires substantially more land than options 3 and 4 require but less than option 2. Due to the larger land requirement, this treatment system will have to be located some distance outside of the planning area and will require pumping of the wastewater to reach the site.

Table ST16.3 summarizes, respectively, (1) the land cost and land value for each option, (2) the capital expenditures for each option, and (3) the O&M costs associated with each option.

The price of land is assumed to be appreciating at an annual rate of 3%, and the equipment installed will require a replacement cycle of 15 years. Its replacement cost will increase at an annual rate

| TABLE ST16.3

Land Cost for Each Option			
Option Number	Land Required (acres)	Land Cost ($)	Land Value (in 20 years)
2	800	$2,400,000	$4,334,600
3	7	49,000	88,500
4	7	49,000	88,500
5	80	400,000	722,400
Capital Expenditures			
Option Number	Equipment	Structure	Pumping	Total
2	$500,000	$700,000	$100,000	$1,300,000
3	500,000	2,100,000	0	2,600,000
4	400,000	2,463,000	0	2,863,000
5	175,000	1,750,000	100,000	2,025,000
Annual O&M costs			
Option Number	Energy	Labor	Repair	Total
2	$200,000	$95,000	$30,000	$325,000
3	125,000	65,000	20,000	210,000
4	100,000	53,000	15,000	168,000
5	50,000	37,000	5,000	92,000

of 5% (over the initial cost), and its salvage value at the end of each replacement cycle will be 50% of the original replacement cost. The structure requires replacement after 40 years and will have a salvage value of 60% of the original cost. The cost of energy and repair will increase at an annual rate of 5% and 2%, respectively. The labor cost will increase at an annual rate of 4%.

With the following sets of assumptions, answer parts (a) and (b).

- Assume an analysis period of 120 years.
- Replacement costs for the equipment, as well as for the pumping facilities, will increase at an annual rate of 5%.
- Replacement cost for the structure will remain constant over the planning period. However, the salvage value of the structure will be 60% of the original cost. (Because it has a 40-year replacement

cycle, any increase in the future replacement cost will have very little impact on the solution.)

- All O&M cost figures are given in today's dollars. For example, the annual energy cost of $200,000 for option 2 means that the actual energy cost during the first operating year will be $200,000(1.05) = $10,000.
- Option 1 is not considered a viable alternative, as its annual operating cost exceeds $3,650,000.

(a) If the interest rate (including inflation) is 10%, which option is the most cost effective?

(b) Suppose a household discharges about 400 gallons of wastewater per day through the facility selected in part (a). What should be the monthly bill assessed for this household?

ST16.4 The Federal Highway Administration predicts that by the year 2015, Americans will be spending 8.1 billion hours per year in traffic jams.

Most traffic experts believe that adding and enlarging highway systems will not alleviate the problem. As a result, current research on traffic management is focusing on three areas: (1) the development of computerized dashboard navigational systems; (2) the development of roadside sensors and signals that monitor and help manage the flow of traffic; and (3) the development of automated steering and speed controls that might allow cars to drive themselves on certain stretches of highway.

In Los Angeles, perhaps the most traffic-congested city in the United States, a Texas Transportation Institute study found that traffic delays cost motorists $8 billion per year. But Los Angeles has already implemented a system of computerized traffic-signal controls that by some estimates has reduced travel time by 13.2%, fuel consumption by 12.5%, and pollution by 10%. And between Santa Monica and downtown Los Angeles, testing of an electronic traffic and navigational system including highway sensors and cars with computerized dashboard maps is being sponsored by federal, state, and local governments and General Motors Corporation. This test program costs $40 million; to install it throughout Los Angeles could cost $2 billion.

On a national scale, the estimates for implementing "smart" roads and vehicles are even more staggering: It would cost $18 billion to build the highways, $4 billion per year to maintain and operate them, $1 billion for research and development of driver-information aids, and $2.5 billion for vehicle-control devices. Advocates say the rewards far outweigh the costs.

(a) On a national scale, how would you identify the users' benefits and disbenefits for this type of public project?

(b) On a national scale, what would be the sponsor's cost?

(c) Suppose that the users' net benefits grow at 3% per year and the sponsor's costs grow at 4% per year. Assuming a social discount rate of 10%, what would be the B/C ratio over a 20-year study period?

FUNDAMENTALS OF ENGINEERING REVIEW QUESTIONS

This appendix is designed primarily to help you develop a working knowledge of the concepts and principles of engineering economics. All questions in this appendix are structured in multiple-choice format as these types of exam questions are used in the Fundamentals of Engineering (FE) exam and, increasingly, in introductory engineering economics courses.

How to Prepare for the Fundamentals of Engineering (FE) Exam

The FE exam consists of 180 multiple-choice questions. During the morning session, all examinees take a general exam common to all disciplines. During the afternoon session, examinees can opt to take a general exam or a discipline-specific (Chemical, Civil, Electrical, Environmental, Industrial, or Mechanical) exam.

The general exam tests four questions related to engineering economics in the morning session and five in the afternoon session. The specific engineering economics topics covered in the FE exam are

- Annual cost
- Breakeven Analysis
- Benefit-Cost Analysis
- Future Worth or Value
- Present Worth
- Valuation and Depreciation

Some examples of questions provided by the National Council of Examiners for Engineering and Surveying (http://www.ncees.org/exams) are shown below. You may develop test-taking skills by adopting the problem-solving format used in our textbook examples. You first highlight the critical data provided by each question, isolate the question being asked, and outline the correct approach in the solution under the headings **Given**, **Find** and **Approach**, respectively. This guidance is intended to stimulate student curiosity to look beyond the mechanics of problem solving to explore "what if" issues, alternative solution methods, and the interpretation of the solution.

Sample Test Question 1:

If $200 is deposited in a savings account at the beginning of each of 15 years and the account draws interest at 8% per compounded annually, the value of the account at the end of 15 years will be nearly:

(a) $6,000   (b) $5,400

(c) $5,900   (d) $6,900

Solution
Given: $A = \$200, N = 15$ years, $i = 8\%$
Find: F

Approach: Note that each deposit is made at the <u>beginning</u> of each year. However, the equal-payment series compound amount factor $(F/A, i, N)$ is based on the end-of-period assumption. To adjust for this timing difference, you may still use the $(F/A, i, N)$ factor, but adjust the resulting F value for the one additional interest-earning period by multiplying it by $(1 + 0.08)$.

$$F' = \$200(F/A, 8\%, 15) = \$5,430.42$$

$$F = \$5,430.42(1.08) = \$5,865$$

Thus, the correct answer is (c).

Sample Test Question 2:

Your county has asked you to analyze the purchase of some dump trucks. Each truck will cost $45,000 and have an operating and maintenance cost that starts at $15,000 the first year and increases by $2,000 per year. Assume the salvage value at the end of 5 years is $9,000 and the interest rate is 12%. The equivalent annual cost of each truck is most nearly

(a) $31,000

(b) $41,200

(c) $26,100

(d) $29,600

Solution

Given: $I = \$45,000$, $S = \$9,000$, O&M cost $= \$15,000$ first year, increasing by $2,000 per year, $N = 5$ years, $i = 12\%$

Find: AE(12%)

Approach: Note that there are two kinds of costs: ownership costs (capital costs) and operating costs. The capital costs can be calculated by using the capital recovery with return formula whereas the O&M cost needs to be annualized, knowing that it takes a linear gradient series with $G = \$2,000$.

$$CR(12\%) = (I - S)(A/P, i, N) + iS$$

$$= (\$45,000 - \$9,000)(A/P, 12\%, 5) + (0.12)(\$9,000)$$

$$= \$11,067$$

$$AEC(12\%)_{O\&M} = \$15,000 + \$2,000(A/G, 12\%, 5)$$

$$= \$15,000 + \$2,000(1.7746)$$

$$= \$18,549$$

$$AEC(12\%) = CR(12\%) + AEC(12\%)_{O\&M}$$

$$= \$11,067 + \$18,549$$

$$= \$29,616$$

The correct answer is (d).

FE Exam Practice Problem Set No. 1

1. How many years would it take an investment to double at 8% annual interest?
   (a) about 5 years
   (b) about 7 years
   (c) about 9 years
   (d) about 12 years

2. If you borrow $10,000 over three years at 10% annual interest, what is the interest payment in the 2nd year? You are making three equal annual payments where the annual payment size is $4,021.15.
   (a) $1,000
   (b) $698
   (c) $366
   (d) $3,323

3. You plan to retire in 15 years. After retirement you will need $20,000 per year for 20 years. If interest remains at 8% compounded annually, how much should you deposit each year until retirement? (Note that you will make a total of 15 end-of-year deposits, and your first withdrawal will be made at the end of 16th year.)
   (a) $13,091
   (b) $7,811
   (c) $6,993
   (d) $7,232

4. Assume that $800 is deposited according to the following cash flow diagram. At a 10% interest compounded annually, determine the future value at the end of year 10?

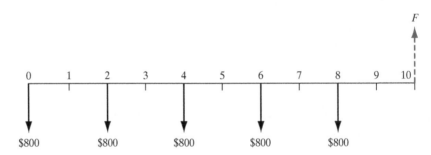

   (a) $7,346
   (b) $4,400
   (c) $ 6,071
   (d) $7,213

5. You will deposit $2,000 per year for the next 10 years. Then, you will start drawing money out of the account which earns 8% annual interest. How much could you draw out if you make equal annual withdrawals for five years? (Note: The first withdrawal will be made at the end of 11 years.)
   (a) $5,795
   (b) $7,257
   (c) $4,320
   (d) $6,988

6. You will deposit $10,000 per year for five years at 9%. Then you will just let the money sit there and continue to earn interest. If the interest rate in this second stage is 10%, how much would you have six years after the second stage starts?
   (a) $142,656
   (b) $95,755
   (c) $100,370
   (d) $106,023

7. You are making $1,000 quarterly deposits into a fund that pays interest at a rate of 8% compounded <u>quarterly</u>. What would be the balance at the end of 10 years?
(a) $60,402
(b) $57,946
(c) $58,611
(d) $45,853

8. How much must you deposit today ($n = 0$) in an account which earns 8% annual interest in order to make 10 equal annual withdrawals of $5,000 if the first withdrawal is made at the end of third year?
(a) $38,433
(b) $28,764
(c) $31,065
(d) $33,550

9. If $500 is deposited in a savings account at the <u>beginning</u> of each quarter for 10 years and the account earns 8% interest compounded quarterly, what will be the balance in the account at the end of 10 years?
(a) $28,933
(b) $30,201
(c) $31,421
(d) $30,805

10. You borrowed $5,000 to finance your office furniture at a 9% APR over 12 months. The furniture store also offers an alternative financing plan with the same terms with a grace period of 12 months (meaning that your first payment occurs at the end of 12 months from purchasing the furniture, but the interest during the grace period is not waived). What would be the monthly payment if you go with the alternative financing with the grace period?
(a) $478.27
(b) $437.26
(c) $249.85
(d) $474.72

11. What is the amount of 10 equal annual deposits that can provide five annual withdrawals, where a first withdrawal of $2,000 is made at the end of year 11 and subsequent withdrawals increase at the rate of $500 year over the previous year's if the interest rate is 10% compounded annually?
(a) $628
(b) $691
(c) $760
(d) $823

12. How many years will it take for the dollar's purchasing power to be <u>three-quarter</u> of what it is now (lose 25% of value), if the general inflation rate is expected to continue at the rate of 5% for an unforeseeable future?
(a) about 6 years
(b) about 7 years
(c) about 9 years
(d) about 14 years

13. You borrowed $20,000 from a bank at an interest rate of 12%, compounded monthly. This loan will be repaid in 60 equal monthly installments over 5 years. Immediately after your 30th payment if you want to pay the remainder of the loan in a single payment, the amount is close to
(a) $12,131
(b) $11,808
(c) $11,482
(d) $10,141

14. Consider the following cash flow series at varying interest rates. What is the equivalent present worth of the cash flow series? How much do you need to

deposit now in an account that provides the following scheme of interest earnings over 5 years, so that you will be able to withdraw the money as indicated?

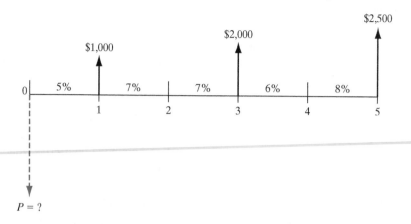

$P = ?$

(a) $P = \$4,922$ (b) $P = \$4,433$

(c) $P = \$4,277$ (d) $P = \$3,944$

15. How many months will it take for an investment to <u>triple</u> if the interest rate is 9% compounded monthly?

(a) less than 140 months

(b) between 140 months and 150 months

(c) between 151 months and 160 months

(d) more than 161 months

16. A series of equal semiannual payment of $5,000 for 10 years is equivalent to what future lump-sum amount at the end of 15 years (from today) at an interest rate of 8% compounded <u>continuously</u>?

(a) $220,394   (b) $150,132

(c) $213,449   (d) $223,997

17. Assume that you deposited $6,000 in a savings account that pays 6% interest compounded monthly. You wish to withdraw $300 at the end of each month. How many months will it take to deplete the balance?

(a) less than 20 months   (b) between 21 and 23 months

(c) between 24 and 26 months   (d) between 27 and 30 months

18. You are considering two options in purchasing an automobile:

Option A: Purchase the vehicle worth $26,200 and pay for the vehicle over 3 years with equal monthly payments at 3% APR financing.

Option B: Purchase the vehicle at a discount price of $24,000 to be paid immediately.

The funds, which would be used to purchase the vehicle, are presently earning 6% annual interest compounded monthly. Which of the following statements is <u>incorrect</u>?

(a) If you go with Option A, the monthly payment would be $761.93.

(b) Option B is a better choice if you have enough money to purchase in cash.

(c) Option B is a better financing plan, as the net cost of financing is cheaper than Option A.

(d) The effective annual yield for your funds is 6.17%.

19. A series of five payments in constant dollars, beginning with $6,000 at the end of the first year, are increasing at the rate of 5% per year. Assume that the average general inflation rate is 4%, and the market interest rate is 11% during this inflationary period. What is the equivalent present worth of the series?

(a) $24,259   (b) $25,892

(c) $27,211   (d) $29,406

20. A father wants to save in advance for his 8-year old daughter's college expenses. The daughter will enter the college 10 years from now. An annual amount of $20,000 in today's dollars (constant dollars) will be required to support the college for 4 years. Assume that these college payments will be made at the *beginning* of each school year. (The first payment occurs at the end of 10 years). The future general inflation rate is estimated to be 5% per year, and the interest rate on the savings account will be 8% compounded quarterly (market interest rate) during this period. If the father has decided to save only $1,000 (actual dollars) at the end of each quarter until she goes to college (the last deposit occurs at the end of 10 years) how much will the daughter have to borrow to cover her sophomore expenses?

(a) $4,120   (b) $4,314

(c) $4,000   (d) $4,090

FE Exam Practice Problem Set No. 2

1. You invested $100,000 in a project and received $40,000 at $n = 1$, $40,000 at $n = 2$, and $30,000 at $n = 3$ years. For some reason, you need to terminate the project at the end of year 3. Your interest rate is 10%; what is the project balance at the time of termination?

(a) gain of $10,000   (b) loss of $8,039

(c) loss of $10,700   (d) just breakeven

2. You borrowed $3,000 at 10% interest compounded annually. The loan was repaid according to the following schedule.

n	Repayment Amount
1	$1,000
2	$1,000
3	X

Find X, the amount that is required to pay off the loan at the end of year 3.

(a) $1,684   (b) $1,398

(c) $950   (d) $1,300

3. J&M Corporation purchased a vibratory finishing machine for $20,000 in year 0. The useful life of the machine is 10 years, at the end of which, the machine is estimated to have a zero salvage value. The machine generates net annual revenues

of $6,000. The annual operating and maintenance expenses are estimated to be $1,000. If J&M's MARR is 15%, how many years does it take before this machine becomes profitable?

(a) 3 years $< n \leq 4$ years   (b) 4 years $< n \leq 5$ years

(c) 5 years $< n \leq 6$ years   (d) 6 years $< n \leq 7$ years

4. Alpha Company is planning to invest in a machine, the use of which will result in the following:

- Annual revenues of $10,000 in the first year and increases of $5,000 each year, up to year 9. From year 10, the revenues will remain constant ($52,000) for an indefinite period.

- The machine is to be overhauled every 10 years. The expense for each overhaul is $40,000.

If Alpha expects a present worth of at least $100,000 at a MARR of 10% for this project, what is the maximum investment that Alpha should be prepared to make?

(a) $250,130   (b) $674,697

(c) $350,100   (d) $509,600

5. A local county government is considering purchasing some dump trucks for the trash pickups. Each truck will cost $55,000 and have an operating and maintenance cost that starts at $18,000 the first year and increases by $3,000 per year thereafter. Assume the salvage value is $12,000 at the end of 5 years and the interest rate is 10%. The equivalent annual cost of owning and operating each truck is most nearly

(a) $35,974   (b) $32,600

(c) $6,956   (d) $37,939

6. Consider the cash flow series for an investment project. If the project breaks even at an interest rate of 10%, what is the required annual revenue (X) in years 2, 3, 5, and 6?

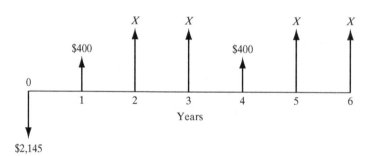

(a) $546   (b) $568

(c) $631   (d) $522

7. Consider a piece of industrial equipment that has an installed cost of $100K. The equipment is expected to generate $30K worth of annual energy savings during its first year of installation. The value of these annual savings is expected to increase at the rate of 3% per year because of increased fuel costs. Assume that the equipment has a service life of 5 years (or 3,000 operating hours per year) with

no appreciable salvage value. Determine the equivalent dollar savings per each operating hour at $i = 14\%$.

(a) $1.300 per hour (b) $1.765 per hour

(c) $0.827 per hour (d) $0.568 per hour

8. Two options are available for painting your house: (1) Oil-based paint, which costs $5,000 and (2) water-based paint, which costs $3,000. The estimated lives are 10 years and 5 years respectively. For either option, no salvage value is expected at the end of respective service lives. Assume that you will keep and maintain the house for 10 years. At an interest rate of 10%, which of the following statement is correct?

(a) On an annual basis, Option 1 will cost about $850

(b) On an annual basis, Option 2 is about $22 cheaper than Option 1

(c) On an annual basis, both options cost about the same

(d) On an annual basis, Option 2 will cost about $820.

9. A new investment project is proposed. The required initial investment is $50,000 and the expected salvage value is $10,000 at the end of 3 years. A 15% return on investment is desired. What should be the minimum annual net cash flows over three years?

Year	Cash Flow
0	−$50,000
1	X
2	X
3	$X + \$10,000$

(a) $13,334 (b) $21,900

(c) $17,520 (d) $19,020

10. You are considering purchasing a new injection molding machine. This machine will have an estimated service life of 10 years with a negligible salvage value. Its annual net operating cash flows are estimated to be $60,000. To expect a 15% rate of return on investment, what would be the maximum amount that should be spent on purchasing the injection molding machine?

(a) $301,126 (b) $234,645

(c) $600,000 (d) $126,450

11. Two types of 150-horsepower (HP) motors are being considered for installation at a municipal sewage-treatment plant. The first motor costs $4,500 and has an operating efficiency of 83% at full load. The second motor costs $3,600 and has an efficiency rating of 80% at full load. Both motors are projected to have zero salvage value after a life of 10 years. The annual operating and maintenance cost (excepting power cost) amounts to 15% of the original cost of each motor. The power costs are a flat 5 cents per kilowatt-hour. What would be the minimum number of hours of full-load operation per year necessary to justify the purchase of the more expensive motor at $i = 6\%$? (1HP = 0.7457 kilowatt)

(a) 851 hours to 900 hours (b) 901 hours to 950 hours

(c) 951 hours to 1,000 hours (d) 1,001 hours to 1,050 hours

12. You just purchased a pin inserting machine to relieve some bottleneck problems that have been experienced in manufacturing a PC board. The machine cost $56,000 and has an estimated service life of 5 years. At that time, the estimated salvage value would be $5,000. The machine is expected to operate 2,500 hours per year. The expected annual operating and maintenance cost would be $6,000. If your firm's interest rate is 15%, what would be the machine cost (owning and operating) per hour?
   (a) $8.79
   (b) $5.89
   (c) $11.85
   (d) $7.85

13. The following information on two mutually exclusive projects is given below:

n	Project A	Project B
0	−3,000	−5,000
1	1,350	1,350
2	1,800	1,800
3	1,500	5,406
IRR	25%	25%

Which of the following statements is *correct?*
   (a) Since the two projects have the same rate of return, they are indifferent.
   (b) Project A would be a better choice, as the required investment is smaller with the same rate of return.
   (c) Project B would be a better choice as long as the investor's MARR is less than 25%.
   (d) Project B is a better choice regardless of the investor's MARR.

14. Consider an investment with the following repeating cash flow cycles. Determine the capitalized equivalent worth of the project at an interest rate of 10%.

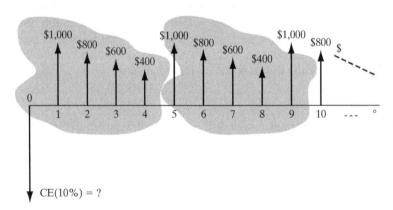

   (a) $7,238
   (b) $8,483
   (c) $7,458
   (d) $7,977

15. A company is currently paying a sales representative $0.58 per mile to drive her car for company business. The company is considering supplying the representative

with a car, which would involve the following: A car costs $24,000, has a service life of 5 years, and a market value of $7,000 at the end of that time. The cost of keeping the car in the garage during the off-hours amounts to $2,500 a year, and the cost of fuel, tires, and maintenance is 30 cents per mile. What annual mileage must a salesman travel by car for the cost of the two methods of providing transportation to be equal if the interest rate is 15%? (Do not consider any income tax effects for the sake of simplicity.)

(a) 22,356 miles
(b) 30,791 miles
(c) 34,226 miles
(d) 35,340 miles

16. You purchased a stamping machine which cost $60,000 five years ago. At that time, the machine was estimated to have a service life of 5 years with salvage value of $5,000. These estimates are still good. The property has been depreciated according to a 7-year MACRS property class. Now (at the end of year 5 from purchase) you are considering selling the machine at $10,000. What book value should you use in determining the taxable gains?

(a) $10,000
(b) $13,386
(c) $16,065
(d) $17,520

17. Lane Construction Ltd. is considering the acquisition of a new 18-wheeler.
   • The truck's base price is $80,000, and it will cost another $20,000 to modify it for special use by the company.
   • This truck falls into the MACRS five-year class.
   • It will be sold after three years (project life) for $30,000.
   • The truck purchase will have no effect on revenues, but it is expected to save the firm $45,000 per year in before-tax operating costs, mainly in leasing expenses.
   • The firm's marginal tax rate (federal plus state) is 40%, and its MARR is 15%.

What is the net present worth of this acquisition?

(a) −$45,158 (loss)
(b) $532
(c) −$1,677 (loss)
(d) $2,742

18. Consider the following investment projects.

Year (n)	Net Cash Flow	
	Project 1	Project 2
0	−$1,200	−$2,000
1	600	1,500
2	1,000	1,000
IRR	19.65%	17.54%

Determine the range of MARR where project 2 would be preferred over project 1.

(a) MARR ≤ 12.5%
(b) 13% ≤ MARR ≤ 15%
(c) 16% ≤ MARR
(d) Not enough information to determine.

19. You are considering the following two options:

- **Option 1:** You continue to use the old machine tool that was bought four years ago for $15,000. It has been fully depreciated but can be sold for $2,000. If kept, it could be used for 3 more years with proper maintenance and with some extra care. No salvage value is expected at the end of 3 years. The maintenance costs would run $10,000 per year for the old machine tool.

- **Option 2:** You purchase a brand-new machine tool at a price of $20,000 to replace the present equipment. Because of the nature of the product manufactured, it also has an expected economic life of 3 years and will have a salvage value of $5,000 at the end of that time. With the new machine tool, the expected operating and maintenance costs (with the scrap savings) amount to $3,000 each year for 3 years.

What is the net before-tax annual equivalent incremental benefit (or loss) associated with replacing the old machine tools now at an interest rate of 15%?

(a) $3,270 (b) $730

(c) $1,270 (d) $556

20. A municipal government is considering two mutually exclusive proposals for a new softball complex on a city-owned lot.

Alternative Design	Seating Capacity	Annual Benefits	Annual Costs	Required Investment
A1	3,000	$194,000	$87,500	$800,000
A2	4,000	224,000	105,000	1,000,000

The complex will be useful for 30 years and has no appreciable salvage value (regardless of seating capacity). Assuming an 8% discount rate, which of the following statements is <u>incorrect</u>?

(a) Select A1 because it has the largest **B/C** ratio.

(b) Select A1 because it has the most benefits per seating capacity.

(c) Select A1 because it has the largest PW.

(d) Select A1 because the incremental benefits generated from A2 are not large enough to offset the additional investment ($200,000 over A1).

Solution to Set No. 1

1. **Given:** $i = 8\%$, $P = 1$, and $F = 2$

Find: N

Method 1:

$$2 = 1(1 + 0.08)^N$$

$$\ln 2 = N \ln 1.08$$

$$N = \frac{\ln 2}{\ln 1.08} = 9.01 \text{ years}$$

Method 2: Using Rule of 72,

$$N \cong \frac{72}{8} = 9 \text{ years}$$

2. **Given:** $P = \$10,000$, $i = 10\%$, $N = 3$ years, and $A = \$4,021.15$
   Find: IP_2

   $n = 0$:
   $B_0 = P = \$10,000$
   $n = 1$:
   $IP_1 = \$10,000(0.10) = \$1,000$
   $PP_1 = \$4,021.15 - \$1,000 = \$3,021.15$
   $B_1 = \$10,000 - \$3,021.15 = \$6,978.85$
   $n = 2$:
   $IP_2 = \$6,978.85(0.10) = \$697.89$

3. **Given:** $i = 8\%$, $N = 15$ years, desired annual withdrawals after retirement $= \$20,000$ for 20 years
   Find: A (required annual contribution)

   - Required savings at $n = 15$:
     $$\$20,000(P/A, 8\%, 20) = \$196,362.95$$

   - Required contributions:
     $$A = \$196,362.95(A/F, 8\%, 15) = \$7,231.96$$

4. **Given:** a deposit schedule, $i = 10\%$, and $N = 10$ years
   Find: F_{10}

   $F_{10} = \$800(F/P, 10\%, 10) + \$800(F/P, 10\%, 8) + \$800(F/P, 10\%, 6)$
   $\quad + \$800(F/P, 10\%, 4) + \$800(F/P, 10\%, 2)$
   $\quad = \$7,346.39$

5. **Given:** annual deposit $= \$2,000$, $N = 10$ years, and $i = 8\%$
   Find: B_{10} and annual withdrawals over 5 years

   $$B_{10} = \$2,000(F/A, 8\%, 10)$$
   $$= \$28,973$$
   $$A = \$28,973(A/P, 8\%, 5)$$
   $$= \$7,256.47$$

6. **Given:** $A = \$10,000$, $N_1 = 5$, $i_1 = 9\%$, $N_2 = 6$, and $i_2 = 10\%$
   Find: F_{11}

   $$B_5 = \$10,000(F/A, 9\%, 5)$$
   $$= \$59,847.11$$
   $$B_{11} = \$59,847.11(F/P, 10\%, 6)$$
   $$= \$106,022.81$$

7. **Given:** $A = \$1,000$, $r = 8\%$ compounded quarterly, and $N = 40$ quarters
   Find: F_{40}

   $$F_{40} = \$1,000(F/A, 2\%, 40)$$
   $$= \$60,401.98$$

8.   **Given:** $i = 8\%$, $A = \$5,000$, $N_1 = 3$ years, and $N_2 = 10$ years
     Find: P

$$P = \$5,000(P/A, 8\%, 10)(P/F, 8\%, 2)$$
$$= \$28,764.06$$

9.   **Given:** $A = \$500$ per quarter, 2% interest per quarter, and $N = 40$ quarters
     Find: F_{40}

$$F = 500(F/A, 2\%, 40)$$
$$= \$30,200.99$$
$$F_{40} = \$30,200.99(1.02)$$
$$= \$30,805.01$$

10. **Given:** $B = \$5,000$, APR $= 9\%$ compounded monthly, $N = 12$ months, and grace period $= 12$ months
     Find: A

$$F = \$5,000(F/P, 0.75\%, 11)$$
$$= \$5,428.32$$
$$A = \$5,428.32(A/P, 0.75\%, 12)$$
$$= \$474.71$$

11. **Given:** $A_1 = \$2,000$, $G = \$500$, $N_1 = 10$, $N_2 = 5$, and $i = 10\%$
     Find: A

$$B_{10} = \$2,000(P/A, 10\%, 5) + \$500(P/G, 10\%, 5)$$
$$= \$11,012.47$$
$$A = \$11,012.47(A/F, 10\%, 10)$$
$$= \$690.98$$

12. **Given:** $F = 0.75P$ and $\overline{f} = 5\%$
     Find: N

$$0.75F = F(1 + 0.05)^{-N}$$
$$\ln 0.75 = -N \ln 1.05$$
$$N = \frac{\ln 0.75}{-\ln 1.05} = 5.90 \text{ years}$$

13. **Given:** $B = \$20,000$, APR $= 12\%$, and $N = 60$ months
     Find: A and B_{30}

$$A = \$20,000(A/P, 1\%, 60)$$
$$= \$444.89$$
$$B_{30} = \$444.89(P/A, 1\%, 30)$$
$$= \$11,481.59$$

14. **Given:** changing interest rates and withdrawal amounts over 5 years
     Find: P

$$P = \$1,000(P/F, 5\%, 1)$$
$$+ \$2,000(P/F, 7\%, 2)(P/F, 5\%, 1)$$
$$+ \$2,500(P/F, 8\%, 1)(P/F, 6\%, 1)(P/F, 7\%, 2)(P/F, 5\%, 1)$$
$$= \$4,432.65$$

15. **Given:** $i = 0.75\%$ per month, and $F = 3P$
    Find: N

$$3 = (1 + 0.0075)^N$$
$$\ln 3 = N \ln 1.0075$$
$$N = \frac{\ln 3}{\ln 1.0075} = 147.03 \text{ months}$$

16. **Given:** $A = \$5,000$ per semiannual, $N = 20$ semiannual, and $r = 8\%$ compounded continuously
    Find: F

$$i = e^{0.08/2} - 1 = 4.0811\% \text{ per semiannual}$$
$$B_{20} = \$5,000(F/A, 4.0811\%, 20)$$
$$= \$150,149.54$$
$$B_{30} = \$150,149.54(F/P, 4.0811\%, 10)$$
$$= \$223,997.28$$

17. **Given:** $P = B_0 = \$6,000$, $i = 0.5\%$ per month, and $A = \$300$ per month
    Find: N

$$\$6,000 = \$300(P/A, 0.5\%, N)$$
$$N = 22 \text{ months}$$

18. **Given:** cash flow transactions for both financing options, $r = 6\%$ compounded monthly
    Find: the incorrect statement

(a) Correct: $A = \$26,200(A/P, 3\%/12, 36) = \$761.93$
(b) Incorrect: It all depends on your opportunity cost rate.
(c) Correct: Option B is cheaper by $\$1,045.41$.

$$P_{\text{Option A}} = \$761.93(P/A, 0.5\%, 36) = \$25,045.41$$
$$P_{\text{Option B}} = \$24,000$$

(d) Correct: $i_a = (1 + 0.005)^{12} - 1 = 6.17\%$

19. **Given:** $A'_1 = \$6,000$ per year, $g = 5\%$, $\overline{f} = 4\%$, $i = 11\%$, and $N = 5$
    Find: P

$$i' = \frac{i - \overline{f}}{1 + \overline{f}} = \frac{0.11 - 0.04}{1 + 0.04} = 6.73\%$$
$$P = \$6,000\left[\frac{1 + (1 + 0.05)^5(1 + 0.0673)^{-5}}{0.0673 + 0.05}\right]$$
$$= \$27,211.74$$

20. **Given:** College expenses − $\$20,000$ (constant $\$$) per year for four years, $N = 10$ years, $\overline{f} = 5\%$, $i = 2\%$ per quarter, and $A = \$1,000$ per quarter
    Find: amount to borrow to meet her sophomore expenses College expenses in actual dollars:
    - Freshman year: $\$20,000(F/P, 5\%, 10) = \$32,577.89$
    - Sophomore year: $\$20,000(F/P, 5\%, 11) = \$34,206.78$

Amount saved at the end of year 10:
$$F = \$1,000(F/A, 2\%, 40) = \$60,401.98$$
Balance after paying the freshman year expense:
$$B_{10} = \$60,401.98 - \$32,577.89 = \$27,824.$$
Amount to borrow to cover the sophomore year expenses:
$$\Delta = \$34,206.78 - \$27,824(F/P, 2\%, 4) = \$4,089.18$$

Solution to Set No. 2

1. **Given:** project cash flows, $n = 3$, and $i = 10\%$
   Find: PB$(10\%)_3$

   $$PB(10\%)_0 = -\$100,000$$
   $$PB(10\%)_1 = -\$100,000(1 + 0.10) + \$40,000 = -\$70,000$$
   $$PB(10\%)_2 = -\$70,000(1 + 0.10) + \$40,000 = -\$37,000$$
   $$PB(10\%)_3 = -\$37,000(1 + 0.10) + \$30,000 = -\$10,700$$

2. **Given:** $B_0 = \$3,000$, $A_1 = A_2 = \$1,000$, and $i = 10\%$
   Find: A_3

   $$\$3,000 = \$1,000(P/A, 10\%, 2) + X(P/F, 10\%, 3)$$
   $$X = \frac{\$3,000 - \$1,000(P/A, 10\%, 2)}{(P/F, 10\%, 3)}$$
   $$= \$1,683$$

3. **Given:** $P = \$20,000$, $N = 10$ years, $S = 0$, $A_n = \$5,000$, and $i = 15\%$
   Find: breakeven n^*

   $$\$20,000 = \$5,000(P/A, 15\%, n)$$
   $$n^* = 7 \text{ years}$$

4. **Given:** $A_1 = \$10,000$, $G = \$5,000$ for $n = 1$ through 9, $A_{10} = A_{11} \ldots A_\infty = \$52,000$, overhaul expenses $= \$40,000$ every 10 years, $i = 10\%$, PW$(10\%) = \$100,000$
   Find: I

   $$PW(10\%) = -I + \$10,000(P/A, 10\%, 9) + \$5,000(P/G, 10\%, 9)$$
   $$+ \frac{\$52,000}{0.10}(P/F, 10\%, 9)$$
   $$- \frac{\$40,000(A/F, 10\%, 10)}{0.10}$$
   $$= \$100,000$$
   $$I = \$250,130$$

5. **Given:** $I = \$55,000$, $S = \$12,000$, $N = 5$ years, $A_1 = \$18,000$, $G = \$3000$, and $i = 10\%$
   Find: AEC(10%)

   $$CR(10\%) = (\$55,000 - \$12,000)(A/P, 10\%, 5) + \$12,000(0.10)$$
   $$= \$12,543.29$$
   $$O\&M(10\%) = \$18,000 + \$3,000(A/G, 10\%, 5)$$
   $$= \$23,430.37$$

$$AEC(10\%) = \$12,543.29 + \$23,430.37$$
$$= \$35,973.66$$

6.   **Given:** Cash flow diagram and $i = 10\%$
     Find: X

$$PW(10\%) = -\$2,145 + X(P/A, 10\%, 6)$$
$$+ (\$400 - X)(P/F, 10\%, 1)$$
$$+ (\$400 - X)(P/F, 10\%, 4)$$
$$= -\$2,145 + 4.3553X + \$363.64 - 0.9090X + \$273.21 - 0.6830X$$
$$= -\$1,508.15 + 2.7633X$$
$$= 0$$
$$X = \$545.78$$

7.   **Given:** $I = \$100,000, S = 0, A_1 = \$30,000, g = 3\%, N = 5$ years, operating hours $= 3,000$ hours per year, $i = 14\%$
     Find: dollar saving per hour

$$-\$100,000(A/P, 14\%, 5)$$
$$+ 30,000(P/A_1, 3\%, 14\%, 5)(A/P, 14\%, 5)$$
$$= \$2,481.95$$

$$\$ \text{ savings per hour} = \frac{\$2,481.95}{3,000}$$
$$= \$0.8273$$

8.   **Given:** Oil-based painting $- I_1 = \$5,000, N_1 = 10$ years, $S_1 = 0$; water-based painting $- I_2 = \$3,000, N_2 = 5$ years, $S_2 = 0$, analysis period $= 10$ years, and $i = 10\%$
     Find: which statement is correct?

   • Oil-based painting:

$$AEC(10\%)_1 = \$5,000(A/P, 10\%, 10) = \$813.72$$

   • Water-based painting:

$$AEC(10\%)_2 = \$3,000(A/P, 10\%, 5) = \$791.39$$

9.   **Given:** $I = \$50,000, S = \$10,000, N = 3$ years, ROR $= 15\%$
     Find: X

$$\$50,000 = X(P/A, 15\%, 3) + \$10,000(P/F, 15\%, 3)$$
$$X = \$19,019.08$$

10. **Given:** $S = 0, N = 10$ years, $A = \$60,000$, and IRR $= 15\%$
     Find: I

$$PW(15\%) = -I + \$60,000(P/A, 15\%, 10)$$
$$= 0$$
$$I = \$301,126$$

11. **Given:** 150 HP motors, $I_1 = \$4,500$, $S_1 = 0$, $N_1 = 10$ years, $j_1 = 83\%$, $I_2 = \$3,600$, $S2 = 0$, $N_2 = 10$ years, $j_2 = 80\%$, power cost $= \$0.05$ per kWh, and $i = 6\%$
    Find: breakeven operating hours

    $$\$4,500(A/P, 6\%, 10) + \frac{150(0.7457)}{0.83}(\omega)(0.05) + 0.15(\$4,500)$$

    $$< \$3,600(A/P, 6\%, 10) + \frac{150(0.7457)}{0.80}(\omega)(0.05) + 0.15(\$3,600)$$

    $\$611.41 + 6.7381\omega + \$675 < \$489.12 + 6.9909w + \$540$
    $\omega^* > 1,017.76$ hours per year

12. **Given:** $I = \$56,000$, $N = 5$ years, $S = \$5,000$, annual operating hours $= \$2,500$ hours, annual O&M $= \$6,000$, and $i = 15\%$
    Find: AEC(15%)

    $$CR(15\%) = (\$56,000 - \$5,000)(A/P, 15\%, 5) + (0.15)(\$5,000)$$
    $$= \$15,964.09$$
    $$O\&M(15\%) = \$6,000$$
    $$AEC(15\%) = \$15,964.09 + \$6,000$$
    $$= \$21,964.09$$
    $$\text{Hourly cost} = \frac{\$21,964.09}{2,500} = \$8.79$$

13. **Given:** project cash flows and IRRs
    Find: which statement is correct?
    Perform incremental analysis (Project B – Project A):

    $$-\$2,000 + \$3,906(P/F, i, 3) = 0$$

    $$IRR_{B-A} = \$25\%$$

    Project B is a better project if MARR $< 25\%$.

14. **Given:** repeating cash flow series, $N = \infty$, and $i = 10\%$
    Find: CE(10%)

    $$AE(10\%) = \$1,000 - \$200(A/G, 10\%, 4)$$
    $$= \$723.76$$
    $$CE(10\%) = \frac{\$723.76}{0.10} = \$7,237.60$$

15. **Given:** reimbursement rate $= \$0.58$ per mile, $I = \$24,000$, $N = 5$ years, $S = \$7,000$, variable O&M costs $= \$0.30$ per mile, fixed O&M $= \$2500$ per year, and $i = 15\%$
    Find: breakeven mileage (M)

    • Option 1: $0.58 per mile

    $$AEC(15\%) = \$0.58(M)$$

    • Option 2:

    $$CR(15\%) = (\$24,000 - \$7,000)(A/P, 15\%, 5) + (0.15)(\$7,000)$$
    $$= \$6,121.36$$

$$O\&M(15\%) = \$2{,}500 + 0.30M$$

$$AEC(15\%) = \$6{,}121.36 + \$2{,}500 + 0.30M$$

Breakeven mileage:

$$0.58M = \$6{,}121.36 + \$2{,}500 + 0.30M$$

$$0.28M = \$8{,}621.36$$

$$M^* = 30{,}790.57 \text{ miles}$$

16. **Given:** $I = \$60{,}000$, $N = 5$ years, $S = \$5{,}000$, 7-year MACRS, and current market value $= \$10{,}000$
    Find: book value

$$B_5 = \$60{,}000 - \$60{,}000\left(0.1429 + 0.2449 + 0.1749 + 0.1249 + \frac{0.0893}{2}\right)$$

$$= \$16{,}065$$

17. **Given:** Financial facts as summarized in the problem statement
    Find: PW(15%) on after-tax basis

Income Statement	0	1	2	3
Revenue:				
Savings		$45,000	$45,000	$45,000
Expenses:				
Depreciation		$20,000	$32,000	$9,600
Taxable income		$25,000	$13,000	$35,400
Income taxes (40%)		$10,000	$5,200	$14,160
Net income		$15,000	$7,800	$21,240
Cash Flow Statement				
Operating activities:				
Net income		$15,000	$7,800	$21,240
Depreciation		$20,000	$32,000	$9,600
Investing activities:				
Investment	−$100,000			
Salvage value				$30,000
Gains tax				$3,360
Net cash flow	−$100,000	$35,000	$39,800	$64,200
PW(15%) =		$2,741.84		

18. **Given:** Cash flows for two mutually exclusive projects
    Find: range of MARR where project 2 is preferred
    Project 2 is preferred as long as MARR is less than 12.50%.

	Net Cash Flow		
Year	Project 1	Project 2	Project 2 – Project 1
0	−$1,200	−$2,000	−$800
1	$600	$1,500	$900
2	$1,000	$1,000	$0
IRR	19.65%	17.54%	12.50%

19. **Given:** Defender − current market value = $2,000, $N = 3$ years, O & M = $10,000 per year; challenger − $I = $20,000, $N = 3$ years, $S = $5,000, O&M = $3,000 per year, and $i = 15\%$

    Find: AEC(15%)$_{\text{Challenger − Defender}}$

    • Defender:
    $$\text{PW}(15\%) = -\$2,000 - \$10,000(P/A, 15\%, 3)$$
    $$= -\$24,832.25$$
    $$\text{AEC}(15\%) = \$24,832.25(A/P, 15\%, 3)$$
    $$= \$10,875.95$$

    • Challenger:
    $$\text{PW}(15\%) = -\$20,000 - \$3,000(P/A, 15\%, 3) + \$5,000(P/F, 15\%, 3)$$
    $$= -\$23,562.09$$
    $$\text{AEC}(15\%) = \$23,562.09(A/P, 15\%, 3)$$
    $$= \$10,319.65$$

    Incremental annual benefit = $\$10,875.95 - \$10,319.65 = \$556.30$

20. **Given:** Financial data for two alternative design proposals, $N = 30$ years, and $i = 8\%$

    Find: which one is an incorrect statement?

    • Present worth analysis:
    $$\text{PW}(8\%)_1 = -\$800,000 + (\$194,000 - \$87,500)(P/A, 8\%, 30) =$$
    $$\$398,953.50$$
    $$\text{PW}(8\%)_2 = -\$1,000,000 + (\$224,000 - \$105,000)(P/A, 8\%, 30) =$$
    $$\$339,675.70$$

    • Benefit-cost analysis:
    $$\text{BC}(8\%)_1 = \frac{\$194,000(P/A, 8\%, 30)}{\$800,000 + \$87,500(P/A, 8\%, 30)} = 1.22$$
    $$\text{BC}(8\%)_2 = \frac{\$224,000(P/A, 8\%, 30)}{\$1,000,000 + \$105,000(P/A, 8\%, 30)} = 1.16$$
    $$\text{BC}(8\%)_{2-1}$$
    $$= \frac{(\$224,000 - \$194,000)(P/A, 8\%, 30)}{(\$1,000,000 - \$800,000) + (\$105,000 - \$87,500)(P/A, 8\%, 30)} = 0.85$$

    Even though the BC ratio for Design option 1 is greater than that of design option 2, a correct procedure is to use an incremental analysis to determine the best decision alternative. In our example, the incremental BC ratio is 0.85, indicating design option 1 is a better choice. The incorrect answer is (a).

INTEREST FACTORS FOR DISCRETE COMPOUNDING

0.25%

	Single Payment		Equal Payment Series				Gradient Series		
N	Compound Amount Factor (F/P,i,N)	Present Worth Factor (P/F,i,N)	Compound Amount Factor (F/A,i,N)	Sinking Fund Factor (A/F,i,N)	Present Worth Factor (P/A,i,N)	Capital Recovery Factor (A/P,i,N)	Gradient Uniform Series (A/G,i,N)	Gradient Present Worth (P/G,i,N)	N
1	1.0025	0.9975	1.0000	1.0000	0.9975	1.0025	0.0000	0.0000	1
2	1.0050	0.9950	2.0025	0.4994	1.9925	0.5019	0.4994	0.9950	2
3	1.0075	0.9925	3.0075	0.3325	2.9851	0.3350	0.9983	2.9801	3
4	1.0100	0.9901	4.0150	0.2491	3.9751	0.2516	1.4969	5.9503	4
5	1.0126	0.9876	5.0251	0.1990	4.9627	0.2015	1.9950	9.9007	5
6	1.0151	0.9851	6.0376	0.1656	5.9478	0.1681	2.4927	14.8263	6
7	1.0176	0.9827	7.0527	0.1418	6.9305	0.1443	2.9900	20.7223	7
8	1.0202	0.9802	8.0704	0.1239	7.9107	0.1264	3.4869	27.5839	8
9	1.0227	0.9778	9.0905	0.1100	8.8885	0.1125	3.9834	35.4061	9
10	1.0253	0.9753	10.1133	0.0989	9.8639	0.1014	4.4794	44.1842	10
11	1.0278	0.9729	11.1385	0.0898	10.8368	0.0923	4.9750	53.9133	11
12	1.0304	0.9705	12.1664	0.0822	11.8073	0.0847	5.4702	64.5886	12
13	1.0330	0.9681	13.1968	0.0758	12.7753	0.0783	5.9650	76.2053	13
14	1.0356	0.9656	14.2298	0.0703	13.7410	0.0728	6.4594	88.7587	14
15	1.0382	0.9632	15.2654	0.0655	14.7042	0.0680	6.9534	102.2441	15
16	1.0408	0.9608	16.3035	0.0613	15.6650	0.0638	7.4469	116.6567	16
17	1.0434	0.9584	17.3443	0.0577	16.6235	0.0602	7.9401	131.9917	17
18	1.0460	0.9561	18.3876	0.0544	17.5795	0.0569	8.4328	148.2446	18
19	1.0486	0.9537	19.4336	0.0515	18.5332	0.0540	8.9251	165.4106	19
20	1.0512	0.9513	20.4822	0.0488	19.4845	0.0513	9.4170	183.4851	20
21	1.0538	0.9489	21.5334	0.0464	20.4334	0.0489	9.9085	202.4634	21
22	1.0565	0.9466	22.5872	0.0443	21.3800	0.0468	10.3995	222.3410	22
23	1.0591	0.9442	23.6437	0.0423	22.3241	0.0448	10.8901	243.1131	23
24	1.0618	0.9418	24.7028	0.0405	23.2660	0.0430	11.3804	264.7753	24
25	1.0644	0.9395	25.7646	0.0388	24.2055	0.0413	11.8702	287.3230	25
26	1.0671	0.9371	26.8290	0.0373	25.1426	0.0398	12.3596	310.7516	26
27	1.0697	0.9348	27.8961	0.0358	26.0774	0.0383	12.8485	335.0566	27
28	1.0724	0.9325	28.9658	0.0345	27.0099	0.0370	13.3371	360.2334	28
29	1.0751	0.9301	30.0382	0.0333	27.9400	0.0358	13.8252	386.2776	29
30	1.0778	0.9278	31.1133	0.0321	28.8679	0.0346	14.3130	413.1847	30
31	1.0805	0.9255	32.1911	0.0311	29.7934	0.0336	14.8003	440.9502	31
32	1.0832	0.9232	33.2716	0.0301	30.7166	0.0326	15.2872	469.5696	32
33	1.0859	0.9209	34.3547	0.0291	31.6375	0.0316	15.7736	499.0386	33
34	1.0886	0.9186	35.4406	0.0282	32.5561	0.0307	16.2597	529.3528	34
35	1.0913	0.9163	36.5292	0.0274	33.4724	0.0299	16.7454	560.5076	35
36	1.0941	0.9140	37.6206	0.0266	34.3865	0.0291	17.2306	592.4988	36
40	1.1050	0.9050	42.0132	0.0238	38.0199	0.0263	19.1673	728.7399	40
48	1.1273	0.8871	50.9312	0.0196	45.1787	0.0221	23.0209	1040.0552	48
50	1.1330	0.8826	53.1887	0.0188	46.9462	0.0213	23.9802	1125.7767	50
60	1.1616	0.8609	64.6467	0.0155	55.6524	0.0180	28.7514	1600.0845	60
72	1.1969	0.8355	78.7794	0.0127	65.8169	0.0152	34.4221	2265.5569	72
80	1.2211	0.8189	88.4392	0.0113	72.4260	0.0138	38.1694	2764.4568	80
84	1.2334	0.8108	93.3419	0.0107	75.6813	0.0132	40.0331	3029.7592	84
90	1.2520	0.7987	100.7885	0.0099	80.5038	0.0124	42.8162	3446.8700	90
96	1.2709	0.7869	108.3474	0.0092	85.2546	0.0117	45.5844	3886.2832	96
100	1.2836	0.7790	113.4500	0.0088	88.3825	0.0113	47.4216	4191.2417	100
108	1.3095	0.7636	123.8093	0.0081	94.5453	0.0106	51.0762	4829.0125	108
120	1.3494	0.7411	139.7414	0.0072	103.5618	0.0097	56.5084	5852.1116	120
240	1.8208	0.5492	328.3020	0.0030	180.3109	0.0055	107.5863	19398.9852	240
360	2.4568	0.4070	582.7369	0.0017	237.1894	0.0042	152.8902	36263.9299	360

0.50%

	Single Payment		Equal Payment Series				Gradient Series		
N	Compound Amount Factor (F/P,i,N)	Present Worth Factor (P/F,i,N)	Compound Amount Factor (F/A,i,N)	Sinking Fund Factor (A/F,i,N)	Present Worth Factor (P/A,i,N)	Capital Recovery Factor (A/P,i,N)	Gradient Uniform Series (A/G,i,N)	Gradient Present Worth (P/G,i,N)	N
1	1.0050	0.9950	1.0000	1.0000	0.9950	1.0050	0.0000	0.0000	1
2	1.0100	0.9901	2.0050	0.4988	1.9851	0.5038	0.4988	0.9901	2
3	1.0151	0.9851	3.0150	0.3317	2.9702	0.3367	0.9967	2.9604	3
4	1.0202	0.9802	4.0301	0.2481	3.9505	0.2531	1.4938	5.9011	4
5	1.0253	0.9754	5.0503	0.1980	4.9259	0.2030	1.9900	9.8026	5
6	1.0304	0.9705	6.0755	0.1646	5.8964	0.1696	2.4855	14.6552	6
7	1.0355	0.9657	7.1059	0.1407	6.8621	0.1457	2.9801	20.4493	7
8	1.0407	0.9609	8.1414	0.1228	7.8230	0.1278	3.4738	27.1755	8
9	1.0459	0.9561	9.1821	0.1089	8.7791	0.1139	3.9668	34.8244	9
10	1.0511	0.9513	10.2280	0.0978	9.7304	0.1028	4.4589	43.3865	10
11	1.0564	0.9466	11.2792	0.0887	10.6770	0.0937	4.9501	52.8526	11
12	1.0617	0.9419	12.3356	0.0811	11.6189	0.0861	5.4406	63.2136	12
13	1.0670	0.9372	13.3972	0.0746	12.5562	0.0796	5.9302	74.4602	13
14	1.0723	0.9326	14.4642	0.0691	13.4887	0.0741	6.4190	86.5835	14
15	1.0777	0.9279	15.5365	0.0644	14.4166	0.0694	6.9069	99.5743	15
16	1.0831	0.9233	16.6142	0.0602	15.3399	0.0652	7.3940	113.4238	16
17	1.0885	0.9187	17.6973	0.0565	16.2586	0.0615	7.8803	128.1231	17
18	1.0939	0.9141	18.7858	0.0532	17.1728	0.0582	8.3658	143.6634	18
19	1.0994	0.9096	19.8797	0.0503	18.0824	0.0553	8.8504	160.0360	19
20	1.1049	0.9051	20.9791	0.0477	18.9874	0.0527	9.3342	177.2322	20
21	1.1104	0.9006	22.0840	0.0453	19.8880	0.0503	9.8172	195.2434	21
22	1.1160	0.8961	23.1944	0.0431	20.7841	0.0481	10.2993	214.0611	22
23	1.1216	0.8916	24.3104	0.0411	21.6757	0.0461	10.7806	233.6768	23
24	1.1272	0.8872	25.4320	0.0393	22.5629	0.0443	11.2611	254.0820	24
25	1.1328	0.8828	26.5591	0.0377	23.4456	0.0427	11.7407	275.2686	25
26	1.1385	0.8784	27.6919	0.0361	24.3240	0.0411	12.2195	297.2281	26
27	1.1442	0.8740	28.8304	0.0347	25.1980	0.0397	12.6975	319.9523	27
28	1.1499	0.8697	29.9745	0.0334	26.0677	0.0384	13.1747	343.4332	28
29	1.1556	0.8653	31.1244	0.0321	26.9330	0.0371	13.6510	367.6625	29
30	1.1614	0.8610	32.2800	0.0310	27.7941	0.0360	14.1265	392.6324	30
31	1.1672	0.8567	33.4414	0.0299	28.6508	0.0349	14.6012	418.3348	31
32	1.1730	0.8525	34.6086	0.0289	29.5033	0.0339	15.0750	444.7618	32
33	1.1789	0.8482	35.7817	0.0279	30.3515	0.0329	15.5480	471.9055	33
34	1.1848	0.8440	36.9606	0.0271	31.1955	0.0321	16.0202	499.7583	34
35	1.1907	0.8398	38.1454	0.0262	32.0354	0.0312	16.4915	528.3123	35
36	1.1967	0.8356	39.3361	0.0254	32.8710	0.0304	16.9621	557.5598	36
40	1.2208	0.8191	44.1588	0.0226	36.1722	0.0276	18.8359	681.3347	40
48	1.2705	0.7871	54.0978	0.0185	42.5803	0.0235	22.5437	959.9188	48
50	1.2832	0.7793	56.6452	0.0177	44.1428	0.0227	23.4624	1035.6966	50
60	1.3489	0.7414	69.7700	0.0143	51.7256	0.0193	28.0064	1448.6458	60
72	1.4320	0.6983	86.4089	0.0116	60.3395	0.0166	33.3504	2012.3478	72
80	1.4903	0.6710	98.0677	0.0102	65.8023	0.0152	36.8474	2424.6455	80
84	1.5204	0.6577	104.0739	0.0096	68.4530	0.0146	38.5763	2640.6641	84
90	1.5666	0.6383	113.3109	0.0088	72.3313	0.0138	41.1451	2976.0769	90
96	1.6141	0.6195	122.8285	0.0081	76.0952	0.0131	43.6845	3324.1846	96
100	1.6467	0.6073	129.3337	0.0077	78.5426	0.0127	45.3613	3562.7934	100
108	1.7137	0.5835	142.7399	0.0070	83.2934	0.0120	48.6758	4054.3747	108
120	1.8194	0.5496	163.8793	0.0061	90.0735	0.0111	53.5508	4823.5051	120
240	3.3102	0.3021	462.0409	0.0022	139.5808	0.0072	96.1131	13415.5395	240
360	6.0226	0.1660	1004.5150	0.0010	166.7916	0.0060	128.3236	21403.3041	360

0.75%

	Single Payment		Equal Payment Series				Gradient Series		
N	Compound Amount Factor (F/P,i,N)	Present Worth Factor (P/F,i,N)	Compound Amount Factor (F/A,i,N)	Sinking Fund Factor (A/F,i,N)	Present Worth Factor (P/A,i,N)	Capital Recovery Factor (A/P,i,N)	Gradient Uniform Series (A/G,i,N)	Gradient Present Worth (P/G,i,N)	N
1	1.0075	0.9926	1.0000	1.0000	0.9926	1.0075	0.0000	0.0000	1
2	1.0151	0.9852	2.0075	0.4981	1.9777	0.5056	0.4981	0.9852	2
3	1.0227	0.9778	3.0226	0.3308	2.9556	0.3383	0.9950	2.9408	3
4	1.0303	0.9706	4.0452	0.2472	3.9261	0.2547	1.4907	5.8525	4
5	1.0381	0.9633	5.0756	0.1970	4.8894	0.2045	1.9851	9.7058	5
6	1.0459	0.9562	6.1136	0.1636	5.8456	0.1711	2.4782	14.4866	6
7	1.0537	0.9490	7.1595	0.1397	6.7946	0.1472	2.9701	20.1808	7
8	1.0616	0.9420	8.2132	0.1218	7.7366	0.1293	3.4608	26.7747	8
9	1.0696	0.9350	9.2748	0.1078	8.6716	0.1153	3.9502	34.2544	9
10	1.0776	0.9280	10.3443	0.0967	9.5996	0.1042	4.4384	42.6064	10
11	1.0857	0.9211	11.4219	0.0876	10.5207	0.0951	4.9253	51.8174	11
12	1.0938	0.9142	12.5076	0.0800	11.4349	0.0875	5.4110	61.8740	12
13	1.1020	0.9074	13.6014	0.0735	12.3423	0.0810	5.8954	72.7632	13
14	1.1103	0.9007	14.7034	0.0680	13.2430	0.0755	6.3786	84.4720	14
15	1.1186	0.8940	15.8137	0.0632	14.1370	0.0707	6.8606	96.9876	15
16	1.1270	0.8873	16.9323	0.0591	15.0243	0.0666	7.3413	110.2973	16
17	1.1354	0.8807	18.0593	0.0554	15.9050	0.0629	7.8207	124.3887	17
18	1.1440	0.8742	19.1947	0.0521	16.7792	0.0596	8.2989	139.2494	18
19	1.1525	0.8676	20.3387	0.0492	17.6468	0.0567	8.7759	154.8671	19
20	1.1612	0.8612	21.4912	0.0465	18.5080	0.0540	9.2516	171.2297	20
21	1.1699	0.8548	22.6524	0.0441	19.3628	0.0516	9.7261	188.3253	21
22	1.1787	0.8484	23.8223	0.0420	20.2112	0.0495	10.1994	206.1420	22
23	1.1875	0.8421	25.0010	0.0400	21.0533	0.0475	10.6714	224.6682	23
24	1.1964	0.8358	26.1885	0.0382	21.8891	0.0457	11.1422	243.8923	24
25	1.2054	0.8296	27.3849	0.0365	22.7188	0.0440	11.6117	263.8029	25
26	1.2144	0.8234	28.5903	0.0350	23.5422	0.0425	12.0800	284.3888	26
27	1.2235	0.8173	29.8047	0.0336	24.3595	0.0411	12.5470	305.6387	27
28	1.2327	0.8112	31.0282	0.0322	25.1707	0.0397	13.0128	327.5416	28
29	1.2420	0.8052	32.2609	0.0310	25.9759	0.0385	13.4774	350.0867	29
30	1.2513	0.7992	33.5029	0.0298	26.7751	0.0373	13.9407	373.2631	30
31	1.2607	0.7932	34.7542	0.0288	27.5683	0.0363	14.4028	397.0602	31
32	1.2701	0.7873	36.0148	0.0278	28.3557	0.0353	14.8636	421.4675	32
33	1.2796	0.7815	37.2849	0.0268	29.1371	0.0343	15.3232	446.4746	33
34	1.2892	0.7757	38.5646	0.0259	29.9128	0.0334	15.7816	472.0712	34
35	1.2989	0.7699	39.8538	0.0251	30.6827	0.0326	16.2387	498.2471	35
36	1.3086	0.7641	41.1527	0.0243	31.4468	0.0318	16.6946	524.9924	36
40	1.3483	0.7416	46.4465	0.0215	34.4469	0.0290	18.5058	637.4693	40
48	1.4314	0.6986	57.5207	0.0174	40.1848	0.0249	22.0691	886.8404	48
50	1.4530	0.6883	60.3943	0.0166	41.5664	0.0241	22.9476	953.8486	50
60	1.5657	0.6387	75.4241	0.0133	48.1734	0.0208	27.2665	1313.5189	60
72	1.7126	0.5839	95.0070	0.0105	55.4768	0.0180	32.2882	1791.2463	72
80	1.8180	0.5500	109.0725	0.0092	59.9944	0.0167	35.5391	2132.1472	80
84	1.8732	0.5338	116.4269	0.0086	62.1540	0.0161	37.1357	2308.1283	84
90	1.9591	0.5104	127.8790	0.0078	65.2746	0.0153	39.4946	2577.9961	90
96	2.0489	0.4881	139.8562	0.0072	68.2584	0.0147	41.8107	2853.9352	96
100	2.1111	0.4737	148.1445	0.0068	70.1746	0.0143	43.3311	3040.7453	100
108	2.2411	0.4462	165.4832	0.0060	73.8394	0.0135	46.3154	3419.9041	108
120	2.4514	0.4079	193.5143	0.0052	78.9417	0.0127	50.6521	3998.5621	120
240	6.0092	0.1664	667.8869	0.0015	111.1450	0.0090	85.4210	9494.1162	240
360	14.7306	0.0679	1830.7435	0.0005	124.2819	0.0080	107.1145	13312.3871	360

1.0%

	Single Payment		Equal Payment Series				Gradient Series		
N	Compound Amount Factor (F/P,i,N)	Present Worth Factor (P/F,i,N)	Compound Amount Factor (F/A,i,N)	Sinking Fund Factor (A/F,i,N)	Present Worth Factor (P/A,i,N)	Capital Recovery Factor (A/P,i,N)	Gradient Uniform Series (A/G,i,N)	Gradient Present Worth (P/G,i,N)	**N**
1	1.0100	0.9901	1.0000	1.0000	0.9901	1.0100	0.0000	0.0000	1
2	1.0201	0.9803	2.0100	0.4975	1.9704	0.5075	0.4975	0.9803	2
3	1.0303	0.9706	3.0301	0.3300	2.9410	0.3400	0.9934	2.9215	3
4	1.0406	0.9610	4.0604	0.2463	3.9020	0.2563	1.4876	5.8044	4
5	1.0510	0.9515	5.1010	0.1960	4.8534	0.2060	1.9801	9.6103	5
6	1.0615	0.9420	6.1520	0.1625	5.7955	0.1725	2.4710	14.3205	6
7	1.0721	0.9327	7.2135	0.1386	6.7282	0.1486	2.9602	19.9168	7
8	1.0829	0.9235	8.2857	0.1207	7.6517	0.1307	3.4478	26.3812	8
9	1.0937	0.9143	9.3685	0.1067	8.5660	0.1167	3.9337	33.6959	9
10	1.1046	0.9053	10.4622	0.0956	9.4713	0.1056	4.4179	41.8435	10
11	1.1157	0.8963	11.5668	0.0865	10.3676	0.0965	4.9005	50.8067	11
12	1.1268	0.8874	12.6825	0.0788	11.2551	0.0888	5.3815	60.5687	12
13	1.1381	0.8787	13.8093	0.0724	12.1337	0.0824	5.8607	71.1126	13
14	1.1495	0.8700	14.9474	0.0669	13.0037	0.0769	6.3384	82.4221	14
15	1.1610	0.8613	16.0969	0.0621	13.8651	0.0721	6.8143	94.4810	15
16	1.1726	0.8528	17.2579	0.0579	14.7179	0.0679	7.2886	107.2734	16
17	1.1843	0.8444	18.4304	0.0543	15.5623	0.0643	7.7613	120.7834	17
18	1.1961	0.8360	19.6147	0.0510	16.3983	0.0610	8.2323	134.9957	18
19	1.2081	0.8277	20.8109	0.0481	17.2260	0.0581	8.7017	149.8950	19
20	1.2202	0.8195	22.0190	0.0454	18.0456	0.0554	9.1694	165.4664	20
21	1.2324	0.8114	23.2392	0.0430	18.8570	0.0530	9.6354	181.6950	21
22	1.2447	0.8034	24.4716	0.0409	19.6604	0.0509	10.0998	198.5663	22
23	1.2572	0.7954	25.7163	0.0389	20.4558	0.0489	10.5626	216.0660	23
24	1.2697	0.7876	26.9735	0.0371	21.2434	0.0471	11.0237	234.1800	24
25	1.2824	0.7798	28.2432	0.0354	22.0232	0.0454	11.4831	252.8945	25
26	1.2953	0.7720	29.5256	0.0339	22.7952	0.0439	11.9409	272.1957	26
27	1.3082	0.7644	30.8209	0.0324	23.5596	0.0424	12.3971	292.0702	27
28	1.3213	0.7568	32.1291	0.0311	24.3164	0.0411	12.8516	312.5047	28
29	1.3345	0.7493	33.4504	0.0299	25.0658	0.0399	13.3044	333.4863	29
30	1.3478	0.7419	34.7849	0.0287	25.8077	0.0387	13.7557	355.0021	30
31	1.3613	0.7346	36.1327	0.0277	26.5423	0.0377	14.2052	377.0394	31
32	1.3749	0.7273	37.4941	0.0267	27.2696	0.0367	14.6532	399.5858	32
33	1.3887	0.7201	38.8690	0.0257	27.9897	0.0357	15.0995	422.6291	33
34	1.4026	0.7130	40.2577	0.0248	28.7027	0.0348	15.5441	446.1572	34
35	1.4166	0.7059	41.6603	0.0240	29.4086	0.0340	15.9871	470.1583	35
36	1.4308	0.6989	43.0769	0.0232	30.1075	0.0332	16.4285	494.6207	36
40	1.4889	0.6717	48.8864	0.0205	32.8347	0.0305	18.1776	596.8561	40
48	1.6122	0.6203	61.2226	0.0163	37.9740	0.0263	21.5976	820.1460	48
50	1.6446	0.6080	64.4632	0.0155	39.1961	0.0255	22.4363	879.4176	50
60	1.8167	0.5504	81.6697	0.0122	44.9550	0.0222	26.5333	1192.8061	60
72	2.0471	0.4885	104.7099	0.0096	51.1504	0.0196	31.2386	1597.8673	72
80	2.2167	0.4511	121.6715	0.0082	54.8882	0.0182	34.2492	1879.8771	80
84	2.3067	0.4335	130.6723	0.0077	56.6485	0.0177	35.7170	2023.3153	84
90	2.4486	0.4084	144.8633	0.0069	59.1609	0.0169	37.8724	2240.5675	90
96	2.5993	0.3847	159.9273	0.0063	61.5277	0.0163	39.9727	2459.4298	96
100	2.7048	0.3697	170.4814	0.0059	63.0289	0.0159	41.3426	2605.7758	100
108	2.9289	0.3414	192.8926	0.0052	65.8578	0.0152	44.0103	2898.4203	108
120	3.3004	0.3030	230.0387	0.0043	69.7005	0.0143	47.8349	3334.1148	120
240	10.8926	0.0918	989.2554	0.0010	90.8194	0.0110	75.7393	6878.6016	240
360	35.9496	0.0278	3494.9641	0.0003	97.2183	0.0103	89.6995	8720.4323	360

1.25%

	Single Payment		Equal Payment Series				Gradient Series		
	Compound Amount Factor	Present Worth Factor	Compound Amount Factor	Sinking Fund Factor	Present Worth Factor	Capital Recovery Factor	Gradient Uniform Series	Gradient Present Worth	
N	(F/P,i,N)	(P/F,i,N)	(F/A,i,N)	(A/F,i,N)	(P/A,i,N)	(A/P,i,N)	(A/G,i,N)	(P/G,i,N)	N
1	1.0125	0.9877	1.0000	1.0000	0.9877	1.0125	0.0000	0.0000	1
2	1.0252	0.9755	2.0125	0.4969	1.9631	0.5094	0.4969	0.9755	2
3	1.0380	0.9634	3.0377	0.3292	2.9265	0.3417	0.9917	2.9023	3
4	1.0509	0.9515	4.0756	0.2454	3.8781	0.2579	1.4845	5.7569	4
5	1.0641	0.9398	5.1266	0.1951	4.8178	0.2076	1.9752	9.5160	5
6	1.0774	0.9282	6.1907	0.1615	5.7460	0.1740	2.4638	14.1569	6
7	1.0909	0.9167	7.2680	0.1376	6.6627	0.1501	2.9503	19.6571	7
8	1.1045	0.9054	8.3589	0.1196	7.5681	0.1321	3.4348	25.9949	8
9	1.1183	0.8942	9.4634	0.1057	8.4623	0.1182	3.9172	33.1487	9
10	1.1323	0.8832	10.5817	0.0945	9.3455	0.1070	4.3975	41.0973	10
11	1.1464	0.8723	11.7139	0.0854	10.2178	0.0979	4.8758	49.8201	11
12	1.1608	0.8615	12.8604	0.0778	11.0793	0.0903	5.3520	59.2967	12
13	1.1753	0.8509	14.0211	0.0713	11.9302	0.0838	5.8262	69.5072	13
14	1.1900	0.8404	15.1964	0.0658	12.7706	0.0783	6.2982	80.4320	14
15	1.2048	0.8300	16.3863	0.0610	13.6005	0.0735	6.7682	92.0519	15
16	1.2199	0.8197	17.5912	0.0568	14.4203	0.0693	7.2362	104.3481	16
17	1.2351	0.8096	18.8111	0.0532	15.2299	0.0657	7.7021	117.3021	17
18	1.2506	0.7996	20.0462	0.0499	16.0295	0.0624	8.1659	130.8958	18
19	1.2662	0.7898	21.2968	0.0470	16.8193	0.0595	8.6277	145.1115	19
20	1.2820	0.7800	22.5630	0.0443	17.5993	0.0568	9.0874	159.9316	20
21	1.2981	0.7704	23.8450	0.0419	18.3697	0.0544	9.5450	175.3392	21
22	1.3143	0.7609	25.1431	0.0398	19.1306	0.0523	10.0006	191.3174	22
23	1.3307	0.7515	26.4574	0.0378	19.8820	0.0503	10.4542	207.8499	23
24	1.3474	0.7422	27.7881	0.0360	20.6242	0.0485	10.9056	224.9204	24
25	1.3642	0.7330	29.1354	0.0343	21.3573	0.0468	11.3551	242.5132	25
26	1.3812	0.7240	30.4996	0.0328	22.0813	0.0453	11.8024	260.6128	26
27	1.3985	0.7150	31.8809	0.0314	22.7963	0.0439	12.2478	279.2040	27
28	1.4160	0.7062	33.2794	0.0300	23.5025	0.0425	12.6911	298.2719	28
29	1.4337	0.6975	34.6954	0.0288	24.2000	0.0413	13.1323	317.8019	29
30	1.4516	0.6889	36.1291	0.0277	24.8889	0.0402	13.5715	337.7797	30
31	1.4698	0.6804	37.5807	0.0266	25.5693	0.0391	14.0086	358.1912	31
32	1.4881	0.6720	39.0504	0.0256	26.2413	0.0381	14.4438	379.0227	32
33	1.5067	0.6637	40.5386	0.0247	26.9050	0.0372	14.8768	400.2607	33
34	1.5256	0.6555	42.0453	0.0238	27.5605	0.0363	15.3079	421.8920	34
35	1.5446	0.6474	43.5709	0.0230	28.2079	0.0355	15.7369	443.9037	35
36	1.5639	0.6394	45.1155	0.0222	28.8473	0.0347	16.1639	466.2830	36
40	1.6436	0.6084	51.4896	0.0194	31.3269	0.0319	17.8515	559.2320	40
48	1.8154	0.5509	65.2284	0.0153	35.9315	0.0278	21.1299	759.2296	48
50	1.8610	0.5373	68.8818	0.0145	37.0129	0.0270	21.9295	811.6738	50
60	2.1072	0.4746	88.5745	0.0113	42.0346	0.0238	25.8083	1084.8429	60
72	2.4459	0.4088	115.6736	0.0086	47.2925	0.0211	30.2047	1428.4561	72
80	2.7015	0.3702	136.1188	0.0073	50.3867	0.0198	32.9822	1661.8651	80
84	2.8391	0.3522	147.1290	0.0068	51.8222	0.0193	34.3258	1778.8384	84
90	3.0588	0.3269	164.7050	0.0061	53.8461	0.0186	36.2855	1953.8303	90
96	3.2955	0.3034	183.6411	0.0054	55.7246	0.0179	38.1793	2127.5244	96
100	3.4634	0.2887	197.0723	0.0051	56.9013	0.0176	39.4058	2242.2411	100
108	3.8253	0.2614	226.0226	0.0044	59.0865	0.0169	41.7737	2468.2636	108
120	4.4402	0.2252	275.2171	0.0036	61.9828	0.0161	45.1184	2796.5694	120
240	19.7155	0.0507	1497.2395	0.0007	75.9423	0.0132	67.1764	5101.5288	240
360	87.5410	0.0114	6923.2796	0.0001	79.0861	0.0126	75.8401	5997.9027	360

1.5%

	Single Payment		Equal Payment Series				Gradient Series		
N	Compound Amount Factor (F/P,i,N)	Present Worth Factor (P/F,i,N)	Compound Amount Factor (F/A,i,N)	Sinking Fund Factor (A/F,i,N)	Present Worth Factor (P/A,i,N)	Capital Recovery Factor (A/P,i,N)	Gradient Uniform Series (A/G,i,N)	Gradient Present Worth (P/G,i,N)	N
1	1.0150	0.9852	1.0000	1.0000	0.9852	1.0150	0.0000	0.0000	1
2	1.0302	0.9707	2.0150	0.4963	1.9559	0.5113	0.4963	0.9707	2
3	1.0457	0.9563	3.0452	0.3284	2.9122	0.3434	0.9901	2.8833	3
4	1.0614	0.9422	4.0909	0.2444	3.8544	0.2594	1.4814	5.7098	4
5	1.0773	0.9283	5.1523	0.1941	4.7826	0.2091	1.9702	9.4229	5
6	1.0934	0.9145	6.2296	0.1605	5.6972	0.1755	2.4566	13.9956	6
7	1.1098	0.9010	7.3230	0.1366	6.5982	0.1516	2.9405	19.4018	7
8	1.1265	0.8877	8.4328	0.1186	7.4859	0.1336	3.4219	25.6157	8
9	1.1434	0.8746	9.5593	0.1046	8.3605	0.1196	3.9008	32.6125	9
10	1.1605	0.8617	10.7027	0.0934	9.2222	0.1084	4.3772	40.3675	10
11	1.1779	0.8489	11.8633	0.0843	10.0711	0.0993	4.8512	48.8568	11
12	1.1956	0.8364	13.0412	0.0767	10.9075	0.0917	5.3227	58.0571	12
13	1.2136	0.8240	14.2368	0.0702	11.7315	0.0852	5.7917	67.9454	13
14	1.2318	0.8118	15.4504	0.0647	12.5434	0.0797	6.2582	78.4994	14
15	1.2502	0.7999	16.6821	0.0599	13.3432	0.0749	6.7223	89.6974	15
16	1.2690	0.7880	17.9324	0.0558	14.1313	0.0708	7.1839	101.5178	16
17	1.2880	0.7764	19.2014	0.0521	14.9076	0.0671	7.6431	113.9400	17
18	1.3073	0.7649	20.4894	0.0488	15.6726	0.0638	8.0997	126.9435	18
19	1.3270	0.7536	21.7967	0.0459	16.4262	0.0609	8.5539	140.5084	19
20	1.3469	0.7425	23.1237	0.0432	17.1686	0.0582	9.0057	154.6154	20
21	1.3671	0.7315	24.4705	0.0409	17.9001	0.0559	9.4550	169.2453	21
22	1.3876	0.7207	25.8376	0.0387	18.6208	0.0537	9.9018	184.3798	22
23	1.4084	0.7100	27.2251	0.0367	19.3309	0.0517	10.3462	200.0006	23
24	1.4295	0.6995	28.6335	0.0349	20.0304	0.0499	10.7881	216.0901	24
25	1.4509	0.6892	30.0630	0.0333	20.7196	0.0483	11.2276	232.6310	25
26	1.4727	0.6790	31.5140	0.0317	21.3986	0.0467	11.6646	249.6065	26
27	1.4948	0.6690	32.9867	0.0303	22.0676	0.0453	12.0992	267.0002	27
28	1.5172	0.6591	34.4815	0.0290	22.7267	0.0440	12.5313	284.7958	28
29	1.5400	0.6494	35.9987	0.0278	23.3761	0.0428	12.9610	302.9779	29
30	1.5631	0.6398	37.5387	0.0266	24.0158	0.0416	13.3883	321.5310	30
31	1.5865	0.6303	39.1018	0.0256	24.6461	0.0406	13.8131	340.4402	31
32	1.6103	0.6210	40.6883	0.0246	25.2671	0.0396	14.2355	359.6910	32
33	1.6345	0.6118	42.2986	0.0236	25.8790	0.0386	14.6555	379.2691	33
34	1.6590	0.6028	43.9331	0.0228	26.4817	0.0378	15.0731	399.1607	34
35	1.6839	0.5939	45.5921	0.0219	27.0756	0.0369	15.4882	419.3521	35
36	1.7091	0.5851	47.2760	0.0212	27.6607	0.0362	15.9009	439.8303	36
40	1.8140	0.5513	54.2679	0.0184	29.9158	0.0334	17.5277	524.3568	40
48	2.0435	0.4894	69.5652	0.0144	34.0426	0.0294	20.6667	703.5462	48
50	2.1052	0.4750	73.6828	0.0136	34.9997	0.0286	21.4277	749.9636	50
60	2.4432	0.4093	96.2147	0.0104	39.3803	0.0254	25.0930	988.1674	60
72	2.9212	0.3423	128.0772	0.0078	43.8447	0.0228	29.1893	1279.7938	72
80	3.2907	0.3039	152.7109	0.0065	46.4073	0.0215	31.7423	1473.0741	80
84	3.4926	0.2863	166.1726	0.0060	47.5786	0.0210	32.9668	1568.5140	84
90	3.8189	0.2619	187.9299	0.0053	49.2099	0.0203	34.7399	1709.5439	90
96	4.1758	0.2395	211.7202	0.0047	50.7017	0.0197	36.4381	1847.4725	96
100	4.4320	0.2256	228.8030	0.0044	51.6247	0.0194	37.5295	1937.4506	100
108	4.9927	0.2003	266.1778	0.0038	53.3137	0.0188	39.6171	2112.1348	108
120	5.9693	0.1675	331.2882	0.0030	55.4985	0.0180	42.5185	2359.7114	120
240	35.6328	0.0281	2308.8544	0.0004	64.7957	0.0154	59.7368	3870.6912	240
360	212.7038	0.0047	14113.5854	0.0001	66.3532	0.0151	64.9662	4310.7165	360

1.75%

	Single Payment		Equal Payment Series				Gradient Series		
N	Compound Amount Factor (F/P,i,N)	Present Worth Factor (P/F,i,N)	Compound Amount Factor (F/A,i,N)	Sinking Fund Factor (A/F,i,N)	Present Worth Factor (P/A,i,N)	Capital Recovery Factor (A/P,i,N)	Gradient Uniform Series (A/G,i,N)	Gradient Present Worth (P/G,i,N)	N
1	1.0175	0.9828	1.0000	1.0000	0.9828	1.0175	0.0000	0.0000	1
2	1.0353	0.9659	2.0175	0.4957	1.9487	0.5132	0.4957	0.9659	2
3	1.0534	0.9493	3.0528	0.3276	2.8980	0.3451	0.9884	2.8645	3
4	1.0719	0.9330	4.1062	0.2435	3.8309	0.2610	1.4783	5.6633	4
5	1.0906	0.9169	5.1781	0.1931	4.7479	0.2106	1.9653	9.3310	5
6	1.1097	0.9011	6.2687	0.1595	5.6490	0.1770	2.4494	13.8367	6
7	1.1291	0.8856	7.3784	0.1355	6.5346	0.1530	2.9306	19.1506	7
8	1.1489	0.8704	8.5075	0.1175	7.4051	0.1350	3.4089	25.2435	8
9	1.1690	0.8554	9.6564	0.1036	8.2605	0.1211	3.8844	32.0870	9
10	1.1894	0.8407	10.8254	0.0924	9.1012	0.1099	4.3569	39.6535	10
11	1.2103	0.8263	12.0148	0.0832	9.9275	0.1007	4.8266	47.9162	11
12	1.2314	0.8121	13.2251	0.0756	10.7395	0.0931	5.2934	56.8489	12
13	1.2530	0.7981	14.4565	0.0692	11.5376	0.0867	5.7573	66.4260	13
14	1.2749	0.7844	15.7095	0.0637	12.3220	0.0812	6.2184	76.6227	14
15	1.2972	0.7709	16.9844	0.0589	13.0929	0.0764	6.6765	87.4149	15
16	1.3199	0.7576	18.2817	0.0547	13.8505	0.0722	7.1318	98.7792	16
17	1.3430	0.7446	19.6016	0.0510	14.5951	0.0685	7.5842	110.6926	17
18	1.3665	0.7318	20.9446	0.0477	15.3269	0.0652	8.0338	123.1328	18
19	1.3904	0.7192	22.3112	0.0448	16.0461	0.0623	8.4805	136.0783	19
20	1.4148	0.7068	23.7016	0.0422	16.7529	0.0597	8.9243	149.5080	20
21	1.4395	0.6947	25.1164	0.0398	17.4475	0.0573	9.3653	163.4013	21
22	1.4647	0.6827	26.5559	0.0377	18.1303	0.0552	9.8034	177.7385	22
23	1.4904	0.6710	28.0207	0.0357	18.8012	0.0532	10.2387	192.5000	23
24	1.5164	0.6594	29.5110	0.0339	19.4607	0.0514	10.6711	207.6671	24
25	1.5430	0.6481	31.0275	0.0322	20.1088	0.0497	11.1007	223.2214	25
26	1.5700	0.6369	32.5704	0.0307	20.7457	0.0482	11.5274	239.1451	26
27	1.5975	0.6260	34.1404	0.0293	21.3717	0.0468	11.9513	255.4210	27
28	1.6254	0.6152	35.7379	0.0280	21.9870	0.0455	12.3724	272.0321	28
29	1.6539	0.6046	37.3633	0.0268	22.5916	0.0443	12.7907	288.9623	29
30	1.6828	0.5942	39.0172	0.0256	23.1858	0.0431	13.2061	306.1954	30
31	1.7122	0.5840	40.7000	0.0246	23.7699	0.0421	13.6188	323.7163	31
32	1.7422	0.5740	42.4122	0.0236	24.3439	0.0411	14.0286	341.5097	32
33	1.7727	0.5641	44.1544	0.0226	24.9080	0.0401	14.4356	359.5613	33
34	1.8037	0.5544	45.9271	0.0218	25.4624	0.0393	14.8398	377.8567	34
35	1.8353	0.5449	47.7308	0.0210	26.0073	0.0385	15.2412	396.3824	35
36	1.8674	0.5355	49.5661	0.0202	26.5428	0.0377	15.6399	415.1250	36
40	2.0016	0.4996	57.2341	0.0175	28.5942	0.0350	17.2066	492.0109	40
48	2.2996	0.4349	74.2628	0.0135	32.2938	0.0310	20.2084	652.6054	48
50	2.3808	0.4200	78.9022	0.0127	33.1412	0.0302	20.9317	693.7010	50
60	2.8318	0.3531	104.6752	0.0096	36.9640	0.0271	24.3885	901.4954	60
72	3.4872	0.2868	142.1263	0.0070	40.7564	0.0245	28.1948	1149.1181	72
80	4.0064	0.2496	171.7938	0.0058	42.8799	0.0233	30.5329	1309.2482	80
84	4.2943	0.2329	188.2450	0.0053	43.8361	0.0228	31.6442	1387.1584	84
90	4.7654	0.2098	215.1646	0.0046	45.1516	0.0221	33.2409	1500.8798	90
96	5.2882	0.1891	245.0374	0.0041	46.3370	0.0216	34.7556	1610.4716	96
100	5.6682	0.1764	266.7518	0.0037	47.0615	0.0212	35.7211	1681.0886	100
108	6.5120	0.1536	314.9738	0.0032	48.3679	0.0207	37.5494	1816.1852	108
120	8.0192	0.1247	401.0962	0.0025	50.0171	0.0200	40.0469	2003.0269	120
240	64.3073	0.0156	3617.5602	0.0003	56.2543	0.0178	53.3518	3001.2678	240
360	515.6921	0.0019	29410.9747	0.0000	57.0320	0.0175	56.4434	3219.0833	360

2.0%

	Single Payment		Equal Payment Series				Gradient Series		
	Compound Amount Factor	Present Worth Factor	Compound Amount Factor	Sinking Fund Factor	Present Worth Factor	Capital Recovery Factor	Gradient Uniform Series	Gradient Present Worth	
N	(F/P,i,N)	(P/F,i,N)	(F/A,i,N)	(A/F,i,N)	(P/A,i,N)	(A/P,i,N)	(A/G,i,N)	(P/G,i,N)	N
1	1.0200	0.9804	1.0000	1.0000	0.9804	1.0200	0.0000	0.0000	1
2	1.0404	0.9612	2.0200	0.4950	1.9416	0.5150	0.4950	0.9612	2
3	1.0612	0.9423	3.0604	0.3268	2.8839	0.3468	0.9868	2.8458	3
4	1.0824	0.9238	4.1216	0.2426	3.8077	0.2626	1.4752	5.6173	4
5	1.1041	0.9057	5.2040	0.1922	4.7135	0.2122	1.9604	9.2403	5
6	1.1262	0.8880	6.3081	0.1585	5.6014	0.1785	2.4423	13.6801	6
7	1.1487	0.8706	7.4343	0.1345	6.4720	0.1545	2.9208	18.9035	7
8	1.1717	0.8535	8.5830	0.1165	7.3255	0.1365	3.3961	24.8779	8
9	1.1951	0.8368	9.7546	0.1025	8.1622	0.1225	3.8681	31.5720	9
10	1.2190	0.8203	10.9497	0.0913	8.9826	0.1113	4.3367	38.9551	10
11	1.2434	0.8043	12.1687	0.0822	9.7868	0.1022	4.8021	46.9977	11
12	1.2682	0.7885	13.4121	0.0746	10.5753	0.0946	5.2642	55.6712	12
13	1.2936	0.7730	14.6803	0.0681	11.3484	0.0881	5.7231	64.9475	13
14	1.3195	0.7579	15.9739	0.0626	12.1062	0.0826	6.1786	74.7999	14
15	1.3459	0.7430	17.2934	0.0578	12.8493	0.0778	6.6309	85.2021	15
16	1.3728	0.7284	18.6393	0.0537	13.5777	0.0737	7.0799	96.1288	16
17	1.4002	0.7142	20.0121	0.0500	14.2919	0.0700	7.5256	107.5554	17
18	1.4282	0.7002	21.4123	0.0467	14.9920	0.0667	7.9681	119.4581	18
19	1.4568	0.6864	22.8406	0.0438	15.6785	0.0638	8.4073	131.8139	19
20	1.4859	0.6730	24.2974	0.0412	16.3514	0.0612	8.8433	144.6003	20
21	1.5157	0.6598	25.7833	0.0388	17.0112	0.0588	9.2760	157.7959	21
22	1.5460	0.6468	27.2990	0.0366	17.6580	0.0566	9.7055	171.3795	22
23	1.5769	0.6342	28.8450	0.0347	18.2922	0.0547	10.1317	185.3309	23
24	1.6084	0.6217	30.4219	0.0329	18.9139	0.0529	10.5547	199.6305	24
25	1.6406	0.6095	32.0303	0.0312	19.5235	0.0512	10.9745	214.2592	25
26	1.6734	0.5976	33.6709	0.0297	20.1210	0.0497	11.3910	229.1987	26
27	1.7069	0.5859	35.3443	0.0283	20.7069	0.0483	11.8043	244.4311	27
28	1.7410	0.5744	37.0512	0.0270	21.2813	0.0470	12.2145	259.9392	28
29	1.7758	0.5631	38.7922	0.0258	21.8444	0.0458	12.6214	275.7064	29
30	1.8114	0.5521	40.5681	0.0246	22.3965	0.0446	13.0251	291.7164	30
31	1.8476	0.5412	42.3794	0.0236	22.9377	0.0436	13.4257	307.9538	31
32	1.8845	0.5306	44.2270	0.0226	23.4683	0.0426	13.8230	324.4035	32
33	1.9222	0.5202	46.1116	0.0217	23.9886	0.0417	14.2172	341.0508	33
34	1.9607	0.5100	48.0338	0.0208	24.4986	0.0408	14.6083	357.8817	34
35	1.9999	0.5000	49.9945	0.0200	24.9986	0.0400	14.9961	374.8826	35
36	2.0399	0.4902	51.9944	0.0192	25.4888	0.0392	15.3809	392.0405	36
40	2.2080	0.4529	60.4020	0.0166	27.3555	0.0366	16.8885	461.9931	40
48	2.5871	0.3865	79.3535	0.0126	30.6731	0.0326	19.7556	605.9657	48
50	2.6916	0.3715	84.5794	0.0118	31.4236	0.0318	20.4420	642.3606	50
60	3.2810	0.3048	114.0515	0.0088	34.7609	0.0288	23.6961	823.6975	60
72	4.1611	0.2403	158.0570	0.0063	37.9841	0.0263	27.2234	1034.0557	72
80	4.8754	0.2051	193.7720	0.0052	39.7445	0.0252	29.3572	1166.7868	80
84	5.2773	0.1895	213.8666	0.0047	40.5255	0.0247	30.3616	1230.4191	84
90	5.9431	0.1683	247.1567	0.0040	41.5869	0.0240	31.7929	1322.1701	90
96	6.6929	0.1494	284.6467	0.0035	42.5294	0.0235	33.1370	1409.2973	96
100	7.2446	0.1380	312.2323	0.0032	43.0984	0.0232	33.9863	1464.7527	100
108	8.4883	0.1178	374.4129	0.0027	44.1095	0.0227	35.5774	1569.3025	108
120	10.7652	0.0929	488.2582	0.0020	45.3554	0.0220	37.7114	1710.4160	120
240	115.8887	0.0086	5744.4368	0.0002	49.5686	0.0202	47.9110	2374.8800	240
360	1247.5611	0.0008	62328.0564	0.0000	49.9599	0.0200	49.7112	2483.5679	360

3.0%

	Single Payment		Equal Payment Series				Gradient Series		
	Compound Amount Factor	Present Worth Factor	Compound Amount Factor	Sinking Fund Factor	Present Worth Factor	Capital Recovery Factor	Gradient Uniform Series	Gradient Present Worth	
N	(*F/P,i,N*)	(*P/F,i,N*)	(*F/A,i,N*)	(*A/F,i,N*)	(*P/A,i,N*)	(*A/P,i,N*)	(*A/G,i,N*)	(*P/G,i,N*)	*N*
1	1.0300	0.9709	1.0000	1.0000	0.9709	1.0300	0.0000	0.0000	1
2	1.0609	0.9426	2.0300	0.4926	1.9135	0.5226	0.4926	0.9426	2
3	1.0927	0.9151	3.0909	0.3235	2.8286	0.3535	0.9803	2.7729	3
4	1.1255	0.8885	4.1836	0.2390	3.7171	0.2690	1.4631	5.4383	4
5	1.1593	0.8626	5.3091	0.1884	4.5797	0.2184	1.9409	8.8888	5
6	1.1941	0.8375	6.4684	0.1546	5.4172	0.1846	2.4138	13.0762	6
7	1.2299	0.8131	7.6625	0.1305	6.2303	0.1605	2.8819	17.9547	7
8	1.2668	0.7894	8.8923	0.1125	7.0197	0.1425	3.3450	23.4806	8
9	1.3048	0.7664	10.1591	0.0984	7.7861	0.1284	3.8032	29.6119	9
10	1.3439	0.7441	11.4639	0.0872	8.5302	0.1172	4.2565	36.3088	110
11	1.3842	0.7224	12.8078	0.0781	9.2526	0.1081	4.7049	43.5330	11
12	1.4258	0.7014	14.1920	0.0705	9.9540	0.1005	5.1485	51.2482	12
13	1.4685	0.6810	15.6178	0.0640	10.6350	0.0940	5.5872	59.4196	13
14	1.5126	0.6611	17.0863	0.0585	11.2961	0.0885	6.0210	68.0141	14
15	1.5580	0.6419	18.5989	0.0538	11.9379	0.0838	6.4500	77.0002	15
16	1.6047	0.6232	20.1569	0.0496	12.5611	0.0796	6.8742	86.3477	16
17	1.6528	0.6050	21.7616	0.0460	13.1661	0.0760	7.2936	96.0280	17
18	1.7024	0.5874	23.4144	0.0427	13.7535	0.0727	7.7081	106.0137	18
19	1.7535	0.5703	25.1169	0.0398	14.3238	0.0698	8.1179	116.2788	19
20	1.8061	0.5537	26.8704	0.0372	14.8775	0.0672	8.5229	126.7987	20
21	1.8603	0.5375	28.6765	0.0349	15.4150	0.0649	8.9231	137.5496	21
22	1.9161	0.5219	30.5368	0.0327	15.9396	0.0627	9.3186	148.5094	22
23	1.9736	0.5067	32.4529	0.0308	16.4436	0.0608	9.7093	159.6566	23
24	2.0328	0.4919	34.4265	0.0290	16.9355	0.0590	10.0954	170.9711	24
25	2.0938	0.4776	36.4593	0.0274	17.4131	0.0574	10.4768	182.4336	25
26	2.1566	0.4637	38.5530	0.0259	17.8768	0.0559	10.8535	194.0260	26
27	2.2213	0.4502	40.7096	0.0246	18.3270	0.0546	11.2255	205.7309	27
28	2.2879	0.4371	42.9309	0.0233	18.7641	0.0533	11.5930	217.5320	28
29	2.3566	0.4243	45.2189	0.0221	19.1885	0.0521	11.9558	229.4137	29
30	2.4273	0.4120	47.5754	0.0210	19.6004	0.0510	12.3141	241.3613	30
31	2.5001	0.4000	50.0027	0.0200	20.0004	0.0500	12.6678	253.3609	31
32	2.5751	0.3883	52.5028	0.0190	20.3888	0.0490	13.0169	265.3993	32
33	2.6523	0.3770	55.0778	0.0182	20.7658	0.0482	13.3616	277.4642	33
34	2.7319	0.3660	57.7302	0.0173	21.1318	0.0473	13.7018	289.5437	34
35	2.8139	0.3554	60.4621	0.0165	21.4872	0.0465	14.0375	301.6267	35
40	3.2620	0.3066	75.4013	0.0133	23.1148	0.0433	15.6502	361.7499	40
45	3.7816	0.2644	92.7199	0.0108	24.5187	0.0408	17.1556	420.6325	45
50	4.3839	0.2281	112.7969	0.0089	25.7298	0.0389	18.5575	477.4803	50
55	5.0821	0.1968	136.0716	0.0073	26.7744	0.0373	19.8600	531.7411	55
60	5.8916	0.1697	163.0534	0.0061	27.6756	0.0361	21.0674	583.0526	60
65	6.8300	0.1464	194.3328	0.0051	28.4529	0.0351	22.1841	631.2010	65
70	7.9178	0.1263	230.5941	0.0043	29.1234	0.0343	23.2145	676.0869	70
75	9.1789	0.1089	272.6309	0.0037	29.7018	0.0337	24.1634	717.6978	75
80	10.6409	0.0940	321.3630	0.0031	30.2008	0.0331	25.0353	756.0865	80
85	12.3357	0.0811	377.8570	0.0026	30.6312	0.0326	25.8349	791.3529	85
90	14.3005	0.0699	443.3489	0.0023	31.0024	0.0323	26.5667	823.6302	90
95	16.5782	0.0603	519.2720	0.0019	31.3227	0.0319	27.2351	853.0742	95
100	19.2186	0.0520	607.2877	0.0016	31.5989	0.0316	27.8444	879.8540	100

4.0%

	Single Payment		Equal Payment Series				Gradient Series		
	Compound Amount Factor	Present Worth Factor	Compound Amount Factor	Sinking Fund Factor	Present Worth Factor	Capital Recovery Factor	Gradient Uniform Series	Gradient Present Worth	
N	(F/P,i,N)	(P/F,i,N)	(F/A,i,N)	(A/F,i,N)	(P/A,i,N)	(A/P,i,N)	(A/G,i,N)	(P/G,i,N)	N
1	1.0400	0.9615	1.0000	1.0000	0.9615	1.0400	0.0000	0.0000	1
2	1.0816	0.9246	2.0400	0.4902	1.8861	0.5302	0.4902	0.9246	2
3	1.1249	0.8890	3.1216	0.3203	2.7751	0.3603	0.9739	2.7025	3
4	1.1699	0.8548	4.2465	0.2355	3.6299	0.2755	1.4510	5.2670	4
5	1.2167	0.8219	5.4163	0.1846	4.4518	0.2246	1.9216	8.5547	5
6	1.2653	0.7903	6.6330	0.1508	5.2421	0.1908	2.3857	12.5062	6
7	1.3159	0.7599	7.8983	0.1266	6.0021	0.1666	2.8433	17.0657	7
8	1.3686	0.7307	9.2142	0.1085	6.7327	0.1485	3.2944	22.1806	8
9	1.4233	0.7026	10.5828	0.0945	7.4353	0.1345	3.7391	27.8013	9
10	1.4802	0.6756	12.0061	0.0833	8.1109	0.1233	4.1773	33.8814	10
11	1.5395	0.6496	13.4864	0.0741	8.7605	0.1141	4.6090	40.3772	11
12	1.6010	0.6246	15.0258	0.0666	9.3851	0.1066	5.0343	47.2477	12
13	1.6651	0.6006	16.6268	0.0601	9.9856	0.1001	5.4533	54.4546	13
14	1.7317	0.5775	18.2919	0.0547	10.5631	0.0947	5.8659	61.9618	14
15	1.8009	0.5553	20.0236	0.0499	11.1184	0.0899	6.2721	69.7355	15
16	1.8730	0.5339	21.8245	0.0458	11.6523	0.0858	6.6720	77.7441	16
17	1.9479	0.5134	23.6975	0.0422	12.1657	0.0822	7.0656	85.9581	17
18	2.0258	0.4936	25.6454	0.0390	12.6593	0.0790	7.4530	94.3498	18
19	2.1068	0.4746	27.6712	0.0361	13.1339	0.0761	7.8342	102.8933	19
20	2.1911	0.4564	29.7781	0.0336	13.5903	0.0736	8.2091	111.5647	20
21	2.2788	0.4388	31.9692	0.0313	14.0292	0.0713	8.5779	120.3414	21
22	2.3699	0.4220	34.2480	0.0292	14.4511	0.0692	8.9407	129.2024	22
23	2.4647	0.4057	36.6179	0.0273	14.8568	0.0673	9.2973	138.1284	23
24	2.5633	0.3901	39.0826	0.0256	15.2470	0.0656	9.6479	147.1012	24
25	2.6658	0.3751	41.6459	0.0240	15.6221	0.0640	9.9925	156.1040	25
26	2.7725	0.3607	44.3117	0.0226	15.9828	0.0626	10.3312	165.1212	26
27	2.8834	0.3468	47.0842	0.0212	16.3296	0.0612	10.6640	174.1385	27
28	2.9987	0.3335	49.9676	0.0200	16.6631	0.0600	10.9909	183.1424	28
29	3.1187	0.3207	52.9663	0.0189	16.9837	0.0589	11.3120	192.1206	29
30	3.2434	0.3083	56.0849	0.0178	17.2920	0.0578	11.6274	201.0618	30
31	3.3731	0.2965	59.3283	0.0169	17.5885	0.0569	11.9371	209.9556	31
32	3.5081	0.2851	62.7015	0.0159	17.8736	0.0559	12.2411	218.7924	32
33	3.6484	0.2741	66.2095	0.0151	18.1476	0.0551	12.5396	227.5634	33
34	3.7943	0.2636	69.8579	0.0143	18.4112	0.0543	12.8324	236.2607	34
35	3.9461	0.2534	73.6522	0.0136	18.6646	0.0536	13.1198	244.8768	35
40	4.8010	0.2083	95.0255	0.0105	19.7928	0.0505	14.4765	286.5303	40
45	5.8412	0.1712	121.0294	0.0083	20.7200	0.0483	15.7047	325.4028	45
50	7.1067	0.1407	152.6671	0.0066	21.4822	0.0466	16.8122	361.1638	50
55	8.6464	0.1157	191.1592	0.0052	22.1086	0.0452	17.8070	393.6890	55
60	10.5196	0.0951	237.9907	0.0042	22.6235	0.0442	18.6972	422.9966	60
65	12.7987	0.0781	294.9684	0.0034	23.0467	0.0434	19.4909	449.2014	65
70	15.5716	0.0642	364.2905	0.0027	23.3945	0.0427	20.1961	472.4789	70
75	18.9453	0.0528	448.6314	0.0022	23.6804	0.0422	20.8206	493.0408	75
80	23.0498	0.0434	551.2450	0.0018	23.9154	0.0418	21.3718	511.1161	80
85	28.0436	0.0357	676.0901	0.0015	24.1085	0.0415	21.8569	526.9384	85
90	34.1193	0.0293	827.9833	0.0012	24.2673	0.0412	22.2826	540.7369	90
95	41.5114	0.0241	1012.7846	0.0010	24.3978	0.0410	22.6550	552.7307	95
100	50.5049	0.0198	1237.6237	0.0008	24.5050	0.0408	22.9800	563.1249	100

5.0%

	Single Payment		Equal Payment Series				Gradient Series		
N	Compound Amount Factor (F/P,i,N)	Present Worth Factor (P/F,i,N)	Compound Amount Factor (F/A,i,N)	Sinking Fund Factor (A/F,i,N)	Present Worth Factor (P/A,i,N)	Capital Recovery Factor (A/P,i,N)	Gradient Uniform Series (A/G,i,N)	Gradient Present Worth (P/G,i,N)	N
1	1.0500	0.9524	1.0000	1.0000	0.9524	1.0500	0.0000	0.0000	1
2	1.1025	0.9070	2.0500	0.4878	1.8594	0.5378	0.4878	0.9070	2
3	1.1576	0.8638	3.1525	0.3172	2.7232	0.3672	0.9675	2.6347	3
4	1.2155	0.8227	4.3101	0.2320	3.5460	0.2820	1.4391	5.1028	4
5	1.2763	0.7835	5.5256	0.1810	4.3295	0.2310	1.9025	8.2369	5
6	1.3401	0.7462	6.8019	0.1470	5.0757	0.1970	2.3579	11.9680	6
7	1.4071	0.7107	8.1420	0.1228	5.7864	0.1728	2.8052	16.2321	7
8	1.4775	0.6768	9.5491	0.1047	6.4632	0.1547	3.2445	20.9700	8
9	1.5513	0.6446	11.0266	0.0907	7.1078	0.1407	3.6758	26.1268	9
10	1.6289	0.6139	12.5779	0.0795	7.7217	0.1295	4.0991	31.6520	10
11	1.7103	0.5847	14.2068	0.0704	8.3064	0.1204	4.5144	37.4988	11
12	1.7959	0.5568	15.9171	0.0628	8.8633	0.1128	4.9219	43.6241	12
13	1.8856	0.5303	17.7130	0.0565	9.3936	0.1065	5.3215	49.9879	13
14	1.9799	0.5051	19.5986	0.0510	9.8986	0.1010	5.7133	56.5538	14
15	2.0789	0.4810	21.5786	0.0463	10.3797	0.0963	6.0973	63.2880	15
16	2.1829	0.4581	23.6575	0.0423	10.8378	0.0923	6.4736	70.1597	16
17	2.2920	0.4363	25.8404	0.0387	11.2741	0.0887	6.8423	77.1405	17
18	2.4066	0.4155	28.1324	0.0355	11.6896	0.0855	7.2034	84.2043	18
19	2.5270	0.3957	30.5390	0.0327	12.0853	0.0827	7.5569	91.3275	19
20	2.6533	0.3769	33.0660	0.0302	12.4622	0.0802	7.9030	98.4884	20
21	2.7860	0.3589	35.7193	0.0280	12.8212	0.0780	8.2416	105.6673	21
22	2.9253	0.3418	38.5052	0.0260	13.1630	0.0760	8.5730	112.8461	22
23	3.0715	0.3256	41.4305	0.0241	13.4886	0.0741	8.8971	120.0087	23
24	3.2251	0.3101	44.5020	0.0225	13.7986	0.0725	9.2140	127.1402	24
25	3.3864	0.2953	47.7271	0.0210	14.0939	0.0710	9.5238	134.2275	25
26	3.5557	0.2812	51.1135	0.0196	14.3752	0.0696	9.8266	141.2585	26
27	3.7335	0.2678	54.6691	0.0183	14.6430	0.0683	10.1224	148.2226	27
28	3.9201	0.2551	58.4026	0.0171	14.8981	0.0671	10.4114	155.1101	28
29	4.1161	0.2429	62.3227	0.0160	15.1411	0.0660	10.6936	161.9126	29
30	4.3219	0.2314	66.4388	0.0151	15.3725	0.0651	10.9691	168.6226	30
31	4.5380	0.2204	70.7608	0.0141	15.5928	0.0641	11.2381	175.2333	31
32	4.7649	0.2099	75.2988	0.0133	15.8027	0.0633	11.5005	181.7392	32
33	5.0032	0.1999	80.0638	0.0125	16.0025	0.0625	11.7566	188.1351	33
34	5.2533	0.1904	85.0670	0.0118	16.1929	0.0618	12.0063	194.4168	34
35	5.5160	0.1813	90.3203	0.0111	16.3742	0.0611	12.2498	200.5807	35
40	7.0400	0.1420	120.7998	0.0083	17.1591	0.0583	13.3775	229.5452	40
45	8.9850	0.1113	159.7002	0.0063	17.7741	0.0563	14.3644	255.3145	45
50	11.4674	0.0872	209.3480	0.0048	18.2559	0.0548	15.2233	277.9148	50
55	14.6356	0.0683	272.7126	0.0037	18.6335	0.0537	15.9664	297.5104	55
60	18.6792	0.0535	353.5837	0.0028	18.9293	0.0528	16.6062	314.3432	60
65	23.8399	0.0419	456.7980	0.0022	19.1611	0.0522	17.1541	328.6910	65
70	30.4264	0.0329	588.5285	0.0017	19.3427	0.0517	17.6212	340.8409	70
75	38.8327	0.0258	756.6537	0.0013	19.4850	0.0513	18.0176	351.0721	75
80	49.5614	0.0202	971.2288	0.0010	19.5965	0.0510	18.3526	359.6460	80
85	63.2544	0.0158	1245.0871	0.0008	19.6838	0.0508	18.6346	366.8007	85
90	80.7304	0.0124	1594.6073	0.0006	19.7523	0.0506	18.8712	372.7488	90
95	103.0347	0.0097	2040.6935	0.0005	19.8059	0.0505	19.0689	377.6774	95
100	131.5013	0.0076	2610.0252	0.0004	19.8479	0.0504	19.2337	381.7492	100

6.0%

	Single Payment		Equal Payment Series				Gradient Series		
N	Compound Amount Factor (F/P,i,N)	Present Worth Factor (P/F,i,N)	Compound Amount Factor (F/A,i,N)	Sinking Fund Factor (A/F,i,N)	Present Worth Factor (P/A,i,N)	Capital Recovery Factor (A/P,i,N)	Gradient Uniform Series (A/G,i,N)	Gradient Present Worth (P/G,i,N)	N
1	1.0600	0.9434	1.0000	1.0000	0.9434	1.0600	0.0000	0.0000	1
2	1.1236	0.8900	2.0600	0.4854	1.8334	0.5454	0.4854	0.8900	2
3	1.1910	0.8396	3.1836	0.3141	2.6730	0.3741	0.9612	2.5692	3
4	1.2625	0.7921	4.3746	0.2286	3.4651	0.2886	1.4272	4.9455	4
5	1.3382	0.7473	5.6371	0.1774	4.2124	0.2374	1.8836	7.9345	5
6	1.4185	0.7050	6.9753	0.1434	4.9173	0.2034	2.3304	11.4594	6
7	1.5036	0.6651	8.3938	0.1191	5.5824	0.1791	2.7676	15.4497	7
8	1.5938	0.6274	9.8975	0.1010	6.2098	0.1610	3.1952	19.8416	8
9	1.6895	0.5919	11.4913	0.0870	6.8017	0.1470	3.6133	24.5768	9
10	1.7908	0.5584	13.1808	0.0759	7.3601	0.1359	4.0220	29.6023	10
11	1.8983	0.5268	14.9716	0.0668	7.8869	0.1268	4.4213	34.8702	11
12	2.0122	0.4970	16.8699	0.0593	8.3838	0.1193	4.8113	40.3369	12
13	2.1329	0.4688	18.8821	0.0530	8.8527	0.1130	5.1920	45.9629	13
14	2.2609	0.4423	21.0151	0.0476	9.2950	0.1076	5.5635	51.7128	14
15	2.3966	0.4173	23.2760	0.0430	9.7122	0.1030	5.9260	57.5546	15
16	2.5404	0.3936	25.6725	0.0390	10.1059	0.0990	6.2794	63.4592	16
17	2.6928	0.3714	28.2129	0.0354	10.4773	0.0954	6.6240	69.4011	17
18	2.8543	0.3503	30.9057	0.0324	10.8276	0.0924	6.9597	75.3569	18
19	3.0256	0.3305	33.7600	0.0296	11.1581	0.0896	7.2867	81.3062	19
20	3.2071	0.3118	36.7856	0.0272	11.4699	0.0872	7.6051	87.2304	20
21	3.3996	0.2942	39.9927	0.0250	11.7641	0.0850	7.9151	93.1136	21
22	3.6035	0.2775	43.3923	0.0230	12.0416	0.0830	8.2166	98.9412	22
23	3.8197	0.2618	46.9958	0.0213	12.3034	0.0813	8.5099	104.7007	23
24	4.0489	0.2470	50.8156	0.0197	12.5504	0.0797	8.7951	110.3812	24
25	4.2919	0.2330	54.8645	0.0182	12.7834	0.0782	9.0722	115.9732	25
26	4.5494	0.2198	59.1564	0.0169	13.0032	0.0769	9.3414	121.4684	26
27	4.8223	0.2074	63.7058	0.0157	13.2105	0.0757	9.6029	126.8600	27
28	5.1117	0.1956	68.5281	0.0146	13.4062	0.0746	9.8568	132.1420	28
29	5.4184	0.1846	73.6398	0.0136	13.5907	0.0736	10.1032	137.3096	29
30	5.7435	0.1741	79.0582	0.0126	13.7648	0.0726	10.3422	142.3588	30
31	6.0881	0.1643	84.8017	0.0118	13.9291	0.0718	10.5740	147.2864	31
32	6.4534	0.1550	90.8898	0.0110	14.0840	0.0710	10.7988	152.0901	32
33	6.8406	0.1462	97.3432	0.0103	14.2302	0.0703	11.0166	156.7681	33
34	7.2510	0.1379	104.1838	0.0096	14.3681	0.0696	11.2276	161.3192	34
35	7.6861	0.1301	111.4348	0.0090	14.4982	0.0690	11.4319	165.7427	35
40	10.2857	0.0972	154.7620	0.0065	15.0463	0.0665	12.3590	185.9568	40
45	13.7646	0.0727	212.7435	0.0047	15.4558	0.0647	13.1413	203.1096	45
50	18.4202	0.0543	290.3359	0.0034	15.7619	0.0634	13.7964	217.4574	50
55	24.6503	0.0406	394.1720	0.0025	15.9905	0.0625	14.3411	229.3222	55
60	32.9877	0.0303	533.1282	0.0019	16.1614	0.0619	14.7909	239.0428	60
65	44.1450	0.0227	719.0829	0.0014	16.2891	0.0614	15.1601	246.9450	65
70	59.0759	0.0169	967.9322	0.0010	16.3845	0.0610	15.4613	253.3271	70
75	79.0569	0.0126	1300.9487	0.0008	16.4558	0.0608	15.7058	258.4527	75
80	105.7960	0.0095	1746.5999	0.0006	16.5091	0.0606	15.9033	262.5493	80
85	141.5789	0.0071	2342.9817	0.0004	16.5489	0.0604	16.0620	265.8096	85
90	189.4645	0.0053	3141.0752	0.0003	16.5787	0.0603	16.1891	268.3946	90
95	253.5463	0.0039	4209.1042	0.0002	16.6009	0.0602	16.2905	270.4375	95
100	339.3021	0.0029	5638.3681	0.0002	16.6175	0.0602	16.3711	272.0471	100

7.0%

	Single Payment		Equal Payment Series				Gradient Series		
N	Compound Amount Factor (F/P,i,N)	Present Worth Factor (P/F,i,N)	Compound Amount Factor (F/A,i,N)	Sinking Fund Factor (A/F,i,N)	Present Worth Factor (P/A,i,N)	Capital Recovery Factor (A/P,i,N)	Gradient Uniform Series (A/G,i,N)	Gradient Present Worth (P/G,i,N)	N
1	1.0700	0.9346	1.0000	1.0000	0.9346	1.0700	0.0000	0.0000	1
2	1.1449	0.8734	2.0700	0.4831	1.8080	0.5531	0.4831	0.8734	2
3	1.2250	0.8163	3.2149	0.3111	2.6243	0.3811	0.9549	2.5060	3
4	1.3108	0.7629	4.4399	0.2252	3.3872	0.2952	1.4155	4.7947	4
5	1.4026	0.7130	5.7507	0.1739	4.1002	0.2439	1.8650	7.6467	5
6	1.5007	0.6663	7.1533	0.1398	4.7665	0.2098	2.3032	10.9784	6
7	1.6058	0.6227	8.6540	0.1156	5.3893	0.1856	2.7304	14.7149	7
8	1.7182	0.5820	10.2598	0.0975	5.9713	0.1675	3.1465	18.7889	8
9	1.8385	0.5439	11.9780	0.0835	6.5152	0.1535	3.5517	23.1404	9
10	1.9672	0.5083	13.8164	0.0724	7.0236	0.1424	3.9461	27.7156	10
11	2.1049	0.4751	15.7836	0.0634	7.4987	0.1334	4.3296	32.4665	11
12	2.2522	0.4440	17.8885	0.0559	7.9427	0.1259	4.7025	37.3506	12
13	2.4098	0.4150	20.1406	0.0497	8.3577	0.1197	5.0648	42.3302	13
14	2.5785	0.3878	22.5505	0.0443	8.7455	0.1143	5.4167	47.3718	14
15	2.7590	0.3624	25.1290	0.0398	9.1079	0.1098	5.7583	52.4461	15
16	2.9522	0.3387	27.8881	0.0359	9.4466	0.1059	6.0897	57.5271	16
17	3.1588	0.3166	30.8402	0.0324	9.7632	0.1024	6.4110	62.5923	17
18	3.3799	0.2959	33.9990	0.0294	10.0591	0.0994	6.7225	67.6219	18
19	3.6165	0.2765	37.3790	0.0268	10.3356	0.0968	7.0242	72.5991	19
20	3.8697	0.2584	40.9955	0.0244	10.5940	0.0944	7.3163	77.5091	20
21	4.1406	0.2415	44.8652	0.0223	10.8355	0.0923	7.5990	82.3393	21
22	4.4304	0.2257	49.0057	0.0204	11.0612	0.0904	7.8725	87.0793	22
23	4.7405	0.2109	53.4361	0.0187	11.2722	0.0887	8.1369	91.7201	23
24	5.0724	0.1971	58.1767	0.0172	11.4693	0.0872	8.3923	96.2545	24
25	5.4274	0.1842	63.2490	0.0158	11.6536	0.0858	8.6391	100.6765	25
26	5.8074	0.1722	68.6765	0.0146	11.8258	0.0846	8.8773	104.9814	26
27	6.2139	0.1609	74.4838	0.0134	11.9867	0.0834	9.1072	109.1656	27
28	6.6488	0.1504	80.6977	0.0124	12.1371	0.0824	9.3289	113.2264	28
29	7.1143	0.1406	87.3465	0.0114	12.2777	0.0814	9.5427	117.1622	29
30	7.6123	0.1314	94.4608	0.0106	12.4090	0.0806	9.7487	120.9718	30
31	8.1451	0.1228	102.0730	0.0098	12.5318	0.0798	9.9471	124.6550	31
32	8.7153	0.1147	110.2182	0.0091	12.6466	0.0791	10.1381	128.2120	32
33	9.3253	0.1072	118.9334	0.0084	12.7538	0.0784	10.3219	131.6435	33
34	9.9781	0.1002	128.2588	0.0078	12.8540	0.0778	10.4987	134.9507	34
35	10.6766	0.0937	138.2369	0.0072	12.9477	0.0772	10.6687	138.1353	35
40	14.9745	0.0668	199.6351	0.0050	13.3317	0.0750	11.4233	152.2928	40
45	21.0025	0.0476	285.7493	0.0035	13.6055	0.0735	12.0360	163.7559	45
50	29.4570	0.0339	406.5289	0.0025	13.8007	0.0725	12.5287	172.9051	50
55	41.3150	0.0242	575.9286	0.0017	13.9399	0.0717	12.9215	180.1243	55
60	57.9464	0.0173	813.5204	0.0012	14.0392	0.0712	13.2321	185.7677	60
65	81.2729	0.0123	1146.7552	0.0009	14.1099	0.0709	13.4760	190.1452	65
70	113.9894	0.0088	1614.1342	0.0006	14.1604	0.0706	13.6662	193.5185	70
75	159.8760	0.0063	2269.6574	0.0004	14.1964	0.0704	13.8136	196.1035	75
80	224.2344	0.0045	3189.0627	0.0003	14.2220	0.0703	13.9273	198.0748	80
85	314.5003	0.0032	4478.5761	0.0002	14.2403	0.0702	14.0146	199.5717	85
90	441.1030	0.0023	6287.1854	0.0002	14.2533	0.0702	14.0812	200.7042	90
95	618.6697	0.0016	8823.8535	0.0001	14.2626	0.0701	14.1319	201.5581	95
100	867.7163	0.0012	12381.6618	0.0001	14.2693	0.0701	14.1703	202.2001	100

8.0%

	Single Payment		Equal Payment Series				Gradient Series		
N	Compound Amount Factor (F/P,i,N)	Present Worth Factor (P/F,i,N)	Compound Amount Factor (F/A,i,N)	Sinking Fund Factor (A/F,i,N)	Present Worth Factor (P/A,i,N)	Capital Recovery Factor (A/P,i,N)	Gradient Uniform Series (A/G,i,N)	Gradient Present Worth (P/G,i,N)	N
1	1.0800	0.9259	1.0000	1.0000	0.9259	1.0800	0.0000	0.0000	1
2	1.1664	0.8573	2.0800	0.4808	1.7833	0.5608	0.4808	0.8573	2
3	1.2597	0.7938	3.2464	0.3080	2.5771	0.3880	0.9487	2.4450	3
4	1.3605	0.7350	4.5061	0.2219	3.3121	0.3019	1.4040	4.6501	4
5	1.4693	0.6806	5.8666	0.1705	3.9927	0.2505	1.8465	7.3724	5
6	1.5869	0.6302	7.3359	0.1363	4.6229	0.2163	2.2763	10.5233	6
7	1.7138	0.5835	8.9228	0.1121	5.2064	0.1921	2.6937	14.0242	7
8	1.8509	0.5403	10.6366	0.0940	5.7466	0.1740	3.0985	17.8061	8
9	1.9990	0.5002	12.4876	0.0801	6.2469	0.1601	3.4910	21.8081	9
10	2.1589	0.4632	14.4866	0.0690	6.7101	0.1490	3.8713	25.9768	10
11	2.3316	0.4289	16.6455	0.0601	7.1390	0.1401	4.2395	30.2657	11
12	2.5182	0.3971	18.9771	0.0527	7.5361	0.1327	4.5957	34.6339	12
13	2.7196	0.3677	21.4953	0.0465	7.9038	0.1265	4.9402	39.0463	13
14	2.9372	0.3405	24.2149	0.0413	8.2442	0.1213	5.2731	43.4723	14
15	3.1722	0.3152	27.1521	0.0368	8.5595	0.1168	5.5945	47.8857	15
16	3.4259	0.2919	30.3243	0.0330	8.8514	0.1130	5.9046	52.2640	16
17	3.7000	0.2703	33.7502	0.0296	9.1216	0.1096	6.2037	56.5883	17
18	3.9960	0.2502	37.4502	0.0267	9.3719	0.1067	6.4920	60.8426	18
19	4.3157	0.2317	41.4463	0.0241	9.6036	0.1041	6.7697	65.0134	19
20	4.6610	0.2145	45.7620	0.0219	9.8181	0.1019	7.0369	69.0898	20
21	5.0338	0.1987	50.4229	0.0198	10.0168	0.0998	7.2940	73.0629	21
22	5.4365	0.1839	55.4568	0.0180	10.2007	0.0980	7.5412	76.9257	22
23	5.8715	0.1703	60.8933	0.0164	10.3711	0.0964	7.7786	80.6726	23
24	6.3412	0.1577	66.7648	0.0150	10.5288	0.0950	8.0066	84.2997	24
25	6.8485	0.1460	73.1059	0.0137	10.6748	0.0937	8.2254	87.8041	25
26	7.3964	0.1352	79.9544	0.0125	10.8100	0.0925	8.4352	91.1842	26
27	7.9881	0.1252	87.3508	0.0114	10.9352	0.0914	8.6363	94.4390	27
78	8.6271	0.1159	95.3388	0.0105	11.0511	0.0905	8.8289	97.5687	28
29	9.3173	0.1073	103.9659	0.0096	11.1584	0.0896	9.0133	100.5738	29
30	10.0627	0.0994	113.2832	0.0088	11.2578	0.0888	9.1897	103.4558	30
31	10.8677	0.0920	123.3459	0.0081	11.3498	0.0881	9.3584	106.2163	31
32	11.7371	0.0852	134.2135	0.0075	11.4350	0.0875	9.5197	108.8575	32
33	12.6760	0.0789	145.9506	0.0069	11.5139	0.0869	9.6737	111.3819	33
34	13.6901	0.0730	158.6267	0.0063	11.5869	0.0863	9.8208	113.7924	34
35	14.7853	0.0676	172.3168	0.0058	11.6546	0.0858	9.9611	116.0920	35
40	21.7245	0.0460	259.0565	0.0039	11.9246	0.0839	10.5699	126.0422	40
45	31.9204	0.0313	386.5056	0.0026	12.1084	0.0826	11.0447	133.7331	45
50	46.9016	0.0213	573.7702	0.0017	12.2335	0.0817	11.4107	139.5928	50
55	68.9139	0.0145	848.9232	0.0012	12.3186	0.0812	11.6902	144.0065	55
60	101.2571	0.0099	1253.2133	0.0008	12.3766	0.0808	11.9015	147.3000	60
65	148.7798	0.0067	1847.2481	0.0005	12.4160	0.0805	12.0602	149.7387	65
70	218.6064	0.0046	2720.0801	0.0004	12.4428	0.0804	12.1783	151.5326	70
75	321.2045	0.0031	4002.5566	0.0002	12.4611	0.0802	12.2658	152.8448	75
80	471.9548	0.0021	5886.9354	0.0002	12.4735	0.0802	12.3301	153.8001	80
85	693.4565	0.0014	8655.7061	0.0001	12.4820	0.0801	12.3772	154.4925	85
90	1018.9151	0.0010	12723.9386	0.0001	12.4877	0.0801	12.4116	154.9925	90
95	1497.1205	0.0007	18701.5069	0.0001	12.4917	0.0801	12.4365	155.3524	95
100	2199.7613	0.0005	27484.5157	0.0000	12.4943	0.0800	12.4545	155.6107	100

9.0%

	Single Payment		Equal Payment Series				Gradient Series		
N	Compound Amount Factor (F/P,i,N)	Present Worth Factor (P/F,i,N)	Compound Amount Factor (F/A,i,N)	Sinking Fund Factor (A/F,i,N)	Present Worth Factor (P/A,i,N)	Capital Recovery Factor (A/P,i,N)	Gradient Uniform Series (A/G,i,N)	Gradient Present Worth (P/G,i,N)	N
1	1.0900	0.9174	1.0000	1.0000	0.9174	1.0900	0.0000	0.0000	1
2	1.1881	0.8417	2.0900	0.4785	1.7591	0.5685	0.4785	0.8417	2
3	1.2950	0.7722	3.2781	0.3051	2.5313	0.3951	0.9426	2.3860	3
4	1.4116	0.7084	4.5731	0.2187	3.2397	0.3087	1.3925	4.5113	4
5	1.5386	0.6499	5.9847	0.1671	3.8897	0.2571	1.8282	7.1110	5
6	1.6771	0.5963	7.5233	0.1329	4.4859	0.2229	2.2498	10.0924	6
7	1.8280	0.5470	9.2004	0.1087	5.0330	0.1987	2.6574	13.3746	7
8	1.9926	0.5019	11.0285	0.0907	5.5348	0.1807	3.0512	16.8877	8
9	2.1719	0.4604	13.0210	0.0768	5.9952	0.1668	3.4312	20.5711	9
10	2.3674	0.4224	15.1929	0.0658	6.4177	0.1558	3.7978	24.3728	10
11	2.5804	0.3875	17.5603	0.0569	6.8052	0.1469	4.1510	28.2481	11
12	2.8127	0.3555	20.1407	0.0497	7.1607	0.1397	4.4910	32.1590	12
13	3.0658	0.3262	22.9534	0.0436	7.4869	0.1336	4.8182	36.0731	13
14	3.3417	0.2992	26.0192	0.0384	7.7862	0.1284	5.1326	39.9633	14
15	3.6425	0.2745	29.3609	0.0341	8.0607	0.1241	5.4346	43.8069	15
16	3.9703	0.2519	33.0034	0.0303	8.3126	0.1203	5.7245	47.5849	16
17	4.3276	0.2311	36.9737	0.0270	8.5436	0.1170	6.0024	51.2821	17
18	4.7171	0.2120	41.3013	0.0242	8.7556	0.1142	6.2687	54.8860	18
19	5.1417	0.1945	46.0185	0.0217	8.9501	0.1117	6.5236	58.3868	19
20	5.6044	0.1784	51.1601	0.0195	9.1285	0.1095	6.7674	61.7770	20
21	6.1088	0.1637	56.7645	0.0176	9.2922	0.1076	7.0006	65.0509	21
22	6.6586	0.1502	62.8733	0.0159	9.4424	0.1059	7.2232	68.2048	22
23	7.2579	0.1378	69.5319	0.0144	9.5802	0.1044	7.4357	71.2359	23
24	7.9111	0.1264	76.7898	0.0130	9.7066	0.1030	7.6384	74.1433	24
25	8.6231	0.1160	84.7009	0.0118	9.8226	0.1018	7.8316	76.9265	25
26	9.3992	0.1064	93.3240	0.0107	9.9290	0.1007	8.0156	79.5863	26
27	10.2451	0.0976	102.7231	0.0097	10.0266	0.0997	8.1906	82.1241	27
28	11.1671	0.0895	112.9682	0.0089	10.1161	0.0989	8.3571	84.5419	28
29	12.1722	0.0822	124.1354	0.0081	10.1983	0.0981	8.5154	86.8422	29
30	13.2677	0.0754	136.3075	0.0073	10.2737	0.0973	8.6657	89.0280	30
31	14.4618	0.0691	149.5752	0.0067	10.3428	0.0967	8.8083	91.1024	31
32	15.7633	0.0634	164.0370	0.0061	10.4062	0.0961	8.9436	93.0690	32
33	17.1820	0.0582	179.8003	0.0056	10.4644	0.0956	9.0718	94.9314	33
34	18.7284	0.0534	196.9823	0.0051	10.5178	0.0951	9.1933	96.6935	34
35	20.4140	0.0490	215.7108	0.0046	10.5668	0.0946	9.3083	98.3590	35
40	31.4094	0.0318	337.8824	0.0030	10.7574	0.0930	9.7957	105.3762	40
45	48.3273	0.0207	525.8587	0.0019	10.8812	0.0919	10.1603	110.5561	45
50	74.3575	0.0134	815.0836	0.0012	10.9617	0.0912	10.4295	114.3251	50
55	114.4083	0.0087	1260.0918	0.0008	11.0140	0.0908	10.6261	117.0362	55
60	176.0313	0.0057	1944.7921	0.0005	11.0480	0.0905	10.7683	118.9683	60
65	270.8460	0.0037	2998.2885	0.0003	11.0701	0.0903	10.8702	120.3344	65
70	416.7301	0.0024	4619.2232	0.0002	11.0844	0.0902	10.9427	121.2942	70
75	641.1909	0.0016	7113.2321	0.0001	11.0938	0.0901	10.9940	121.9646	75
80	986.5517	0.0010	10950.5741	0.0001	11.0998	0.0901	11.0299	122.4306	80
85	1517.9320	0.0007	16854.8003	0.0001	11.1038	0.0901	11.0551	122.7533	85
90	2335.5266	0.0004	25939.1842	0.0000	11.1064	0.0900	11.0726	122.9758	90
95	3593.4971	0.0003	39916.6350	0.0000	11.1080	0.0900	11.0847	123.1287	95
100	5529.0408	0.0002	61422.6755	0.0000	11.1091	0.0900	11.0930	123.2335	100

10.0%

	Single Payment		Equal Payment Series				Gradient Series		
	Compound Amount Factor	Present Worth Factor	Compound Amount Factor	Sinking Fund Factor	Present Worth Factor	Capital Recovery Factor	Gradient Uniform Series	Gradient Present Worth	
N	(F/P,i,N)	(P/F,i,N)	(F/A,i,N)	(A/F,i,N)	(P/A,i,N)	(A/P,i,N)	(A/G,i,N)	(P/G,i,N)	N
1	1.1000	0.9091	1.0000	1.0000	0.9091	1.1000	0.0000	0.0000	1
2	1.2100	0.8264	2.1000	0.4762	1.7355	0.5762	0.4762	0.8264	2
3	1.3310	0.7513	3.3100	0.3021	2.4869	0.4021	0.9366	2.3291	3
4	1.4641	0.6830	4.6410	0.2155	3.1699	0.3155	1.3812	4.3781	4
5	1.6105	0.6209	6.1051	0.1638	3.7908	0.2638	1.8101	6.8618	5
6	1.7716	0.5645	7.7156	0.1296	4.3553	0.2296	2.2236	9.6842	6
7	1.9487	0.5132	9.4872	0.1054	4.8684	0.2054	2.6216	12.7631	7
8	2.1436	0.4665	11.4359	0.0874	5.3349	0.1874	3.0045	16.0287	8
9	2.3579	0.4241	13.5795	0.0736	5.7590	0.1736	3.3724	19.4215	9
10	2.5937	0.3855	15.9374	0.0627	6.1446	0.1627	3.7255	22.8913	10
11	2.8531	0.3505	18.5312	0.0540	6.4951	0.1540	4.0641	26.3963	11
12	3.1384	0.3186	21.3843	0.0468	6.8137	0.1468	4.3884	29.9012	12
13	3.4523	0.2897	24.5227	0.0408	7.1034	0.1408	4.6988	33.3772	13
14	3.7975	0.2633	27.9750	0.0357	7.3667	0.1357	4.9955	36.8005	14
15	4.1772	0.2394	31.7725	0.0315	7.6061	0.1315	5.2789	40.1520	15
16	4.5950	0.2176	35.9497	0.0278	7.8237	0.1278	5.5493	43.4164	16
17	5.0545	0.1978	40.5447	0.0247	8.0216	0.1247	5.8071	46.5819	17
18	5.5599	0.1799	45.5992	0.0219	8.2014	0.1219	6.0526	49.6395	18
19	6.1159	0.1635	51.1591	0.0195	8.3649	0.1195	6.2861	52.5827	19
20	6.7275	0.1486	57.2750	0.0175	8.5136	0.1175	6.5081	55.4069	20
21	7.4002	0.1351	64.0025	0.0156	8.6487	0.1156	6.7189	58.1095	21
22	8.1403	0.1228	71.4027	0.0140	8.7715	0.1140	6.9189	60.6893	22
23	8.9543	0.1117	79.5430	0.0126	8.8832	0.1126	7.1085	63.1462	23
24	9.8497	0.1015	88.4973	0.0113	8.9847	0.1113	7.2881	65.4813	24
25	10.8347	0.0923	98.3471	0.0102	9.0770	0.1102	7.4580	67.6964	25
26	11.9182	0.0839	109.1818	0.0092	9.1609	0.1092	7.6186	69.7940	26
27	13.1100	0.0763	121.0999	0.0083	9.2372	0.1083	7.7704	71.7773	27
28	14.4210	0.0693	134.2099	0.0075	9.3066	0.1075	7.9137	73.6495	28
29	15.8631	0.0630	148.6309	0.0067	9.3696	0.1067	8.0489	75.4146	29
30	17.4494	0.0573	164.4940	0.0061	9.4269	0.1061	8.1762	77.0766	30
31	19.1943	0.0521	181.9434	0.0055	9.4790	0.1055	8.2962	78.6395	31
32	21.1138	0.0474	201.1378	0.0050	9.5264	0.1050	8.4091	80.1078	32
33	23.2252	0.0431	222.2515	0.0045	9.5694	0.1045	8.5152	81.4856	33
34	25.5477	0.0391	245.4767	0.0041	9.6086	0.1041	8.6149	82.7773	34
35	28.1024	0.0356	271.0244	0.0037	9.6442	0.1037	8.7086	83.9872	35
40	45.2593	0.0221	442.5926	0.0023	9.7791	0.1023	9.0962	88.9525	40
45	72.8905	0.0137	718.9048	0.0014	9.8628	0.1014	9.3740	92.4544	45
50	117.3909	0.0085	1163.9085	0.0009	9.9148	0.1009	9.5704	94.8889	50
55	189.0591	0.0053	1880.5914	0.0005	9.9471	0.1005	9.7075	96.5619	55
60	304.4816	0.0033	3034.8164	0.0003	9.9672	0.1003	9.8023	97.7010	60
65	490.3707	0.0020	4893.7073	0.0002	9.9796	0.1002	9.8672	98.4705	65
70	789.7470	0.0013	7887.4696	0.0001	9.9873	0.1001	9.9113	98.9870	70
75	1271.8954	0.0008	12708.9537	0.0001	9.9921	0.1001	9.9410	99.3317	75
80	2048.4002	0.0005	20474.0021	0.0000	9.9951	0.1000	9.9609	99.5606	80
85	3298.9690	0.0003	32979.6903	0.0000	9.9970	0.1000	9.9742	99.7120	85
90	5313.0226	0.0002	53120.2261	0.0000	9.9981	0.1000	9.9831	99.8118	90
95	8556.6760	0.0001	85556.7605	0.0000	9.9988	0.1000	9.9889	99.8773	95
100	13780.6123	0.0001	137796.1234	0.0000	9.9993	0.1000	9.9927	99.9202	100

11.0%

	Single Payment		Equal Payment Series				Gradient Series		
N	Compound Amount Factor (F/P,i,N)	Present Worth Factor (P/F,i,N)	Compound Amount Factor (F/A,i,N)	Sinking Fund Factor (A/F,i,N)	Present Worth Factor (P/A,i,N)	Capital Recovery Factor (A/P,i,N)	Gradient Uniform Series (A/G,i,N)	Gradient Present Worth (P/G,i,N)	N
1	1.1100	0.9009	1.0000	1.0000	0.9009	1.1100	0.0000	0.0000	1
2	1.2321	0.8116	2.1100	0.4739	1.7125	0.5839	0.4739	0.8116	2
3	1.3676	0.7312	3.3421	0.2992	2.4437	0.4092	0.9306	2.2740	3
4	1.5181	0.6587	4.7097	0.2123	3.1024	0.3223	1.3700	4.2502	4
5	1.6851	0.5935	6.2278	0.1606	3.6959	0.2706	1.7923	6.6240	5
6	1.8704	0.5346	7.9129	0.1264	4.2305	0.2364	2.1976	9.2972	6
7	2.0762	0.4817	9.7833	0.1022	4.7122	0.2122	2.5863	12.1872	7
8	2.3045	0.4339	11.8594	0.0843	5.1461	0.1943	2.9585	15.2246	8
9	2.5580	0.3909	14.1640	0.0706	5.5370	0.1806	3.3144	18.3520	9
10	2.8394	0.3522	16.7220	0.0598	5.8892	0.1698	3.6544	21.5217	10
11	3.1518	0.3173	19.5614	0.0511	6.2065	0.1611	3.9788	24.6945	11
12	3.4985	0.2858	22.7132	0.0440	6.4924	0.1540	4.2879	27.8388	12
13	3.8833	0.2575	26.2116	0.0382	6.7499	0.1482	4.5822	30.9290	13
14	4.3104	0.2320	30.0949	0.0332	6.9819	0.1432	4.8619	33.9449	14
15	4.7846	0.2090	34.4054	0.0291	7.1909	0.1391	5.1275	36.8709	15
16	5.3109	0.1883	39.1899	0.0255	7.3792	0.1355	5.3794	39.6953	16
17	5.8951	0.1696	44.5008	0.0225	7.5488	0.1325	5.6180	42.4095	17
18	6.5436	0.1528	50.3959	0.0198	7.7016	0.1298	5.8439	45.0074	18
19	7.2633	0.1377	56.9395	0.0176	7.8393	0.1276	6.0574	47.4856	19
20	8.0623	0.1240	64.2028	0.0156	7.9633	0.1256	6.2590	49.8423	20
21	8.9492	0.1117	72.2651	0.0138	8.0751	0.1238	6.4491	52.0771	21
22	9.9336	0.1007	81.2143	0.0123	8.1757	0.1223	6.6283	54.1912	22
23	11.0263	0.0907	91.1479	0.0110	8.2664	0.1210	6.7969	56.1864	23
24	12.2392	0.0817	102.1742	0.0098	8.3481	0.1198	6.9555	58.0656	24
25	13.5855	0.0736	114.4133	0.0087	8.4217	0.1187	7.1045	59.8322	25
26	15.0799	0.0663	127.9988	0.0078	8.4881	0.1178	7.2443	61.4900	26
27	16.7386	0.0597	143.0786	0.0070	8.5478	0.1170	7.3754	63.0433	27
28	18.5799	0.0538	159.8173	0.0063	8.6016	0.1163	7.4982	64.4965	28
29	20.6237	0.0485	178.3972	0.0056	8.6501	0.1156	7.6131	65.8542	29
30	22.8923	0.0437	199.0209	0.0050	8.6938	0.1150	7.7206	67.1210	30
31	25.4104	0.0394	221.9132	0.0045	8.7331	0.1145	7.8210	68.3016	31
32	28.2056	0.0355	247.3236	0.0040	8.7686	0.1140	7.9147	69.4007	32
33	31.3082	0.0319	275.5292	0.0036	8.8005	0.1136	8.0021	70.4228	33
34	34.7521	0.0288	306.8374	0.0033	8.8293	0.1133	8.0836	71.3724	34
35	38.5749	0.0259	341.5896	0.0029	8.8552	0.1129	8.1594	72.2538	35
40	65.0009	0.0154	581.8261	0.0017	8.9511	0.1117	8.4659	75.7789	40
45	109.5302	0.0091	986.6386	0.0010	9.0079	0.1110	8.6763	78.1551	45
50	184.5648	0.0054	1668.7712	0.0006	9.0417	0.1106	8.8185	79.7341	50
55	311.0025	0.0032	2818.2042	0.0004	9.0617	0.1104	8.9135	80.7712	55
60	524.0572	0.0019	4755.0658	0.0002	9.0736	0.1102	8.9762	81.4461	60

12.0%

	Single Payment		Equal Payment Series				Gradient Series		
N	Compound Amount Factor (F/P,i,N)	Present Worth Factor (P/F,i,N)	Compound Amount Factor (F/A,i,N)	Sinking Fund Factor (A/F,i,N)	Present Worth Factor (P/A,i,N)	Capital Recovery Factor (A/P,i,N)	Gradient Uniform Series (A/G,i,N)	Gradient Present Worth (P/G,i,N)	N
1	1.1200	0.8929	1.0000	1.0000	0.8929	1.1200	0.0000	0.0000	1
2	1.2544	0.7972	2.1200	0.4717	1.6901	0.5917	0.4717	0.7972	2
3	1.4049	0.7118	3.3744	0.2963	2.4018	0.4163	0.9246	2.2208	3
4	1.5735	0.6355	4.7793	0.2092	3.0373	0.3292	1.3589	4.1273	4
5	1.7623	0.5674	6.3528	0.1574	3.6048	0.2774	1.7746	6.3970	5
6	1.9738	0.5066	8.1152	0.1232	4.1114	0.2432	2.1720	8.9302	6
7	2.2107	0.4523	10.0890	0.0991	4.5638	0.2191	2.5515	11.6443	7
8	2.4760	0.4039	12.2997	0.0813	4.9676	0.2013	2.9131	14.4714	8
9	2.7731	0.3606	14.7757	0.0677	5.3282	0.1877	3.2574	17.3563	9
10	3.1058	0.3220	17.5487	0.0570	5.6502	0.1770	3.5847	20.2541	10
11	3.4785	0.2875	20.6546	0.0484	5.9377	0.1684	3.8953	23.1288	11
12	3.8960	0.2567	24.1331	0.0414	6.1944	0.1614	4.1897	25.9523	12
13	4.3635	0.2292	28.0291	0.0357	6.4235	0.1557	4.4683	28.7024	13
14	4.8871	0.2046	32.3926	0.0309	6.6282	0.1509	4.7317	31.3624	14
15	5.4736	0.1827	37.2797	0.0268	6.8109	0.1468	4.9803	33.9202	15
16	6.1304	0.1631	42.7533	0.0234	6.9740	0.1434	5.2147	36.3670	16
17	6.8660	0.1456	48.8837	0.0205	7.1196	0.1405	5.4353	38.6973	17
18	7.6900	0.1300	55.7497	0.0179	7.2497	0.1379	5.6427	40.9080	81
19	8.6128	0.1161	63.4397	0.0158	7.3658	0.1358	5.8375	42.9979	19
20	9.6463	0.1037	72.0524	0.0139	7.4694	0.1339	6.0202	44.9676	20
21	10.8038	0.0926	81.6987	0.0122	7.5620	0.1322	6.1913	46.8188	21
22	12.1003	0.0826	92.5026	0.0108	7.6446	0.1308	6.3514	48.5543	22
23	13.5523	0.0738	104.6029	0.0096	7.7184	0.1296	6.5010	50.1776	23
24	15.1786	0.0659	118.1552	0.0085	7.7843	0.1285	6.6406	51.6929	24
25	17.0001	0.0588	133.3339	0.0075	7.8431	0.1275	6.7708	53.1046	25
26	19.0401	0.0525	150.3339	0.0067	7.8957	0.1267	6.8921	54.4177	26
27	21.3249	0.0469	169.3740	0.0059	7.9426	0.1259	7.0049	55.6369	27
28	23.8839	0.0419	190.6989	0.0052	7.9844	0.1252	7.1098	56.7674	28
29	26.7499	0.0374	214.5828	0.0047	8.0218	0.1247	7.2071	57.8141	29
30	29.9599	0.0334	241.3327	0.0041	8.0552	0.1241	7.2974	58.7821	30
31	33.5551	0.0298	271.2926	0.0037	8.0850	0.1237	7.3811	59.6761	31
32	37.5817	0.0266	304.8477	0.0033	8.1116	0.1233	7.4586	60.5010	32
33	42.0915	0.0238	342.4294	0.0029	8.1354	0.1229	7.5302	61.2612	33
34	47.1425	0.0212	384.5210	0.0026	8.1566	0.1226	7.5965	61.9612	34
35	52.7996	0.0189	431.6635	0.0023	8.1755	0.1223	7.6577	62.6052	35
40	93.0510	0.0107	767.0914	0.0013	8.2438	0.1213	7.8988	65.1159	40
45	163.9876	0.0061	1358.2300	0.0007	8.2825	0.1207	8.0572	66.7342	45
50	289.0022	0.0035	2400.0182	0.0004	8.3045	0.1204	8.1597	67.7624	50
55	509.3206	0.0020	4236.0050	0.0002	8.3170	0.1202	8.2251	68.4082	55
60	897.5969	0.0011	7471.6411	0.0001	8.3240	0.1201	8.2664	68.8100	60

13.0%

	Single Payment		Equal Payment Series				Gradient Series		
N	Compound Amount Factor (F/P,i,N)	Present Worth Factor (P/F,i,N)	Compound Amount Factor (F/A,i,N)	Sinking Fund Factor (A/F,i,N)	Present Worth Factor (P/A,i,N)	Capital Recovery Factor (A/P,i,N)	Gradient Uniform Series (A/G,i,N)	Gradient Present Worth (P/G,i,N)	N
1	1.1300	0.8850	1.0000	1.0000	0.8850	1.1300	0.0000	0.0000	1
2	1.2769	0.7831	2.1300	0.4695	1.6681	0.5995	0.4695	0.7831	2
3	1.4429	0.6931	3.4069	0.2935	2.3612	0.4235	0.9187	2.1692	3
4	1.6305	0.6133	4.8498	0.2062	2.9745	0.3362	1.3479	4.0092	4
5	1.8424	0.5428	6.4803	0.1543	3.5172	0.2843	1.7571	6.1802	5
6	2.0820	0.4803	8.3227	0.1202	3.9975	0.2502	2.1468	8.5818	6
7	2.3526	0.4251	10.4047	0.0961	4.4226	0.2261	2.5171	11.1322	7
8	2.6584	0.3762	12.7573	0.0784	4.7988	0.2084	2.8685	13.7653	8
9	3.0040	0.3329	15.4157	0.0649	5.1317	0.1949	3.2014	16.4284	9
10	3.3946	0.2946	18.4197	0.0543	5.4262	0.1843	3.5162	19.0797	10
11	3.8359	0.2607	21.8143	0.0458	5.6869	0.1758	3.8134	21.6867	11
12	4.3345	0.2307	25.6502	0.0390	5.9176	0.1690	4.0936	24.2244	12
13	4.8980	0.2042	29.9847	0.0334	6.1218	0.1634	4.3573	26.6744	13
14	5.5348	0.1807	34.8827	0.0287	6.3025	0.1587	4.6050	29.0232	14
15	6.2543	0.1599	40.4175	0.0247	6.4624	0.1547	4.8375	31.2617	15
16	7.0673	0.1415	46.6717	0.0214	6.6039	0.1514	5.0552	33.3841	16
17	7.9861	0.1252	53.7391	0.0186	6.7291	0.1486	5.2589	35.3876	17
18	9.0243	0.1108	61.7251	0.0162	6.8399	0.1462	5.4491	37.2714	18
19	10.1974	0.0981	70.7494	0.0141	6.9380	0.1441	5.6265	39.0366	19
20	11.5231	0.0868	80.9468	0.0124	7.0248	0.1424	5.7917	40.6854	20
21	13.0211	0.0768	92.4699	0.0108	7.1016	0.1408	5.9454	42.2214	21
22	14.7138	0.0680	105.4910	0.0095	7.1695	0.1395	6.0881	43.6486	22
23	16.6266	0.0601	120.2048	0.0083	7.2297	0.1383	6.2205	44.9718	23
24	18.7881	0.0532	136.8315	0.0073	7.2829	0.1373	6.3431	46.1960	24
25	21.2305	0.0471	155.6196	0.0064	7.3300	0.1364	6.4566	47.3264	25
26	23.9905	0.0417	176.8501	0.0057	7.3717	0.1357	6.5614	48.3685	26
27	27.1093	0.0369	200.8406	0.0050	7.4086	0.1350	6.6582	49.3276	27
28	30.6335	0.0326	227.9499	0.0044	7.4412	0.1344	6.7474	50.2090	28
29	34.6158	0.0289	258.5834	0.0039	7.4701	0.1339	6.8296	51.0179	29
30	39.1159	0.0256	293.1992	0.0034	7.4957	0.1334	6.9052	51.7592	30
31	44.2010	0.0226	332.3151	0.0030	7.5183	0.1330	6.9747	52.4380	31
32	49.9471	0.0200	376.5161	0.0027	7.5383	0.1327	7.0385	53.0586	32
33	56.4402	0.0177	426.4632	0.0023	7.5560	0.1323	7.0971	53.6256	33
34	63.7774	0.0157	482.9034	0.0021	7.5717	0.1321	7.1507	54.1430	34
35	72.0685	0.0139	546.6808	0.0018	7.5856	0.1318	7.1998	54.6148	35
40	132.7816	0.0075	1013.7042	0.0010	7.6344	0.1310	7.3888	56.4087	40
45	244.6414	0.0041	1874.1646	0.0005	7.6609	0.1305	7.5076	57.5148	45
50	450.7359	0.0022	3459.5071	0.0003	7.6752	0.1303	7.5811	58.1870	50
55	830.4517	0.0012	6380.3979	0.0002	7.6830	0.1302	7.6260	58.5909	55
60	1530.0535	0.0007	11761.9498	0.0001	7.6873	0.1301	7.6531	58.8313	60

14.0%

	Single Payment		Equal Payment Series				Gradient Series		
N	Compound Amount Factor (F/P,i,N)	Present Worth Factor (P/F,i,N)	Compound Amount Factor (F/A,i,N)	Sinking Fund Factor (A/F,i,N)	Present Worth Factor (P/A,i,N)	Capital Recovery Factor (A/P,i,N)	Gradient Uniform Series (A/G,i,N)	Gradient Present Worth (P/G,i,N)	N
1	1.1400	0.8772	1.0000	1.0000	0.8772	1.1400	0.0000	0.0000	1
2	1.2996	0.7695	2.1400	0.4673	1.6467	0.6073	0.4673	0.7695	2
3	1.4815	0.6750	3.4396	0.2907	2.3216	0.4307	0.9129	2.1194	3
4	1.6890	0.5921	4.9211	0.2032	2.9137	0.3432	1.3370	3.8957	4
5	1.9254	0.5194	6.6101	0.1513	3.4331	0.2913	1.7399	5.9731	5
6	2.1950	0.4556	8.5355	0.1172	3.8887	0.2572	2.1218	8.2511	6
7	2.5023	0.3996	10.7305	0.0932	4.2883	0.2332	2.4832	10.6489	7
8	2.8526	0.3506	13.2328	0.0756	4.6389	0.2156	2.8246	13.1028	8
9	3.2519	0.3075	16.0853	0.0622	4.9464	0.2022	3.1463	15.5629	9
10	3.7072	0.2697	19.3373	0.0517	5.2161	0.1917	3.4490	17.9906	10
11	4.2262	0.2366	23.0445	0.0434	5.4527	0.1834	3.7333	20.3567	11
12	4.8179	0.2076	27.2707	0.0367	5.6603	0.1767	3.9998	22.6399	12
13	5.4924	0.1821	32.0887	0.0312	5.8424	0.1712	4.2491	24.8247	13
14	6.2613	0.1597	37.5811	0.0266	6.0021	0.1666	4.4819	26.9009	14
15	7.1379	0.1401	43.8424	0.0228	6.1422	0.1628	4.6990	28.8623	15
16	8.1372	0.1229	50.9804	0.0196	6.2651	0.1596	4.9011	30.7057	16
17	9.2765	0.1078	59.1176	0.0169	6.3729	0.1569	5.0888	32.4305	17
18	10.5752	0.0946	68.3941	0.0146	6.4674	0.1546	5.2630	34.0380	18
19	12.0557	0.0829	78.9692	0.0127	6.5504	0.1527	5.4243	35.5311	19
20	13.7435	0.0728	91.0249	0.0110	6.6231	0.1510	5.5734	36.9135	20
21	15.6676	0.0638	104.7684	0.0095	6.6870	0.1495	5.7111	38.1901	21
22	17.8610	0.0560	120.4360	0.0083	6.7429	0.1483	5.8381	39.3658	22
23	20.3616	0.0491	138.2970	0.0072	6.7921	0.1472	5.9549	40.4463	23
24	23.2122	0.0431	158.6586	0.0063	6.8351	0.1463	6.0624	41.4371	24
25	26.4619	0.0378	181.8708	0.0055	6.8729	0.1455	6.1610	42.3441	25
26	30.1666	0.0331	208.3327	0.0048	6.9061	0.1448	6.2514	43.1728	26
27	34.3899	0.0291	238.4993	0.0042	6.9352	0.1442	6.3342	43.9289	27
28	39.2045	0.0255	272.8892	0.0037	6.9607	0.1437	6.4100	44.6176	28
29	44.6931	0.0224	312.0937	0.0032	6.9830	0.1432	6.4791	45.2441	29
30	50.9502	0.0196	356.7868	0.0028	7.0027	0.1428	6.5423	45.8132	30
31	58.0832	0.0172	407.7370	0.0025	7.0199	0.1425	6.5998	46.3297	31
32	66.2148	0.0151	465.8202	0.0021	7.0350	0.1421	6.6522	46.7979	32
33	75.4849	0.0132	532.0350	0.0019	7.0482	0.1419	6.6998	47.2218	33
34	86.0528	0.0116	607.5199	0.0016	7.0599	0.1416	6.7431	47.6053	34
35	98.1002	0.0102	693.5727	0.0014	7.0700	0.1414	6.7824	47.9519	35
40	188.8835	0.0053	1342.0251	0.0007	7.1050	0.1407	6.9300	49.2376	40
45	363.6791	0.0027	2590.5648	0.0004	7.1232	0.1404	7.0188	49.9963	45
50	700.2330	0.0014	4994.5213	0.0002	7.1327	0.1402	7.0714	50.4375	50

15.0%

	Single Payment		Equal Payment Series				Gradient Series		
N	Compound Amount Factor (F/P,i,N)	Present Worth Factor (P/F,i,N)	Compound Amount Factor (F/A,i,N)	Sinking Fund Factor (A/F,i,N)	Present Worth Factor (P/A,i,N)	Capital Recovery Factor (A/P,i,N)	Gradient Uniform Series (A/G,i,N)	Gradient Present Worth (P/G,i,N)	N
1	1.1500	0.8696	1.0000	1.0000	0.8696	1.1500	0.0000	0.0000	1
2	1.3225	0.7561	2.1500	0.4651	1.6257	0.6151	0.4651	0.7561	2
3	1.5209	0.6575	3.4725	0.2880	2.2832	0.4380	0.9071	2.0712	3
4	1.7490	0.5718	4.9934	0.2003	2.8550	0.3503	1.3263	3.7864	4
5	2.0114	0.4972	6.7424	0.1483	3.3522	0.2983	1.7228	5.7751	5
6	2.3131	0.4323	8.7537	0.1142	3.7845	0.2642	2.0972	7.9368	6
7	2.6600	0.3759	11.0668	0.0904	4.1604	0.2404	2.4498	10.1924	7
8	3.0590	0.3269	13.7268	0.0729	4.4873	0.2229	2.7813	12.4807	8
9	3.5179	0.2843	16.7858	0.0596	4.7716	0.2096	3.0922	14.7548	9
10	4.0456	0.2472	20.3037	0.0493	5.0188	0.1993	3.3832	16.9795	10
11	4.6524	0.2149	24.3493	0.0411	5.2337	0.1911	3.6549	19.1289	11
12	5.3503	0.1869	29.0017	0.0345	5.4206	0.1845	3.9082	21.1849	12
13	6.1528	0.1625	34.3519	0.0291	5.5831	0.1791	4.1438	23.1352	13
14	7.0757	0.1413	40.5047	0.0247	5.7245	0.1747	4.3624	24.9725	14
15	8.1371	0.1229	47.5804	0.0210	5.8474	0.1710	4.5650	26.6930	15
16	9.3576	0.1069	55.7175	0.0179	5.9542	0.1679	4.7522	28.2960	16
17	10.7613	0.0929	65.0751	0.0154	6.0472	0.1654	4.9251	29.7828	17
18	12.3755	0.0808	75.8364	0.0132	6.1280	0.1632	5.0843	31.1565	18
19	14.2318	0.0703	88.2118	0.0113	6.1982	0.1613	5.2307	32.4213	19
20	16.3665	0.0611	102.4436	0.0098	6.2593	0.1598	5.3651	33.5822	20
21	18.8215	0.0531	118.8101	0.0084	6.3125	0.1584	5.4883	34.6448	21
22	21.6447	0.0462	137.6316	0.0073	6.3587	0.1573	5.6010	35.6150	22
23	24.8915	0.0402	159.2764	0.0063	6.3988	0.1563	5.7040	36.4988	23
24	28.6252	0.0349	184.1678	0.0054	6.4338	0.1554	5.7979	37.3023	24
25	32.9190	0.0304	212.7930	0.0047	6.4641	0.1547	5.8834	38.0314	25
26	37.8568	0.0264	245.7120	0.0041	6.4906	0.1541	5.9612	38.6918	26
27	43.5353	0.0230	283.5688	0.0035	6.5135	0.1535	6.0319	39.2890	27
28	50.0656	0.0200	327.1041	0.0031	6.5335	0.1531	6.0960	39.8283	28
29	57.5755	0.0174	377.1697	0.0027	6.5509	0.1527	6.1541	40.3146	29
30	66.2118	0.0151	434.7451	0.0023	6.5660	0.1523	6.2066	40.7526	30
31	76.1435	0.0131	500.9569	0.0020	6.5791	0.1520	6.2541	41.1466	31
32	87.5651	0.0114	577.1005	0.0017	6.5905	0.1517	6.2970	41.5006	32
33	100.6998	0.0099	664.6655	0.0015	6.6005	0.1515	6.3357	41.8184	33
34	115.8048	0.0086	765.3654	0.0013	6.6091	0.1513	6.3705	42.1033	34
35	133.1755	0.0075	881.1702	0.0011	6.6166	0.1511	6.4019	42.3586	35
40	267.8635	0.0037	1779.0903	0.0006	6.6418	0.1506	6.5168	43.2830	40
45	538.7693	0.0019	3585.1285	0.0003	6.6543	0.1503	6.5830	43.8051	45
50	1083.6574	0.0009	7217.7163	0.0001	6.6605	0.1501	6.6205	44.0958	50

16.0%

	Single Payment		Equal Payment Series				Gradient Series		
	Compound Amount Factor	Present Worth Factor	Compound Amount Factor	Sinking Fund Factor	Present Worth Factor	Capital Recovery Factor	Gradient Uniform Series	Gradient Present Worth	
N	(*F/P,i,N*)	(*P/F,i,N*)	(*F/A,i,N*)	(*A/F,i,N*)	(*P/A,i,N*)	(*A/P,i,N*)	(*A/G,i,N*)	(*P/G,i,N*)	*N*
1	1.1600	0.8621	1.0000	1.0000	0.8621	1.1600	0.0000	0.0000	1
2	1.3456	0.7432	2.1600	0.4630	1.6052	0.6230	0.4630	0.7432	2
3	1.5609	0.6407	3.5056	0.2853	2.2459	0.4453	0.9014	2.0245	3
4	1.8106	0.5523	5.0665	0.1974	2.7982	0.3574	1.3156	3.6814	4
5	2.1003	0.4761	6.8771	0.1454	3.2743	0.3054	1.7060	5.5858	5
6	2.4364	0.4104	8.9775	0.1114	3.6847	0.2714	2.0729	7.6380	6
7	2.8262	0.3538	11.4139	0.0876	4.0386	0.2476	2.4169	9.7610	7
8	3.2784	0.3050	14.2401	0.0702	4.3436	0.2302	2.7388	11.8962	8
9	3.8030	0.2630	17.5185	0.0571	4.6065	0.2171	3.0391	13.9998	9
10	4.4114	0.2267	21.3215	0.0469	4.8332	0.2069	3.3187	16.0399	10
11	5.1173	0.1954	25.7329	0.0389	5.0286	0.1989	3.5783	17.9941	11
12	5.9360	0.1685	30.8502	0.0324	5.1971	0.1924	3.8189	19.8472	12
13	6.8858	0.1452	36.7862	0.0272	5.3423	0.1872	4.0413	21.5899	13
14	7.9875	0.1252	43.6720	0.0229	5.4675	0.1829	4.2464	23.2175	14
15	9.2655	0.1079	51.6595	0.0194	5.5755	0.1794	4.4352	24.7284	15
16	10.7480	0.0930	60.9650	0.0164	5.6685	0.1764	4.6086	26.1241	16
17	12.4677	0.0802	71.6730	0.0140	5.7487	0.1740	4.7676	27.4074	17
18	14.4625	0.0691	84.1407	0.0119	5.8178	0.1719	4.9130	28.5828	18
19	16.7765	0.0596	98.6032	0.0101	5.8775	0.1701	5.0457	29.6557	19
20	19.4608	0.0514	115.3797	0.0087	5.9288	0.1687	5.1666	30.6321	20
21	22.5745	0.0443	134.8405	0.0074	5.9731	0.1674	5.2766	31.5180	21
22	26.1864	0.0382	157.4150	0.0064	6.0113	0.1664	5.3765	32.3200	22
23	30.3762	0.0329	183.6014	0.0054	6.0442	0.1654	5.4671	33.0442	23
24	35.2364	0.0284	213.9776	0.0047	6.0726	0.1647	5.5490	33.6970	24
25	40.8742	0.0245	249.2140	0.0040	6.0971	0.1640	5.6230	34.2841	25
26	47.4141	0.0211	290.0883	0.0034	6.1182	0.1634	5.6898	34.8114	26
27	55.0004	0.0182	337.5024	0.0030	6.1364	0.1630	5.7500	35.2841	27
28	63.8004	0.0157	392.5028	0.0025	6.1520	0.1625	5.8041	35.7073	28
29	74.0085	0.0135	456.3032	0.0022	6.1656	0.1622	5.8528	36.0856	29
30	85.8499	0.0116	530.3117	0.0019	6.1772	0.1619	5.8964	36.4234	30
31	99.5859	0.0100	616.1616	0.0016	6.1872	0.1616	5.9356	36.7247	31
32	115.5196	0.0087	715.7475	0.0014	6.1959	0.1614	5.9706	36.9930	32
33	134.0027	0.0075	831.2671	0.0012	6.2034	0.1612	6.0019	27.2318	33
34	155.4432	0.0064	965.2698	0.0010	6.2098	0.1610	6.0299	37.4441	34
35	180.3141	0.0055	1120.7130	0.0009	6.2153	0.1609	6.0548	37.6327	35
40	378.7212	0.0026	2360.7572	0.0004	6.2335	0.1604	6.1441	38.2992	40
45	795.4438	0.0013	4965.2739	0.0002	6.2421	0.1602	6.1934	38.6598	45
50	1670.7038	0.0006	10435.6488	0.0001	6.2463	0.1601	6.2201	38.8521	50

18.0%

	Single Payment		Equal Payment Series				Gradient Series		
N	Compound Amount Factor (F/P,i,N)	Present Worth Factor (P/F,i,N)	Compound Amount Factor (F/A,i,N)	Sinking Fund Factor (A/F,i,N)	Present Worth Factor (P/A,i,N)	Capital Recovery Factor (A/P,i,N)	Gradient Uniform Series (A/G,i,N)	Gradient Present Worth (P/G,i,N)	N
1	1.1800	0.8475	1.0000	1.0000	0.8475	1.1800	0.0000	0.0000	1
2	1.3924	0.7182	2.1800	0.4587	1.5656	0.6387	0.4587	0.7182	2
3	1.6430	0.6086	3.5724	0.2799	2.1743	0.4599	0.8902	1.9354	3
4	1.9388	0.5158	5.2154	0.1917	2.6901	0.3717	1.2947	3.4828	4
5	2.2878	0.4371	7.1542	0.1398	3.1272	0.3198	1.6728	5.2312	5
6	2.6996	0.3704	9.4420	0.1059	3.4976	0.2859	2.0252	7.0834	6
7	3.1855	0.3139	12.1415	0.0824	3.8115	0.2624	2.3526	8.9670	7
8	3.7589	0.2660	15.3270	0.0652	4.0776	0.2452	2.6558	10.8292	8
9	4.4355	0.2255	19.0859	0.0524	4.3030	0.2324	2.9358	12.6329	9
10	5.2338	0.1911	23.5213	0.0425	4.4941	0.2225	3.1936	14.3525	10
11	6.1759	0.1619	28.7551	0.0348	4.6560	0.2148	3.4303	15.9716	11
12	7.2876	0.1372	34.9311	0.0286	4.7932	0.2086	3.6470	17.4811	12
13	8.5994	0.1163	42.2187	0.0237	4.9095	0.2037	3.8449	18.8765	13
14	10.1472	0.0985	50.8180	0.0197	5.0081	0.1997	4.0250	20.1576	14
15	11.9737	0.0835	60.9653	0.0164	5.0916	0.1964	4.1887	21.3269	15
16	14.1290	0.0708	72.9390	0.0137	5.1624	0.1937	4.3369	22.3885	16
17	16.6722	0.0600	87.0680	0.0115	5.2223	0.1915	4.4708	23.3482	17
18	19.6733	0.0508	103.7403	0.0096	5.2732	0.1896	4.5916	24.2123	18
19	23.2144	0.0431	123.4135	0.0081	5.3162	0.1881	4.7003	24.9877	19
20	27.3930	0.0365	146.6280	0.0068	5.3527	0.1868	4.7978	25.6813	20
21	32.3238	0.0309	174.0210	0.0057	5.3837	0.1857	4.8851	26.3000	21
22	38.1421	0.0262	206.3448	0.0048	5.4099	0.1848	4.9632	26.8506	22
23	45.0076	0.0222	244.4868	0.0041	5.4321	0.1841	5.0329	27.3394	23
24	53.1090	0.0188	289.4945	0.0035	5.4509	0.1835	5.0950	27.7725	24
25	62.6686	0.0160	342.6035	0.0029	5.4669	0.1829	5.1502	28.1555	25
26	73.9490	0.0135	405.2721	0.0025	5.4804	0.1825	5.1991	28.4935	26
27	87.2598	0.0115	479.2211	0.0021	5.4919	0.1821	5.2425	28.7915	27
28	102.9666	0.0097	566.4809	0.0018	5.5016	0.1818	5.2810	29.0537	28
29	121.5005	0.0082	669.4475	0.0015	5.5098	0.1815	5.3149	29.2842	29
30	143.3706	0.0070	790.9480	0.0013	5.5168	0.1813	5.3448	29.4864	30
31	169.1774	0.0059	934.3186	0.0011	5.5227	0.1811	5.3712	29.6638	31
32	199.6293	0.0050	1103.4960	0.0009	5.5277	0.1809	5.3945	29.8191	32
33	235.5625	0.0042	1303.1253	0.0008	5.5320	0.1808	5.4149	29.9549	33
34	277.9638	0.0036	1538.6878	0.0006	5.5356	0.1806	5.4328	30.0736	34
35	327.9973	0.0030	1816.6516	0.0006	5.5386	0.1806	5.4485	30.1773	35
40	750.3783	0.0013	4163.2130	0.0002	5.5482	0.1802	5.5022	30.5269	40
45	1716.6839	0.0006	9531.5771	0.0001	5.5523	0.1801	5.5293	30.7006	45
50	3927.3569	0.0003	21813.0937	0.0000	5.5541	0.1800	5.5428	30.7856	50

20.0%

	Single Payment		Equal Payment Series				Gradient Series		
N	Compound Amount Factor (F/P,i,N)	Present Worth Factor (P/F,i,N)	Compound Amount Factor (F/A,i,N)	Sinking Fund Factor (A/F,i,N)	Present Worth Factor (P/A,i,N)	Capital Recovery Factor (A/P,i,N)	Gradient Uniform Series (A/G,i,N)	Gradient Present Worth (P/G,i,N)	N
1	1.2000	0.8333	1.0000	1.0000	0.8333	1.2000	0.0000	0.0000	1
2	1.4400	0.6944	2.2000	0.4545	1.5278	0.6545	0.4545	0.6944	2
3	1.7280	0.5787	3.6400	0.2747	2.1065	0.4747	0.8791	1.8519	3
4	2.0736	0.4823	5.3680	0.1863	2.5887	0.3863	1.2742	3.2986	4
5	2.4883	0.4019	7.4416	0.1344	2.9906	0.3344	1.6405	4.9061	5
6	2.9860	0.3349	9.9299	0.1007	3.3255	0.3007	1.9788	6.5806	6
7	3.5832	0.2791	12.9159	0.0774	3.6046	0.2774	2.2902	8.2551	7
8	4.2998	0.2326	16.4991	0.0606	3.8372	0.2606	2.5756	9.8831	8
9	5.1598	0.1938	20.7989	0.0481	4.0310	0.2481	2.8364	11.4335	9
10	6.1917	0.1615	25.9587	0.0385	4.1925	0.2385	3.0739	12.8871	10
11	7.4301	0.1346	32.1504	0.0311	4.3271	0.2311	3.2893	14.2330	11
12	8.9161	0.1122	39.5805	0.0253	4.4392	0.2253	3.4841	15.4667	12
13	10.6993	0.0935	48.4966	0.0206	4.5327	0.2206	3.6597	16.5883	13
14	12.8392	0.0779	59.1959	0.0169	4.6106	0.2169	3.8175	17.6008	14
15	15.4070	0.0649	72.0351	0.0139	4.6755	0.2139	3.9588	18.5095	15
16	18.4884	0.0541	87.4421	0.0114	4.7296	0.2114	4.0851	19.3208	16
17	22.1861	0.0451	105.9306	0.0094	4.7746	0.2094	4.1976	20.0419	17
18	26.6233	0.0376	128.1167	0.0078	4.8122	0.2078	4.2975	20.6805	18
19	31.9480	0.0313	154.7400	0.0065	4.8435	0.2065	4.3861	21.2439	19
20	38.3376	0.0261	186.6880	0.0054	4.8696	0.2054	4.4643	21.7395	20
21	46.0051	0.0217	225.0256	0.0044	4.8913	0.2044	4.5334	22.1742	21
22	55.2061	0.0181	271.0307	0.0037	4.9094	0.2037	4.5941	22.5546	22
23	66.2474	0.0151	326.2369	0.0031	4.9245	0.2031	4.6475	22.8867	23
24	79.4968	0.0126	392.4842	0.0025	4.9371	0.2025	4.6943	23.1760	24
25	95.3962	0.0105	471.9811	0.0021	4.9476	0.2021	4.7352	23.4276	25
26	114.4755	0.0087	567.3773	0.0018	4.9563	0.2018	4.7709	23.6460	26
27	137.3706	0.0073	681.8528	0.0015	4.9636	0.2015	4.8020	23.8353	27
28	164.8447	0.0061	819.2233	0.0012	4.9697	0.2012	4.8291	23.9991	28
29	197.8136	0.0051	984.0680	0.0010	4.9747	0.2010	4.8527	24.1406	29
30	237.3763	0.0042	1181.8816	0.0008	4.9789	0.2008	4.8731	24.2628	30
31	284.8516	0.0035	1419.2579	0.0007	4.9824	0.2007	4.8908	24.3681	31
32	341.8219	0.0029	1704.1095	0.0006	4.9854	0.2006	4.9061	24.4588	32
33	410.1863	0.0024	2045.9314	0.0005	4.9878	0.2005	4.9194	24.5368	33
34	492.2235	0.0020	2456.1176	0.0004	4.9898	0.2004	4.9308	24.6038	34
35	590.6682	0.0017	2948.3411	0.0003	4.9915	0.2003	4.9406	24.6614	35
40	1469.7716	0.0007	7343.8578	0.0001	4.9966	0.2001	4.9728	24.8469	40
45	3657.2620	0.0003	18281.3099	0.0001	4.9986	0.2001	4.9877	24.9316	45

25.0%

	Single Payment		Equal Payment Series				Gradient Series		
N	Compound Amount Factor (F/P,i,N)	Present Worth Factor (P/F,i,N)	Compound Amount Factor (F/A,i,N)	Sinking Fund Factor (A/F,i,N)	Present Worth Factor (P/A,i,N)	Capital Recovery Factor (A/P,i,N)	Gradient Uniform Series (A/G,i,N)	Gradient Present Worth (P/G,i,N)	N
1	1.2500	0.8000	1.0000	1.0000	0.8000	1.2500	0.0000	0.0000	1
2	1.5625	0.6400	2.2500	0.4444	1.4400	0.6944	0.4444	0.6400	2
3	1.9531	0.5120	3.8125	0.2623	1.9520	0.5123	0.8525	1.6640	3
4	2.4414	0.4096	5.7656	0.1734	2.3616	0.4234	1.2249	2.8928	4
5	3.0518	0.3277	8.2070	0.1218	2.6893	0.3718	1.5631	4.2035	5
6	3.8147	0.2621	11.2588	0.0888	2.9514	0.3388	1.8683	5.5142	6
7	4.7684	0.2097	15.0735	0.0663	3.1611	0.3163	2.1424	6.7725	7
8	5.9605	0.1678	19.8419	0.0504	3.3289	0.3004	2.3872	7.9469	8
9	7.4506	0.1342	25.8023	0.0388	3.4631	0.2888	2.6048	9.0207	9
10	9.3132	0.1074	33.2529	0.0301	3.5705	0.2801	2.7971	9.9870	10
11	11.6415	0.0859	42.5661	0.0235	3.6564	0.2735	2.9663	10.8460	11
12	14.5519	0.0687	54.2077	0.0184	3.7251	0.2684	3.1145	11.6020	12
13	18.1899	0.0550	68.7596	0.0145	3.7801	0.2645	3.2437	12.2617	13
14	22.7374	0.0440	86.9495	0.0115	3.8241	0.2615	3.3559	12.8334	14
15	28.4217	0.0352	109.6868	0.0091	3.8593	0.2591	3.4530	13.3260	15
16	35.5271	0.0281	138.1085	0.0072	3.8874	0.2572	3.5366	13.7482	16
17	44.4089	0.0225	173.6357	0.0058	3.9099	0.2558	3.6084	14.1085	17
18	55.5112	0.0180	218.0446	0.0046	3.9279	0.2546	3.6698	14.4147	18
19	69.3889	0.0144	273.5558	0.0037	3.9424	0.2537	3.7222	14.6741	19
20	86.7362	0.0115	342.9447	0.0029	3.9539	0.2529	3.7667	14.8932	20
21	108.4202	0.0092	429.6809	0.0023	3.9631	0.2523	3.8045	15.0777	21
22	135.5253	0.0074	538.1011	0.0019	3.9705	0.2519	3.8365	15.2326	22
23	169.4066	0.0059	673.6264	0.0015	3.9764	0.2515	3.8634	15.3625	23
24	211.7582	0.0047	843.0329	0.0012	3.9811	0.2512	3.8861	15.4711	24
25	264.6978	0.0038	1054.7912	0.0009	3.9849	0.2509	3.9052	15.5618	25
26	330.8722	0.0030	1319.4890	0.0008	3.9879	0.2508	3.9212	15.6373	26
27	413.5903	0.0024	1650.3612	0.0006	3.9903	0.2506	3.9346	15.7002	27
28	516.9879	0.0019	2063.9515	0.0005	3.9923	0.2505	3.9457	15.7524	28
29	646.2349	0.0015	2580.9394	0.0004	3.9938	0.2504	3.9551	15.7957	29
30	807.7936	0.0012	3227.1743	0.0003	3.9950	0.2503	3.9628	15.8316	30
31	1009.7420	0.0010	4034.9678	0.0002	3.9960	0.2502	3.9693	15.8614	31
32	1262.1774	0.0008	5044.7098	0.0002	3.9968	0.2502	3.9746	15.8859	32
33	1577.7218	0.0006	6306.8872	0.0002	3.9975	0.2502	3.9791	15.9062	33
34	1972.1523	0.0005	7884.6091	0.0001	3.9980	0.2501	3.9828	15.9229	34
35	2465.1903	0.0004	9856.7613	0.0001	3.9984	0.2501	3.9858	15.9367	35
40	7523.1638	0.0001	30088.6554	0.0000	3.9995	0.2500	3.9947	15.9766	40

30.0%

	Single Payment		Equal Payment Series				Gradient Series		
	Compound Amount Factor	Present Worth Factor	Compound Amount Factor	Sinking Fund Factor	Present Worth Factor	Capital Recovery Factor	Gradient Uniform Series	Gradient Present Worth	
N	(F/P,i,N)	(P/F,i,N)	(F/A,i,N)	(A/F,i,N)	(P/A,i,N)	(A/P,i,N)	(A/G,i,N)	(P/G,i,N)	N
1	1.3000	0.7692	1.0000	1.0000	0.7692	1.3000	0.0000	0.0000	1
2	1.6900	0.5917	2.3000	0.4348	1.3609	0.7348	0.4348	0.5917	2
3	2.1970	0.4552	3.9900	0.2506	1.8161	0.5506	0.8271	1.5020	3
4	2.8561	0.3501	6.1870	0.1616	2.1662	0.4616	1.1783	2.5524	4
5	3.7129	0.2693	9.0431	0.1106	2.4356	0.4106	1.4903	3.6297	5
6	4.8268	0.2072	12.7560	0.0784	2.6427	0.3784	1.7654	4.6656	6
7	6.2749	0.1594	17.5828	0.0569	2.8021	0.3569	2.0063	5.6218	7
8	8.1573	0.1226	23.8577	0.0419	2.9247	0.3419	2.2156	6.4800	8
9	10.6045	0.0943	32.0150	0.0312	3.0190	0.3312	2.3963	7.2343	9
10	13.7858	0.0725	42.6195	0.0235	3.0915	0.3235	2.5512	7.8872	10
11	17.9216	0.0558	56.4053	0.0177	3.1473	0.3177	2.6833	8.4452	11
12	23.2981	0.0429	74.3270	0.0135	3.1903	0.3135	2.7952	8.9173	12
13	30.2875	0.0330	97.6250	0.0102	3.2233	0.3102	2.8895	9.3135	13
14	39.3738	0.0254	127.9125	0.0078	3.2487	0.3078	2.9685	9.6437	14
15	51.1859	0.0195	167.2863	0.0060	3.2682	0.3060	3.0344	9.9172	15
16	66.5417	0.0150	218.4722	0.0046	3.2832	0.3046	3.0892	10.1426	16
17	86.5042	0.0116	285.0139	0.0035	3.2948	0.3035	3.1345	10.3276	17
18	112.4554	0.0089	371.5180	0.0027	3.3037	0.3027	3.1718	10.4788	18
19	146.1920	0.0068	483.9734	0.0021	3.3105	0.3021	3.2025	10.6019	19
20	190.0496	0.0053	630.1655	0.0016	3.3158	0.3016	3.2275	10.7019	20
21	247.0645	0.0040	820.2151	0.0012	3.3198	0.3012	3.2480	10.7828	21
22	321.1839	0.0031	1067.2796	0.0009	3.3230	0.3009	3.2646	10.8482	22
23	417.5391	0.0024	1388.4635	0.0007	3.3254	0.3007	3.2781	10.9009	23
24	542.8008	0.0018	1806.0026	0.0006	3.3272	0.3006	3.2890	10.9433	24
25	705.6410	0.0014	2348.8033	0.0004	3.3286	0.3004	3.2979	10.9773	25
26	917.3333	0.0011	3054.4443	0.0003	3.3297	0.3003	3.3050	11.0045	26
27	1192.5333	0.0008	3971.7776	0.0003	3.3305	0.3003	3.3107	11.0263	27
28	1550.2933	0.0006	5164.3109	0.0002	3.3312	0.3002	3.3153	11.0437	28
29	2015.3813	0.0005	6714.6042	0.0001	3.3317	0.3001	3.3189	11.0576	29
30	2619.9956	0.0004	8729.9855	0.0001	3.3321	0.3001	3.3219	11.0687	30
31	3405.9943	0.0003	11349.9811	0.0001	3.3324	0.3001	3.3242	11.0775	31
32	4427.7926	0.0002	14755.9755	0.0001	3.3326	0.3001	3.3261	11.0845	32
33	5756.1304	0.0002	19183.7681	0.0001	3.3328	0.3001	3.3276	11.0901	33
34	7482.9696	0.0001	24939.8985	0.0000	3.3329	0.3000	3.3288	11.0945	34
35	9727.8604	0.0001	32422.8681	0.0000	3.3330	0.3000	3.3297	11.0980	35

35.0%

	Single Payment		Equal Payment Series				Gradient Series		
N	Compound Amount Factor (F/P,i,N)	Present Worth Factor (P/F,i,N)	Compound Amount Factor (F/A,i,N)	Sinking Fund Factor (A/F,i,N)	Present Worth Factor (P/A,i,N)	Capital Recovery Factor (A/P,i,N)	Gradient Uniform Series (A/G,i,N)	Gradient Present Worth (P/G,i,N)	N
1	1.3500	0.7407	1.0000	1.0000	0.7407	1.3500	0.0000	0.0000	1
2	1.8225	0.5487	2.3500	0.4255	1.2894	0.7755	0.4255	0.5487	2
3	2.4604	0.4064	4.1725	0.2397	1.6959	0.5897	0.8029	1.3616	3
4	3.3215	0.3011	6.6329	0.1508	1.9969	0.5008	1.1341	2.2648	4
5	4.4840	0.2230	9.9544	0.1005	2.2200	0.4505	1.4220	3.1568	5
6	6.0534	0.1652	14.4384	0.0693	2.3852	0.4193	1.6698	3.9828	6
7	8.1722	0.1224	20.4919	0.0488	2.5075	0.3988	1.8811	4.7170	7
8	11.0324	0.0906	28.6640	0.0349	2.5982	0.3849	2.0597	5.3515	8
9	14.8937	0.0671	39.6964	0.0252	2.6653	0.3752	2.2094	5.8886	9
10	20.1066	0.0497	54.5902	0.0183	2.7150	0.3683	2.3338	6.3363	10
11	27.1439	0.0368	74.6967	0.0134	2.7519	0.3634	2.4364	6.7047	11
12	36.6442	0.0273	101.8406	0.0098	2.7792	0.3598	2.5205	7.0049	12
13	49.4697	0.0202	138.4848	0.0072	2.7994	0.3572	2.5889	7.2474	13
14	66.7841	0.0150	187.9544	0.0053	2.8144	0.3553	2.6443	7.4421	14
15	90.1585	0.0111	254.7385	0.0039	2.8255	0.3539	2.6889	7.5974	15
16	121.7139	0.0082	344.8970	0.0029	2.8337	0.3529	2.7246	7.7206	16
17	164.3138	0.0061	466.6109	0.0021	2.8398	0.3521	2.7530	7.8180	17
18	221.8236	0.0045	630.9247	0.0016	2.8443	0.3516	2.7756	7.8946	18
19	299.4619	0.0033	852.7483	0.0012	2.8476	0.3512	2.7935	2.9547	19
20	404.2736	0.0025	1152.2103	0.0009	2.8501	0.3509	2.8075	8.0017	20
21	545.7693	0.0018	1556.4838	0.0006	2.8519	0.3506	2.8186	8.0384	21
22	736.7886	0.0014	2102.2532	0.0005	2.8533	0.3505	2.8272	8.0669	22
23	994.6646	0.0010	2839.0418	0.0004	2.8543	0.3504	2.8340	8.0890	23
24	1342.7973	0.0007	3833.7064	0.0003	2.8550	0.3503	2.8393	8.1061	24
25	1812.7763	0.0006	5176.5037	0.0002	2.8556	0.3502	2.8433	8.1194	25
26	2447.2480	0.0004	6989.2800	0.0001	2.8560	0.3501	2.8465	8.1296	26
27	3303.7848	0.0003	9436.5280	0.0001	2.8563	0.3501	2.8490	8.1374	27
28	4460.1095	0.0002	12740.3128	0.0001	2.8565	0.3501	2.8509	8.1435	28
29	6021.1478	0.0002	17200.4222	0.0001	2.8567	0.3501	2.8523	8.1481	29
30	8128.5495	0.0001	23221.5700	0.0000	2.8568	0.3500	2.8535	8.1517	30

40.0%

	Single Payment		Equal Payment Series				Gradient Series		
N	Compound Amount Factor (F/P,i,N)	Present Worth Factor (P/F,i,N)	Compound Amount Factor (F/A,i,N)	Sinking Fund Factor (A/F,i,N)	Present Worth Factor (P/A,i,N)	Capital Recovery Factor (A/P,i,N)	Gradient Uniform Series (A/G,i,N)	Gradient Present Worth (P/G,i,N)	N
1	1.4000	0.7143	1.0000	1.0000	0.7143	1.4000	0.0000	0.0000	1
2	1.9600	0.5102	2.4000	0.4167	1.2245	0.8167	0.4167	0.5102	2
3	2.7440	0.3644	4.3600	0.2294	1.5889	0.6294	0.7798	1.2391	3
4	3.8416	0.2603	7.1040	0.1408	1.8492	0.5408	1.0923	2.0200	4
5	5.3782	0.1859	10.9456	0.0914	2.0352	0.4914	1.3580	2.7637	5
6	7.5295	0.1328	16.3238	0.0613	2.1680	0.4613	1.5811	3.4278	6
7	10.5414	0.0949	23.8534	0.0419	2.2628	0.4419	1.7664	3.9970	7
8	14.7579	0.0678	34.3947	0.0291	2.3306	0.4291	1.9185	4.4713	8
9	20.6610	0.0484	49.1526	0.0203	2.3790	0.4203	2.0422	4.8585	9
10	28.9255	0.0346	69.8137	0.0143	2.4136	0.4143	2.1419	5.1696	10
11	40.4957	0.0247	98.7391	0.0101	2.4383	0.4101	2.2215	5.4166	11
12	56.6939	0.0176	139.2348	0.0072	2.4559	0.4072	2.2845	5.6106	12
13	79.3715	0.0126	195.9287	0.0051	2.4685	0.4051	2.3341	5.7618	13
14	111.1201	0.0090	275.3002	0.0036	2.4775	0.4036	2.3729	5.8788	14
15	155.5681	0.0064	386.4202	0.0026	2.4839	0.4026	2.4030	5.9688	15
16	217.7953	0.0046	541.9883	0.0018	2.4885	0.4018	2.4262	6.0376	16
17	304.9135	0.0033	759.7837	0.0013	2.4918	0.4013	2.4441	6.0901	17
18	426.8789	0.0023	1064.6971	0.0009	2.4941	0.4009	2.4577	6.1299	18
19	597.6304	0.0017	1491.5760	0.0007	2.4958	0.4007	2.4682	6.1601	19
20	836.6826	0.0012	2089.2064	0.0005	2.4970	0.4005	2.4761	6.1828	20
21	1171.3556	0.0009	2925.8889	0.0003	2.4979	0.4003	2.4821	6.1998	21
22	1639.8978	0.0006	4097.2445	0.0002	2.4985	0.4002	2.4866	6.2127	22
23	2295.8569	0.0004	5737.1423	0.0002	2.4989	0.4002	2.4900	6.2222	23
24	3214.1997	0.0003	8032.9993	0.0001	2.4992	0.4001	2.4925	6.2294	24
25	4499.8796	0.0002	11247.1990	0.0001	2.4994	0.4001	2.4944	6.2347	25
26	6299.8314	0.0002	15747.0785	0.0001	2.4996	0.4001	2.4959	6.2387	26
27	8819.7640	0.0001	22046.9099	0.0000	2.4997	0.4000	2.4969	6.2416	27
28	12347.6696	0.0001	30866.6739	0.0000	2.4998	0.4000	2.4977	6.2438	28
29	17286.7374	0.0001	43214.3435	0.0000	2.4999	0.4000	2.4983	6.2454	29
30	24201.4324	0.0000	60501.0809	0.0000	2.4999	0.4000	2.4988	6.2466	30

50.0%

	Single Payment		Equal Payment Series				Gradient Series		
N	Compound Amount Factor (F/P,i,N)	Present Worth Factor (P/F,i,N)	Compound Amount Factor (F/A,i,N)	Sinking Fund Factor (A/F,i,N)	Present Worth Factor (P/A,i,N)	Capital Recovery Factor (A/P,i,N)	Gradient Uniform Series (A/G,i,N)	Gradient Present Worth (P/G,i,N)	N
1	1.5000	0.6667	1.0000	1.0000	0.6667	1.5000	0.0000	0.0000	1
2	2.2500	0.4444	2.5000	0.4000	1.1111	0.9000	0.4000	0.4444	2
3	3.3750	0.2963	4.7500	0.2105	1.4074	0.7105	0.7368	1.0370	3
4	5.0625	0.1975	8.1250	0.1231	1.6049	0.6231	1.0154	1.6296	4
5	7.5938	0.1317	13.1875	0.0758	1.7366	0.5758	1.2417	2.1564	5
6	11.3906	0.0878	20.7813	0.0481	1.8244	0.5481	1.4226	2.5953	6
7	17.0859	0.0585	32.1719	0.0311	1.8829	0.5311	1.5648	2.9465	7
8	25.6289	0.0390	49.2578	0.0203	1.9220	0.5203	1.6752	3.2196	8
9	38.4434	0.0260	74.8867	0.0134	1.9480	0.5134	1.7596	3.4277	9
10	57.6650	0.0173	113.3301	0.0088	1.9653	0.5088	1.8235	3.5838	10
11	86.4976	0.0116	170.9951	0.0058	1.9769	0.5058	1.8713	3.6994	11
12	129.7463	0.0077	257.4927	0.0039	1.9846	0.5039	1.9068	3.7842	12
13	194.6195	0.0051	387.2390	0.0026	1.9897	0.5026	1.9329	3.8459	13
14	291.9293	0.0034	581.8585	0.0017	1.9931	0.5017	1.9519	3.8904	14
15	437.8939	0.0023	873.7878	0.0011	1.9954	0.5011	1.9657	3.9224	15
16	656.8408	0.0015	1311.6817	0.0008	1.9970	0.5008	1.9756	3.9452	16
17	985.2613	0.0010	1968.5225	0.0005	1.9980	0.5005	1.9827	3.9614	17
18	1477.8919	0.0007	2953.7838	0.0003	1.9986	0.5003	1.9878	3.9729	18
19	2216.8378	0.0005	4431.6756	0.0002	1.9991	0.5002	1.9914	3.9811	19
20	3325.2567	0.0003	6648.5135	0.0002	1.9994	0.5002	1.9940	3.9868	20

APPENDIX

C

VALUES OF THE STANDARD NORMAL DISTRIBUTION FUNCTION

PART A: The Cumulative Standardized Normal Distribution Function

z	0.00	0.01	0.02	0.03	0.04	0.05	0.06	0.07	0.08	0.09
−3	0.00135	0.00131	0.00126	0.00122	0.00118	0.00114	0.00111	0.00107	0.00104	0.00100
−2.9	0.00187	0.00181	0.00175	0.00169	0.00164	0.00159	0.00154	0.00149	0.00144	0.00139
−2.8	0.00256	0.00248	0.00240	0.00233	0.00226	0.00219	0.00212	0.00205	0.00199	0.00193
−2.7	0.00347	0.00336	0.00326	0.00317	0.00307	0.00298	0.00289	0.00280	0.00272	0.00264
−2.6	0.00466	0.00453	0.00440	0.00427	0.00415	0.00402	0.00391	0.00379	0.00368	0.00357
−2.5	0.00621	0.00604	0.00587	0.00570	0.00554	0.00539	0.00523	0.00508	0.00494	0.00480
−2.4	0.00820	0.00798	0.00776	0.00755	0.00734	0.00714	0.00695	0.00676	0.00657	0.00639
−2.3	0.01072	0.01044	0.01017	0.00990	0.00964	0.00939	0.00914	0.00889	0.00866	0.00842
−2.2	0.01390	0.01355	0.01321	0.01287	0.01255	0.01222	0.01191	0.01160	0.01130	0.01101
−2.1	0.01786	0.01743	0.01700	0.01659	0.01618	0.01578	0.01539	0.01500	0.01463	0.01426
−2	0.02275	0.02222	0.02169	0.02118	0.02068	0.02018	0.01970	0.01923	0.01876	0.01831
−1.9	0.02872	0.02807	0.02743	0.02680	0.02619	0.02559	0.02500	0.02442	0.02385	0.02330
−1.8	0.03593	0.03515	0.03438	0.03362	0.03288	0.03216	0.03144	0.03074	0.03005	0.02938
−1.7	0.04457	0.04363	0.04272	0.04182	0.04093	0.04006	0.03920	0.03836	0.03754	0.03673
−1.6	0.05480	0.05370	0.05262	0.05155	0.05050	0.04947	0.04846	0.04746	0.04648	0.04551
−1.5	0.06681	0.06552	0.06426	0.06301	0.06178	0.06057	0.05938	0.05821	0.05705	0.05592
−1.4	0.08076	0.07927	0.07780	0.07636	0.07493	0.07353	0.07215	0.07078	0.06944	0.06811
−1.3	0.09680	0.09510	0.09342	0.09176	0.09012	0.08851	0.08691	0.08534	0.08379	0.08226
−1.2	0.11507	0.11314	0.11123	0.10935	0.10749	0.10565	0.10383	0.10204	0.10027	0.09853
−1.1	0.13567	0.13350	0.13136	0.12924	0.12714	0.12507	0.12302	0.12100	0.11900	0.11702
−1	0.15866	0.15625	0.15386	0.15151	0.14917	0.14686	0.14457	0.14231	0.14007	0.13786
−0.9	0.18406	0.18141	0.17879	0.17619	0.17361	0.17106	0.16853	0.16602	0.16354	0.16109
−0.8	0.21186	0.20897	0.20611	0.20327	0.20045	0.19766	0.19489	0.19215	0.18943	0.18673
−0.7	0.24196	0.23885	0.23576	0.23270	0.22965	0.22663	0.22363	0.22065	0.21770	0.21476
−0.6	0.27425	0.27093	0.26763	0.26435	0.26109	0.25785	0.25463	0.25143	0.24825	0.24510
−0.5	0.30854	0.30503	0.30153	0.29806	0.29460	0.29116	0.28774	0.28434	0.28096	0.27760
−0.4	0.34458	0.34090	0.33724	0.33360	0.32997	0.32636	0.32276	0.31918	0.31561	0.31207
−0.3	0.38209	0.37828	0.37448	0.37070	0.36693	0.36317	0.35942	0.35569	0.35197	0.34827
−0.2	0.42074	0.41683	0.41294	0.40905	0.40517	0.40129	0.39743	0.39358	0.38974	0.38591
−0.1	0.46017	0.45620	0.45224	0.44828	0.44433	0.44038	0.43644	0.43251	0.42858	0.42465
0	0.50000	0.50399	0.50798	0.51197	0.51595	0.51994	0.52392	0.52790	0.53188	0.53586

PART B: The Cumulative Standardized Normal Distribution Function

z	0.00	0.01	0.02	0.03	0.04	0.05	0.06	0.07	0.08	0.09
0	0.50000	0.50399	0.50798	0.51197	0.51595	0.51994	0.52392	0.52790	0.53188	0.53586
0.1	0.53983	0.54380	0.54776	0.55172	0.55567	0.55962	0.56356	0.56749	0.57142	0.57535
0.2	0.57926	0.58317	0.58706	0.59095	0.59483	0.59871	0.60257	0.60642	0.61026	0.61409
0.3	0.61791	0.62172	0.62552	0.62930	0.63307	0.63683	0.64058	0.64431	0.64803	0.65173
0.4	0.65542	0.65910	0.66276	0.66640	0.67003	0.67364	0.67724	0.68082	0.68439	0.68793
0.5	0.69146	0.69497	0.69847	0.70194	0.70540	0.70884	0.71226	0.71566	0.71904	0.72240
0.6	0.72575	0.72907	0.73237	0.73565	0.73891	0.74215	0.74537	0.74857	0.75175	0.75490
0.7	0.75804	0.76115	0.76424	0.76730	0.77035	0.77337	0.77637	0.77935	0.78230	0.78524
0.8	0.78814	0.79103	0.79389	0.79673	0.79955	0.80234	0.80511	0.80785	0.81057	0.81327
0.9	0.81594	0.81859	0.82121	0.82381	0.82639	0.82894	0.83147	0.83398	0.83646	0.83891
1	0.84134	0.84375	0.84614	0.84849	0.85083	0.85314	0.85543	0.85769	0.85993	0.86214
1.1	0.86433	0.86650	0.86864	0.87076	0.87286	0.87493	0.87698	0.87900	0.88100	0.88298
1.2	0.88493	0.88686	0.88877	0.89065	0.89251	0.89435	0.89617	0.89796	0.89973	0.90147
1.3	0.90320	0.90490	0.90658	0.90824	0.90988	0.91149	0.91309	0.91466	0.91621	0.91774
1.4	0.91924	0.92073	0.92220	0.92364	0.92507	0.92647	0.92785	0.92922	0.93056	0.93189
1.5	0.93319	0.93448	0.93574	0.93699	0.93822	0.93943	0.94062	0.94179	0.94295	0.94408
1.6	0.94520	0.94630	0.94738	0.94845	0.94950	0.95053	0.95154	0.95254	0.95352	0.95449
1.7	0.95543	0.95637	0.95728	0.95818	0.95907	0.95994	0.96080	0.96164	0.96246	0.96327
1.8	0.96407	0.96485	0.96562	0.96638	0.96712	0.96784	0.96856	0.96926	0.96995	0.97062
1.9	0.97128	0.97193	0.97257	0.97320	0.97381	0.97441	0.97500	0.97558	0.97615	0.97670
2	0.97725	0.97778	0.97831	0.97882	0.97932	0.97982	0.98030	0.98077	0.98124	0.98169
2.1	0.98214	0.98257	0.98300	0.98341	0.98382	0.98422	0.98461	0.98500	0.98537	0.98574
2.2	0.98610	0.98645	0.98679	0.98713	0.98745	0.98778	0.98809	0.98840	0.98870	0.98899
2.3	0.98928	0.98956	0.98983	0.99010	0.99036	0.99061	0.99086	0.99111	0.99134	0.99158
2.4	0.99180	0.99202	0.99224	0.99245	0.99266	0.99286	0.99305	0.99324	0.99343	0.99361
2.5	0.99379	0.99396	0.99413	0.99430	0.99446	0.99461	0.99477	0.99492	0.99506	0.99520
2.6	0.99534	0.99547	0.99560	0.99573	0.99585	0.99598	0.99609	0.99621	0.99632	0.99643
2.7	0.99653	0.99664	0.99674	0.99683	0.99693	0.99702	0.99711	0.99720	0.99728	0.99736
2.8	0.99744	0.99752	0.99760	0.99767	0.99774	0.99781	0.99788	0.99795	0.99801	0.99807
2.9	0.99813	0.99819	0.99825	0.99831	0.99836	0.99841	0.99846	0.99851	0.99856	0.99861
3	0.99865	0.99869	0.99874	0.99878	0.99882	0.99886	0.99889	0.99893	0.99896	0.99900

Note 1: The standard normal distribution is the function of $f(z) = \dfrac{1}{\sqrt{2\pi}} e^{\frac{-z^2}{2}}$. Its graph is a bell curve above a region of area 1, with inflection points at $z = -1$ and $z = +1$.

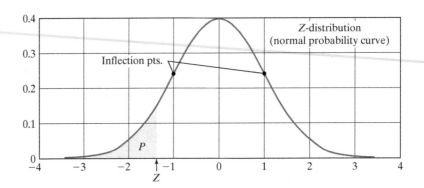

Note 2: If a normal variable X is not "standard," its value must be standardized: $z = \dfrac{x - \mu}{\sigma}$. To find the area **P** under the normal probability curve, $P(X \le x) = \Phi\left(z = \dfrac{x - \mu}{\sigma}\right)$ and $\Phi(-z) = 1 - \Phi(z)$. For example, to find the probability that the NPW would be negative where the NPW distribution is normally distributed with mean of \$20 and variance of 400, we evaluate the following:

$$P(\text{NPW} \le 0) = \Phi\left(z = \frac{0 - 20}{\sqrt{400}} = -1\right) = 0.15866$$

Or if we want to find the probability that the NPW exceeds \$50, we calculate

$$P(\text{NPW} \ge 50) = 1 - P(\text{NPW} \le 50) = 1 - \Phi\left(z = \frac{50 - 20}{\sqrt{400}} = 1.5\right)$$

$$= 1 - 0.93319 = 0.0668$$

INDEX

A

Accelerated depreciation
methods, 466
Accounting, 46–83, 457–458, 460
accrual-basis, 60
asset depreciation and, 460
business financial status, 49–64
depreciation, 457–458, 460
decision making, 46–83
equation, 51–52
financial statements, 46–83
matching concept, 460
net worth, 47, 53
ratio analysis, 64–75
uses of information from,
48–49
Accrual-basis accounting, 60
Accumulated cash-flow sign test,
353–355
Actual dollars (A_n), 87, 575–577,
579–583
adjusted-discount method of
analysis, 581–583
analysis for inflation, 579–583
conversion from constant
dollars, 575–576
conversion to constant dollars,
576–577
deflation method of analysis,
580–581
mixed-dollar analysis, 583
Add-on interest loans, 197–198
Adjustable rate mortgage (ARM),
162–163, 205–206
AE, *see* Annual equivalent-worth
(AE) analysis
American option, 694, 717
Amortization expense, 488
Amortized loans, 190–196, 203
Analysis period, 265–279,
300–302
annual equivalence (AE)
analysis, 300–302
present worth (PW) analysis,
265–279

differing from project lives,
269–276, 301–302
matching project lives, 266–269,
301
required service period,
265–266
unspecified, 276–279
Annual effective yield, 164.
See also Effective
interest rate
Annual equivalent cost (AEC),
296, 303–305, 764–765
Annual equivalent-worth (AE)
analysis, 294–345
analysis periods and project
lives, 300–302
applications of, 306–312
benefits of, 306
capital recovery (CR) cost,
303–305
design economics, 320–330
Excel calculations for, 325
fundamental decision rule,
296–302
life-cycle cost analysis (LCCA),
306, 312–320
make-or-buy (outsourcing)
decisions, 308–311
mutually exclusive alternatives,
296, 300–302
net present worth for, 296, 298
operating costs, 303–305
pricing asset use, 311–312
repeating cash flow cycles,
296–299
robot cost reduction, 296–304
single-project evaluation,
296–298
unit profit (cost) calculation,
306–308
Annual percentage rate (APR),
164, 191
Annual percentage yield (APY),
164. *See also* Effective
interest rate

Annual production budget,
442–444
Annual sales budget, 441–442
Annual worth (*AW*), 235
Annuity factor (*A/P*), 119
Arbitrage opportunity, 708
Arbitrage (risk-free) portfolio, 706
Asset depreciation ranges
(ADRs), 463
Asset management analysis, 68–70
Assets, 52, 209–220, 311–312,
458–482, 486–488, 539–542,
763–780, 824–825. *See also*
Investments
accounting depreciation, 460
annual equivalent-worth (AE)
analysis, 311–312
bonds, 212–220
book depreciation methods,
464–474, 486
cash flow statements for,
539–542
cost basis, 461–462
current, 52
depreciation, 458–464, 486–488
economic depreciation, 459–460
expected return, 209–212
financial management of,
208–220
fixed, 458
long-term (fixed), 52
multiple project requirements,
539–542
pricing use of, 311–312
property depreciation, 460–461
rate of return on, 824–825
recovery periods, 474–475
repair or improvement
adjustments, 486–488
replacement analysis, 763–780
salvage value, 463–464
service life, 763–780
tax depreciation methods,
474–482
trade-in allowance, 462

Assets (*Continued*)
  unrecognized gain or loss, 462
  useful life, 463–464
Automobile design, risk and,
  608–609
Average collection period, 69–70
Average inflation rate (*f*), 572–573
Average unit cost, 420–424

B

Balance sheets, 51–55
  accounting equation for, 51–52
  assets, 52
  liabilities, 53
  reading, 53–55
  stockholder's equity (net worth),
  53
Base period, 570
Benefit–cost analysis (BCA), 313,
  863–871
  decision making use of, 863
  framework of, 863–864
  public project difficulties, 871
  quantifying benefits and costs,
  866–871
  social discount rate, 865–866
  sponsor's costs, 865, 868, 870
  sponsor's revenues, 868, 870
  user's benefits, 864, 867–869
  user's disbenefits, 868, 869–870
  valuation of benefits and costs,
  864–866
Benefit–cost (*B/C*) ratios,
  872–881
  calculations for, 872–875
  incremental analysis using,
  877–881
  mutually exclusive alternatives
  and, 877–881
  net-present-worth (NPW) and,
  876–877
  profitability index (net *B/C*
  ratio), 875–876
Binomial lattice, 712–716
  multiperiod model, 713–716
  option pricing using, 712–716
  two-period option valuation,
  712–713

Black–Scholes option model,
  716–718
Bonds, 65, 212–220
  current yield, 216
  fixed-income securities, 510
  loans compared to, 212–213
  long-term debt financing, 65
  prices of, 214–220
  rating by financial risk, 216
  safety of, 215
  types of, 213–214
  yield to maturity (YTM),
  215–220
Book depreciation methods,
  464–474, 486
  declining balance (DB),
  466–473
  straight-line (SL), 465–466
  units-of-production, 473–474
Book value per share, 73
Borrowed funds, 251–252, 522,
  533–535, 589–592
  cash inflow from, 522
  cash outflow from repayment
  of, 522
  inflation effects on, 589–592
  net-present-worth (NPW)
  method and, 251–252
  project financing, 533–535
Bottom line, 50
Break-even analysis, 616–620
Break-even interest rate (*i**), 248,
  249, 349–350, 353–355,
  366–369. *See also* Internal
  rate of return (IRR)
  accumulated cash-flow sign test,
  353–355
  benchmark, as a, 366
  calculations for, 349–350
  internal rate of return (IRR),
  366–369
  mixed investments, 368–369
  multiple rates-of-return,
  net cash flow rule of signs,
  353
  present-worth (PW) and, 248,
  249, 349–350
  pure investments, 366–368

rate-of-return (ROR) analysis,
  349–350, 353–355
Break-even sales, 426–429
Budgeted income statement,
  447–448
Budgets, 441–447, 810–857.
  See also Profit estimation
  annual sales, 441–442
  annual production, 442–444
  capital, 810–857
  cost of goods-sold, 444–444
  nonmanufacturing cost budget,
  445–447
Business, 33–40, 46–47, 49–75,
  429–433
  corporations, 34–35
  costs for decisions, 429–433
  differential cost and revenue,
  429–433
  economic versus design
  decisions, 36
  engineer's role in, 33–36
  engineering economic
  decisions, 35
  financial statements, 40, 49–64
  large-scale projects, 38–40
  make-or-buy (outsourcing)
  decisions, 433
  managers, 48
  methods changes, 430–431
  net worth, 46–47
  operations planning, 431–433
  partnerships, 34
  personal economic decisions, 36
  proprietorships, 34
  ratio analysis, 64–75
  risks of large-scale projects,
  39–40

C

Call options, 693, 695–697, 699–
  702, 705–708, 716, 719–720
  buying to reduce capital at risk,
  699–702
  buying when price expected to
  rise, 695–696
  European, 716
  real, 719–720

replicating-portfolio approach
 with, 705–708
Callable bonds, 214
Capital, 35, 53, 55, 59–60, 65, 67
 debt, 65
 equity, 65
 expenditure, 35
 net change in working, 59–60
 paid-in (surplus), 53
 working, 55, 59–60, 67
Capital budgeting, 810–857
 contingent projects, 838–839
 cost of capital, 820–829
 cost of debt, 826–827
 cost of equity, 821–826
 debt financing, 812, 814–815
 debt ratio (capital structure),
  816–820
 dependent projects, 837
 equity financing, 812–813
 independent projects, 837, 838
 investment alternative
  evaluation, 837–846
 limited budgets, 839–846
 minimum attractive rate of
  return (MARR), 829–837
 mutually exclusive projects, 838
Capital rationing, 833–837
Capital recovery (CR) cost,
 303–305
Capital-recovery factor (*A/P*),
 119–121
Capitalized-equivalent (CE)
 method, 256–260
Cash flow, 94–95, 101–111,
 149–150, 189–190, 298–299,
 501–505, 516–566,
 568–607
 composite, 141–147
 composite series (two), 123–124
 compound growth, 129
 conversions for equivalence
  calculations, 99–101
 cost–benefit estimation,
  518–520
 discrete compounding formulas,
  140
 economic analysis and, 501–505

equal-payment (uniform) series,
 101–102, 111–112
 equivalent, 94–95
 future-worth factor (*F*), 101, 108
 generalized approach,
  543–553
 geometric gradient series, 103,
  129–134, 140
 incremental, 520–524
 independent projects and, 239
 inflation impact on, 568–607
 interest rates and, 108–111,
  189–190
 irregular (mixed) series, 103,
  135–140
 lease-or-buy decision, 547–553
 linear-gradient series, 102–103,
  123–129, 140
 loans, 236–239
 mutually exclusive projects and,
  239
 net equations, 543–544
 net income versus, 502–505
 noncash expenses, 502–503
 present-worth factor (*P*), 101,
  106–108, 122–123, 124–126,
  130–132
 projects, 236–239, 516–566,
  583–592
 repeating cycles, 298–299
 series of, 189–190
 simple borrowing, 351
 single, 101, 103–111, 140
 skipping pattern, 149–150
 solving for time and interest
  rates, 108–111
 tabular formats, 544–545
 unconventional calculations,
  141–150
 uneven-payment series, 103,
  135–140
Cash flow diagrams, 89–90
Cash flow statements, 57–64,
 524–542
 borrowed funds for project
  financing, 533–535
 cash sources and uses, 58
 development of, 524–542

estimating from operations,
 61–64
 financing activities and, 59
 investing activities and, 59
 multiple assets project
  requirements, 539–542
 negative taxable income,
  535–539
 operating activities and, 59
 operating and investing project
  requirements, 524–528
 project development and,
  524–542
 reading, 59–61
 working-capital project
  requirements, 528–533
Challenger, replacement decisions
 by, 758, 760–762
Coefficient of variation, 632–633
Collectible cars: return on
 investment, 346–347
Commercial loans, 190–198
 add-on interest, 197–198
 amortized, 190–196
 remaining-balance method,
  195–196
 tabular method, 192–194
Common (capital) stock, 65–66,
 83, 822
Complex projects, cash flow for,
 519–520
Composite cash flows, 141–147
Composite series, 123–124
Compound-amount factor, 103,
 105, 111–113, 140
 discrete compounding formulas,
  140
 equal-payment series, 111–113
 single cash flows, 103, 105
Compound growth, 129
Compound interest, 91–94,
 164–187
 calculations for, 91–94
 continuous compounding,
  169–170, 180–187
 continuous-funds flow,
  182–187
 discrete compounding, 180–181

Compound interest *(Continued)*
  effective annual interest rates, 164–179
  equal to payment period, 171–172
  equivalence calculations for, 171–187
  less frequent than payments, 176–179
  more frequent than payments, 172–175
  payment periods and, 167–168
  periods of, 164–167
  quarterly, 164–165
  various intervals of, 165–166
Compound options, 733–739
Compounding process, 103
Computer simulations, 649–650
Conditional probabilities, 630–632, 673–674
Constant-dollar prices, 279
Constant dollars (A'_n), 575–577, 578–579, 583
  analysis for inflation, 578–579
  conversion from actual dollars, 576–577
  conversion to actual dollars, 575–576
  mixed-dollar analysis, 583
Consumer price index (CPI), 568–569, 570–572
  annual calculation of, 571–572
  base period for, 570
  calculating, 570
  industrial goods, 571
  inflation measurement using, 570
  purchasing power and, 568–569
Contingent projects, capital budgeting for, 838–839
Continuous compounding, 169–170, 180–187
Continuous-funds flow, 182–187
Contribution margin, 426–427
Corporate bonds, 214
Corporate taxes, 488–515
  calculating gains or losses, 491–496

depreciation and, 455–515
depreciation recapture, 493
economic analysis, 497–505
effective (average) tax rate, 489
income tax rate, 497–501
incremental income tax rate, 497–499
MACRS property disposal and, 491
marginal tax rate, 489
operating income, 488–491
state income taxes, 500–501
Corporations, 34–35
Cost basis, 461–462, 491, 493–494
Cost–benefit estimation, 518–520
Cost depletion, 482–483
Cost-effectiveness, 881–899
  cost-utility analysis, 896–899
  discounting, 892
  health-care services, 890–899
  independent interventions, 892–893
  life-cycle costs, 887–890
  macro analysis, 886–887
  micro analysis, 886
  mutually exclusive interventions, 892, 894–896
  public projects, 881–890
  ratio, 892
Cost of capital, 820–829
  calculating, 827–829
  capital budgeting and, 820–829
  cost of debt, 826–827
  cost of equity, 821–826
  marginal, 828–829
  weighted-average, 827–828
Cost of goods-sold budget, 444–444
Cost-only projects, 536
Costs, 42–44, 56, 85–94, 244–245, 252–253, 262, 296, 303–308, 410–455, 522, 759–768, 887–890
  annual equivalent (AEC), 296, 303–305, 764–765
  average unit, 420–424
  business decisions and, 429–433

capital (ownership), 252–253, 763
capital recovery (CR), 303–305
concept of, 412–413
cost–volume–profit analysis, 424–429
differential cost and revenue, 429–433
economic service life and, 763–768
financial statement classification of, 415–418
fixed (capacity), 419
funds, payback and, 244–245
incremental, 262
iPhone manufacture, 410–411
life-cycle, 887–890
make-or-buy (outsourcing) decisions, 433
manufacturing, 413–414
marginal, 43–44, 435–440
matching principle, 415
mixed (semi-variable), 420
money, time value of, 85–94
nonmanufacturing, 414–415
operating, 303–305, 760, 763–764
operations planning, 431–433
opportunity cost, 433–435, 760–762
period, 415
predicting behavior, 418–429
product, 415–418
profit estimation, 441–449
reduction, 42
replacement analysis, 759–768
revenue, 56, 429–431, 435–437
savings as cash inflow, 522
sunk, 435, 759–760
unit profit, 306–308
variable operating, 419
volume index, 419
Cost-utility analysis, 896–899
Cost–volume chart, 420–424
Covariance, 632–633
Creditors, 48–49
Cumulative distribution, 625–627
Current market value, 758–759

Current ratio, 68
Current yield, 216

D

Day's sales outstanding (DSO),
    69–70
Debt, 65, 533, 812, 814–815,
    826–827
  capital, 65
  cost of, 826–827
  financing, 812, 814–815
Debt financing, 812, 814–815
Debt management, 66–67,
    190–208
  amortized loans, 190
  analysis, 66–67
  commercial loans, 190–198
  discount rate, 199–201
  installments, 190
  lease financing and loans,
      comparison of, 198–201
  loans, 190–208
  money management, 190–208
  mortgages, 202–208
Debt ratio (capital structure), 66,
    533, 816–820
Decision making, 35–42, 46–83,
    261–279, 308–311, 429–433,
    547–553, 622–633, 664–678,
    756–809, 810–857, 863–871.
    See also Costs
  accounting, 46–83
  benefit–cost analysis, 863–871
  capital budgeting, 810–857
  capital expenditure, 35
  decision trees for risk
      evaluation, 664–678
  design versus economic, 36
  difficulty of, 37–38
  "do nothing" option, 261–262
  economic, 35–38, 40–42
  engineering economic, 35, 40–42
  financial statements, 46–83
  future costs for, 429–433
  large-scale projects and, 38–40
  lease-or-buy, 547–553
  make-or-buy (outsourcing),
      308–311, 433

method changed and, 430–431
  mutually exclusive project
      alternatives, 261–279
  operations planning for,
      431–433
  personal, 36
  probabilities for investments,
      622–633
  replacement analysis for,
      756–809
  sequential investment decisions,
      664–678
  strategic, 40–42
Decision rules, 366–374, 380–381,
    645–649
  expected-value criterion,
      645–646
  incremental investment analysis,
      380–381
  internal rate of return (IRR),
      366–369
  mean-and-variance criterion,
      646
  mixed investments, 368–369
  probability distribution,
      645–649
  pure investments, 366–368
  risk-adjusted discount rate
      method, 646
Decision trees, 664–678
  conditional profit or loss
      updates, 671–674
  imperfect information,
      673–678
  net after-tax cash flow
      relevance, 665
  perfect information, 669–671
  probabilities from information,
      673–676
  risk evaluation using, 664–678
  rollback procedure, 665–666
  sequential investment decisions
      and, 664–678
  structure of, 665–669
  value of information, 669–678
Declining balance (DB)
    depreciation method,
    466–473

Defender, replacement decisions
    by, 758, 760–762
Defer investment option, 724–727
Deferred annuity, 121
Deflation, inflation compared to,
    570
Dependent projects, capital
    budgeting for, 837
Depletion, 482–485
  cost, 482–483
  percentage, 482–485
Depreciation, 419, 456–515
  accounting, 457–458, 460
  assets, 458–482, 486–488
  book methods, 464–474, 486
  cash flow needs, 501–505
  corporate taxes and, 488–515
  cost basis, 461–462, 491,
      493–494
  declining-balance (DB) method,
      466–473
  depletion, 482–485
  economic, 459–460
  economic analysis for, 501–505
  Excel calculations for, 466, 471,
      473, 478, 482
  functional, 459
  modified accelerated cost
      recovery system (MACRS),
      474–482
  net income and, 502–505
  personal property, 476–478
  physical, 459
  property, 460–461
  real property, 479–482
  recapture, 493
  repair or improvement
      adjustments, 486–488
  salvage value, 463–464, 491,
      493–494
  straight-line (SL) method,
      465–466
  tax methods, 464, 474–482, 486
  units-of-production method,
      473–474
  useful life, 463–464
Design economics, 320–330
Differential analysis, 43

Differential cash flows, *see* Incremental cash flows
Differential cost and revenue, 429–433
Direct-solution method, rate-of-return (ROR) analysis, 356–357
Disbursements, 88
Discount rate, 199–201, 865–866
Discounted cash flow techniques (DCFs), 245–253
  minimum attractive rate of return (MARR), 246, 252–253
  net-present-worth (NPW), 245–252
  project balance diagrams, 207–210
Discounted payback period, 240, 243–244
Discounting factor (e^{-rt}), 180, 182
Discounting process, 106–107
Discounts for health-care services, 892
Discrete compounding, 140, 180–181, 931–960
  cash flow formulas, 140
  compound interest and, 180–181
  interest factors for, 931–960
Discrete payment or receipt (A_n), 88
Distribution functions, 961–964
Dividends, 66
"Do nothing" option, 261–262
Double-declining balance (DDB) depreciation, 467

E

Earning power, 86–87
Earnings per share (EPS), 56
Economic analysis, 497–505, 858–909
  benefit–cost analysis, 863–872
  benefit–cost ratios, 872–881
  cash flow needs, 501–505
  cost-effectiveness, 881–899
  cost-utility analysis, 896–899
  depreciation and, 497–505

health-care services, 890–899
  income tax rate and, 497–501
  incremental income tax rate for, 497–499
  motor vehicle crash costs, 858–859
  net income versus cash flow, 502–505
  noncash expenses, 502–504
  pricing difficulties, 861–862
  public sector projects, 862–872, 881–890
  service sectors, 858–909
  state income taxes, 500–501
Economic depreciation, 459–460
Economic equivalence, *see* Equivalence
Economic service life, 763–768
Economic versus design decisions, 36
Effective interest rates (i_a), 164–179
  compounding periods of, 165–167
  continuous compounding, 169–170
  equivalence calculations with, 171–179
  nominal interest rate compared to, 164–166
  payment periods of, 167–168
  quarterly compounding, 164–165
Effective (average) tax rate, 489
End-of-period convention, 90–91
End-of-period payment (A), 88
Engineering economics, 32–45
  business organizations, 34–35
  decision making, 35–38, 40–42
  difficulty of decisions, 37–38
  engineer's role in business, 33–36
  large-scale projects, 38–40
  motivation and, 32–33
  principles of, 42–45
  strategic decisions, 40–42
Equal-payment series, 101–102, 111–123, 140
  annuity factor (A/P), 119

cash flow transactions, 107–102
  discrete compounding formulas, 140
  capital-recovery factor (A/P), 119–121
  compound-amount factor, 111–113
  deferred annuity, 121
  present-worth factor (P), 122–123
  sinking-fund factor, 114–118
  uniform factors, 112, 115
Equipment, 41–42, 520–521
  cash outflow from, 520–521
  replacement, 41–42
  selection, 41
  strategic decisions and, 41–42
Equity, 65, 533, 812–813, 821–826, 829–831
  borrowed funds from, 533
  capital budgeting and, 812–813, 821–826
  capital, 65
  common stock, 822
  cost of, 821–826
  financing, 812–813
  net flow, 829–831
  preferred stock, 822–823
  rate of return on risky assets, 824–825
  retained earnings, 821–822
Equivalence, 84–161, 171–187, 578–583
  actual dollar (A_n) analysis, 579–583
  adjusted-discount method of analysis, 581–583
  annual worth (AW), 235
  calculations, 95–101
  cash flow and, 141–147, 149–150
  cash flow conversions and, 99–101
  common time basis for, 96–98
  composite cash flows, 141–147
  compounding interest, 171–187
  constant dollar (A'_n) analysis, 578–579
  continuous compounding, 180–187

deflation method of analysis, 580–581

development of formulas for, 101–140

earning power, 86–87

economic, 84–101

effective interest rates using, 171–179

Excel financial functions, 106, 107, 109, 110, 113, 120, 134, 140

future worth (FW), 96, 235

inflation and calculations of, 578–583

inflation-free (real) interest rate (*i'*), 578

interest rate and, 95–104, 147–149

market interest rate (*i*), 578

mixed-dollar analysis, 583

present worth (PW), 96, 235

purchasing power, 86–87

skipping pattern in cash flow, 149–150

unconventional calculations, 141–150

Equivalent cash flow, 94–95

European option, 694, 716–717

Excel, 106, 107, 109, 110, 113, 120, 134, 140, 174, 194, 218, 268, 325, 362, 466, 471, 473, 478, 482, 552, 586, 588, 591, 593, 595, 597, 620, 662, 718, 775, 782–783, 785–786, 788

after-tax interest rate calculations, 782–783, 785–786, 788

annual equivalent-worth (AE) calculations, 325

Black–Scholes formula evaluation, 718

cash flow statements under inflation, 586, 588, 591

depreciation calculations, 466, 471, 473, 478, 482

economic service life calculations, 775, 788

financial functions, 106, 107, 109, 110, 113, 120, 134, 140

interest rate calculations, 174, 194, 218, 782–783

loan repayment schedule, 194

net present-worth (NPW) calculations, 268

project cash flow tables, 552

rate-of-return (ROR) calculations, 362, 593, 595

risk analysis worksheets, 620, 662

working-capital spreadsheet, 597

Excel-based simulation software, 661–664

Expected return, 209–212

Expected value, 627–628, 645–646

F

Factor notation, 105

Financial options, 693–699

Financial statements, 40, 46–83, 415–418

accounting and, 46–83

balance sheets, 51–55

business status and, 40, 46–83

cash flow statement, 57–64

cost classification for, 415–418

engineering project impact on, 40

income statement, 56–57

matching principle, 415

period costs, 415

product costs, 415–418

Financing, 59, 65, 162–163, 191, 198–201, 523–524, 812–816. *See also* Interest; Loans

capital budgeting and, 812–816

cash flow statements for, 59

debt, 812, 814–815

discount rate, 199–201

equity, 812–813

incremental cash flow from, 523–524

lease or loan, comparison of, 198–201

loan charges, 191

long-term debt, 65

mixed, 816

money management and, 65, 162–163, 191, 198–201

mortgages, 162–163

short-term debt, 65

Fixed assets, 458

Fixed (capacity) costs, 419

Fixed-rate mortgage, 203–205

Floating-rate bonds, 214

Flotation costs, 813

Functional depreciation, 459

Fundamental decision rule, 296–302

Fundamentals of Engineering review questions, 911–930

Future worth (FW), 88, 96, 235, 253–256. *See also* Net-future-worth (NFW) method

Future-worth factor (*F*), 101, 105–106

G

Geometric-gradient series, 103, 129–134, 140

Gradient (*G*) series, 102–103, 123–134, 140

cash flow transactions, 102–103

composite series as, 123–124

compound growth as, 129

discrete compounding formulas, 140

geometric-gradient series, 103, 129–134

gradient-to-equal-payment series conversion factor, 126–129

linear, 102–103, 123–129

linear-gradient, 102–103, 123–129

present-worth factor (*P*), 124–126, 130–132

single-payment present-worth factor, 130

strict, 123

Graphical method, rate-of-return (ROR) analysis, 360–362

Gross margin, 56, 447–448

Growth (expansion) option, 728–730

H

Half-year convention (MACRS depreciation), 479
Health-care services analysis, 890–899
Hedging (insuring) stock, 702–704

I

Income statements, 56–57, 447–448
Income tax rate, 497–501
Incremental cash flows, 520–524
    classification of elements, 522–524
    financing activities, 523–524
    inflow elements, 522
    investing activities, 523–524
    operating activities, 523–524
    outflow elements, 520–522
Incremental cost-utility ratios, 897
Incremental costs, 262
Incremental income tax rate, 497–499
Incremental investment analysis, 263, 380–387, 877–881
    benefit–cost (B/C) ratios for, 877–881
    rate-of-return (ROR) analysis using, 263, 380–387
Independent projects, 239, 837, 838, 892–893
    cost-effectiveness and intervention programs, 892–893
    capital budgeting for, 837, 838
Inflation, 568–607
    actual dollars (A_n), 575–577, 579–583
    average rate (f), 572–573
    borrowed funds, effects of on, 589–592
    cash flows, impact on, 568–607
    constant dollars (A'_n), 575–577, 578–579
    consumer price index (CPI), 568–569, 570–572
    deflation compared to, 570
    dollar conversions, 575–577

equivalence calculations under, 578–583
Excel cash flow statements under, 586, 588, 591
general rate (f), 578
interest rates, 578
internal rate of return (IRR), effects of on, 592–595
measure of, 570–577
mixed-dollar analysis, 583
multiple rates of, 587–589
producer price index (PPI), 571
project cash flows, effects of on, 583–592
rate-of-return (ROR) analysis under, 592–598
specific rate (f_i), 573–574
working capital, effects of on, 596–598
Inflation-free interest rate (i), 578
Interest, 85–94, 522
    actual dollar transactions, 87
    calculating methods, 91–94
    cash flow diagrams, 89–90
    cash outflow from, 522
    compound, 91–94
    cost of money, 85–94
    discrete payment or receipt (A_n), 88
    earning power, 86–87
    end-of-period convention, 90–91
    end-of-period payment (A), 88
    future amount (F), 88
    market interest rate, 86, 87
    number of periods (N), 88, 90–91
    period (n), 88, 90–91
    purchasing power, 86–87
    rate (i), 86
    repayment of borrowed funds, 522
    simple, 91
    transaction elements, 87–91
Interest factors, 931–960
Interest-only mortgage, 203
Interest rate cap, 206

Interest rates (i), 84–161, 164–179, 246, 248, 252–253, 348–350, 353–362, 578, 780–794. *See also* Time and value of money
    after-tax rates, 780–794
    annual effective yield, 164
    annual percentage rate (APR), 164
    break-even (i^*), 248, 349–350, 353–362
    cash flow solutions for, 108–111
    changing, 187–190
    compounding period, 165–167
    continuous compounding, 169–170, 180–187
    continuous funds flow, 182–187
    discounting, 180, 182
    discrete compounding, 180–181
    effective, 164–179
    equivalence calculations for, 95–101, 147–149, 171–187
    Excel calculations for, 174, 194, 218
    inflation and, 578
    inflation-free (i'), 578
    internal rate of return (IRR), 248
    market, 86, 87, 578
    minimum attractive rate of return (MARR), 246, 252–253
    money management and, 164–190
    nominal, 164
    payment periods, 167–168
    periodic, 191
    periods (n) for, 88, 90–91
    rate of return, 348–349
    series of cash flows, 189–190
    single sums, 187–188
    time and, 108–111
Interest tables, 103–104
Internal rate of return (IRR), 248, 348, 350–351, 363–379, 387–390, 592–595
    break-even interest rate (i^*), 366–369

decision rules, 366–374, 380–381
inflation effects on, 592–595
minimum acceptable rate of return (MARR), 366–376
mixed investments, 363, 368–369
modified internal rate of return (MIRR), 368, 377–379
mutually exclusive alternatives and, 379
net-investment test, 363–365
net-present-worth (NPW) method, 248
present worth (PW) analysis and, 248, 363
project ranking flaws, 379
pure investments, 363, 366–368
rate-of-return (ROR) analysis, 348, 350–351, 363–379, 387–390, 592–595
relative (percentage) measure of, 379
return on invested capital (RIC), 350–351, 369–374
trial-and-error method, 374–376
true, 369
unequal service lines using, 387–390
Intervention programs, 892–896
Inventory, manufacturing costs, 416–417
Inventory turnover ratio, 69
Investing activities, 59
Investment pool, 246, 249–251, 833–834
Investments, 209–220, 262–265, 350–390, 523–524, 622–649, 664–678, 719–739, 837–846
absolute (dollar) measures, 379
aggregating risk over time, 639–645
alternative evaluation of, 837–846
bonds, 212–220
break-even interest rate (i^*), 353–355, 366–369
capital budgeting, 837–846

compound options, 733–739
contingent projects, 838–839
decision trees for risk evaluation, 664–678
dependent projects, 837
diversification,
expected return, 209–212
expected-value criterion, 645–646
growth (expansion) option, 728–730
incremental cash flows from, 523–524
incremental-investment analysis, 263, 380–387
independent projects, 837, 838
internal rate of return (IRR) for, 350–351, 363–379
limited budgets, 839–846
liquidity, 209
mean-and-variance criterion, 646
mixed, 363, 368–376
mutually exclusive projects, 838
net-investment test, 363–365
nonsimple, 351–353
option to defer, 724–727
present-worth (PW) analysis and, 262–265
probabilities for decision making, 622–633
probability distribution, 634–649
project risk evaluation, 645–649, 664–678
pure, 363, 366–368
rate-of-return (ROR) analysis, 262–265, 350–390
real-options analysis, 719–739
relative (percentage) measures, 379
return on invested capital (RIC), 350–351, 369–374
return on investment over time, 210–212
return, 209–212
risk-adjusted discount rate method, 646

risk of, 209
scale of, 263–265
scale-up option, 730–733
sequential decisions, 664–678
simple, 351–353
total-investment approach, 263
unequal service lines, 387–390
yield, 215–220
Investors, 48–49
iPhone manufacturing costs, 410–411
IRR, *see* Internal rate of return (IRR)
Irregular (mixed) payment series, 103, 135–139

J

Joint probabilities, 630–632, 674–676

L

Labor cost, 414
Law of large numbers, 646
Lease financing and loans, comparison of, 198–201
Lease-or-buy decision, 547–553
Leasing expenses, cash outflow from, 521
Lending rate (l), 834
Liabilities, 53
Life-cycle costs, 887–890
Life-cycle cost analysis (LCCA), 306, 312–320
Linear-gradient series, 102–103, 123–129, 140
Liquidity, 209
Liquidity analysis, 67–68
Loans, 190–208, 212–213, 236–239
amortized, 190–196
annual percentage rate (APR), 191
bonds compared to, 212–213
cash flow, 236–239
commercial, 190–198
fees, 191
financing charges, 191
home mortgages, 202–208

Loans *(Continued)*
 installments, 190
 lease financing compared to,
  198–201
 periodic interest rate, 191
 project cash flow versus,
  236–239
 repayment schedule, Excel, 194
 terms of, 191
Long-term debt financing, 65
Lowest common multiple of
 project lives, 276–279

M

Macro analysis, 886–887
Make-or-buy (outsourcing)
 decisions, 308–311, 433
Manufacturing, 413–418, 441–
 449, 521
 annual production budget,
  442–444
 cash outflow from, 521
 costs, 413–418, 441–449
 budgeted income statement,
  447–448
 cost-of-goods-sold budget,
  444–445
 direct labor cost, 414
 direct raw material cost, 414
 profit estimation, 441–449
 finished-goods inventory, 417
 overhead, 414
 product costs, 415–418
 raw-materials inventory, 416
 work-in-process inventory,
  416–417
Marginal cost and revenue, 43–44,
 435–440
Marginal contribution rate, 427
Marginal cost of capital, 828–829
Marginal probabilities, 630, 632,
 674–676
Marginal tax rate, 489
Marginal yield, 321
Market basket, 570
Market interest rate (i), 86–87, 578
Market value (current), 758–759
Market value analysis, 72–73

Matching concept, 460
Matching principle, 415
Maturity of option (T), 694
Mean-and-variance criterion, 646
Micro analysis, 886
Midmonth convention (MACRS
 depreciation), 479
Minimum attractive rate of return
 (MARR), 246, 252–253,
 366–376, 829–837
 capital budgeting and, 829–837
 capital rationing and, 833–837
 internal rate of return (IRR) and,
  366–376
 net-present worth (NPW) and,
  246, 252–253
 project financing known,
  829–831
 project financing unknown,
  831–833
 selection of, 252–253
Mixed (semi-variable) costs, 420
Mixed-dollar analysis, 583
Mixed financing, 816
Mixed investments, 363, 368–369
Model simulations, 650–653
Modified accelerated cost
  recovery system (MACRS),
  474–482, 491
 depreciation rules, 476–573
 half-year convention, 479
 midmonth convention, 479
 property classification, 475
 recovery allowance percentages,
  476
 recovery periods, 474–475
 straight-line method switched
  from, 479
 taxes on disposal of property,
  491
Modified internal rate of return
 (MIRR), 368, 377–379
Money management, 64–75, 84–
 161, 162–232
 annual percentage rate (APR),
  164
 asset management analysis,
  68–70

 compound interest, 164–187
 debt management, 66–67,
  190–208
 effective interest rates (i_a),
  164–179
 equivalence calculations for,
  171–187
 interest rates, 164–190
 investments, 209–220
 mortgage financing, 162–163
 nominal interest rate, 164
 payment periods, 167–179,
  203–206
 payment series, 135–140
 ratio analysis, 64–75
 series of cash flows, 189–190
 single sums, 187–188
 time value of money, 84–161
Monte Carlo sampling, 654–659
Mortgages, 162–163, 202–208
 adjustable rate (ARM), 205–206
 adjustment period, 206
 cost of, 202
 financing, 162–163
 fixed-rate, 203–205
 fully amortizing, 203
 index, 205–206
 interest rate cap, 206
 interest-only, 203
 margin, 206
 monthly payment calculation,
  203–206
 payment cap, 207
 points, 202, 203
Motivation and, 32–33
Motor vehicle crash costs,
 858–859
Multiple inflation rates, 587–589
Municipal bonds, 214
Mutually exclusive project
  alternatives, 236, 261–279,
  296, 300–302, 379–390,
  645–649, 838–846, 877–881,
  892, 894–896
 analysis period and project lives,
  265–279, 300–302
 annual equivalence (AE)
  analysis, 296, 300–302

benefit–cost (*B/C*) ratios for, 877–881

capital budgeting for, 838–846

cost-effective interventions, 892, 894–896

"do nothing" option, 261–262

expected-value criterion, 645–646

incremental investment analysis, 263, 380–387, 877–881

independent projects compared to, 239

internal rate of return (IRR) and, 379, 387–390

mean-and-variance criterion, 646

present-worth (PW) analysis, 261–279

probability distribution, 645–649

project ranking and, 379

rate-of-return (ROR) analysis, 379–390

replacement chain approach, revenue projects, 262, 263

risk-adjusted discount rate method, 646

risky alternative comparisons, 645–649

scale of investment, 263–265

service projects, 262, 263, 301–302

total investment approach, 263

unequal-service-life problems, IRR approach, 387–390

N

Negative rate of return, 350

Negative taxable income, 535–539

Net cash-flow equations, 543–544

Net cash-flow rule of signs, 353

Net change in working capital, 59

Net equity flow, 829–831

Net-future-worth (NFW) method, 253–256

Net income, 56

Net-investment test, 363–365

Net margin, 56

Net-present-worth (NPW) method, 245–252, 609, 634–649, 876–877

benefit–cost (*B/C*) ratios and, 876–877

borrowed-funds concept, 251–252

cost of capital, 252–253

internal rate of return (IRR), 248

investment pool, 246, 249–251

minimum attractive rate of return (MARR), 246, 252–253

multiple project alternatives, 247–249

probability distribution of, 609, 634–649

single project evaluation, 246

Net profit margin, 448

Net revenue, 56

Net worth, 46–47

Nominal interest rate, 164–166

Nonmanufacturing costs, 414–415, 445–447

NPW, *see* Net-present-worth (NPW) method

Number of periods (*N*), 88, 90–91

O

Operating activities, 59, 523–524

Operating costs, 303–305, 760, 763–764

Operating income, 488–491

Operating margin, 448

Operating revenue, cash inflow from, 522

Opportunity cost, 433–435, 760–762

Options, 692–753

American, 694, 717

Black–Scholes model, 716–718

call, 693, 695–697, 699–702, 705–708, 716, 719–720

compound, 733–739

defer investment, 724–727

European, 694, 716–717

financial, 693–699

growth (expansion), 728–730

hedging (insuring) stock, 702–704

multiperiod binomial lattice model, 713–716

payoffs, 695

positions, 695

pricing (value), 705–718

put, 693, 697–699, 702–704, 711, 716–717, 720

put-call parity equation, 695

real-options analysis, 719–753

reducing capital at risk, 699–702

replicating-portfolio approach, 705–708

risk management, 693–699

risk-free financing approach, 708–709

risk-neutral probability approach, 709–711

scale-up, 730–733

strategies, 699–704

two-period binomial lattice, 712–713

valuation, 711–718, 727–728

Oracle Crystal Ball software, 661–664

Output analysis, simulation results, 659–661

Overhead cost, 414

P

Paid-in capital (capital surplus), 53

Partnerships, 34

Patent and lease valuation, 727–728

Payback methods, 240–245

conventional, 240

cost of funds and, 244–245

discounted, 240, 243–244

payback period for, 240–242

screening, 240, 243

Payback period, 240–242

Payment cap, 207

Payment periods, 167–179, 203–206

compounding interest and, 172–179

Payment periods *(Continued)*
  continuous compounding and,
    169–170
  effective interest rates and,
    167–168, 170–179
  interest rates, 167–168
  monthly mortgage calculation,
    203–206
Payoffs, 695
Percentage depletion, 482–485
Period costs, 415
Periodic interest rate, 191
Periods, 88, 90–91, 164–168, 240–
    242, 265–279, 474–475
  analysis, 265–279
  base, for inflation, 570
  compounding interest, 164–167
  interest (n), 88, 90–91
  payback, 240–242
  payment, 167–168
  recovery (assets), 474–475
  required service, 265
  study, 265
Perpetual service life, 256–258
Physical depreciation, 459
Plan for receipts, 89
Planning horizon, 769, 771–779
Points, 202, 203
Preferred stock, 65–66, 822–823
Premium of option (C), 694
Present economic studies, 412
Present value, 88
Present worth (PW), 88, 96, 235
Present-worth (PW) analysis,
    234–293
  analysis period and project lives,
    265–279
  capitalized-equivalent (CE)
    method, 256–260
  discounted cash flow techniques
    (DCFs), 245–253
  incremental-investment
    approach, 263
  internal rate of return (IRR) and,
    248, 363
  minimum attractive rate of
    return (MARR), 246,
    252–253

mutually exclusive alternatives,
    261–279
  net-future-worth (NFW)
    method, 253–256
  net-present-worth (NPW)
    method, 245–252
  payback methods, 240–245
  project and loan cash flows,
    236–239
  scale of investment, 263–265
  service life and, 256–260
  service versus revenue projects,
    262
  stadium expansion plan,
    234–235
  total-investment approach, 263
Present-worth factor (P), 101,
    106–108, 122–123, 124–126,
    130–132, 140
  discrete compounding with, 140
  discounting process, 106–107
  equal-payment series, 122–123
  geometric-gradient series,
    130–132
  gradient series (G), 124–126,
    130–132
  linear-gradient series, 124–126
  single cash flows, 101, 106–108
  single-payment factor, 130
  time and, 108
Price index, 571. *See also*
    Consumer price index (CPI)
Price-to-earnings ratio (P/E), 73
Principle, 87
Probabilities, 622–633, 673–676,
    710
  assessment of, 622–627
  coefficient of variation, 632–633
  conditional, 630–632, 673–674
  covariance, 632–633
  cumulative distribution,
    625–627
  decision tree information and,
    673–676
  distribution of, 623–625
  expected value, 627–628
  investment decision making and,
    622–633

joint, 630–632, 674–676
  marginal, 630, 632, 674–676
  measurement of, 627–630
  random variables, 623
  revised (posterior), 676
  risk analysis, 622
  risk-neutral, 710
  standard deviation, 629
  variance, 628–630
Probability distribution, 623–625,
    634–649
  aggregating risk over time,
    639–645
  expected-value criterion,
    645–646
  mean-and-variance criterion, 646
  mutually exclusive alternatives,
    645–649
  net-present worth (NPW) and,
    609, 634–649
  procedure for, 634–639
  risk-adjusted discount rate
    method, 646
Producer price index (PPI), 571
Product costs, 415–418
Product expansion, 42
Profit estimation, 441–449
  annual production budget,
    442–444
  annual sales budget, 441–442
  budgeted income statement,
    447–448
  cost-of-goods-sold budget,
    444–444
  economic effects on, 449
  gross margin, 447–448
  net profit margin, 448
  nonmanufacturing cost budget,
    445–447
  operating income calculation,
    441
  operating margin, 448
  project expenses, 441
  project revenue, 441
Profit margin on sales, 70–71
Profitability analysis, 70–72
Profitability index (net B/C ratio),
    875–876

Project balance (PB), 368
Project lives, 256–260, 265–279
  analysis period and, 265–279
  differing from analysis period,
    269–276
  lowest common multiple of,
    276–279
  matching analysis period,
    266–269
  present-worth (PW) analysis,
    265–279
  salvage value, 269
  service life, 256–260
Project risk, 609, 610–622
  break-even analysis, 616–620
  origins of, 610
  scenario analysis, 620–622
  sensitivity (what-if) analysis,
    611–616
Projects, 38–42, 236–239, 262,
    441–449, 516–566, 583–592,
    608–691, 739–745, 837–846,
    858–909
  budgeted income statement,
    447–448
  benefit–cost analysis,
    863–871
  budgets for, 441–447
  business risks, 39–40
  capital budgeting for, 837–846
  cash flow statements, 524–542
  cash flows, 236–239, 516–566,
    583–592
  complex, 519–520
  contingent, 838–839
  cost–benefit estimation,
    518–520
  cost-effectiveness of, 881–890
  cost-only, 536
  decision-making for, 38–40
  dependent, 837
  economic analysis of, 858–909
  economical benefits, 38–39
  estimating volatility of cash
    flow, 739–745
  expenses, 441
  impact on financial statements,
    40

incremental cash flows,
    520–524
independent, 837, 838
independent, 239
inflation effects on, 583–592
large-scale, 38–40
lease-or-buy decision, 547–553
limited budgets, 839–846
loan cash flow versus, 236–239
mutually exclusive, 236, 838
operating income calculation,
    441
profit estimation, 441–449
public (service sector), 858–909
realism of, 38
real-option analysis, 739–745
revenue, 262, 441
risk and uncertainty, 608–691
selection of, 41
service, 262, 536
simple, 518–519
strategic economic decisions for,
    40–42
Property depreciation, 460–461
Public sector projects, economic
    analysis in, 862–871. *See
    also* Service sector
Purchasing power, 86–87, 568–569
Pure borrowing, 363
Pure investments, 363, 366–368
Put-call parity equation, 695
Put options, 693, 697–699,
    702–704, 711, 716–717, 720
  buying when price expected to
    decrease, 697–699
  European, 716–717
  protective, as a hedge, 702–704
  real, 720
PW, *see* Present-worth (PW)
    analysis

Q

Quality improvement, 42
Quick (acid-test) ratio, 68

R

Random numbers, simulation
    sampling with, 654

Random variables, 623
Rate-of-return (ROR) analysis,
    346–407, 592–598
  break-even interest rate (i^*),
    349–350, 353–362
  collectible cars: return on
    investment, 346–347
  computational methods,
    356–362
  decision rules, 366–374,
    380–381
  direct-solution method, 356–357
  Excel calculations for, 362, 593,
    595
  graphical method, 360–362
  incremental investment analysis,
    263, 380–387
  inflation effects on, 592–598
  interest rate earned, 348–349
  internal rate of return (IRR),
    348, 350–351, 363–379,
    592–595
  marginal efficiency of capital,
    348
  methods for finding, 351–362
  minimum acceptable rate of
    return (MARR), 366–376
  modified internal rate of return
    (MIRR), 368, 377–379
  mutually exclusive alternatives,
    379–390
  negative rate of return, 350
  nonsimple investments,
    351–353
  return on invested capital (RIC),
    350–351, 369–374
  return on investment, 346–347,
    348–350
  simple investments, 351–353
  trial-and-error method, 357–
    360, 374–376
  unequal service lines, 387–390
  yield, 348, 350
  working capital and, 596–598
Ratio analysis, 64–75
  asset management analysis,
    68–70
  business decisions using, 64–75

Time value of money *(Continued)*
 U.S. lottery: Sharing the
  jackpot, 84–85
 uneven-payment series, 103,
  135–140
 unconventional calculations,
  141–150
Times-interest-earned ratio, 66–67
Total assets turnover ratio, 70
Total double-declining balance
 (TDB) depreciation, 467
Total-investment approach, 263
Trade-in allowance, 462
Treasury notes and bonds, 214
Treasury stock, 53
Trend analysis, 71
Trial-and-error method, 357–360,
 374–376
True rate of return, 369

U

U.S. lottery: Sharing the jackpot,
 84–85
Underlying asset (S), 694
Unequal service lines, 387–390
Uneven-payment series, 103,
 135–140
Uniform series, *see*
 Equal-payment series
Unit of use, 311

Unit profit (cost) calculation,
 306–308
Units-of-production depreciation
 method, 473–474
Unrecognized gain or loss, 462
Useful life (assets), 463–464
User's benefits and disbenefits,
 864, 867–870

V

Value, 705–718, 864–866,
 864–866, 868–870
 benefits and costs, 864–866,
  868–870
 Black-Scholes option model,
  716–718
 calculations for benefit–cost
  analysis, 868–870
 multiperiod binomial lattice
  model, 713–716
 option pricing and, 705–718
 put-option valuation, 711
 replicating-portfolio approach,
  705–708
 risk-free financing approach,
  708–709
 risk-neutral probability
  approach, 709–711
 social discount rate, 865–866
 sponsor's costs, 865, 870

two-period binomial lattice
 valuation, 712–713
 user's benefits, 864, 868–869
 valuation, 711–718, 727–728
Variable operating costs, 419
Variance, 628–630
Volume index, 419

W

Weighted-average cost of capital,
 827–828
Working capital, 55, 59–60, 67,
 521–522, 528–533, 596–598
 cash flow statements and,
  528–533
 Excel spreadsheet for, 597
 inflation effects on, 596–598
 investments, cash outflow
  from, 521
 net change in, 59–60
 project requirements, 528–533
 release, cash inflow from, 522
 rate-of-return (ROR) analysis
  and, 596–598

Y

Yield to maturity (YTM), 215–220

Z

Zero-coupon bonds, 214

Summary of Useful Excel's Financial Functions (Part A)

	Description	Excel Function	Example	Solution
Single-Payment Cash Flows	**Find:** F **Given:** P	$= \text{FV}(i\%, N, 0, P)$	Find the future worth of $500 in 5 years at 8%.	$= \text{FV}(8\%, 5, 0, -500)$ $= \$734.66$
	Find: P **Given:** F	$= \text{PV}(i\%, N, 0, F)$	Find the present worth of $1,300 due in 10 years at a 16% interest rate.	$= \text{PV}(16\%, 10, 0, 1300)$ $= (\$294.69)$
	Find: F **Given:** A	$= \text{FV}(i\%, N, A)$	Find the future worth of a payment series of $200 per year for 12 years at 6%	$= \text{FV}(6\%,12, -200)$ $= \$3,373.99$
Equal-Payment-Series	**Find:** P **Given:** A	$= \text{PV}(i\%, N, A)$	Find the present worth of a payment series of $900 per year for 5 years at 8% interest rate.	$= \text{PV}(8\%, 5, 900)$ $= (\$3,593.44)$
	Find: A **Given:** P	$= \text{PMT}(i\%, N, P)$	What equal-annual-payment series is required to repay $25,000 in 5 years at 9% interest rate?	$= \text{PMT}(9\%, 5, -25000)$ $= \$6,427.31$
	Find: A **Given:** F	$= \text{PMT}(i\%, N, 0, F)$	What is the required annual savings to accumulate $50,000 in 3 years at 7% interest rate?	$= \text{PMT}(7\%, 3, 0, 50000)$ $= (\$15,552.58)$
Measures of Investment Worth	**Find:** NPW **Given:** Cash flow series	$= \text{NPV}(i\%, \text{series})$	Consider a project with the following cash flow series at 12% ($n = 0, -\$200; n = 1, \$150, n = 2, \$300, n = 3, 250)$?	$= \text{NPV}(12\%, \text{B3:B5}) + \text{B2}$ $= \$351.03$
	Find: IRR **Given:** Cash flow series	$= \text{IRR (values, guess)}$		$= \text{IRR(B2:B4, 10\%)}$ $= 89\%$

	A	B
	Period	Cash Flow
1	0	-200
2	1	150
3	2	300
4	3	250

	Description	Excel Function		Solution
	Find: AE **Given:** Cash flow series	$= \text{PMT}(i\%, N, \text{NPW})$		$= \text{PMT}(12\%, 3, -351.03)$ $= \$146.15$

Note: When specifying the cash flow input parameters in any Excel function, an outflow must be a negative number.

Summary of Useful Excel's Financial Functions (Part B)

	Description	Excel Function	Example	Solution
Loan Analysis Functions	Loan payment size	= PMT($i\%$, N, P)	Suppose you borrow $10,000 at 9% interest over 48 months. Find the size of the equal monthly payment.	= PMT(9%/12,48,1000) = ($248.45)
	Interest payment	= IPMT($i\%$, n, N, P)	Find the portion of interest payment for the 10th payment.	= IPMT(9%/12,10,48,10000) = ($62.91)
	Principal payment	= PPMT($i\%$, n, N, P)	Find the portion of principal payment for the 10th payment.	= PPMT(9%/12,10,48,10000) = ($185.94)
	Cumulative interest payment	= CUMIPMT(($i\%$, N, P, start_period, end_period, type)	Find the total interest payment over 48 months.	= CUMIPMT(9%/12,48,10000,1,48,0) = $1,944.82 (type "0 → payment at the end of period)
	Interest rate	= RATE(N, A, P, F)	What nominal interest rate is being paid on the following financing arrangement? Loan amount: $10,000, loan period: 60 months, and monthly payment: $207.58.	= RATE(60,207.58, −10000) = 0.7499% APR = 0.7499% × 12 = 9%
	Number of payments	= NPER($i\%$, A, P, F)	Find the number of months required to pay off a loan of $10,000 with 12% APR where you can afford a monthly payment of $200.	= NPER(12%/12,200, −10000) = 69.66 months
Depreciation functions	Straight-line	= SLN(cost, salvage, life)	Cost = $100,000, S = $20,000, life = 5 years	= SLN(100000,20000,5) = $16,000
	Declining balance (fixed rate)	= DB(cost, salvage, life, period)	Find the depreciation amount in period 3 with fixed rate = $1 - (\text{salvage/cost})^{(1/\text{life})} = 0.2752$	= DB(100000,20000,5,3) = $14,455
	Double declining balance	= DDB(cost, salvage, life, period, factor)	Find the depreciation amount in period 3 with α = 150%.	= DDB(100000,20000,5,3,1.5) = $14,700
	Declining balance with switching to straight-line	= VDB(cost, salvage, life, start_period, end_period, factor)	Find the depreciation amount in period 3 with α = 150%, with switching allowed.	= VDB(100000,20000,5,3,4,1.5) = $10,290